BAYESIAN
THEORY

BAYESIAN THEORY

José M. Bernardo
Professor of Statistics
Universidad de Valencia, Spain

Adrian F. M. Smith
Professor of Statistics
Imperial College of Science, Technology and Medicine, London, UK

JOHN WILEY & SONS, LTD
Chichester · New York · Weinheim · Brisbane · Singapore · Toronto

Telephone (+44) 1243 779777

Email (for orders and customer service enquiries): cs-books@wiley.co.uk
Visit our Home Page on www.wileyeurope.com or www.wiley.com

First published in paperback 1994 (ISBN 0 471 92416 4)

Reprinted September 2001, May 2002, June 2003, July 2004

Other Wiley Editorial Offices

John Wiley & Sons Inc., 111 River Street, Hoboken, NJ 07030, USA

Jossey-Bass, 989 Market Street, San Francisco, CA 94103-1741, USA

Wiley-VCH Verlag GmbH, Boschstr. 12, D-69469 Weinheim, Germany

John Wiley & Sons Australia Ltd, 33 Park Road, Milton, Queensland 4064, Australia

John Wiley & Sons (Asia) Pte Ltd, 2 Clementi Loop #02-01, Jin Xing Distripark, Singapore 129809

John Wiley & Sons Canada Ltd, 22 Worcester Road, Etobicoke, Ontario, Canada M9W 1L1

British Library Cataloguing in Publication Data

A catalogue record for this book is available from the British Library

ISBN 0 471 49464 X

Produced from postscript files supplied by the author

To MARINA and DANIEL

Preface

This volume, first published in hardback in 1994, presents an overview of the foundations and key theoretical concepts of Bayesian Statistics. Our original intention had been to produce further volumes on computation and methods. However, these projects have been shelved as tailored Markov chain Monte Carlo methods have emerged and are being refined as the standard Bayesian computational tools. We have taken the opportunity provided by this reissue in paperback form to make a number of typographical corrections.

The original motivation for this enterprise stemmed from the impact and influence of de Finetti's two-volume *Theory of Probability*, which one of us helped translate into English from the Italian in the early 1970's. This was widely acknowledged as the definitive exposition of the operationalist, subjectivist approach to uncertainty, and provided further impetus at that time to a growth in activity and interest in Bayesian ideas.

From a philosophical, foundational perspective, the de Finetti volumes provide —in the words of the author's dedication to his friend Segre—

> . . . a necessary document for clarifying one point of view in its entirety.

From a statistical, methodological perspective, however, the de Finetti volumes end abruptly, with just the barest introduction to the mechanics of Bayesian inference.

Some years ago, we decided to try to write a series of books which would take up the story where de Finetti left off, with the grandiose objective of "clarifying in its entirety" the world of Bayesian statistical theory and practice.

It is now clear that this was a hopeless undertaking. The world of Bayesian Statistics has been changing shape and growing in size rapidly and unpredictably—most notably in relation to developments in computational methods and the subsequent opening up of new application horizons. We are greatly relieved that we were too incompetent to finish our books a few years ago!

And, of course, these changes and developments continue. There is no static world of Bayesian Statistics to describe in a once-and-for-all way. Moreover, we are dealing with a field of activity where, even among those whose intellectual perspectives fall within the broad paradigm, there are considerable differences of view at the level of detail and nuance of interpretation.

This volume on *Bayesian Theory* attempts to provide a fairly complete and up-to-date overview of what we regard as the key concepts, results and issues. However, it necessarily reflects the prejudices and interests of its authors—as well as the temporal constraints imposed by a publisher whose patience has been sorely tested for far too long. We can but hope that our sins of commission and omission are not too grievous.

Too many colleagues have taught us too many things for it to be practical to list everyone to whom we are beholden. However, Dennis Lindley has played a special role, not least in supervising us as Ph.D. students, and we should like to record our deep gratitude to him. We also shared many enterprises with Morrie DeGroot and continue to miss his warmth and intellectual stimulation. For detailed comments on earlier versions of material in this volume, we are indebted to our colleagues M. J. Bayarri, J. O. Berger, J. de la Horra, P. Diaconis, F. J. Girón, M. A. Gómez-Villegas, D. V. Lindley, M. Mendoza, J. Muñoz, E. Moreno, L. R. Pericchi, A. van der Linde, C. Villegas and M. West.

We are also grateful, in more ways than one, to the State of Valencia. It has provided a beautiful and congenial setting for much of the writing of this book. And, in the person of the Governor, Joan Lerma, it has been wonderfully supportive of the celebrated series of *Valencia International Meetings on Bayesian Statistics*. During the secondment of one of us as scientific advisor to the Governor, it also provided resources to enable the writing of this book to continue.

This volume has been produced directly in TEX and we are grateful to María Dolores Tortajada for all her efforts.

Finally, we thank past and present editors at John Wiley & Sons for their support of this project: Jamie Cameron for saying "Go!" and Helen Ramsey for saying "Stop!"

Valencia, Spain
January 26, 2000

J. M. Bernardo
A. F. M. Smith

Contents

Chapter 1

Introduction

Summary

A brief historical introduction to Bayes' theorem and its author is given, as a prelude to a statement of the perspective adopted in this volume regarding Bayesian Statistics. An overview is provided of the material to be covered in successive chapters and appendices, and a Bayesian reading list is provided.

1.1 THOMAS BAYES

According to contemporary journal death notices and the inscription on his tomb in Bunhill Fields cemetery in London, Thomas Bayes died on 7th April, 1761, at the age of 59. The inscription on top of the tomb reads:

> Rev. Thomas Bayes. Son of the said Joshua and Ann Bayes (59). 7 April 1761. In recognition of Thomas Bayes's important work in probability. The vault was restored in 1969 with contributions received from statisticians throughout the world.

Definitive records of Bayes' birth do not seem to exist, but, allowing for the calendar reform of 1752 and accepting that he died at the age of 59, it seems likely that he was born in 1701 (an argument attributed to Bellhouse in the *Inst. Math.*

Statist. Bull. **26**, 1992). Some background on the life and the work of Bayes may be found in Barnard (1958), Holland (1962), Pearson (1978), Gillies (1987), Dale (1990, 1991) and Earman (1990). See, also, Stigler (1986a).

That his name lives on in the characterisation of a modern statistical methodology is a consequence of the publication of *An essay towards solving a problem in the doctrine of chances*, attributed to Bayes and communicated to the Royal Society after Bayes' death by Richard Price in 1763 (*Phil. Trans. Roy. Soc.* **53**, 370–418).

The technical result at the heart of the essay is what we now know as *Bayes' theorem*. However, from a purely formal perspective there is no obvious reason why this essentially trivial probability result should continue to excite interest.

In its simplest form, if H denotes an hypothesis and D denotes data, the theorem states that

$$P(H \mid D) = P(D \mid H) \times P(H)/P(D).$$

With $P(H)$ regarded as a probabilistic statement of belief about H before obtaining data D, the left-hand side $P(H \mid D)$ becomes a probabilistic statement of belief about H after obtaining D. Having specified $P(D \mid H)$ and $P(D)$, the mechanism of the theorem provides a solution to the problem of how to learn from data.

Actually, Bayes only stated his result for a uniform prior. According to Stigler (1986b), it was Laplace (1774/1986)—apparently unaware of Bayes' work—who stated the theorem in its general (discrete) form.

Like any theorem in probability, at the technical level Bayes' theorem merely provides a form of "uncertainty accounting", which asserts that the left-hand side of the equation must equal the right-hand side. The interest and controversy, of course, lie in the interpretation and assumed scope of the formal inputs to the two sides of the equation—and it is here that past and present commentators part company in their responses to the idea that Bayes' theorem can or should be regarded as a central feature of the statistical learning process. At the heart of the controversy is the issue of the philosophical interpretation of probability—objective or subjective?—and the appropriateness and legitimacy of basing a scientific theory on the latter.

What Thomas Bayes—from the tranquil surroundings of Bunhill Fields, where he lies in peace with Richard Price for company—has made of all the fuss over the last 233 years we shall never know. We would like to think that he is a subjectivist fellow-traveller but, in any case, he is in no position to complain at the liberties we are about to take in his name.

1.2 THE SUBJECTIVIST VIEW OF PROBABILITY

Throughout this work, we shall adopt a wholehearted subjectivist position regarding the interpretation of probability. The definitive account and defence of this position are given in de Finetti's two-volume *Theory of Probability* (1970/1974, 1970/1975) and the following brief extract from the Preface to that work perfectly encapsulates the essence of the case.

> The only relevant thing is uncertainty—the extent of our own knowledge and ignorance. The actual fact of whether or not the events considered are in some sense *determined*, or known by other people, and so on, is of no consequence.
>
> The numerous, different, opposed attempts to put forward particular points of view which, in the opinion of their supporters, would endow Probability Theory with a 'nobler' status, or a 'more scientific' character, or 'firmer' philosophical or logical foundations, have only served to generate confusion and obscurity, and to provoke well-known polemics and disagreements—even between supporters of essentially the same framework.
>
> The main points of view that have been put forward are as follows.
>
> The *classical* view, based on physical considerations of symmetry, in which one should be *obliged* to give the same probability to such 'symmetric' cases. But which symmetry? And, in any case, why? The original sentence becomes meaningful if reversed: the symmetry is probabilistically significant, in someone's opinion, if it leads him to assign the same probabilities to such events.
>
> The *logical* view is similar, but much more superficial and irresponsible inasmuch as it is based on similarities or symmetries which no longer derive from the facts and their actual properties, but merely from the sentences which describe them, and from their formal structure or language.
>
> The *frequentist* (or *statistical*) view presupposes that one accepts the classical view, in that it considers an *event* as a class of *individual events*, the latter being 'trials' of the former. The individual events not only have to be 'equally probable', but also 'stochastically independent' ... (these notions when applied to individual events are virtually impossible to define or explain in terms of the frequentist interpretation). In this case, also, it is straightforward, by means of the subjective approach, to obtain, under the appropriate conditions, in a perfectly valid manner, the result aimed at (but unattainable) in the statistical formulation. It suffices to make use of the notion of exchangeability. The result, which acts as a bridge connecting the new approach with the old, has often been referred to by the objectivists as "de Finetti's representation theorem".
>
> It follows that all the three proposed definitions of 'objective' probability, although useless *per se*, turn out to be useful and good as valid auxiliary devices when included as such in the subjectivist theory.
>
> (de Finetti, 1970/1974, Preface, xi–xii)

1.3 BAYESIAN STATISTICS IN PERSPECTIVE

The theory and practice of Statistics span a range of diverse activities, which are motivated and characterised by varying degrees of formal intent. Activity in the context of initial data exploration is typically rather informal; activity relating to concepts and theories of evidence and uncertainty is somewhat more formally structured; and activity directed at the mathematical abstraction and rigorous analysis of these structures is intentionally highly formal.

What is the nature and scope of Bayesian Statistics within this spectrum of activity?

Bayesian Statistics offers a rationalist theory of personalistic beliefs in contexts of uncertainty, with the central aim of characterising how an individual should act in order to avoid certain kinds of undesirable behavioural inconsistencies. The theory establishes that expected utility maximisation provides the basis for rational decision making and that Bayes' theorem provides the key to the ways in which beliefs should fit together in the light of changing evidence. The goal, in effect, is to establish rules and procedures for individuals concerned with disciplined uncertainty accounting. The theory is not descriptive, in the sense of claiming to model actual behaviour. Rather, it is prescriptive, in the sense of saying "if you wish to avoid the possibility of these undesirable consequences you must act in the following way".

From the very beginning, the development of the theory necessarily presumes a rather formal frame of discourse, within which uncertain events and available actions can be described and axioms of rational behaviour can be stated. But this formalism is preceded and succeeded in the scientific learning cycle by activities which, in our view, cannot readily be seen as part of the formalism.

In any field of application, a prerequisite for arriving at a structured frame of discourse will typically be an informal phase of exploratory data analysis. Also, it can happen that evidence arises which discredits a previously assumed and accepted formal framework and necessitates a rethink. Part of the process of realising that a change is needed can take place within the currently accepted framework using Bayesian ideas, but the process of rethinking is again outside the formalism. Both these phases of initial structuring and subsequent restructuring might well be guided by "Bayesian thinking"—by which we mean keeping in mind the objective of creating or re-creating a formal framework for uncertainty analysis and decision making—but are not themselves part of the Bayesian formalism. That said, there is, of course, often a pragmatic ambiguity about the boundaries of the formal and the informal.

The emphasis in this book is on ideas and we have sought throughout to keep the level of the mathematical treatment as simple as is compatible with giving what we regard as an honest account. However, there are sections where the full story would require a greater level of abstraction than we have adopted, and we have drawn attention to this whenever appropriate.

1.4 AN OVERVIEW OF BAYESIAN THEORY

1.4.1 Scope

This volume on *Bayesian Theory* focuses on the basic concepts and theory of Bayesian Statistics, with chapters covering elementary *Foundations*, mathematical *Generalisations* of the Foundations, *Modelling, Inference* and *Remodelling*. In addition, there are two appendices providing a *Summary of Basic Formulae* and a review of *Non-Bayesian Theories*. The emphasis throughout is on general ideas—the *Why?*—of Bayesian Statistics. A detailed treatment of analytical and numerical techniques for implementing Bayesian procedures—the *How?*—will be provided in the volume *Bayesian Computation*. A systematic study of the methods of analysis for a wide range of commonly encountered model and problem types—the *What?*—will be provided in the volume *Bayesian Methods*.

The selection of topics and the details of approach adopted in this volume necessarily reflect our own preferences and prejudices. Where we hold strong views, these are, for the most part, rather clearly and forcefully stated, while, hopefully, avoiding too dogmatic a tone. We acknowledge, however, that even colleagues who are committed to the Bayesian paradigm will disagree with at least some points of detail and emphasis in our account. For this reason, and to avoid complicating the main text with too many digressionary asides and references, each of Chapters 2 to 6 concludes with a *Discussion and Further References* section, in which some of the key issues in the chapter are critically re-examined.

In most cases, the omission of a topic, or its abbreviated treatment in this volume, reflects the fact that a detailed treatment will be given in one or other of the volumes *Bayesian Computation* and *Bayesian Methods*. Topics falling into this category include *Design of Experiments*, *Image Analysis*, *Linear Models*, *Multivariate Analysis*, *Nonparametric Inference*, *Prior Elicitation*, *Robustness*, *Sequential Analysis*, *Survival Analysis* and *Time Series*. However, there are important topics, such as *Game Theory* and *Group Decision Making*, which are omitted simply because a proper treatment seemed to us to involve too much of a digression from our central theme. For a convenient source of discussion and references at the interface of Decision Theory and Game Theory, see French (1986).

1.4.2 Foundations

In Chapter 2, the concept of rationality is explored in the context of representing beliefs or choosing actions in situations of uncertainty. We introduce a formal framework for decision problems and an axiom system for the foundations of decision theory, which we believe to have considerable intuitive appeal and to be an improvement on the many such systems that have been previously proposed. Here, and throughout this volume, we stress the importance of a decision-oriented framework in providing a disciplined setting for the discussion of issues relating to uncertainty and rationality.

The dual concepts of probability and utility are formally defined and analysed within this decision making context and the criterion of maximising expected utility is shown to be the only decision criterion which is compatible with the axiom system. The analysis of sequential decision problems is shown to reduce to successive applications of the methodology introduced.

A key feature of our approach is that statistical inference is viewed simply as a particular form of decision problem; specifically, a decision problem where an action corresponds to reporting a probability belief distribution for some unknown quantity of interest. Thus defined, the inference problem can be analysed within the general decision theory framework, rather than requiring a separate "theory of inference".

An important special feature of what we shall call a pure inference problem is the form of utility function to be adopted. We establish that the logarithmic utility function—more often referred to as a score function in this context—plays a special role as the natural utility function for describing the preferences of an individual faced with a pure inference problem.

Within this framework, measures of the discrepancy between probability distributions and the amount of information contained in a distribution are naturally defined in terms of expected loss and expected increase, respectively, in logarithmic utility. These measures are mathematically closely related to well-known information-theoretic measures pioneered by Shannon (1948) and employed in statistical contexts by Kullback (1959/1968). A resulting characteristic feature of our approach is therefore the systematic appearance of these information-theoretic quantities as key elements in the Bayesian analysis of inference and general decision problems.

1.4.3 Generalisations

In Chapter 3, the ideas and results of Chapter 2 are extended to a much more general mathematical setting. An additional postulate concerning the comparison of a countable collection of events is appended to the axiom system of Chapter 2, and is shown to provide a justification for restricting attention to countably additive probability as the basis for representing beliefs. The elements of mathematical probability theory required in our subsequent development are then reviewed.

The notions of actions and utilities, introduced in a simple discrete setting in Chapter 2, are extended in a natural way to provide a very general mathematical framework for our development of decision theory. A further additional mathematical postulate regarding preferences is introduced and, within this more general framework, the criterion of maximising expected utility is shown to be the only decision making criterion compatible with the extended axiom system.

In this generalised setting, inference problems are again considered simply as special cases of decision problems and generalised definitions of score functions and measures of information and discrepancy are given.

1.4.4 Modelling

In Chapter 4, we examine in detail the role of familiar mathematical forms of statistical models and the possible justifications—from a subjectivist perspective—for their use as representations of actual beliefs about observable random quantities. A feature of our approach is an emphasis on the primacy of observables and the notion of a model as a (probabilistic) prediction device for such observables. From this perspective, the role of conventional parametric statistical modelling is problematic, and requires fundamental re-examination.

The problem is approached by considering simple structural characteristics —such as symmetry with respect to the labelling of individual counts or measurements, a feature common to many individual beliefs about sequences of observables. The key concept here is that of exchangeability, which we motivate, formalise and then use to establish a version of de Finetti's celebrated representation theorem. This demonstrates that judgements of exchangeability lead to general mathematical representations of beliefs that justify and clarify the use and interpretations of such familiar statistical concepts as parameters, random samples, likelihoods and prior distributions.

Going beyond simple exchangeability, we show that beliefs which have certain additional invariance properties—for example, to rotation of the axes of measurements, or translation of the origin—can lead to mathematical representations involving other familiar specific forms of parametric distributions, such as normals and exponentials.

A further approach to characterising belief distributions is considered, based on data reduction. The concept of a sufficient statistic is introduced and related to representations involving the exponential family of distributions.

Various forms of partial exchangeability judgements about data structures are then discussed in a number of familiar contexts and links are established with a number of other commonly used statistical models. Structures considered include those of several samples, multiway layouts, problems involving covariates, and hierarchies.

1.4.5 Inference

In Chapter 5, the key role of Bayes' theorem in the updating of beliefs about observables in the light of new information is identified and related to conventional mechanisms of predictive and parametric inference. The roles of sufficiency, ancillarity and stopping rules in such inference processes are also examined.

Various standard forms of statistical problems, such as point and interval estimation and hypothesis testing, are re-examined within the general Bayesian decision framework and related to formal and informal inference summaries.

The problems of implementing Bayesian procedures are discussed at length. The mathematical convenience and elegance of conjugate analysis are illustrated in detail, as are the mathematical approximations available under the assumption of the validity of large-sample asymptotic analysis. A particular feature of this volume is the extended account of so-called reference analysis, which can be viewed as a Bayesian formalisation of the idea of "letting the data speak for themselves". An alternative, closely related idea is that of how to represent "vague beliefs" or "ignorance". We provide a detailed historical review of attempts that have been made to solve this problem and compare and contrast some of these with the reference analysis approach. A brief account is given of recent analytic approximation strategies derived from Laplace-type methods, together with outline accounts of numerical quadrature, importance sampling, sampling-importance-resampling, and Markov chain Monte Carlo methods.

1.4.6 Remodelling

In Chapter 6, it is argued that, whether viewed from the perspective of a sensitive individual modeller or from that of a group of modellers, there are good reasons for systematically entertaining a range of possible belief models, rather than predicating all analysis on a single assumed model.

A variety of decision problems are examined within this framework, some involving model choice only, some involving model choice followed by a terminal action, such as prediction, others involving only a terminal action.

A feature of our treatment of this topic is that, throughout, a clear distinction is drawn among three rather different perspectives on the comparison of and choice from among a range of competing models. The first perspective arises when the range of models under consideration is assumed to include the "true" model. The second perspective arises when the range of models is assumed to be under consideration in order to provide a more conveniently implemented proxy for an actual, but intractable, belief model. The third perspective arises when the range of models is under consideration because the models are "all there is available", in the absence of any specification of an actual belief model. Our discussion relates and links these ideas with aspects of hypothesis testing, significance testing and cross-validation.

1.4.7 Basic Formulae

In Appendix A, we collect together for convenience, in tabular format, summaries of the main univariate and multivariate probability distributions that appear in the text, together with summaries of the prior/posterior/predictive forms corresponding to these distributions in the context of conjugate and reference analyses.

1.4.8 Non-Bayesian Theories

In Appendix B, we review what we perceive to be the main alternatives to the Bayesian approach; namely, classical decision theory, frequentist procedures, likelihood theory, and fiducial and related theories.

We compare and contrast these alternatives in the context of "stylised" inference problems such as point and interval estimation, hypothesis and significance testing. Through counter-examples and general discussion, we indicate why we find all these alternatives seriously deficient as formal inference theories.

1.5 A BAYESIAN READING LIST

As we have already remarked, this work is—necessarily—a selective account of Bayesian theory, reflecting our own interests and perspectives. The following is a list of other Bayesian books—by no means exhaustive—whose contents would provide a significant complement to the material in this volume.

In those cases where there are several editions, or when the original is not in English, we quote both the original date and the date of the most recent English edition. Thus, Jeffreys (1939/1961) refers to Jeffreys' *Theory of Probability*, first published in 1939, and to its most recent (3rd) edition, published in 1961; similarly, de Finetti (1970/1974) refers to the original (1970) Italian version of de Finetti's *Teoria delle Probabilità* vol. 1 and to its English translation (published in 1974).

Pioneering Bayesian books include Laplace (1812), Keynes (1921/1929), Jeffreys (1939/1961), Good (1950, 1965), Savage (1954/1972, 1962), Schlaifer (1959, 1961), Raiffa and Schlaifer (1961), Mosteller and Wallace (1964/1984), Dubins and Savage (1965/1976), Lindley (1965, 1972), Pratt *et al.* (1965), Tribus (1969), DeGroot (1970), de Finetti (1970/1974, 1970/1975, 1972) and Box and Tiao (1973).

Elementary and intermediate Bayesian textbooks include those of Savage (1968), Schmitt (1969), Lavalle (1970), Lindley (1971/1985), Winkler (1972), Kleiter (1980), Bernardo (1981b), Daboni and Wedlin (1982), Iversen (1984), O'Hagan (1988a), Cifarelli and Muliere (1989), Lee (1989), Press (1989), Scozzafava (1989), Wichmann (1990), Borovcnik (1992), and Berry (1996).

More advanced Bayesian monographs include Hartigan (1983), Regazzini (1983), Berger (1985a), Savchuk (1989), Florens *et al.* (1990), Robert (1992) and O'Hagan (1994a). Polson and Tiao (1995) is a two volume collection of classic papers in Bayesian inference.

Special topics have also been examined from a Bayesian point of view; these include *Actuarial Science* (Klugman, 1992), *Biostatistics* (Girelli-Bruni, 1981;

Lecoutre, 1984; Barrai *et al.*, 1992; Berry and Stangl, 1996), *Control Theory* (Aoki, 1967; Sawagari *et al.*, 1967), *Decision Analysis* (Duncan and Raiffa, 1957; Chernoff and Moses, 1959; Grayson, 1960; Fellner, 1965; Roberts, 1966; Edwards and Tversky, 1967; Hadley, 1967; Martin, 1967; Morris, 1968; Raiffa, 1968; Lusted, 1968; Schlaifer, 1969; Aitchison, 1970; Fishburn, 1970, 1982; Halter and Dean, 1971; Lindgren, 1971; Keeney and Raiffa, 1976; Ríos, 1977; Lavalle, 1978; Roberts, 1979; French *et al.* 1983; French, 1986, 1989; Marinell and Seeber, 1988; Smith, 1988a), *Dynamic Forecasting* (Spall, 1988; West and Harrison, 1989; Pole *et al.*, 1994), *Economics and Econometrics* (Morales, 1971; Zellner, 1971; Richard, 1973; Bauwens, 1984; Boyer and Kihlstrom, 1984; Cyert and DeGroot, 1987), *Educational and Psychological Research* (Novick and Jackson, 1974; Pollard, 1986), *Foundations* (Fishburn, 1964, 1970, 1982, 1987, 1988a; Berger and Wolpert, 1984/1988; Brown, 1985), *History* (Dale, 1991), *Information Theory* (Yaglom and Yaglom, 1960/1983; Osteyee and Good, 1974), *Law and Forensic Science* (DeGroot *et al.*, 1986; Aitken and Stoney, 1991), *Linear Models* (Lempers, 1971; Leamer, 1978; Pilz, 1983/1991; Broemeling, 1985), *Logic and Philosophy of Science* (Jeffrey, 1965/1983, 1981; Rosenkranz, 1977; Seidenfeld, 1979; Howson and Urbach 1989; Verbraak, 1990; Rivadulla, 1991), *Maximum Entropy* (Levine and Tribus, 1978; Smith and Grandy, 1985; Justice, 1987; Smith and Erickson, 1987; Erickson and Smith, 1988; Skilling, 1989; Fougère, 1990; Grandy and Schick, 1991; Kapur and Kesavan,1992; Mohammad-Djafari and Demoment, 1993), *Multivariate Analysis* (Press, 1972/1982; Berger and DasGupta, 1991), *Optimisation* (Mockus, 1989), *Pattern Recognition* (Simon, 1984), *Prediction* (Aitchison and Dunsmore, 1975; Geisser, 1993), *Probability Assessment* (Stäel von Holstein, 1970; Stäel von Holstein and Matheson, 1979; Cooke, 1991), *Reliability* (Martz and Waller, 1982; Claroti and Lindley, 1988), *Sample Surveys* (Rubin, 1987), *Social Science* (Phillips, 1973) and *Spectral Analysis* (Bretthorst, 1988).

A number of collected works also include a wealth of Bayesian material. Among these, we note particularly; Kyburg and Smokler (1964/1980), Meyer and Collier (1970), Godambe and Sprott (1971), Fienberg and Zellner (1974), White and Bowen (1975), Aykaç and Brumat (1977), Parenti (1978), Zellner (1980), Savage (1981), Box *et al.* (1983), Dawid and Smith (1983), Florens *et al.* (1983, 1985), Good (1983), Jaynes (1983), Kadane (1984), Box (1985), Goel and Zellner (1986), Smith and Dawid (1987), Viertl (1987), Gärdenfors and Sahlin (1988), Gupta and Berger (1988, 1994), Geisser *et al.* (1990), Hinkelmann (1990), Oliver and Smith (1990), Ghosh and Pathak (1992), Goel and Iyengar (1992), de Finetti (1993), Fearn and O'Hagan (1993), Gatsonis *et al.* (1993), Freeman and Smith (1994), and last, but not least, the *Proceedings of the Valencia International Meetings on Bayesian Statistics* (Bernardo *et al.* 1980, 1985, 1988, 1992, 1996 and 1999).

General discussions of Bayesian Statistics may be found in review papers and encyclopedia articles. Among these, we note de Finetti (1951), Lindley (1953, 1976, 1978, 1982b, 1982c, 1984, 1990, 1992), Anscombe (1961), Savage (1961,

1970), Edwards *et al.* (1963), Bartholomew (1965), Cornfield (1969), Good (1976, 1992), Roberts (1978), DeGroot (1982), Dawid (1983a), Smith (1984), Zellner (1985, 1987, 1988a), Pack (1986a, 1986b), Cifarelli (1987), Bernardo (1989), Ghosh (1991) and Berger (1993).

For discussion of important specific topics, see Luce and Suppes (1965), Birnbaum (1968, 1978), de Finetti (1968), Press (1980a, 1985a), Fishburn (1981, 1986, 1988b), Dickey (1982), Geisser (1982, 1986), Good (1982, 1985, 1987, 1988a, 1988b), Dawid (1983b, 1986a, 1992), Joshi (1983), LaMotte (1985), Genest and Zidek (1986), Goldstein (1986c), Racine-Poon *et al.* (1986), Hodges (1987), Ericson (1988), Zellner (1988c), Trader (1989), Breslow (1990), Lindley (1991), Smith (1991), Barlow and Irony (1992), Ferguson *et al.* (1992), Arnold (1993), Kadane (1993), Bartholomew (1994), Berger (1994) and Hill (1994).

Chapter 2

Foundations

Summary

The concept of rationality is explored in the context of representing beliefs or choosing actions in situations of uncertainty. An axiomatic basis, with intuitive operational appeal, is introduced for the foundations of decision theory. The dual concepts of probability and utility are formally defined and analysed within this context. The criterion of maximising expected utility is shown to be the only decision criterion which is compatible with the axiom system. The analysis of sequential decision problems is shown to reduce to successive applications of the methodology introduced. Statistical inference is viewed as a particular decision problem which may be analysed within the framework of decision theory. The logarithmic score is established as the natural utility function to describe the preferences of an individual faced with a pure inference problem. Within this framework, the concept of discrepancy between probability distributions and the quantification of the amount of information in new data are naturally defined in terms of expected loss and expected increase in utility, respectively.

2.1 BELIEFS AND ACTIONS

We spend a considerable proportion of our lives, both private and professional, in a state of uncertainty. This uncertainty may relate to past situations, where direct knowledge or evidence is not available, or has been lost or forgotten; or to present and future developments which are not yet completed. Whatever the circumstances, there is a sense in which all states of uncertainty may be described in the same way: namely, an individual feeling of incomplete knowledge in relation to a specified situation (a feeling which may, of course, be shared by other individuals). And yet it is obvious that we do not attempt to treat all our individual uncertainties with the same degree of interest or seriousness.

Many feelings of uncertainty are rather insubstantial and we neither seek to analyse them, nor to order our thoughts and opinions in any kind of responsible way. This typically happens when we feel no actual or practical involvement with the situation in question. In other words, when we feel that we have no (or only negligible) capacity to influence matters, or that the possible outcomes have no (or only negligible) consequences so far as we are concerned. In such cases, we are not motivated to think carefully about our uncertainty either because nothing depends on it, or the potential effects are trivial in comparison with the effort involved in carrying out a conscious analysis.

On the other hand, we all regularly encounter uncertain situations in which we at least aspire to behave "rationally" in some sense. This might be because we face the direct practical problem of choosing from among a set of possible actions, where each involves a range of uncertain consequences and we are concerned to avoid making an "illogical" choice. Alternatively, we might be called upon to summarise our beliefs about the uncertain aspects of the situation, bearing in mind that others may subsequently use this summary as the basis for choosing an action. In this case, we are concerned that our summary be in a form which will enable a "rational" choice to be made at some future time. More specifically, we might regard the summary itself, i.e., the choice of a particular mode of representing and communicating our beliefs, as being a form of action to which certain criteria of "rationality" might be directly applied.

Our basic concern in this chapter is with exploring the concept of "rationality" in the context of representing beliefs or choosing actions in situations of uncertainty. To choose the best among a set of actions would, in principle, be immediate if we had perfect information about the consequences to which they would lead. So far as this work is concerned, interesting decision problems are those for which such perfect information is not available, and we must take *uncertainty* into account as a major feature of the problem.

It might be argued that there are complex situations where we *do* have complete information and yet still find it difficult to take the best decision. Here, however, the difficulty is *technical*, not conceptual. For example, even though we have, in principle, complete information, it is typically not easy to decide what is the optimal strategy to rebuild a Rubik cube or which is the cheapest diet fulfilling specified nutritional requirements. We take the view that such problems are purely technical. In the first case, they result from the large number of possible strategies;

in the second, they reduce to the mathematical problem of finding a minimum under certain constraints. But in neither case is there any doubt about the decision criterion to be used. In this work we shall not consider these kinds of combinatorial or mathematical programming problems, and we shall assume that in the presence of complete information we can, in principle, always choose the best alternative.

Our concern, instead, is with the *logical process of decision making in situations of uncertainty*. In other words, with the decision criterion to be adopted when we do not have complete information and are thus faced with, at least some, elements of uncertainty.

To avoid any possible confusion, we should emphasise that we do not interpret "actions in situations of uncertainty" in a narrow, directly "economic" sense. For example, within our purview we include the situation of an individual scientist summarising his or her own current beliefs following the results of an experiment; or trying to facilitate the task of others seeking to decide upon their beliefs in the light of the experimental results.

It is assumed in our approach to such problems that the notion of "rational belief" cannot be considered separately from the notion of "rational action". Either a statement of beliefs in the light of available information is, actually or potentially, an input into the process of choosing some practical course of action,

> ... it is not asserted that a belief ... does actually lead to action, but would lead to action in suitable circumstances; just as a lump of arsenic is called poisonous not because it actually has killed or will kill anyone, but because it would kill anyone if he ate it (Ramsey, 1926).

or, alternatively, a statement of beliefs might be regarded as an end in itself, in which case the choice of the form of statement to be made constitutes an action,

> Frequently, it is a question of providing a convenient summary of the data ... In such cases, the emphasis is on the inference rather than the decision aspect of problem, although formally it can still be considered a decision problem if the inferential statement itself is interpreted as the decision to be taken (Lehmann, 1959/1986).

We can therefore explore the notion of "rationality" for both beliefs and actions by concentrating on the latter and asking ourselves what kinds of rules should govern preference patterns among sets of alternative actions in order that choices made in accordance with such rules commend themselves to us as "rational", in that they cannot lead us into forms of behavioural inconsistency which we specifically wish to avoid.

In Section 2.2, we describe the general structure of problems involving choices under uncertainty and introduce the idea of preferences between options. In Section 2.3, we make precise the notion of "rational" preferences in the form of axioms.

We describe these as *principles of quantitative coherence* because they specify the ways in which preferences need to be made *quantitatively* precise and fit together, or *cohere*, if "illogical" forms of behaviour are to be avoided. In Sections 2.4 and 2.5, we prove that, in order to conform with the principles of quantitative coherence, degrees of belief about uncertain events should be described in terms of a (finitely additive) *probability measure*, relative values of individual possible consequences should be described in terms of a *utility function*, and the rational choice of an action is to select one which has the *maximum expected utility*.

In Section 2.6, we discuss sequential decision problems and show that their analysis reduces to successive applications of the maximum expected utility methodology; in particular, we identify the design of experiments as a particular case of a sequential decision problem. In Section 2.7, we make precise the sense in which choosing a form of a statement of beliefs can be viewed as a special case of a decision problem. This identification of inference as decision provides the fundamental justification for beginning our development of Bayesian Statistics with the discussion of decision theory. Finally, a general review of ideas and references is given in Section 2.8.

2.2 DECISION PROBLEMS

2.2.1 Basic Elements

We shall describe any situation in which choices are to be made among alternative courses of action with uncertain consequences as a *decision problem*, whose structure is determined by three basic elements:

(i) a set $\{a_i, i \in I\}$ of available *actions*, one of which is to be selected;

(ii) for each action a_i, a set $\{E_j, j \in J\}$ of *uncertain events*, describing the uncertain outcomes of taking action a_i;

(iii) corresponding to each set $\{E_j, j \in J\}$, a set of *consequences* $\{c_j, j \in J\}$.

The idea is as follows. Suppose we choose action a_i; then one and only one of the uncertain events E_j, $j \in J$, occurs and leads to the corresponding consequence c_j, $j \in J$. Each set of events $\{E_j, j \in J\}$ forms a *partition* (an exclusive and exhaustive decomposition) of the total set of possibilities. Naturally, both the set of consequences and the partition which labels them may depend on the particular action considered, so that a more precise notation would be $\{E_{ij}, j \in J_i\}$ and $\{c_{ij}, j \in J_i\}$ for each action a_i. However, to simplify notation, we shall omit this dependence, while remarking that it should always be borne in mind. We shall come back to this point in Section 2.6.

In practical problems, the labelling sets, I and J (for each i), are typically finite. In such cases, the decision problem can be represented schematically by means of a decision tree as shown in Figure 2.1.

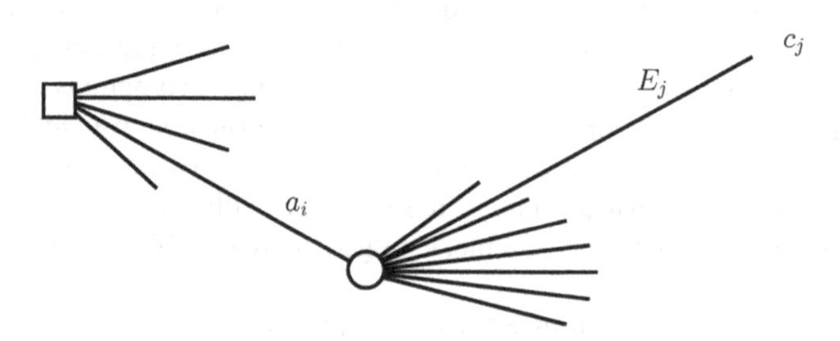

Figure 2.1 *Decision tree*

The square represents a decision node, where the choice of an action is required. The circle represents an uncertainty node where the outcome is beyond our control. Following the choice of an action and the occurrence of a particular event, the branch leads us to the corresponding consequence.

Of course, most practical problems involve sequential considerations but, as shown in Section 2.6, these reduce, essentially, to repeated analyses based on the above structure.

It is clear, either from our general discussion, or from the decision tree representation, that we can formally identify any a_i, $i \in I$, with the combination of $\{E_j,\ j \in J\}$ and $\{c_j,\ j \in J\}$ to which it leads. In other words, to choose a_i is to opt for the uncertain scenario labelled by the pairs (E_j, c_j), $j \in J$. We shall write $a_i = \{c_j \mid E_j,\ j \in J\}$ to denote this identification, where the notation $c_j \mid E_j$ signifies that event E_j leads to consequence c_j, i.e., that $a_i(E_j) = c_j$.

An individual's perception of the state of uncertainty resulting from the choice of any particular a_i is very much dependent on the *information currently available*. In particular, $\{E_j,\ j \in J\}$ forms a partition of the total set of relevant possibilities as the individual decision-maker now perceives them to be. Further information, of a kind which leads to a restriction on what can be regarded as the total set of possibilities, will change the perception of the uncertainties, in that some of the E_j's may become very implausible (or even logically impossible) in the light of the new information, whereas others may become more plausible. It is therefore of considerable importance to bear in mind that a representation such as Figure 2.1 only captures the structure of a decision problem as perceived at a particular point in time. Preferences about the uncertain scenarios resulting from the choices of actions depend on *attitudes to the consequences involved and assessments of the uncertainties attached to the corresponding events*. The latter are clearly subject to change as new information is acquired and this may well change overall preferences among the various courses of action.

The notion of *preference* is, of course, very familiar in the everyday context of actual or potential choice. Indeed, an individual decision-maker often prefaces an actual choice (from a menu, an investment portfolio, a range of possible forms of medical treatment, a textbook of statistical methods, etc.) with the phrase "I prefer... " (caviar, equities, surgery, Bayesian procedures, etc.). To prefer action a_1 to action a_2 means that if these were the only two options available a_1 would be chosen (conditional, of course, on the information available at the time). In everyday terms, the idea of indifference between two courses of action also has a clear operational meaning. It signifies a willingness to accept an externally determined choice (for example, letting a disinterested third party choose, or tossing a coin).

In addition to *representing the structure* of a decision problem using the three elements discussed above, we must also be able to *represent the idea of preference* as applied to the comparison of some or all of the pairs of available options. We shall therefore need to consider a fourth basic element of a decision problem:

(iv) the relation $\leq$, which expresses the individual decision-maker's preferences between pairs of available actions, so that $a_1 \leq a_2$ signifies that a_1 *is not preferred to* a_2.

These four basic elements have been introduced in a rather informal manner. In order to study decision problems in a precise way, we shall need to reformulate these concepts in a more formal framework. The development which follows, here and in Section 3.3, is largely based on Bernardo, Ferrándiz and Smith (1985).

2.2.2 Formal Representation

When considering a particular, concrete decision problem, we do not usually confine our thoughts to *only* those outcomes and options explicitly required for the specification of that problem. Typically, we *expand our horizons* to encompass analogous problems, which we hope will aid us in ordering our thoughts by providing suggestive points of reference or comparison. The collection of uncertain scenarios defined by the original concrete problem is therefore implicitly embedded in a somewhat wider framework of actual and hypothetical scenarios. We begin by describing this *wider frame of discourse* within which the comparisons of scenarios are to be carried out. It is to be understood that the initial specification of any such particular frame of discourse, together with the preferences among options within it, are dependent on the decision-maker's overall state of information at that time. Throughout, we shall denote this initial state of mind by M_0.

We now give a formal definition of a decision problem. This will be presented in a rather compact form; detailed elaboration is provided in the remarks following the definition.

Definition 2.1. (***Decision problem***). *A decision problem is defined by the elements* $(\mathcal{E}, \mathcal{C}, \mathcal{A}, \leq)$, *where:*

(i) $\mathcal{E}$ *is an algebra of relevant events,* E_j;

(ii) *$\mathcal{C}$ is a set of possible consequences, c_j;*

(iii) *$\mathcal{A}$ is a set of options, or potential acts, consisting of functions which map finite partitions of Ω, the certain event in $\mathcal{E}$, to compatibly-dimensioned, ordered sets of elements of $\mathcal{C}$;*

(iv) *$\leq$ is a preference order, taking the form of a binary relation between some of the elements of $\mathcal{A}$.*

We now discuss each of these elements in detail. Within this wider frame of discourse, an individual decision-maker will wish to consider the uncertain events judged to be *relevant* in the light of the initial state of information M_0. However, it is natural to assume that if $E_1 \in \mathcal{E}$ and $E_2 \in \mathcal{E}$ are judged to be relevant events then it may also be of interest to know about their joint occurrence, or whether at least one of them occurs. This means that $E_1 \cap E_2$ and $E_1 \cup E_2$ should also be assumed to belong to $\mathcal{E}$. Repetition of this argument suggests that $\mathcal{E}$ should be closed under the operations of arbitrary finite intersections and unions. Similarly, it is natural to require $\mathcal{E}$ to be closed under complementation, so that $E^c \in \mathcal{E}$. In particular, these requirements ensure that the *certain event* Ω and the *impossible event* $\emptyset$, both belong to $\mathcal{E}$. Technically, we are assuming that the class of relevant events has the structure of an *algebra*. (However, it can certainly be argued that this is too rigid an assumption. We shall provide further discussion of this and related issues in Section 2.8.4.)

As we mentioned when introducing the idea of a wider frame of discourse, the algebra $\mathcal{E}$ will consist of what we might call the *real-world* events (that is, those occurring in the structure of any concrete, actual decision problem that we may wish to consider), together with any other *hypothetical* events, which it may be convenient to bring to mind as an aid to thought. The class $\mathcal{E}$ will simply be referred to as the *algebra of (relevant) events*.

We denote by $\mathcal{C}$ the set of all consequences that the decision-maker wishes to take into account; preferences among such consequences will later be assumed to be independent of the state of information concerning relevant events. The class $\mathcal{C}$ will simply be referred to as the *set of (possible) consequences*.

In our introductory discussion we used the term *action* to refer to each potential act available as a choice at a decision node. Within the wider frame of discourse, we prefer the term *option*, since the general, formal framework may include hypothetical scenarios (possibly rather far removed from potential concrete actions).

So far as the definition of an option as a function is concerned, we note that this is a rather natural way to view options from a mathematical point of view: an option consists precisely of the linking of a partition of Ω, $\{E_j,\ j \in J\}$, with a corresponding set of consequences, $\{c_j,\ j \in J\}$. To represent such a mapping we shall adopt the notation $\{c_j \mid E_j,\ j \in J\}$, with the interpretation that event E_j leads to consequence c_j, $j \in J$.

It follows immediately from the definition of an option that the ordering of the labels within J is irrelevant, so that, for example, the options $\{c_1 \mid E,\ c_2 \mid E^c\}$, and $\{c_2 \mid E^c,\ c_1 \mid E\}$ are identical, and forms such as $\{c \mid E_1,\ c \mid E_2,\ c_j \mid E_j, j \in J\}$ and $\{c \mid E_1 \cup E_2,\ c_j \mid E_j,\ j \in J\}$ are completely equivalent. Which form is used in any particular context is purely a matter of convenience. Sometimes, the interpretation of an option with a rather cumbersome description is clarified by an appropriate reformulation. For example, $a = \{c_1 \mid E \cap G,\ c_2 \mid E^c \cap G,\ c_3 \mid G^c\}$ may be more compactly written as $a = \{a_1 \mid G, c_3 \mid G^c\}$, with $a_1 = \{c_1 \mid E, c_2 \mid E^c\}$. Thus, if

$$a = \{c_{k(j)} \mid E_{k(j)} \cap F_j, k(j) \in K_j, j \in J\}, \quad a_j = \{c_{k(j)} \mid E_{k(j)}, k(j) \in K_j\},$$

we shall use the *composite function notation* $a = \{a_j \mid F_j, j \in J\}$. In all cases, the ordering of the labels is irrelevant. The class $\mathcal{A}$ of options, or potential actions, will simply be referred to as the *action space*.

In defining options, the assumption of a *finite* partition into events of $\mathcal{E}$ seems to us to correspond most closely to the structure of practical problems. However, an extension to admit the possibility of *infinite* partitions has certain mathematical advantages and will be fully discussed, together with other mathematical extensions, in Chapter 3.

In introducing the preference binary relation $\leq$, we are not assuming that all pairs of options $(a_1, a_2) \in \mathcal{A} \times \mathcal{A}$ can necessarily be related by $\leq$. If the relation *can* be applied, in the sense that either $a_1 \leq a_2$ or $a_2 \leq a_1$ (or both), we say that a_1 is not preferred to a_2, or a_2 is not preferred to a_1 (or both). From $\leq$, we can derive a number of other useful binary relations.

Definition 2.2. (***Induced binary relations***).

(i) $a_1 \sim a_2 \iff a_1 \leq a_2$ *and* $a_2 \leq a_1$.

(ii) $a_1 < a_2 \iff a_1 \leq a_2$ *and it is not true that* $a_2 \leq a_1$.

(iii) $a_1 \geq a_2 \iff a_2 \leq a_1$.

(iv) $a_1 > a_2 \iff a_2 < a_1$.

Definition 2.2 is to be understood as referring to any options a_1, a_2 in $\mathcal{A}$. To simplify the presentation we shall omit such universal quantifiers when there is no danger of confusion. The induced binary relations are to be interpreted to mean that a_1 is equivalent to a_2 if and only if $a_1 \sim a_2$, and a_1 is strictly preferred to a_2 if and only if $a_1 > a_2$. Together with the interpretation of $\leq$, these suffice to describe all cases where pairs of options can be compared.

We can identify individual consequences as special cases of options by writing $c = \{c \mid \Omega\}$, for any $c \in \mathcal{C}$. Without introducing further notation, we shall simply regard c as denoting either an element of $\mathcal{C}$, or the element $\{c \mid \Omega\}$ of $\mathcal{A}$. There will be no danger of any confusion arising from this identification. Thus, we shall write $c_1 \leq c_2$ if and only if $\{c_1 \mid \Omega\} \leq \{c_2 \mid \Omega\}$ and say that consequence c_1 is not preferred to consequence c_2. Strictly speaking, we should introduce a new symbol to replace $\leq$ when referring to a preference relation over $\mathcal{C} \times \mathcal{C}$, since $\leq$ is defined over $\mathcal{A} \times \mathcal{A}$. In fact, this parsimonious abuse of notation creates no danger of confusion and we shall routinely adopt such usage in order to avoid a proliferation of symbols. We shall proceed similarly with the binary relations $\sim$ and $<$ introduced in Definition 2.2. To avoid triviality, we shall later formally assume that there exist at least two consequences c_1 and c_2 such that $c_1 < c_2$.

The basic preference relation between options, $\leq$, conditional on the initial state of information M_0, can also be used to define a binary relation on $\mathcal{E} \times \mathcal{E}$, the collection of all pairs of *relevant events*. This binary relation will capture the intuitive notion of one event being "more likely" than another. Since, once again, there is no danger of confusion, we shall further economise on notation and also use the symbol $\leq$ to denote this new uncertainty binary relation between events.

Definition 2.3. (***Uncertainty relation***).

$$E \leq F \iff \text{for all } c_1 < c_2,\ \{c_2 \mid E, c_1 \mid E^c\} \leq \{c_2 \mid F, c_1 \mid F^c\};$$

*we then say that E is **not more likely** than F.*

The intuitive content of the definition is clear. If we compare two dichotomised options, involving the same pair of consequences and differing only in terms of their uncertain events, we will prefer the option under which we feel it is "more likely" that the preferred consequence will obtain. Clearly, the force of this argument applies independently of the choice of the particular consequences c_1 and c_2, provided that our preferences between the latter are assumed independent of any considerations regarding the events E and F.

Continuing the (convenient and harmless) abuse of notation, we shall also use the derived binary relations given in Definition 2.2 to describe uncertainty relations between events. Thus, $E \sim F$ if and only if E and F are equally likely, and $E > F$ if and only if E is strictly more likely than F. Since, for all $c_1 < c_2$,

$$c_1 \equiv \{c_2 \mid \emptyset, c_1 \mid \Omega\} < \{c_2 \mid \Omega, c_1 \mid \emptyset\} \equiv c_2,$$

it is always true, as one would expect, that $\emptyset < \Omega$.

It is worth stressing once again at this point that *all* the order relations over $\mathcal{A} \times \mathcal{A}$, and hence over $\mathcal{C} \times \mathcal{C}$ and $\mathcal{E} \times \mathcal{E}$, are to be understood as *personal*, in the sense that, given an agreed structure for a decision problem, each individual is free to express his or her own personal preferences, in the light of his or her initial state of information M_0. Thus, for a given individual, a statement such as $E > F$ is to be interpreted as "*this individual, given the state of information described by M_0, considers event E to be more likely than event F*". Moreover,

Definition 2.3 provides such a statement with an *operational meaning* since for all $c_1 < c_2$, $E > F$ is equivalent to an agreement to choose option $\{c_2 \mid E, c_1 \mid E^c\}$ in preference to option $\{c_2 \mid F, c_1 \mid F^c\}$.

To complete our discussion of basic ideas and definitions, we need to consider one further important topic. Throughout this section, we have stressed that preferences, initially defined among options but inducing binary relations among consequences and events, are *conditional* on the current state of information. The *initial* state of information, taking as an arbitrary "origin" the first occasion on which an individual thinks systematically about the problem, has been denoted by M_0. Subsequently, however, we shall need to take into account *further information*, obtained by considering the occurrence of real-world events. Given the assumed occurrence of a possible event G, preferences between options will be described by a new binary relation $\leq_G$, taking into account both the initial information M_0 and the additional information provided by G. The obvious relation between $\leq$ and $\leq_G$ is given by the following:

Definition 2.4. **(*Conditional preference*).** *For any* $G > \emptyset$,

(i) $a_1 \leq_G a_2 \iff$ *for all* $a\{a_1 \mid G, a \mid G^c\} \leq \{a_2 \mid G, a \mid G^c\}$;

(ii) $E \leq_G F \iff$ *for* $c_1 \leq_G c_2, \{c_2 \mid E, c_1 \mid E^c\} \leq_G \{c_2 \mid F, c_1 \mid F^c\}$.

The intuitive content of the definition is clear. If we do not prefer a_1 to a_2, given G, then this preference obviously carries over to any pair of options leading, respectively, to a_1 or a_2 if G occurs, and defined identically if G^c occurs. Conversely, comparison of options which are identical if G^c occurs depends entirely on consideration of what happens if G occurs. Naturally, the induced binary relations set out in Definition 2.2 have their obvious counterparts, denoted by $\sim_G$ and $<_G$.

The induced binary relation between consequences is obviously defined by

$$c_1 \leq_G c_2 \iff \{c_1 \mid \Omega\} \leq_G \{c_2 \mid \Omega\}.$$

However, when we come, in Section 2.3, to discuss the desirable properties of $\leq$ and $\leq_G$ we shall make formal assumptions which imply that, as one would expect, $c_1 \leq_G c_2$ if and only if $c_1 \leq c_2$, so that preferences between pure consequences are not affected by additional information regarding the uncertain events in $\mathcal{E}$.

The definition of the conditional uncertainty relation $\leq_G$ is a simple translation of Definition 2.3 to a conditional preference setting. The conditional uncertainty relation $\leq_G$ induced between events is of fundamental importance. This relation, with its derived forms $\sim_G$ and $<_G$, provides the key to investigating the way in which uncertainties about events should be modified in the light of new information. Obviously, if $G = \Omega$, all conditional relations reduce to their unconditional counterparts. Thus, it is only when $\emptyset < G < \Omega$ that conditioning on G may yield new preference patterns.

2.3 COHERENCE AND QUANTIFICATION

2.3.1 Events, Options and Preferences

The formal representation of the decision-maker's "wider frame of discourse" includes an algebra of events $\mathcal{E}$, a set of consequences $\mathcal{C}$, and a set of options $\mathcal{A}$, whose generic element has the form $\{c_j \mid E_j, j \in J\}$, where $\{E_j, j \in J\}$ is a finite partition of the certain event Ω, $E_j \in \mathcal{E}$, $c_j \in \mathcal{C}$, $j \in J$. The set $\mathcal{A} \times \mathcal{A}$ is equipped with a collection of binary relations $\leq_G$, $G > \emptyset$, representing the notion that one option is not preferred to another, given the assumed occurrence of a possible event G. In addition, all preferences are assumed conditional on an initial state of information, M_0, with the binary relation $\leq$ (i.e., $\leq_\Omega$) representing the preference relation on $\mathcal{A} \times \mathcal{A}$ conditional on M_0 alone.

We now wish to make precise our assumptions about these elements of the formal representation of a decision problem. Bearing in mind the overall objective of developing a rational approach to choosing among options, our assumptions, presented in the form of a series of *axioms*, can be viewed as responses to the questions: "what rules should preference relations obey?" and "what events should be included in $\mathcal{E}$?"

Each formal axiom will be accompanied by a detailed discussion of the intuitive motivation underlying it.

It is important to recognise that the axioms we shall present are *prescriptive*, not *descriptive*. Thus, they do not purport to describe the ways in which individuals actually *do* behave in formulating problems or making choices, neither do they assert, on some presumed "ethical" basis, the ways in which individuals *should* behave. The axioms simply prescribe constraints which it seems to us imperative to acknowledge in those situations where an individual aspires to choose among alternatives in such a way as to avoid certain forms of behavioural inconsistency.

2.3.2 Coherent Preferences

We shall begin by assuming that problems represented within the formal framework are non-trivial and that we are able to compare any pair of simple *dichotomised* options.

Axiom 1. (***Comparability of consequences and dichotomised options***).

(i) *There exist consequences* c_1, c_2 *such that* $c_1 < c_2$.

(ii) *For all consequences* c_1, c_2, *and events* E, F,
either $\{c_2 \mid E, c_1 \mid E^c\} \leq \{c_2 \mid F, c_1 \mid F^c\}$
or $\{c_2 \mid E, c_1 \mid E^c\} \geq \{c_2 \mid F, c_1 \mid F^c\}$.

Discussion of Axiom 1. Condition (i) is very natural. If all consequences were equivalent, there would not be a decision problem in any real sense, since all choices would certainly lead to precisely equivalent outcomes. We have already noted that, in any given decision problem, C can be defined as simply the set of consequences required for that problem. Condition (ii) does *not* therefore assert that we should be able to compare *any* pair of conceivable options, however bizarre or fantastic. In most *practical* problems, there will typically be a high degree of similarity in the form of the consequences (e.g. all monetary), although it is easy to think of examples where this form is complex (e.g. combinations of monetary, health and industrial relations elements). We are trying to capture the essence of what is required for an orderly and systematic approach to comparing alternatives of genuine interest. We are not, at this stage, making the direct assumption that *all options*, however complex, can be compared. But there could be no possibility of an orderly and systematic approach if we were unwilling to express preferences among simple dichotomised options and hence (with $E = F = \Omega$) among the consequences themselves. Condition (ii) is therefore to be interpreted in the following sense: "*If we aspire to make a rational choice between alternative options, then we must at least be willing to express preferences between simple dichotomised options.*"

There are certainly many situations where we find the task of comparing simple options, and even consequences, very difficult. Resource allocation among competing health care programmes involving different target populations and morbidity and mortality rates is one obvious such example. However, the difficulty of comparing options in such cases does not, of course, obviate the *need* for such comparisons if we are to aspire to responsible decision making.

We shall now state our assumptions about the ways in which preferences should fit together or *cohere* in terms of the order relation over $\mathcal{A} \times \mathcal{A}$.

Axiom 2. (***Transitivity of preferences***).

(i) $a \leq a$.

(ii) *If* $a_1 \leq a_2$ *and* $a_2 \leq a_3$, *then* $a_1 \leq a_3$.

Discussion of Axiom 2. Condition (i) has obvious intuitive support. It would make little sense to assert that an option was strictly preferred to itself. It would also seem strangely perverse to claim to be unable to compare an option with itself! We note that, from Definition 2.2 (i), if $a \leq a$, then $a \sim a$. Condition (ii) requires preferences to be *transitive*. The intuitive basis for such a requirement is perhaps best illustrated by considering the consequences of *intransitive* preferences. Suppose, therefore, that we found ourselves expressing the preferences $a_1 < a_2$, $a_2 < a_3$ and $a_3 < a_1$ among three options a_1, a_2 and a_3. The assertion of strict preference rules out equivalence between any pair of the options, so that our expressed preferences reveal that we perceive *some actual difference in value* (no matter how small) between the two options in each case. Let us now examine the behavioural implications of these expressed preferences. If we consider, for example, the preference $a_1 < a_2$, we are implicitly stating that there exists a "price", say x, that we would be willing to pay in order to move from a position of having to accept option a_1 to one where we have, instead, to accept option a_2. Let y and z denote the corresponding "prices" for switching from a_2 to a_3 and from a_3 to a_1, respectively. Suppose now that we are confronted with the prospect of having to accept option a_1. By virtue of the expressed preference $a_1 < a_2$ and the above discussion, we are willing to pay x in order to exchange option a_1 for option a_2. But now, by virtue of the preference $a_2 < a_3$, we are willing to pay y in order to exchange a_2 for a_3. Repeating the argument once again, since $a_3 < a_1$ we are willing to pay z in order to avoid a_3 and have, instead, the prospect of option a_1. *We would thus have paid $x + y + z$ in order to find ourselves in precisely the same position as we started from!* What is more, we could find ourselves arguing through this cycle over and over again. Willingness to act on the basis of intransitive preferences is thus seen to be equivalent to a willingness to suffer unnecessarily the *certain loss* of something to which one attaches positive value. We regard this as inherently inconsistent behaviour and recall that the purpose of the axioms is to impose rules of coherence on preference orderings that will exclude the possibility of such inconsistencies. Thus, Axiom 2(ii) is to be understood in the following sense: "*If we aspire to avoid expressing preferences whose behavioural implications are such as to lead us to the certain loss of something we value, then we must ensure that our preferences fit together in a transitive manner.*"

Our discussion of this axiom is, of course, informal and appeals to directly intuitive considerations. At this stage, it would therefore be inappropriate to become involved in a formal discussion of terms such as "value" and "price". It is intuitively clear that if we assert strict preference there must be some amount of money (or grains of wheat, or beads, or whatever), however small, having a "value" less than the perceived difference in "value" between the two options. We should therefore be willing to pay this amount to switch from the less preferred to the more preferred option.

The following consequences of Axiom 2 are easily established and will prove useful in our subsequent development.

Proposition 2.1. (***Transitivity of uncertainties***).

(i) $E \sim E$.

(ii) $E_1 \leq E_2$ *and* $E_2 \leq E_3$ *imply* $E_1 \leq E_3$.

Proof. This is immediate from Definition 2.3 and Axiom 2. $\triangleleft$

Proposition 2.2. (***Derived transitive properties***).

(i) *If* $a_1 \sim a_2$ *and* $a_2 \sim a_3$ *then* $a_1 \sim a_3$.
If $E_1 \sim E_2$ *and* $E_2 \sim E_3$ *then* $E_1 \sim E_3$.

(ii) *If* $a_1 < a_2$ *and* $a_2 \sim a_3$ *then* $a_1 < a_3$.
If $E_1 < E_2$ *and* $E_2 \sim E_3$ *then* $E_1 < E_3$.

Proof. To prove (i), let $a_1 \sim a_2$ and $a_2 \sim a_3$ so that, by Definition 2.2, $a_1 \leq a_2$, $a_2 \leq a_1$ and $a_2 \leq a_3$, $a_3 \leq a_2$. Then, by Axiom 2(ii), $a_1 \leq a_3$ and $a_3 \leq a_1$, and thus $a_1 \sim a_3$. A similar argument applies to events using Proposition 2.1. Again, part (ii) follows rather similarly. $\triangleleft$

Axiom 3. (***Consistency of preferences***).

(i) *If* $c_1 \leq c_2$ *then, for all* $G > \emptyset$, $c_1 \leq_G c_2$.

(ii) *If, for some* $c_1 < c_2$, $\{c_2 \,|\, E, c_1 \,|\, E^c\} \leq \{c_2 \,|\, F, c_1 \,|\, F^c\}$, *then* $E \leq F$.

(iii) *If, for some* c *and* $G > \emptyset$, $\{a_1 \,|\, G, c \,|\, G^c\} \leq \{a_2 \,|\, G, c \,|\, G^c\}$, *then* $a_1 \leq_G a_2$.

Discussion of Axiom 3. Condition (i) formalises the idea that preferences between pure consequences should not be affected by the acquisition of further information regarding the uncertain events in $\mathcal{E}$. Conditions (ii) and (iii) ensure that Definitions 2.3 and 2.4 have operational content. Indeed, (ii) asserts that if we have $\{c_2 \,|\, E, c_1 \,|\, E^c\} \leq \{c_2 \,|\, F, c_1 \,|\, F^c\}$ for *some* $c_1 < c_2$ then we should have this preference for *any* $c_1 < c_2$. This formalises the intuitive idea that the stated preference should only depend on the "relative likelihood" of E and F and should *not* depend on the particular consequences used in constructing the options. Similarly, (iii) asserts that if we have the preference $\{a_1 \,|\, G, c \,|\, G^c\} \leq \{a_2 \,|\, G, c \,|\, G^c\}$ for *some* c then, given G, a_1 should not be preferred to a_2, so that, for *any* a, $\{a_1 \,|\, G, a \,|\, G^c\} \leq \{a_2 \,|\, G, a \,|\, G^c\}$. This latter argument is a version of what might be called the ***sure-thing principle***: if two situations are such that whatever the outcome of the first there is a preferable corresponding outcome of the second, then the second situation is preferable overall.

An important implication of Axiom 3 is that preferences between consequences are invariant under changes in the information "origin" regarding events in $\mathcal{E}$.

Proposition 2.3. (***Invariance of preferences between consequences***).
$c_1 \leq c_2$ *if and only if there exist* $G > \emptyset$ *such that* $c_1 \leq_G c_2$.

Proof. If $c_1 \leq c_2$ then, by Axiom 3(i), $c_1 \leq_G c_2$ for any event G. Conversely, by Definition 2.4(i), for any $G > \emptyset$, $c_1 \leq_G c_2$ implies that for *any* option a, one has $\{c_1 \,|\, G, a \,|\, G^c\} \leq \{c_2 \,|\, G, a \,|\, G^c\}$. Taking $a = \{c_1 \,|\, G, c_2 \,|\, G^c\}$, this implies that $\{c_1 \,|\, G, c_2 \,|\, G^c\} \leq \{c_1 \,|\, \emptyset, c_2 \,|\, \Omega\}$. If $c_1 > c_2$ this implies, by Axiom 3(ii), that $G \leq \emptyset$, thus contradicting $G > \emptyset$. Hence, by Axiom 1(ii), $c_1 \leq c_2$. $\triangleleft$

Another important consequence of Axiom 3 is that uncertainty orderings of events respect logical implications, in the sense that if E logically implies F, i.e., if $E \subseteq F$, then F cannot be considered less likely than E.

Proposition 2.4. **(*Monotonicity*).** *If $E \subseteq F$ then $E \leq F$.*

Proof. For any $c_1 < c_2$, define

$$\begin{aligned} a_1 &= \{c_2 \,|\, E, c_1 \,|\, E^c\} = \{c_1 \,|\, F - E, \{c_2 \,|\, E, c_1 \,|\, E^c\} \,|\, (F - E)^c\}, \\ a_2 &= \{c_2 \,|\, F, c_1 \,|\, F^c\} = \{c_2 \,|\, F - E, \{c_2 \,|\, E, c_1 \,|\, E^c\} \,|\, (F - E)^c\}. \end{aligned}$$

By Axiom 3(i) with $G = F - E = F \cap E^c$, $a_1 \leq a_2$. It now follows immediately from Definition 2.2 that $E \leq F$. ◁

This last result is an example of how coherent *qualitative* comparisons of uncertain events in terms of the "not more likely" relation conform to intuitive requirements.

If follows from Proposition 2.4 that, as one would expect, for any event E, $\emptyset \leq E \leq \Omega$. We shall mostly work, however, with "significant" events, for which this ordering is strict.

Definition 2.5. **(*Significant events*).** *An event E is significant given $G > \emptyset$ if $c_1 <_G c_2$ implies that $c_1 <_G \{c_2 \,|\, E, c_1 \,|\, E^c\} <_G c_2$. If $G = \Omega$, we shall simply say that E is significant.*

Intuitively, significant events given G are those operationally perceived by the decision-maker as "practically possible but not certain" given the information provided by G. Thus, given $G > \emptyset$ and assuming $c_1 <_G c_2$, if E is judged to be significant given G, one would strictly prefer the option $\{c_2 \,|\, E, c_1 \,|\, E^c\}$ to c_1 for sure, since it provides an additional perceived possibility of obtaining the more desirable consequence c_2. Similarly, one would strictly prefer c_2 for sure to the stated option.

Proposition 2.5. **(*Characterisation of significant events*).** *An event E is significant given $G > \emptyset$, if and only if $\emptyset < E \cap G < G$. In particular, E is significant if and only if $\emptyset < E < \Omega$.*

Proof. Using Definitions 2.4 and 2.5, if E is significant given G then, for all $c_1 \leq_G c_2$ and for any option a,

$$\{c_1 \,|\, G, a \,|\, G^c\} < \{c_2 \,|\, E \cap G, c_1 \,|\, E^c \cap G, a \,|\, G^c\} < \{c_2 \,|\, G, a \,|\, G^c\}.$$

Taking $a = c_1$, we have

$$c_1 = \{c_2 \,|\, \emptyset, c_1 \,|\, \Omega\} < \{c_2 \,|\, E \cap G, c_1 \,|\, (E \cap G)^c\} < \{c_2 \,|\, G, c_1 \,|\, G^c\}$$

and hence, by Definition 2.3, $\emptyset < E \cap G < G$. Conversely, if $\emptyset < E \cap G < G$,

$$\{c_1 \,|\, G, c_1 \,|\, G^c\} < \{c_2 \,|\, E \cap G, c_1 \,|\, E^c \cap G, c_1 \,|\, G^c\} < \{c_2 \,|\, G, c_1 \,|\, G^c\}$$

and hence, by Axiom 3(iii), $c_1 <_G \{c_2 \,|\, E, c_1 \,|\, E^c\} <_G c_2$. If, in particular, $G = \Omega$ then E is significant if and only if $\emptyset < E < \Omega$. $\triangleleft$

The operational essence of "learning from experience" is that a decision-maker's preferences may change in passing from one state of information to a new state brought about by the acquisition of further information regarding the occurrence of events in $\mathcal{E}$, which leads to changes in assessments of uncertainty. There are, however, too many complex ways in which such changes in assessments can take place for us to be able to capture the idea in a simple form. On the other hand, the very special case in which preferences do *not* change is easy to describe in terms of the concepts thus far available to us.

Definition 2.6. ***(Pairwise independence of events)***.
We say that E *and* F *are (pairwise) independent, denoted by* $E \perp F$, *if, and only if, for all* c, c_1, c_2

(i) $c \bullet \{c_2 \,|\, E, c_1 \,|\, E^c\} \Rightarrow c \bullet_F \{c_2 \,|\, E, c_1 \,|\, E^c\}$,

(ii) $c \bullet \{c_2 \,|\, F, c_1 \,|\, F^c\} \Rightarrow c \bullet_E \{c_2 \,|\, F, c_1 \,|\, F^c\}$,

where $\bullet$ *is any one of the relations* $<$, $\sim$ *or* $>$.

The definition is given for the simple situation of preferences between pure consequences and dichotomised options. Since by Proposition 2.3 preferences regarding pure consequences are unaffected by additional information, the condition stated captures, in an operational form, the notion that uncertainty judgements about E, say, are unaffected by the additional information F. We interpret $E \perp F$ as "E is independent of F". An alternative characterisation will be given in Proposition 2.13.

2.3.3 Quantification

The notion of preference between options, formalised by the binary relation $\leq$, provides a *qualitative* basis for comparing options and, by extension, for comparing consequences and events. The *coherence axioms* (Axioms 1 to 3) then provide a minimal set of rules to ensure that qualitative comparisons based on $\leq$ cannot have intuitively undesirable implications.

We shall now argue that this purely qualitative framework is inadequate for serious, systematic comparisons of options. An illuminating analogy can be drawn between $\leq$ and a number of qualitative relations in common use both in an everyday setting and in the physical sciences.

Consider, for example, the relations *not heavier than, not longer than, not hotter than.* It is abundantly clear that these cannot suffice, as they stand, as an adequate basis for the physical sciences. Instead, we need to introduce in each case some form of *quantification* by setting up a *standard unit of measurement*, such as the *kilogram*, the *metre*, or the *centigrade interval*, together with an (implicitly) continuous scale such as *arbitrary decimal fractions* of a kilogram, a metre, a centigrade interval. This enables us to assign a *numerical value*, representing *weight*, *length*, or *temperature*, to any given physical or chemical entity.

This can be achieved by carrying out, implicitly or explicitly, a series of qualitative pairwise comparisons of the feature of interest with appropriately chosen points on the standard scale. For example, in quantifying the length of a stick, we place one end against the origin of a metre scale and then use a series of qualitative comparisons, based on "not longer than" (and derived relations, such as "strictly longer than"). If the stick is "not longer than" the scale mark of 2.5 metres, but is "strictly longer than" the scale mark of 2.4 metres, we might lazily report that the stick is "2.45 metres long". If we needed to, we could continue to make qualitative comparisons of this kind with finer subdivisions of the scale, thus extending the number of decimal places in our answer. The example is, of course, a trivial one, but the general point is extremely important. *Precision, through quantification, is achieved by introducing some form of numerical standard into a context already equipped with a coherent qualitative ordering relation.*

We shall regard it as essential to be able to *aspire* to some kind of quantitative precision in the context of comparing options. It is therefore necessary that we have available some form of *standard options*, whose definitions have close links with an easily understood numerical scale, and which will play a role analogous to the standard metre or standard kilogram. As a first step towards this, we make the following assumption about the algebra of events, $\mathcal{E}$.

Axiom 4. (***Existence of standard events***). *There exists a subalgebra $\mathcal{S}$ of $\mathcal{E}$ and a function $\mu : \mathcal{S} \to [0,1]$ such that:*

(i) $S_1 \leq S_2$ *if, and only if,* $\mu(S_1) \leq \mu(S_2)$;

(ii) $S_1 \cap S_2 = \emptyset$ *implies that* $\mu(S_1 \cup S_2) = \mu(S_1) + \mu(S_2)$;

(iii) *for any number α in $[0,1]$, and events E, F, there is a standard event S such that $\mu(S) = \alpha$, $E \perp S$ and $F \perp S$;*

(iv) $S_1 \perp S_2$ *implies that* $\mu(S_1 \cap S_2) = \mu(S_1)\mu(S_2)$.

(v) *if $E \perp S$, $F \perp S$ and $E \perp F$, then $E \sim S \Rightarrow E \sim_F S$.*

Discussion of Axiom 4. A family of events satisfying conditions (i) and (ii) is easily identified by imagining an idealised roulette wheel of unit circumference. We suppose that no point on the circumference is "favoured" as a resting place for the ball (considered as a point) in the sense that given any c_1, c_2 and events S_1, S_2 corresponding to the ball landing within specified connected arcs, or finite unions and intersections of such arcs, $\{c_1 \mid S_1, c_2 \mid S_1^c\}$ and $\{c_1 \mid S_2, c_2 \mid S_2^c\}$ are considered equivalent if and only if $\mu(S_1) = \mu(S_2)$, where μ is the function mapping the "arc-event" to its total length. Conditions (i) and (ii) are then intuitively obvious, as is the fact, in (iii), that for any $\alpha \in [0, 1]$ we can construct an S with $\mu(S) = \alpha$. Note that $\mathcal{S}$ is required to be an algebra and thus both $\emptyset$ and Ω are standard events. It follows from Proposition 2.4 and Axiom 4(i) that $\mu(\emptyset) = 0$ and $\mu(\Omega) = 1$. The remainder of (iii) is intuitively obvious; we note first that the basic idea of an idealised roulette wheel does assume that each "play" on such a wheel is "independent", in the sense of Definition 2.6, of any other events, including previous "plays" on the same wheel. Thus, for any events E, F in $\mathcal{E}$, we can always think of an "independent" play which generates independent events S in $\mathcal{S}$ with $\mu(S) = \alpha$ for any specified α in $[0, 1]$. In this extended setting, if we think of the circumferences for two independent plays as unravelled to form the sides of a unit square, with μ mapping events to the areas they define, condition (iv) is clearly satisfied. Finally, (v) encapsulates an obviously desirable consequence of independence; namely, that if E is independent of F and S, and F is independent of S, a judgement of equivalence between E and S should not be affected by the occurrence of F.

We will refer to $\mathcal{S}$ as a *standard family* of events in $\mathcal{E}$ and will think of $\mathcal{E}$ as the algebra generated by the relevant events in the decision problem together with the elements of $\mathcal{S}$. Other forms of standard family satisfying (i) to (v) are easily imagined. For example, it is obvious that a roulette wheel of unit circumference could be imagined cut at some point and "unravelled" to form a unit interval. The underlying image would then be that of a point landing in the unit interval and an event S such that $\mu(S) = p$ would denote a subinterval of length p; alternatively, we could imagine a point landing in the unit square, with S denoting a region of area p. The obvious intuitive content of conditions (i) to (v) can clearly be similarly motivated in these cases, the discussion for the unit interval being virtually identical to that given for the roulette wheel. It is important to emphasise that we do *not* require the assumption that standard families of events actually, physically exist, or could be *precisely* constructed in accordance with conditions (i) to (v). We only require that we can invoke such a set up as a mental image.

There is, of course, an element of mathematical idealisation involved in thinking about *all* $p \in [0, 1]$, rather than, for example, some subset of the rationals, corresponding to binary expansions consisting of zeros from some specified location onwards, reflecting the inherent limits of accuracy in any actual procedure for determining arc lengths or areas. The same is true, however, of *all* scientific discourse in which measurements are taken, in principle, to be real numbers, rather than a subset of the rationals chosen to reflect the limits of accuracy in the physical measurement procedure being employed. Our argument for accepting this degree of mathematical idealisation in setting up our formal system is the same as would apply in the physical sciences. Namely, that no serious conceptual distortion is introduced, while many irrelevant technical difficulties are avoided; in particular, those concerning the non-closure of a set of numbers with respect to operations of interest. This argument is not universally accepted, however, and further, related discussion of the issue is provided in Section 2.8.

Our view is that, from the perspective of the foundations of decision-making, the step from the finite to the infinite implicit in making use of real numbers is simply a pragmatic convenience, whereas the step from comparing a finite set of possibilities to comparing an infinite set has more substantive implications. We have emphasised this latter point by postponing infinite extensions of the decision framework until Chapter 3.

Proposition 2.6. (***Collections of disjoint standard events***).
For any finite collection $\{\alpha_1, \ldots, \alpha_n\}$ *of real numbers such that* $\alpha_i > 0$ *and* $\alpha_1 + \cdots + \alpha_n \leq 1$ *there exists a corresponding collection* $\{S_1, \ldots, S_n\}$ *of disjoint standard events such that* $\mu(S_i) = \alpha_i$, $i = 1, \ldots, n$.

Proof. By Axiom 4(iii) there exists S_1 such that $\mu(S_1) = \alpha_1$. For $1 < j \leq n$, suppose inductively that $S_1, \ldots, S_{j-1}$ are disjoint, $B_j = S_1 \cup \cdots \cup S_{j-1}$ and define $\beta_j = \alpha_1 + \cdots + \alpha_{j-1} = \mu(B_j)$. By Axiom 4 (iii, iv), there exists T_j in $\mathcal{S}$ such that $\mu(B_j \cap T_j) = \mu(B_j)\{\alpha_j/(1 - \beta_j)\}$. Define $S_j = T_j \cap B_j^c$, so that $S_j \cap S_i = \emptyset$, $i = 1, \ldots, j-1$. Then, $T_j = S_j \cup (T_j \cap B_j)$ and hence, using Axiom 4(ii), $\mu(T_j) = \mu(S_j) + \mu(T_j \cap B_j)$. Thus, $\mu(S_j) = \alpha_j/(1 - \beta_j) - \alpha_j\beta_j/(1 - \beta_j) = \alpha_j$ and the result follows. $\triangleleft$

Axiom 5. (***Precise measurement of preferences and uncertainties***).

(i) *If* $c_1 \leq c \leq c_2$, *there exists a standard event* S *such that* $c \sim \{c_2 \mid S, c_1 \mid S^c\}$.

(ii) *For each event* E, *there exists a standard event* S *such that* $E \sim S$.

Discussion of Axiom 5. In the introduction to this section, we discussed the idea of precision through quantification and pointed out, using analogies with other measurement systems such as weight, length and temperature, that the process is based on successive comparisons with a standard. Let S_q denote a standard event such that $\mu(S_q) = q$. We start with the obvious preferences, $\{c_2 \mid S_0, c_1 \mid S_0^c\} \leq c \leq \{c_2 \mid S_1, c_1 \mid S_1^c\}$, for any $c_1 \leq c \leq c_2$, and then begin to explore comparisons with standard options based on S_x, S_y with $0 < x < y < 1$. In this way, by gradually increasing x away from 0 and decreasing y away from 1, we arrive at comparisons such as $\{c_2 \mid S_x, c_1 \mid S_x^c\} \leq c \leq \{c_2 \mid S_y, c_1 \mid S_y^c\}$, with the difference $y - x$ becoming increasingly small. Intuitively, as we increase x, $\{c_2 \mid S_x, c_1 \mid S_x^c\}$ becomes more and more "attractive" as an option, and as we decrease y, $\{c_2 \mid S_y, c_1 \mid S_y^c\}$ becomes less "attractive". Any given consequence c, such that $c_1 \leq c \leq c_2$, can therefore be "sandwiched" arbitrarily tightly and, in the limit, be judged equivalent to one of the standard options defined in terms of c_1, c_2. The essence of Axiom 5(i) is that we can proceed to a common limit, α, say, approached from below by the successive values of x and above by the successive values of y. The standard family of options is thus assumed to provide a continuous scale against which any consequence can be *precisely* compared.

Condition (ii) extends the idea of precise comparison to include the assumption that, for any event E and for all consequences c_1, c_2 such that $c_1 < c_2$, the option $\{c_2 \mid E, c_1 \mid E^c\}$ can be compared precisely with the family of standard options $\{c_2 \mid S_x, c_1 \mid S_x^c\}, x \in [0, 1]$, defined by c_1 and c_2. The underlying idea is similar to that motivating condition (i). Indeed, given the intuitive content of the relation "not more likely than", we can begin with the obvious ordering $\{c_2 \mid S_0, c_1 \mid S_0^c\} \leq \{c_2 \mid E, c_1 \mid E^c\} \leq \{c_2 \mid S_1, c_1 \mid S_1^c\}$ for any event E, and then consider refinements of this of the form $\{c_2 \mid S_x, c_1 \mid S_x^c\} \leq \{c_2 \mid E, c_1 \mid E^c\} \leq \{c_2 \mid S_y, c_1 \mid S_y^c\}$, with x increasing gradually from 0, y decreasing gradually from 1, and $y - x$ becoming increasingly small, so that, in terms of the ordering of the events, $S_x \leq E \leq S_y$. Again, the essence of the axiom is that this "sandwiching" can be refined arbitrarily closely by an increasing sequence of x's and a decreasing sequence of y's tending to a common limit.

The preceding argument certainly again involves an element of mathematical idealisation. In practice, there might, in fact, be some *interval of indifference*, in the sense that we judge $\{c_2 \mid S_x, c_1 \mid S_x^c\} \leq c \leq \{c_2 \mid S_y, c_1 \mid S_y^c\}$ for some (possibly rational) x and y but feel unable to express a more precise form of preference. This is analogous to the situation where a physical measuring instrument has inherent limits, enabling one to conclude that a reading is in the range 3.126 to 3.135, say, but not permitting a more precise statement. In this case, we would typically report the measurement to be 3.13 and proceed *as if* this were a precise measurement. We formulate the theory on the prescriptive assumption that we aspire to exact measurement (exact comparisons in our case), whilst acknowledging that, in practice, we have to make do with the best level of precision currently available (or devote some resources to improving our measuring instruments!).

In the context of measuring beliefs, several authors have suggested that this imprecision be *formally incorporated into the axiom system*. For many applications, this would seem to be an unnecessary confusion of the *prescriptive* and the *descriptive*. Every physicist or chemist knows that there are inherent limits of accuracy in any given laboratory context but, so far as we know, no one has suggested developing the structures of theoretical physics or chemistry on the assumption that quantities appearing in fundamental equations should be constrained to take values in some subset of the rationals. However, it may well be that there are situations where imprecision in the context of comparing consequences is too basic and problematic a feature to be adequately dealt with by an approach based on theoretical precision, tempered with pragmatically acknowledged approximation. We shall return to this issue in Section 2.8.

The particular standard option to which c is judged equivalent will, of course, depend on c, but we have implicitly assumed that it does *not* depend on any information we might have concerning the occurrence of real-world events. Indeed, Proposition 2.3 implies that our "attitudes" or "values" regarding consequences are *fixed* throughout the analysis of any particular decision problem. It is intuitively obvious that, if the time-scale on which values change were not rather long compared with the time-scale within which individual problems are analysed, there would be little hope for rational analysis of any kind.

2.4 BELIEFS AND PROBABILITIES

2.4.1 Representation of Beliefs

It is clear that an individual's preferences among options in any decision problem should depend, at least in part, on the "degrees of belief" which that individual attaches to the uncertain events forming part of the definitions of the options.

The principles of *coherence* and *quantification* by comparison with a standard, expressed in axiomatic form in the previous section, will enable us to give a formal definition of *degree of belief*, thus providing a numerical measure of the uncertainty attached to each event.

The conceptual basis for this numerical measure will be seen to derive from the formal rules governing quantitative, coherent preferences, irrespective of the nature of the uncertain events under consideration. This is in vivid contrast to what are sometimes called the *classical* and *frequency* approaches to defining numerical measures of uncertainty (see Section 2.8), where the existence of *symmetries* and the possibility of *indefinite replication*, respectively, play fundamental roles in defining the concepts for restricted classes of events.

We cannot emphasise strongly enough the important distinction between *defining a general concept* and *evaluating a particular case*. Our *definition* will depend only on the logical notions of quantitative, coherent preferences; our practical *evaluations* will often make use of perceived symmetries and observed frequencies.

We begin by establishing some basic results concerning the uncertainty relation between events.

Proposition 2.7. (***Complete comparability of events***).
Either $E_1 > E_2$, *or* $E_1 \sim E_2$, *or* $E_2 > E_1$.

Proof. By Axiom 5(ii), there exist S_1 and S_2 such that $E_1 \sim S_1$ and $E_2 \sim S_2$; the complete ordering now follows from Axiom 4(i) and Proposition 2.1. $\triangleleft$

We see from Proposition 2.7 that, although the order relation $\leq$ between options was not assumed to be complete (i.e., not *all* pairs of options were assumed to be comparable), it turns out, as a consequence of Axiom 5 (the axiom of precise measurement), that the uncertainty relation induced between events *is* complete. A similar result concerning the comparability of all options will be established in Section 2.5.

Proposition 2.8. (***Additivity of uncertainty relations***). *If $A \leq B, C \leq D$ and $A \cap C = B \cap D = \emptyset$, then $A \cup C \leq B \cup D$. Moreover, if $A < B$ or $C < D$, then $A \cup C < B \cup D$.*

Proof. We first show that, for any G, if $A \cap G = B \cap G = \emptyset$ then $A \leq B \iff A \cup G \leq B \cup G$. For any $c_2 > c_1$, $A \cap G = B \cap G = \emptyset$, define:

$$\begin{aligned}
a_1 &= \{c_2 \,|\, A, c_1 \,|\, A^c\} = \{c_1 \,|\, G, \{c_2 \,|\, A, c_1 \,|\, A^c\} \,|\, G^c\} \\
a_2 &= \{c_2 \,|\, B, c_1 \,|\, B^c\} = \{c_1 \,|\, G, \{c_2 \,|\, B, c_1 \,|\, B^c\} \,|\, G^c\} \\
a_3 &= \{c_2 \,|\, A \cup G, c_1 \,|\, (A \cup G)^c\} = \{c_2 \,|\, G, \{c_2 \,|\, A, c_1 \,|\, A^c\} \,|\, G^c\} \\
a_4 &= \{c_2 \,|\, B \cup G, c_1 \,|\, (B \cup G)^c\} = \{c_2 \,|\, G, \{c_2 \,|\, B, c_1 \,|\, B^c\} \,|\, G^c\}.
\end{aligned}$$

Then, by Definition 2.3, $A \leq B \iff a_1 \leq a_2$; by Axiom 3, $a_1 \leq a_2 \iff a_3 \leq a_4$; and using again Definition 2.3, $a_3 \leq a_4 \iff A \cup G \leq B \cup G$. Thus,

$$A \cup (C - B) \leq B \cup (C - B) = B \cup C = C \cup (B - C) \leq D \cup (B - C),$$

$$A \cup C = A \cup (C - B) \cup (C \cap B) \leq D \cup (B - C) \cup (C \cap B) = B \cup D.$$

The final statement follows from essentially the same argument. ◁

We now make the key definition which enables us to move to a *quantitative* notion of degree of belief.

Definition 2.7. (***Measure of degree of belief***). *Given an uncertainty relation $\leq$, the **probability** $P(E)$ of an event E is the real number $\mu(S)$ associated with any standard event S such that $E \sim S$.*

This definition provides a natural, operational extension of the qualitative uncertainty relation encapsulated in Definition 2.3, by linking the equivalence of any $E \in \mathcal{E}$ to some $S \in \mathcal{S}$ and exploiting the fact that the nature of the construction of $\mathcal{S}$ provides a direct obvious quantification of the uncertainty regarding S.

With our operational definition, the *meaning* of a probability statement is clear. For instance, the statement $P(E) = 0.5$ precisely means that E is judged to be equally likely as a standard event of 'measure' 0.5, maybe a conceptual perfect coin falling heads, or a computer generated 'random' integer being an odd number.

It should be emphasised that, according to Definition 2.7, probabilities are always *personal degrees of belief*, in that they are a numerical representation of the decision-maker's personal uncertainty relation $\leq$ between events. Moreover, probabilities are always conditional on the information currently available. It makes no sense, within the framework we are discussing, to qualify the word probability with adjectives such as "objective", "correct" or "unconditional".

Since probabilities are obviously conditional on the initial state of information M_0, a more precise and revealing notation in Definition 2.7 would have been $P(E \mid M_0)$. In order to avoid cumbersome notation, we shall stick to the shorter version, but the implicit conditioning on M_0 should always be borne in mind.

Proposition 2.9. (***Existence and uniqueness***). *Given an uncertainty relation $\leq$, there exists a unique probability $P(E)$ associated with each event E.*

Proof. Existence follows from Axiom 5(ii). For uniqueness, if $E \sim S_1$ and $E \sim S_2$ then by Proposition 2.2(ii), $S_1 \sim S_2$. The result now follows from Axiom 4(i). $\triangleleft$

Definition 2.8. (***Compatibility***). *A function $f : \mathcal{E} \to \Re$ is said to be compatible with an order relation $\leq$ on $\mathcal{E} \times \mathcal{E}$ if, for all events,*

$$E \leq F \iff f(E) \leq f(F).$$

Proposition 2.10. (***Compatibility of probability and degrees of belief***). *The probability function $P(.)$ is compatible with the uncertainty relation $\leq$.*

Proof. By Axiom 5(ii) there exist standard events S_1 and S_2 such that $E \sim S_1$ and $F \sim S_2$. Then, by Proposition 2.2(ii) , $E \leq F$ iff $S_1 \leq S_2$ and hence, by Axiom 4(i), iff $\mu(S_1) \leq \mu(S_2)$. The result follows from Definition 2.7. $\triangleleft$

The following proposition is of fundamental importance. It establishes that coherent, quantitative degrees of belief have the structure of a finitely additive probability measure over $\mathcal{E}$. Moreover, it establishes that significant events, i.e., events which are "practically possible but not certain", should be assigned probability values in the *open* interval $(0, 1)$.

Proposition 2.11. (***Probability structure of degrees of belief***).

(i) $P(\emptyset) = 0$ *and* $P(\Omega) = 1$.

(ii) *If* $E \cap F = \emptyset$, *then* $P(E \cup F) = P(E) + P(F)$.

(iii) E *is significant if, and only if,* $0 < P(E) < 1$.

Proof. (i) By Definition 2.7, $0 \leq P(E) \leq 1$. Moreover, by Axiom 4(iii) there exist S_* and S^* such that $\mu(S_*) = 0$ and $\mu(S^*) = 1$. By Proposition 2.4, $\emptyset \leq S_*$ and, by Proposition 2.10 $P(\emptyset) \leq 0$; hence, $P(\emptyset) = 0$; similarly, $S^* \leq \Omega$ implies that $P(\Omega) = 1$.

(ii) If $E = \emptyset$ or $F = \emptyset$, or both, the result is trivially true. If $E > \emptyset$ and $F > \emptyset$, then, by Proposition 2.8, $E \cup F > E$; thus, if $\alpha = P(E)$ and $\beta = P(E \cup F)$, we have $\alpha < \beta$ and, by Proposition 2.6, there exist events S_1, S_2 such that $S_1 \cap S_2 = \emptyset$, $P(S_1) = \alpha$ and $P(S_2) = \beta - \alpha$. By Proposition 2.7, $F > S_2$ or $F \sim S_2$ or $F < S_2$. If $F > S_2$, then, by Proposition 2.8, $E \cup F > S_1 \cup S_2$ and hence $P(E \cup F) > \beta$, which is impossible; similarly, if $F < S_2$ then $E \cup F < S_1 \cup S_2$ and $P(E \cup F) < \beta$ which, again, is impossible. Hence, $F \sim S_2$ and therefore $P(F) = \beta - \alpha$, so that $P(E \cup F) = P(E) + P(F)$, as stated.

(iii) By Proposition 2.5, E is significant iff $\emptyset < E < \Omega$. The result then follows immediately from Proposition 2.10. ◁

Corollary. ***(Finitely additive structure of degrees of belief).***

(i) *If $\{E_j, j \in J\}$ is a finite collection of disjoint events, then*

$$P\left(\bigcup_{j \in J} E_j\right) = \sum_{j \in J} P(E_j).$$

(ii) *For any event E, $P(E^c) = 1 - P(E)$.*

Proof. The first part follows by induction from Proposition 2.11(iii); the second part is a special case of (i) since if $\cup_j E_j = \Omega$ then, by Proposition 2.11(i), $\Sigma_j P(E_j) = 1$. ◁

Proposition 2.11 is crucial. It establishes formally that coherent, quantitative measures of uncertainty about events must take the form of probabilities, therefore justifying the nomenclature adopted in Definition 2.6 for this measure of degree of belief. In short, *coherent degrees of belief are probabilities.*

It will often be convenient for us to use probability terminology, without explicit reference to the fact that the mathematical structure is merely serving as a representation of (personal) degrees of belief. The latter fact should, however, be constantly borne in mind.

Definition 2.9. ***(Probability distribution).*** *If $\{E_j, j \in J\}$ form a finite partition of Ω, with $P(E_j) = p_j, j \in J$, then $\{p_j, j \in J\}$ is said to be a probability distribution over the partition.*

This terminology will prove useful in later discussions. The idea is that total belief (in Ω, having measure 1) is *distributed* among the events of the partition, $\{E_j, j \in J\}$, according to the relative degrees of belief $\{p_j, j \in J\}$, with $\Sigma_j p_j = \Sigma_j P(E_j) = 1$.

Starting from the qualitative ordering among events, we have derived a quantitative measure, $P(.) \equiv P(. \mid M_0)$, over $\mathcal{E}$ and shown that, expressed in conventional mathematical terminology, it has the form of a *finitely additive probability measure*, compatible with the qualitative ordering $\leq$. We now establish that this is the only probability measure over $\mathcal{E}$ compatible with $\leq$.

Proposition 2.12. (***Uniqueness of the probability measure***). *P is the only probability measure compatible with the uncertainty relation $\leq$.*

Proof. If P' were another compatible measure, then by Proposition 2.8 we would always have $P'(E) \leq P'(F) \iff P(E) \leq P(F)$; hence, there exists a monotonic function f of $[0,1]$ into itself such that $P'(E) = f\{P(E)\}$. By Proposition 2.6, for all non-negative α, β such that $\alpha + \beta \leq 1$, there exist disjoint standard events S_1 and S_2, such that $P(S_1) = \alpha$ and $P(S_2) = \beta$. Hence, by Axiom 4(ii), $f(\alpha + \beta) = P'(S_1 \cup S_2) = P'(S_1) + P'(S_2) = f(\alpha) + f(\beta)$ and so (Eichhorn, 1978, Theorem 2.63), $f(\alpha) = k\alpha$ for all α in $[0,1]$. But, by Proposition 2.9, $P'(\Omega) = 1$ and hence, $k = 1$, so that we have $P'(E) = P(E)$ for all E. $\triangleleft$

We shall now establish that our operational definition of (pairwise) independence of events is compatible with its more standard, *ad hoc*, product definition.

Proposition 2.13. (***Characterisation of independence***).

$$E \perp F \iff P(E \cap F) = P(E)P(F).$$

Proof. Suppose $E \perp F$. By Axiom 4(iii), there exists S_1 such that $P(S_1) = P(E)$, $E \perp S_1$ and $F \perp S_1$. Hence, by Axiom 4(v), $E \sim_F S_1$, so that, for any consequences $c_1 < c_2$, and any option a,

$$\{c_2 \mid E \cap F, c_1 \mid E^c \cap F, a \mid F^c\} \sim \{c_2 \mid S_1 \cap F, c_1 \mid S_1^c \cap F, a \mid F^c\}.$$

Taking $a = c_1$, we have

$$\{c_2 \mid E \cap F, c_1 \mid (E \cap F)^c\} \sim \{c_2 \mid S_1 \cap F, c_1 \mid (S_1 \cap F)^c\},$$

so that $E \cap F \sim S_1 \cap F$. Again by Axiom 4(iii), given F, S_1, there exists S_2 such that $P(S_2) = P(F)$, $F \perp S_2$ and $S_1 \perp S_2$. Hence, by an identical argument to the above, and noting from Definition 2.6 the symmetry of $\perp$, we have

$$S_1 \cap F \sim S_1 \cap S_2.$$

By Propositions 2.1, 2.10, and Axiom 4(iv),

$$P(E \cap F) = P(S_1 \cap S_2) = P(S_1)P(S_2),$$

and hence $P(E \cap F) = P(E)P(F)$.

Suppose $P(E \cap F) = P(E)P(F)$. By Axiom 4(iii), there exists S such that $P(S) = P(F)$ and $F \perp S$, $E \perp S$. Hence, by the first part of the proof,

$$P(E \cap S) = P(E)P(S) = P(E)P(F) = P(E \cap F),$$

so that $E \cap F \sim E \cap S$. Now suppose, without loss of generality, that $c \leq \{c_2 \,|\, E, c_1 \,|\, E^c\}$. Then, by Definition 2.6,

$$\{c \,|\, S, c_1 \,|\, S^c\} \leq \{c_2 \,|\, E \cap S, c_1 \,|\, (E \cap S)^c\}.$$

But $\{c \,|\, S, c_1 \,|\, S^c\} \sim \{c \,|\, F, c_1 \,|\, F^c\}$ and

$$\{c_2 \,|\, E \cap S, c_1 \,|\, (E \cap S)^c\} \sim \{c_2 \,|\, E \cap F, c_1 \,|\, (E \cap F)^c\};$$

hence by Proposition 2.2,

$$\{c \,|\, F, c_1 \,|\, F^c\} \leq \{c_2 \,|\, E \cap F, c_1 \,|\, (E \cap F)^c\},$$

so that $c \leq_F \{c_2 \,|\, E, c_1 \,|\, E^c\}$. A similar argument can obviously be given reversing the roles of E and F, hence establishing that $E \perp F$. $\triangleleft$

2.4.2 Revision of Beliefs and Bayes' Theorem

The assumed occurrence of a real-world event will typically modify preferences between options by modifying the degrees of belief attached, by an individual, to the events defining the options. In this section, we use the assumptions of Section 2.3 in order to identify the precise way in which coherent modification of initial beliefs should proceed.

The starting point for analysing order relations between events, given the assumed occurrence of a possible event G, is the uncertainty relation $\leq_G$ defined between events. Given the assumed occurrence of $G > \emptyset$, the ordering $\leq$ between acts is replaced by $\leq_G$. Analogues of Propositions 2.1 and 2.2 are trivially established and we recall (Proposition 2.3) that, for any $G > \emptyset$, $c_2 \leq c_1$ iff $c_2 \leq_G c_1$.

Proposition 2.14. ***(Properties of conditional beliefs).***

(i) $E \leq_G F \iff E \cap G \leq F \cap G$.

(ii) *If there exist* $c_1 < c_2$ *such that* $\{c_2 \,|\, E, c_1 \,|\, E^c\} \leq_G \{c_2 \,|\, F, c_1 \,|\, F^c\}$, *then* $E \leq_G F$.

Proof. By Definition 2.4 and Proposition 2.3, $E \leq_G F$ iff, for all $c_2 \geq c_1$,

$$\{c_2 \,|\, E, c_1 \,|\, E^c\} \leq_G \{c_2 \,|\, F, c_1 \,|\, F^c\},$$

i.e., if, and only if, for all a,

$$\{c_2 \,|\, E \cap G, c_1 \,|\, E^c \cap G, a \,|\, G^c\} \leq \{c_2 \,|\, F \cap G, c_1 \,|\, F^c \cap G, a \,|\, G^c\}.$$

Taking $a = c_1$,

$$E \leq_G F \iff \{c_2 \,|\, E \cap G, c_1 \,|\, (E \cap G)^c\} \leq \{c_2 \,|\, F \cap G, c_1 \,|\, (F \cap G)^c\},$$

and this is true iff $E \cap G \leq F \cap G$.

Moreover, if there exist $c_2 > c_1$ such that $\{c_2 \,|\, E, c_1 \,|\, E^c\} \leq_G \{c_2 \,|\, F, c_1 \,|\, F^c\}$ then, by Definition 2.4, with $a = c_1$,

$$\{c_2 \,|\, E \cap G, c_1 \,|\, E^c \cap G, c_1 \,|\, G^c\} \leq \{c_2 \,|\, F \cap G, c_1 \,|\, F^c \cap G, c_1 \,|\, G^c\},$$

so that

$$\{c_2 \,|\, E \cap G, c_1 \,|\, (E \cap G)^c\} \leq \{c_2 \,|\, F \cap G, c_1 \,|\, (F \cap G)^c\}$$

and the result follows from Axiom 3(ii) and part (i) of this proposition. ◁

Definition 2.10. ***(Conditional measure of degree of belief).*** *Given a conditional uncertainty relation $\leq_G, G > \emptyset$, the conditional probability $P(E \,|\, G)$ of an event E given the assumed occurrence of G is the real number $\mu(S)$ such that $E \sim_G S$, where S is an standard event independent of G.*

Generalising the idea encapsulated in Definition 2.7, $P(E \,|\, G)$ provides a quantitative operational measure of the uncertainty attached to E given the assumed occurrence of the event G. The following fundamental result provides the key to the process of revising beliefs in a coherent manner in the light of new information. It relates the *conditional* measure of degree of belief $P(. \,|\, G)$ to the *initial* measure of degree of belief $P(.)$.

We have, of course, already stressed that *all* degrees of belief are conditional. The intention of the terminology used above is to emphasise the additional conditioning resulting from the occurrence of G; the initial state of information, M_0, is always present as a conditioning factor, although omitted throughout for notational convenience.

Proposition 2.15. ***(Conditional probability).*** *For any $G > \emptyset$,*

$$P(E \,|\, G) = \frac{P(E \cap G)}{P(G)}\,.$$

Proof. By Axiom 4(iii) and Proposition 2.13, there exists $S \bot G$ such that $\mu(S) = P(E \cap G)/P(G)$. By Proposition 2.13,

$$P(S \cap G) = P(S)P(G) = \mu(S)P(G) = P(E \cap G).$$

Thus, by Proposition 2.10, $S \cap G \sim E \cap G$ and, by Proposition 2.14, $S \sim_G E$. Thus, by Definition 2.10, $P(E \,|\, G) = \mu(S) = P(E \cap G)/P(G)$. ◁

Note that, in our formulation, $P(E \mid G) = P(E \cap G)/P(G)$ is a logical derivation from the axioms, *not* an *ad hoc* definition. In fact, this is the simplest version of *Bayes' theorem*. An extended form is given later in Proposition 2.19.

Proposition 2.16. (***Compatibility of conditional probability and conditional degrees of belief***).

$$E \leq_G F \iff P(E \mid G) \leq P(F \mid G).$$

Proof. By Proposition 2.14(i), $E \leq_G F$ iff $E \cap G \leq F \cap G$, which, by Proposition 2.10, holds if and only if $P(E \cap G) \leq P(F \cap G)$; the result now follows from Proposition 2.15. ◁

We now extend Proposition 2.11 to degrees of belief conditional on the occurrence of significant events.

Proposition 2.17. (***Probability structure of conditional degrees of belief***).
For any event $G > \emptyset$,

(i) $P(\emptyset \mid G) = 0 \leq P(E \mid G) \leq P(\Omega \mid G) = 1;$
(ii) *if* $E \cap F \cap G = \emptyset$, *then* $P(E \cup F \mid G) = P(E \mid G) + P(F \mid G);$
(iii) E *is significant given* $G \iff 0 < P(E \mid G) < 1.$

Proof. By Proposition 2.15, $P(E \mid G) \geq 0$ and $P(\emptyset \mid G) = 0$; moreover, since $E \cap G \leq G$, Proposition 2.10 implies that $P(E \cap G) \leq P(G)$, so that, by Proposition 2.15, $P(E \mid G) \leq 1$. Finally, $\Omega \cap G = G$, so that, using again Proposition 2.15 , $P(\Omega \mid G) = 1$.

By Proposition 2.15,

$$P(E \cup F \mid G) = \frac{P((E \cap G) \cup (F \cap G))}{P(G)}$$

$$= \frac{P(E \cap G)}{P(G)} + \frac{P(F \cap G)}{P(G)} = P(E \mid G) + P(F \mid G).$$

Finally, by Proposition 2.5, E is significant given G iff $\emptyset < E \cap G < G$. Thus, by Proposition 2.10, E is significant given G iff $0 < P(E \cap G) < P(G)$. The result follows from Proposition 2.15. ◁

Corollary. (***Finitely additive structure of conditional degrees of belief***).
For all $G > \emptyset$,

(i) *if* $\{E_j \cap G, j \in J\}$ *is a finite collection of disjoint events, then*

$$P\left(\bigcup_{j \in J} E_j \,\middle|\, G\right) = \sum_{j \in J} P(E_j \mid G);$$

(ii) *for any event* E, $P(E^c \mid G) = 1 - P(E \mid G)$.

Proof. This parallels the proof of the Corollary to Proposition 2.11. ◁

Proposition 2.18. ***(Uniqueness of the conditional probability measure).*** *$P(.\,|\,G)$ is the only probability measure compatible with the conditional uncertainty relation $\leq_G$.*

Proof. This parallels the proof of Proposition 2.12. ◁

Example 2.1. ***(Simpson's paradox).*** The following example provides an instructive illustration of the way in which the formalism of conditional probabilities provides a coherent resolution of an otherwise seemingly paradoxical situation.

Suppose that the results of a clinical trial involving 800 sick patients are as shown in Table 2.1, where T, T^c denote, respectively, that patients did or did not receive a certain treatment, and R, R^c denote, respectively, that the patients did or did not recover.

Table 2.1 *Trial results for all patients*

	R	R^c	*Total*	*Recovery rate*
T	200	200	400	50%
T^c	160	240	400	40%

Intuitively, it seems clear that the treatment is beneficial, and were one to base probability judgements on these reported figures, it would seem reasonable to specify

$$P(R\,|\,T) = 0.5, \qquad P(R\,|\,T^c) = 0.4,$$

where recovery and the receipt of treatment by individuals are now represented, in an obvious notation, as events. Suppose now, however, that one became aware of the trial outcomes for male and female patients separately, and that these have the summary forms described in Tables 2.2 and 2.3.

Table 2.2 *Trial results for male patients*

	R	R^c	*Total*	*Recovery rate*
T	180	120	300	60%
T^c	70	30	100	70%

The results surely seem paradoxical. Tables 2.2 and 2.3 tell us that the treatment is neither beneficial for males nor for females; but Table 2.1 tells us that overall it is beneficial! How are we to come to a coherent view in the light of this apparently conflicting evidence?

Table 2.3 *Trial results for female patients*

	R	R^c	*Total*	*Recovery rate*
T	20	80	100	20%
T^c	90	210	300	30%

The seeming paradox is easily resolved by an appeal to the logic of probability which, after all, we have just demonstrated to be the prerequisite for the coherent treatment of uncertainty. With M, M^c denoting, respectively, the events that a patient is either male or female, were one to base probability judgements on the figures reported in Tables 2.2 and 2.3, it would seem reasonable to specify

$$P(R \mid M \cap T) = 0.6, \qquad P(R \mid M \cap T^c) = 0.7,$$

$$P(R \mid M^c \cap T) = 0.2, \qquad P(R \mid M^c \cap T^c) = 0.3.$$

To see that these judgements do indeed cohere with those based on Table 2.1, we note, from the Corollary to Proposition 2.11, Proposition 2.15 and the Corollary to Proposition 2.17, that

$$P(R \mid T) = P(R \mid M \cap T)P(M \mid T) + P(R \mid M^c \cap T)P(M^c \mid T)$$
$$P(R \mid T^c) = P(R \mid M \cap T^c)P(M \mid T^c) + P(R \mid M^c \cap T^c)P(M^c \mid T^c),$$

where

$$P(M \mid T) = 0.75, \qquad P(M \mid T^c) = 0.25.$$

The probability formalism reveals that the seeming paradox has arisen from the confounding of sex with treatment as a consequence of the unbalanced trial design. See Simpson (1951), Blyth (1972, 1973) and Lindley and Novick (1981) for further discussion.

Proposition 2.19. ***(Bayes' theorem).***
For any finite partition $\{E_j, j \in J\}$ *of* Ω *and* $G > \emptyset$,

$$P(E_i \mid G) = \frac{P(G \mid E_i)P(E_i)}{\sum_{j \in J} P(G \mid E_j)P(E_j)}.$$

Proof. By Proposition 2.15,

$$P(E_i \mid G) = \frac{P(E_i \cap G)}{P(G)} = \frac{P(G \mid E_i)P(E_i)}{P(G)}.$$

The result now follows from the Corollary to Proposition 2.11 when applied to $G = \cup_j (G \cap E_j)$. $\triangleleft$

Bayes' theorem is a simple mathematical consequence of the fact that quantitative coherence implies that degrees of belief should obey the rules of probability. From another point of view, it may also be established (Zellner, 1988b) that, under some reasonable desiderata, Bayes' theorem is an optimal information processing system.

Since the $\{E_j, j \in J\}$ form a partition and hence, by the Corollary to Proposition 2.17, $\sum_j P(E_j \mid G) = 1$, Bayes' theorem may be written in the form

$$P(E_j \mid G) \propto P(G \mid E_j)P(E_j), \quad j \in J,$$

since the missing proportionality constant is $[P(G)]^{-1} = [\Sigma_j P(G \mid E_j)P(E_j)]^{-1}$, and thus it is always possible to normalise the products by dividing by their sum. This form of the theorem is often very useful in applications.

Bayes' theorem acquires a particular significance in the case where the uncertain events $\{E_j, j \in J\}$ correspond to an exclusive and exhaustive set of *hypotheses* about some aspect of the world (for example, in a medical context, the set of possible diseases from which a patient may be suffering) and the event G corresponds to a relevant piece of *evidence*, or *data* (for example, the outcome of a clinical test). If we adopt the more suggestive notation, $E_j = H_j, j \in J, G = D$, and, as usual, we omit explicit notational reference to the initial state of information M_0, Proposition 2.17 leads to Bayes' theorem in the form $P(H_j \mid D) = P(D \mid H_j)P(H_j)/P(D), j \in J$, where $P(D) = \Sigma_j P(D \mid H_j)P(H_j)$, characterizing the way in which initial beliefs about the hypotheses, $P(H_j)$, $j \in J$, are modified by the data, D, into a revised set of beliefs, $P(H_j \mid D)$, $j \in J$. This process is seen to depend crucially on the specification of the quantities $P(D \mid H_j)$, $j \in J$, which reflect how beliefs about obtaining the given data, D, vary over the different underlying hypotheses, thus defining the "relative likelihoods" of the latter. The four elements, $P(H_j)$, $P(D \mid H_j)$, $P(H_j \mid D)$ and $P(D)$, occur, in various guises, throughout Bayesian statistics and it is convenient to have a standard terminology available.

Definition 2.11. (***Prior, posterior, and predictive probabilities***).
If $\{H_j, j \in J\}$ are exclusive and exhaustive events (hypotheses), then for any event (data) D,

(i) *$P(H_j)$, $j \in J$, are called the* ***prior probabilities*** *of the H_j, $j \in J$;*

(ii) *$P(D \mid H_j)$, $j \in J$, are called the* ***likelihoods*** *of the H_j, $j \in J$, given D;*

(iii) *$P(H_j \mid D)$, $j \in J$, are called the* ***posterior probabilities*** *of the H_j, $j \in J$;*

(iv) *$P(D)$ is called the* ***predictive probability*** *of D implied by the likelihoods and the prior probabilities.*

It is important to realise that the terms "prior" and "posterior" only have significance *given* an initial state of information and *relative* to an additional piece of information. Thus, $P(H_j)$, which could be more properly be written as $P(H_j \mid M_0)$, represents beliefs prior to conditioning on data D, but posterior to conditioning on whatever history led to the state of information described by M_0. Similarly, $P(H_j \mid D)$, or, more properly, $P(H_j \mid M_0 \cap D)$, represents beliefs posterior to conditioning on M_0 and D, but prior to conditioning on any further data which may be obtained subsequent to D.

The predictive probability $P(D)$, logically implied by the likelihoods and the prior probabilities, provides a basis for assessing the compatibility of the data D with our beliefs (see Box, 1980). We shall consider this in more detail in Chapter 6.

Example 2.2. ***(Medical diagnosis)***. In simple problems of medical diagnosis, Bayes' theorem often provides a particularly illuminating form of analysis of the various uncertainties involved. For simplicity, let us consider the situation where a patient may be characterised as belonging either to state H_1, or to state H_2, representing the presence or absence, respectively, of a specified disease. Let us further suppose that $P(H_1)$ represents *the prevalence rate* of the disease in the population to which the patient is assumed to belong, and that further information is available in the form of the result of a single clinical test, whose outcome is either positive (suggesting the presence of the disease and denoted by $D = T$), or negative (suggesting the absence of the disease and denoted by $D = T^c$).

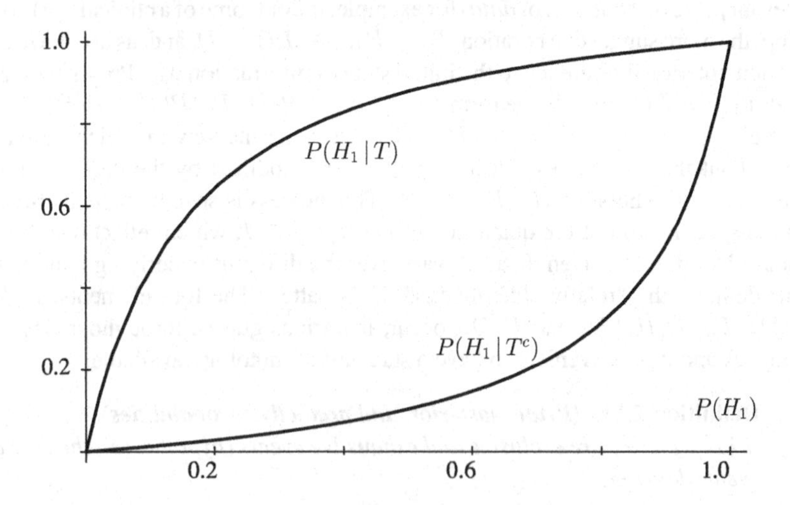

Figure 2.2 *$P(H_1 \mid T)$ and $P(H_1 \mid T^c)$ as functions of $P(H_1)$*

The quantities $P(T \mid H_1)$ and $P(T^c \mid H_2)$ represent the *true positive* and *true negative rates* of the clinical test (often referred to as the *test sensitivity* and *test specificity*, respectively) and the systematic use of Bayes' theorem then enables us to understand the manner in which these characteristics of the test combine with the prevalence rate to produce varying degrees of diagnostic discriminatory power. In particular, for a given clinical test of known sensitivity and specificity, we can investigate the range of underlying prevalence rates for which the test has worthwhile diagnostic value.

As an illustration of this process, let us consider the assessment of the diagnostic value of stress thallium-201 scintigraphy, a technique involving analysis of Gamma camera image data as an indicator of coronary heart disease. On the basis of a controlled experimental study, Murray *et al.* (1981) concluded that $P(T \mid H_1) = 0.900$, $P(T^c \mid H_2) = 0.875$ were reasonable orders of magnitude for the sensitivity and specificity of the test.

Insight into the diagnostic value of the test can be obtained by plotting values of $P(H_1 \mid T)$, $P(H_1 \mid T^c)$ against $P(H_1)$, where

$$P(H_1 \mid D) = \frac{P(D \mid H_1)P(H_1)}{P(D \mid H_1)P(H_1) + P(D \mid H_2)P(H_2)} ,$$

for $D = T$ or $D = T^c$, as shown in Figure 2.2.

As a single, overall measure of the discriminatory power of the test, one may consider the difference $P(H_1 \mid T) - P(H_1 \mid T^c)$. In cases where $P(H_1)$ has very low or very high values (e.g. for large population screening or following individual patient referral on the basis of suspected coronary disease, respectively), there is limited diagnostic value in the test. However, in clinical situations where there is considerable uncertainty about the presence of coronary heart disease, for example, $0.25 \leq P(H_1) \leq 0.75$, the test may be expected to provide valuable diagnostic information.

One further point about the terms prior and posterior is worth emphasising. *They are not necessarily to be interpreted in a chronological sense*, with the assumption that "prior" beliefs are specified first and then later modified into "posterior" beliefs. Propositions 2.15 and 2.17 do not involve any such chronological notions. They merely indicate that, *for coherence*, specifications of degrees of belief must satisfy the given relationships. Thus, for example, in Proposition 2.15 one might first specify $P(G)$ and $P(E \mid G)$ and then use the relationship stated in the theorem to arrive at coherent specification of $P(E \cap G)$. In any given situation, the particular order in which we specify degrees of belief and check their coherence is a pragmatic one; thus, some assessments seem straightforward and we feel comfortable in making them directly, while we are less sure about other assessments and need to approach them indirectly via the relationships implied by coherence. It is true that the natural order of assessment does coincide with the "chronological" order in a number of practical applications, but it is important to realise that this is a pragmatic issue and not a requirement of the theory.

2.4.3 Conditional Independence

An important special case of Proposition 2.15 arises when E and G are such that $P(E \mid G) = P(E)$, so that beliefs about E are *unchanged* by the assumed occurrence of G. Not surprisingly, this is directly related to our earlier operational definition of (pairwise) independence.

Proposition 2.20. *For all* $F > \emptyset$, $E \perp F \iff P(E \mid F) = P(E)$.

Proof. $E \perp F \iff P(E \cap F) = P(E)P(F)$ and, by Proposition 2.15, we have $P(E \cap F) = P(E \mid F)P(F)$. $\triangleleft$

In the case of three events, E, F and G, the situation is somewhat more complicated in that, from an intuitive point of view, we would regard our degree of belief for E as being "independent" of knowledge of F and G if and only if $P(E \mid H) = P(E)$, for any of the four possible forms of H,

$$\{F \cap G,\ F^c \cap G,\ F \cap G^c,\ F^c \cap G^c\},$$

describing the combined occurrences, or otherwise, of F and G (and, of course, similar conditions must hold for the "independence" of F from E and G, and of G from E and F). These considerations motivate the following formal definition, which generalises Definition 2.6 and can be shown (see e.g. Feller, 1950/1968, pp. 125–128) to be necessary and sufficient for encapsulating, in the general case, the intuitive conditions discussed above.

Definition 2.12. (*Mutual independence*).
Events $\{E_j, j \in J\}$ are said to be mutually independent if, for any $I \subseteq J$,

$$P\left(\bigcap_{i \in I} E_i\right) = \prod_{i \in I} P(E_i).$$

An important consequence of the fact that coherent degrees of belief combine in conformity with the rules of (finitely additive) mathematical probability theory is that the task of specifying degrees of belief for complex combinations of events is often greatly simplified. Instead of being forced into a direct specification, we can attempt to represent the complex event in terms of simpler events, for which we feel more comfortable in specifying degrees of belief. The latter are then recombined, using the probability rules, to obtain the desired specification for the complex event. Definition 2.12 makes clear that the judgement of independence for a collection of events leads to considerable additional simplification when complex intersections of events are to be considered. Note that Proposition 2.20 derives from the uncertainty relation $\leq_F$ and therefore reflects an inherently personal judgement (although coherence may rule out some events from being judged independent: for example, any E, F such that $\emptyset \subset E \subseteq F \subset \Omega$).

There is a sense, however, in which the judgement of independence (given M_0) for large classes of events of interest reflects a rather extreme form of belief, in that scope for learning from experience is very much reduced. This motivates consideration of the following weaker form of independence judgement.

Definition 2.13. (***Conditional independence***). *The events $\{E_j, j \in J\}$ are said to be conditionally independent given $G > \emptyset$ if, for any $I \subseteq J$,*

$$P\left(\bigcap_{i \in I} E_i \,|\, G\right) = \prod_{i \in I} P(E_i \,|\, G).$$

For any subalgebra $\mathcal{F}$ of $\mathcal{E}$, the events $\{E_j, j \in J\}$ are said to be conditionally independent given $\mathcal{F}$ if and only if they are conditionally independent given any $G > \emptyset$ in $\mathcal{F}$.

Definitions 2.12 and 2.13 could, of course, have been stated in primitive terms of choices among options, as in Definition 2.6. However, having seen in detail the way in which the latter leads to the standard "product definition", it will be clear that a similar equivalence holds in these more general cases, but that the algebraic manipulations involved are somewhat more tedious.

The form of degree of belief judgement encapsulated in Definition 2.13 is one which is utilised in some way or another in a wide variety of practical contexts and statements of scientific theories. Indeed, a detailed discussion of the kinds of circumstances in which it may be reasonable to structure beliefs on the basis of such judgements will be a main topic of Chapter 4. Thus, for example, in the practical context of sampling, with or without replacement, from large dichotomised populations (of voters, manufactured items, or whatever), successive outcomes (voting intention, marketable quality, ...) may very often be judged independent, *given exact knowledge of the proportional split in the dichotomised population.* Similarly, in simple Mendelian theory, the genotypes of successive offspring are typically judged to be independent events, *given the knowledge of the two genotypes forming the mating*. In the absence of such knowledge, however, in neither case would the judgement of independence for successive outcomes be intuitively plausible, since earlier outcomes provide information about the unknown population or mating composition and this, in turn, influences judgements about subsequent outcomes. For a detailed analysis of the concept of conditional independence, see Dawid (1979a, 1979b, 1980b).

2.4.4 Sequential Revision of Beliefs

Bayes' theorem characterises the way in which current beliefs about a set of mutually exclusive and exhaustive hypotheses, H_j, $j \in J$, are revised in the light of new data, D. In practice, of course, we typically receive data in successive stages, so that the process of revising beliefs is sequential.

As a simple illustration of this process, let us suppose that data are obtained in two stages, which can be described by real-world events D_1 and D_2. Omitting, for convenience, explicit conditioning on M_0, revision of beliefs on the basis of the first piece of data D_1 is described by $P(H_j \mid D_1) = P(D_1 \mid H_j)P(H_j)/P(D_1)$, $j \in J$. When it comes to the further, subsequent revision of beliefs in the light of D_2, the likelihoods and prior probabilities to be used in Bayes' theorem are now $P(D_2 \mid H_j \cap D_1)$ and $P(H_j \mid D_1)$, $j \in J$, respectively, since all judgements are now conditional on D_1. We thus have, for all $j \in J$,

$$P(H_j \mid D_1 \cap D_2) = \frac{P(D_2 \mid H_j \cap D_1)P(H_j \mid D_1)}{P(D_2 \mid D_1)},$$

where $P(D_2 \mid D_1) = \sum_j P(D_2 \mid H_j \cap D_1)P(H_j \mid D_1)$.

From an intuitive standpoint, we would obviously anticipate that coherent revision of initial belief in the light of the combined data, $D_1 \cap D_2$, should not depend on whether D_1, D_2 were analysed successively or in combination. This is easily verified by substituting the expression for $P(H_j \mid D_1)$ into the expression for $P(H_j \mid D_1 \cap D_2)$, whereupon we obtain

$$\frac{P(D_2 \mid H_j \cap D_1)P(D_1 \mid H_j)P(H_j)}{P(D_2 \mid D_1)P(D_1)} = \frac{P(D_1 \cap D_2 \mid H_j)P(H_j)}{P(D_1 \cap D_2)},$$

the latter being the direct expression for $P(H_j \mid D_1 \cap D_2)$ from Bayes' theorem when $D_1 \cap D_2$ is treated as a single piece of data.

The generalisation of this sequential revision process to any number of stages, corresponding to data, $D_1, D_2, \ldots, D_n, \ldots$, proceeds straightforwardly. If we write $D^{(k)} = D_1 \cap D_2 \cap \cdots \cap D_k$ to denote all the data received up to and including stage k, then, for all $j \in J$,

$$P(H_j \mid D^{(k+1)}) = \frac{P(D_{k+1} \mid H_j \cap D^{(k)})P(H_j \mid D^{(k)})}{P(D_{k+1} \mid D^{(k)})},$$

which provides a recursive algorithm for the revision of beliefs.

There is, however, a potential practical difficulty in implementing this process, since there is an implicit need to specify the successively *conditioned likelihoods*, $P(D_{k+1} \mid H_j \cap D^{(k)})$, $j \in J$, a task which, in the absence of simplifying assumptions, may appear to be impossibly complex if k is at all large. One possible form of simplifying assumption is the judgement of conditional independence for $D_1, D_2, \ldots, D_n$, given any $H_j, j \in J$, since, by Definition 2.13, we then only need the evaluations $P(D_{k+1} \mid H_j \cap D^{(k)}) = P(D_{k+1} \mid H_j)$, $j \in J$. Another possibility might be to assume a rather weak form of dependence by making the judgement that a (Markov) property such as $P(D_{k+1} \mid H_j \cap D^{(k)}) = P(D_{k+1} \mid H_j \cap D_k)$, $j \in J$, holds for all k. As we shall see later, these kinds of simplifying structural assumptions play a fundamental role in statistical modelling and analysis.

In the case of two hypotheses, H_1, H_2, the judgement of conditional independence for $D_1, D_2, \ldots, D_n, \ldots$, given H_1 or H_2, enables us to provide an alternative description of the process of revising beliefs by noting that, in this case,

$$\frac{P(H_1 \mid D^{(k+1)})}{P(H_2 \mid D^{(k+1)})} = \frac{P(H_1 \mid D^{(k)})}{P(H_2 \mid D^{(k)})} \times \frac{P(D_{k+1} \mid H_1)}{P(D_{k+1} \mid H_2)} \,.$$

With due regard to the relative nature of the terms prior and posterior, we can thus summarise the learning process (in "favour" of H_1) as follows:

$$\textit{posterior odds} = \textit{prior odds} \times \textit{likelihood ratio}.$$

In Section 2.6, we shall examine in more detail the key role played by the sequential revision of beliefs in the context of complex, sequential decision problems.

2.5 ACTIONS AND UTILITIES

2.5.1 Bounded Sets of Consequences

At the beginning of Section 2.4, we argued that choices among options are governed, in part, by the relative degrees of belief that an individual attaches to the uncertain events involved in the options. It is equally clear that choices among options should depend on the relative values that an individual attaches to the consequences flowing from the events. The measurement framework of Axiom 5(i) provides us with a direct, intuitive way of introducing a *numerical measure of value for consequences*, in such a way that the latter has a coherent, operational basis. Before we do this, we need to consider a little more closely the nature of the set of consequences $\mathcal{C}$. The following special case provides a useful starting point for our development of a measure of value for consequences.

Definition 2.14. (***Extreme consequences***). *The pair of consequences c_* and c^* are called, respectively, the* ***worst*** *and the* ***best*** *consequences in a decision problem if, for any other consequence $c \in \mathcal{C}$, $c_* \leq c \leq c^*$.*

It could be argued that *all* real decision problems actually have extreme consequences. Indeed, we recall that all consequences are to be thought of as relevant consequences in the context of the decision problem. This eliminates pathological, mathematically motivated choices of $\mathcal{C}$, which could be constructed in such a way as to rule out the existence of extreme consequences. For example, in *mathematical* modelling of decision problems involving monetary consequences, $\mathcal{C}$ is often taken to be the real line $\Re$ or, in a no-loss situation with current assets k, to be the interval $[k, \infty)$. Such $\mathcal{C}$'s would not contain *both* a best and a worst consequence but, on the other hand, they clearly do not correspond to concrete, practical problems. In the next section, we shall consider the solution to decision problems for which extreme consequences are assumed to exist.

Nevertheless, despite the force of the pragmatic argument that extreme consequences always exist, it must be admitted that insisting upon problem formulations which satisfy the assumption of the existence of extreme consequences can sometimes lead to rather tedious complications of a conceptual or mathematical nature.

Consider, for example, a medical decision problem for which the consequences take the form of different numbers of years of remaining life for a patient. Assuming that more value is attached to longer survival, it would appear rather difficult to justify any *particular* choice of realistic upper bound, even though we believe there to be one. To choose a particular c^* would be tantamount to putting forward c^* years as a realistic possible survival time, but regarding $c^* + 1$ years as impossible! In such cases, it is attractive to have available the possibility, for conceptual and mathematical convenience, of dealing with sets of consequences *not* possessing extreme elements (and the same is true of many problems involving monetary consequences). For this reason, we shall also deal (in Section 2.5.3) with the situation in which extreme consequences are not assumed to exist.

2.5.2 Bounded Decision Problems

Let us consider a decision problem $(\mathcal{E}, \mathcal{C}, \mathcal{A}, \leq)$ for which extreme consequences $c_* < c^*$ are assumed to exist. We shall refer to such decision problems as *bounded.*

Definition 2.15. ***(Canonical utility function for consequences).*** *Given a preference relation $\leq$, the utility $u(c) = u(c \,|\, c_*, c^*)$ of a consequence c, relative to the extreme consequences $c_* < c^*$, is the real number $\mu(S)$ associated with any standard event S such that $c \sim \{c^* \,|\, S, c_* \,|\, S^c\}$. The mapping $u : \mathcal{C} \to \Re$ is called the* ***utility function.***

It is important to note that the definition of utility only involves comparison among consequences and options constructed with standard events. Since the preference patterns among consequences is unaffected by additional information, we would expect the utility of a consequence to be uniquely defined and to remain unchanged as new information is obtained. This is indeed the case.

Proposition 2.21. ***(Existence and uniqueness of bounded utilities).*** *For any bounded decision problem $(\mathcal{E}, \mathcal{C}, \mathcal{A}, \leq)$ with extreme consequences $c_* < c^*$,*

(i) *for all c, $u(c \,|\, c_*, c^*)$ exists and is unique;*

(ii) *the value of $u(c \,|\, c_*, c^*)$ is unaffected by the assumed occurrence of an event $G > \emptyset$;*

(iii) $0 = u(c_* \,|\, c_*, c^*) \leq u(c \,|\, c_*, c^*) \leq u(c^* \,|\, c_*, c^*) = 1.$

Proof. (i) Existence follows immediately from Axiom 5(i). For uniqueness, note that if $c \sim \{c^* \mid S_1, c_* \mid S_1^c\}$ and $c \sim \{c^* \mid S_2, c_* \mid S_2^c\}$ then, by transitivity and Axiom 3(ii), $\{c^* \mid S_1, c_* \mid S_1^c\} \sim \{c^* \mid S_2, c_* \mid S_2^c\}$ and $S_1 \sim S_2$; the result now follows from Axiom 4(i).

(ii) To establish this, let $c \sim \{c^* \mid S_1, c_* \mid S_1^c\}$, so that $u(c \mid c_*, c^*) = \mu(S_1)$; using Axiom 4(iii), for any $G > \emptyset$ choose S_2 such that $G \perp S_2$ and $\mu(S_2) = \mu(S_1)$. Then, by Definition 2.6, $c \sim_G \{c^* \mid S_2, c_* \mid S_2^c\}$ and so the utility of c given G is just the original value $\mu(S_2)$.

(iii) Finally, since $c^* \equiv \{c^* \mid \emptyset, c_* \mid \Omega\}$, $c_* \equiv \{c^* \mid \Omega, c_* \mid \emptyset\}$, and both $\emptyset$ and Ω belong to the algebra of standard events, we have $u(c_* \mid c_*, c^*) = \mu(\emptyset) = 0$ and $u(c^* \mid c_*, c^*) = \mu(\Omega) = 1$. It then follows, from Definition 2.15 and Axiom 4(i), that $0 \leq u(c \mid c_*, c^*) \leq 1$. $\triangleleft$

It is interesting to note that $u(c \mid c_*, c^*)$, which we shall often simply denote by $u(c)$, can be given an operational interpretation in terms of degrees of belief. Indeed, if we consider a choice between the fixed consequence c and the option $\{c^* \mid E, c_* \mid E^c\}$, for some event E, then the utility of c can be thought of as defining a threshold value for the degree of belief in E, in the sense that values greater than u would lead an individual to prefer the uncertain option, whereas values less than u would lead the individual to prefer c for certain. The value u itself corresponds to indifference between the two options and is the degree of belief in the occurrence of the best, rather than worst, consequence.

This suggests one possible technique for the experimental *elicitation of utilities*, a subject which has generated a large literature (with contributions from economists and psychologists, as well as from statisticians). We shall illustrate the ideas in Example 2.3.

Using the coherence and quantification principles set out in Section 2.3, we have seen how numerical measures can be assigned to two of the elements of a decision problem in the form of *degrees of belief for events* and *utilities for consequences*. It remains now to investigate how an *overall numerical measure of value* can be attached to an *option*, whose form depends both on the events of a finite partition of the certain event Ω and on the particular consequences to which these events lead.

Definition 2.16. (***Conditional expected utility***).
For any $c_* < c^*, G > \emptyset$, and $a \equiv \{c_j \mid E_j, j \in J\}$,

$$\overline{u}(a \mid c_*, c^*, G) = \sum_{j \in J} u(c_j \mid c_*, c^*) P(E_j \mid G)$$

is the expected utility of the option a, given G, with respect to the extreme consequences c_*, c^*. If $G = \Omega$, we shall simply write $\overline{u}(a \mid c_*, c^*)$ in place of $\overline{u}(a \mid c_*, c^*, \Omega)$.

In the language of mathematical probability theory (see Chapter 3), if the utility value of a is considered as a "random quantity", contingent on the occurrence of a particular event E_j, then $\overline{u}$ is simply the *expected value* of that utility when the probabilities of the events are considered conditional on G.

Proposition 2.22. (***Decision criterion for a bounded decision problem***). *For any bounded decision with extreme consequences $c_* < c^*$, and $G > \emptyset$,*

$$a_1 \leq_G a_2 \iff \overline{u}(a_1 \,|\, c_*, c^*, G) \leq \overline{u}(a_2 \,|\, c_*, c^*, G).$$

Proof. Let $a_i = \{c_{ij} \,|\, E_{ij}, j = 1, \ldots, n_i\}$, $i = 1, 2$. By Axioms 5(ii), 4(iii), and Proposition 2.13, for all (i, j) there exist S_{ij} and S'_{ij} such that

$$c_{ij} \sim \{c^* \,|\, S'_{ij}, c_* \,|\, S'^c_{ij}\}, \quad S_{ij} \perp (E_{ij} \cap G), \quad P(S'_{ij}) = P(S_{ij}).$$

Hence, by Proposition 2.10, $c_{ij} \sim \{c^* \,|\, S_{ij}, c_* \,|\, S^c_{ij}\}$ with $S_{ij} \perp (E_{ij} \cap G)$ and $P(S_{ij}) = u(c_{ij} \,|\, c_*, c^*)$. By Definition 2.6, for $i = 1, 2$ and any option a,

$$\{[c_{ij} \,|\, E_{ij} \cap G], j = 1, \ldots, n_i, \; a \,|\, G^c\}$$
$$\sim \{[(c^* \,|\, S_{ij}, c_* \,|\, S^c_{ij}) \,|\, E_{ij} \cap G], j = 1, \ldots, n_i, \; a \,|\, G^c\},$$

which may be written as $\{c^* \,|\, A_i, c_* \,|\, B_i, a \,|\, G^c\}$, where $A_i = \cup_j (E_{ij} \cap G \cap S_{ij})$ and $B_i = \cup_j (E_{ij} \cap G \cap S^c_{ij})$. By Propositions 2.14(ii) and 2.16, and using Definition 2.5, $a_1 \leq_G a_2 \Rightarrow A_1 \leq_G A_2 \Rightarrow P(A_1 \,|\, G) \leq P(A_2 \,|\, G)$. But, by Proposition 2.15, $P(E_{ij} \cap G \cap S_{ij}) = P(E_{ij} \cap G) P(S_{ij}) = P(S_{ij}) P(E_{ij} \,|\, G) P(G)$. Hence,

$$P(A_i \,|\, G) = \sum_{j=1}^{n_i} u(c_{ij} \,|\, c_*, c^*) P(E_{ij} \,|\, G) = \overline{u}(a_i \,|\, c_*, c^*, G)$$

and so $a_1 \leq_G a_2 \Leftrightarrow \overline{u}(a_1 \,|\, c_*, c^*, G) \leq \overline{u}(a_2 \,|\, c, c^*, G)$. ◁

The result just established is sometimes referred to as the *principle of maximising expected utility*. In our development, this is clearly not an independent "principle", but rather an implication of our assumptions and definitions. In summary form, the resulting prescription for quantitative, coherent decision-making is: *choose the option with the greatest expected utility*.

Technically, of course, Proposition 2.22 merely establishes, for each $\leq_G$, a complete ordering of the options considered and does not guarantee the *existence* of an optimal option for which the expected utility is a maximum. However, in most (if not all) concrete, practical problems the set of options considered will be finite and so a *best option* (not necessarily unique) will exist. In more abstract mathematical formulations, the existence of a maximum will depend on analytic features of the set of options and on the utility function $u : C \to \Re$.

Example 2.3. ***(Utilities of oil wildcatters).*** One of the earliest reported systematic attempts at the quantification of utilities in a practical decision-making context was that of Grayson (1960), whose decision-makers were oil wildcatters engaged in exploratory searches for oil and gas. The consequences of drilling decisions and their outcomes are ultimately changes in the wildcatters' monetary assets, and Grayson's work focuses on the assessment of utility functions for this latter quantity.

For the purposes of illustration, suppose that we restrict attention to changes in monetary assets ranging, in units of one thousand dollars, from -150 (the *worst consequence*) to $+825$ (the *best consequence*). Assuming $u(-150) = 0$, $u(825) = 1$, the above development suggests ways in which we might try to elicit an individual wildcatter's values of $u(c)$ for various c in the range $-150 < c < 825$. For example, one could ask the wildcatter, using a series of values of c, which option he or she would prefer out of the following:

(i) c for sure,
(ii) entry into a venture having outcome 825 with probability p and an outcome -150 with probability $1 - p$, for some specified p.

If c_p emerges from such interrogation as an approximate "indifference" value, the theory developed above suggests that, for a coherent individual,

$$u(c_p) = p\,u(825) + (1-p)\,u(-150) = p.$$

Repeating this exercise for a range of values of p, provides a series of (c_p, p) pairs, from which a "picture" of $u(c)$ over the range of interest can be obtained. An alternative procedure, of course, would be to fix c, perform an interrogation for various p until an "indifference" value, p_c is found, and then repeat this procedure for a range of values of c to obtain a series of (c, p_c) pairs.

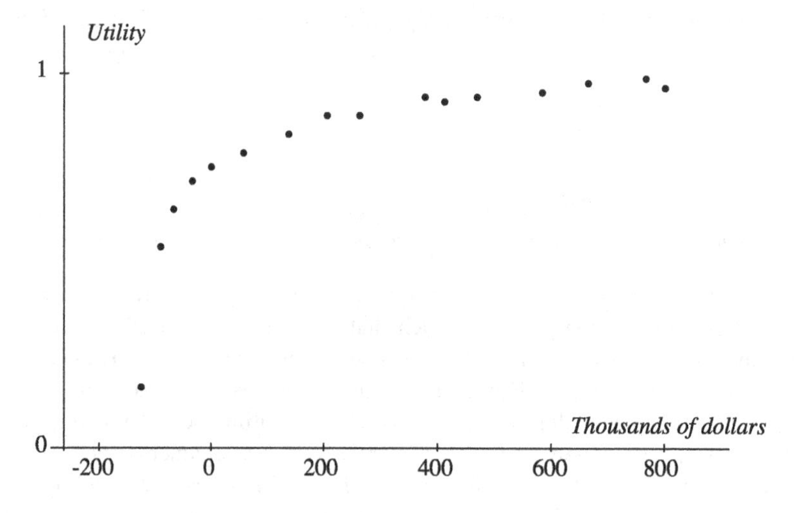

Figure 2.3 *William Beard's utility function for changes in monetary assets*

Figure 2.3 shows the results obtained by Grayson using procedures of this kind to interrogate oil company executive, W. Beard, on October 23, 1957. A "picture" of Beard's utility function clearly emerges from the empirical data. In particular, over the range concerned, the utility function reflects considerable risk aversion, in the sense that even quite small asset losses lead to large (negative) changes in utility compared with the (positive) changes associated with asset gains.

Since the expected utility $\overline{u}$ is a linear combination of values of the utility function, Proposition 2.22 guarantees that preferences among options are invariant under changes in the origin and scale of the utility measure used; i.e., invariant with respect to transformations of the form $Au(.) + B$, provided we take $A > 0$, so that the orientation of "best" and "worst" is not changed. In general, therefore, such an origin and scale can be chosen for convenience in any given problem, and we can simply refer to the expected utility of an option without needing to specify the (positive linear) transformation of the utility function which has been used. However, there may be bounded decision problems where the probabilistic interpretation discussed above makes it desirable to work in terms of *canonical* utilities, derived by referring to the best and worst consequences.

In the next section, we shall provide an extension of these ideas to more general decision problems where extreme consequences are not assumed to exist.

2.5.3 General Decision Problems

We begin with a more general definition of the utility of a consequence which preserves the linear combination structure and the invariance discussed above.

Definition 2.17. (***General utility function***). *Given a preference relation $\leq$, the utility $u(c \mid c_1, c_2)$ of a consequence c, relative to the consequences $c_1 < c_2$, is defined to be the real number u such that*

if $c < c_1$ and $c_1 \sim \{c_2 \mid S_x, c \mid S_x^c\}$, then $u = -x/(1-x)$;

if $c_1 \leq c \leq c_2$ and $c \sim \{c_2 \mid S_x, c_1 \mid S_x^c\}$, then $u = x$;

if $c > c_2$ and $c_2 \sim \{c \mid S_x, c_1 \mid S_x^c\}$, then $u = 1/x$

where $x = \mu(S_x)$ is the measure associated with the standard event S_x.

Our restricted definition of utility (Definition 2.15) relied on the existence of extreme consequences c_*, c^*, such that $c_* \leq c \leq c^*$ for all $c \in \mathcal{C}$. In the absence of this assumption, we have to select some *reference consequences*, c_1, c_2 to play the role of c_*, c^*. However, we cannot then assume that $c_1 \leq c \leq c_2$ for all c, and this means that if c_1, c_2 are to define a utility scale by being assigned values $0, 1$, respectively, we shall require *negative* assignments for $c < c_1$ and assignments *greater than one* for $c > c_2$. The definition is motivated by a desire to maintain the linear features of the utility function obtained in the case where extreme consequences exist. It can be checked straightforwardly that if $c_{(1)}$, $c_{(2)}$, $c_{(3)}$ denote any permutation of c, c_1, c_2, where $c_1 < c_2$ and $c_{(1)} \leq c_{(2)} \leq c_{(3)}$, the definition given ensures that for any $G > \emptyset$, $c_{(2)} \sim_G \{c_{(3)} \mid S_x, c_{(1)} \mid S_x^c\}$ implies that

$$u(c_{(2)} \mid c_1, c_2) = x\, u(c_{(3)} \mid c_1, c_2) + (1 - x)\, u(c_{(1)} \mid c_1, c_2).$$

The following result extends Proposition 2.21 to the general utility function defined above.

Proposition 2.23. (***Existence and uniqueness of utilities***). *For any decision problem, and for any pair of consequences $c_1 < c_2$,*

(i) *for all c, $u(c \mid c_1, c_2)$ exists and is unique;*

(ii) *the value of $u(c \mid c_1, c_2)$ is unaffected by the occurrence of an event $G > \emptyset$;*

(iii) $u(c_1 \mid c_1, c_2) = 0$ *and* $u(c_2 \mid c_1, c_2) = 1$.

Proof. This is virtually identical to the proof of Proposition 2.21. $\triangleleft$

The following results guarantee that the utilities of consequences are linearly transformed if the pair of consequences chosen as a reference is changed.

Proposition 2.24. (***Linearity***). *For all $c_1 < c_2$ and $c_3 < c_4$ there exist $A > 0$ and B such that, for all c, $u(c \mid c_1, c_2) = Au(c \mid c_3, c_4) + B$.*

Proof. Suppose first that $c_3 \geq c_1$, $c_4 \leq c_2$, and $c_1 \leq c \leq c_2$. By Axiom 5(ii), $c_3 \leq c \leq c_4$ implies that there exists a standard event S_x such that $c \sim \{c_4 \mid S_x, c_3 \mid S_x^c\}$. Hence, by Proposition 2.22,

$$u(c \mid c_1, c_2) = xu(c_4 \mid c_1, c_2) + (1 - x)u(c_3 \mid c_1, c_2),$$

where $x = P(S_x)$ and, by Definition 2.17, $u(c \mid c_3, c_4) = x$. Hence, $u(c \mid c_1, c_2) = Au(c \mid c_3, c_4) + B$, where $A = u(c_4 \mid c_1, c_2) - u(c_3 \mid c_1, c_2)$ and $B = u(c_3 \mid c_1, c_2)$.

By Axiom 5(ii), if $c_3 > c$ there exists S_y such that $c_3 \sim \{c_4 \mid S_y, c \mid S_y^c\}$. Hence, by Proposition 2.22,

$$u(c_3 \mid c_1, c_2) = yu(c_4 \mid c_1, c_2) + (1 - y)u(c \mid c_1, c_2),$$

where $y = P(S_y)$ and, by Definition 2.17, $u(c \mid c_3, c_4) = -y/(1 - y)$. Hence, $u(c \mid c_1, c_2) = Au(c \mid c_3, c_4) + B$, with A and B as above. Similarly, if $c > c_4$ there exists S_z such that $c_4 \sim \{c \mid S_z, c_3 \mid S_z^c\}$ and

$$u(c_4 \mid c_1, c_2) = yu(c \mid c_1, c_2) + (1 - y)u(c_3 \mid c_1, c_2),$$

where $y = P(S_y)$ and, by Definition 2.17, $u(c \mid c_3, c_4) = 1/y$. Hence, we have $u(c \mid c_1, c_2) = Au(c \mid c_3, c_4) + B$, with A and B as above.

Now suppose that the c's have arbitrary order, subject to $c_2 > c_1$, $c_4 > c_3$. Let c_*, c^* be the minimum and maximum, respectively, of $\{c_1, c_2, c_3, c_4, c\}$. Then, by the above, there exist A_1, B_1, A_2, B_2 such that, for $c_{(i)} \in \{c_1, c_2, c_3, c_4, c\}$, $u(c_{(i)} \mid c_*, c^*) = A_1 u(c_{(i)} \mid c_1, c_2) + B_1$ and $u(c_{(i)} \mid c_*, c^*) = A_2 u(c_{(i)} \mid c_3, c_4) + B_2$; hence, $u(c_{(i)} \mid c_1, c_2) = (A_2/A_1)u(c_{(i)} \mid c_3, c_4) + (B_2 - B_1)/A_1$. $\triangleleft$

Finally, we generalise Proposition 2.22 to unbounded decision problems;

Proposition 2.25. (***General decision criterion***).
For any decision problem, pair of consequences $c_1 < c_2$, *and event* $G > \emptyset$,

$$a_1 \leq_G a_2 \iff \overline{u}(a_1 \,|\, c_1, c_2, G) \leq \overline{u}(a_2 \,|\, c_1, c_2, G).$$

Proof. Suppose $a_i = \{c_{ij} \,|\, E_{ij}, j = 1, \ldots, n_i\}, i = 1, 2$, and let c_*, c^* be such that for all c_{ij}, $c_* \leq c_{ij} \leq c^*$. Then, by Proposition 2.22, $a_2 \leq_G a_1$ iff $\overline{u}(a_2 \,|\, c_*, c^*, G) \leq \overline{u}(a_1 \,|\, c_*, c^*, G)$. But, by Proposition 2.24, there exists $A > 0$ and B such that $u(c \,|\, c_*, c^*) = Au(c \,|\, c_1, c_2) + B$, and so the result follows. ◁

An immediate implication of Proposition 2.25 is that all options can be compared among themselves. We recall that we did *not* directly assume that comparisons could be made between all pair of options (an *assumption* which is often criticised as unjustified; see, for example, Fine 1973, p. 221). Instead, we merely assumed that all consequences could be compared among themselves and with the (very simply structured) standard dichotomised options, and that the latter could be compared among themselves.

This completes our elaboration of the axiom system set out in Section 2.3. Starting from the primitive notion of preference, $\leq$, we have shown that quantitative, coherent comparisons of options must proceed *as if* a utility function has been assigned to consequences, probabilities to events and the choice of an option made on the basis of maximising expected utility.

If we *begin* by defining a utility function over $u : \mathcal{C} \to \Re$, this *induces* in turn a preference ordering which is necessarily coherent. Any function can serve as a utility function (subject only to the existence of the expected utility for each option, a problem which does not arise in the case of finite partitions) and the choice is a personal one. In some contexts, however, there are further formal considerations which may delimit the form of function chosen. An important special case is discussed in detail in Section 2.7.

2.6 SEQUENTIAL DECISION PROBLEMS

2.6.1 Complex Decision Problems

Many real decision problems would appear to have a more complex structure than that encapsulated in Definition 2.1. For instance, in the fields of market research and production engineering investigators often consider first whether or not to run a pilot study and only then, in the light of information obtained (or on the basis of initial information if the study is not undertaken), are the major options considered. Such a two-stage process provides a simple example of a *sequential* decision problem, involving successive, interdependent decisions. In this section, we shall demonstrate that complex problems of this kind can be solved with the tools already at our disposal, thus substantiating our claim that the principles of quantitative coherence suffice to provide a prescriptive solution to *any* decision problem.

Before explicitly considering sequential problems, we shall review, using a more detailed notation, some of our earlier developments.

Let $\mathcal{A} = \{a_i, i \in I\}$ be the set of alternative actions we are willing to consider. For each a_i, there is a class $\{E_{ij}, j \in J_i\}$ of exhaustive and mutually exclusive events, which label the possible consequences $\{c_{ij}, j \in J_i\}$ which may result from action a_i. Note that, with this notation, we are merely emphasising the obvious dependence of both the consequences and the events on the action from which they result. If M_0 is our initial state of information and $G > \emptyset$ is additional information obtained subsequently, the main result of the previous section (Proposition 2.25) may be restated as follows.

For behaviour consistent with the principles of quantitative coherence, action a_1 is to be preferred to action a_2, given M_0 and G, if and only if

$$\overline{u}(a_1 \mid G) > \overline{u}(a_2 \mid G),$$

where

$$\overline{u}(a_i \mid G) = \sum_{j \in J_i} u(c_{ij}) P(E_{ij} \mid a_i, M_0, G),$$

$u(c_{ij})$ is the value attached to the consequence foreseen if action a_i is taken and the event E_{ij} occurs, and $P(E_{ij} \mid a_i, M_0, G)$ is the degree of belief in the occurrence of event E_{ij}, conditional on action a_i having been taken, and the state of information being (M_0, G).

We recall that the probability measure used to compute the expected utility is taken to be a representation of the decision-maker's degree of belief conditional on the total information available. By using the extended notation $P(E_{ij} \mid a_i, G, M_0)$, rather than the more economical $P(E_j \mid G)$ used previously, we are emphasising that (i) the actual events considered may depend on the particular action envisaged, (ii) the information available certainly includes the initial information together with $G > \emptyset$, and (iii) degrees of belief in the occurrence of events such as E_{ij} are understood to be conditional on action a_i having been assumed to be taken, so that the possible influence of the decision-maker on the real world is taken into account.

For any action a_i, it is sometimes convenient to describe the relevant events E_{ij}, $j \in J$, in a sequential form. For example, in considering the relevant events which label the consequences of a surgical intervention for cancer, one may first think of whether the patient will survive the operation and then, conditional on survival, whether or not the tumour will eventually reappear were this particular form of surgery to be performed.

These situations are most easily described diagrammatically using decision trees, such as that shown in Figure 2.4, with as many successive random nodes as necessary. Obviously, this does not represent any formal departure from our previous structure, since the problem can be restated with a single random node where relevant events are defined in terms of appropriate intersections, such as $E_{ij} \cap F_{ijk}$ in the example shown. It is also usually the case, in practice, that it is easier to elicit the relevant degrees of belief conditionally, so that, for example, $P(E_{ij} \cap F_{ijk} \mid a_i, G, M_0)$ would often be best assessed by combining the separately assessed terms $P(F_{ijk} \mid E_{ij}, a_i, G, M_0)$ and $P(E_{ij} \mid a_i, G, M_0)$.

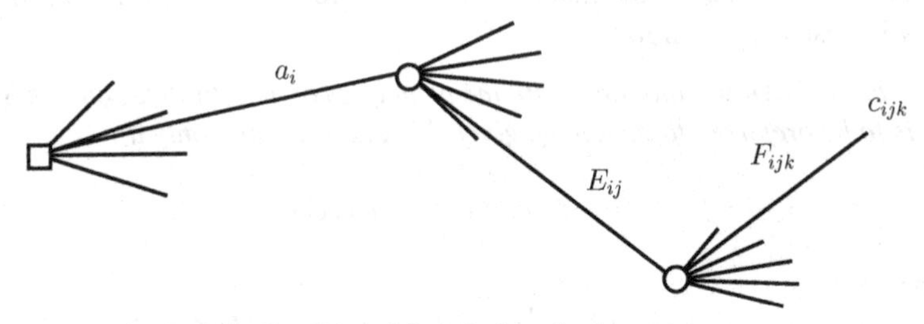

Figure 2.4 *Conditional description of relevant events*

Conditional analysis of this kind is usually necessary in order to understand the structure of complicated situations. Consider, for instance, the problem of placing a bet on the result of a race after which the total amount bet is to be divided up among those correctly guessing the winner. Clearly, if we bet on the favourite we have a higher probability of winning; but, if the favourite wins, many people will have guessed correctly and the prize will be small. It may appear at first sight that this is a decision problem where the utilities involved in an action (the possible prizes to be obtained from a bet) depend on the probabilities of the corresponding uncertain events (the possible winning horses), a possibility *not* contemplated in our structure. A closer analysis reveals, however, that the structure of the problem is similar to that of Figure 2.4. The prize received depends on the bet you place (a_i) the related betting behaviour of other people (E_{ij}) and the outcome of the race (F_{ijk}). It is only natural to assume that our degree of belief in the possible outcomes of the race may be influenced by the betting behaviour of other people. This conditional analysis straightforwardly resolves the initial, apparent complication.

We now turn to considering *sequences* of decision problems. We shall consider situations where, after an action has been taken and its consequences observed, a new decision problem arises, conditional on the new circumstances. For example, when the consequences of a given medical treatment have been observed, a physician has to decide whether to continue the same treatment, or to change to an alternative treatment, or to declare the patient cured.

If a decision problem involves a succession of decision nodes, it is intuitively obvious that the optimal choice at the first decision node depends on the optimal choices at the subsequent decision nodes. In colloquial terms, we typically cannot decide what to do today without thinking first of what we might do tomorrow, and that, of course, will typically depend on the possible consequences of today's actions. In the next section, we consider a technique, *backward induction*, which makes it possible to solve these problems *within* the framework we have already established.

2.6.2 Backward Induction

In any actual decision problem, the number of scenarios which may be contemplated at any given time is necessarily finite. Consequently, and bearing in mind that the analysis is only strictly valid under certain fixed general assumptions and we cannot seriously expect these to remain valid for an indefinitely long period, the number of decision nodes to be considered in any given sequential problem will be assumed to be finite. Thus, we should be able to define a *finite horizon*, after which no further decisions are envisaged in the particular problem formulation. If, at each node, the possibilities are finite in number, the situation may be diagrammatically described by means of a decision tree like that of Figure 2.5.

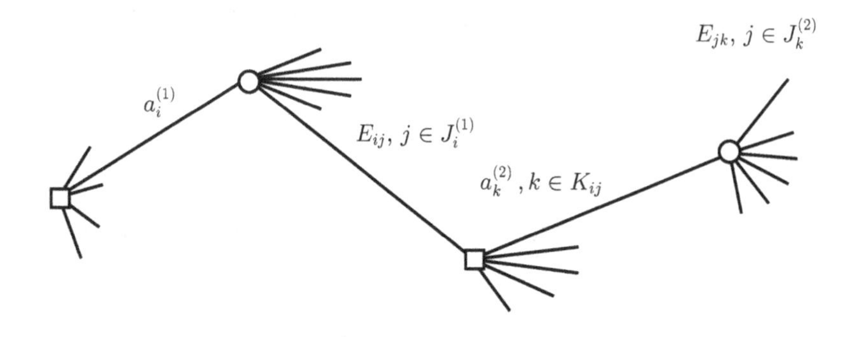

Figure 2.5 *Decision tree with several decision nodes*

Let n be the number of decision stages considered and let $a^{(m)}$ denote an action being considered at the mth stage. Using the notation for composite options introduced in Section 2.2, all first-stage actions may be compactly described in the form

$$a_i^{(1)} = \left\{ \max_{k \in K_{ij}} a_k^{(2)} \,|\, E_{ij},\ j \in J_i^{(1)} \right\},$$

where $\{E_{ij}, j \in J_i^{(1)}\}$ is the partition of relevant events which corresponds to $a_i^{(1)}$ and the notation "max $a_k^{(2)}$" refers to the *most preferred* of the set of options $\{a_k^{(2)}, k \in K_{ij}\}$ which we would be confronted with were the event E_{ij} to occur. The "maximisation" is naturally to be understood in the sense of our conditional preference ordering among the available second-stage options, given the occurrence of E_{ij}. Indeed, the "consequence" of choosing $a_i^{(1)}$ and having E_{ij} occur is that we are confronted with a set of options $\{a_k^{(2)}, k \in K_{ij}\}$ from which *we can choose* that option which is preferred on the basis of our pattern of preferences at that stage. Similarly, second-stage options may be written in terms of third-stage options, and the process continued until we reach the nth stage, consisting of "ordinary" options defined in terms of the events and consequences to which they may lead. Formally, we have

$$a_i^{(m)} = \left\{ \max_{k \in K_{ij}} a_k^{(m+1)} \,|\, E_{ij},\ j \in J_i^{(m)} \right\}, \qquad m = 1, 2, \ldots, n-1,$$

$$a_i^{(n)} = \left\{ c_{ij} \,|\, E_{ij}, j \in J_i^{(n)} \right\}.$$

It is now apparent that sequential decision problems are a special case of the general framework which we have developed.

It follows from Proposition 2.25 that, at each stage m, if G_m is the relevant information available, and $u(.)$ is the (generalised) utility function, we may write

$$a_i^{(m)} \leq_{G_m} a_j^{(m)} \iff \overline{u}\left\{a_i^{(m)} \,|\, G_m\right\} \leq \overline{u}\left\{a_j^{(m)} \,|\, G_m\right\},$$

where

$$\overline{u}\left\{a_i^{(m)} \,|\, G_m\right\} = \sum_{j \in J_i^{(m)}} \left[\max_{k \in K_{ij}} \overline{u}\left\{a_k^{(m+1)} \,|\, G_{m+1}\right\} \right] P(E_{ij} \,|\, G_m),$$

$$\overline{u}\left\{a_i^{(n)} \,|\, G_n\right\} = \sum_{j \in J_i^{(n)}} u(c_{ij}) P(E_{ij} \,|\, G_n).$$

This means that one has to first solve the final (nth) stage, by maximising the appropriate expected utility; then one has to solve the $(n-1)$th stage by maximizing the expected utility conditional on making the optimal choice at the nth stage; and so on, working backwards progressively, until the optimal first stage option has been obtained, a procedure often referred to as *dynamic programming*.

This process of *backward induction* satisfies the requirement that, at any stage of the procedure, the mth, say, the continuation of the procedure must be identical to the optimal procedure starting at the mth stage with information G_m. This requirement is usually known as *Bellman's optimality principle* (Bellman, 1957). As with the "principle" of maximising expected utility, we see that this is not required as a further assumed "principle" in our formulation, but is simply a consequence of the principles of quantitative coherence.

Example 2.4. ***(An optimal stopping problem)***. We now consider a famous problem, which is usually referred to in the literature as the "marriage problem" or the "secretary problem". Suppose that a specified number of objects $n \geq 2$ are to be inspected sequentially, one at a time, in order to select one of them. Suppose further that, at any stage r, $1 \leq r \leq n$, the inspector has the option of either stopping the inspection process, receiving, as a result, the object currently under inspection, or of continuing the inspection process with the next object. No backtracking is permitted and if the inspection process has not terminated before the nth stage the outcome is that the nth object is received. At each stage, r, the only information available to the inspector is the relative rank (1=best, r=worst) of the current object among those inspected so far, and the knowledge that the n objects are being presented in a completely random order.

When should the inspection process be terminated? Intuitively, if the inspector stops too soon there is a good chance that objects more preferred to those seen so far will remain uninspected. However, if the inspection process goes on too long there is a good chance that the overall preferred object will already have been encountered and passed over.

This kind of dilemma is inherent in a variety of practical problems, such as property purchase in a limited seller's market when a bid is required immediately after inspection, or staff appointment in a skill shortage area when a job offer is required immediately after interview. More exotically—and assuming a rather egocentric inspection process, again with no backtracking possibilities—this stopping problem has been suggested as a model for choosing a mate. Potential partners are encountered sequentially; the proverb "marry in haste, repent at leisure" warns against settling down too soon; but such hesitations have to be balanced against painful future realisations of missed golden opportunities.

Less romantically, let c_i, $i = 1, \ldots, n$, denote the possible consequences of the inspection process, with $c_i = i$ if the eventual object chosen has rank i out of all n objects. We shall denote by $u(c_i) = u(i)$, $i = 1, \ldots, n$, the inspector's utility for these consequences.

Now suppose that $r < n$ objects have been inspected and that the relative rank among these of the object under current inspection is x, where $1 \leq x \leq r$. There are two actions available at the rth stage: a_1 = stop, a_2 = continue (where, to simplify notation, we have dropped the superscript, r). The information available at the rth stage is $G_r = (x, r)$; the information available at the $(r+1)$th stage would be $G_{r+1} = (y, r+1)$, where y, $1 \leq y \leq r+1$, is the rank of the next object relative to the $r+1$ then inspected, all values of y being, of course, equally likely since the n objects are inspected in a random order. If we denote the expected utility of stopping, given G_r, by $\overline{u}_s(x, r)$ and the expected utility of acting optimally, given G_r, by $\overline{u}_0(x,r)$, the general development given above establishes that

$$\overline{u}_0(x,r) = \max\left\{\overline{u}_s(x,r), \frac{1}{r+1}\sum_{y=1}^{r+1}\overline{u}_0(y,r+1)\right\},$$

where

$$\overline{u}_s(x,r) = \sum_{z=x}^{n-r+x} u(z)\frac{\binom{z-1}{x-1}\binom{n-z}{r-x}}{\binom{n}{r}},$$

$$\overline{u}_0(x,n) = \overline{u}_s(x,n) = u(x), \quad x = 1,\ldots,n.$$

Values of $\overline{u}_0(x,r)$ can be found from the final condition and the technique of backwards induction. The optimal procedure is then seen to be:

(i) continue if $\overline{u}_0(x,r) > \overline{u}_s(x,r)$,

(ii) stop if $\overline{u}_0(x,r) = \overline{u}_s(x,r)$.

For illustration, suppose that the inspector's preference ordering corresponds to a "nothing but the best" utility function, defined by $u(1) = 1$, $u(x) = 0$, $x = 2,\ldots,n$. It is then easy to show that

$$\overline{u}_s(1,r) = \frac{r}{n},$$

$$\overline{u}_s(x,r) = 0, \quad x = 2,\ldots,n;$$

thus, if $x > 1$,

$$\overline{u}_0(x,r) > \overline{u}_s(x,r), \quad r = 1,\ldots,n-1.$$

This implies that *inspection should never be terminated if the current object is not the best seen so far*. The decision as to whether to stop if $x = 1$ is determined from the equation

$$\overline{u}_0(x,r) = \max\left\{\frac{r}{n}, \frac{r}{n}\left(\frac{1}{n-1}+\cdots+\frac{1}{r}\right)\right\},$$

which is easily verified by induction. If r^* is the smallest positive integer for which

$$\frac{1}{n-1} + \frac{1}{n-2} + \cdots + \frac{1}{r^*} \leq 1,$$

the optimal procedure is defined as follows:

(i) continue until at least r^* objects have been inspected;

(ii) if the r^*th object is the best so far, stop;

(iii) otherwise, continue until the object under inspection is the best so far, then stop (stopping in any case if the nth stage is reached).

If n is large, approximation of the sum in the above inequality by an integral readily yields the approximation $r^* \approx n/e$. For further details, see DeGroot (1970, Chapter 13), whose account is based closely on Lindley (1961a). For reviews of further, related work on this fascinating problem, see Freeman (1983) and Ferguson (1989).

Applied to the problem of "choosing a mate", and assuming that potential partners are encountered uniformly over time between the ages of 16 and 60, the above analysis suggests delaying a choice until one is at least 32 years old, thereafter ending the search as soon as one encounters someone better than anyone encountered thus far. Readers who are suspicious of putting this into practice have the option, of course, of staying at home and continuing their study of this volume.

Sequential decision problems are now further illustrated by considering the important special case of situations involving an initial choice of experimental design.

2.6.3 Design of Experiments

A simple, very important example of a sequential problem is provided by the situation where we have available a class of experiments, one of which is to be performed in order to provide information for use in a subsequent decision problem. We want to choose the "best" experiment. The structure of this problem, which embraces the topic usually referred to as the problem of *experimental design*, may be diagrammatically described by means of a sequential decision tree such as that shown in Figure 2.6.

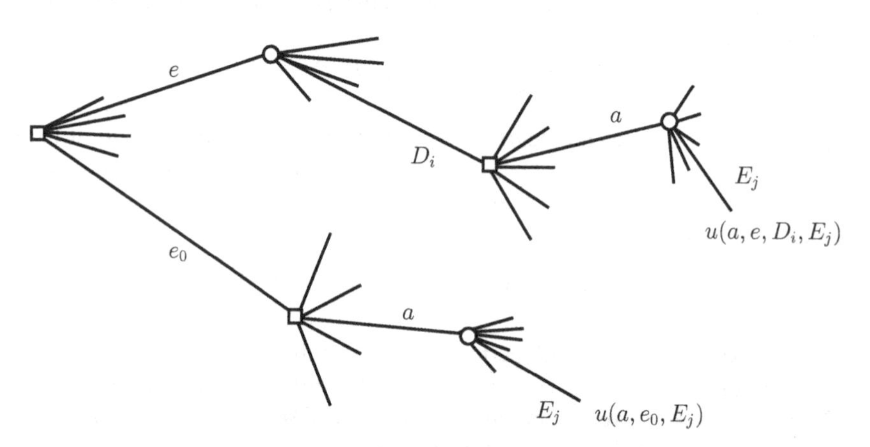

Figure 2.6 *Decision tree for experimental design*

We must first choose an experiment e and, in light of the data D obtained, take an action a, which, were event E to occur, would produce a consequence having utility which, modifying earlier notation in order to be explicit about the elements involved, we denote by $u(a, e, D, E)$. Usually, we also have available the possibility, denoted by e_0 and referred to as the *null experiment*, of directly choosing an action without performing any experiment.

Within the general structure for sequential decision problems developed in the previous section, we note that the possible sets of data obtainable may depend on the particular experiment performed, the set of available actions may depend on the results of the experiment performed, and the sets of consequences and labelling events may depend on the particular combination of experiment and action chosen. However, in our subsequent development we will use a simplified notation which suppresses these possible dependencies in order to centre attention on other, more important, aspects of the problem.

We have seen, in Section 2.6.2, that to solve a sequential decision problem we start at the last stage and work backwards. In this case, the expected utility of option a, given the information available at the stage when the action is to be taken, is

$$\overline{u}(a, e, D_i) = \sum_{j \in J} u(a, e, D_i, E_j) P(E_j \mid e, D_i, a).$$

For each pair (e, D_i) we can therefore choose the best possible continuation; namely, that action a_i^* which maximises the expression given above. Thus, the expected utility of the pair (e, D_i) is given by

$$\overline{u}(e, D_i) = \overline{u}(a_i^*, e, D_i) = \max_a \overline{u}(a, e, D_i).$$

We are now in a position to determine the best possible experiment. This is that e which maximises, in the class of available experiments, the unconditional expected utility

$$\overline{u}(e) = \sum_{i \in I} \overline{u}(a_i^*, e, D_i) P(D_i \mid e),$$

where $P(D_i \mid e)$ denotes the degree of belief attached to the occurrence of data D_i if e were the experiment chosen. On the other hand, the expected utility of performing no experiment and choosing that action a_0^* which maximises the (prior) expected utility is

$$\overline{u}(e_0) = \overline{u}(a_0^*, e_0) = \max_a \sum_{j \in J} u(a, e_0, E_j,) P(E_j \mid e_0, a),$$

so that an experiment e is worth performing if and only if $\overline{u}(e) > \overline{u}(e_0)$.

Naturally, $\overline{u}(a, e, D_i), \overline{u}(e, D_i)$ and $\overline{u}(e)$ are different functions defined on different spaces. However, to simplify the notation and without danger of confusion we shall always use $\overline{u}$ to denote an expected utility.

Proposition 2.26. (***Optimal experimental design***). *The optimal action is to perform the experiment* e^* *if* $\overline{u}(e^*) > \overline{u}(e_0)$ *and* $\overline{u}(e^*) = \max_e \overline{u}(e)$; *otherwise, the optimal action is to perform no experiment.*

Proof. This is immediate from Proposition 2.25. $\triangleleft$

It is often interesting to determine the value which additional information might have in the context of a given decision problem.

The expected value of the information provided by new data may be computed as the (posterior) expected difference between the utilities which correspond to optimal actions after and before the data have been obtained.

Definition 2.18. (***The value of additional information***).

(i) *The* ***expected value of the data*** D_i *provided by an experiment* e *is*

$$v(e, D_i) = \sum_{j \in J} \{u(a_i^*, e, D_i, E_j) - u(a_0^*, e_0, E_j)\} \, P(E_j \mid e, D_i, a_i^*);$$

where a_i^*, a_0^* *are, respectively, the optimal actions given* D_i, *and with no data.*

(ii) *the* ***expected value of an experiment*** e *is given by*

$$v(e) = \sum_{i \in I} v(e, D_i) p(D_i \mid e).$$

It is sometimes convenient to have an upper bound for the expected value $v(e)$ of an experiment e. Let us therefore consider the optimal actions which would be available with perfect information, i.e., were we to know the particular event E_j which will eventually occur, and let $a_{(j)}^*$ be the optimal action given E_j, i.e., such that, for all E_j,

$$u(a_{(j)}^*, e_0, E_j) = \max_a u(a, e_0, E_j).$$

Then, given E_j, the loss suffered by choosing any other action a will be

$$u(a_{(j)}^*, e_0, E_j) - u(a, e_0, E_j).$$

For $a = a_0^*$, the optimal action under prior information, this difference will measure, conditional on E_j, the value of perfect information and, under appropriate conditions, its expected value will provide an upper bound for the increase in utility which additional data about the E_j's could be expected to provide.

Definition 2.19. (***Expected value of perfect information***). *The* ***opportunity loss*** *which would be suffered if action* a *were taken and event* E_j *occurred is*

$$l(a, E_j) = \max_{a_i} u(a_i, e_0, E_j) - u(a, e_0, E_j);$$

the expected value of perfect information is then given by

$$v^*(e_0) = \sum_{j \in J} l(a_0^*, E_j) P(E_j \mid a_0^*).$$

It is important to bear in mind that the functions $v(D_i), v(e)$ and the number $v^*(e_0)$, all crucially depend on the (prior) probability distributions $\{(P(E_j \mid a), a \in \mathcal{A}\}$ although, for notational convenience, we have not made this dependency explicit.

In many situations, the utility function $u(a, e, D_i, E_j)$ may be thought of as made up of two separate components. One is the (experimental) *cost* of performing e and obtaining D_i; the other is the (terminal) *utility* of directly choosing a and then finding that E_j occurs. Often, the latter component does not actually depend on the preceding e and D_i, so that, assuming additivity of the two components, we may write $u(a, e, D_i, E_j) = u(a, e_0, E_j) - c(e, D_i)$ where $c(e, D_i) \geq 0$. Moreover, the probability distributions over the events are often independent of the action taken. When these conditions apply, we can establish a useful upper bound for the expected value of an experiment in terms of the difference between the expected value of complete data and the expected cost of the experiment itself.

Proposition 2.27. (***Additive decomposition***). *If the utility function has the form*

$$u(a, e, D_i, E_j) = u(a, e_0, E_j) - c(e, D_i),$$

with $c(e, D_i) \geq 0$, *and the probability distributions are such that*

$$p(E_j \mid e, D_i, a) = p(E_j \mid e, D_i), \quad p(E_j \mid e_0, a) = p(E_j \mid e_0),$$

then, for any available experiment e,

$$v(e) \leq v^*(e_0) - \overline{c}(e),$$

where

$$\overline{c}(e) = \sum_{i \in I} c(e, D_i) P(D_i \mid e)$$

is the expected cost of e.

Proof. Using Definitions 2.18 and 2.19, $v(e)$ may be written as

$$\sum_{i \in I} \Big[\sum_{j \in J} \Big\{ u(a_i^*, e_0, E_j) - c(e, D_i) - u(a_0^*, e_0, E_j) \Big\} P(E_j \mid e, D_i) \Big] P(D_i \mid e)$$

$$= \sum_{i \in I} \Big[\max_a \sum_{j \in J} \Big\{ u(a, e_0, E_j) - u(a_0^*, e_0, E_j) \Big\} P(E_j \mid e, D_i) \Big] P(D_i \mid e) - \overline{c}(e)$$

$$\leq \sum_{i \in I} \sum_{j \in J} \Big[\max_a u(a, e_0, E_j) - u(a_0^*, e_0, E_j) \Big] P(E_j \cap D_i \mid e) - \overline{c}(e)$$

and, hence,

$$
\begin{aligned}
v(e) &\leq \Big\{ \sum_{j \in J} l(a, E_j) P(E_j \mid e_0) \Big\} \Big\{ \sum_{i \in I} P(D_i \mid E_j, e) \Big\} - \bar{c}(e) \\
&= \Big\{ \sum_{j \in J} l(a, E_j) P(E_j \mid e_0) \Big\} - \bar{c}(e) \\
&= v^*(e_0) - \bar{c}(e),
\end{aligned}
$$

as stated. $\triangleleft$

In Section 2.7, we shall study in more detail the special case of experimental design in situations where data are being collected for the purpose of pure inference, rather than as an input into a directly practical decision problem.

We have shown that the simple decision problem structure introduced in Section 2.2, and the tools developed in Sections 2.3 to 2.5, suffice for the analysis of complex, sequential problems which, at first sight, appear to go beyond that simple structure. In particular, we have seen that the important problem of experimental design can be analysed within the sequential decision problem framework. We shall now use this framework to analyse the very special form of decision problem posed by *statistical inference*, thus establishing the fundamental relevance of these foundational arguments for statistical theory and practice.

2.7 INFERENCE AND INFORMATION

2.7.1 Reporting Beliefs as a Decision Problem

The results on quantitative coherence (Sections 2.2 to 2.5) establish that if we aspire to analyse a given decision problem, $\{\mathcal{E}, \mathcal{C}, \mathcal{A}, \leq\}$, in accordance with the axioms of quantitative coherence, we must represent degrees of belief about uncertain events in the form of a finite probability measure over $\mathcal{E}$ and values for consequences in the form of a utility function over $\mathcal{C}$. Options are then to be compared on the basis of expected utility.

The probability measure represents an individual's beliefs conditional on his or her current state of information. Given the initial state of information described by M_0 and further information in the form of the assumed occurrence of a significant event G, we previously denoted such a measure by $P(. \mid G)$. We now wish to specialise our discussion somewhat to the case where G can be thought of as a description of the outcome of an investigation (typically a survey, or an experiment) involving the deliberate collection of data (usually, in numerical form). The event G will then be defined directly in terms of the counts or measurements obtained, either as a precise statement, or involving a description of intervals within which readings lie. To emphasise the fact that G characterises the actual *data* collected, we shall denote the event which describes the new information obtained by D. An individual's degree of belief measure over $\mathcal{E}$ will then be denoted $P(. \mid D)$ representing the individual's current beliefs in the light of the data obtained (where, again, we have suppressed, for notational convenience, the explicit dependence on M_0). So far as uncertainty about the events of $\mathcal{E}$ is concerned, $P(. \mid D)$ constitutes a complete encapsulation of the information provided by D, given the initial state of information M_0. Moreover, in conjunction with the specification of a utility function, $P(. \mid D)$ provides all that is necessary for the calculation of the expected utility of any option and, hence, for the solution of *any decision problem* defined in terms of the frame of reference adopted.

Starting from the decision problem framework, we thus have a formal justification for the main topic of this book; namely, *the study of models and techniques for analysing the ways in which beliefs are modified by data.* However, many eminent writers have argued that basic problems of reporting scientific inferences do not fall within the framework of decision problems as defined in earlier sections:

> Statistical inferences involve the data, a specification of the set of possible populations sampled and a question concerning the true population... Decisions are based on not only the considerations listed for inferences, but also on an assessment of the losses resulting from wrong decisions... (Cox, 1958);
>
> ... a considerable body of doctrine has attempted to explain, or rather to reinterpret these (significance) tests on the basis of quite a different model, namely as means to making decisions in an acceptance procedure. The differences between these two situations seem to the author many and wide, ... (Fisher, 1956/1973).

If views such as these were accepted, they would, of course, undermine our conclusion that *problems concerning uncertainty are to be solved by revising degrees of belief in the light of new data in accordance with Bayes' theorem.* Our main purpose in this section is therefore to demonstrate that *the problem of reporting inferences is essentially a special case of a decision problem.*

By way of preliminary clarification, let us recall from Section 2.1 that we distinguished two, possibly distinct, reasons for trying to think rationally about uncertainty. On the one hand, quoting Ramsey (1926), we noted that, even if an immediate decision problem does not appear to exist, we know that our statements of uncertainty may be used by others in contexts representable within the decision framework. In such situations, our conclusion holds. On the other hand, quoting Lehmann (1959/1986), we noted that the inference, or inference statement, may sometimes be regarded *as an end in itself,* to be judged independently of any "practical" decision problem. It is this case that we wish to consider in more detail in this section, establishing that, indeed, it can be regarded as falling within the general framework of Sections 2.2 to 2.5.

Formalising the first sentence of the remark of Cox, given above, a pure inference problem may be described as one in which we seek to learn which of a set of mutually exclusive "hypotheses" ("theories", "states of nature", or "model parameters") is true. From a strictly realistic viewpoint, there is always, implicitly, a finite set of such hypotheses, say $\{H_j,\, j \in J\}$, although it may be mathematically convenient to work as if this were not the case. We shall regard this set of hypotheses as equivalent to a finite partition of the certain event into events $\{E_j,\, j \in J\}$, having the interpretation $E_j \equiv$ "the hypothesis H_j is true". The actions available to an individual are the various inference statements that might be made about the events $\{E_j,\, j \in J\}$, the latter constituting the uncertain events corresponding to each action. To complete the basic decision problem framework, we need to acknowledge that, corresponding to each *inference statement* and each E_j, there will be a *consequence*; namely, the record of what the individual put forward as an appropriate inference statement, together with what actually turned out to be the case.

If we aspire to quantitative coherence in such a framework, we know that our uncertainty about the $\{E_j,\, j \in J\}$ should be represented by $\{P(E_j \mid D),\, j \in J\}$, where $P(.\mid D)$ denotes our current degree of belief measure, given data D in addition to the initial information M_0. It is natural, therefore, to regard the set of possible *inference statements* as the class of probability distributions over $\{E_j,\, j \in J\}$ compatible with the information D. The inference reporting problem can thus be viewed as one of choosing a probability distribution to serve as an inference statement. *But there is nothing (so far) in this formulation which leads to the conclusion that the best action is to state one's actual beliefs.* Indeed, we know from our earlier development that options cannot be ordered without an (implicit or explicit) specification of utilities for the consequences. We shall consider this specification and its implications in the following sections. A particular form of utility function for inference statements will be introduced and it will then be seen that the idea of *inference as decision* leads to rather natural interpretations of commonly used information measures in terms of expected utility. In the discussion which follows, we shall only consider the case of finite partitions $\{E_j,\, j \in J\}$. Mathematical extensions will be discussed in Chapter 3.

2.7.2 The Utility of a Probability Distribution

We have argued above that the provision of a statistical inference statement about a class of exclusive and exhaustive "hypotheses" $\{E_j,\, j \in J\}$, conditional on some relevant data D, may be precisely stated as a decision problem, where the set of "hypotheses" $\{E_j,\, j \in J\}$ is a partition consisting of elements of $\mathcal{E}$, and the action space $\mathcal{A}$ relates to the class $\mathcal{Q}$ of conditional probability distributions over $\{E_j,\, j \in J\}$; thus,

$$\mathcal{Q} = \Big\{\boldsymbol{q} \equiv (q_j,\, j \in J); \quad q_j \geq 0, \quad \sum\nolimits_{j \in J} q_j = 1\Big\},$$

where q_j is assumed to be the probability which, conditional on the available data D, an individual *reports* as the probability of $E_j \equiv H_j$ being true. The set of consequences $\mathcal{C}$, consists of all pairs $(\boldsymbol{q}, E_j)$, representing the conjunctions of reported beliefs and true hypotheses. The action corresponding to the choice of $\boldsymbol{q}$ is defined as $\{(\boldsymbol{q}, E_j) \mid E_j,\ j \in J\}$.

To avoid triviality, we assume that none of the hypotheses is certain and that, without loss of generality, all are compatible with the available data; i.e., that all the E_j's are significant given D, so that (Proposition 2.5) $\emptyset < E_j \cap D < D$ for all $j \in J$. If this were not so, we could simply discard any incompatible hypotheses. It then follows from Proposition 2.17(iii) that each of the personal degrees of belief attached by the individual to the conflicting hypotheses given the data must be strictly positive. Throughout this section, we shall denote by

$$\boldsymbol{p} \equiv (p_j = P(E_j \mid D),\ j \in J)\,, \quad p_j > 0, \quad \sum\nolimits_{j \in J} p_j = 1,$$

the probability distribution which describes, conditional again on the available data D, the individual's *actual* beliefs about the alternative "hypotheses".

We emphasise again that, in the structure described so far, there is no logical requirement which forces an individual to *report* the probability distribution $\boldsymbol{p}$ which describes his or her personal *beliefs*, in preference to any other probability distribution $\boldsymbol{q}$ in $\mathcal{Q}$.

We complete the specification of this decision problem by inducing the preference ordering through direct specification of a utility function $u(.)$, which describes the "value" $u(\boldsymbol{q}, E_j)$ of reporting the probability distribution $\boldsymbol{q}$ as the final inferential summary of the investigation, were E_j to turn out to be the true "state of nature". Our next task is to investigate the properties which such a function should possess in order to describe a preference pattern which accords with what a scientific community ought to demand of an inference statement. This special class of utility functions is often referred to as the class of score functions (see also Section 2.8) since the functions describe the possible "scores" to be awarded to the individual as a "prize" for his or her "prediction".

Definition 2.20. (***Score function***). *A score function u for probability distributions $\boldsymbol{q} = \{q_j,\ j \in J\}$ defined over a partition $\{E_j,\ j \in J\}$ is a mapping which assigns a real number $u\{\boldsymbol{q}, E_j\}$ to each pair $(\boldsymbol{q}, E_j)$. This function is said to be* ***smooth*** *if it is continuously differentiable as a function of each q_j.*

It seems natural to assume that score functions should be smooth (in the intuitive sense), since one would wish small changes in the reported distribution to produce only small changes in the obtained score. The mathematical condition imposed is a simple and convenient representation of such smoothness.

We have characterised the problem faced by an individual reporting his or her beliefs about conflicting "hypotheses" as a problem of choice among probability distributions over $\{E_j, j \in J\}$, with preferences described by a score function. This is a well specified problem, whose solution, in accordance with our development based on quantitative coherence, is to report that distribution q which maximises the expected utility

$$\sum_{j \in J} u(\boldsymbol{q}, E_j)\, P(E_j \mid D).$$

In order to ensure that a coherent individual is also *honest*, we need a form of $u(.)$ which guarantees that the expected utility is maximised if, and only if, $q_j = p_j = P(E_j \mid D)$, for each j; otherwise, the individual's best policy could be to report something other than his or her true beliefs. This motivates the following definition:

Definition 2.21. (***Proper score function***). *A score function u is proper if, for each strictly positive probability distribution $\boldsymbol{p} = \{p_j, j \in J\}$ defined over a partition $\{E_j, j \in J\}$,*

$$\sup_{\boldsymbol{q} \in \mathcal{Q}} \left\{ \sum_{j \in J} u(\boldsymbol{q}, E_j) p_j \right\} = \sum_{j \in J} u(\boldsymbol{p}, E_j) p_j \, ,$$

where the supremum, taken over the class $\mathcal{Q}$ of all probability distributions over $\{E_j, j \in J\}$, is attained if, and only if, $\boldsymbol{q} = \boldsymbol{p}$.

It would seem reasonable that, in a scientific inference context, one should require a score function to be proper. Whether a scientific report presents the inference of a single scientist or a range of inferences, purporting to represent those that might be made by some community of scientists, we should wish to be reassured that any reported inference could be justified as a *genuine* current belief.

Smooth, proper score functions have been successfully used in practice in the following contexts: (i) to determine an appropriate fee to be paid to meteorologists in order to encourage them to report reliable predictions (Murphy and Epstein, 1967); (ii) to score multiple choice examinations so that students are encouraged to assign, over the possible answers, probability distributions which truly describe their beliefs (de Finetti, 1965; Bernardo, 1981b, Section 3.6); (iii) to devise general procedures to elicit personal probabilities and expectations (Savage, 1971); (iv) to select best subsets of variables for prediction purposes in political or medical contexts (Bernardo and Bermúdez, 1985).

The simplest proper score function is the quadratic function (Brier, 1950; de Finetti, 1962) defined as follows.

Definition 2.22. (***Quadratic score function***). *A quadratic score function for probability distributions* $\boldsymbol{q} = \{q_j, j \in J\}$ *defined over a partition* $\{E_j,\ j \in J\}$ *is any function of the form*

$$u\{\boldsymbol{q}, E_j\} = A\left\{2q_j - \sum_{i \in J} q_i^2\right\} + B_j, \quad A > 0,$$

where $\boldsymbol{q} = \{q_j,\ j \in J\}$ *is any probability distribution over* $\{E_j,\ j \in J\}$.

Using the indicator function for E_j, 1_{Ej}, an alternative expression for the quadratic score function is given by

$$u\{\boldsymbol{q}, E_j\} = A\left\{1 - \sum_{i \in J} (q_i - 1_{Ej})^2\right\} + B_j, \quad A > 0,$$

which makes explicit the role of a 'penalty' equal to the squared euclidean distance from $\boldsymbol{q}$ to a perfect prediction.

Proposition 2.28. *A quadratic score function is proper.*

Proof. We have to maximise, over $\boldsymbol{q}$, the expected score

$$\sum_{j \in J} u\{\boldsymbol{q}, E_j\}\, p_j = \sum_{j \in J}\left\{A\left(2q_j - \sum_{i \in J} q_i^2\right) + B_j\right\} p_j\,.$$

Taking derivatives with respect to the q_j's and equating them to zero, we have the system of equations $2p_j - 2q_j\{\sum_k p_k\} = 0$, $j \in J$, and since $\sum_i p_i = 1$, we have $q_j = p_j$ for all j. It is easily checked that this gives a maximum. ◁

Note that in the proof of Proposition 2.28 we did *not* need to use the condition $\sum_j q_j = 1$; this is a rather special feature of the quadratic score function.

A further condition is required for score functions in contexts, which we shall refer to as "pure inference problems", where the value of a distribution, $\boldsymbol{q}$, is only to be assessed in terms of the probability it assigned to the actual outcome.

Definition 2.23. (***Local score function***). *A score function* u *is local if, for each element* $\boldsymbol{q} = \{q_j,\ j \in J\}$ *of the class* $\mathcal{Q}$ *of probability distributions defined over a partition* $\{E_j,\ j \in J\}$, *there exist functions* $\{u_j(.),\ j \in J\}$ *such that* $u\{\boldsymbol{q}, E_j\} = u_j(q_j)$.

It is intuitively clear that the preferences of an individual scientist faced with a pure inference problem should correspond to the ordering induced by a local score function. The reason for this is that, by definition, in a "pure" inference problem we are solely concerned with "the truth". It is therefore natural that if E_j, say, turns out to be true, the individual scientist should be assessed (i.e., scored) only on the basis of his or her reported judgement about the plausibility of E_j.

This can be contrasted with the forms of "score" function that would typically be appropriate in more directly practical contexts. In stock control, for example, probability judgements about demand would usually be assessed in the light of the relative seriousness of under- or over-stocking, rather than by just concentrating on the belief previously attached to what turned out to be the actual level of demand.

Note that, in Definition 2.23, the functional form $u_j(p_j)$ of the dependence of the score on the probability attached to the true E_j is allowed to vary with the particular E_j considered. By permitting different $u_j(.)$'s for each E_j, we allow for the possibility that "bad predictions" regarding some "truths" may be judged more harshly than others.

The situation described by a local score function is, of course, an idealised, limit situation, but one which seems, at least approximately, appropriate in reporting pure scientific research. In addition, later in this section we shall see that certain well-known criteria for choosing among experimental designs are optimal if, and only if, preferences are described by a smooth, proper, local score function.

Proposition 2.29. (***Characterisation of proper local score functions***). *If u is a smooth, proper, local score function for probability distributions $\boldsymbol{q} = \{q_j, j \in J\}$ defined over a partition $\{E_j,\ j \in J\}$ which contains more than two elements, then it must be of the form $u\{\boldsymbol{q}, E_j\} = A \log q_j + B_j$, where $A > 0$ and the B_j's are arbitrary constants.*

Proof. Since $u(.)$ is local and proper, then for some $\{u_j(.), j \in J\}$, we must have

$$\sup_{\boldsymbol{q}} \sum_{j \in J} u(\boldsymbol{q}, E_j)\, p_j = \sup_{\boldsymbol{q}} \sum_{j \in J} u_j(q_j)\, p_j = \sum_{j \in J} u_j(p_j)\, p_j,$$

where $p_j > 0,\ \sum_j p_j = 1$ and the supremum is taken over the class of probability distributions $\boldsymbol{q} = (q_j, j \in J),\ q_j \geq 0,\ \sum_j q_j = 1$.

Writing $\boldsymbol{p} = \{p_1, p_2, \ldots\}$ and $\boldsymbol{q} = \{q_1, q_2, \ldots\}$, with

$$p_1 = 1 - \sum_{j>1} p_j, \quad q_1 = 1 - \sum_{j>1} q_j,$$

we seek $\{u_j(.)\,, j \in J\}$, giving an extremal of

$$F\{q_2, q_3, \ldots\} = \left(1 - \sum_{j>1} p_j\right) u_1\left(1 - \sum_{j>1} q_j\right) + \sum_{j>1} p_j u_j(q_j)\,,$$

For $\{q_2, q_3, \ldots\}$ to make F stationary it is necessary (see e.g. Jeffreys and Jeffreys, 1946, p. 315) that

$$\frac{\partial}{\partial \alpha} F\{q_2 + \alpha\varepsilon_2, q_3 + \alpha\varepsilon_3, \ldots\}\bigg|_{\alpha=0} = 0$$

for any $\varepsilon = \{\varepsilon_2, \varepsilon_3, \ldots\}$ such that all the ε_j are sufficiently small. Calculating this derivative, the condition is seen to reduce to

$$\sum_{j>1} \left\{ \left(1 - \sum_{i>1} p_i\right) u_1' \left(1 - \sum_{j>1} q_j\right) - p_j u_j'(q_j) \right\} \varepsilon_j = 0$$

for all ε_j's sufficiently small, where u' stands for the derivative of u. Moreover, since u is proper, $\{p_2, p_3, \ldots\}$ must be an extremal of F and thus we have the system of equations

$$p_j \, u_j'(p_j) = \left(1 - \sum_{i>1} p_i\right) u_1' \left(1 - \sum_{i>1} p_i\right), \quad j = 2, 3, \ldots$$

so that all the functions $u_j, j = 1, 2, \ldots$ satisfy the same functional equation, namely

$$p_j \, u_j'(p_j) = p_1 \, u_1'(p_1), \quad j = 2, 3, \ldots,$$

for all $\{p_2, p_3, \ldots\}$ and, hence,

$$p \, u_j'(p) = A, \quad 0 < p \leq 1, \qquad \text{for all } j = 1, 2, \ldots$$

so that $u_j(p) = A \log p + B_j$. The condition $A \geq 0$ suffices to guarantee that the extremal found is indeed a maximum. $\triangleleft$

Definition 2.24. (***Logarithmic score function***). *A logarithmic score function for strictly positive probability distributions* $\boldsymbol{q} = \{q_j, j \in J\}$ *defined over a partition* $\{E_j, \; j \in J\}$ *is any function of the form*

$$u\{\boldsymbol{q}, E_j\} = A \log \, q_j + B_j, \quad A > 0.$$

If the partition $\{E_j, j \in J\}$ only contains two elements, so that the partition is simply $\{H, H^c\}$, the locality condition is, of course, vacuous. In this case, $u\{\boldsymbol{q}, E_j\} = u\{(q_1, 1 - q_1), 1_H\} = f(q_1, 1_H)$, say, where 1_H is the indicator function for H, and the score function only depends on the probability q_1 attached to H, whether or not H occurs.

For $u\{(q_1, 1 - q_1), \; 1_H\}$ to be proper we must have

$$\sup_{q_1 \in [0,1]} \{p_1 \, f(q_1, 1) + (1 - p_1) \, f(q_1, 0)\} = p_1 \, f(p_1, 1) + (1 - p_1) \, f(p_1, 0)$$

so that, if the score function is smooth, then f must satisfy the functional equation

$$x \, f'(x, 1) + (1 - x) \, f'(x, 0) = 0.$$

The logarithmic function $f(x,1) = A \log x + B_1, f(x,0) = A \log(1-x) + B_2$ is then just one of the many possible solutions (see Good, 1952).

We have assumed that the probability distributions to be considered as options assign strictly positive q_j to each E_j. This means that, given any particular $\boldsymbol{q} \in \mathcal{Q}$, we have no problem in calculating the expected utility arising from the logarithmic score function. It is worth noting, however, that since we place no (strictly positive) lower bound on the possible q_j, we have an example of an unbounded decision problem; i.e., a decision problem without extreme consequences.

2.7.3 Approximation and Discrepancy

We have argued that the optimal solution to an inference reporting problem (either for an individual, or for each of several individuals) is to state the appropriate actual beliefs, $\boldsymbol{p}$, say. From a technical point of view, however, particularly within the mathematical extensions to be considered in Chapter 3, the precise computation of $\boldsymbol{p}$ may be difficult and we may choose instead to report an approximation to our beliefs, $\boldsymbol{q}$, say, on the grounds that $\boldsymbol{q}$ is "close" to $\boldsymbol{p}$, but much easier to calculate. The justification of such a procedure requires a study of the notion of "closeness" between two distributions.

Proposition 2.30. (***Expected loss in probability reporting***). *If preferences are described by a logarithmic score function, the expected loss of utility in reporting a probability distribution $\boldsymbol{q} = \{q_j, j \in J\}$ defined over a partition $\{E_j, j \in J\}$, rather than the distribution $\boldsymbol{p} = \{p_j, j \in J\}$ representing actual beliefs, is given by*

$$\delta\{\boldsymbol{q} \mid \boldsymbol{p}\} = A \sum_{j \in J} p_j \log (p_j / q_j), \quad A > 0.$$

Moreover, $\delta\{\boldsymbol{q} \mid \boldsymbol{p}\} \geq 0$ with equality if and only if $\boldsymbol{q} = \boldsymbol{p}$.

Proof. Using Definition 2.24, the expected utility of reporting $\boldsymbol{q}$ when $\boldsymbol{p}$ is the actual distribution of beliefs is $\overline{u}(\boldsymbol{q}) = \sum_j \{A \log q_j + B_j\} p_j$, and thus

$$\begin{aligned} \delta\{\boldsymbol{q} \mid \boldsymbol{p}\} &= \overline{u}(\boldsymbol{p}) - \overline{u}(\boldsymbol{q}) \\ &= \sum_{j \in J} \{(A \log p_j + B_j) - (A \log q_j + B_j)\} \, p_j = A \sum_{j \in J} p_j \log \frac{p_j}{q_j} \, . \end{aligned}$$

The final statement in the theorem is a consequence of Proposition 2.29 since, because the logarithmic score function is proper, the expected utility of reporting $\boldsymbol{q}$ is maximised if, and only if, $\boldsymbol{q} = \boldsymbol{p}$, so that $\overline{u}(\boldsymbol{p}) \geq \overline{u}(\boldsymbol{q})$, with equality if, and only if, $p = q$. An immediate direct proof is obtained using the fact that for all $x > 0$, $\log x \leq x - 1$ with equality if, and only if, $x = 1$. Indeed, we then have

$$\begin{aligned} -\delta\{q\,|\,p\} &= \sum_{j\in J} p_j \log \frac{q_j}{p_j} \\ &\leq \sum_j p_j\{(q_j/p_j) - 1\} = \sum_{j\in J} q_j - \sum_{j\in J} p_j = 1 - 1 = 0, \end{aligned}$$

with equality if, and only if, $q_j = p_j$ for all j. $\triangleleft$

The quantity $\delta\{q\,|\,p\}$, which arises here as a difference between two expected utilities, was introduced by Kullback and Leibler (1951) as an *ad hoc* measure of (directed) *divergence* between two probability distributions.

Combining Propositions 2.29 and 2.30, it is clear that an individual with preferences approximately described by a proper local score function should beware of approximating by zero. This reflects the fact that the "tails" of the distribution are, generally speaking, extremely important in pure inference problems. This is in contrast to many practical decision problems where the form of the utility function often makes the solution robust with respect to changes in the "tails" of the distribution assumed.

Proposition 2.30 suggests a natural, general measure of "lack of fit", or *discrepancy*, between a distribution and an approximation, when preferences are described by a logarithmic score function.

Definition 2.25. ***(Discrepancy of an approximation).*** *The discrepancy between a strictly positive probability distribution* $p = \{p_j,\, j \in J\}$ *over a partition* $\{E_j,\, j \in J\}$ *and an approximation* $\hat{p} = \{\hat{p}_j,\, j \in J\}$ *is defined by*

$$\delta\{\hat{p}\,|\,p\} = \sum_{j\in J} p_j \log \frac{p_j}{\hat{p}_j}\, .$$

Example 2.5. ***(Poisson approximation to a binomial distribution).*** The behaviour of $\delta\{\hat{p}\,|\,p\}$ is well illustrated by a familiar, elementary example. Consider the binomial distribution

$$\begin{aligned} p_j &= \binom{n}{j}\theta^j(1-\theta)^{n-j}, \quad j = 0, 1, \ldots, n, \\ &= 0\,, \ \text{otherwise} \end{aligned}$$

and let

$$\hat{p}_j = \exp\{-n\theta\}\frac{(n\theta)^j}{j!}, \quad j = 0, 1, \ldots$$

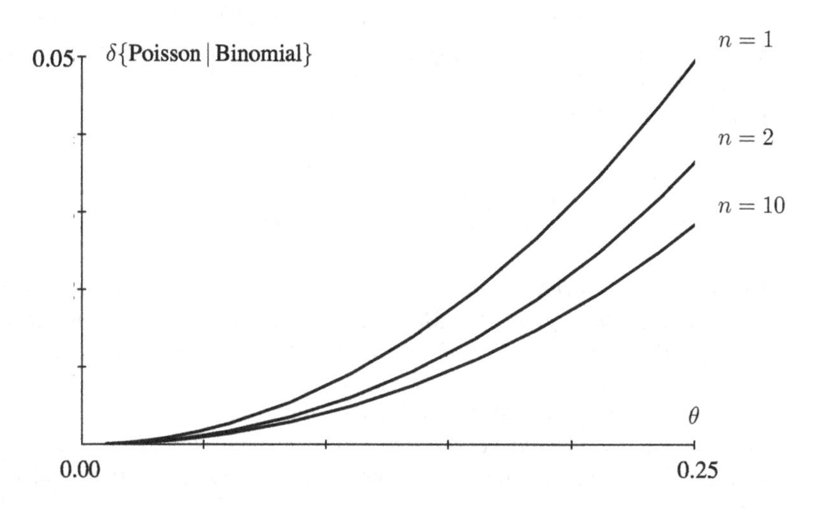

Figure 2.7 *Discrepancy between a binomial distribution and its Poisson approximation (logarithms to base 2).*

be its Poisson approximation. It is apparent from Figure 2.7 that $\delta\{\hat{p} \mid p\}$ decreases as either n increases or θ decreases, or both, and that the second factor is far more important than the first. However, it follows from our previous discussion that it would not be a good idea to reverse the roles and try to approximate a Poisson distribution by a binomial distribution.

When, as in Figure 2.7, logarithms to base 2 are used, the utility and discrepancy are measured on the well-known scale of bits of information (or *entropy*), which can be interpreted in terms of the expected number of yes-no questions required to identify the true event in the partition (see, for example, de Finetti, 1970/1974, p. 103, or Renyi, 1962/1970, p. 564).

Clearly, Definition 2.25 provides a systematic approach to approximation in pure inference contexts. The best *approximation* within a given family will be that which *minimises the discrepancy.*

2.7.4 Information

In Section 2.4.2, we showed that, for quantitative coherence, any new information D should be incorporated into the analysis by updating beliefs via Bayes' theorem, so that the initial representation of beliefs $P(.)$ is updated to the conditional probability measure $P(. \mid D)$. In Section 2.7.2, we showed that, within the context of the pure inference reporting problem, utility is defined in terms of the logarithmic score function.

Proposition 2.31. ***(Expected utility of data).*** *If preferences are described by a logarithmic score function for the class of probability distributions defined over a partition* $\{E_j,\ j \in J\}$*, then the expected increase in utility provided by data* D*, when the initial probability distribution* $\{P(E_j), j \in J\}$ *is strictly positive, is given by*

$$A \sum_{j \in J} P(E_j \mid D) \log \frac{P(E_j \mid D)}{P(E_j)},$$

where $A > 0$ *is arbitrary, and* $\{P(E_j \mid D), j \in J\}$ *is the conditional probability distribution, given* D*. Moreover, this expected increase in utility is non-negative and is zero if, and only if,* $P(E_j \mid D) = P(E_j)$ *for all* j*.*

Proof. By Definition 2.24, the utilities of reporting $P(.)$ or $P(. \mid D)$, were E_j known to be true, would be $A \log P(E_j) + B_j$ and $A \log P(E_j \mid D) + B_j$, respectively. Thus, conditional on D, the expected increase in utility provided by D is given by

$$\sum_{j \in J} \{(A \log P(E_j \mid D) + B_j) - (A \log P(E_j) + B_j)\} P(E_j \mid D)$$

$$= A \sum_{j \in J} P(E_j \mid D) \log \frac{P(E_j \mid D)}{P(E_j)},$$

which, by Proposition 2.30, is non-negative and is zero if and only if, for all j, $P(E_j \mid D) = P(E_j)$. $\triangleleft$

In the context of pure inference problems, we shall find it convenient to underline the fact that, because of the use of the logarithmic score function, utility assumes a special form and establishes a link between utility theory and classical information theory. This motivates Definitions 2.26 and 2.27.

Definition 2.26. ***(Information from data).*** *The amount of information about a partition* $\{E_j,\ j \in J\}$ *provided by the data* D*, when the initial distribution over* $\{E_j,\ j \in J\}$ *is* $\mathbf{p}_0 = \{P(E_j), j \in J\}$*, is defined to be*

$$I(D \mid \mathbf{p}_0) = \sum_{j \in J} P(E_j \mid D) \log \frac{P(E_j \mid D)}{P(E_j)},$$

where $\{P(E_j \mid D),\ j \in J\}$ *is the conditional probability distribution given the data* D*.*

It follows from Definition 2.26 that the amount of information provided by data D is equal to $\delta(\boldsymbol{p}_0 \mid \boldsymbol{p}_D)$, the discrepancy measure if $\boldsymbol{p}_0 = \{P(E_j), j \in J\}$ is considered as an approximation to $\boldsymbol{p}_D = \{P(E_j \mid D), j \in J\}$. Another interesting interpretation of $I(D \mid \boldsymbol{p}_0)$ arises from the following analysis. Conditional on E_j, $\log P(E_j)$ and $\log P(E_j \mid D)$ measure, respectively, how good the initial and the conditional distributions are in "predicting" the "true hypothesis" $E_j = H_j$, so that $\log P(E_j \mid D) - \log P(E_j)$ is a measure of the value of D, were E_j known to be true; $I(D \mid \boldsymbol{p}_0)$ is simply the expected value of that difference calculated with respect to $\boldsymbol{p}_D$.

It should be clear from the preceding discussion that $I(D \mid \boldsymbol{p}_0)$ measures indirectly the information provided by the data in terms of the changes produced in the probability distribution of interest. The amount of information is thus seen to be a relative measure, which obviously depends on the initial distribution. Attempts to define absolute measures of information have systematically failed to produce concepts of lasting value.

In the finite case, the *entropy* of the distribution $\boldsymbol{p} = \{p_1, \ldots, p_n\}$, defined by

$$H\{\boldsymbol{p}\} = -\sum_{j=1}^{n} p_j \log p_j,$$

has been proposed and widely accepted as an absolute measure of uncertainty. The recognised fact that its apparently natural extension to the continuous case does not make sense (if only because it is heavily dependent on the particular parametrisation used) should, however, have raised doubts about the universality of this concept. The fact that, in the finite case, $H\{\boldsymbol{p}\}$ as a measure of uncertainty (and $-H\{\boldsymbol{p}\}$ as a measure of "absolute" information) seems to work correctly is explained (from our perspective) by the fact that

$$\sum_{j=1}^{n} p_j \log \frac{p_j}{n^{-1}} = \log n - H\{\boldsymbol{p}\},$$

so that, in terms of the above discussion, $-H\{\boldsymbol{p}\}$ may be interpreted, apart from an unimportant additive constant, as the amount of information which is necessary to obtain $\boldsymbol{p} = \{p_1, \ldots, p_n\}$ from an *initial discrete uniform distribution* (see Section 3.2.2), which acts as an "origin" or "reference" measure of uncertainty. As we shall see in detail later, the problem of extending the entropy concept to continuous distributions is closely related to that of defining an "origin" or "reference" measure of uncertainty in the continuous case, a role unambiguously played by the uniform distribution in the finite case. For detailed discussion of $H\{\boldsymbol{p}\}$ and other proposed entropy measures, see Renyi (1961).

We shall on occasion wish to consider the idea of the amount of information which may be expected from an experiment e, the expectation being calculated before the results of the experiment are actually available.

Definition 2.27. ***(Expected information from an experiment).*** *The expected information to be provided by an experiment* e *about a partition* $\{E_j,\, j \in J\}$*, when the initial distribution over* $\{E_j,\, j \in J\}$ *is* $\mathbf{p}_0 = \{P(E_j),\, j \in J\}$*, is given by*

$$I(e \,|\, \mathbf{p}_0) = \sum_{i \in I} I(D_i \,|\, \mathbf{p}_0)\, P(D_i),$$

where the possible results of the experiment e*,* $\{D_i,\, i \in I\}$*, occur with probabilities* $\{P(D_i), i \in I\}$*.*

Proposition 2.32. *An alternative expression for the expected information is*

$$I(e \,|\, \mathbf{p}_0) = \sum_{i \in I} \sum_{j \in J} P(E_j \cap D_i) \log \frac{P(E_j \cap D_i)}{P(E_j) P(D_i)},$$

where $P(E_j \cap D_i) = P(D_i)\; P(E_j \,|\, D_i)$*, and* $\{P(E_j \,|\, D_i), j \in J\}$ *is the conditional distribution, given the occurrence of* D_i*, corresponding to the initial distribution* $\mathbf{p}_0 = \{P(E_j), j \in J\}$*. Moreover,* $I(e \,|\, \mathbf{p}_0) \geq 0$*, with equality if and only if, for all* E_i *and* D_j*,* $P(E_j \cap D_i) = P(E_j)P(D_i)$*.*

Proof. Let $q_i = P(D_i)$, $p_j = P(E_j)$ and $p_{ji} = P(E_j \,|\, D_i)$. Then, by Definition 2.27,

$$I(e \,|\, \mathbf{p}_0) = \sum_{i \in I} \left\{ \sum_{j \in J} p_{ji} \log \frac{p_{ji}}{p_j} \right\} q_i \;=\; \sum_{i \in I} \sum_{j \in J} p_{ji} q_i \log \frac{p_{ji} q_i}{p_j q_i}$$

and the result now follows from the fact that, by Bayes' theorem,

$$P(E_j \cap D_i) = P(E_j \,|\, D_i) P(D_i) = p_{ji} q_i.$$

Since, by Proposition 2.31, $I(D_i \,|\, \mathbf{p}_0) \geq 0$ with equality iff, $P(E_j \,|\, D_i) = P(E_j)$, it follows from Definition 2.27 that $I(e \,|\, \mathbf{p}_0) \geq 0$ with equality if, and only if, for all E_j and D_i, $P(E_j \cap D_i) = P(E_j)P(D_i)$. ◁

The expression for $I(e \,|\, \mathbf{p}_0)$ given by Proposition 2.32 is Shannon's (1948) measure of expected information. We have thus found, in a decision theoretical framework, a natural interpretation of this famous measure of expected information: *Shannon's expected information is the expected utility provided by an experiment in a pure inference context, when an individual's preferences are described by a smooth, proper, local score function.*

In conclusion, we have suggested that the problem of reporting inferences can be viewed as a particular decision problem and thus should be analysed within the framework of decision theory. We have established that, with a natural characterisation of an individual's utility function when faced with a pure inference problem, preferences should be described by a logarithmic score function. We have also seen that, within this framework, discrepancy and amount of information are naturally defined in terms of expected loss of utility and expected increase in utility, respectively, and that maximising expected Shannon information is a particular instance of maximising expected utility. We shall see in Section 3.4 how these results, established here for finite partitions, extend straightforwardly to the continuous case.

2.8 DISCUSSION AND FURTHER REFERENCES

2.8.1 Operational Definitions

In everyday conversation, the way in which we use language is typically rather informal and unselfconscious, and we tolerate each other's ambiguities and vacuities for the most part, occasionally seeking an *ad hoc* clarification of a particular statement or idea if the context seems to justify the effort required in trying to be a little more precise. (For a detailed account of the ambiguities which plague qualitative probability expressions in English, see Mosteller and Youtz, 1990.)

In the context of scientific and philosophical discourse, however, there is a paramount need for statements which are meaningful and unambiguous. The everyday, tolerant, *ad hoc* response will therefore no longer suffice. More rigorous habits of thought are required, and we need to be selfconsciously aware of the precautions and procedures to be adopted if we are to arrive at statements which make sense.

A prerequisite for "making sense" is that the fundamental concepts which provide the substantive content of our statements should themselves be defined in an essentially unambiguous manner. We are thus driven to seek for definitions of fundamental notions which can be reduced ultimately to the touchstone of actual or potential personal experience, rather than remaining at the level of mere words or phrases.

This kind of approach to definitions is closely related to the philosophy of *pragmatism*, as formulated in the second half of the nineteenth century by Peirce, who insisted that clarity in thinking about concepts could only be achieved by concentrating attention on the conceivable practical effects associated with a concept, or the practical consequence of adopting one form of definition rather than another. In Peirce (1878), this point of view was summarised as follows:

> Consider what effects, that might conceivably have practical bearings, we conceive the object of our conception to have. Then, our conception of these effects is the whole of our conception of the object.

In some respects, however, this position is not entirely satisfactory in that it fails to go far enough in elaborating what is to be understood by the term "practical". This crucial elaboration was provided by Bridgman (1927) in a book entitled *The Logic of Modern Physics*, where the key idea of an *operational* definition is introduced and illustrated by considering the concept of "length":

> ... what do we mean by the length of an object? We evidently know what we mean by length if we can tell what the length of any and every object is and for the physicist nothing more is required. To find the length of an object, we have to perform certain physical operations. The concept of length is therefore fixed when the operations by which length is measured are fixed: that is, the concept of length involves as much as, and nothing more, than the set of operations by which length is determined. In general, we mean by any concept nothing more than a set of operations; the concept is synonymous with the corresponding set of operations. If the concept is physical, ... the operations are actual physical measurements ... ; or if the concept is mental, ... the operations are mental operations...

Throughout this work, we shall seek to adhere to the operational approach to defining concepts in order to arrive at meaningful and unambiguous statements in the context of representing beliefs and taking actions in situations of uncertainty. Indeed, we have stressed this aspect of our thinking in Sections 2.1 to 2.7, where we made the practical, operational idea of preference between options the fundamental starting point and touchstone for all other definitions.

We also noted the inevitable element of idealisation, or approximation, implicit in the operational approach to our concepts, and we remarked on this at several points in Section 2.3. Since many critics of the personalistic Bayesian viewpoint claim to find great difficulty with this feature of the approach, often suggesting that it undermines the entire theory, it is worth noting Bridgman's very explicit recognition that *all* experience is subject to error and that all we can do is to take sufficient precautions when specifying sets of operations to ensure that remaining unspecified variations in procedure have negligible effects on the results of interest. This is well illustrated by Bridgman's account of the operational concept of length and its attendant idealisations and approximations:

> ...we take a measuring rod, lay it on the object so that one of its ends coincides with one end of the object, mark on the object the position of the rod, then move the rod along in a straight line extension of its previous position until the first end coincides with the previous position of the second end, repeat this process as often as we can, and call the length the total number of times the rod was applied. This procedure, apparently so simple, is in practice exceedingly complicated, and doubtless a full description of all the precautions that must be taken would fill a large treatise. We must, for example, be sure that the temperature of the rod is the standard temperature at which its length is defined, or else we must make a

> correction for it; or we must correct for the gravitational distortion of the rod if we measure a vertical length; or we must be sure that the rod is not a magnet or is not subject to electrical forces ... we must go further and specify all the details by which the rod is moved from one position to the next on the object, its precise path through space and its velocity and acceleration in getting from one position to another. Practically, of course, precautions such as these are not taken, but the justification is in our experience that variations of procedure of this kind are without effect on the final result...

This pragmatic recognition that there are inevitable limitations in any concrete application of a set of operational procedures is precisely the spirit of our discussion of Axioms 4 and 5 in Section 2.3. In practical terms, we have to stop somewhere, even though, in principle, we could indefinitely refine our measurement operations. What matters is to be able to achieve sufficient accuracy to avoid unacceptable distortion in any analysis of interest.

2.8.2 Quantitative Coherence Theories

In a comprehensive review of normative decision theories leading to the expected utility criterion, Fishburn (1981) lists over thirty different axiomatic formulations of the principles of coherence, reflecting a variety of responses to the underlying conflict between axiomatic simplicity and structural flexibility in the representation of decision problems. Fishburn sums up the dilemma as follows:

> On the one hand, we would like our axioms to be simple, interpretable, intuitively clear, and capable of convincing others that they are appealing criteria of coherency and consistency in decision making under uncertainty, but to do this it seems essential to invoke strong structural conditions. On the other hand, we would like our theory to adhere to the loose structures that often arise in realistic decision situations, but if this is done then we will be faced with fairly complicated axioms that accommodate these loose structures.

In addition, we should like the definitions of the basic concepts of probability and utility to have strong and direct links with practical assessment procedures, in conformity with the operational philosophy outlined above.

With these considerations in mind, our purpose here is to provide a brief historical review of the foundational writings which seem to us the most significant. This will serve in part to acknowledge our general intellectual indebtedness and orientation, and in part to explain and further motivate our own particular choice of axiom system.

The earliest axiomatic approach to the problem of decision making under uncertainty is that of Ramsey (1926), who presented the outline of a formal system. The key postulate in Ramsey's theory is the existence of a so-called *ethically neutral* event E, say, which, expressed in terms of our notation for options, has the property that $\{c_1 \,|\, E, c_2 \,|\, E^c\} \sim \{c_1 \,|\, E^c, c_2 \,|\, E\}$, for any consequences c_1, c_2. It is then rather natural to define the degree of belief in such an event to be $1/2$ and, from this quantitative basis, it is straightforward to construct an operational measure of utility for consequences. This, in turn, is used to extend the definition of degree of belief to general events by means of an expected utility model.

From a conceptual point of view, Ramsey's theory seems to us, as indeed it has to many other writers, a revolutionary landmark in the history of ideas. From a mathematical point of view, however, the treatment is rather incomplete and it was not until 1954, with the publication of Savage's (1954) book *The Foundations of Statistics* that the first complete formal theory appeared. No mathematical completion of Ramsey's theory seems to have been published, but a closely related development can be found in Pfanzagl (1967, 1968).

Savage's major innovation in structuring decision problems is to define what he calls acts (options, in our terminology) as functions from the set of uncertain possible outcomes into the set of consequences. His key coherence assumption is then that of a complete, transitive order relation among acts and this is used to define qualitative probabilities. These are extended into quantitative probabilities by means of a "continuously divisible" assumption about events. Utilities are subsequently introduced using ideas similar to those of von Neumann and Morgenstern (1944/1953), who had, ten years earlier, presented an axiom system for utility alone, assuming the prior existence of probabilities.

The Savage axiom system is a great historical achievement and provides the first formal justification of the personalistic approach to probability and decision making; for a modern appraisal see Shafer (1986) and lively ensuing discussion. See, also, Hens (1992). Of course, many variations on an axiomatic theme are possible and other Savage-type axiom systems have been developed since by Stigum (1972), Roberts (1974), Fishburn (1975) and Narens (1976). Suppes (1956) presented a system which combined elements of Savage's and Ramsey's approaches. See, also, Suppes (1960, 1974) and Savage (1970). There are, however, two major difficulties with Savage's approach, which impose severe limitations on the range of applicability of the theory.

The first of these difficulties stems from the "continuously divisible" assumption about events, which Savage uses as the basis for proceeding from qualitative to quantitative concepts. Such an assumption imposes severe constraints on the allowable forms of structure for the set of uncertain outcomes: in fact, it even prevents the theory from being directly applicable to situations involving a finite or countably infinite set of possible outcomes.

One way of avoiding this embarrassing structural limitation is to introduce a quantitative element into the system by a device like that of Ramsey's ethically neutral event. This is directly defined to have probability $1/2$ and thus enables Ramsey to get the quantitative ball rolling without imposing undue constraints on the structure. All he requires is that (at least) one such event be included in the representation of the uncertain outcomes. In fact, a generalisation of Ramsey's idea re-emerges in the form of canonical lotteries, introduced by Anscombe and Aumann (1963) for defining degrees of belief, and by Pratt, Raiffa and Schlaifer (1964, 1965) as a basis for simultaneously quantifying personal degrees of belief and utilities in a direct and intuitive manner.

The basic idea is essentially that of a standard measuring device, in some sense external to the real-world events and options of interest. It seems to us that this idea ties in perfectly with the kind of operational considerations described above, and the standard events and options that we introduced in Section 2.3 play this fundamental operational role in our own system. Other systems using standard measuring devices (sometimes referred to as external scaling devices) are those of Fishburn (1967b, 1969) and Balch and Fishburn (1974). A theory which, like ours, combines a standard measuring device with a fundamental notion of *conditional* preference is that of Luce and Krantz (1971).

The second major difficulty with Savage's theory, and one that also exists in many other theories (see Table I in Fishburn, 1981), is that the Savage axioms imply the boundedness of utility functions (an implication of which Savage was apparently unaware when he wrote *The Foundations of Statistics*, but which was subsequently proved by Fishburn, 1970). The theory does not therefore justify the use of many mathematically convenient and widely used utility functions; for example, those implicit in forms such as "quadratic loss" and "logarithmic score".

We take the view, already hinted at in our brief discussion of medical and monetary consequences in Section 2.5, that it is often conceptually and mathematically convenient to be able to use structural representations going beyond what we perceive to be the essentially finitistic and bounded characteristics of real-world problems. And yet, in presenting the basic quantitative coherence axioms it is important not to confuse the primary definitions and coherence principles with the secondary issues of the precise forms of the various sets involved. For this reason, we have so far always taken options to be defined by finite partitions; indeed, within this simple structure, we hope that the essence of the quantitative coherence theory has already been clearly communicated, uncomplicated by structural complexities. Motivated by considerations of mathematical convenience, however, we shall, in Chapter 3, relax the constraint imposed on the form of the action space. We shall then arrive at a sufficiently general setting for all our subsequent developments and applications.

2.8.3 Related Theories

Our previous discussion centred on complete axiomatic approaches to decision problems, involving a unified development of both probability and utility concepts. In our view, a unified treatment of the two concepts is inescapable if operational considerations are to be taken seriously. However, there have been a number of attempted developments of probability ideas separate from utility considerations, as well as separate developments of utility ideas presupposing the existence of probabilities. In addition, there is a considerable literature on information-theoretic ideas closely related to those of Section 2.7. In this section, we shall provide a summary overview of a number of these related theories, grouped under the following subheadings: (i) *Monetary Bets and Degrees of Belief*, (ii) *Scoring Rules and Degrees of Belief*, (iii) *Axiomatic Approaches to Degrees of Belief*, (iv) *Axiomatic Approaches to Utilities* and (v) *Information Theories*.

For the most part, we shall simply give what seem to us the most important historical references, together with some brief comments. The first two topics will, however, be treated at greater length; partly because of their close relation with the main concerns of this book, and partly because of their connections with the important practical topic of the assessment of beliefs.

Monetary Bets and Degrees of Belief

An elegant demonstration that coherent degrees of belief satisfy the rules of (finitely additive) probability was given by de Finetti (1937/1964), without explicit use of the utility concept. Using the notation for options introduced in Section 2.3, de Finetti's approach can be summarised as follows.

If consequences are assumed to be monetary, and if, given an arbitrary monetary sum m and uncertain event E, an individual's preferences among options are such that $\{pm \mid \Omega\} \sim \{m \mid E, 0 \mid E^c\}$, then the individual's degree of belief in E is defined to be p.

This definition is virtually identical to Bayes' own definition of probability (see our later discussion under the heading of *Axiomatic Approaches to Degrees of Belief*). In modern economic terminology, probability can be considered to be a marginal rate of substitution or, more simply, a kind of "price".

Given that an individual has specified his or her degrees of belief for some collection of events by repeated use of the above definition, either it is possible to arrange a form of monetary bet in terms of these events which is such that the individual will certainly lose, a so-called "Dutch book", or such an arrangement is impossible. In the latter case, the individual is said to have specified a *coherent* set of degrees of belief. It is now straightforward to verify that coherent degrees of belief have the properties of finitely additive probabilities.

To demonstrate that $0 \leq p \leq 1$, for any E and m, we can argue as follows. An individual who assigns $p > 1$ is implicitly agreeing to pay a stake larger than m to enter a gamble in which the maximum prize he or she can win is m; an individual who assigns $p < 0$ is implicitly agreeing to offer a gamble in which he or she will pay out either m or nothing in return for a negative stake, which is equivalent to paying an opponent to enter such a gamble. In either case, a bet can be arranged which will result in a certain loss to the individual and avoidance of this possibility requires that $0 \leq p \leq 1$.

To demonstrate the additive property of degrees of belief for exclusive and exhaustive events, $E_1, E_2, \ldots, E_n$, we proceed as follows. If an individual specifies $p_1, p_2, \ldots, p_n$, to be his or her degrees of belief in those events, this is an implicit agreement to pay a total stake of $p_1 m_1 + p_2 m_2 + \cdots + p_n m_n$ in order to enter a gamble resulting in a prize of m_i if E_i occurs and thus a "gain", or "net return", of $g_i = m_i - \sum_j p_j m_j$, which could, of course, be negative. In order to avoid the possibility of the m_j's being chosen in such a way as to guarantee the negativity of the g_i's for fixed p_j's in this system of linear equations, it is necessary that the determinant of the matrix relating the m_j's to the g_i's be zero so that the linear system cannot be solved; this turns out to require that $p_1 + p_2 + \cdots + p_n = 1$. Moreover, it is easy to check that this is also a sufficient condition for coherence: it implies $\sum_j p_j g_j = 0$, for any choice of the m_j's, and hence the impossibility of all the returns being negative.

The extension of these ideas to cover the revision of degrees of belief conditional on new information proceeds in a similar manner, except that an individual's degree of belief in an event E conditional on an event F is defined to be the number q such that, given any monetary sum m, we have the equivalence $\{qm \mid \Omega\} \sim \{m \mid E \cap F, 0 \mid E^c \cap F, qm \mid F^c\}$, according to the individual's preference ordering among options. The interpretation of this definition is straightforward: having paid a stake of qm, if F occurs we are confronted with a gamble with prizes m if E occurs, and nothing otherwise; if F does not occur the bet is "called off" and the stake returned.

However, despite the intuitive appeal of this simple and neat approach, it has two major shortcomings from an operational viewpoint.

In the first place, it is clear that the definitions cannot be taken seriously in terms of arbitrary monetary sums: the "perceived value" of a stake or a return is not equivalent to its monetary value and the missing "utility" concept is required in order to overcome the difficulty. This point was later recognised by de Finetti (see Kyburg and Smokler, 1964/1980, p. 62, footnote (a)), but has its earlier origins in the celebrated *St. Petersburg paradox* (first discussed in terms of utility by Daniel Bernoulli, 1730/1954). For further discussion of possible forms of "utility for money", see, for example, Pratt (1964), Lavalle (1968), Lindley (1971/1985, Chapter 5) and Hull *et al.* (1973). Additionally, one may explicitly recognise that some people have a positive utility for gambling (see, for instance, Conlisk, 1993).

An *ad hoc* modification of de Finetti's approach would be to confine attention to "small" stakes (thus, in effect, restricting attention to a range of outcomes over which the "utility" can be taken as approximately linear) and the argument, thus modified, has considerable pedagogical and, perhaps, practical use, despite its rather informal nature. A more formal argument based on the avoidance of certain losses in betting formulations has been given by Freedman and Purves (1969). Related arguments have also been used by Cornfield (1969), Heath and Sudderth (1972) and Buehler (1976) to expand on de Finetti's concept of coherent systems of bets.

In addition to the problem of "non-linearity in the face of risk", alluded to above, there is also the difficulty that unwanted game-theoretic elements may enter the picture if we base a theory on ideas such as "opponents" choosing the levels of prizes in gambles. For this reason, de Finetti himself later preferred to use an approach based on scoring rules, a concept we have already introduced in Section 2.7.

Scoring Rules and Degrees of Belief

The scoring rule approach to the definition of degrees of belief and the derivation of their properties when constrained to be coherent is due to de Finetti (1963, 1964), with important subsequent generalisations by Savage (1971) and Lindley (1982a).

In terms of the quadratic scoring rule, the development proceeds as follows. Given an uncertain event E, an individual is asked to select a number, p, with the understanding that if E occurs he or she is to suffer a penalty (or loss) of $L = (1 - p)^2$, whereas if E does not occur he or she is to suffer a penalty of $L = p^2$. Using the indicator function for E, the penalty can be written in the general form, $L = (1_E - p)^2$. The number, p, which the individual chooses is defined to be his or her degree of belief in E.

Suppose now that $E_1, E_2, \ldots, E_n$ are an exclusive and exhaustive collection of uncertain events for which the individual, using the quadratic scoring rule scheme, has to specify degrees of belief $p_1, p_2, \ldots, p_n$, respectively, subject now to the penalty

$$L = (1_{E_1} - p_1)^2 + (1_{E_2} - p_2)^2 + \cdots + (1_{E_n} - p_n)^2.$$

Given a specification, $p_1, p_2, \ldots, p_n$, either it is possible to find an alternative specification, $q_1, q_2, \ldots, q_n$, say, such that

$$\sum_{i=1}^{n} (1_{E_i} - q_i)^2 < \sum_{i=1}^{n} (1_{E_i} - p_i)^2,$$

for any assignment of the value 1 to one of the E_i's and 0 to the others, or it is not possible to find such $q_1, q_2, \ldots, q_n$. In the latter case, the individual is said to have specified a *coherent* set of degrees of belief. The underlying idea in this development is clearly very similar to that of de Finetti's (1937/1964) approach where the avoidance of a "Dutch book" is the basic criterion of coherence.

A simple geometric argument now establishes that, for coherence we must have $0 \leq p_i \leq 1$, for $i = 1, 2, \ldots, n$, and $p_1 + p_2 + \cdots + p_n = 1$. To see this, note that the n logically compatible assignments of values 1 and 0 to the E_i's define n points in $\Re^n$. Thinking of $p_1, p_2, \ldots, p_n$ as defining a further point in $\Re^n$, the coherence condition can be reinterpreted as requiring that this latter point cannot be moved in such a way as to reduce the distance from all the other n points. This means that $p_1, p_2, \ldots, p_n$ must define a point in the convex hull of the other n points, thus establishing the required result.

The extension of this approach to cover the revision of degrees of belief conditional on new information proceeds as follows. An individual's degree of belief in an event E conditional on the occurrence of an event F is defined to be the number q, which he or she chooses when confronted with a penalty defined by $L = 1_F\,(1_E - q)^2$. The interpretation of this penalty is straightforward. Indeed, if F occurs, the specification of q proceeds according to the penalty $(1_E - q)^2$; if F does not occur, there is no penalty, a formulation which is clearly related to the idea of "called-off" bets used in de Finetti's 1937 approach. Suppose now that, in addition to the conditional degree of belief q, the numbers p and r are the individual's degrees of belief, respectively, for the events $E \cap F$ and F, specified subject to the penalty

$$L = 1_F\,(1_E - q)^2 + (1_E\,1_F - p)^2 + (1_F - r)^2.$$

To derive the constraints on p, q and r imposed by coherence, which demands that no other choices will lead to a strictly smaller L, whatever the logically compatible outcomes of the events are, we argue as follows.

If u, v, w, respectively, are the values which L takes in the cases where $E \cap F$, $E^c \cap F$ and F^c occur, then p, q, r satisfy the equations

$$\begin{aligned} u &= (1-q)^2 + (1-p)^2 + (1-r)^2 \\ v &= \quad\; q^2 + \quad\;\; p^2 + (1-r)^2 \\ w &= \qquad\qquad\quad\;\; p^2 + \qquad r^2. \end{aligned}$$

If p, q, r defined a point in $\Re^3$ where the Jacobian of the transformation defined by the above equations did not vanish, it would be possible to move from that point in a direction which simultaneously reduced the values u, v and w. Coherence therefore requires that the Jacobian be zero. A simple calculation shows that this reduces to the condition $q = p/r$, which is, again, Bayes' theorem.

De Finetti's 'penalty criterion' and related ideas have been critically re-examined by a number of authors. Relevant additional references are Myerson (1979), Regazzini (1983), Gatsonis (1984), Eaton (1992) and Gilio (1992a). See, also, Piccinato (1986).

Axiomatic Approaches to Degrees of Belief

Historically, the idea of probability as "degree of belief" has received a great deal of distinguished support, including contributions from James Bernoulli (1713/1899), Laplace (1774/1986, 1814/1952), De Morgan (1847) and Borel (1924/1964). However, so far as we know, none of these writers attempted an axiomatic development of the idea.

The first recognisably "axiomatic" approach to a theory of degrees of belief was that of Bayes (1763) and the magnitude of his achievement has been clearly recognised in the two centuries following his death by the adoption of the adjective *Bayesian* as a description of the philosophical and methodological developments which have been inspired, directly or indirectly, by his essay.

By present day standards, Bayes' formulation is, of course, extremely informal, and a more formal, modern approach only began to emerge a century and a half later, in a series of papers by Wrinch and Jeffreys (1919, 1921). Formal axiom systems which whole-heartedly embrace the principle of revising beliefs through systematic use of Bayes' theorem, are discussed in detail by Jeffreys (1931/1973, 1939/1961), whose profound philosophical and methodological contributions to Bayesian statistics are now widely recognised; see for example, the evaluations of his work by Geisser (1980a), by Good (1980a) and by Lindley (1980a), in the volume edited by Zellner (1980).

From a *foundational perspective*, however, the flavour of Jeffreys' approach seems to us to place insufficient emphasis on the inescapably personal nature of degrees of belief, resulting in an over-concentration on "conventional" representations of degrees of belief derived from "logical" rather than operational considerations (despite the fact that Jeffreys was highly motivated by real world applications!). Similar criticisms seem to us to apply to the original and elegant formal development given by Cox (1946, 1961) and Jaynes (1958), who showed that the probability axioms constitute the only consistent extension of ordinary (Aristotelian) logic in which degrees of belief are represented by real numbers.

We should point out, however, that our emphasis on operational considerations and the subjective character of degrees of belief would, in turn, be criticised by many colleagues who, in other respects, share a basic commitment to the Bayesian approach to statistical problems. See Good (1965, Chapter 2) for a discussion of the variety of attitudes to probability compatible with a systematic use of the Bayesian paradigm.

There are, of course, many other examples of axiomatic approaches to quantifying uncertainty in some form or another. In the finite case, this includes work by Kraft *et al.* (1959), Scott (1964), Fishburn (1970, Chapter 4), Krantz *et al.* (1971), Domotor and Stelzer (1971), Suppes and Zanotti (1976, 1982), Heath and Sudderth (1978) and Luce and Narens (1978). The work of Keynes (1921/1929) and Carnap (1950/1962) deserves particular mention and will be further discussed later in Section 2.8.4. Fishburn (1986) provided an authoritative review of the axiomatic foundations of subjective probability, which is followed by a long, stimulating discussion. See, also, French (1982) and Chuaqui and Malitz (1983).

Axiomatic Approaches to Utilities

Assuming the prior existence of probabilities, von Neumann and Morgenstern (1944/1953) presented axioms for coherent preferences which led to a justification of utilities as numerical measures of value for consequences and to the optimality criterion of maximising expected utility. Much of Savage's (1954/1972) system was directly inspired by this seminal work of von Neumann and Morgenstern and the influence of their ideas extends into a great many of the systems we have mentioned. Other early developments which concentrate on the utility aspects of the decision problem include those of Friedman and Savage (1948, 1952), Marschak (1950), Arrow (1951a), Herstein and Milnor (1953), Edwards (1954) and Debreu (1960). Seminal references are reprinted in Page (1968). General accounts of utility are given in the books by Blackwell and Girshick (1954), Luce and Raiffa (1957), Chernoff and Moses (1959) and Fishburn (1970). Extensive bibliographies are given in Savage (1954/1972) and Fishburn (1968, 1981).

Discussions of the experimental measurement of utility are provided by Edwards (1954), Davison *et al.* (1957), Suppes and Walsh (1959), Becker *et al.* (1963), DeGroot (1963), Becker and McClintock (1967), Savage (1971) and Hull *et al.* (1973). DeGroot (1970, Chapter 7) presents a general axiom system for utilities which imposes rather few mathematical constraints on the underlying decision problem structure. Multiattribute utility theory is discussed, among others, by Fishburn (1964) and Keeney and Raiffa (1976). Other discussions of utility theory include Fishburn (1967a, 1988b) and Machina (1982, 1987). See, also, Schervish *et al.* (1990).

Information Theories

Measures of information are closely related to ideas of uncertainty and probability and there is a considerable literature exploring the connections between these topics.

The logarithmic information measure was proposed independently by Shannon (1948) and Wiener (1948) in the context of communication engineering; Lindley (1956) later suggested its use as a statistical criterion in the design of experiments. The logarithmic divergency measure was first proposed by Kullback and Leibler (1951) and was subsequently used as the basis for an information-theoretic approach to statistics by Kullback (1959/1968). A formal axiomatic approach to measures of information in the context of uncertainty was provided by Good (1966), who has made numerous contributions to the literature of the foundations of decision making and the evaluation of evidence. Other relevant references on information concepts are Renyi (1964, 1966, 1967) and Särndal (1970).

The mathematical results which lead to the characterisation of the logarithmic scoring rule for reporting probability distributions have been available for some considerable time. Logarithmic scores seem to have been first suggested by Good (1952), but he only dealt with dichotomies, for which the uniqueness result is not applicable. The first characterisation of the logarithmic score for a finite distribution was attributed to Gleason by McCarthy (1956); Aczel and Pfanzagl (1966), Arimoto (1970) and Savage (1971) have also given derivations of this form of scoring rule under various regularity conditions.

By considering the inference reporting problem as a particular case of a decision problem, we have provided (in Section 2.7) a natural, unifying account of the fundamental and close relationship between information-theoretic ideas and the Bayesian treatment of "pure inference" problems. Based on work of Bernardo (1979a), this analysis will be extended, in Chapter 3, to cover continuous distributions.

2.8.4 Critical Issues

We shall conclude this chapter by providing a summary overview of our position in relation to some of the objections commonly raised against the foundations of Bayesian statistics. These will be dealt with under the following subheadings: (i) *Dynamic Frame of Discourse*, (ii) *Updating Subjective Probability*, (iii) *Relevance of an Axiomatic Approach*, (iv) *Structure of the Set of Relevant Events*, (v) *Prescriptive Nature of the Axioms*, (vi) *Precise, Complete, Quantitative Preference*, (vii) *Subjectivity of Probability*, (viii) *Statistical Inference as a Decision Problem* and (ix) *Communication and Group Decision Making*.

Dynamic Frame of Discourse

As we indicated in Chapter 1, our concern in this volume is with coherent beliefs and actions in relation to a limited set of specified possibilities, currently assumed necessary and sufficient to reflect key features of interest in the problem under study. In the language of Section 2.2, we are operating in terms of a fixed frame of discourse, defined in the light of our current knowledge and assumptions, M_0. However, as many critics have pointed out, this activity constitutes only one static phase of the wider, evolving, scientific learning and decision process. In the more general, dynamic, context, this activity has to be viewed, either potentially or actually, as sandwiched between two other vital processes. On the one hand, the creative generation of the set of possibilities to be considered; on the other hand, the critical questioning of the adequacy of the currently entertained set of possibilities (see, for example, Box, 1980). We accept that the mode of reasoning encapsulated within the quantitative coherence theory as presented here is ultimately conditional, and thus not directly applicable to every phase of the scientific process. But we do not accept, as Box (1980) appears to, that alternative *formal* statistical theories have a convincing, complementary role to play.

The problem of generating the frame of discourse, i.e., inventing new models or theories, seems to us to be one which currently lies outside the purview of any "statistical" formalism, although some limited formal clarification is actually possible within the Bayesian framework, as we shall see in Chapter 4. Substantive subject-matter inputs would seem to be of primary importance, although informal, exploratory data analysis is no doubt a necessary adjunct and, particularly in the context of the possibilities opened up by modern computer graphics, offers considerable intellectual excitement and satisfaction in its own right.

The problem of criticising the frame of discourse also seems to us to remain essentially unsolved by any "statistical" theory. In the case of a "revolution", or even "rebellion", in scientific paradigm (Kuhn, 1962), the issue is resolved for us as statisticians by the consensus of the subject-matter experts, and we simply begin again on the basis of the frame of discourse implicit in the new paradigm. However, in the absence of such "externally" directed revision or extension of the current frame of discourse, it is not clear what questions one should pose in order to arrive at an "internal" assessment of adequacy in the light of the information thus far available.

On the one hand, exploratory diagnostic probing would seem to have a role to play in confirming that specific forms of local elaboration of the frame of discourse should be made. The logical catch here, however, is that such specific diagnostic probing can only stem from the prior realisation that the corresponding specific elaborations might be required. The latter could therefore be incorporated *ab initio* into the frame of discourse and a fully coherent analysis carried out. The issue here is one of pragmatic convenience, rather than of circumscribing the scope of the coherent theory.

On the other hand, the issue of assessing adequacy in relation to a *total absence of any specific suggested elaborations* seems to us to remain an open problem. Indeed, it is not clear that the "problem" as usually posed is well-formulated. For example, is the key issue that of "surprise"; or is some kind of extension of the notion of a decision problem required in order to give an operational meaning to the concept of "adequacy"?

Readers interested in this topic will find in Box (1980), and the ensuing discussion, a range of reactions. We shall return to these issues in Chapter 6. Related issues arise in discussions of the general problem of assessing, or "calibrating", the external, empirical performance of an internally coherent individual; see, for example, Dawid (1982a).

Overall, our responses to critics who question the relevance of the coherent approach based on a fixed frame of reference can be summarised as follows. So far as the scope and limits of Bayesian theory are concerned: (i) we acknowledge that the mode of reasoning encapsulated within the quantitative coherence theory is ultimately *conditional*, and thus not directly applicable to every phase of the scientific process; (ii) informal, *exploratory* techniques are an essential part of the process of generating ideas; there can be no purely "statistical" theory of model formulation; this aspect of the scientific process is not part of the foundational debate, although the process of passing from such ideas to their mathematical *representation* can often be subjected to formal analysis; (iii) we *all* lack a decent theoretical formulation of and solution to the problem of global model criticism in the absence of concrete suggested alternatives.

However, critics of the Bayesian approach should recognise that: (i) an enormous amount of current theoretical and applied statistical activity is concerned with the analysis of uncertainty in the context of models which are accepted, for the purposes of the analysis, as working frames of discourse, subject only to local probing of specific potential elaborations, and (ii) our arguments thus far, and those to follow, are an attempt to convince the reader that *within this latter context* there are compelling reasons for adopting the Bayesian approach to statistical theory and practice.

Updating Subjective Probability

An issue related to the topic just discussed is that of the mechanism for updating subjective probabilities.

In Section 2.4.2, we defined, in terms of a conditional uncertainty relation, $\leq_G$, the notion of the conditional probability, $P(E \mid G)$, of an event E given the *assumed* occurrence of an event G. From this, we derived Bayes' theorem, which establishes that $p(E \mid G) = P(G \mid E)P(E)/P(G)$. If we actually *know for certain* that G has occurred, $P(E \mid G)$ becomes our actual degree of belief in E. The *prior* probability $P(E)$, has been updated to the *posterior* probability $P(E \mid G)$.

However, a number of authors have questioned whether it is justified to identify assessments made conditional on the *assumed* occurrence of G with actual beliefs once G is *known*. We shall not pursue this issue further, although we acknowledge its interest and potential importance. Detailed discussion and relevant references can be found in Diaconis and Zabell (1982), who discuss, in particular, *Jeffrey's rule* (Jeffrey, 1965/1983), and Goldstein (1985), who examines the role of *temporal coherence*. See, also, Good (1977).

Relevance of the Axiomatic Approach

Arguments against over-concern with foundational issues come in many forms. At one extreme, we have heard Bayesian colleagues argue that the mechanics and flavour of the Bayesian inference process have their own sufficient, direct, intuitive appeal and do not need axiomatic reinforcement. Another form of this argument asserts that developments from axiom systems are "pointless" because the conclusions are, tautologically, contained in the premises. Although this is literally true, we simply do not accept that the methodological imperatives which flow from the assumptions of quantitative coherence are in any way "obvious" to someone contemplating the axioms. At the other extreme, we have heard proponents of supposedly "model-free" exploratory methodology proclaim that we can evolve towards "good practice" by simply giving full encouragement to the creative imagination and then "seeing what works".

Our objection to both these attitudes is that they each implicitly assume, albeit from different perspectives, the existence of a commonly agreed notion of what constitutes "desirable statistical practice". This does not seem to us a reasonable assumption at all, and to avoid potential confusion, an operational definition of the notion is required. The quantitative coherence approach is based on the assumption that, *within the structured framework set out in Section 2.2*, desirable practice requires, at least, to avoid Dutch-book inconsistencies, an assumption which leads to the Bayesian paradigm for the revision of belief.

Structure of the Set of Relevant Events

But is the structure assumed for the set of relevant events too rigid? In particular, is it reasonable to assume that, in each and every context involving uncertainty, the logical description of the possibilities should be forced into the structure of an algebra (or σ-algebra), in which each event has the same logical status? It seems to us that this may not always be reasonable and that there is a potential need for further research into the implications of applying appropriate concepts of quantitative coherence to event structures other than simple algebras. For example, this problem has already been considered in relation to the foundations of quantum mechanics, where the notion of "sample space" has been generalised to allow for the simultaneous representation of the outcomes of a set of "related" experiments (see, for example, Randall and Foulis, 1975). In that context, it has been established that there exists a natural extension of the Bayesian paradigm to the more general setting.

Another area where the applicability of the standard paradigm has been questioned is that of so-called "knowledge-based expert systems", which often operate on knowledge representations which involve complex and loosely structured spaces of possibilities, including hierarchies and networks. Proponents of such systems have argued that (Bayesian) probabilistic reasoning is incapable of analysing these structures and that novel forms of quantitative representations of uncertainty are required (see Spiegelhalter and Knill-Jones, 1984, and ensuing discussion, for references to these ideas). However, alternative proposals, which include "fuzzy logic", "belief functions" and "confirmation theory", are, for the most part, *ad hoc* and the challenge to the probabilistic paradigm seems to us to be elegantly answered by Lauritzen and Spiegelhalter (1988). We shall return to this topic later in this section.

Finally, another form of query relating to the logical status of events is sometimes raised (see, for example, Barnard, 1980a). This draws attention to the interpretational asymmetry between a statement like "the underlying distribution is normal" and its negation. This raises questions about their implicitly symmetric treatment within the framework given in Section 2.2. Choices of the elements to be included in $\mathcal{E}$ are, of course, bound up with general questions of "modelling" and the issue here seems to us to be one concerning sensible modelling strategies. We shall return to this topic in Chapters 4 and 6.

Prescriptive Nature of the Axioms

When introducing our formal development, we emphasised that the Bayesian foundational approach is prescriptive and not descriptive. We are concerned with understanding how we *ought* to proceed, *if* we wish to avoid a specified form of behavioural inconsistency. We are *not* concerned with sociological or psychological description of actual behaviour. For the latter, see, for example, Wallsten (1974), Kahneman and Tversky (1979), Kahneman *et al.* (1982), Machina (1987), Bordley (1992), Luce (1992) and Yilmaz (1992). See, also, Savage (1980).

Despite this, many critics of the Bayesian approach have somehow taken comfort from the fact that there is empirical evidence, from experiments involving hypothetical gambles, which suggests that people often do not act in conformity with the coherence axioms; see, for example, Allais (1953) and Ellsberg (1961).

Allais' criticism is based on a study of the actual preferences of individuals in contexts where they are faced with pairs of hypothetical situations, like those described in Figure 2.8, in each of which a choice has to be made between the two options where C stands for current assets and the numbers describe thousands of units of a familiar currency.

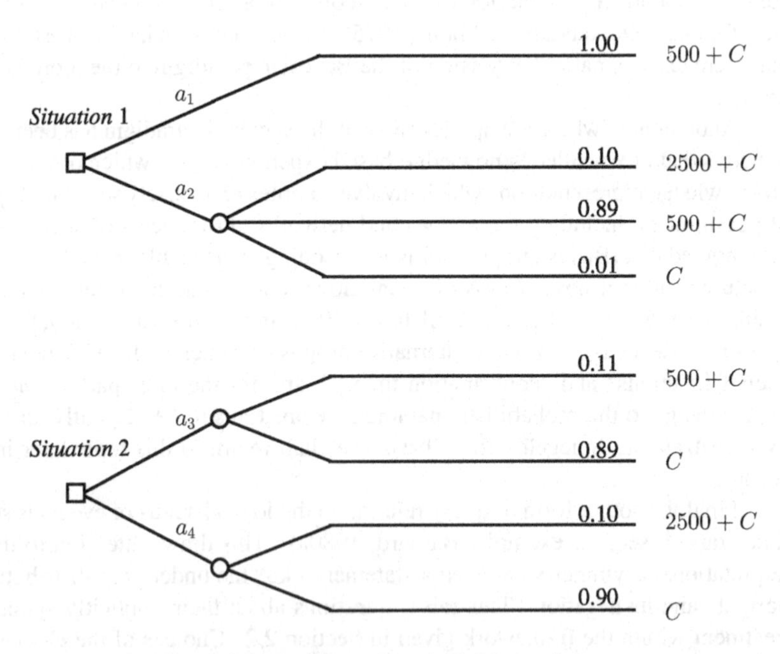

Figure 2.8 *An illustration of Allais' paradox*

It has been found (see, for example, Allais and Hagen, 1979) that there are a great many individuals who prefer option 1 to option 2 in the first situation, and at the same time prefer option 4 to option 3 in the second situation.

To examine the coherence of these two revealed preferences, we note that, if they are to correspond to a consistent utility ordering, there must exist a utility function $u(.)$, defined over consequences (in this case, total assets in thousands of monetary units), satisfying the inequalities

$$u(500 + C) > 0.10\, u(2,500 + C) + 0.89\, u(500 + C) + 0.01\, u(C)$$

$$0.10\, u(2,500 + C) + 0.90\, u(C) > 0.11\, u(500 + C) + 0.89\, u(C).$$

But simple rearrangement reveals that these inequalities are logically incompatible for any function $u(.)$, and, therefore, the stated preferences are incoherent.

How should one react to this conflict between the compelling intuitive attraction (for many individuals) of the originally stated preferences, and the realisation that they are not in accord with the prescriptive requirements of the formal theory? Allais and his followers would argue that the force of examples of this kind is so powerful that it undermines the whole basis of the axiomatic approach set out in Section 2.3. This seems to us a very peculiar argument. It is as if one were to argue for the abandonment of ordinary logical or arithmetic rules, on the grounds that individuals can often be shown to perform badly at deduction or long division.

The conclusion to be drawn is surely the opposite: namely, the more liable people are to make mistakes, the more need there is to have the formal prescription available, both as a reference point, to enable us to discover the kinds of mistakes and distortions to which we are prone in *ad hoc* reasoning, and also as a suggestive source of improved strategies for thinking about and structuring problems.

Table 2.4 *Savage's reformulation of Allais' example*

	Ticket number	1	2–11	12–100
situation 1	*option* 1	$500 + C$	$500 + C$	$500 + C$
	option 2	C	$2500 + C$	$500 + C$
situation 2	*option* 3	$500 + C$	$500 + C$	C
	option 4	C	$2500 + C$	C

In the case of Allais' example, Savage (1954/1972, Chapter 5) pointed out that a concrete realisation of the options described in the two situations could be achieved by viewing the outcomes as prizes from a lottery involving one hundred numbered tickets, as shown in Table 2.4. Indeed, when the problem is set out in this form, it is clear that if any of the tickets numbered from 12 to 100 is chosen it will not matter, in either situation, which of the options is selected. Preferences in both situations should therefore only depend on considerations relating to tickets in the range from 1 to 11. But, for this range of tickets, situations 1 and 2 are identical in structure, so that preferring option 1 to option 2 and at the same time preferring option 4 to option 3 is now seen to be indefensible.

Viewed in this way, Allais' problem takes on the appearance of a decision-theoretic version of an "optical illusion" achieved through the distorting effects of "extreme" consequences, which go far beyond the ranges of our normal experience. The lesson of Savage's analysis is that, when confronted with complex or tricky problems, we must be prepared to shift our angle of vision in order to view the structure in terms of more concrete and familiar images with which we feel more comfortable.

Ellsberg's (1961) criticism is of a similar kind to Allais', but the "distorting" elements which are present in his hypothetical gambles stem from the rather vague nature of the uncertainty mechanisms involved, rather than from the extreme nature of the consequences. In such cases, where confusion is engendered by the probabilities rather than the utilities, the perceived incoherence may, in fact, disappear if one takes into account the possibility that the experimental subjects' utility may be a function of more than one attribute. In particular, we may need to consider the attribute "avoidance of looking foolish", often as a result of thinking that there is a "right answer" if the problem seems predominantly to do with sorting out "experimentally assigned" probabilities, in addition to the monetary consequences specified in the hypothetical gambles. Even without such refinements, however, and arguing solely in terms of the gambles themselves, Raiffa (1961) and Roberts (1963) have provided clear and convincing rejoinders to the Ellsberg criticism. Indeed, Roberts presents a particularly lucid and powerful defence of the axioms, also making use of the analogy with "optical" and "magical" illusions. The form of argument used is similar to that in Savage's rejoinder to Allais, and we shall not repeat the details here. For a recent discussion of both the Allais and Ellsberg phenomena, see Kadane (1992).

Precise, Complete, Quantitative Preferences

In our axiomatic development we have not made the a priori assumption that all options can be compared directly using the preference relation. We have, however, assumed, in Axiom 5, that all consequences and certain general forms of dichotomised options can be compared with dichotomised options involving standard events. This latter assumption then turns out to imply a quantitative basis for all preferences, and hence for beliefs and values.

The view has been put forward by some writers (e.g. Keynes, 1921/1929, and Koopman, 1940) that not all degrees of belief are quantifiable, or even comparable. However, beginning with Jeffreys' review of Keynes' *Treatise* (see also Jeffreys, 1931/1973) the general response to this view has been that some form of quantification is essential if we are to have an operational, scientifically useful theory. Other references, together with a thorough review of the mathematical consequences of these kind of assumptions, are given by Fine (1973, Chapter 2).

Nevertheless, there has been a widespread feeling that the demand for precise quantification, implicit in "standard" axiom systems, is rather severe and certainly ought to be questioned. We should consider, therefore, some of the kinds of suggestions that have been put forward from this latter perspective.

Among the attempts to present formal alternatives to the assumption of precise quantification are those of Good (1950, 1962), Kyburg (1961), Smith (1961), Dempster (1967, 1985), Walley and Fine (1979), Girón and Ríos (1980), DeRobertis and Hartigan (1981), Walley (1987, 1991) and Nakamura (1993). In essence, the suggestion in relation to probabilities is to replace the usual representation of a degree of belief in terms of a single number, by an interval defined by two numbers, to be interpreted as "upper" and "lower" probabilities. So far as decisions are concerned, such theories lead to the identification of a class of "would-be" actions, but provide no operational guidance as to how to choose from among these. Particular ideas, such as Dempster's (1968) generalization of the Bayesian inference mechanism, have been shown to be suspect (see, for example, Aitchison, 1968), but have led on themselves to further generalizations, such as Shafer's (1976, 1982a) theory of "belief functions". This has attracted some interest (see e.g., Wasserman (1990a, 1990b), but its operational content has thus far eluded us.

In general, we accept that the assumption of precise quantification, i.e., that comparisons with standard options can be successively refined without limit, is clearly absurd if taken literally and interpreted in a *descriptive* sense. We therefore echo our earlier detailed commentary on Axiom 5 in Section 2.3, to the effect that these kinds of proposed extension of the axioms seem to us to be based on a confusion of the descriptive and the prescriptive and to be largely unnecessary. It is rather as though physicists and surveyors were to feel the need to rethink their practices on the basis of a physical theory incorporating explicit concepts of upper and lower lengths. We would not wish, however, to be dogmatic about this. Our basic commitment is to quantitative coherence. The question of whether this should be precise, or allowed to be imprecise, is certainly an open, debatable one, and it might well be argued that "measurement" of beliefs and values is not totally analogous to that of physical "length". An obvious, if often technically involved solution, is to consider simultaneously all probabilities which are compatible with elicited comparisons. This and other forms of "robust Bayesian" approaches will be reviewed in Section 5.6.3. In this work, we shall proceed on the basis of a *prescriptive* theory which assumes precise quantification, but then pragmatically acknowledges that, in practice, all this should be taken with a large pinch of salt and a great deal of systematic sensitivity analysis. For a related practical discussion, see Hacking (1965). See, also, Chateaneuf and Jaffray (1984).

Subjectivity of Probability

As we stressed in Section 2.2, the notion of preference between options, the primitive operational concept which underlies all our other definitions, is to be understood as personal, in the sense that it derives from the response of a particular individual to a decision making situation under uncertainty. A particular consequence of this is that the concept which emerges is personal degree of belief, defined in Section 2.4 and subsequently shown to combine for compound events in conformity with the properties of a finitely additive probability measure.

The "individual" referred to above could, of course, be some kind of group, such as a committee, provided the latter had agreed to "speak with a single voice", in which case, to the extent that we ignore the processes by which the group arrives at preferences, it can conveniently be regarded as a "person". Further comments on the problem of individuals versus groups will be given later under the heading *Communication and Group Decision Making*.

This idea that personal (or subjective) probability should be the key to the "scientific" or "rational" treatment of uncertainty has proved decidedly unpalatable to many statisticians and philosophers (although in some application areas, such as actuarial science, it has met with a more favourable reception; see Clarke, 1954). At the very least, it appears to offend directly against the general notion that the methods of science should, above all else, have an "objective" character. Nevertheless, bitter though the subjectivist pill may be, and admittedly difficult to swallow, the alternatives are either inert, or have unpleasant and unexpected side-effects or, to the extent that they appear successful, are found to contain subjectivist ingredients.

From the objectivistic standpoint, there have emerged two alternative kinds of approach to the definition of "probability" both seeking to avoid the subjective degree of belief interpretation. The first of these retains the idea of probability as measurement of partial belief, but rejects the subjectivist interpretation of the latter, regarding it, instead, as a unique degree of partial *logical* implication between one statement and another. The second approach, by far the most widely accepted in some form or another, asserts that the notion of probability should be related in a fundamental way to certain "objective" aspects of physical reality, such as *symmetries or frequencies*.

The logical view was given its first explicit formulation by Keynes (1921/1929) and was later championed by Carnap (1950/1962) and others; it is interesting to note, however, that Keynes seems subsequently to have changed his view and acknowledged the primacy of the subjectivist interpretation (see Good, 1965, Chapter 2). Brown (1993) proposes the related concept of "impersonal" probability.

From an historical point of view, the first systematic foundation of the frequentist approach is usually attributed to Venn (1886), with later influential contributions from von Mises (1928) and Reichenbach (1935). The case for the subjectivist approach and against the objectivist alternatives can be summarised as follows.

The *logical* view is entirely lacking in operational content. Unique probability values are simply assumed to exist as a measure of the degree of implication between one statement and another, to be intuited, in some undefined way, from the formal structure of the language in which these statements are presented.

The *symmetry* (or classical) view asserts that physical considerations of symmetry lead directly to a primitive notion of "equally likely cases". But any uncertain situation typically possesses many plausible "symmetries": a truly "objective" theory would therefore require a procedure for choosing a particular symmetry and for justifying that choice. The subjectivist view explicitly recognises that regarding a specific symmetry as probabilistically significant is itself, inescapably, an act of *personal* judgement.

The *frequency* view can only attempt to assign a measure of uncertainty to an individual event by embedding it in an infinite class of "similar" events having certain "randomness" properties, a "collective" in von Mises' (1928) terminology, and then identifying "probability" with some notion of limiting relative frequency. But an individual event can be embedded in many different "collectives" with no guarantee of the same resulting limiting relative frequencies: a truly "objective" theory would therefore require a procedure for justifying the choice of a particular embedding sequence. Moreover, there are obvious difficulties in defining the underlying notions of "similar" and "randomness" without lapsing into some kind of circularity. The subjectivist view explicitly recognises that any assertion of "similarity" among different, individual events is itself, inescapably, an act of *personal* judgement, requiring, in addition, an operational definition of which is meant by "similar".

In fact, this latter requirement finds natural expression in the concept of an *exchangeable* sequence of events, which we shall discuss at length in Chapter 4. This concept, via the celebrated de Finetti representation theorem, provides an elegant and illuminating explanation, from an entirely subjectivistic perspective, of the fundamental role of symmetries and frequencies in the structuring and evaluation of personal beliefs. It also provides a meaningful operational interpretation of the word "objective" in terms of "intersubjective consensus".

The identification of probability with frequency or symmetry seems to us to be profoundly misguided. It is of paramount importance to maintain the distinction between the *definition of a general concept* and the *evaluation of a particular case*. In the subjectivist approach, the definition derives from logical notions of quantitative coherent preferences: practical evaluations in particular instances often derive from perceived symmetries and observed frequencies, and it is only in this evaluatory process that the latter have a role to play.

The subjectivist point of view outlined above is, course, not new and has been expounded at considerable length and over many years by a number of authors. The idea of probability as individual "degree of confidence" in an event whose outcome is uncertain seems to have been first put forward by James Bernoulli (1713/1899). However, it was not until Thomas Bayes' (1763) famous essay that it was explicitly used as a definition:

> The probability of any event is the ratio between the value at which an expectation depending on the happening of the event ought to be computed, and the value of the thing expected upon its happening.

Not only is this directly expressed in terms of operational comparisons of certain kinds of simple options on the basis of expected values, but the style of Bayes' presentation strongly suggests that these expectations were to be interpreted as personal evaluations.

A number of later contributions to the field of subjective probability are collected together and discussed in the volume edited by Kyburg and Smokler (1964/1980), which includes important seminal papers by Ramsey (1926) and de Finetti (1937/1964). An exhaustive and profound discussion of all aspects of subjective probability is given in de Finetti's magisterial *Theory of Probability* (1970/1974, 1970/1975). Other interpretations of probability are discussed in Renyi (1955), Good (1959), Kyburg (1961, 1974), Fishburn (1964), Fine (1973), Hacking (1975), de Finetti (1978), Walley and Fine (1979) and Shafer (1990).

Statistical Inference as a Decision Problem

Stylised statistical problems have often been approached from a decision-theoretical viewpoint; see, for instance, the books by Ferguson (1967), DeGroot (1970), Barnett (1973/1982), Berger (1985a) and references therein. However, we have already made clear that, in our view, the supposed dichotomy between inference and decision is illusory, since any report or communication of beliefs following the receipt of information inevitably itself constitutes a form of action. In Section 2.7, we formalised this argument and characterised the utility structure that is typically appropriate for consequences in the special case of a "pure inference" problem. The expected utility of an "experiment" in this context was then seen to be identified with expected information (in the Shannon sense), and a number of information-theoretic ideas and their applications were given a unified interpretation within a purely subjectivist Bayesian framework.

Many approaches to statistical inference do not, of course, assign a primary role to reporting probability distributions, and concentrate instead on stylised estimation and hypothesis testing formulations of the problem (see Appendix B, Section 3). We shall deal with these topics in more detail in Chapters 5 and 6.

Communication and Group Decision Making

The Bayesian approach which has been presented in this chapter is predicated on the primitive notion of *individual* preference. A seemingly powerful argument against the use of the Bayesian paradigm is therefore that it provides an inappropriate basis for the kinds of interpersonal communication and reporting processes which characterise both public debate about beliefs regarding scientific and social issues, and also "cohesive-small-group" decision making processes. We believe that the two contexts, "public" and "cohesive-small-group", pose rather different problems, requiring separate discussion.

In the case of the revision and communication of beliefs in the context of general scientific and social debate, we feel that criticism of the Bayesian paradigm is largely based on a misunderstanding of the issues involved, and on an over-simplified view of the paradigm itself, and the uses to which it can be put. So far as the issues are concerned, we need to distinguish two rather different activities: on the one hand, the prescriptive processes by which we ought individually to revise our beliefs in the light of new information if we aspire to coherence; on the other hand, the pragmatic processes by which we seek to report to and share perceptions with others. The first of these processes leads us inescapably to the conclusion that beliefs should be handled using the Bayesian paradigm; the second reminds us that a "one-off" application of the paradigm to summarise a single individual's revision of beliefs is inappropriate in this context.

But, so far as we are aware, no Bayesian statistician has ever argued that the latter would be appropriate. Indeed, the whole basis of the subjectivist philosophy predisposes Bayesians to seek to report a rich *range* of the possible belief mappings induced by a data set, the range being chosen both to reflect (and even to challenge) the initial beliefs of a range of interested parties. Some discussion of the Bayesian reporting process may be found in Dickey (1973), Dickey and Freeman (1975) and Smith (1978). Further discussion is given in Smith (1984), together with a review of the connections between this issue and the role of models in facilitating communication and consensus. This latter topic will be further considered in Chapter 4.

We concede that much remains to be done in developing Bayesian reporting technology, and we conjecture that modern interactive computing and graphics will have a major role to play. Some of the literature on expert systems is relevant here; see, for instance, Lindley (1987), Spiegelhalter (1987) and Gaul and Schader (1988). On the broader issue, however, one of the most attractive features of the Bayesian approach is its recognition of the legitimacy of the plurality of (coherently constrained) responses to data. Any approach to scientific inference which seeks to legitimise *an answer* in response to complex uncertainty seems to us a totalitarian parody of a would-be rational human learning process.

On the other hand, in the "cohesive-small-group" context there may be an imposed need for *group* belief and decision. A variety of problems can be isolated within this framework, depending on whether the emphasis is on combining probabilities, or utilities, or both; and on how the group is structured in relation to such issues as "democracy", "information-sharing", "negotiation" or "competition". It is not yet clear to us whether the analyses of these issues will impinge directly on the broader controversies regarding scientific inference methodology, and so we shall not attempt a detailed review of the considerable literature that is emerging.

Useful introductions to the extensive literature on amalgamation of beliefs or utilities, together with most of the key references, are provided by Arrow (1951b), Edwards (1954), Luce and Raiffa (1957), Luce (1959), Stone (1961), Blackwell and Dubins (1962), Fishburn (1964, 1968, 1970, 1987), Kogan and Wallace (1964), Wilson (1968), Winkler (1968, 1981), Sen (1970), Kranz *et al.* (1971), Marschak and Radner (1972), Cochrane and Zeleny (1973), DeGroot (1974, 1980), Morris (1974), White and Bowen (1975), White (1976a, 1976b), Press (1978, 1980b, 1985b), Lindley *et al.* (1979), Roberts (1979), Hogarth (1980), Saaty (1980), Berger (1981), French (1981, 1985, 1986, 1989), Hylland and Zeckhauser (1981), Weerahandi and Zidek (1981, 1983), Brown and Lindley (1982, 1986), Chankong and Haimes (1982), Edwards and Newman (1982), DeGroot and Feinberg (1982, 1983, 1986), Raiffa (1982), French *et al.* (1983), Lindley (1983, 1985, 1986), Bunn (1984), Caro *et al.* (1984), Genest (1984a, 1984b), Yu (1985), De Waal *et al.* (1986), Genest and Zidek (1986), Arrow and Raynaud (1987), Clemen and Winkler (1987, 1993), Kim and Roush (1987), Barlow *et al.* (1988), Bayarri and DeGroot (1988, 1989, 1991), Huseby (1988), West (1988, 1992a), Clemen (1989, 1990), Ríos *et al.* (1989), Seidenfeld *et al.* (1989), Ríos (1990), DeGroot and Mortera (1991), Kelly (1991), Lindley and Singpurwalla (1991, 1993), Goel *et al.* (1992), Goicoechea *et al.* (1992), Normand and Tritchler (1992) and Gilardoni and Clayton (1993). Important, seminal papers are reproduced in Gärdenfors and Sahlin (1968). For related discussion in the context of policy analysis, see Hodges (1987).

References relating to the Bayesian approach to game theory include Harsany (1967), DeGroot and Kadane (1980), Eliashberg and Winkler (1981), Kadane and Larkey (1982, 1983), Raiffa (1982), Wilson (1986), Aumann (1987), Smith (1988b), Nau and McCardle (1990), Young and Smith (1991), Kadane and Seidenfeld (1992) and Keeney (1992).

A recent review of related topics, followed by an informative discussion, is provided by Kadane (1993).

Chapter 3

Generalisations

Summary

The ideas and results of Chapter 2 are extended to a much more general mathematical setting. An additional postulate concerning the comparison of a countable collection of events is introduced and is shown to provide a justification for restricting attention to countably additive probability as the basis for representing beliefs. The elements of mathematical probability theory are reviewed. The notions of options and utilities are extended to provide a very general mathematical framework for decision theory. A further additional postulate regarding preferences is introduced, and is shown to justify the criterion of maximising expected utility within this more general framework. In the context of inference problems, generalised definitions of score functions and of measures of information and discrepancy are given.

3.1 GENERALISED REPRESENTATION OF BELIEFS

3.1.1 Motivation

The developments of Chapter 2, based on Axioms 1 to 5, led to the fundamental result that quantitatively coherent degrees of belief for events belonging to the *algebra* $\mathcal{E}$ should fit together in conformity with the mathematical rules of *finitely additive probability*. From a directly practical and intuitive point of view, there seems no compelling reason to require anything beyond this finitistic framework, a view argued forcefully, and in great detail, by de Finetti (1970/1974, 1970/1975).

However, there are many situations where the implied necessity of choosing a particular finitistic representation of a problem can lead to annoying conceptual and mathematical complications, as we remarked in Section 2.5.1 when discussing bounded sets of consequences. The example given in that context involved the problem of representing the length of remaining life of a medical patient. Most people would accept that there is an implicit upper bound, but find it difficult to justify any particular choice of its value. Similar problems obviously arise in representing other forms of survival time (of equipment, transplanted organs, or whatever) and further difficulties occur in representing the possible outcomes of many other measurement processes, since these are generally regarded as being on a *continuous* scale. For these reasons, and provided we do not feel that in so doing we are distorting essential features of our beliefs, it is certainly attractive, from the point of view of descriptive and mathematical convenience, to consider the possibility of extending our ideas beyond the finite, discrete framework.

In Section 3.1.2, we shall provide a formal extension of the quantitative coherence theory to the infinite domain. Our fundamental conclusion about beliefs in this setting will be that quantitatively coherent degrees of belief for events belonging to a *σ-algebra* $\mathcal{E}$ should fit together in conformity with the mathematical rules of *countably additive probability*.

The major mathematical advantage of this generalised framework is that all the standard manipulative tools and results of mathematical probability theory then become available to us; convenient references are, for example, Kingman and Taylor (1966) or Ash (1972). A selection of these tools and results will be reviewed in Section 3.2, and then used in Section 3.3 to develop natural extensions of our finitistic definitions of actions and utilities, thus establishing an extremely general mathematical setting for the representation and analysis of decision problems. The important special case of inference as a decision problem will be considered in Section 3.4, which extends to the general mathematical framework the discussion of the finite, discrete case, given in Section 2.7. Finally, a discussion of some particular issues is given in Section 3.5.

3.1.2 Countable Additivity

In Definition 2.1 and the subsequent discussion, we assumed that the collection $\mathcal{E}$ of events included in the underlying frame of discourse should be closed under the operations of arbitrary *finite* intersections and unions. As the first step in providing a mathematical extension to the infinite domain, we shall now assume that we allow arbitrary *countable* intersections and unions in $\mathcal{E}$, so that the latter is taken to be a *σ-algebra*. Within this extended structure, Axioms 1 to 5 will continue to encapsulate the requirements of quantitative coherence for preferences, and hence for degrees of belief, provided that only finite combinations of events of $\mathcal{E}$ are involved. However, if we wish to deal with countable combinations of events, we shall need an extension of the existing requirements for quantitative coherence. One possible such extension is encapsulated in the following postulate.

Postulate 1. (***Monotone continuity***).
If, for all j, $E_j \supseteq E_{j+1}$ *and* $E_j \geq F$, *then* $\bigcap_{j=1}^{\infty} E_j \geq F$.

Discussion of Postulate 1. If the relation $E_j \geq F$ holds for every member of a decreasing sequence of events $E_1 \supseteq E_2 \ldots$, and if we accept the limit event $\bigcap_j E_j$ into our frame of discourse, then it would seem very natural in terms of "continuity" that the relation should "carry over". The operational justification for considering such a countable sequence of comparisons is certainly open to doubt. However, if, for descriptive and mathematical convenience, we admit the possibility, then this form of continuity would seem to be a minimal requirement for coherence.

Proposition 3.1. (***Continuity at*** $\emptyset$). *If, for all* j, $E_j \supseteq E_{j+1}$ *and* $\bigcap_{j=1}^{\infty} E_j = \emptyset$, *then, for any* $G > \emptyset$, $\lim_{j \to \infty} P(E_j \,|\, G) = 0$.

Proof. We note first that the condition encapsulated in Postulate 1 carries over to conditional preferences. By Proposition 2.14(i), if $E_j \geq_G F$ then we have $E_j \cap G \geq F \cap G$; moreover, $E_j \cap G \supseteq E_{j+1} \cap G$, for all $j = 1, 2, \ldots$, and thus, by Postulate 1, $\left(\bigcap_j E_j\right) \cap G \geq F \cap G$. It now follows from Proposition 2.14(i) that $\left(\bigcap_j E_j\right) \geq_G F$. By Proposition 2.16, $P(E_j \,|\, G) \geq P(E_{j+1} \,|\, G) \geq 0$, and so there exists a number $p \geq 0$ such that $\lim_{j \to \infty} P(E_j \,|\, G) = p$, and, for all j, $P(E_j \,|\, G) \geq p$. By Axiom 4(iii) and Proposition 2.16, there exists a standard event S such that $\mu(S) = p$, $G \perp S$ and, for all j, $E_j \geq_G S$. Hence, by the above, we have $\left(\bigcap_j E_j\right) = \emptyset \geq_G S$, which implies that $S \sim \emptyset$ and thus, by Propositions 2.10 and 2.11, that $p = 0$. $\triangleleft$

Since we have already established in Proposition 2.17 that $P(. \,|\, G)$ is a finitely additive probability measure, the above result, *based on the postulate of monotone continuity*, enables us to establish immediately that, in this extended setting, $P(. \,|\, G)$ is a *countably additive* probability measure.

Proposition 3.2. (***Countably additive structure of degrees of belief***).
If $\{E_j,\ j = 1, 2, \ldots\}$ *are disjoint events in* $\mathcal{E}$, *and* $G > \emptyset$, *then*

$$P\left(\bigcup_{j=1}^{\infty} E_j \,|\, G\right) = \sum_{j=1}^{\infty} P(E_j \,|\, G).$$

Proof. Since $P(.\,|\,G)$ is finitely additive we have, for any $n \geq 1$,

$$P\left(\bigcup_{j=1}^{\infty} E_j \,|\, G\right) = \sum_{j=1}^{n} P(E_j \,|\, G) + P(F_n \,|\, G)$$

where

$$F_n = \bigcup_{j=n+1}^{\infty} E_j.$$

It follows that $F_n \supseteq F_{n+1}$ with $\bigcap_n F_n = \emptyset$; hence, by Proposition 3.1,

$$\lim_{n \to \infty} P(F_n \,|\, G) = 0.$$

The result follows by taking limits in the last expression for $P(\bigcup_j E_j \,|\, G)$. ◁

We shall consider the finite versus countable additivity debate in a little more detail in Section 3.5.2. For the present, we simply note that philosophical allegiance to the finitistic framework is in no way incompatible with the systematic adoption and use of countably additive probability measures for the overwhelming majority of applications. The debate centres on whether this particular restriction to a subclass of the finitely additive measures should be considered as a *necessary feature* of quantitative coherence, or whether it is a *pragmatic option*, outside the quantitative coherence framework encapsulated in Axioms 1 to 5. From a philosophical point of view, we identify strongly with this latter viewpoint; but in almost all the developments which follow we shall rarely feel discomfited by implicitly working within a countably additive framework.

We have established (in Propositions 2.4 and 2.10) that if E and F are events in $\mathcal{E}$ with $E \subseteq F$, then $P(E) \leq P(F)$, so that if $P(F) = 0$ then $P(E) = 0$. However, in general, not all subsets of an event of probability zero (a so-called *null* event) will belong to $\mathcal{E}$ and so we cannot logically even refer to their probabilities, let alone infer that they are zero. In some circumstances it may be desirable, as well as mathematically convenient, to be able to do this. If so, this can be done "automatically" by simply agreeing that $\mathcal{E}$ be replaced by the smallest *σ-algebra*, $\mathcal{F}$, which contains $\mathcal{E}$ and all the subsets of the null events of $\mathcal{E}$ (the so-called *completion* of $\mathcal{E}$). The induced probability measure over $\mathcal{F}$ is unique and has the property that all subsets of null events are themselves null events. It is called a *complete* probability measure.

Definition 3.1. ***(Probability space)***. *A probability space is defined by the elements $\{\Omega, \mathcal{F}, P\}$ where $\mathcal{F}$ is a σ-algebra of Ω and P is a complete, σ-additive probability measure on $\mathcal{F}$.*

From now on, our mathematical development will take place within the assumed structure of a probability space. We do not anticipate encountering situations where these mathematical assumptions lead to conceptual distortions beyond the usual, inevitable element of mathematical idealisation which enters *any* formal analysis. However, as de Finetti (1970/1974, 1970/1975) has so eloquently warned, one must always be on guard and aware that distortions *might* occur.

The material which follows (in Section 3.2) will differ in flavour somewhat from our preceding discussion of the *general foundations* of coherent beliefs and actions (Chapter 2 and Section 3.1) and our subsequent discussions of generalised decision problems (Section 3.3) and of the link between beliefs about observables and the structure of *specific models* for representing such beliefs (Chapter 4). These developments systematically invoke the subjectivist, operationalist philosophy as a basic motivation and guiding principle. In the next section, we shall concentrate instead on reviewing, from a purely mathematical standpoint, the concepts and results from mathematical probability theory which will provide the *technical* underpinning of our theory.

3.2 REVIEW OF PROBABILITY THEORY

3.2.1 Random Quantities and Distributions

In the framework we have been discussing, the constituent possibilities and probabilities of any decision problem are encapsulated in the structure of the probability space $\{\Omega, \mathcal{F}, P\}$. Now, in a certain abstract sense, we might think of Ω as the "primitive" collection of all possible outcomes in a situation of interest; for example, that surrounding the birth of an infant, or the state of international commodity markets at a particular time point. However, we are not really interested in a "complete description", even if such were possible, but rather in some numerical summary of the outcomes, in the forms of counts or measurements.

Recalling the discussion of Section 2.8, it might be argued that "measurements" are always, in fact, "counts". However, when convenient, we shall distinguish the two in the usual pragmatic (fuzzy) way: "counts" will typically mean integer-valued data; "measurements" will typically mean data which we *pretend* are real-valued.

We move, therefore, from $\{\Omega, \mathcal{F}, P\}$ to a more explicitly numerical setting by invoking a mapping,

$$x : \Omega \to X \subseteq \Re,$$

which associates a real number $x(\omega)$ with each elementary outcome ω of Ω (our initial exposition will be in terms of a single-valued x; the vector extension will be made in Section 3.2.4). Subsets of Ω are thus mapped into subsets of $\Re$ and the probability measure P defined on $\mathcal{F}$ will induce a probability measure, P_x, say, over appropriate subsets of $\Re$. However, we shall wish to ensure that P_x is defined on certain special subsets of $\Re$, for example, intervals such as $(-\infty, a], a \in \Re$, in which case we shall want sets of the form $\{\omega;\ -\infty < x(\omega) \leq a\}, a \in \Re$, to belong to $\mathcal{F}$, and this will constrain the class of functions x which we would wish to use to define the numerical mapping. The standard requirement is that P_x be definable on the σ-algebra of *Borel sets*, $\mathcal{B}$, of $\Re$, the smallest σ-algebra containing intervals of the form $(-\infty, a], a \in \Re$, and hence all forms of interval, since the latter can be generated by appropriate countable unions and intersections of the intervals $(-\infty, a], a \in \Re$.

Definition 3.2. (***Random quantity***). *A random quantity on a probability space $\{\Omega, \mathcal{F}, P\}$ is a function $x : \Omega \to X \subseteq \Re$ such that $x^{-1}(B) \in \mathcal{F}$, for all $B \in \mathcal{B}$.*

Following de Finetti (1970/1974, 1970/1975), we use the term *random quantity* to signify a numerical entity whose value is uncertain, rather than use the traditional, but potentially confusing, term *random variable*, which might suggest a restriction to contexts involving repeated "trials" over which the quantity may vary. Notationally, we shall use the same symbol for both a random quantity and its value. Thus, for example, x may denote a function, or a particular value of the function $x(\omega) = x$, say. The interpretation will always be clear from the context.

For a random quantity x, the induced measure P_x is defined in the natural way by

$$P_x(B) = P(x^{-1}(B)), B \in \mathcal{B}.$$

The function P_x is easily seen to be a *probability* measure, and describes the way in which probability is "distributed" over the possible values $x \in X$. This information can also be encapsulated in a single real-valued function.

Definition 3.3. (***Distribution function***). *The distribution function of a random quantity $x : \Omega \to X \subseteq \Re$ on $\{\Omega, \mathcal{F}, P\}$ is the function $F_x : \Re \to [0, 1]$ defined by*

$$F_x(x) = P\{\omega;\ x(\omega) \leq x\} = P_x\{(-\infty, x]\}, \quad x \in \Re.$$

If the probability distribution concentrates on a countable set of values, so that $X = \{x_1, x_2, \ldots\}$, x is called a *discrete* random quantity and the function $p_x : \Re \to [0, 1]$ such that

$$p_x(x) = P\{\omega;\ x(\omega) = x\}$$

is called its *probability (mass) function.* The distribution function is then a step function with jumps $p_x(x_i)$ at each x_i.

If the probability distribution is such that there exists a real, non-negative (measurable) function p_x such that

$$P_x(x \in B) = \int_B p_x(t)dt = \int_B dF_x(t), \quad B \in \mathcal{B},$$

then x is called an ***(absolutely) continuous*** random quantity and p_x is called its *density function.* In addition, of course, we might have a mixture of both discrete and continuous elements. No use of *singular* distributions will be made in this volume (for discussion of such distributions, see, for example, Ash, 1972, Section 2.2).

We shall use the same notation, p_x, for both the mass function of a discrete random quantity and the density function of a continuous random quantity. In measure-theoretic terms, both are, of course, special cases of the Radon-Nikodym derivative. In general, we shall use the notation and results of Lebesgue and Lebesgue-Stieltjes integration theory as and when it suits us. Readers unfamiliar with these concepts need not worry: virtually none of the machinery will be visible, and the meanings of integrals will rarely depend on the niceties of the interpretation adopted. Moreover, when there is no danger of confusion, we shall often omit the suffix x in $p_x(x)$, using $p(x)$ both to represent the density or mass function $p_x(\cdot)$ and its value $p_x(x)$ at a particular $x \in X$. Also, to avoid tedious repetition of phrases like "almost everywhere", we shall, when appropriate, simply state that densities are equal, leaving it to be understood that, with respect to the relevant measure, this means "equal, except possibly on a set of measure zero".

If x is a random quantity defined on $\{\Omega, \mathcal{F}, P\}$ such that $x : \Omega \to X \subseteq \Re$, and if $g : \Re \to Y \subseteq \Re$ is a function such that $(g \circ x)^{-1}(B) \in \mathcal{F}$ for all $B \in \mathcal{B}$, then $g \circ x$ is also a random quantity. We shall typically denote $g \circ x$ by $g(x)$, and, whenever we refer to such functions of a random quantity x, it is to be understood that the composite function is indeed a random quantity. Writing $y = g(x)$, the random quantity y induces a probability space $\{\Re, \mathcal{B}, P_y\}$, where

$$P_y(B) = P_x(\{g^{-1}(B)\}) = P(\{(g \circ x)^{-1}(B)\}), \quad B \in \mathcal{B}.$$

Functions such as F_y and p_y are defined in the obvious way. These forms are easily related to those of F_x and p_x. In particular, if g^{-1} exists and is strictly monotonic increasing we have

$$F_y(y) = P_x(g(x) \le y) = P_x(x \le g^{-1}(y)) = F_x(g^{-1}(y))$$

and, in the continuous case, if g is monotonic and differentiable, the density p_y is given by

$$p_y(y) = F_y'(y) = F_x'(g^{-1}(y)) = p_x(g^{-1}(y)) \left| \frac{\partial g^{-1}(y)}{\partial y} \right| .$$

Some examples of this relationship are given at the end of Section 3.2.2.

Definition 3.4. (Expectation). *If x, y are random quantities with $y = g(x)$, the expectation of y, $E[y] = E[g(x)]$, is defined by either*

$$\sum_{y \in Y} y\, p_y(y) = \sum_{x \in X} g(x)\, p_x(x), \quad or$$

$$\int_Y y\, p_y(y) dy = \int_X g(x)\, p_x(x) dx,$$

for the discrete and continuous cases, respectively, where the equality is to be interpreted in the sense that if either side exists so does the other and they are equal.

As in Definition 3.4, most sums or integrals over possible values of random quantities will involve the complete set of possible values (for example, X and Y). To simplify notation we shall usually omit the range of summation or integration, assuming it to be understood from the context. To avoid tiresome duplication, we shall also typically use the integral form to represent both the continuous and the discrete cases.

It is useful to be able to summarise the main features of a probability distribution by quantities defined to encapsulate its location, spread, or shape, often in terms of special cases of Definition 3.4. Assuming, in each case, the right-hand side to exist, such summary quantities include:

(i) $E[x]$, the *mean* of the distribution of the random quantity x;

(ii) $E[x^k]$, the kth *(absolute) moment*;

(iii) $V[x] = E[(x - E[x])^2] = E[x^2] - E^2[x]$, the *variance*;

(iv) $D[x] = V[x]^{1/2}$, the *standard deviation*;

(v) $M[x]$, a *mode* of the distribution of x, such that

$$p_x(M[x]) = \sup_{x \in X} p_x(x);$$

(vi) $Q_\alpha[x]$, an α-*quantile* of the distribution of x, such that

$$F_x(Q_\alpha[x]) = P_x(x \leq Q_\alpha[x]) = \alpha;$$

(vi) $Me[x] = Q_{0.5}[x]$, a *median*;

(vii) $(Q_{(1-p)/2}[x], Q_{(1+p)/2}[x])$, a p-*interquantile range*.

The expectation operator of Definition 3.4 is linear, so that if x_1, x_2 are two random quantities and c_1, c_2 are finite real numbers then

$$E[c_1 x_1 + c_2 x_2] = c_1 E[x_1] + c_2 E[x_2].$$

In the special case of the transformation $g(x) = cx$, for some real constant c, we clearly have

$$E[g(x)] = c\,E[x] = g(E[x]),$$
$$D[g(x)] = \left(V[g(x)]\right)^{1/2} = \left(c^2 V[x]\right)^{1/2} = g\big(D[x]\big).$$

For general transformations $g(x)$, $E[g(x)] \neq g(E[x])$ and the moments of a transformed random quantity $g(x)$ do not exactly relate in any straightforward manner to those of x. However, for suitably well-behaved $g(x)$ the following result, which we shall illustrate at the end of Section 3.2.2, often provides useful approximations.

Proposition 3.3. **(*Approximate mean and variance*).** *If x is a random quantity with $E[x] = \mu$, $V[x] = \sigma^2$ and $y = g(x)$ then, subject to conditions on the distribution of x and the smoothness of g,*

$$E[y] \approx g(\mu) + \tfrac{1}{2}\sigma^2 g''(\mu),$$
$$V[y] \approx \sigma^2 \left[g'(\mu)\right]^2.$$

Outline proof. Expanding $g(x)$ in a Taylor series about μ, we obtain

$$g(x) \approx g(\mu) + (x-\mu)g'(\mu) + \tfrac{1}{2}(x-\mu)^2 g''(\mu),$$

where we are assuming regularity conditions sufficient to ensure the adequacy of this approximation in what follows. Taking expectations immediately yields the approximate form for $E[y]$; subtracting the latter approximation from both sides, squaring, taking expectations and ignoring higher order terms, yields the result for $V[y]$. Clearly, more refined approximations are easily obtained by including higher order terms. $\triangleleft$

In Definition 2.6, we introduced the notion of the independence of two events with subsequent generalisations to mutual independence (Definition 2.12) and conditional independence (Definition 2.13). These notions can be extended to random quantities in the following way.

Definition 3.5. **(*Mutual independence*).** *The random quantities $x_1, \ldots, x_n$ are mutually independent if, for any $t_i \in \Re$, the events $\{\omega; x_i(\omega) \leq t_i\}$, for $i = 1, \ldots, n$, are mutually independent.*

We note that for independent random quantities $x_1, \ldots, x_n$,

$$E\left[\prod_{i=1}^{n} x_i\right] = \prod_{i=1}^{n} E[x_i] \quad \text{and} \quad V\left[\sum_{i=1}^{n} x_i\right] = \sum_{i=1}^{n} V[x_i].$$

Definition 3.6. **(*Conditional independence*).** *For any random quantity y, the random quantities $x_1, \ldots, x_n$ are conditionally independent given y if, for any $t_i \in \Re$, $i = 1, \ldots, n$, the events $\{\omega;\ x_i(\omega) \leq t_i\}$ are conditionally independent given the event $\{\omega;\ y(\omega) \leq y\}$, for all y.*

Conditional independence will play a major role in our later discussion of modelling in Chapter 4.

Many forms of technical manipulation of probability distributions are greatly facilitated by working with some suitable transformation of the original density or distribution function. One of the most useful such transforms is the following.

Definition 3.7. (***Characteristic function***). *The characteristic function of a random quantity x is the function ϕ_x, mapping $\Re$ to the complex plane, given by*

$$\phi_x(t) = E[e^{itx}], \quad t \in \Re.$$

Among the most important properties of the characteristic function, we note the following.

(i) $|\phi_x(t)| \leq 1$ and $\phi_x(0) = 1$.

(ii) ϕ_x is a uniformly continuous function of t.

(iii) If $x_1, \ldots, x_n$ are independent random quantities, and $s = \sum_{i=1}^n x_i$, then $\phi_s(t) = \prod_{i=1}^n \phi_{x_i}(t)$.

(iv) Two random quantities have the same distribution if and only if they have the same characteristic function.

(v) If $E[x^k] < \infty$, then $\phi_x(t) = \sum_{j=1}^{k} \frac{(it)^j}{j!} E[x^j] + o(t^k)$.

Many similar properties hold for the closely related alternative transforms $E[e^{tx}]$, the *moment generating function*, and $E[t^x]$, the *probability generating function*.

3.2.2 Some Particular Univariate Distributions

In this section, we shall review a number of particular univariate distributions which are frequently used in applications, and list some of their properties and characteristics. We shall assume that the reader is familiar with most of this material, and detailed discussion and derivations are therefore not given. The books by Johnson and Kotz (1969, 1970) provide a mass of detail on these and other distributions.

One important initial warning is required! These distributions provide the building blocks for *statistical models* and are typically defined in terms of "parameters". The role and interpretation of "*models*" and "*parameters*" within the general subjectivist, operationalist framework are extremely important issues, which will be discussed at length in Chapter 4. For the present, " parameters" should simply be regarded as "labels" of the various mathematical functions we shall be considering, although, as we shall see, these "labelling parameters" often relate closely to one or other of the characteristics of the distribution.

The Binomial Distribution

A discrete random quantity x has a *binomial* distribution with parameters θ and n $(0 < \theta < 1,\ n = 1, 2, \ldots)$ if its probability function $\text{Bi}(x \mid n, \theta)$ is

$$\text{Bi}(x \mid \theta, n) = \binom{n}{x} \theta^x (1-\theta)^{n-x}, \quad x = 0, 1, \ldots, n.$$

The mean and variance are $E[x] = n\theta$, and $V[x] = n\theta(1-\theta)$. A mode is attained at the greatest integer $M[x]$ which does not exceed $x_m = (n+1)\theta$; if x_m is an integer, then both x_m and $x_m - 1$ are modes.

If $n = 1$, x is said to have a *Bernoulli* distribution, with probability function denoted by $\text{Br}(x \mid \theta)$. The sum of k independent binomial random quantities with parameters (θ, n_i), $i = 1, \ldots k$, is a binomial random quantity with parameters θ and $n_1 + \cdots + n_k$.

The Hypergeometric Distribution

A discrete random quantity x has an *hypergeometric* distribution with integer parameters N, M and n $(n \le N + M)$ if its probability function $\text{Hy}(x \mid N, M, n)$ is

$$\text{Hy}(x \mid N, M, n) = c \binom{N}{x} \binom{M}{n-x}, \quad \max\{0, n-M\} \le x \le \min\{n, N\},$$

where

$$c = \binom{N+M}{n}^{-1}.$$

The mean and variance are given by

$$E[x] = \frac{nN}{N+M} \quad \text{and} \quad V[x] = \frac{nMN}{(N+M)^2} \frac{(N+M-n)}{(N+M-1)}.$$

A mode is attained at the greatest integer $M[x]$ which does not exceed

$$x_m = \frac{(n+1)(N+1)}{M+N+2};$$

if x_m is an integer, then both x_m and $x_m - 1$ are modes.

The Negative-Binomial Distribution

A discrete random quantity x has a *negative-binomial* distribution with parameters θ and r $(0 < \theta < 1,\ r = 1, 2, \ldots)$ if its probability function $\text{Nb}(x \mid r, \theta)$ is

$$\text{Nb}(x \mid \theta, r) = c \binom{r+x-1}{r-1} (1-\theta)^x, \quad x = 0, 1, 2, \ldots$$

where $c = \theta^r$. The mean and variance are $E[x] = r\theta$ and $V[x] = r(1-\theta)/\theta^2$. If $r(1-\theta) > 1$ the mode $M[x]$ is the least integer not less than $[r(1-\theta)]/\theta$; if $r(1-\theta) = 1$, there are two modes at 0 and 1; if $r(1-\theta) < 1$, $M[x] = 0$.

If $r = 1$, x is said to have a *geometric* or *Pascal* distribution. Moreover, the sum of k independent negative binomial random quantities with parameters (θ, r_i), $i = 1, \ldots, k$, is a negative binomial random quantity with parameters θ and $r_1 + \cdots + r_k$.

The Poisson Distribution

A discrete random quantity x has a *Poisson* distribution with parameter λ $(\lambda > 0)$ if its probability function $\text{Pn}(x \mid \lambda)$ is

$$\text{Pn}(x \mid \lambda) = c \frac{\lambda^x}{x!}, \quad x = 0, 1, 2, \ldots$$

where $c = e^{-\lambda}$. The mean and variance are given by $E[x] = V[x] = \lambda$. A mode $M[x]$ is attained at the greatest integer which does not exceed λ. If λ is an integer, both λ and $\lambda - 1$ are modes.

The sum of k independent Poisson random quantities with parameters λ_i, $i = 1, \ldots, k$, is a Poisson random quantity with parameter $\lambda_1 + \cdots + \lambda_k$.

The Beta Distribution

A continuous random quantity x has a *beta* distribution with parameters α and β $(\alpha > 0, \beta > 0)$ if its density function $\text{Be}(x \mid \alpha, \beta)$ is

$$\text{Be}(x \mid \alpha, \beta) = c\, x^{\alpha-1}(1-x)^{\beta-1}, \quad 0 < x < 1\,,$$

where

$$c = \frac{\Gamma(\alpha+\beta)}{\Gamma(\alpha)\Gamma(\beta)}$$

and $\Gamma(x) = \int_0^\infty t^{x-1} e^{-t} dt$; integer and half-integer values of the gamma function are easily found from the recursive relation $\Gamma(x+1) = x\Gamma(x)$, and the values $\Gamma(1) = 1$ and $\Gamma(1/2) = \sqrt{\pi} \approx 1.7725$.

Systematic application of the beta integral,

$$\int_0^1 x^{\alpha-1}(1-x)^{\beta-1}dx = \frac{\Gamma(\alpha)\Gamma(\beta)}{\Gamma(\alpha+\beta)},$$

gives

$$E[x] = \frac{\alpha}{\alpha+\beta} \quad \text{and} \quad V[x] = \frac{\alpha\beta}{(\alpha+\beta)^2(\alpha+\beta+1)}.$$

If $\alpha > 1$ and $\beta > 1$, there is a unique mode at $(\alpha-1)/(\alpha+\beta-2)$. If x has a $\text{Be}(x \mid \alpha, \beta)$ density, then $y = 1 - x$ has a $\text{Be}(y \mid \beta, \alpha)$ density. If $\alpha = \beta = 1$, x is said to have a *uniform* distribution $\text{Un}(x \mid 0, 1)$ on $(0, 1)$.

By considering the transformed random quantity $y = a + x(b-a)$, where x has a $\text{Be}(x \mid \alpha, \beta)$ density, the beta distribution can be generalised to any finite interval (a, b). In particular, the *uniform* distribution $\text{Un}(y \mid a, b)$ on (a, b),

$$\text{Un}(y \mid a, b) = (b-a)^{-1}, \quad a < y < b,$$

has mean $E[y] = (a+b)/2$ and variance $V[y] = (b-a)^2/12$.

The Binomial-Beta Distribution

A discrete random quantity x has a *binomial-beta* distribution with parameters α, β and n $(\alpha > 0, \beta > 0, n = 1, 2, \ldots)$ if its probability function $\text{Bb}(x \mid \alpha, \beta, n)$ is

$$\text{Bb}(x \mid \alpha, \beta, n) = c\binom{n}{x}\Gamma(\alpha+x)\Gamma(\beta+n-x), \quad x = 0, \ldots, n,$$

where

$$c = \frac{\Gamma(\alpha+\beta)}{\Gamma(\alpha)\Gamma(\beta)\Gamma(\alpha+\beta+n)}.$$

The distribution is generated by the mixture

$$\text{Bb}(x \mid \alpha, \beta, n) = \int_0^1 \text{Bi}(x \mid \theta, n)\, \text{Be}(\theta \mid \alpha, \beta)\, d\theta.$$

The mean and variance are given by

$$E[x] = n\frac{\alpha}{\alpha+\beta} \quad \text{and} \quad V[x] = \frac{n\alpha\beta}{(\alpha+\beta)^2}\frac{(\alpha+\beta+n)}{(\alpha+\beta+1)}.$$

A mode is attained at the greatest integer $M[x]$ which does not exceed

$$x_m = \frac{(n+1)(\alpha-1)}{\alpha+\beta-2};$$

if x_m is an integer, both x_m and $x_m - 1$ are modes. If $\alpha = \beta = 1$ we obtain the *discrete uniform* distribution, assigning mass $(n+1)^{-1}$ to each possible x.

The Negative-Binomial-Beta Distribution

A discrete random quantity x has a *negative-binomial-beta* distribution with parameters α, β and r $(\alpha > 0, \beta > 0, r = 1, 2, \ldots)$ if its probability function $\text{Nbb}(x \mid \alpha, \beta, r)$ is

$$\text{Nbb}(x \mid \alpha, \beta, r) = c \binom{r+x-1}{r-1} \frac{\Gamma(\beta + x)}{\Gamma(\alpha + \beta + x + r)}, \quad x = 0, 1, \ldots$$

where

$$c = \frac{\Gamma(\alpha+\beta)\Gamma(\alpha+r)}{\Gamma(\alpha)\Gamma(\beta)}.$$

The distribution is generated by the mixture

$$\text{Nbb}(x \mid \alpha, \beta, r) = \int_0^1 \text{Nb}(x \mid \theta, r)\, \text{Be}(\theta \mid \alpha, \beta)\, d\theta.$$

The mean is $E[x] = r\beta(\alpha - 1)^{-1}, \alpha > 1$, and the variance is given by

$$V[x] = \frac{r\beta}{\alpha - 1}\left[\frac{\alpha + \beta + r - 1}{\alpha - 2} + \frac{r\beta}{(\alpha-1)(\alpha-2)}\right], \quad \alpha > 2.$$

The Gamma Distribution

A continuous random quantity x has a *gamma* distribution with parameters α and β $(\alpha > 0, \beta > 0)$ if its density function $\text{Ga}(x \mid \alpha, \beta)$ is

$$\text{Ga}(x \mid \alpha, \beta) = c\, x^{\alpha-1} e^{-\beta x}, \quad x > 0,$$

where $c = \beta^\alpha / \Gamma(\alpha)$. Systematic application of the gamma integral

$$\int_0^\infty x^{\alpha-1} e^{-\beta x} dx = \frac{\Gamma(\alpha)}{\beta^\alpha}$$

gives $E[x] = \alpha/\beta$ and $V[x] = \alpha/\beta^2$. If $\alpha > 1$, there is a unique mode at $(\alpha - 1)/\beta$; if $\alpha < 1$ there are no modes (the density is unbounded).

If $\alpha = 1$, x is said to have an *exponential* $\text{Ex}(x \mid \beta)$ distribution with parameter β and density

$$\text{Ex}(x \mid \beta) = \beta e^{-\beta x}, \quad x \geq 0.$$

The mode of an exponential distribution is located at zero. If $\beta = 1$, x is said to have an *Erlang* distribution with parameter α. If $\alpha = \nu/2$, $\beta = 1/2$, x is said to have a *(central) chi-squared* (χ^2) distribution with parameter ν (often referred to as *degrees of freedom*) and density denoted by $\chi^2(x \mid \nu)$ or $\chi^2_\nu(x)$.

By considering the transformed random quantities $y = a + x$ or $z = b - x$, where x has a $\text{Ga}(x \mid \alpha, \beta)$ density, the gamma distribution can be generalised to the ranges (a, ∞) or $(-\infty, b)$. Moreover, the sum of k independent gamma random quantities with parameters (α_i, β), $i = 1, \ldots, k$, is a gamma random quantity with parameters $\alpha_1 + \cdots + \alpha_k$ and β.

The Inverted-Gamma Distribution

A continuous random quantity x has an *inverted-gamma* distribution with parameters α and β ($\alpha > 0$, $\beta > 0$) if its density function $\text{Ig}(x \mid \alpha, \beta)$ is

$$\text{Ig}(x \mid \alpha, \beta) = c\, x^{-(\alpha+1)} e^{-\beta/x}, \quad x > 0,$$

where $c = \beta^\alpha / \Gamma(\alpha)$. Systematic application of the gamma integral gives

$$E[x] = \frac{\beta}{(\alpha - 1)}, \qquad \alpha > 1,$$

$$V[x] = \frac{\beta^2}{(\alpha - 1)^2(\alpha - 2)}, \qquad \alpha > 2.$$

There is a unique mode at $\beta/(\alpha + 1)$. The term inverted-gamma derives from the easily established fact that if y has a $\text{Ga}(y \mid \alpha, \beta)$ density then $x = y^{-1}$ has an $\text{Ig}(x \mid \alpha, \beta)$ density.

If x has an inverted-gamma distribution with $\alpha = \nu/2$, $\beta = 1/2$, then x is said to have an *inverted-*χ^2_ν distribution.

A continuous random quantity y has a *square-root inverted-gamma density*, $\text{Ga}^{-(1/2)}(y \mid \alpha, \beta)$, if $x = y^{-2}$ has a $\text{Ga}(x \mid \alpha, \beta)$ density.

The Poisson-Gamma Distribution

A discrete random quantity x has a *Poisson-gamma* distribution with parameters α, β and ν $(\alpha > 0, \beta > 0, \nu > 0)$ if its probability function $\text{Pg}(x \mid \alpha, \beta, \nu)$ is

$$\text{Pg}(x \mid \alpha, \beta, \nu) = c\, \frac{\Gamma(\alpha + x)}{x!} \frac{\nu^x}{(\beta + \nu)^{\alpha + x}}, \quad x = 0, 1, 2, \ldots$$

where $c = \beta^\alpha / \Gamma(\alpha)$. The distribution is generated by the mixture

$$\text{Pg}(x \mid \alpha, \beta, \nu) = \int_0^\infty \text{Pn}(x \mid \nu\lambda)\, \text{Ga}(\lambda \mid \alpha, \beta)\, d\lambda.$$

This compound Poisson distribution is, in fact, a generalisation of the negative binomial distribution $\text{Nb}\big(x \mid \alpha, \beta/(\beta + \nu)\big)$, previously defined only for integer α. The mean is $E[x] = \nu\alpha/\beta$, and the variance is $V[x] = \nu\alpha(\beta + \nu)/\beta^2$. Moreover, if $\alpha\nu > \beta + \nu$, there is a mode at the least integer not less than $(\nu(\alpha - 1)/\beta) - 1$; if $\alpha\nu = \beta + \nu$, there are two modes at 0 and 1; if $\alpha\nu < \beta + \nu$, $M[x] = 0$.

The Gamma-Gamma Distribution

A continuous random quantity x has a *gamma-gamma* distribution with parameters α, β and n $(\alpha > 0, \beta > 0, n = 1, 2, \ldots)$ if its probability function $\mathrm{Gg}(x \mid \alpha, \beta, n)$ is

$$\mathrm{Gg}(x \mid \alpha, \beta, n) = \frac{\beta^\alpha}{\Gamma(\alpha)} \frac{\Gamma(\alpha + n)}{\Gamma(n)} \frac{x^{n-1}}{(\beta + x)^{\alpha+n}}, \quad x > 0.$$

The distribution is generated by the mixture

$$\mathrm{Gg}(x \mid \alpha, \beta, n) = \int_0^\infty \mathrm{Ga}(x \mid n, \lambda)\, \mathrm{Ga}(\lambda \mid \alpha, \beta)\, d\lambda.$$

The mean and variance are given by

$$E[x] = n \frac{\beta}{\alpha - 1}, \qquad \alpha > 1,$$

$$V[x] = \frac{\beta^2 (n^2 + n(\alpha - 1))}{(\alpha - 1)^2 (\alpha - 2)}, \qquad \alpha > 2.$$

The Pareto Distribution

A continuous random quantity x has a *Pareto* distribution with parameters α and β $(\alpha > 0, \beta > 0)$ if its density function $\mathrm{Pa}(x \mid \alpha, \beta)$ is

$$\mathrm{Pa}(x \mid \alpha, \beta) = c\, x^{-(\alpha+1)}, \quad x \geq \beta,$$

where $c = \alpha\beta^\alpha$. The mean and variance are given by

$$E[x] = \frac{\alpha\beta}{\alpha - 1}, \quad \text{if } \alpha > 1$$

$$V[x] = \frac{\alpha\beta^2}{(\alpha - 1)^2 (\alpha - 2)}, \quad \text{if } \alpha > 2.$$

The mode is $M[x] = \beta$. The distribution is generated by the mixture

$$\mathrm{Pa}(x \mid \alpha, \beta) = \int_0^\infty \mathrm{Ex}(x - \beta \mid \theta)\, \mathrm{Ga}(\theta \mid \alpha, \beta)\, d\theta.$$

A continuous random quantity y has an *inverted-Pareto* density $\mathrm{Ip}(y \mid \alpha, \beta)$ if $x = y^{-1}$ has a $Pa(x \mid \alpha, \beta)$ density.

The Normal Distribution

A continuous random quantity x has a *normal* distribution with parameters μ and λ ($\mu \in \Re, \lambda > 0$) if its density function $\mathrm{N}(x \mid \mu, \lambda)$ is

$$\mathrm{N}(x \mid \mu, \lambda) = c \exp\left\{-\frac{\lambda}{2}(x-\mu)^2\right\}, \quad x \in \Re\,,$$

where

$$c = \left(\frac{\lambda}{2\pi}\right)^{1/2}.$$

The distribution is symmetrical about $x = \mu$. The mean and mode are $E[x] = M[x] = \mu$ and the variance is $V[x] = \lambda^{-1}$, so that λ here represents the *precision* of the distribution. Alternatively, $\mathrm{N}(x \mid \mu, \lambda)$ is denoted by $\mathrm{N}(x \mid \mu, \sigma^{-2})$, where $\sigma^2 = V[x]$ is the *variance*. If $\mu = 0$, $\lambda = 1$, x is said to have a *standard normal* distribution, with distribution function Φ given by

$$\Phi(x) = \frac{1}{\sqrt{2\pi}} \int_{-\infty}^{x} \exp\{-\tfrac{1}{2}t^2\}\, dt.$$

If $y = \lambda^{1/2}(x-\mu) = (x-\mu)/\sigma$, where x has a normal density $\mathrm{N}(x \mid \mu, \lambda)$, then y has a $\mathrm{N}(y \mid 0, 1)$ (standard) density. In general, if $y = a + \sum_{i=1}^{k} b_i x_i$, where the x_i are independent with $\mathrm{N}(x_i \mid \mu_i, \lambda_i)$ densities, then y has a normal density, $N(y \mid a + \sum_{i=1}^{k} b_i \mu_i, \lambda)$ where $\lambda = (\sum_{i=1}^{k} b_i^2/\lambda_i)^{-1}$, a weighted harmonic mean of the individual precisions.

If $x_1, \ldots, x_k$ are mutually independent standard normal random quantities, then $z = \sum_{i=1}^{k} x_i^2$ has a (central) χ_k^2 distribution.

The Non-central χ^2 Distribution

A continuous random quantity x has a *non-central* χ^2 distribution with parameters ν (degrees of freedom) and λ (non-centrality) ($\nu > 0, \lambda > 0$) if its density function $\chi^2(x \mid \nu, \lambda)$ is

$$\chi^2(x \mid \nu, \lambda) = \sum_{i=0}^{\infty} \mathrm{Pn}\,(i \mid \lambda/2)\, \chi^2(x \mid \nu + 2i),$$

i.e., a mixture of central χ^2 distributions with Poisson weights. It reduces to a central $\chi^2(\nu)$ when $\lambda = 0$. The mean and variance are $E[x] = \nu + \lambda$ and $V[x] = 2(\nu + 2\lambda)$. The distribution is unimodal; the mode occurs at the value $M[x]$ such that $\chi^2(M[x] \mid \nu, \lambda) = \chi^2(M[x] \mid \nu - 2, \lambda)$.

If $x_1, \ldots, x_k$ are mutually independent normal random quantities $N(x_i \mid \mu_i, 1)$, then $z = \sum_{i=1}^{k} x_i^2$ has a non-central χ^2 distribution,

$$\chi^2(z \mid k, \textstyle\sum_{i=1}^{k} \mu_i^2).$$

The sum of k independent non-central χ^2 distributions with parameters (ν_i, λ_i) is a non-central χ^2 with parameters $\nu_1 + \cdots + \nu_k$ and $\lambda_1 + \cdots + \lambda_k$.

The Logistic Distribution

A continuous random quantity x has a *logistic* distribution with parameters α and β ($\alpha \in \Re$, $\beta > 0$) if its density function $\text{Lo}(x \mid \alpha, \beta)$ is

$$\text{Lo}(x \mid \alpha, \beta) = c \, \frac{\exp\left\{-\left(\frac{x-\alpha}{\beta}\right)\right\}}{\left[1+\exp\left\{-\left(\frac{x-\alpha}{\beta}\right)\right\}\right]^2}, \quad x \in \Re,$$

where $c = \beta^{-1}$. An alternative expression for the density function is

$$\frac{1}{4\beta} \operatorname{sech}^2 \left\{\frac{1}{2}\left(\frac{x-\alpha}{\beta}\right)\right\},$$

so that the logistic is sometimes called the *sech-squared* distribution.

The logistic distribution is most simply expressed in terms of its distribution function,

$$F_x(x) = \left[1+\exp\left\{-\left(\frac{x-\alpha}{\beta}\right)\right\}\right]^{-1}.$$

The distribution is symmetrical about $x = \alpha$. The mean and mode are given by $E[x] = M[x] = \alpha$, and the variance is $V[x] = \beta^2\pi^2/3$.

The Student (t) Distribution

A continuous random quantity x has a *Student* distribution with parameters μ, λ and α ($\mu \in \Re, \lambda > 0, \alpha > 0$) if its density $\text{St}(x \mid \mu, \lambda, \alpha)$ is

$$\text{St}(x \mid \mu, \lambda, \alpha) = c \left[1 + \frac{\lambda}{\alpha}(x-\mu)^2\right]^{-(\alpha+1)/2}, \quad x \in \Re,$$

where

$$c = \frac{\Gamma\left((\alpha+1/2)\right)}{\Gamma(\alpha/2)\Gamma(1/2)} \left(\frac{\lambda}{\alpha}\right)^{1/2}.$$

The distribution is symmetrical about $x = \mu$, and has a unique mode $M[x] = \mu$. The mean and variance are

$$E[x] = \mu, \quad \text{if } \alpha > 1,$$

$$V[x] = \frac{1}{\lambda} \frac{\alpha}{(\alpha-2)}, \quad \text{if } \alpha > 2.$$

The parameter α is usually referred to as the *degrees of freedom* of the distribution.

The distribution is generated (Dickey, 1968) by the mixture

$$\text{St}(x \mid \mu, \lambda, \alpha) = \int_0^\infty \text{N}(x \mid \mu, \lambda y) \text{ Ga}\left(y \mid \frac{\alpha}{2}, \frac{\alpha}{2}\right) dy,$$

and includes the normal distribution as a limiting case, since

$$\text{N}(x \mid \mu, \lambda) = \lim_{\alpha \to \infty} \text{St}(x \mid \mu, \lambda, \alpha).$$

If $y = \lambda^{1/2}(x - \mu)$, where x has a $\text{St}(x \mid \mu, \lambda, \alpha)$ density, then y has a (*standard*) student density $\text{St}(y \mid 0, 1, \alpha)$. If $\alpha = 1$, x is said to have a *Cauchy* distribution, with density $\text{Ca}(x \mid \mu, \lambda)$.

If x has a standard normal distribution, y has a χ^2_ν distribution, and x and y are mutually independent, then

$$z = \frac{x}{(y/\nu)^{1/2}}$$

has a standard Student density $\text{St}(z \mid 0, 1, \nu)$.

The Snedecor (F) Distribution

A continuous random quantity x has a *Snedecor*, or *Fisher*, distribution with parameters α and β (*degrees of freedom*) $(\alpha > 0,\ \beta > 0)$ if its density $\text{Fs}\,(x \mid \alpha, \beta)$ is

$$\text{Fs}\,(x \mid \alpha, \beta) = c\, \frac{x^{(\alpha/2)-1}}{(\beta + \alpha x)^{(\alpha+\beta)/2}}, \quad x > 0,$$

where

$$c = \frac{\Gamma\left((\alpha+\beta)/2\right)}{\Gamma(\alpha/2)\Gamma(\beta/2)} \alpha^{\alpha/2} \beta^{\beta/2}.$$

If $\beta > 2$, $E[x] = \beta/(\beta - 2)$ and there is a unique mode at $[\beta/(\beta+2)][(\alpha-2)/\alpha]$; moreover, if $\beta > 4$,

$$V[x] = 2\frac{\beta^2}{\alpha(\beta-4)} \frac{(\alpha+\beta-2)}{(\beta-2)^2}.$$

If x and y are independent random quantities with central χ^2 distributions, with, respectively, ν_1 and ν_2 degrees of freedom, then

$$z = \frac{(x/\nu_1)}{(y/\nu_2)}$$

has a Snedecor distribution with ν_1 and ν_2 degrees of freedom.

Relationships between some of the distributions described above can be established using the techniques described in Section 3.2.1. For a geometrical interpretation of some of these relations, see Bailey (1992).

Example 3.1. ***(Gamma and*** χ^2 ***distributions).*** Suppose that x has a $\text{Ga}(x \mid \alpha, \beta)$ density and let $y = 2\beta x$. Then, for $y > 0$, we have

$$p_y(y) = p_x\left(\frac{y}{2\beta}\right)\left|\frac{\partial(y/2\beta)}{\partial y}\right| = \frac{1}{2^\alpha \Gamma(\alpha)} y^{\alpha-1} \exp\{-\frac{y}{2}\} = \text{Ga}(y \mid \alpha, \tfrac{1}{2}),$$

so that y has a $\chi^2(y \mid 2\alpha)$ density. Since χ^2 distributions are extensively tabulated, this relationship provides a useful basis for numerical work involving any gamma distribution.

Example 3.2. ***(Beta, Binomial and Snedecor (F) distributions).*** Suppose that x has a density $\text{Be}(x \mid \alpha, \beta)$ and let $y = \beta x[\alpha(1-x)]^{-1}$. Then, for $y > 0$, and noting that $x = \alpha y[\beta + \alpha y]^{-1}$, we have

$$\begin{aligned} p_y(y) &= p_x(\alpha y[\beta + \alpha y]^{-1})\left|\frac{\partial\{\alpha y[\beta + \alpha y]^{-1}\}}{\partial y}\right| \\ &= \frac{\Gamma(\alpha+\beta)}{\Gamma(\alpha)\Gamma(\beta)}\left(\frac{\alpha y}{\beta + \alpha y}\right)^{\alpha-1}\left(\frac{\beta}{\beta+\alpha y}\right)^{\beta-1}\frac{\alpha\beta}{(\beta+\alpha y)^2} \\ &= \frac{\Gamma(\alpha+\beta)\,\alpha^\alpha\beta^\beta}{\Gamma(\alpha)\Gamma(\beta)}\frac{y^{\alpha-1}}{(\beta+\alpha y)^{\alpha+\beta}} = \text{Fs}(y \mid 2\alpha, 2\beta), \end{aligned}$$

so that y has the stated Snedecor (F) density. Binomial probabilities may also be obtained from the F distribution using the exact relation between their distribution functions given by Peizer and Pratt (1968)

$$F_{\text{Bi}}(x \mid \theta, n) = F_F\left(\frac{(x+1)(1-\theta)}{\theta(n-x)}\,\middle|\, 2(n-x),\ 2(x+1)\right), \quad x = 0, 1, \ldots, n.$$

Since F distributions are extensively tabulated, these relationships provide a useful basis for numerical work involving any beta or binomial distribution.

Example 3.3. ***(Approximate moments for transformed random quantities).*** Suppose that x has a $\text{Be}(x \mid \alpha, \beta)$ density, $0 < x < 1$, but that we are interested in the means and variances of the transformed random quantities

$$y_1 = g_1(x) = \log\left(\frac{x}{1-x}\right), \quad y_2 = g_2(x) = 2\sin^{-1}\sqrt{x}.$$

Recalling that

$$E[x] = \mu = \frac{\alpha}{\alpha+\beta}, \quad V[x] = \sigma^2 = \frac{\mu(1-\mu)}{\alpha+\beta+1},$$

and noting that

$$g_1'(x) = \frac{1}{x(1-x)}, \quad g_1''(x) = \frac{x-(1-x)}{x^2(1-x)^2},$$

$$g_2'(x) = \frac{1}{x^{1/2}(1-x)^{1/2}}, \quad g_1''(x) = \frac{x-(1-x)}{x^{3/2}(1-x)^{3/2}},$$

application of Proposition 3.3 immediately yields the following approximations:

$$E[y_1] \approx \log\left(\frac{\mu}{1-\mu}\right) + \frac{1}{2}\sigma^2\left[\frac{\mu-(1-\mu)}{\mu^2(1-\mu)^2}\right],$$

$$V[y_1] \approx \frac{\sigma^2}{\mu^2(1-\mu)^2},$$

$$E[y_2] \approx 2\sin^{-1}\sqrt{\mu} + \frac{1}{2}\sigma^2\left[\frac{\mu-(1-\mu)}{\mu^{3/2}(1-\mu)^{3/2}}\right],$$

$$V[y_2] \approx \frac{\sigma^2}{\mu(1-\mu)} = \frac{1}{\alpha+\beta+1}.$$

We note, in particular, that if $\mu \approx \frac{1}{2}$ the second (correction) terms in the mean approximations will be small, and that, for all μ, the variance is "stabilised" (i.e., does not depend on μ) under the second transformation.

3.2.3 Convergence and Limit Theorems

Within the countably additive framework for probability which we are currently reviewing, much of the powerful resulting mathematical machinery rests on various notions of limit process. We shall summarise a few of the main ideas and results, beginning with the four most widely used notions of convergence for random quantities.

Definition 3.8. **(*Convergence*).** *A sequence $x_1, x_2, \ldots$, of random quantities:*

(i) *converges **in mean square** to a random quantity x if and only if*

$$\lim_{i\to\infty} E[(x_i - x)^2] = 0;$$

(ii) *converges **almost surely** to a random quantity x if and only if*

$$P\left(\left\{\omega; \lim_{i\to\infty} x_i(\omega) = x(\omega)\right\}\right) = 1;$$

in other words, if $x_i(\omega)$ tends to $x(\omega)$ for all ω except those lying in a set of P-measure zero;

(iii) *converges **in probability** to a random quantity x if and only if*

$$\text{for all } \varepsilon > 0, \lim_{i\to\infty} P(\{\omega;\ |x_i(\omega) - x(\omega)| > \varepsilon\}) = 0;$$

(iv) *converges **in distribution** to a random quantity x if and only if the corresponding distribution functions are such that*

$$\lim_{i\to\infty} F_i(t) = F(t)$$

at all continuity points t of F in $\Re$; we denote this by $F_i \to F$.

Convergence in mean square implies convergence in probability; for finite random quantities, almost sure convergence also implies convergence in probability. Convergence in probability implies convergence in distribution; the converse is false. Convergence in distribution is completely determined by the distribution functions: the corresponding random quantities need not be defined on the same probability space. Moreover,

(i) $F_i \to F$ if and only if, for every bounded continuous function g, the sequence of the expected values $E[g(x_i)]$ with respect to F_i converges to the expected value $E[g(x)]$ with respect to F;

(ii) if $F_i \to F$ and $\phi_i(t)$ and $\phi(t)$ are the corresponding characteristic functions, then $\phi_i(t) \to \phi(t)$ for all $t \in \Re$; the converse also holds, provided that $\phi(t)$ is continuous at $t = 0$;

(iii) (*Helly's theorem*) Given a sequence $\{F_1, F_2, \ldots\}$ of distributions functions such that for each $\varepsilon > 0$ there exists an a such that for all i sufficiently large $F_i(a) - F_i(-a) > 1 - \varepsilon$, there exists a distribution function F and a subsequence $F_{i_1}, F_{i_2}, \ldots,$ such that $F_{i_j} \to F$.

An important class of limit results, the so-called *laws of large numbers*, link the limiting behaviour of averages of (independent) random quantities with their expectations. Some of the most basic of these are the following:

(i) If $x_1, x_2, \ldots,$ are independent, identically distributed random quantities with $E[x_i^2] < \infty$ and $E[x_i] = \mu$, then the sequence of random quantities $\overline{x}_n = n^{-1}\sum_{i=1}^n x_i$, $n = 1, 2, \ldots,$ converges in mean square (and hence in probability) to μ; that is to say, to a degenerate, discrete random quantity which assigns probability one to μ.

(ii) *The weak law of large numbers.* If $x_1, x_2, \ldots$ are independent, identically distributed random quantities with $E[x_i] = \mu < \infty$, then the sequence of random quantities $\overline{x}_n$, $n = 1, 2, \ldots,$ converges in probability to μ.

(iii) *The strong law of large numbers.* Under the same conditions as in (ii), $\overline{x}_n$, $n = 1, 2, \ldots,$ converges almost surely to μ.

In addition, there is a further class of limit results which characterises in more detail the properties of the distance between the sequence and the limit values. Two important examples are the following:

(i) *The central limit theorem.* If $x_1, x_2, \ldots$ are independent identically distributed random quantities with $E[x_i] = \mu$ and $V[x_i] = \sigma^2 < \infty$, for all i, then the sequence of standardised random quantities

$$z_n = \frac{\overline{x}_n - \mu}{\sigma/\sqrt{n}}, \quad n = 1, 2, \ldots$$

converges in distribution to the standard normal distribution.

(ii) *The law of the iterated logarithm.* Under the conditions assumed for the central limit theorem,

$$\limsup_{n\to\infty} \frac{\overline{x}_n - \mu}{\sigma/\sqrt{n}} \left(2 \log\log n\right)^{-1/2} = 1.$$

There are enormously wide-ranging variations and generalisations of these results, but we shall rarely need to go beyond the above in our subsequent discussion.

3.2.4 Random Vectors, Bayes' Theorem

A random quantity represents a numerical summary of the potential outcomes in an uncertain situation. However, in general each outcome has many different numerical summaries which may be associated with it. For example, a description of the state of the international commodity market would typically involve a whole complex of price information; the birth of an infant might be recorded in terms of weight and heart-rate measurements, as well as an encoding (for example, using a 0-1 convention) of its sex. It is necessary therefore to have available the mathematical apparatus for handling a *vector* of numerical information.

Formally, we wish to define a mapping

$$\boldsymbol{x} : \Omega \to X \subseteq \Re^k$$

which associates a vector $\boldsymbol{x}(\omega)$ of k real numbers with each elementary outcome ω of Ω. As in the case of (univariate) random quantities, we move the focus of attention from the underlying probability space $\{\Omega, \mathcal{F}, P\}$ to the context of $\Re^k$ and an induced probability measure $P_{\boldsymbol{x}}$. However, we shall again wish to ensure that $P_{\boldsymbol{x}}$ is well-defined for particular subsets of $\Re^k$ and this puts mathematical constraints on the form of the function $\boldsymbol{x}$. Generalising our earlier discussion given in Section 3.2.1, we shall take this class of subsets to be the smallest σ-algebra, $\mathcal{B}$, containing all forms of k-dimensional interval (the so-called Borel sets of $\Re^k$). This then prompts the following definition.

Definition 3.9. (***Random vector***). *A random vector $\boldsymbol{x}$ on a probability space $\{\Omega, \mathcal{F}, P\}$ is a function $\boldsymbol{x} : \Omega \to X \subseteq \Re^k$ such that*

$$\boldsymbol{x}^{-1}(B) \in \mathcal{F}, \quad \textit{for all} \quad B \in \mathcal{B}.$$

For a random vector $\boldsymbol{x}$, the induced probability measure $P_{\boldsymbol{x}}$ is defined in the natural way by

$$P_{\boldsymbol{x}}(B) = P\big(\boldsymbol{x}^{-1}(B)\big), \quad B \in \mathcal{B}.$$

The possible forms of *distribution* for $\boldsymbol{x}$, $P_{\boldsymbol{x}}$, are potentially much more complicated for a random vector than in the case of a single random quantity, in that they not only describe the uncertainty about each of the individual component random quantities in the vector, but also the dependencies among them.

As in the one-dimensional case, we can distinguish *discrete distributions*, where x takes only a countable number of possible values and the distribution can be described by the *probability (mass) function*

$$p_x(x) = P(\{\omega;\ x(\omega) = x\}),$$

and *(absolutely) continuous distributions*, where the distribution may be described by a *density function* $p_x(x)$ such that

$$P_x(B) = \int_B p_x(x)\, dx = \int_B p_x(x_1, \ldots, x_k)\, dx_1 \cdots dx_k, \qquad B \in \mathcal{B}.$$

The *distribution function* of a random vector x is the real-valued function $F_x : \Re^k \to [0,1]$ defined by

$$F_x(x) = F_x(x_1, \ldots, x_k) = P_x\{(-\infty, x_1] \cap \ldots \cap (-\infty, x_k]\}.$$

In addition, we could have cases where some of the components are discrete and others are continuous. Some components, of course, might themselves be a mixture of the two types. In what follows, we shall usually present our discussion using the notation for the continuous case. It will always be clear from the context how to reinterpret things in the discrete (or mixed) cases.

The density $p_x(x) = p_x(x_1, \ldots, x_k)$ of the random vector x is often referred to as the *joint density* of the random quantities $x_1, \ldots, x_k$. If the random vector x is partitioned into $x = (y, z)$, say, where $y = (x_1, \ldots, x_t)$, $z = (x_{t+1}, \ldots, x_k)$, the *marginal density* for the vector y is given by

$$p_y(y) = \int_{\Re^{k-t}} p_x(y, z)\, dz,$$

or alternatively, dropping the subscripts without danger of confusion,

$$p(x_1, \ldots, x_t) = \int_{\Re^{k-t}} p(x_1, \ldots, x_k)\, dx_{t+1} \cdots dx_k.$$

This operation of passing from a joint to a marginal density occurs so commonly in Bayesian inference contexts (see Chapter 5, in particular) that it is useful to have available a simple alternative notation, emphasising the operation itself, rather than the technical integration required. To denote the *marginalisation operation* we shall therefore write

$$p_x(y, z) \underset{y}{\rightarrow} p_y(y).$$

The *conditional density* for the random vector z, given that $y(\omega) = y$, is defined by

$$p_{z|y}(z \mid y) = \frac{p_x(y, z)}{p_y(y)},$$

or, alternatively, again dropping the subscripts for convenience,

$$p(x_{t+1},\ldots,x_k \mid x_1,\ldots,x_t) = \frac{p(x_1,\ldots,x_k)}{p(x_1,\ldots,x_t)}\,.$$

We shall almost always use the generic subscript-free notation for densities. It is therefore important to remember that the functional forms of the various marginal and conditional densities will typically differ.

Proposition 3.4. (***Generalised Bayes' theorem***).

$$p_{\mathbf{y}|\mathbf{z}}(\boldsymbol{y} \mid \boldsymbol{z}) = \frac{p_{\mathbf{z}|\mathbf{y}}(\boldsymbol{z} \mid \boldsymbol{y})p_{\mathbf{y}}(\boldsymbol{y})}{p_{\mathbf{z}}(\boldsymbol{z})}$$

Proof. Exchanging the roles of $\boldsymbol{y}$ and $\boldsymbol{z}$ in the above, it is obvious that

$$p_{\mathbf{x}}(\boldsymbol{x}) = p_{\mathbf{z}|\mathbf{y}}(\boldsymbol{z} \mid \boldsymbol{y})p_{\mathbf{y}}(\boldsymbol{y}) = p_{\mathbf{y}|\mathbf{z}}(\boldsymbol{y} \mid \boldsymbol{z})p_{\mathbf{z}}(\boldsymbol{z}),$$

which immediately yields the result. $\triangleleft$

It is often convenient to re-express Bayes' theorem in the simple proportionality form

$$p_{\mathbf{y}|\mathbf{z}}(\boldsymbol{y} \mid \boldsymbol{z}) \propto p_{\mathbf{z}|\mathbf{y}}(\boldsymbol{z} \mid \boldsymbol{y})p_{\mathbf{y}}(\boldsymbol{y}),$$

since the right-hand side contains all the information required to reconstruct the normalising constant,

$$[p_{\mathbf{z}}(\boldsymbol{z})]^{-1} = \left[\int p_{\mathbf{z}|\mathbf{y}}(\boldsymbol{z} \mid \boldsymbol{y})p_{\mathbf{y}}(\boldsymbol{y})\, d\boldsymbol{y}\right]^{-1},$$

should the latter be needed explicitly. In many cases, however, it is not explicitly required since the "shape" of $p_{\mathbf{y}|\mathbf{z}}(\boldsymbol{y} \mid \boldsymbol{z})$ is all that one needs to know.

In fact, this latter observation is often extremely useful for avoiding unnecessary detail when carrying out manipulations involving Bayes' theorem. More generally, we note that if a density function $p(\boldsymbol{x})$ can be expressed in the form $c\, q(\boldsymbol{x})$, where q is a function and c is a constant, not depending on $\boldsymbol{x}$, then

$$\int_{\Re^k} q(\boldsymbol{x})\, d\boldsymbol{x} = c^{-1}, \quad \text{since} \quad \int_{\Re^k} p(\boldsymbol{x})\, d\boldsymbol{x} = P(\Re^k) = 1\,.$$

Any such $q(\boldsymbol{x})$ will be referred to as a *kernel* of the density $p(\boldsymbol{x})$. The proportionality form of Bayes' theorem then makes it clear that, up to the final stage of calculating the normalising constant, we can always just work with kernels of densities.

For further technical discussion of the generalised Bayes' theorem, see Mouchart (1976), Hartigan (1983, Chapter 3) and Wasserman and Kadane (1990).

As with marginalising, the *Bayes' theorem operation* of passing from a conditional and a marginal density to the "other" conditional density is also fundamental to Bayesian inference and, again, it is useful to have available an alternative notation. To denote the Bayes' theorem operation, we shall therefore write

$$p_{z|y}(z \mid y) \otimes p_y(y) \equiv p_{y|z}(y|z).$$

In more explicit terms, and dropping the subscripts on densities, Bayes' theorem can be written in the form

$$p(x_1, \ldots, x_t \mid x_{t+1}, \ldots, x_k) = \frac{p(x_{t+1}, \ldots, x_k \mid x_1, \ldots, x_t)\, p(x_1, \ldots, x_t)}{\int_{\Re^t} p(x_{t+1}, \ldots, x_k \mid x_1, \ldots, x_t)\, p(x_1, \ldots, x_t)\, dx_1 \cdots dx_t}.$$

Manipulations based on this form will underlie the greater part of the ideas and results to be developed in subsequent chapters.

In particular, extending the use of the terms given in Chapter 2, we shall typically interpret densities such as $p(x_1, \ldots, x_t)$ as describing beliefs for the random quantities $x_1, \ldots, x_t$ before (i.e., *prior* to) observing the random quantities $x_{t+1}, \ldots, x_k$, and $p(x_1, \ldots, x_t \mid x_{t+1}, \ldots, x_k)$ as describing beliefs after (i.e., *posterior* to) observing $x_{t+1}, \ldots, x_k$.

Often, manipulation is simplified if *independence* or *conditional independence* assumptions can be made. For example, if $x_1, \ldots, x_t$ were independent we would have

$$p(x_1, \ldots, x_t) = \prod_{i=1}^{t} p(x_i)\ ;$$

if $x_{t+1}, \ldots, x_k$ were conditionally independent, given $x_1, \ldots, x_t$, we would have

$$p(x_{t+1}, \ldots, x_k \mid x_1, \ldots, x_t) = \prod_{i=t+1}^{k} p(x_i \mid x_1, \ldots, x_t)\ .$$

If x is a random vector defined on $\{\Omega, \mathcal{F}, P\}$ such that $x : \Omega \to X \subseteq \Re^k$ and if $g : \Re^k \to Y \subseteq \Re^h$ $(h \leq k)$ is a function such that $(g \circ x)^{-1}(B) \in \mathcal{F}$ for all $B \in \mathcal{B}$, then $g \circ x$ is also a random vector. We shall typically denote $g \circ x$ by $g(x)$, and, whenever we refer to such vector functions of a random vector x, it is to be understood that the composite function is indeed a random vector. Writing y=$g(x)$, the random vector y induces a probability space $\{\Re^h, \mathcal{B}, P_y\}$ where

$$P_y(B) = P_x(g^{-1}(B)) = P((g \circ x)^{-1}(B)), \quad B \in \mathcal{B},$$

and distribution and density functions $F_{\boldsymbol{y}}, p_{\boldsymbol{y}}$ are defined in the obvious way. These forms are easily related to $F_{\boldsymbol{x}}, p_{\boldsymbol{x}}$. In particular, if $\boldsymbol{g}$ is a one-to-one differentiable function with inverse $\boldsymbol{g}^{-1}$, we have, for each $\boldsymbol{y} \in Y$,

$$p_{\boldsymbol{y}}(\boldsymbol{y}) = p_{\boldsymbol{x}}(\boldsymbol{g}^{-1}(\boldsymbol{y}))\left|\boldsymbol{J}_{\boldsymbol{g}^{-1}}(\boldsymbol{y})\right|,$$

where

$$\boldsymbol{J}_{\boldsymbol{g}^{-1}}(\boldsymbol{y}) = \frac{\partial \boldsymbol{g}^{-1}(\boldsymbol{y})}{\partial \boldsymbol{y}}$$

is the Jacobian of the transformation $\boldsymbol{g}^{-1}$, defined by

$$\left[\boldsymbol{J}_{\boldsymbol{g}^{-1}}(\boldsymbol{y})\right]_{ij} = \frac{\partial h_i(\boldsymbol{y})}{\partial y_j}, \quad i,j = 1,\ldots,k\,,$$

where

$$h_i(\boldsymbol{y}) = \left[\boldsymbol{g}^{-1}(\boldsymbol{y})\right]_i \,.$$

If $h < k$, we are usually able to define an appropriate $\boldsymbol{z}$, with dimension $k - h$, such that $\boldsymbol{w} = \boldsymbol{f}(\boldsymbol{x}) = (\boldsymbol{y}, \boldsymbol{z}) = \big(\boldsymbol{g}(\boldsymbol{x}), \boldsymbol{z}\big)$ is a one-to-one function with inverse $\boldsymbol{f}^{-1}$, and then proceed in two steps to obtain $p_{\boldsymbol{y}}(\boldsymbol{y})$ by first obtaining

$$p_{\boldsymbol{w}}(\boldsymbol{\omega}) = p_{\boldsymbol{y},\boldsymbol{z}}(\boldsymbol{y}, \boldsymbol{z}) = p_{\boldsymbol{x}}\big(\boldsymbol{f}^{-1}(\boldsymbol{w})\big)\left|\boldsymbol{J}_{\boldsymbol{f}^{-1}}(\boldsymbol{\omega})\right|,$$

and then marginalising to

$$p_{\boldsymbol{y}}(\boldsymbol{y}) = \int_{\Re^{k-h}} p_{\boldsymbol{y},\boldsymbol{z}}(\boldsymbol{y}, \boldsymbol{z})\, d\boldsymbol{z}.$$

The *expectation* concept generalises to the case of random vectors in an obvious way.

Definition 3.10. (***Expectation of a random vector***). *If $\boldsymbol{x}, \boldsymbol{y}$ are random vectors such that $\boldsymbol{y} = \boldsymbol{g}(\boldsymbol{x})$, $\boldsymbol{x} : \Omega \to X \subseteq \Re^k$, $\boldsymbol{g} : \Re^k \to Y \subseteq \Re^h$ $(h \leq k)$, the expectation of $\boldsymbol{y}$, $E[\boldsymbol{y}] = E[\boldsymbol{g}(\boldsymbol{x})]$, is a vector whose ith component,*

$$E[y_i] = E[\boldsymbol{g}(\boldsymbol{x})]_i = E\big[g_i(\boldsymbol{x})\big], \quad i = 1,\ldots,h,$$

is defined by either

$$\sum_{\boldsymbol{y}\in Y} y_i\, p_{\boldsymbol{y}}(\boldsymbol{y}) = \sum_{y_i} y_i p_{y_i}(y_i) = \sum_{\boldsymbol{x}\in X} g_i(\boldsymbol{x}) p_{\boldsymbol{x}}(\boldsymbol{x}),$$

or

$$\int_{\Re^h} y_i\, p_{\boldsymbol{y}}(\boldsymbol{y})\, d\boldsymbol{y} = \int_{\Re} y_i p_{y_i}(y_i)\, dy_i = \int_{\Re^k} g_i(\boldsymbol{x}) p_{\boldsymbol{x}}(\boldsymbol{x})\, d\boldsymbol{x},$$

for the discrete and absolutely continuous cases, respectively, where all the equalities are to be interpreted in the sense that if either side exists, so does the other and they are equal.

In particular, the forms defined by $E\left[\Pi_{i=1}^{k} x_i^{n_i}\right]$ are called the *moments* of x of order $n = n_1 + \cdots + n_k$. Important special cases include the first-order moments, $E[x_i], i = 1, \ldots, k$, and the second-order moments, $E[x_i^2], (i = 1, \ldots, k), E[x_i x_j]$, $(1 \leq i \neq j \leq k)$. If $x_1, \ldots, x_k$ are independent, then $E\left[\Pi_i x_i^{n_i}\right] = \Pi_i \, E[x_i^{n_i}]$. The *covariance* between x_i and x_j is defined by

$$C[x_i, x_j] = E\big[\big(x_i - E[x_i]\big)\big(x_j - E[x_j]\big)\big] = E[x_i x_j] - E[x_i]\; E[x_j]\,,$$

and the *correlation* by

$$R[x_i, x_j] = \frac{C[x_i, x_j]}{V[x_i]^{1/2} V[x_j]^{1/2}}\,.$$

The Cauchy-Schwarz inequality establishes that $|\, R[x_i, x_j]\,| \leq 1$. The expectation vector with components $E[x_1], \ldots, E[x_k]$ is also called the *mean vector*, $E[\boldsymbol{x}]$, of $\boldsymbol{x}$; the $k \times k$ matrix with (i, j)th element $C[x_i, x_j]$ is called the *covariance matrix*, $V[\boldsymbol{x}]$, of $\boldsymbol{x}$. If the components of $\boldsymbol{x}$ are independent random quantities, $V[\boldsymbol{x}]$ reduces to a diagonal matrix with (i, i)th entry given by $V[x_i]$.

As in the case of a single random quantity, exact forms for moments of an arbitrary transformation, $\boldsymbol{y} = \boldsymbol{g}(\boldsymbol{x})$, are not available. We shall not need very general results in this area, but the following will occasionally prove useful.

Proposition 3.5. (***Approximate mean and covariance***). *If $\boldsymbol{x}$ is a random vector in $\Re^k$, with $E[\boldsymbol{x}] = \boldsymbol{\mu}$, $V[\boldsymbol{x}] = \Sigma$ and $\boldsymbol{y} = \boldsymbol{g}(\boldsymbol{x})$ is a one-to-one transformation of $\boldsymbol{x}$ such that $\boldsymbol{g}^{-1}$ exists, then, subject to conditions on the distribution of $\boldsymbol{x}$ and on the smoothness of $\boldsymbol{g}$,*

$$E\big[\big(\boldsymbol{g}(\boldsymbol{x})\big)_i\big] = E\big[g_i(\boldsymbol{x})\big] \approx g_i(\boldsymbol{\mu}) + \tfrac{1}{2}\,\mathrm{tr}\left[\Sigma\, \nabla^2 g_i(\boldsymbol{\mu})\right]$$

$$V\big[\boldsymbol{g}(\boldsymbol{x})\big] \approx \boldsymbol{J}_{\boldsymbol{g}}(\boldsymbol{\mu})\; \Sigma\; \boldsymbol{J}_{\boldsymbol{g}}^t(\boldsymbol{\mu}),$$

where, for $i = 1, \ldots, k$,

$$\big(\nabla^2 g_i(\boldsymbol{\mu})\big)_{jl} = \left.\frac{\partial^2 g_i(\boldsymbol{x})}{\partial x_j \partial x_l}\right|_{\boldsymbol{x}=\boldsymbol{\mu}},$$

$$\big(\boldsymbol{J}_{\boldsymbol{g}}(\boldsymbol{\mu})\big)_{jl} = \left.\frac{\partial g_j(\boldsymbol{x})}{\partial x_l}\right|_{\boldsymbol{x}=\boldsymbol{\mu}},$$

where $\mathrm{tr}[.]$ *denotes the trace of a matrix argument.*

Proof. This follows straightforwardly from a multivariate Taylor expansion; the details are tedious and we will omit them here. ◁

A note on measure theory. Readers familiar with measure theory will be aware that there are many subtle steps in passing to a density representation of the probability measure $P_{\mathbf{x}}$. In particular, a detailed rigorous treatment of densities (Radon-Nikodym derivatives) requires statements about dominating measures and comments on the versions assumed for such densities. Readers unfamiliar with measure theory will already have assumed—correctly!—that we shall almost always be dealing with the "standard" versions of probability mass functions and densities (corresponding to counting and Lebesgue dominating measures and "smoothly" defined). Only occasionally, in Chapter 4, do we refer to general, i.e., non-density, forms.

3.2.5 Some Particular Multivariate Distributions

We conclude our review of probability theory with a selection of the more frequently used multivariate probability distributions; that is to say, distributions for random vectors. As in Section 3.2.3, no very detailed discussion will be given: see, for example Wilks (1962), Johnson and Kotz (1969, 1972) and DeGroot (1970) for further information.

The Multinomial Distribution

A discrete random vector $\boldsymbol{x} = (x_1, \ldots, x_k)$ has a *multinomial* distribution of dimension k, with parameters $\boldsymbol{\theta} = (\theta_1, \ldots, \theta_k)$ and n $(0 < \theta_i < 1, \sum_i \theta_i < 1,$ $n = 1, 2, \ldots)$ if its probability function $\text{Mu}_k(\boldsymbol{x} \mid \boldsymbol{\theta}, n)$, for $x_i = 0, 1, 2, \ldots$, with $\sum_{i=1}^k x_i \leq n$, is

$$\text{Mu}_k(\boldsymbol{x} \mid \boldsymbol{\theta}, n) = \frac{n!}{\prod_{i=1}^k x_i!\,(n - \sum_{i=1}^k x_i)!} \prod_{i=1}^k \theta_i^{x_i} (1 - \textstyle\sum_{i=1}^k \theta_i)^{n - \Sigma_{i=1}^k x_i}.$$

The mean vector and covariance matrix are given by

$$E[x_i] = n\theta_i, \quad V[x_i] = n\theta_i(1 - \theta_i), \quad C[x_i, x_j] = -n\theta_i\theta_j.$$

The mode(s) of the distribution is (are) located near $E[\boldsymbol{x}]$, satisfying

$$n\theta_i < M[x_i] \leq (n + k - 1)\theta_i, \quad i = 1, \ldots, k\ ;$$

these inequalities, with the condition $\sum_{i=1}^k x_i \leq n$, restrict the possible modes to a relatively few points.

The marginal distribution of $\boldsymbol{x}^{(m)} = (x_1, \ldots, x_m)$, $m < k$, is the multinomial $\text{Mu}_m(\boldsymbol{x}^{(m)} \,|\, \theta_1, \ldots, \theta_m, n)$. The conditional distribution of $\boldsymbol{x}^{(m)}$ given the remaining x_i's is also multinomial, and it depends on the remaining x_i's only through their sum $s = \sum_{i=m+1}^{k} x_i$; specifically,

$$p(\boldsymbol{x}^{(m)} \,|\, x_{m+1}, \ldots, x_k) = \text{Mu}_k \left(\boldsymbol{x}^{(m)} \middle| \frac{\theta_1}{\sum_{j=1}^{m} \theta_j}, \ldots, \frac{\theta_m}{\sum_{j=1}^{m} \theta_j}, n - s \right).$$

If $\boldsymbol{x} = (x_1, \ldots, x_k)$ has density $\text{Mu}_k(\boldsymbol{x} \,|\, \boldsymbol{\theta}, n)$ then $\boldsymbol{y} = (y_1 \ldots, y_t)$ where

$$y_1 = x_1 + \cdots + x_{i_1}, \;\; \ldots, \;\; y_t = x_{i_{t-1}+1} + \ldots + x_{i_k}, \quad 1 \leq t < k,$$

has density $\text{Mu}_t(\boldsymbol{y} \,|\, \boldsymbol{\phi}, n)$, where

$$\phi_1 = \theta_1 + \cdots + \theta_{i_1}, \;\; \ldots, \;\; \phi_t = \theta_{i_{t-1}+1} + \cdots + \theta_{i_k}.$$

If $\boldsymbol{z}$ is the sum of m independent random vectors having multinomial densities with parameters $(\boldsymbol{\theta}, n_i)$, $i = 1, \ldots, m$, then $\boldsymbol{z}$ also has a multinomial density with parameters $\boldsymbol{\theta}$ and $(n_1 + \cdots + n_m)$. If $k = 1$, $\text{Mu}_k(\boldsymbol{x} \,|\, \boldsymbol{\theta}, n)$ reduces to the binomial density $\text{Bi}(x \,|\, \theta, n)$.

If $x_1, \ldots, x_k$ are k independent Poisson random quantities with densities $\text{Pn}(x_i \,|\, \lambda_i)$, then the joint distribution of $\boldsymbol{x} = (x_1, \ldots, x_k)$ given $\sum_{j=1}^{k} x_i = n$ is multinomial $\text{Mu}(x \,|\, \boldsymbol{\theta}, n)$, with $\theta_i = \lambda_i / \sum_{j=1}^{k} \lambda_j$.

The Dirichlet Distribution

A continuous random vector $\boldsymbol{x} = (x_1, \ldots, x_k)$ has a *Dirichlet* distribution of dimension k, with parameters $\boldsymbol{\alpha} = (\alpha_1, \ldots, \alpha_{k+1})$ $(\alpha_i > 0, i = 1, \ldots, k+1)$ if its probability density $\text{Di}_k(\boldsymbol{x} \,|\, \boldsymbol{\alpha})$, $0 < x_i < 1$ and $x_1 + \cdots + x_k < 1$, is

$$\text{Di}_k(\boldsymbol{x} \,|\, \boldsymbol{\alpha}) = c\, x_1^{\alpha_1 - 1} \cdots x_k^{\alpha_k - 1} \left(1 - \textstyle\sum_{i=1}^{k} x_i\right)^{\alpha_{k+1} - 1},$$

where

$$c = \frac{\Gamma\left(\sum_{i=1}^{k+1} \alpha_i\right)}{\prod_{i=1}^{k+1} \Gamma(\alpha_i)}.$$

If $k = 1$, $\text{Di}_k(\boldsymbol{x} \,|\, \boldsymbol{\alpha})$ reduces to the beta density $\text{Be}(x \,|\, \alpha_1, \alpha_2)$. In the general case, the mean vector and covariance matrix are given by

$$E[x_i] = \frac{\alpha_i}{\sum_{j=1}^{k+1} \alpha_j}; \quad V[x_i] = \frac{E[x_i](1 - E[x_i])}{1 + \sum_{j=1}^{k+1} \alpha_j}; \quad C[x_i, x_j] = \frac{-E[x_i]E[x_j]}{1 + \sum_{j=1}^{k+1} \alpha_j}.$$

If $\alpha_i > 1$, $i = 1, \ldots, k$, there is a mode given by

$$M[x_i] = \frac{\alpha_i - 1}{\sum_{j=1}^{k+1} \alpha_j \ - k - 1}\,.$$

The marginal distribution of $\boldsymbol{x}^{(m)} = (x_1, \ldots, x_m)$, $m < k$, is the Dirichlet

$$p(\boldsymbol{x}^{(m)}) = \text{Di}_m(\boldsymbol{x}^{(m)} \,|\, \alpha_1, \ldots, \alpha_m, \textstyle\sum_{j=m+1}^{k+1} \alpha_j).$$

The conditional distribution, given $x_{m+1}, \ldots, x_k$, of

$$x_i' = \frac{x_i}{1 - \sum_{j=m+1}^{k} x_j}\,, \quad i = 1, \ldots, m$$

is also Dirichlet, $\text{Di}_m(x_1', \ldots, x_m' \,|\, \alpha_1, \ldots, \alpha_m, \alpha_{k+1})$. In particular,

$$p(x_i' \,|\, x_{m+1}, \ldots, x_k) = \text{Be}(x_i' \,|\, \alpha_i, \textstyle\sum_{j=1}^{m} \alpha_j + \alpha_{k+1} - \alpha_i), \quad i = 1, \ldots, m\,.$$

Moreover, if $\boldsymbol{x} = (x_1, \ldots, x_k)$ has density $\text{Di}_k(\boldsymbol{x} \,|\, \boldsymbol{\alpha})$, then $\boldsymbol{y} = (y_1, \ldots, y_t)$ where

$$y_1 = x_1 + \cdots + x_{i_1}, \ \ldots, \ y_t = x_{i_{t-1}+1} + \cdots + x_k, \quad 1 \le t < k,$$

has density $\text{Di}_t(\boldsymbol{y} \,|\, \boldsymbol{\beta})$, where

$$\beta_1 = \alpha_1 + \cdots + \alpha_{i_1}, \ \ldots, \ \beta_t = \alpha_{i_{t-1}+1} + \cdots + \alpha_k, \beta_{t+1} = \alpha_{k+1}.$$

The Multinomial-Dirichlet Distribution

A discrete random vector $\boldsymbol{x} = (x_1, \ldots, x_k)$ has a *multinomial-Dirichlet* distribution of dimension k, with parameters $\boldsymbol{\alpha} = (\alpha_1, \ldots, \alpha_{k+1})$ and n where $\alpha_i > 0$, and $n = 1, 2, \ldots$, if its probability function $\text{Md}_k(\boldsymbol{x} \,|\, \boldsymbol{\alpha}, n)$, for $x_i = 0, 1, 2, \ldots$, with $\sum_{i=1}^{k} x_i \le n$, is

$$\text{Md}_k(\boldsymbol{x} \,|\, \boldsymbol{\alpha}, n) = c \prod_{j=1}^{k+1} \left(\frac{\alpha_j^{[x_j]}}{x_j!} \right)$$

where $\alpha^{[s]} = \prod_{j=1}^{s} (\alpha + j - 1)$ defines the *ascending factorial* function, with $x_{k+1} = n - \sum_{j=1}^{k} x_j$ and

$$c = \frac{n!}{\sum_{j=1}^{k+1} \alpha_j^{[n]}}\,.$$

The mean vector and covariance matrix are given by

$$E[x_i] = np_i, \quad p_i = \frac{\alpha_i}{\sum_{j=1}^{k+1} \alpha_j}$$

$$V[x_i] = \frac{n + \sum_{j=1}^{k+1} \alpha_j}{1 + \sum_{j=1}^{k+1} \alpha_j} np_i(1 - p_i)$$

$$C[x_i, x_j] = -\frac{n + \sum_{j=1}^{k+1} \alpha_j}{1 + \sum_{j=1}^{k+1} \alpha_j} np_i p_j \,.$$

The marginal distribution of the subset $\{x_1, \ldots, x_s\}$ is a multinomial-Dirichlet with parameters $\{\alpha_1, \ldots, \alpha_s, \sum_{j=1}^{k+1} \alpha_j - \sum_{j=1}^{s} \alpha_j\}$ and n. In particular, the marginal distribution of x_i is the binomial-beta $\text{Bb}(x_i \mid \alpha_i, \sum_{j=1}^{k+1} \alpha_j - \alpha_i)$. Moreover, the conditional distribution of $\{x_{s+1}, \ldots, x_k\}$ given $\{x_1, \ldots, x_s\}$ is also multinomial-Dirichlet, with parameters $\{\alpha_{s+1}, \ldots, \alpha_k, \sum_{j=1}^{k+1} \alpha_j - \sum_{j=s+1}^{k} \alpha_j\}$ and $n - \sum_{j=1}^{s} x_j$. For an interesting characterization of this distribution, see Basu and Pereira (1983).

The Normal-Gamma Distribution

A continuous bivariate random vector (x, y) has a *normal-gamma* distribution, with parameters μ, λ, α and β, $(\mu \in \Re, \lambda > 0, \alpha > 0, \beta > 0)$ if its density $\text{Ng}(x, y \mid \mu, \lambda, \alpha, \beta)$ is

$$\text{Ng}(x, y \mid \mu, \lambda, \alpha, \beta) = \text{N}(x \mid \mu, \lambda y)\text{Ga}(y \mid \alpha, \beta), \quad x \in \Re, y > 0,$$

where the normal and gamma densities are defined in Section 3.2.2. It is clear from the definition, that the conditional density of x given y is $\text{N}(x \mid \mu, \lambda y)$ and that the marginal density of y is $\text{Ga}(y \mid \alpha, \beta)$. Moreover, the marginal density of x is $\text{St}(x \mid \mu, \lambda\alpha/\beta, 2\alpha)$.

The shape of a normal-gamma distribution is illustrated in Figure 3.1, where the probability density of $\text{Ng}(x, y \mid 0, 1, 5, 5)$ is displayed both as a surface and in terms of equal density contours.

The Multivariate Normal Distribution

A continuous random vector $\boldsymbol{x} = (x_1, \ldots, x_k)$ has a *multivariate normal* distribution of dimension k, with parameters $\boldsymbol{\mu} = (\mu_1, \ldots, \mu_k)$ and $\boldsymbol{\lambda}$, where $\boldsymbol{\mu} \in \Re^k$ and $\boldsymbol{\lambda}$ is a $k \times k$ symmetric positive-definite matrix, if its probability density $\text{N}_k(\boldsymbol{x} \mid \boldsymbol{\mu}, \boldsymbol{\lambda})$ is

$$\text{N}_k(\boldsymbol{x} \mid \boldsymbol{\mu}, \boldsymbol{\lambda}) = c \exp\{-\tfrac{1}{2}(\boldsymbol{x} - \boldsymbol{\mu})^t \boldsymbol{\lambda}(\boldsymbol{x} - \boldsymbol{\mu})\}, \quad \boldsymbol{x} \in \Re^k,$$

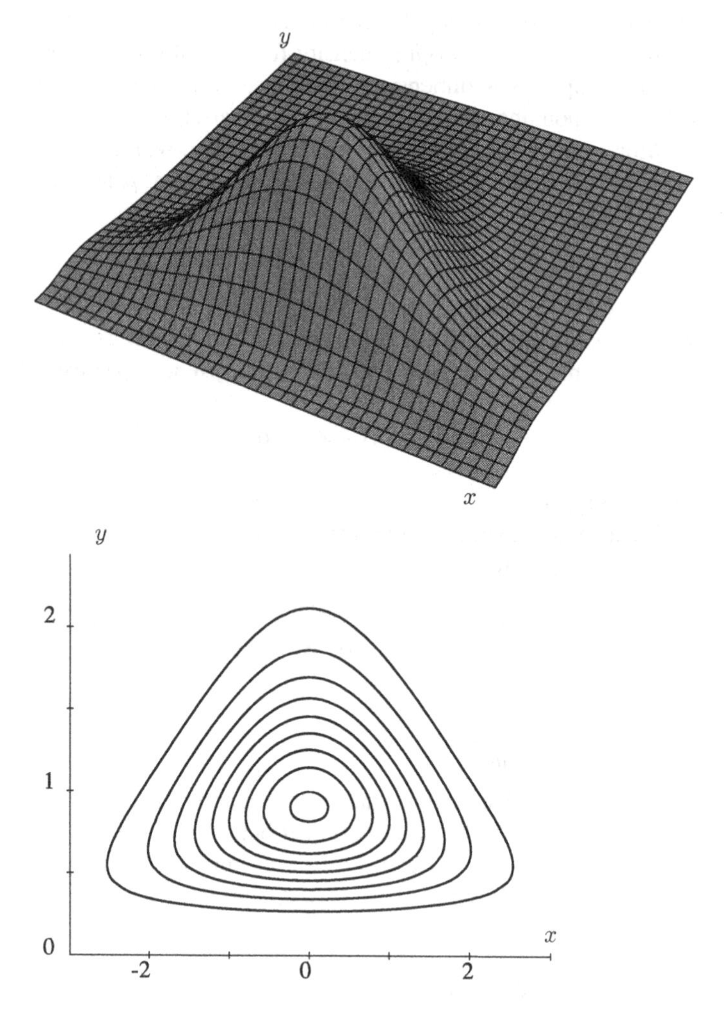

Figure 3.1 *The Normal-gamma density* $\text{Ng}(x, y \mid 0, 1, 5, 5)$

where $c = (2\pi)^{-k/2}|\boldsymbol{\lambda}|^{1/2}$.

If $k = 1$, so that $\boldsymbol{\lambda}$ is a scalar, λ, $\text{N}_k(\boldsymbol{x} \mid \boldsymbol{\mu}, \boldsymbol{\lambda})$ reduces to the univariate normal density $\text{N}(x \mid \mu, \lambda)$.

In the general case, $E[x_i] = \mu_i$, and, with $\Sigma = \boldsymbol{\lambda}^{-1}$ of general element σ_{ij}, $V[x_i] = \sigma_{ii}$ and $C[x_i, x_j] = \sigma_{ij}$, so that $V[\boldsymbol{x}] = \boldsymbol{\lambda}^{-1}$. The parameter $\boldsymbol{\mu}$ therefore labels the *mean vector* and the parameter $\boldsymbol{\lambda}$ the *precision matrix* (the inverse of the *covariance matrix*, Σ). If $\boldsymbol{y} = \boldsymbol{A}\boldsymbol{x}$, where $\boldsymbol{A}$ is an $m \times k$ matrix of real numbers such that $\boldsymbol{A}\Sigma\boldsymbol{A}^t$ is non-singular, then $\boldsymbol{y}$ has density $\mathrm{N}_m(\boldsymbol{y} \mid \boldsymbol{A}\boldsymbol{\mu}, (\boldsymbol{A}\Sigma\boldsymbol{A}^t)^{-1})$.

In particular, the *marginal* density for any subvector of $\boldsymbol{x}$ is (multivariate) normal, of appropriate dimension, with mean vector and covariance matrix given by the corresponding subvector of $\boldsymbol{\mu}$ and submatrix of $\boldsymbol{\lambda}^{-1}$.

Moreover, if $\boldsymbol{x} = (\boldsymbol{x}_1, \boldsymbol{x}_2)$ is a partition of $\boldsymbol{x}$, with $\boldsymbol{x}_i$ having dimension k_i, and $k_1 + k_2 = k$, and if the corresponding partitions of $\boldsymbol{\mu}$ and $\boldsymbol{\lambda}$ are

$$\boldsymbol{\mu} = \begin{pmatrix} \boldsymbol{\mu}_1 \\ \boldsymbol{\mu}_2 \end{pmatrix}, \quad \boldsymbol{\lambda} = \begin{pmatrix} \boldsymbol{\lambda}_{11} & \boldsymbol{\lambda}_{12} \\ \boldsymbol{\lambda}_{21} & \boldsymbol{\lambda}_{22} \end{pmatrix},$$

then the *conditional* density of $\boldsymbol{x}_1$ given $\boldsymbol{x}_2$ is also (multivariate) normal, of dimension k_1 with mean vector and precision matrix given, respectively, by

$$\boldsymbol{\mu}_1 - \boldsymbol{\lambda}_{11}^{-1}\boldsymbol{\lambda}_{12}(\boldsymbol{x}_2 - \boldsymbol{\mu}_2) \quad \text{and} \quad \boldsymbol{\lambda}_{11}.$$

The random quantity $y = (\boldsymbol{x} - \boldsymbol{\mu})^t\boldsymbol{\lambda}(\boldsymbol{x} - \boldsymbol{\mu})$ has a $\chi^2(y \mid k)$ density.

We also note that, from the form of the multivariate normal density, we can deduce the integral formula

$$\int_{\Re^k} \exp\{-\tfrac{1}{2}(\boldsymbol{x} - \boldsymbol{\mu})^t\boldsymbol{\lambda}(\boldsymbol{x} - \boldsymbol{\mu})\}\, d\boldsymbol{x} = \frac{(2\pi)^{k/2}}{|\boldsymbol{\lambda}|^{1/2}}.$$

The Wishart Distribution

A symmetric, positive-definite matrix $\boldsymbol{x}$ of random quantities $x_{ij} = x_{ji}$, for $i = 1, \ldots, k$, $j = 1, \ldots, k$, has a *Wishart* distribution of dimension k, with parameters α and $\boldsymbol{\beta}$ (with $2\alpha > k - 1$ and $\boldsymbol{\beta}$ a $k \times k$ symmetric, nonsingular matrix), if the density $\mathrm{Wi}_k(\boldsymbol{x} \mid \alpha, \boldsymbol{\beta})$ of the $k(k+1)/2$ dimensional random vector of the distinct entries of $\boldsymbol{x}$ is

$$\mathrm{Wi}_k(\boldsymbol{x} \mid \alpha, \boldsymbol{\beta}) = c\, |\boldsymbol{x}|^{\alpha-(k+1)/2} \exp\{-\operatorname{tr}(\boldsymbol{\beta}\boldsymbol{x})\},$$

where $c = |\boldsymbol{\beta}|^{\alpha}/\Gamma_k(\alpha)$,

$$\Gamma_k(\alpha) = \pi^{k(k-1)/4} \prod_{i=1}^{k} \Gamma\left(\frac{2\alpha + 1 - i}{2}\right)$$

is the *generalised gamma function* and $\operatorname{tr}(.)$, as before, denotes the *trace* of a matrix argument. If $k = 1$, so that $\boldsymbol{\beta}$ is a scalar β, then $W_k(\boldsymbol{x} \mid \alpha, \boldsymbol{\beta})$ reduces to the gamma density $\mathrm{Ga}(x \mid \alpha, \beta)$.

If $\{\boldsymbol{x}_1, \ldots, \boldsymbol{x}_n\}$ is a random sample of size $n > 1$ from a multivariate normal $\mathrm{N}_k(\boldsymbol{x}_i \mid \boldsymbol{\mu}, \boldsymbol{\lambda})$, and $\overline{\boldsymbol{x}} = n^{-1} \sum \boldsymbol{x}_i$, then $\overline{\boldsymbol{x}}$ is $\mathrm{N}_k(\overline{\boldsymbol{x}} \mid \boldsymbol{\mu}, n\boldsymbol{\lambda})$, and

$$S = \sum_{i=1}^{n} (\boldsymbol{x}_i - \overline{\boldsymbol{x}})(\boldsymbol{x}_i - \overline{\boldsymbol{x}})^t$$

is independent of $\overline{\boldsymbol{x}}$, and has a Wishart distribution $\mathrm{Wi}_k(S \mid \frac{1}{2}(n-1), \frac{1}{2}\boldsymbol{\lambda})$.

The following properties of the Wishart distribution are easily established: $E[\boldsymbol{x}] = \alpha\boldsymbol{\beta}^{-1}$ and $E[\boldsymbol{x}^{-1}] = (\alpha - (k+1)/2)^{-1}\boldsymbol{\beta}$; if $\boldsymbol{y} = \boldsymbol{A}\boldsymbol{x}\boldsymbol{A}^t$ where $\boldsymbol{A}$ is an $m \times k$ matrix ($m \leq k$) of real numbers, then $\boldsymbol{y}$ has a Wishart distribution of dimension m with parameters α and $(\boldsymbol{A}\boldsymbol{\beta}^{-1}\boldsymbol{A}^t)^{-1}$, if the latter exists; in particular, if $\boldsymbol{x}$ and $\boldsymbol{\beta}^{-1}$ conformably partition into

$$\boldsymbol{x} = \begin{pmatrix} \boldsymbol{x}_{11} & \boldsymbol{x}_{12} \\ \boldsymbol{x}_{21} & \boldsymbol{x}_{22} \end{pmatrix}, \quad \boldsymbol{\beta}^{-1} = \begin{pmatrix} \boldsymbol{\sigma}_{11} & \boldsymbol{\sigma}_{12} \\ \boldsymbol{\sigma}_{21} & \boldsymbol{\sigma}_{22} \end{pmatrix},$$

where $\boldsymbol{x}_{11}$, $\boldsymbol{\sigma}_{11}$ are square $h \times h$ matrices ($1 \leq h < k$), then $\boldsymbol{x}_{11}$ has a Wishart distribution of dimension h with parameters α and $(\boldsymbol{\sigma}_{11})^{-1}$. Moreover, if $\boldsymbol{x}_1, \ldots, \boldsymbol{x}_s$ are independent $k \times k$ random matrices, each with a Wishart distribution, with parameters $\alpha_i, \boldsymbol{\beta}, i = 1, \ldots, s$, then $\boldsymbol{x}_1 + \cdots + \boldsymbol{x}_s$ also has a Wishart distribution, with parameters $\alpha_1 + \cdots + \alpha_s$ and $\boldsymbol{\beta}$.

We note that, from the form of the Wishart density, we can deduce the integral formula

$$\int |\boldsymbol{x}|^{\alpha-(k+1)/2} \exp\{-\mathrm{tr}(\boldsymbol{\beta}\boldsymbol{x})\}\, d\boldsymbol{x} = c^{-1},$$

the integration being understood to be with respect to the $k(k+1)/2$ distinct elements of the matrix $\boldsymbol{x}$.

The Multivariate Student Distribution

A continuous random vector $\boldsymbol{x} = (x_1, \ldots, x_k)$ has a *multivariate Student* distribution of dimension k, with parameters $\boldsymbol{\mu} = (\mu_1, \ldots, \mu_k)$, $\boldsymbol{\lambda}$ and α ($\boldsymbol{\mu} \in \Re^k$, $\boldsymbol{\lambda}$ a symmetric, positive-definite $k \times k$ matrix, $\alpha > 0$) if its probability density $\mathrm{St}_k(\boldsymbol{x} \mid \boldsymbol{\mu}, \boldsymbol{\lambda}, \alpha)$ is

$$\mathrm{St}_k(\boldsymbol{x} \mid \boldsymbol{\mu}, \boldsymbol{\lambda}, \alpha) = c\left[1 + \frac{1}{\alpha}(\boldsymbol{x} - \boldsymbol{\mu})^t\boldsymbol{\lambda}(\boldsymbol{x} - \boldsymbol{\mu})\right]^{-(\alpha+k)/2}, \quad \boldsymbol{x} \in \Re^k,$$

where

$$c = \frac{\Gamma((\alpha + k)/2)}{\Gamma(\alpha/2)(\alpha\pi)^{k/2}}|\boldsymbol{\lambda}|^{1/2}.$$

If $k = 1$, so that $\boldsymbol{\lambda}$ is a scalar, λ, then $\mathrm{St}_k(\boldsymbol{x} \mid \boldsymbol{\mu}, \boldsymbol{\lambda}, \alpha)$ reduces to the univariate Student density $\mathrm{St}(x \mid \mu, \lambda, \alpha)$. In the general case, $E[\boldsymbol{x}] = \boldsymbol{\mu}$ and $V[\boldsymbol{x}] = \boldsymbol{\lambda}^{-1}(\alpha/(\alpha - 2))$. Although not exactly equal to the inverse of the covariance matrix, the parameter $\boldsymbol{\lambda}$ is often referred to as the *precision matrix* of the distribution.

If $y = Ax$, where A is an $m \times k$ matrix ($m \leq k$) of real numbers such that $A\lambda^{-1}A^t$ is non-singular, then y has density $\mathrm{St}_m(y \mid A\mu, (A\lambda^{-1}A^t)^{-1}, \alpha)$. In particular, the *marginal* density for any subvector of x is (multivariate) Student, of appropriate dimension, with mean vector and inverse of the precision matrix given by the corresponding subvector of μ and submatrix of λ^{-1}. Moreover, if $x = (x_1, x_2)$ is a partition of x and the corresponding partitions of μ and λ are given by

$$\mu = \begin{pmatrix} \mu_1 \\ \mu_2 \end{pmatrix}, \quad \lambda = \begin{pmatrix} \lambda_{11} & \lambda_{12} \\ \lambda_{21} & \lambda_{22} \end{pmatrix},$$

then the *conditional* density of x_1, given x_2 is also (multivariate) Student, of dimension k_1, with $\alpha + k_2$ degrees of freedom, and mean vector and precision matrix, respectively, given by

$$\mu_1 - \lambda_{11}^{-1}\lambda_{12}(x_2 - \mu_2),$$

$$\lambda_{11}\left[\frac{\alpha + k_2}{\alpha + (x_2 - \mu_2)^t(\lambda_{22} - \lambda_{21}\lambda_{11}^{-1}\lambda_{12})(x_2 - \mu_2)}\right].$$

The random quantity $y = (x - \mu)^t\lambda(x - \mu)$ has an $\mathrm{Fs}(y \mid k, \alpha)$ density.

The Multivariate Normal-Gamma Distribution

A continuous random vector $x = (x_1, \ldots, x_k)$ and a random quantity y have a joint *multivariate normal-gamma* distribution of dimension k, with parameters $\mu, \lambda, \alpha, \beta$ ($\mu \in \Re^k$, λ a $k \times k$ symmetric, positive-definite matrix, $\alpha > 0$ and $\beta > 0$) if the joint probability density of x and y, $\mathrm{Ng}_k(x, y \mid \mu, \lambda, \alpha, \beta)$ is

$$\mathrm{Ng}_k(x, y \mid \mu, \lambda, \alpha, \beta) = \mathrm{N}_k(x \mid \mu, \lambda y)\mathrm{Ga}(y \mid \alpha, \beta),$$

where the multivariate normal and gamma densities have already been defined. From the definition, the conditional density of x given y is $\mathrm{N}_k(x \mid \mu, \lambda y)$ and the marginal density of y is $\mathrm{Ga}(y \mid \alpha, \beta)$. Moreover, the marginal density of x is $\mathrm{St}_k(x \mid \mu, \alpha^{-1}\beta\lambda, 2\alpha)$.

The Multivariate Normal-Wishart Distribution

A continuous random vector x and a symmetric, positive-definite matrix of random quantities y have a joint *Normal-Wishart* distribution of dimension k, with parameters $\mu, \lambda, \alpha, \beta$ ($\mu \in \Re^k$, $\lambda > 0$, integer $2\alpha > k - 1$, and β a $k \times k$ symmetric, non-singular matrix), if the probability density of x and the $k(k+1)/2$ distinct elements of y, $\mathrm{Nw}_k(x, y \mid \mu, \lambda, \alpha, \beta)$ is

$$\mathrm{Nw}_k(x, y \mid \mu, \lambda, \alpha, \beta) = \mathrm{N}_k(x \mid \mu, \lambda y)\,\mathrm{Wi}_k(y \mid \alpha, \beta),$$

where the multivariate normal and Wishart densities are as defined above.

From the definition, the conditional density of x given y is $\mathrm{N}_k(x \mid \mu, \lambda y)$ and the marginal density of y is $\mathrm{Wi}_k(y \mid \alpha, \beta)$. Moreover, the marginal density of x is $\mathrm{St}_k(x \mid \mu, \lambda\alpha\beta^{-1}, 2\alpha)$.

The Bilateral Pareto Distribution

A continuous bivariate random vector (x, y) has a *bilateral Pareto* distribution with parameters β_0, β_1, and α $(\{\beta_0, \beta_1\} \in \Re^2,\ \beta_0 < \beta_1,\ \alpha > 0)$ if its density function $\text{Pa}_2(x, y \mid \alpha, \beta_0, \beta_1)$ is

$$\text{Pa}_2(x, y \mid \alpha, \beta_0, \beta_1) = c\,(y - x)^{-(\alpha+2)}, \quad x \leq \beta_0,\ y \geq \beta_1,$$

where $c = \alpha(\alpha + 1)(\beta_1 - \beta_0)^{\alpha}$. The mean and variance are given by

$$E[x] = \frac{\alpha\beta_0 - \beta_1}{\alpha - 1}, \quad E[y] = \frac{\alpha\beta_1 - \beta_0}{\alpha - 1}, \quad \text{if } \alpha > 1,$$

$$V[x] = V[y] = \frac{\alpha(\beta_1 - \beta_0)^2}{(\alpha - 1)^2(\alpha - 2)}, \quad \text{if } \alpha > 2,$$

and the correlation between x and y is $-\alpha^{-1}$. The marginal distributions of $t_1 = \beta_1 - x$ and $t_2 = y - \beta_0$ are both $\text{Pa}(t \mid \beta_1 - \beta_0, \alpha)$.

3.3 GENERALISED OPTIONS AND UTILITIES

3.3.1 Motivation and Preliminaries

For reasons of mathematical or descriptive convenience, it is common in statistical decision problems to consider sets of options which consist of part or all of the real line (as in problems of point estimation) or are part of some more general space. It is therefore desirable to extend the concepts and results of Chapter 2 to a much more general mathematical setting, going beyond finite, or even countable, frameworks, first by taking $\mathcal{E}$ to be a σ-algebra and then suitably extending the fundamental notion of an option.

In the finite case, an option was denoted by $a = \{c_j \mid E_j,\ j \in J\}$, with the straightforward interpretation that, if option a is chosen, c_j is the consequence of the occurrence of the event E_j. The extension of this function definition to infinite settings clearly requires some form of constructive limit process, analogous to that used in Lebesgue measure and integration theory in passing from simple (i.e., "step") functions to more general functions. Since the development given in Chapter 2 led to the assessment of options in terms of their expected utilities, the "natural" definition of limit that suggests itself is one based fundamentally on the expected utility idea (Bernardo, Ferrándiz and Smith, 1985).

Let us therefore consider a decision problem $\{\mathcal{A}, \mathcal{E}, \mathcal{C}, \leq\}$, which is described by a probability space $\{\Omega, \mathcal{F}, P\}$ and utility function $u : \mathcal{C} \to \Re$, and let

$$\mathcal{D} = \{d : \Omega \to \mathcal{C};\ \overline{u}(d) = \int_{\Omega} u\big(d(\omega)\big)\, dP(\omega) < \infty\}.$$

In other words, $\mathcal{D}$ consists of those functions (soon to be called decisions) $d : \Omega \to \mathcal{C}$ for which $u \circ d \equiv u(d(.))$ is a random quantity whose expectation exists. In the case of the particular subset $\mathcal{A}$ of $\mathcal{D}$, $\overline{u}(a) = \overline{u}(a \mid \Omega)$ is precisely the expected utility of the *simple option* a (see Definition 2.16 with $G = \Omega$; we shall return later to the case of a general conditioning event G and corresponding probability measure $P(\cdot \mid G)$). In all the definitions and propositions in this section, $\mathcal{A}$ and $\{\Omega, \mathcal{F}, P\}$ are to be understood as fixed background specifications.

Definition 3.11. (***Convergence in expected utility***). *For a given utility function, $u : \mathcal{C} \to \Re$, a sequence of functions $d_1, d_2, \ldots$ in $\mathcal{D}$ is said to u-converge to a function d in $\mathcal{D}$, written $d_i \to_u d$, if and only if*

(i) *$u \circ d_i$ converges to $u \circ d$ almost surely (with respect to P),*

(ii) $\overline{u}(d_i) \to \overline{u}(d)$.

Definition 3.12. (***Decisions***). *For a given utility function, $u : \mathcal{C} \to \Re$, a function $d \in \mathcal{D}$ is a decision (generalised option) if and only if there exists a sequence $a_1, a_2, \ldots$ of simple options such that $a_i \to_u d$; the value of*

$$\overline{u}(d) = \lim_i \overline{u}(a_i)$$

is then called the expected utility of the decision d.

Discussion of Definitions 3.11 and 3.12. In abstract mathematical terms, the extension from simple functions, mapping $\mathcal{C}$ to $\Re$, to more general functions requires some form of limit process. However, the fundamental coherence result of Proposition 2.25 was that simple options should be compared in terms of their expected utilities. In order for this to carry over smoothly to decisions (generalised options), it is natural to require a constructive definition of the latter in terms of a limit concept directly expressed in terms of expected utilities.

As it stands, however, this constructive definition does not provide a straightforward means of checking whether or not, given a specified utility function, u, a function $d \in \mathcal{D}$ is or is not a generalised option. However, we can prove that any $d \in \mathcal{D}$ such that $u \circ d$ is essentially bounded (i.e., $u \circ d$ is bounded except on a subset of Ω of P-measure zero) *is* a decision. More specifically, we can prove the following.

Proposition 3.6. *Given a utility function $u : \mathcal{C} \to \Re$, for any function $d \in \mathcal{D}$ such that $u \circ d$ is essentially bounded, there exist sequences $a_1, a_2, \ldots$ and $a'_1, a'_2, \ldots$ of simple options such that $a_i \to_u d$, $a'_i \to_u d$ and, for all i, $\overline{u}(a_i) \leq \overline{u}(d) \leq \overline{u}(a'_i)$.*

Proof. We prove first that if $u \circ d$ is essentially bounded above then there exists a sequence of simple acts $a_1, a_2, \ldots,$ such that $a_i \to_u d$ and $\overline{u}(a_i) \geq \overline{u}(d)$, for all i. (An exactly parallel proof exists if "above" is replaced by "below" and $\geq$ by $\leq$.) We begin by defining the partitions $\{E_{ij},\, j \in J_i\}$, $i = 1, 2, \ldots$, where

$$
\begin{aligned}
E_{ij} &= \{\omega \in \Omega;\ u \circ d(\omega) < -i\} && \text{if } j = -i2^i - 1\\
&= \{\omega \in \Omega;\ u \circ d(\omega) \in \left[j2^{-i}, (j+1)2^{-i}\right]\} && \text{if } j = -i2^i, \ldots, i2^i - 1\\
&= \{\omega \in \Omega;\ u \circ d(\omega) \geq i\} && \text{if } j = i2^i.
\end{aligned}
$$

For each i, this establishes a partition of Ω into $2(i2^i + 1)$ events, in such a way that two extreme events contain outcomes with values of $u \circ d(\omega) < -i$ or $\geq i$, whereas the other events contain outcomes whose values of $u \circ d(\omega)$ do not differ by more than 2^{-i}.

We now define a sequence $\{a_i\}$, of simple options $a_i = \{c_{ij} \mid E_{ij},\, j \in J_i\}$ such that:

(i) if $P(E_{ij}) = 0$ then c_{ij} is an arbitrary element of $\mathcal{C}$;

(ii) if $P(E_{ij}) > 0$ then $c_{ij} \in d(E_{ij})$ and

$$u(c_{ij})P(E_{ij}) \geq \int_{E_{ij}} u(d(\omega))dP(\omega).$$

To see that the c_{ij} exist and are well defined, note that, since $\overline{u}(d) < \infty$, there exists $\overline{u}_{ij}(d) < \infty$, defined by

$$\overline{u}_{ij}(d) = \frac{1}{P(E_{ij})} \int_{E_{ij}} u(d(\omega))dP(\omega);$$

but, if $u(d(\omega)) < \overline{u}_{ij}(d)$ for all $\omega \in E_{ij}$, then we would have

$$\int_{E_{ij}} u(d(\omega))dP(\omega) < \overline{u}_{ij}(d)P(E_{ij}),$$

thus contradicting the definition of $\overline{u}_{ij}(d)$.

By construction, $a_i \to d$ almost surely. Hence, for all $\varepsilon > 0$, there exists i such that $u \circ d(\omega) \in [-i, i)$, with $2^{-i} < \varepsilon$ and, for this i, $|a_i(\omega) - d(\omega)| < \varepsilon$. In addition, for all i,

$$\overline{u}(a_i) = \sum_j u(c_{ij})P(E_{ij}) \geq \sum_{j \in J_i} \int_{E_{ij}} u(d(\omega))dP(\omega) = \overline{u}(d).$$

To show that $a_i \to_u d$, it remains to prove that

$$\int_{\Omega} |\, u(d(\omega)) - u(a_i(\omega))\,|\, dP(\omega) \to 0, \quad \text{as} \quad i \to \infty.$$

Writing $\int_{\Omega} = \int_{A_i} + \int_{A_i^c}$, where $A_i = \{\omega \in \Omega \mid u(d(\omega)) < -i\}$, we note that for sufficiently large i (larger than the essential supremum of $u \circ d$),

$$\int_{A_i^c} |\, u(d(\omega)) - u(a_i(\omega))\,|\, dP(\omega) \leq 2^{-i} P(A_i^c),$$

which converges to zero as $i \to \infty$; moreover, since $\overline{u}(a_i) \geq \overline{u}(d)$, for all i, and $\overline{u}(d) < \infty$,

$$\int_{A_i} |u(d(\omega)) - u(a_i(\omega))| dP(\omega) \leq -2 \int_{A_i} u(d(\omega)) dP(\omega)$$

and, since $A_i \to \emptyset$ as $i \to \infty$, this also converges to zero. $\triangleleft$

In fact, we can show that any decision, whether or not $u \circ d$ is essentially bounded, can be obtained as the limit of "bounding sequences of simple options" in the sense made precise in the following.

Proposition 3.7. *Given a utility function $u : \mathcal{C} \to \Re$, for any decision $d \in \mathcal{D}$, there exist sequences $a_1, a_2, \ldots$, and $a_1', a_2', \ldots$, of simple options such that $a_i \to_u d$, $a_i' \to_u d$ and, for all i, $\overline{u}(a_i) < \overline{u}(d) < \overline{u}(a_i')$.*

Proof. We shall show that there exists a sequence of simple options $a_1, a_2, \ldots$ such that $a_i \to_u d$ and, for all i, $\overline{u}(a_i) < \overline{u}(d)$. An obviously parallel proof exists for the other inequality.

We first note that either $u \circ d$ is essentially bounded above or it is not. In the former case, let K denote the essential supremum and define $A_0 = \{\omega \in \Omega; u \circ d(\omega) = K\}$; in the latter case, define $K = \infty$ and $A_0 = \emptyset$.

If $P(A_0) > 0$, choose a decreasing sequence of real numbers $\alpha_j \in [0, 1]$ such that $\alpha_j \to 0$. Then by Axiom 4 and Proposition 2.6 there exists a sequence of standard events $S_1, S_2, \ldots$ such that $S_{j+1} \supset S_j$, $\mu(S_j) = \alpha_j$ and $\Pr(A_0 \cap S_j) = P(A_0)\alpha_j$, for all j; then, define $A_j = A_0 \cap S_j$ and choose a consequence $c \in \mathcal{C}$ such that $u(c) < K$.

If $P(A_0) = 0$, choose a consequence $c \in \mathcal{C}$ and an increasing sequence of real numbers β_j, such that $\beta_j \to \infty$ and $u(c) < \beta_1$; let $A_j = \{\omega \in \Omega; u \circ d(\omega) > \beta_j\}$. In either case, define

$$d_j(\omega) = \begin{cases} d(\omega), & \text{if } \omega \in A_j^c \\ c, & \text{if } \omega \in A_j. \end{cases}$$

Since d is a decision, there exists a sequence $a_1, a_2, \ldots$ of simple options such that $a_i \rightarrow_u d$. If, for each a_i, we now define a new sequence of simple options a^*_{ij}, by

$$a^*_{ij}(\omega) = \begin{cases} a_i(\omega), & \text{if } \omega \in A^c_j \\ c, & \text{if } \omega \in A_j, \end{cases}$$

we clearly have $a^*_{ij} \rightarrow_u d_i$, so, that, for all i, d_i is a decision. Moreover, by construction $d_i \rightarrow_u d$, $\overline{u}(d_i) < \overline{u}(d)$ and $u \circ d_i$ is bounded above, for all i. Hence, by Proposition 3.6, there exist sequences of simple options

$$a^{(i)}_1, a^{(i)}_2, \ldots, \quad \text{such that} \quad a^{(i)}_j \rightarrow_u d_i \quad \text{and} \quad \overline{u}(a^{(i)}_j) \geq \overline{u}(d_i).$$

If we now choose a subsequence

$$a^{(i)}_{j_1}, a^{(i)}_{j_2}, \ldots, \quad \text{such that} \quad \overline{u}\left(a^{(i)}_{j_k}\right) - \overline{u}(d_i) < \overline{u}(d) - \overline{u}(d_i),$$

the required result follows, since for all k,

$$\overline{u}(d_i) \leq \overline{u}\left(a^{(i)}_{j_k}\right) < \overline{u}(d)$$

and $d_i \rightarrow_u d$ implies that $a^{(i)}_{j_k} \rightarrow d$. $\triangleleft$

3.3.2 Generalised Preferences

Given the adoption, for mathematical or descriptive convenience, of the extended framework developed in the previous section, it is natural to require that preferences among simple options should "carry over", under the limit process we have introduced, to the corresponding decisions. This is made precise in the following.

Postulate 2. (***Extension of the preference relation***). *Given a utility function $u : \mathcal{C} \rightarrow \Re$, for any decisions d_1,d_2, and sequences of simple options $\{a_i\}$ and $\{a'_i\}$ such that $\{a_i\} \rightarrow_u d_1$, and $\{a'_i\} \rightarrow_u d_2$, we have:*

(i) *if, for all $i > i_0$, for some i_0, $a_i \geq a'_i$, then $d_1 \geq d_2$;*

(ii) *if, for all $i > i_0$, for some i_0, $a_i > a'_i$, then $d_1 > d_2$;*

The first part of the postulate simply captures the notion of the carry-over of preferences in the limit; the second part of the postulate is an obvious necessary condition for strict preference.

Together with our previous axioms, this postulate enables us to establish a very general statement of the identification of quantitative coherence with the principle of maximising expected utility.

Proposition 3.8. ***(Maximisation of expected utility for decisions).***
Given any two decisions d_1, d_2,

$$d_1 \geq d_2 \iff \overline{u}(d_1) \geq \overline{u}(d_2).$$

Proof. We first establish that $\overline{u}(d_2) > \overline{u}(d_1)$ implies that $d_2 > d_1$. By Proposition 3.7, there exist sequences of simple options $a_1, a_2, \ldots$, and $a'_1, a'_2, \ldots$, such that $a_i \to_u d_1$, $a'_i \to_u d_2$ and, for all i, $\overline{u}(a_i) > \overline{u}(d_1)$ and $\overline{u}(a'_i) < \overline{u}(d_2)$. With $\varepsilon = [\overline{u}(d_2) - \overline{u}(d_1)]/3$, we can choose i_1, i_2 such that, for all $j > \max\{i_1, i_2\}$,

$$\overline{u}(a_j) - \overline{u}(d_1) < \varepsilon, \quad \overline{u}(d_2) - \overline{u}(a'_j) < \varepsilon,$$

and we can choose i'_1, i'_2 such that, for all $j > i'_1$,

$$\overline{u}(a_j) - \overline{u}(d_1) \leq \overline{u}(a_{i_1}) - \overline{u}(d_1)$$

and, for all $j > i'_2$,

$$\overline{u}(d_2) - \overline{u}(a'_j) \leq \overline{u}(d_2) - \overline{u}(a'_{i_2}).$$

It follows from Proposition 2.25 that, for all $j > \max\{i'_1, i'_2\}$,

$$a'_j \geq a'_{i_2} > a_{i_1} \geq a_j$$

and so, by Postulate 2, $d_2 > d_1$.

To complete the proof ($d_1 \sim d_2 \Rightarrow \overline{u}(d_1) = \overline{u}(d_2)$ being obvious), we must show that $\overline{u}(d_1) = \overline{u}(d_2)$ implies that $d_1 \sim d_2$. By Proposition 3.7, there exist sequences of simple options $(a_i^{(k)}, k = 1, 2, 3, 4)$ such that $a_i^{(k)} \to_u d_1$ for $k = 1, 2$, $a_i^{(k)} \to_u d_2$ for $k = 3, 4$, and, for all i,

$$\overline{u}(a_i^{(1)}) \leq \overline{u}(d_1) \leq \overline{u}(a_i^{(2)}), \quad \overline{u}(a_i^{(3)}) \leq \overline{u}(d_2) \leq \overline{u}(a_i^{(4)}).$$

Since we have $\overline{u}(d_1) = \overline{u}(d_2)$, this implies, by Proposition 2.25, that $a_i^{(4)} \geq a_i^{(1)}$, and $a_i^{(2)} \geq a_i^{(3)}$, for all i, and hence, by Postulate 2, $d_2 \geq d_1$ and $d_1 \geq d_2$, so that $d_1 \sim d_2$. ◁

Proposition 3.9. *For any* $G > \emptyset$,

$$d_1 \geq_G d_2 \iff \overline{u}(d_1 \,|\, G) \geq \overline{u}(d_2 \,|\, G).$$

Proof. Throughout the above, the probability measure $P(\cdot)$ can be replaced by $P(\cdot \,|\, G)$ without any basic modifications to the proofs and results. Writing

$$\overline{u}_G(d) = \int_\Omega u \circ d(\omega) dP(\omega \,|\, G),$$

it is easily verified that $P(G)\overline{u}_G(d) = \overline{u}(1_G \circ d)$, where 1_G is the indicator function of G. It follows that if $u \circ d$ is integrable with respect to $P(\cdot)$ then it is integrable with respect to $P(\cdot \,|\, G)$. ◁

This establishes in full generality that the decision criterion of maximising the expected utility is the *only* criterion which is compatible with an intuitive set of quantitative coherence axioms and the natural mathematical extensions encapsulated in Postulates 1 and 2. Specifically, we have shown that, given a general decision problem, where $\omega \in \Omega$ labels the uncertain outcomes associated with the problem, $u(d(\omega))$ describes the *current* preferences among consequences and $p(\omega)$, the probability density of P with respect to the appropriate dominating measure, describes *current* beliefs about ω, the optimal action is that d_0^* which maximises the expected utility,

$$\overline{u}(d) = \int_{\Omega} u(d(\omega))\, p(\omega)\, d\omega.$$

As we saw in Section 2.6.3, it is natural before making a decision to consider trying to reduce the current uncertainty by obtaining further information by experimentation. Whether or not this is sensible obviously depends on the relative costs and benefits of such additional information, and we shall now extend the notions related to the value of information, introduced in Section 2.6.3, into the more general mathematical framework established in this chapter.

3.3.3 The Value of Information

For the general decision problem, the decision tree for experimental design, given originally in Figure 2.6, now takes the form given in Figure 3.2, where, as in Section 2.6.3, the utility notation is extended to make explicit the possible dependence on the experiment performed e (or e_0 if no data are collected) and the data obtained, x. If d_0^* is the optimal decision corresponding to e_0, the expected utility from an optimal decision with no additional information is defined by

$$\overline{u}(e_0) = \overline{u}(d_0^*, e_0) = \sup_d \int u(d, e_0, \omega)\, p(\omega \mid e_0, d)\, d\omega.$$

Let $d_{\boldsymbol{x}}^*$ be the optimal decision after experiment e has been performed and data x have been obtained, so that $\overline{u}(d_{\boldsymbol{x}}^*, e, x)$, the expected utility from the optimal decision given e and x, is

$$\overline{u}(d_{\boldsymbol{x}}^*, e, \boldsymbol{x}) = \sup_d \int_{\Omega} u(d, e, \boldsymbol{x}, \omega) p(\omega \mid e, \boldsymbol{x}, d) d\omega,$$

and, hence, the expected utility from the optimal decision following e is

$$\overline{u}(e) = \int_X \overline{u}(d_{\boldsymbol{x}}^*, e, \boldsymbol{x})\, p(\boldsymbol{x} \mid e)\, d\boldsymbol{x},$$

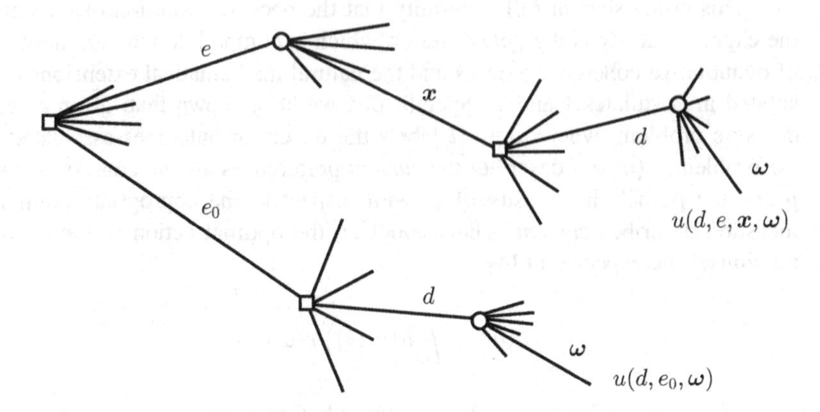

Figure 3.2 *Generalised decision tree for experimental design*

where $p(\boldsymbol{x} \mid e)$ describes beliefs about the occurrence of $\boldsymbol{x}$, were e to be performed.

Proposition 3.10. ***(Optimal experimental design).*** *The optimal decision is to perform experiment e^* if $\overline{u}(e^*) = \max_e \overline{u}(e)$ and $\overline{u}(e^*) > \overline{u}(e_0)$, and to perform no experiment otherwise.*

Proof. This follows immediately. $\triangleleft$

The expected value of the information provided by additional data $\boldsymbol{x}$ may be computed as the (posterior) expected difference between the utilities which correspond to optimal decisions after and before the data. Thus,

Definition 3.13. ***(The value of additional information).***

(i) *The expected value of the information provided by $\boldsymbol{x}$, is*

$$v(e, \boldsymbol{x}) = \int_{\Omega} \{u(d^*_{\boldsymbol{x}}, e, \boldsymbol{x}, \boldsymbol{\omega}) - u(d^*_0, e_0, \boldsymbol{\omega})\} p(\boldsymbol{\omega} \mid e, \boldsymbol{x}, d^*_{\boldsymbol{x}})\, d\boldsymbol{\omega} \; ;$$

(ii) *the expected value of the experiment e is*

$$v(e) = \int_X v(e, \boldsymbol{x})\, p(\boldsymbol{x} \mid e)\, d\boldsymbol{x}.$$

Let us now consider the optimal decisions which would be available to us if we knew the value of $\boldsymbol{\omega}$. Thus, let d^*_{ω} be the optimal decision given $\boldsymbol{\omega}$; i.e., such that, for all d,

$$u(d^*_{\omega}, e_0, \boldsymbol{\omega}) \geq u(d, e_0, \boldsymbol{\omega}), \quad \boldsymbol{\omega} \in \Omega.$$

Then, given ω, the loss suffered by choosing another decision $d \neq d^*_\omega$ would be

$$u(d^*_\omega, e_0, \boldsymbol{\omega}) - u(d, e_0, \boldsymbol{\omega}).$$

For $d = d^*_0$, the optimal decision with no additional data, this utility difference measures (conditional on ω) the value of perfect information. Its expected value with respect to $p(\omega)$ will define, under certain conditions, an upper bound on the increase in utility which additional information about ω could be expected to provide.

Definition 3.14. **(*Expected value of perfect information*).** *The* ***opportunity loss*** *of choosing* d *is defined to be*

$$l(d, \boldsymbol{\omega}) = u(d^*_\omega, e_0, \boldsymbol{\omega}) - u(d, e_0, \boldsymbol{\omega}),$$

and the expected value of perfect information about ω *is defined by*

$$v^*\{e_0\} = \int_\Omega l(d^*_0, \boldsymbol{\omega})\, p(\boldsymbol{\omega} \mid e_0, d^*_0)\, d\boldsymbol{\omega},$$

where d^*_0 *is the optimal decision with no additional information.*

As we remarked in Section 2.6.3, in many situations the utility function may often be thought of as made up of two separate components: the experimental cost of performing e and obtaining x, and the utility of directly taking decision d and finding ω to be the state of the world. Given such an (additive) decomposition, we can establish a useful upper bound for the expected value of an experiment.

Proposition 3.11. **(*Additive decomposition*).** *If the utility function has the form*

$$u(d, e, \boldsymbol{x}, \boldsymbol{\omega}) = u(d, e_0, \boldsymbol{\omega}) - c(e, \boldsymbol{x}),$$

with $c(e, \boldsymbol{x}) \geq 0$, *and the probability distributions are such that*

$$p(\boldsymbol{\omega} \mid e, \boldsymbol{x}, d) = p(\boldsymbol{\omega} \mid e, \boldsymbol{x}), \quad p(\boldsymbol{\omega} \mid e_0, d) = p(\boldsymbol{\omega} \mid e_0),$$

then, for any available experiment e,

$$v(e) \leq v^*(e_0) - \overline{c}(e),$$

where $\overline{c}(e) = \int c(e, \boldsymbol{x})\, p(\boldsymbol{x} \mid e)\, d\boldsymbol{x}$ *is the expected cost of* e.

Proof. This closely parallels the proof, given in Proposition 2.27, for the finite case. $\triangleleft$

This concludes the mathematical extension of the basic framework and associated axioms. In the next section, we reconsider the important special problem of statistical inference, previously discussed in detail in its finitistic setting in Section 2.7.

3.4 GENERALISED INFORMATION MEASURES

3.4.1 The General Problem of Reporting Beliefs

In Section 2.7, we argued that the problem of reporting a degree of belief distribution for a (finite) class of exclusive and exhaustive "hypotheses" $\{H_j,\ j \in J\}$, conditional on some relevant data D and initial state of information M_0, could be formulated as a decision problem $\{\mathcal{E}, \mathcal{C}, \mathcal{A}, \leq\}$. Here, $\{E_j,\ j \in J\}$ is a partition of Ω, consisting of elements of $\mathcal{E}$, with the interpretation $E_j \equiv$ "hypothesis H_j is true", $\mathcal{A}$ relates to

$$\mathcal{Q} = \{\boldsymbol{q} \equiv (q_j,\ j \in J);\quad q_j \geq 0\quad \textstyle\sum_{j\in J} q_j = 1\},$$

where q_j is the probability which, conditional on D, an individual *reports* as the probability of E_j being true, and the set of consequences $\mathcal{C}$ consists of all pairs $(\boldsymbol{q}, E_j)$, representing the possible conjunction of reported beliefs and true hypotheses. In the previous finitistic setting, we denoted by

$$\boldsymbol{p} = \{p_j = P(E_j \,|\, D),\ j \in J\},\quad p_j > 0,\quad \textstyle\sum_{j\in J} p_j = 1,$$

the probability measure describing an individual's *actual* beliefs, conditional on D. We then proceeded to consider a special class of utility functions (score functions) appropriate to this reporting problem and to examine the resulting forms of implied decisions and the links with information theory. In this section, we shall generalise these concepts and results to the extended framework developed in the previous sections.

The first generalisation consists in noting that the set of alternative "hypotheses" now corresponds to the set of possible values of a (possibly continuous) random vector, ω, say, labelling the "unknown states of the world", so that the relevant uncertain events are $E_\omega = \{\omega\}, \omega \in \Omega$, with the interpretation $E_\omega =$ "the hypothesis ω is true". Quantitative coherence requires that any particular individual's uncertainty about ω, given data D and initial state of information M_0, should be represented by a probability distribution P over a σ-algebra of subsets of Ω, which we shall assume can be described by a density (to be understood as a mass function in the discrete case)

$$p_\omega(\cdot \,|\, D) = \left\{p(\omega \,|\, D),\ \omega \in \Omega,\ p(\omega \,|\, D) > 0,\ \int_\Omega p(\omega \,|\, D)\, d\omega = 1\right\}.$$

We shall take the set of possible inference statements to be the set of probability distributions for ω, compatible with D. We denote by $\mathcal{D}$ the set of functions d_p, one for each $p_\omega(\cdot \,|\, D)$, which map ω to the pair $(p_\omega(\cdot \,|\, D), \omega)$.

3.4.2 The Utility of a General Probability Distribution

In this general setting, the problem of providing an inference statement about a class of exclusive and exhaustive "hypotheses" $\{\omega,\ \omega \in \Omega\}$, conditional on data D, is a decision problem, which we can conveniently denote by $\{\mathcal{D}, \Omega, u, P\}$, where Ω is the set of possible values of the random quantity ω, $\mathcal{D}$ relates to the class of probability densities for $\omega \in \Omega$ compatible with D,

$$\mathcal{Q} \equiv \left\{ q_\omega(\cdot \mid D); \quad q(\omega \mid D) \geq 0,\ \omega \in \Omega \text{ and } \int_\Omega q(\omega \mid D)\, d\omega = 1 \right\},$$

where $q_\omega(\cdot \mid D)$ is the density which an individual *reports* as the basis for describing beliefs about ω conditional on D. The set of consequences $\mathcal{C}$ consists of all pairs (d_q, ω), representing the conjunction of *reported* beliefs and true "states of nature". Throughout this section, we shall denote an individual's *actual* belief density by $p_\omega(\cdot \mid D)$. The decision space $\mathcal{D}$ consists of d_q's corresponding to choosing to report $q_\omega(\cdot \mid D)$'s and defined by $d_q(\omega) = (q_\omega(\cdot \mid D), \omega)$. We shall assume the individual to be coherent, so that $d_p \in \mathcal{D}$. Without loss of generality, we shall assume that $p_\omega(\cdot \mid D)$ and the $q_\omega(\cdot \mid D) \in \mathcal{Q}$ are strictly positive probability densities, so that, for all $\omega \in \Omega$, $p(\omega \mid D) > 0$ and $q(\omega \mid D) > 0$ for all $d_q \in \mathcal{D}$.

We complete the specification of this decision problem, by inducing the preference ordering through direct specification of a utility function u, which describes the "value" $u\big(q_\omega(\cdot \mid D), \omega\big)$ of reporting the probability density $q_\omega(\cdot \mid D)$ were ω to turn out to be the true "state of nature". For this purpose and with the same motivation, we generalise the notion of score function introduced in Definition 2.20.

Definition 3.15. (***Score function***). *A score function for probability densities $q_\omega(\cdot \mid D)$ defined on Ω, is a mapping $u : \mathcal{Q} \times \Omega \to \Re$. A score function is said to be **smooth** if it is continuously differentiable as a function of $q(\omega \mid D)$ for each $\omega \in \Omega$.*

The solution to the decision problem is then to report the density $q_\omega(\cdot \mid D)$ which maximises the expected utility

$$\overline{u}(d_q) = \int_\Omega u\big(q_\omega(\cdot \mid D), \omega\big)\, p(\omega \mid D)\, d\omega.$$

As in our earlier development in Chapter 2, we shall wish to restrict utility functions for the reporting problem in such a way as to encourage a coherent individual to be honest, given data D, in the sense that his or her expected utility is maximised if and only if d_q is chosen such that, for each $\omega \in \Omega$, $q(\omega \mid D) = p(\omega \mid D)$. The appropriate generalisation of Definition 2.21 is the following.

Definition 3.16. (***Proper score function***). *A score function u is proper if, for each strictly positive probability density $p_\omega(\cdot \mid D)$,*

$$\sup_{\mathcal{Q}} \int u(q_\omega(\cdot \mid D), \omega)\; p(\omega \mid D)\; d\omega = \int u(p_\omega(\cdot \mid D), \omega)\; p(\omega \mid D)\; d\omega,$$

where the supremum, taken over the class $\mathcal{Q}$ of all distribution for ω compatible with D, is attained if and only if $q_\omega(\cdot \mid D) = q_\omega(\cdot \mid D)$, up to sets of zero measure.

As in the finite case (see Definition 2.22), the simplest proper score function in the general case is the quadratic.

Definition 3.17. (***Quadratic score function***). *A quadratic score function for probability densities $q_\omega(\cdot \mid D) \in \mathcal{Q}$ defined on Ω is a mapping $u : \mathcal{Q} \times \Omega \to \Re$ of the form*

$$u(q_\omega(\cdot \mid D), \omega) = A\left\{2\, q(\omega \mid D) - \int q^2(\omega \mid D)\; d\omega\right\} + B(\omega), \;\; A > 0,$$

such that the otherwise arbitrary function, $B(\cdot)$, ensures the existence of $\overline{u}(d_q)$ for all $d_q \in \mathcal{D}$.

Proposition 3.12. *A quadratic score function is proper.*

Proof. Given data D, we must choose $q_\omega(\cdot \mid D) \in \mathcal{Q}$ to maximise

$$\begin{aligned}\overline{u}(d_q) &= \int_\Omega u(q_\omega(\cdot \mid D), \omega)\; p(\omega \mid D)\; d\omega \\ &= \int_\Omega \left[A\left\{2q(\omega \mid D) - \int q^2(\omega \mid D) d\omega\right\} + B(\omega)\right]\; p(\omega \mid D)\; d\omega\;,\end{aligned}$$

subject to $\int q(\omega \mid D) d\omega = 1$. Rearranging, it is easily seen that this is equivalent to maximising

$$-\int_\Omega (p(\omega \mid D) - q(\omega \mid D))^2\; d\omega,$$

from which it follows that we require $q(\omega \mid D) = p(\omega \mid D)$ for almost all $\omega \in \Omega$. We note again (cf. Proposition 2.28) that the constraint $\int q(\omega \mid D) d\omega = 1$ has not been needed in establishing this result for the quadratic scoring rule. ◁

In fact, as we argued in Section 2.7, for the problem of reporting *pure inference* statements it is natural to restrict further the class of appropriate utility functions. The following generalises Definition 2.23.

Definition 3.18. (***Local score function***). *A score function is local if for each $q_\omega(\cdot \mid D) \in \mathcal{Q}$ there exist functions u_ω, $\omega \in \Omega$, defined on $\Re^+$ such that*

$$u\big(q_\omega(\cdot \mid D), \omega\big) = u_\omega\big(q(\omega \mid D)\big).$$

Note that, as in Definition 2.23, the functional form, $u_\omega(\cdot)$, of the dependence of the score function on the density value $q(\omega \mid D)$ which d_q assigns to ω is allowed to vary with the particular ω in question. Intuitively, this enables us to incorporate the possibility that "bad predictions", i.e., values of $q(\omega \mid D)$, for some "true states of nature", ω, may be judged more harshly than others.

The next result generalises Proposition 2.29 and characterises the form of a smooth, proper, local score function.

Proposition 3.13. (***Characterisation of proper local score functions***). *If $u : \mathcal{Q} \times \Omega \to \Re$ is a smooth, proper, local score function, then it must be of the form*

$$u\big(q_\omega(\cdot \mid D), \omega\big) = A \log q(\omega \mid D) + B(\omega)$$

where $A > 0$ is an arbitrary constant and $B(\cdot)$ is an arbitrary function of ω, subject to the existence of $\bar{u}(d_q)$ for all $d_q \in \mathcal{D}$.

Proof. Given data D, we need to maximise, with respect to $q(\cdot \mid D)$, the expected utility

$$\overline{u}(d_q) = \int u\big(q_\omega(\cdot \mid D), \omega\big)\ p(\omega \mid D)\ d\omega$$

subject to $\int_\Omega q(\omega \mid D) d\omega = 1$. Since u is local, this reduces to finding an extremal of

$$F\big(q_\omega(\cdot \mid D)\big) = \int u_\omega\big(q(\omega \mid D)\big)\ p(\omega \mid D)\ d\omega - A\left[\int q(\omega \mid D)\ d\omega - 1\right].$$

However, for $q_\omega(\cdot \mid D)$ to give a stationary value of $F\big(q_\omega(\cdot \mid D)\big)$ it is necessary that

$$\frac{\partial}{\partial \alpha} F\big(q(\omega \mid D) + \alpha \tau(\omega)\big)\Big|_{\alpha=0} = 0,$$

for any function $\tau : \Omega \to \Re$ of sufficiently small norm (see, for example, Jeffreys and Jeffreys, 1946, Chapter 10). This condition reduces to the differential equation

$$D_1 u_\omega\big(q(\omega \mid D)\big)\ p(\omega \mid D) - A = 0,$$

where $D_1 u_\omega$ denotes the first derivative of u_ω. But, since u_ω is proper, the maximum of $F\big(q_\omega(\cdot \mid D)\big)$ must be attained at $q_\omega(\cdot \mid D) = p_\omega(\cdot \mid D)$, so that a smooth, proper, local utility function must satisfy the differential equation

$$D_1 u_\omega\big(p(\omega \mid D)\big)\, p(\omega \mid D) - A = 0,$$

whose solution is given by

$$u_\omega\big(p(\omega \mid D)\big) = A \log p(\omega \mid D) + B(\omega),$$

as stated. $\triangleleft$

This result prompts us to make the following formal definition.

Definition 3.19. (***Logarithmic score function***). *A logarithmic score function for probability densities $q_\omega(\cdot \mid D) \in \mathcal{Q}$ defined on Ω is a mapping $u : \mathcal{Q} \times \Omega \to \Re$ of the form*

$$u\big(q_\omega(\cdot \mid D), \omega\big) = A \log q(\omega \mid D) + B(\omega),$$

$A > 0$, $B(\cdot)$ *arbitrary, subject to the existence of* $\overline{u}(d_q)$ *for all* $d_q \in \mathcal{D}$.

For additional discussion of generalised score functions see Good(1969) and Buehler (1971).

3.4.3 Generalised Approximation and Discrepancy

As we remarked in Section 2.7.3, although the optimal solution to an inference problem under the above conditions is to state one's actual beliefs, there may be technical reasons why the computation of this "optimal" density $p_\omega(\cdot \mid D)$ is difficult. In such cases, we may need to seek a tractable approximation, $q_\omega(\cdot \mid D)$, say, which is in some sense "close" to $p_\omega(\cdot \mid D)$, but much easier to specify. As in the previous discussion of this idea, we shall need to examine carefully this notion of "closeness". The next result generalises Proposition 2.30.

Proposition 3.14. (***Expected loss in probability reporting***). *If preferences are described by a logarithmic score function, the expected loss of utility in reporting a probability density $q(\omega \mid D)$, rather than the density $p(\omega \mid D)$ representing actual beliefs, is given by*

$$\delta\big(q_\omega(\cdot \mid D) \mid p_\omega(\cdot \mid D)\big) = A \int p(\omega \mid D) \log \frac{p(\omega \mid D)}{q(\omega \mid D)}\, d\omega.$$

Moreover, $\delta\big(q_\omega(\cdot \mid D) \mid p_\omega(\cdot \mid D)\big)$ is non-negative and is zero if, and only if, $q_\omega(\cdot \mid D) = p_\omega(\cdot \mid D)$.

Proof. From Definition 3.19 the expected utility of reporting $q_\omega(\cdot \mid D)$ when $p_\omega(\cdot \mid D)$ is the actual belief distribution is given by

$$\overline{u}(d_q) = \int \left[A \log q(\boldsymbol{\omega} \mid D) + B(\boldsymbol{\omega})\right] p(\boldsymbol{\omega} \mid D)\, d\boldsymbol{\omega},$$

so that

$$\delta\big(q_\omega(\cdot \mid D) \mid p_\omega(\cdot \mid D)\big) = \overline{u}(d_p) - \overline{u}(d_q) = A \int p(\boldsymbol{\omega} \mid D) \log \frac{p(\boldsymbol{\omega} \mid D)}{q(\boldsymbol{\omega} \mid D)}\, d\boldsymbol{\omega}.$$

The final condition follows either from the fact that u is proper, so that $\overline{u}(d_p) \geq \overline{u}(d_q)$ with equality if and only if $q_\omega(\cdot \mid D) = p_\omega(\cdot \mid D)$; or directly from the fact that for all $x > 0$, $\log x \leq x - 1$ with equality if and only if $x = 1$ (cf. Proposition 2.30). $\triangleleft$

As in the finitistic discussion of Section 2.7, the above result suggests a natural, general measure of "lack of fit", or *discrepancy*, between a distribution and an approximation, when preferences are described by a logarithmic score function.

Definition 3.20. ***(Discrepancy of an approximation).***
The discrepancy between a strictly positive probability density $p_\omega(\cdot)$ and an approximation $\hat{p}_\omega(\cdot)$, $\boldsymbol{\omega} \in \Omega$, is defined by

$$\delta\big(\hat{p}_\omega(\cdot) \mid p_\omega(\cdot)\big) = \int p(\boldsymbol{\omega}) \log \frac{p(\boldsymbol{\omega})}{\hat{p}(\boldsymbol{\omega})}\, d\boldsymbol{\omega}.$$

Example 3.4. ***(General normal approximations).*** Suppose that $p(\omega) > 0$, $\omega \in \Re$, is an arbitrary density on the real line, with finite first two moments given by

$$\int_{-\infty}^{\infty} \omega\, p(\omega)\, d\omega = m, \qquad \int_{-\infty}^{\infty} (\omega - m)^2 p(\omega)\, d\omega = t^{-1},$$

and that we wish to approximate $p(\cdot)$ by a $\hat{p}(\cdot)$ corresponding to a normal density, $N(\omega \mid \mu, \lambda)$, with labelling parameters μ, λ chosen to minimise the discrepancy measure given in Definition 3.20. It is easy to see that, subject to the given constraints, minimising $\delta(\hat{p} \mid p)$ with respect to μ and λ is equivalent to minimising

$$-\int_{\infty}^{\infty} p(\omega) \log \mathbf{N}(\omega \mid \mu, \lambda)\, d\omega,$$

and hence to minimising

$$-\tfrac{1}{2} \log \lambda + \frac{\lambda}{2} \int_{-\infty}^{\infty} p(\omega)(\omega - \mu)^2 d\omega.$$

Invoking the two moment constraints, and writing $(\omega - \mu)^2 = (\omega - m + m - \mu)^2$, this reduces to minimising, with respect to μ and λ, the expression

$$-\log\lambda + \frac{\lambda}{t} + \lambda(m-\mu)^2.$$

It follows that the optimal choice is $\mu = m$, $\lambda = t$. In other words, *for the reporting problem with a logarithmic score function*, the *best normal approximation* to a distribution on the real line (whose mean and variance exist) is the normal distribution having the *same mean and variance.*

Example 3.5. ***(Normal approximations to Student distributions).*** Suppose that we wish to approximate the density $\mathrm{St}(x \mid \mu, \lambda, \alpha)$, $\alpha > 2$, by a normal density. We have just shown that the best normal approximation to *any* distribution is that with the same first two moments (assuming the latter to exist, corresponding here to the restriction $\alpha > 2$). Thus, recalling from Section 3.2.2 that the mean and precision of $\mathrm{St}(x \mid \mu, \lambda, \alpha)$ are given by μ and $\lambda(\alpha-2)/\alpha$, respectively, it follows that the best normal approximation to $\mathrm{St}(x \mid \mu, \lambda, \alpha)$ is provided by $\mathrm{N}(x \mid \mu, \lambda(\alpha-2)/\alpha)$. From Definition 3.20, the corresponding discrepancy will be

$$\delta(\mathrm{N} \mid \mathrm{St}) = \int \mathrm{St}(x \mid \mu, \lambda, \alpha) \log \frac{\mathrm{St}(x \mid \mu, \lambda, \alpha)}{\mathrm{N}\big(x \mid \mu, \lambda(\alpha-2)/\alpha\big)}\, dx.$$

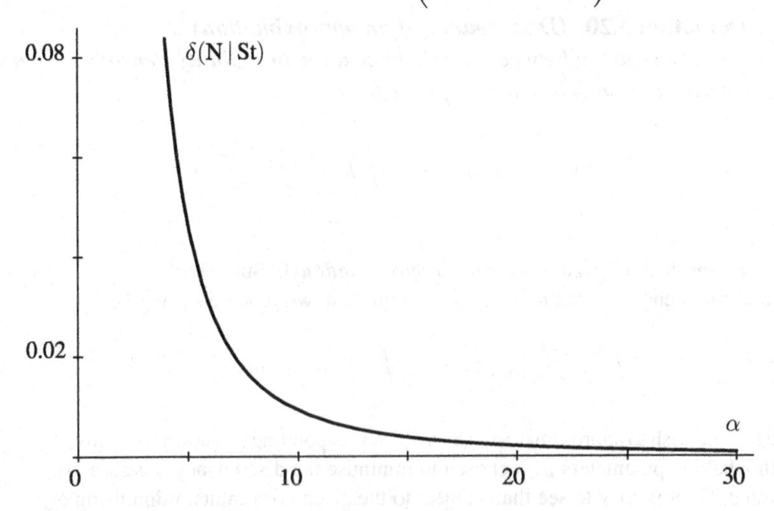

Figure 3.3 *Discrepancy between Student and normal densities*

This is easily evaluated (see, for example, Bernardo, 1978a) using the fact that the entropy of a Student distribution is given by

$$\begin{aligned} H\{\mathrm{St}(x \mid \mu, \lambda, \alpha)\} &= -\int \mathrm{St}(x \mid \mu, \lambda, \alpha) \log \mathrm{St}(x \mid \mu, \lambda, \alpha)\, dx \\ &= \log \frac{\Gamma((\alpha+1)/2)}{\Gamma(\alpha/2)\Gamma(1/2)} + \frac{1}{2}\log\frac{\lambda}{\alpha} + \left(\frac{\alpha+1}{2}\right)\left[\psi\left(\frac{\alpha}{2}\right) - \psi\left(\frac{\alpha+1}{2}\right)\right], \end{aligned}$$

where $\psi(z) = \Gamma'(z)/\Gamma(z)$ denotes the *digamma function* (see, for example, Abramowitz and Stegun, 1964), from which it follows that $\delta(\mathrm{N} \mid \mathrm{St})$ may be written as

$$\log \frac{\Gamma((\alpha+1)/2)}{\Gamma(\alpha/2)} + \left(\frac{\alpha+1}{2}\right)\left[\psi\left(\frac{\alpha}{2}\right) - \psi\left(\frac{\alpha+1}{2}\right)\right] + \frac{1}{2}\left[1 - \log\left(\frac{\alpha}{2} - 1\right)\right],$$

which only depends on the degrees of freedom, α, of the Student distribution. Figure 3.3 shows a plot of $\delta(\mathrm{N} \mid \mathrm{St})$ against α.

Using Stirling's approximation,

$$\log \Gamma(z) \sim \left(z - \frac{1}{2}\right) \log z - z + \frac{1}{2} \log(2\pi),$$

we obtain, for moderate to large values of α,

$$\delta(\mathrm{N} \mid \mathrm{St}) \approx [\alpha(\alpha - 2)]^{-1} = O(1/\alpha^2),$$

so that $[\alpha(\alpha-2)]^{-1}$ provides a simple, approximate measure of the departure from normality of a Student distribution.

3.4.4 Generalised Information

In Section 2.7.4, we examined, in the finitistic context, the increase in expected utility provided by given data D. We now extend this analysis to the general setting, writing x to denote observed data D.

Proposition 3.15. ***(Expected utility of data).*** *If preferences are described by a logarithmic score function for the class of probability densities $p(\omega \mid x)$ defined on Ω, then the expected increase in utility provided by data x, when the prior probability density is $p(\omega)$, is given by*

$$A \int p(\omega \mid x) \log \frac{p(\omega \mid x)}{p(\omega)} \, d\omega,$$

where $p(\omega \mid x)$ is the density of the posterior distribution for ω, given x. This expected increase in utility is non-negative, and zero if, and only if, $p(\omega \mid x)$ is identical to $p(\omega)$.

Proof. Using Definition 3.19, the expected increase in utility provided by x is given by

$$\int \Big\{ \big[A \log p(\omega \mid x) + B(\omega)\big] - \big[A \log p(\omega) + B(\omega)\big] \Big\} \, p(\omega \mid x) \, d\omega$$

$$= A \int p(\omega \mid x) \log \frac{p(\omega \mid x)}{p(\omega)} \, d\omega,$$

which, by Proposition 3.14, is non-negative with equality if and only if $p(\omega \mid x)$ and $p(\omega)$ are identical. $\triangleleft$

The following natural definition of the amount of information provided by the data extends that given in Definition 2.26.

Definition 3.21. (***Information from data***). *The amount of information about $\omega \in \Omega$ provided by data x when the prior density is $p(\omega)$ is given by*

$$I\{x \mid p_\omega(\cdot)\} = \int p(\omega \mid x) \log \frac{p(\omega \mid x)}{p(\omega)}\, d\omega,$$

where $p(\omega \mid x)$ is the corresponding posterior density.

As in the finite case, it is interesting to note that the amount of information provided by x is equivalent to the discrepancy measure if the prior is considered as an approximation to the posterior. Alternatively, we see that $\log p(\omega)$, $\log p(\omega \mid x)$, respectively, measure how "good", on a logarithmic scale, the prior and posterior are at "predicting" the "true state of nature" ω, so that $\log p(\omega \mid x) - \log p(\omega)$ is a measure of the usefulness of x were ω known to be the true value. Thus $I\{x \mid p_\omega(\cdot)\}$ is simply the expected value of that utility difference with respect to the posterior density, given x.

The functional $\int p(\omega) \log p(\omega)\, d\omega$ has been used (see e.g., Lindley, 1956, and references therein) as a measure of the 'absolute' information about ω contained in the probability density $p(\omega)$. The increase in utility from observing x is then

$$\int p(\omega \mid x) \log p(\omega \mid x)\, d\omega - \int p(\omega) \log p(\omega)\, d\omega,$$

instead of our Definition 3.21. However, this expression is *not* invariant under one-to-one transformations of ω, a property which seems to us to be essential. Note, however, that both expressions have the same expectation with respect to the distribution of x. Draper and Guttman (1969) put forward yet another non-invariant definition of information.

Additional references on statistical information concepts are Renyi (1964, 1966, 1967), Goel and DeGroot (1979) and De Waal and Groenewald (1989).

More generally, we may wish to step back to the situation before data become available, and consider the idea of the amount of information to be expected from an experiment e. We therefore generalise Definition 2.27.

Definition 3.22. (***Expected information from an experiment***). *The expected information to be provided by an experiment e about $\omega \in \Omega$, when the prior density is $p(\omega)$ is given by*

$$I\{e \mid p_\omega(\cdot)\} = \int_X I\{x \mid p_\omega(\cdot)\}\, p(x \mid e)\, dx,$$

where the distribution of the possible data outcomes $x \in X$ resulting from the experiment e is described by $p(x \mid e)$.

The following result, which is a generalisation of Proposition 2.32, provides an alternative expression for $I\{e \mid p_\omega(\cdot)\}$.

Proposition 3.16. *An alternative expression for the expected information is*

$$I\{e \mid p_\omega(\cdot)\} = \int\int p(\omega, x \mid e) \log \frac{p(\omega, x \mid e)}{p(\omega)p(x \mid e)} \, d\omega \, dx,$$

where $p(\omega, x \mid e) = p(\omega \mid x, e)p(x \mid e)$ *and* $p(\omega \mid x, e)$ *is the posterior density for* ω *given data* x *and prior density* $p(\omega)$*. Moreover,* $I\{e \mid p_\omega(\cdot)\} \geq 0$*, with equality if and only if* x *and* ω *are independent random quantities, so that* $p(\omega, x \mid e) = p(\omega)p(x \mid e)$ *for all* ω *and* x*.*

Proof.

$$\begin{aligned} I\{e \mid p_\omega(\cdot)\} &= \int \left\{ \int p(\omega \mid x, e) \log \frac{p(\omega \mid x, e)}{p(\omega)} \, d\omega \right\} p(x \mid e) \, dx \\ &= \int\int p(\omega \mid x, e) \, p(x \mid e) \log \frac{p(\omega \mid x, e)p(x \mid e)}{p(\omega)p(x \mid e)} \, d\omega \, dx \end{aligned}$$

and the result now follows from the fact that $p(\omega \mid x, e) = p(\omega \mid x, e)p(x \mid e)$. Moreover, since, by Proposition 3.14, $I\{e \mid p_\omega(\cdot)\} \geq 0$ with equality if and only if $p(\omega \mid x, e) = p(\omega)$, it follows from Definition 3.19 that $I\{e \mid p_\omega(\cdot)\} \geq 0$ with equality if and only if, for all ω and x, $p(\omega, x \mid e) = p(\omega)p(x \mid e)$. $\triangleleft$

Maximisation of the expected Shannon information was proposed by Lindley (1956) as a "reasonable" *ad hoc* criterion for choosing among alternative experiments. Fedorov (1972) proved later that certain classical design criteria (in particular, D-optimality) are special cases of this when normal distributions are assumed. We have shown that maximising expected information is just a particular (albeit important) case of the general criterion, implied by quantitative coherence, of maximising the expected utility in the case of pure inference problems. See Polson (1992) for a closely related argument.

It follows from Proposition 2.31 and the last remark, that someone who adopts the classical D-optimality criterion of optimal design under standard normality assumptions should, for consistency, have preferences which are described by a logarithmic scoring rule; otherwise, such designs are not optimal with respect to his or her underlying preferences.

There is a considerable literature on the Bayesian design of experiments, which we will not attempt to review here. A detailed discussion will be given in the volume *Bayesian Methods*. We note that important references include Blackwell (1951,

1953), Lindley (1956), Chernoff (1959), Stone (1959), DeGroot (1962, 1970), Duncan and DeGroot (1976), Bandemer (1977), Smith and Verdinelli (1980), Pilz (1983/1991), Chaloner (1984), Sugden (1985), Mazloum and Meeden (1987), Felsenstein (1988, 1992), DasGupta and Studden (1991), El-Krunz and Studden (1991), Pardo *et al.* (1991), Mitchell and Morris (1992), Pham-Gia and Turkkan (1992), Verdinelli (1992), Verdinelli and Kadane (1992), Lindley and Deely (1993), Lad and Deely (1994) and Parmigiani and Berry (1994).

3.5 DISCUSSION AND FURTHER REFERENCES

3.5.1 The Role of Mathematics

The translation of any substantive theory into a precise mathematical formalism necessarily involves an element of idealisation.

We have already had occasion to remark on aspects of this problem in Chapter 2, in the context of using real numbers rather than subsets of the rationals to represent actual measurements (necessarily "finitised" by inherent accuracy limits of the uncertainty apparatus). Similar remarks are obviously called for in the context of using, for example, probability densities to represent belief distributions for real-valued observables.

In some situations, as we shall see in Chapter 4, the adoption of specific forms of density may follow from simple, structural assumptions about the form of the belief distribution. In other situations, however, if we really try to think of such a density as being practically identified by expressions of preference among, say, standard options, we would encounter the obvious operational problem that, implicitly, an infinite number of revealed preferences would be required.

Clearly, in such situations the precise mathematical form of a density is likely to have arisen as an approximation to a "rough shape" obtained from some finite elicitation or observation process, and has been chosen, arbitrarily, for reasons of mathematical convenience, from an available mathematical tool-kit. Similar remarks apply to the choice, for descriptive or mathematical convenience, of infinite sets to represent consequences or decisions, with the attendant problems of defining appropriate concepts of expected utility.

There are obvious dangers, therefore, in accepting too uncritically any orientation, or would-be insightful mathematical analysis, that flows from arbitrary, idealized mathematical inputs into the general quantitative coherence theory. However, given an awareness of the dangers involved, we can still systematically make use of the power and elegance of the (idealised) mathematics by simultaneously asserting, as a central tenet of our approach, a concern with the *robustness* and *sensitivity* of the output of an analysis to the form of input assumed (see Section 5.6.3). Of course, we shall later have to make precise the sense in which these terms are to be interpreted and the actual forms of procedures to be adopted. That being understood, our approach, as with the earlier formalism of Chapter 2, will be to work with the mathematical idealization, in order to exploit its potential power and insight, while constantly bearing in mind the need for a large pinch of salt and a repertoire of sensitivity diagnostics.

3.5.2 Critical Issues

We shall comment further on three aspects of the general mathematical structure we have developed and will be using throughout the remainder of this volume. These will be dealt with under the following subheadings: (i) *Finite versus Countable Additivity*; (ii) *Measure versus Linear Theory*; (iii) *Proper versus Improper Probabilities*; (iv) *Abstract versus Concrete Mathematics*.

Finite versus Countable Additivity

In Chapter 2, we developed, from a directly intuitive and operational perspective, a minimal mathematical framework for a theory of quantitative coherence. The role of the mathematics employed in this development was simply that of a tool to capture the essentials of the substantive concepts and theory; within the resulting finitistic framework we then established that uncertainties should be represented in terms of *finitely additive* probabilities.

The generalisations and extensions of the theory given in the present chapter lead, instead, to the mathematical framework of *countable additivity*, within which we have available the full panoply of analytic tools from mathematical probability theory. The latter is clearly highly desirable from the point of view of mathematical convenience, but it is important to pause and consider whether the development of a more convenient mathematical framework has been achieved at the expense of a distortion of the basic concepts and ideas.

First, let us emphasise that, from a philosophical perspective, the monotone continuity postulate introduced in Section 3.1.2 does not have the fundamental status of the axioms presented in Chapter 2. We regard the latter as encapsulating the essence of what is required for a theory of quantitative coherence. The former is an "optional extra" assumption that one might be comfortable with in specific contexts, but should in no way be obliged to accept as a prerequisite for quantitative coherence.

Secondly, we note that the effect of accepting that preferences should conform to the monotone continuity postulate is to restrict one's available (in the sense of coherent) belief specifications to a *subset* of the finitely additive uncertainty measures; namely, those that are *also* countably additive. This is, of course, potentially disturbing from a subjectivist perspective, since a key feature of the theory is that the only constraints on belief specifications should be that they are coherent. For some such representations to be ruled out a priori, as a consequence of a postulate adopted purely for mathematical convenience, would indeed be a distortion of the theory. This is why we regard such a postulate as different from the basic axioms. However, provided one is aware of, and not concerned about, the implicit restriction of the available belief representations, its adoption may be very natural in contexts where one is, in any case, prepared to work in an extended mathematical setting.

Throughout this work, we shall, in fact, make systematic use of concepts and tools from mathematical probability theory, without further concern or debate about this issue. However, to underline what we already said in Section 3.1.2, it is important to be on guard and to be aware that distortions might occur. To this end, we draw attention to some key references to which the reader may wish to refer in order to heighten such awareness and to study in detail the issues involved.

De Finetti (1970/1974, pp. 116–133, 173–177 and 228–241; 1970/1975, pp. 267–276 and 340–361), provides a wealth of detailed analysis, illustration and comment on the issues surrounding finite versus countable (and other) additivity assumptions, his own analysis being motivated throughout by the guiding principle that

> ... mathematics is an instrument which should conform itself strictly to the exigencies of the field in which it is to be applied. (1970/1974, p. 3)

Further technical and philosophical discussion is given in de Finetti (1972, Chapters 5 and 6); see, also, Stone (1986). Systematic use of finite additivity in decision-related contexts is exemplified in Dubins and Savage (1965/1976), Heath and Sudderth (1978, 1989), Stone (1979b), Hill (1980), Sudderth (1980), Seidenfeld and Schervish (1983), Hill and Lane (1984), Regazzini (1987) and Regazzini and Petris (1993). A discussion of the statistical implications of finitely additive probability is given by Kadane *et al.* (1986).

In Section 2.8.3 we discussed , within a finitistic framework, several "betting" approaches to establishing probability as the only coherent measure of degree of belief. These ideas may be extended to the general case. Dawid and Stone (1972, 1973) introduce the concept of "*expectation consistency*", and show the necessity of using Bayes' theorem to construct probability distributions corresponding to fair bets made with additional information. Other generalised discussions on coherence of inference in terms of gambling systems include Lane and Sudderth (1983) and Brunk (1991).

Measure versus Linear Theory

Mathematical probability theory can be developed, equivalently, starting either from the usual Kolmogorov axioms for a set function defined over a σ-field of events (see, for example, Ash, 1972), or from axioms for a linear operator defined over a linear space of random quantities (see, for example, Whittle, 1976). The former deals directly with *probability measure*; the latter with an *expectation operator* (or a *prevision*, in de Finetti's terminology).

In our development of a quantitative coherence theory, the axiomatic approach to preferences among options has led us more naturally towards probability measures as the primary probabilistic element, with expectation (prevision) defined subsequently. In the approach to coherence put forward in de Finetti (1972, 1970/1974, 1970/1975), prevision is the primary element, with probability subsequently emerging as a special case for 0-1 random quantities. The case for adopting the linear rather than the measure theory approach is argued at length by de Finetti, there being many points of contact with the argument regarding finite versus countable additivity, particularly the need to avoid, in the mathematical formulation, going beyond those aspects required for the problem in hand. In the specific context of statistical modelling and inference, Goldstein (1981, 1986a, 1986b, 1987a, 1987b, 1988, 1991, 1994) has systematically developed the linear approach advocated by de Finetti, showing that a version of a subjectivist programme for revising beliefs in the light of data can be implemented without recourse to the full probabilistic machinery developed in this chapter. Lad *et al.* (1990) provide further discussion on the concept of prevision.

We view these and related developments with great interest and with no dogmatic opinion concerning the ultimate relative usefulness and acceptance of "linear" versus "probabilistic" Bayesian statistical concepts and methods. That said, the present volume is motivated by our conviction that, currently, there remains a need for a detailed exposition of the Bayesian approach within the, more or less, conventional framework of full probabilistic descriptions.

Proper versus Improper Probabilities

Whether viewed in terms of finite or countable additivity, we have taken probability to be a measure with values in the interval $[0, 1]$. However, it is possible to adopt axiomatic approaches which allow for infinite (or improper) probabilities: see, for example, Renyi (1955, 1962/1970, Chapter 2, and references therein), who uses conditional arguments to derive proper probabilities form improper distributions, and Hartigan (1983, Chapter 3), who directly provides an axiomatic foundation for improper or, as he terms them, *non-unitary*, probabilities. We shall not review such axiomatic theories in detail, but note that we shall encounter improper distributions systematically in Section 5.4.

Abstract versus Concrete Mathematics

When probabilistic mathematics is being used as a tool for the representation and analysis of substantive non-mathematical problems, rather than as a direct mathematical concern in its own right, there is always a dilemma regarding the appropriate level of mathematics to be used. Specifically, there are basic decisions to be made about how much measure-theoretical machinery should be invoked. The introduction of too much abstract mathematics can easily make the substantive content seem totally opaque to the very reader at whom it is most aimed. On the other hand, too little machinery may prove inadequate to provide a complete mathematical treatment, requiring the omission of certain topics, or the provision of just a partial, non-rigorous treatment, with insight and illustration attempted only by concrete examples.

Thus far, we have tried to provide a complete, rigorous treatment of the Foundations and Generalisations of the theory of quantitative coherence, within the mathematical framework of Chapters 2 and 3. This chapter essentially defines the upper limit of mathematical machinery we shall be using and, in fact, most of our subsequent development will be much more straightforward. However, it will be the case, for example in Chapter 4, that some results of interest to us require rather more sophisticated mathematical tools than we have made available. Our response to this problem will be to try to make it clear to the reader when this is the case, and to provide references to a complete treatment of such results, together with (hopefully) sufficient concrete discussion and illustration to illuminate the topic.

For more sophisticated mathematical treatments of Bayesian theory, the reader is referred to Hartigan (1983) and Florens *et al.* (1990).

Chapter 4

Modelling

Summary

The relationship between beliefs about observable random quantities and their representation using conventional forms of statistical models is investigated. It is shown that judgements of exchangeability lead to representations that justify and clarify the use and interpretation of such familiar concepts as parameters, random samples, likelihoods and prior distributions. Beliefs which have certain additional invariance properties are shown to lead to representations involving familiar specific forms of parametric distributions, such as normals and exponentials. The concept of a sufficient statistic is introduced and related to representations involving the exponential family of distributions. Various forms of partial exchangeability judgements about data structures involving several samples, structured layouts, covariates and designed experiments are investigated, and links established with a number of other commonly used statistical models.

4.1 STATISTICAL MODELS

4.1.1 Beliefs and Models

The subjectivist, operationalist viewpoint has led us to the conclusion that, if we aspire to quantitative coherence, individual degrees of belief, expressed as probabilities, are inescapably the starting point for descriptions of uncertainty. There can be no theories without theoreticians; no learning without learners; in general, no science without scientists. It follows that learning processes, whatever their particular concerns and fashions at any given point in time, are necessarily reasoning processes which take place in the minds of individuals. To be sure, the object of attention and interest may well be an assumed external, objective reality: but the actuality of the learning process consists in the evolution of individual, subjective beliefs about that reality. However, it is important to emphasise, as in our earlier discussion in Section 2.8, that the primitive and fundamental notions of *individual* preference and belief will typically provide the starting point for *interpersonal* communication and reporting processes. In what follows, both here, and more particularly in Chapter 5, we shall therefore often be concerned to identify and examine features of the individual learning process which relate to interpersonal issues, such as the conditions under which an approximate consensus of beliefs might occur in a population of individuals.

In Chapters 2 and 3, we established a very general foundational framework for the study of degrees of belief and their evolution in the light of new information. We now turn to the detailed development of these ideas for the broad class of problems of primary interest to statisticians; namely, those where the events of interest are defined explicitly in terms of *random quantities*, $x_1, \ldots, x_n$ (discrete or continuous, and possibly vector-valued) representing observed or experimental data.

In such cases, we shall assume that an individual's degrees of belief for events of interest are derived from the specification of a joint distribution function $P(x_1, \ldots, x_n)$, which we shall typically assume, without systematic reference to measure-theoretic niceties, to be representable in terms of a joint density function $p(x_1, \ldots, x_n)$ (to be understood as a mass function in the discrete case).

Of course, any such specification implicitly defines a number of other degrees of belief specifications of possible interest: for example, for $1 \leq m < n$,

$$p(x_1, \ldots, x_m) = \int p(x_1, \ldots, x_n) dx_{m+1} \ldots dx_n$$

provides the *marginal* joint density for $x_1, \ldots, x_m$, and

$$p(x_{m+1}, \ldots, x_n \mid x_1, \ldots, x_m) = p(x_1, \ldots, x_n)/p(x_1, \ldots, x_m)$$

gives the joint density for the as yet unobserved $x_{m+1}, \ldots, x_n$, conditional on having observed $x_1 = x_1, \ldots, x_m = x_m$. Within the Bayesian framework, this latter conditional form is the key to "learning from experience".

We recall that, throughout, we shall use notation such as P and p in a *generic* sense, rather than as specifying particular functions. In particular, P may sometimes refer to an underlying probability measure, and sometimes refer to implied distribution functions, such as $P(x_1), P(x_1, \ldots, x_n)$ or $P(x_{m+1}, \ldots, x_n \mid x_1, \ldots, x_m)$. Similarly, we may write $p(x_1), p(x_1, \ldots, x_n)$, etc, so that, for example,

$$p(x_{m+1}, \ldots, x_n \mid x_1, \ldots, x_m) = p(x_1, \ldots, x_n)/p(x_1, \ldots, x_m)$$

simply indicates that the conditional density for $x_{m+1}, \ldots, x_n$ given $x_1, \ldots, x_m$ is given by the ratio of the specified joint densities. Such usage avoids notational proliferation, and the context will always ensure that there is no confusion of meaning.

Thus far, however, our discussion is rather "abstract". In actual applications we shall need to choose specific, concrete forms for joint distributions. This is clearly a somewhat daunting task, since direct contemplation and synthesis of the many complex marginal and conditional judgements implicit in such a specification are almost certainly beyond our capacity in all but very simple situations. We shall therefore need to examine rather closely this process of choosing a specific form of probability measure to represent degrees of belief.

Definition 4.1. (***Predictive probability model***). *A predictive model for a sequence of random quantities $x_1, x_2, \ldots$ is a probability measure P, which mathematically specifies the form of the joint belief distribution for any subset of $x_1, x_2, \ldots$.*

In some cases, we shall find that we are able to identify general types of belief structure which "pin down", in some sense, the mathematical representation strategy to be adopted. In other cases, this "formal" approach will not take us very far towards solving the representation problem and we shall have to fall back on rather more pragmatic modelling strategies.

At this stage, a word of warning is required. In much statistical writing, the starting point for formal analysis is the *assumption* of a mathematical model form, typically involving "unknown parameters", the main object of the study being to infer something about the values of these parameters. From our perspective, this is all somewhat premature and mysterious! We are seeking to represent degrees of belief about observables: nothing in our previous development justifies or gives any insight into the choice of particular "models", and thus far we have no way of attaching any operational meaning to the "parameters" which appear in conventional models. However, as we shall soon see, the subjectivist, operationalist approach *will* provide considerable insight into the nature and status of these conventional assumptions.

4.2 EXCHANGEABILITY AND RELATED CONCEPTS

4.2.1 Dependence and Independence

Consider a sequence of random quantities $x_1, x_2, \ldots$, and suppose that a predictive model is assumed which specifies that, for all n, the joint density can be written in the form

$$p(x_1, \ldots, x_n) = \prod_{i=1}^{n} p(x_i),$$

so that the x_i are *independent* random quantities. It then follows straightforwardly that, for any $1 \leq m < n$,

$$p(x_{m+1}, \ldots, x_n \mid x_1, \ldots, x_m) = p(x_{m+1}, \ldots, x_n),$$

so that *no learning from experience* can take place within this sequence of observations. In other works, past data provide us with no additional information about the possible outcomes of future observations in the sequence.

A predictive model specifying such an independence structure is clearly inappropriate in contexts where we believe that the successive accumulation of data will provide increasing information about future events. In such cases, the structure of the joint density $p(x_1, \ldots, x_n)$ must encapsulate some form of *dependence* among the individual random quantities. In general, however, there are a vast number of possible subjective assumptions about the form such dependencies might take and there can be no all-embracing theoretical discussion. Instead, what we can do is to concentrate on some particular simple forms of judgement about dependence structures which might correspond to actual judgements of individuals in certain situations.

There is no suggestion that the structures we are going to discuss in subsequent subsections have any special status, or *ought* to be adopted in most cases, or whatever. They simply represent forms of judgement which may often be felt to be appropriate and whose detailed analysis provides illuminating insight into the specification and interpretation of certain classes of predictive models.

4.2.2 Exchangeability and Partial Exchangeability

Suppose that, in thinking about $P(x_1, \ldots, x_n)$, his or her joint degree of belief distribution for a sequence of random quantities $x_1, \ldots, x_n$, an individual makes the judgement that the subscripts, the "labels" identifying the individual random quantities, are "uninformative", in the sense that he or she would specify all the marginal distributions for the individual random quantities identically, and similarly for all the marginal joint distributions for all possible pairs, triples, etc., of the random quantities. It is easy to see that this implies that the form of the joint distribution must be such that

$$P(x_1, \ldots, x_n) = P(x_{\pi(1)}, \ldots, x_{\pi(n)}),$$

for any possible permutation π of the subscripts $\{1, \ldots, n\}$. We formalise this notion of "symmetry" of beliefs for the individual random quantities as follows.

Definition 4.2. ***(Finite exchangeability).*** *The random quantities $x_1, \ldots, x_n$ are said to be judged (finitely) exchangeable under a probability measure P if the implied joint degree of belief distribution satisfies*

$$P(x_1, \ldots, x_n) = P(x_{\pi(1)}, \ldots, x_{\pi(n)})$$

for all permutations π defined on the set $\{1, \ldots, n\}$. In terms of the corresponding density or mass function, the condition reduces to

$$p(x_1, \ldots, x_n) = p(x_{\pi(1)}, \ldots, x_{\pi(n)}).$$

Example 4.1. ***(Tossing a thumb tack).*** Consider a sequence of tosses of a standard metal drawing pin (or thumb tack), and let $x_i = 1$ if the pin lands point uppermost on the ith toss, $x_i = 0$ otherwise, $i = 1, \ldots, n$. If the tosses are performed in such a way that time order appears to be irrelevant and the conditions of the toss appear to be essentially held constant throughout, it would seem to be the case that, whatever precise *quantitative* form their beliefs take, most observers would judge the outcomes of the sequence of tosses $x_1, x_2, \ldots$ to be exchangeable in the above sense.

In general, the exchangeability assumption captures, for a subjectivist interested in belief distributions for observables, the essence of the idea of a so-called "random sample". This latter notion is, of course, of no direct use to us at this stage, since it (implicitly) involves the idea of "conditional independence, given the value of the underlying parameter", a meaningless phrase thus far within our framework.

The notion of exchangeability involves a judgement of *complete symmetry* among all the observables $x_1, \ldots, x_n$ under consideration. Clearly, in many situations this might be too restrictive an assumption, even though a partial judgement of symmetry is present.

Example 4.1. ***(cont.).*** Suppose that the sequence of tosses of a drawing pin are not all made with the same pin, but that the even and odd numbered tosses are made with different pins: an all metal one for the odd tosses; a plastic-coated one for the even tosses. Alternatively, suppose that the same pin were used throughout, but that the odd tosses are made by a different person, using a completely different tossing mechanism from that used for the even tosses. In such cases, many individuals would retain an exchangeable form of belief distribution within the sequences of odd and even tosses separately, but might be reluctant to make a judgement of symmetry for the combined sequence of tosses.

Example 4.2. ***(Laboratory measurements).*** Suppose that $x_1, x_2, \ldots$ are real-valued measurements of a physical or chemical property of a given substance, all made on the same sample with the same measurement procedure. Under such conditions, many individuals might judge the complete sequence of measurements to be exchangeable.

Suppose, however, that sequences of such measurements are combined from k different laboratories, the substance being identical but the measurement procedures varying from laboratory to laboratory. In this case, judgements of exchangeability for each laboratory sequence separately might be appropriate, whereas such a judgement for the combined sequence might not be.

Example 4.3. ***(Physiological responses).*** Suppose that $\{x_1, x_2, \ldots, \}$ are real-valued measurements of a specific physiological response in human subjects when a particular drug is administered. If the drug is administered at more than one dose level and if there are both male and female subjects, spanning a wide age range, most individuals would be very reluctant to make a judgement of exchangeability for the entire sequence of results. However, within each combination of dose-level, sex and appropriately defined age-group, a judgement of exchangeability might be regarded as reasonable.

Judgements of the kind suggested in the above examples correspond to forms of *partial exchangeability*. Clearly, there are many possible forms of departure from overall judgements of exchangeability to those of partial exchangeability and so a formal definition of the term does not seem appropriate. In general, it simply signifies that there may be additional "labels" on the random quantities (for example, odd and even, or the identification of the tossing mechanism in Example 4.1) with exchangeable judgements made separately for each group of random quantities having the same additional labels. A detailed discussion of various possible forms of partial exchangeability will be given in Section 4.6.

We shall now return to the simple case of exchangeability and examine in detail the form of representation of $p(x_1, \ldots, x_n)$ which emerges in various special cases. As a preliminary, we shall generalise our previous definition of exchangeability to allow for "potentially infinite" sequences of random quantities. In practice, it should, at least in principle, always be possible to give an upper bound to the number of observables to be considered. However, specifying an actual upper bound may be somewhat difficult or arbitrary and so, for mathematical and descriptive purposes, it is convenient to be able to proceed as if we were contemplating an infinite sequence of potential observables. Of course, it will be important to establish that working within the infinite framework does not cause any fundamental conceptual distortion. These and related issues of finite versus infinite exchangeability will be considered in more detail in Section 4.7.1. For the time being, we shall concentrate on the "potentially infinite" case.

Definition 4.3. ***(Infinite exchangeability).*** *The infinite sequence of random quantities $x_1, x_2, \ldots$ is said to be judged (infinitely) exchangeable if every finite subsequence is judged exchangeable in the sense of Definition 4.2.*

One might be tempted to wonder whether *every* finite sequence of exchangeable random quantities could be embedded in or extended to an infinitely exchangeable sequence of similarly defined random quantities. However, this is certainly not the case as the following example shows.

Example 4.4. ***(Non-extendible exchangeability).*** Suppose that we define the three random quantities x_1, x_2, x_3 such that either $x_i = 1$ or $x_i = 0$, $i = 1, 2, 3$, with joint probability function given by

$$\begin{aligned} p(x_1 = 0, x_2 = 1, x_3 = 1) &= p(x_1 = 1, x_2 = 0, x_3 = 1) \\ &= p(x_1 = 1, x_2 = 1, x_3 = 0) \\ &= 1/3, \end{aligned}$$

with all other combinations of x_1, x_2, x_3 having probability zero, so that x_1, x_2, x_3 are clearly exchangeable. We shall now try to identify an x_4, taking only values 0 and 1, such that $x_1, \ldots, x_4$ are exchangeable. For this to be possible, we require, for example,

$$p(x_1 = 0, x_2 = 1, x_3 = 1, x_4 = 0) = p(x_1 = 0, x_2 = 0, x_3 = 1, x_4 = 1).$$

But

$$\begin{aligned} &p(x_1 = 0, x_2 = 1, x_3 = 1, x_4 = 0) \\ &\quad = p(x_1 = 0, x_2 = 1, x_3 = 1) - p(x_1 = 0, x_2 = 1, x_3 = 1, x_4 = 1) \\ &\quad = 1/3 - p(x_1 = 0, x_2 = 1, x_3 = 1, x_4 = 1) \\ &\quad = 1/3 - p(x_1 = 1, x_2 = 1, x_3 = 1, x_4 = 0), \end{aligned}$$

where

$$p(x_1 = 1, x_2 = 1, x_3 = 1, x_4 = 0) \leq p(x_1 = 1, x_2 = 1, x_3 = 1) = 0,$$

so that

$$p(x_1 = 0, x_2 = 1, x_3 = 1, x_4 = 0) = 1/3.$$

However, we also have

$$p(x_1 = 0, x_2 = 0, x_3 = 1, x_4 = 1) \leq p(x_1 = 0, x_2 = 0, x_3 = 1) = 0$$

and so

$$p(x_1 = 0, x_2 = 1, x_3 = 1, x_4 = 0) \neq p(x_1 = 0, x_2 = 0, x_3 = 1, x_4 = 1).$$

It follows that a finitely exchangeable sequence cannot even necessarily be embedded in a larger finitely exchangeable sequence, let alone an infinitely exchangeable sequence.

4.3 MODELS VIA EXCHANGEABILITY

4.3.1 The Bernoulli and Binomial Models

We consider first the case of an infinitely exchangeable sequence of 0–1 random quantities, $x_1, x_2, \ldots$, with $x_i = 0$ or $x_i = 1$, for all $i = 1, 2, \ldots$. Without loss of generality, we shall derive a representation result for the joint mass function, $p(x_1, \ldots, x_n)$, of the first n random quantities $x_1, \ldots, x_n$.

Proposition 4.1. (***Representation theorem for 0–1 random quantities***). *If $x_1, x_2, \ldots$ is an infinitely exchangeable sequence of 0–1 random quantities with probability measure P, there exists a distribution function Q such that the joint mass function $p(x_1, \ldots, x_n)$ for $x_1 \ldots, x_n$ has the form*

$$p(x_1, \ldots, x_n) = \int_0^1 \prod_{i=1}^n \theta^{x_i}(1-\theta)^{1-x_i}\, dQ(\theta),$$

where,

$$Q(\theta) = \lim_{n\to\infty} P[y_n/n \leq \theta],$$

with $y_n = x_1 + \cdots + x_n$, and $\theta = \lim_{n\to\infty} y_n/n$.

Proof. (De Finetti, 1930, 1937/1964; here we follow closely the proof given by Heath and Sudderth, 1976; see also Barlow, 1991). Suppose $x_1 + \cdots + x_n = y_n$, then, by exchangeability, for any $0 \leq y_n \leq n$,

$$p(x_1 + \cdots + x_n = y_n) = \binom{n}{y_n} p(x_{\pi(1)}, \ldots, x_{\pi(n)})$$

for any permutation π of $\{1, \ldots, n\}$ such that $x_{\pi(1)} + \cdots + x_{\pi(n)} = y_n$. Moreover, for arbitrary $N \geq n \geq y_n \geq 0$, and with the summations below taken over the range $y_N = y_n$ to $y_N = N - (n - y_n)$, we see that

$$\begin{aligned}
&p(x_1 + \cdots + x_n = y_n) \\
&\quad = \sum p(x_1 + \cdots + x_n = y_n \,|\, x_1 + \cdots + x_N = y_N)\, p(x_1 + \cdots + x_N = y_N), \\
&\quad = \sum \binom{N}{n}^{-1} \binom{y_N}{y_n} \binom{N - y_N}{n - y_n} p(x_1 + \cdots + x_N = y_N), \quad 0 \leq y_n \leq n \leq N, \\
&\quad = \binom{n}{y_n} \sum \frac{(y_N)_{y_n}(N - y_N)_{n-y_n}}{(N)_n} p(x_1 + \cdots + x_N = y_N),
\end{aligned}$$

where $(y_N)_{y_n} = y_N(y_N - 1)\cdots[y_N - (y_n - 1)]$, etc. (Intuitively, we can imagine sampling n items without replacement from an urn of N items containing y_N 1's and $N - y_N$ 0's, corresponding to the hypergeometric distribution of Section 3.2.2.)

If we now define $Q_N(\theta)$ on $\Re$ to be the step function which is 0 for $\theta < 0$ and has jumps of $p(x_1 + \cdots + x_N = y_N)$ at $\theta = y_N/N, y_N = 0, \ldots, N$, we see that

$$p(x_1 + \cdots + x_n = y_n) = \binom{n}{y_n} \int_0^1 \frac{(\theta N)_{y_n}[(1-\theta)N]_{n-y_n}}{(N)_n} \, dQ_N(\theta).$$

As $N \to \infty$,

$$\frac{(\theta N)_{y_n}[(1-\theta)N]_{n-y_n}}{(N)_n} \to \theta^{y_n}(1-\theta)^{n-y_n}$$

uniformly in θ. Moreover, by Helly's theorem (see, for example, Section 3.2.3 and Ash, 1972, Section 8.2), there exists a subsequence $Q_{N_1}, Q_{N_2}, \ldots$ such that

$$\lim_{N_j \to \infty} Q_{N_j} = Q,$$

where Q is a distribution function. The result follows. $\triangleleft$

The interpretation of this representation theorem is of profound significance from the point of view of subjectivist modelling philosophy. It is *as if*:

(i) the x_i are judged to be independent, Bernoulli random quantities (see Section 3.2.2) conditional on a random quantity θ;
(ii) θ is itself assigned a probability distribution Q;
(iii) by the strong law of large numbers, $\theta = \lim_{n\to\infty}(y_n/n)$, so that Q may be interpreted as "beliefs about the limiting relative frequency of 1's".

In more conventional notation and language, it is as if, conditional on θ, $x_1, \ldots, x_n$ are a *random sample* from a Bernoulli distribution with *parameter* θ, generating a parametrised *joint sampling distribution*

$$p(x_1, \ldots, x_n \mid \theta) = \prod_{i=1}^{n} p(x_i \mid \theta) = \prod_{i=1}^{n} \theta^{x_i}(1-\theta)^{1-x_i},$$

where the parameter is assigned a *prior distribution* $Q(\theta)$. The operational content of this prior distribution derives from the fact that it is *as if* we are assessing beliefs about what we would anticipate observing as the *limiting relative frequency* from a "very large number" of observations. Thought of as a function of θ, we shall refer to the joint sampling distribution as the *likelihood* function.

In terms of Definition 4.1, the assumption of exchangeability for the infinite sequence of 0–1 random quantities $x_1, x_2, \ldots$ places a strict limitation on the family of probability measures P which can serve as predictive probability models for the sequence. Any such P must correspond to the mixture form given in Proposition 4.1, for some choice of prior distribution $Q(\theta)$. As we range over all possible choices of this latter distribution, we generate all possible predictive probability models compatible with the assumption of infinite exchangeability for the 0–1 random quantities.

Thus, "at a stroke", we establish a justification for the conventional model building procedure of combining a likelihood and a prior. The likelihood is defined in terms of an assumption of conditional independence of the observations given a parameter; the latter, and its associated prior distribution, acquire an operational interpretation in terms of a limiting average of observables (in this case a limiting frequency).

In many applications involving 0–1 random quantities, we may be more interested in a summary random quantity, such as $y_n = x_1 + \cdots + x_n$, than in the individual sequences of x_i's. The representation of $p(x_1 + \cdots + x_n = y_n)$ is straightforwardly obtained from Proposition 4.1.

Corollary 1. *Given the conditions of Proposition 4.1,*

$$p(x_1 + \cdots + x_n = y_n) = \int_0^1 \binom{n}{y_n} \theta^{y_n}(1-\theta)^{n-y_n} dQ(\theta).$$

Proof. This follows immediately from Proposition 4.1 and the fact that

$$p(x_1 + \cdots + x_n = y_n) = \binom{n}{y_n} p(x_1, \ldots, x_n)$$

for all $x_1, \ldots, x_n$ such that $x_1 + \cdots + x_n = y_n$. $\triangleleft$

This provides a justification, when expressing beliefs about y_n, for acting *as if* we have a binomial likelihood, defined by $\text{Bi}(y_n \mid \theta, n)$, with a prior distribution $Q(\theta)$ for the binomial parameter θ.

The formal learning process for models such as this will be developed systematically and generally in Chapter 5. However, this simple example provides considerable insight into the learning process, showing how, in a sense, the key step is a straightforward consequence of the representation theorem.

Corollary 2. *If $x_1, x_2, \ldots$ is an infinitely exchangeable sequence of 0–1 random quantities with probability measure P, the conditional probability function $p(x_{m+1}, \ldots, x_n \mid x_1, \ldots, x_m)$, for $x_{m+1}, \ldots, x_n$ given $x_1, \ldots, x_m$, has the form*

$$\int_0^1 \prod_{i=m+1}^{n} \theta^{x_i}(1-\theta)^{1-x_i} dQ(\theta \mid x_1, \ldots, x_m), \quad 1 \le m < n,$$

where

$$dQ(\theta \mid x_1, \ldots, x_m) = \frac{\prod_{i=1}^{m} \theta^{x_i}(1-\theta)^{1-x_i} dQ(\theta)}{\int_0^1 \prod_{i=1}^{m} \theta^{x_i}(1-\theta)^{1-x_i}\, dQ(\theta)}$$

and

$$Q(\theta) = \lim_{n\to\infty} P(y_n/n \le \theta).$$

Proof. Clearly,

$$p(x_{m+1}, \ldots, x_n \mid x_1, \ldots, x_m) = \frac{p(x_1, \ldots, x_n)}{p(x_1, \ldots, x_m)},$$

and the result follows by applying Proposition 4.1 to both $p(x_1, \ldots, x_n)$ and $p(x_1, \ldots, x_m)$ and rearranging the resulting expression. ◁

We thus see that the basic form of representation of beliefs does not change. All that has happened, expressed in conventional terminology, is that the *prior* distribution $Q(\theta)$ for θ has been revised, via *Bayes' theorem*, into the *posterior* distribution $Q(\theta \mid x_1, \ldots, x_m)$.

The conditional probability function $p(x_{m+1}, \ldots, x_n \mid x_1, \ldots, x_m)$ is called the (*conditional, or posterior*) *predictive* probability function for $x_{m+1}, \ldots, x_n$ given $x_1, \ldots, x_m$, and this, of course, also provides the basis for deriving the conditional predictive distribution of any other random quantity defined in terms of the future observations. For example, given $x_1, \ldots, x_m$, the predictive probability function $p(y_{n-m} \mid x_1, \ldots, x_m)$ for y_{n-m}, i.e., the total number of 1's in $x_{m+1}, \ldots, x_n$, has the form

$$\int_0^1 \binom{n-m}{y_{n-m}} \theta^{y_{n-m}}(1-\theta)^{(n-m)-y_{n-m}} dQ(\theta \mid x_1, \ldots, x_m).$$

A particularly important random quantity defined in terms of future observations is the frequency of 1's in a large sample. But, by Proposition 4.1 and its Corollary 2,

$$\lim_{(n-m)\to\infty} P\left(\frac{y_{n-m}}{(n-m)} \le \theta \,\middle|\, x_1, \ldots, x_m\right) = Q(\theta \mid x_1, \ldots, x_m).$$

Thus, *a posterior distribution for a parameter is seen to be a limiting case of a posterior (conditional) predictive distribution for an observable.*

4.3.2 The Multinomial Model

An alternative way of viewing the 0–1 random quantities discussed in Section 3.1 is as defining category membership (given two exclusive and exhaustive categories), in the sense that $x_i = 1$ signifies that the ith observation belongs to category 1 and $x_i = 0$ signifies membership of category 2. We can extend this idea in an obvious way by considering k-dimensional random vectors $\boldsymbol{x}_i$ whose jth component, x_{ij}, takes the value 1 to indicate membership of the jth of $k+1$ categories. At most one of the k components can take the value 1; if they all take the value 0 this signifies membership of the $(k+1)$th category. In what follows, we shall refer to such $\boldsymbol{x}_i$ as "0–1 random vectors". If $\boldsymbol{x}_1, \boldsymbol{x}_2, \ldots$ is an infinitely exchangeable sequence of 0–1 random vectors, we can extend Proposition 4.1 in an obvious way.

Proposition 4.2. (***Representation theorem for 0–1 random vectors***).
If $\boldsymbol{x}_1, \boldsymbol{x}_2, \ldots$ is an infinitely exchangeable sequence of 0–1 random vectors with probability measure P, there exists a distribution function Q such that the joint mass function $p(\boldsymbol{x}_1, \ldots, \boldsymbol{x}_n)$ for $\boldsymbol{x}_1, \ldots, \boldsymbol{x}_n$ has the form

$$p(\boldsymbol{x}_1, \ldots, \boldsymbol{x}_n) = \int_{\Theta^*} \prod_{i=1}^{n} \theta_1^{x_{i1}} \theta_2^{x_{i2}} \cdots \theta_k^{x_{ik}} \left(1 - \sum_{j=1}^{k} \theta_j\right)^{1-\Sigma x_{ij}} dQ(\boldsymbol{\theta}),$$

where

$$\Theta^* = \left\{ \boldsymbol{\theta} = (\theta_1, \ldots, \theta_k); \quad 0 \leq \theta_i \leq 1, \ \sum_{i=1}^{k} \theta_i \leq 1 \right\}$$

and

$$Q(\boldsymbol{\theta}) = \lim_{n \to \infty} P\left[(\overline{x}_{1n} \leq \theta_1,) \cup \ldots \cup (\overline{x}_{kn} \leq \theta_k) \right],$$

with $\overline{x}_{in} = n^{-1}(x_{1i} + \cdots + x_{ni})$, *and* $\theta_i = \lim_{n \to \infty} \overline{x}_{in}$.

Proof. This is a straightforward, albeit algebraically cumbersome, generalisation of the proof of Proposition 4.1. ◁

As in the previous case, we are often most interested in the summary random vector $\boldsymbol{y}_n = \boldsymbol{x}_1 + \cdots + \boldsymbol{x}_n$ whose jth component y_{nj} is the random quantity corresponding to the total number of occurrences of category j in the n observations. We shall give the representation of $p(\boldsymbol{x}_1 + \cdots + \boldsymbol{x}_n = \boldsymbol{y}_n) = p(y_{n1}, \ldots, y_{nk})$, generalising Corollary 1 to Proposition 4.1, and then comment on the interpretation of these results.

Corollary. *Given the conditions of Proposition 4.2, the joint mass function* $p(y_{n1}, \ldots, y_{nk})$ *may be represented as*

$$\int_{\Theta^*} \binom{n}{y_{n1} \cdots y_{nk}} [\theta_1^{y_{n1}} \theta_2^{y_{n2}} \cdots \theta_k^{y_{nk}}](1 - \Sigma\theta_i)^{n - \Sigma y_{ni}} dQ(\boldsymbol{\theta})$$

where

$$\binom{n}{y_{n1} \cdots y_{nk}} = \frac{n!}{y_{n1}! y_{n2}! \cdots y_{nk}! (n - \Sigma y_{ni})!},$$

Proof. This follows immediately from the generalisation of the argument used in proving Corollary 1 to Proposition 4.1. $\triangleleft$

Thus, we see in Proposition 4.2 that it is *as if* we have a likelihood corresponding to the joint sampling distribution of a random sample $\boldsymbol{x}_1, \ldots, \boldsymbol{x}_n$, where each $\boldsymbol{x}_i$ has a multinomial distribution with probability function $\text{Mu}_k(\boldsymbol{x}_i \mid \boldsymbol{\theta}, 1)$, together with a prior distribution Q over the multinomial parameter $\boldsymbol{\theta}$, where the components $\boldsymbol{\theta}_j$ of the latter can be thought of as the limiting relative frequency of membership of the jth category. In the corollary, it is *as if* we assume a multinomial likelihood, $\text{Mu}_k(\boldsymbol{y}_{n}, \mid \boldsymbol{\theta}, n)$, with a prior $Q(\boldsymbol{\theta})$ for $\boldsymbol{\theta}$.

4.3.3 The General Model

We now consider the case of an infinitely exchangeable sequence of real-valued random quantities $x_1, x_2, \ldots$. As one might expect, the mathematical technicalities of establishing a representation theorem in the real-valued case are somewhat more complicated than in the 0–1 cases, and a rigorous treatment involves the use of measure-theoretic tools beyond the general mathematical level at which this volume is aimed. For this reason, we shall content ourselves with providing an *outline proof* of a form of the representation theorem, having no pretence at mathematical rigour but, hopefully, providing some intuitive insight into the result, as well as the key ideas underlying a form of proper proof.

Proposition 4.3. (***General representation theorem***).
If $x_1, x_2, \ldots$, *is an infinitely exchangeable sequence of real-valued random quantities with probability measure* P, *there exists a probability measure* Q *over* $\Im$, *the space of all distribution functions on* $\Re$, *such that the joint distribution function of* $x_1, \ldots, x_n$ *has the form*

$$P(x_1, \ldots, x_n) = \int_{\Im} \prod_{i=1}^{n} F(x_i) dQ(F),$$

where

$$Q(F) = \lim_{n \to \infty} P(F_n)$$

and F_n *is the empirical distribution function defined by* $x_1, \ldots, x_n$.

Outline proof. (See Chow and Teicher, 1978/1988). Since

$$F_n(x) = \frac{1}{n}\sum_{i=1}^{n} I_{(x_i \le x)}$$

we have, by exchangeability,

$$\mathrm{E}(F_n(x) - F_N(x))^2 = \frac{|N-n|}{Nn}\Big\{P(x_1 < x) - P[(x_1 < x)\cap(x_2 < x)]\Big\}.$$

To see this, writing I_i in place of $I_{[x_i \le x]}$ and noting that $I_i^2 = I_i$, we have

$$\begin{aligned}\left[F_n(x) - F_N(x)\right]^2 &= \left(\frac{1}{n^2}\sum_{i=1}^{n} - \frac{1}{N^2}\sum_{i=1}^{N} \frac{2}{nN}\sum_{i=1}^{n}\right)(I_i)\\ &\quad + 2\left(\frac{1}{n^2}\sum_{j}^{n}\sum_{i<j}^{n} - N^2\sum_{j}^{N}\sum_{i<j}^{N}\frac{1}{nN}\sum_{j}^{n}\sum_{i<j}^{N}\right)(I_iI_j).\end{aligned}$$

Note also that $\mathrm{E}(I_i) = P(x_1 < x)$ and $\mathrm{E}(I_iI_j) = P[(x_1 < x)\cap(x_2 < x)]$, for all i, j, by exchangeability. A straightforward count of the numbers of terms involved in the summations then gives the required result.

The right-hand side tends to zero as $N, n \to \infty$, and hence the random quantity $F_n(x)$ tends in probability to some random quantity, $F(x)$, say, which implies that

$$\prod_{j=1}^{n} F_N(x_j) \to \prod_{j=1}^{n} F(x_j) \qquad (*)$$

in probability as $N \to \infty$, for fixed n.

Suppose we now let $\alpha_1, \ldots, \alpha_n$ denote positive integers and set

$$A = \{\alpha = (\alpha_1, \ldots, \alpha_n);\quad 1 \le \alpha_i \le N \text{ for } 1 \le i \le n\}$$

and

$$A^* = I(\alpha) = I[(x_{\alpha_1} \le x_1)\cap\cdots\cap(x_{\alpha_n} \le x_n)].$$

For $N > n$, it then follows that

$$\begin{aligned}\prod_{j=1}^{n} F_N(x_j) &= N^{-n}\prod_{j=1}^{n}\sum_{i=1}^{N} I_{(x_i < x_j)} = N^{-n}\sum_{\alpha\in A} I(\alpha)\\ &= N^{-n}\left(\sum_{\alpha\in A-A^*} + \sum_{\alpha\in A^*}\right) I(\alpha).\end{aligned}$$

However, as $N \to \infty$,

$$N^{-n} \sum_{\alpha \in A - A^*} I(\alpha) \leq N^{-n} \sum_{\alpha \in A - A^*} 1 = [N^n - N(N-1) \cdots (N-n+1)]/N^n \to 0$$

so that,

$$\prod_{j=1}^{n} F_N(x_j) \approx N^{-n} \sum_{\alpha \in A^*} I(\alpha)$$

But, by exchangeability,

$$\begin{aligned} \int I(\alpha)\, dP &= \int I_{[(x \leq x_1) \cap \cdots \cap (x \leq x_n)]}\, dP \\ &= P[(x \leq x_1) \cap \cdots \cap (x \leq x_n)] = P(x_1, \ldots, x_n) \end{aligned}$$

and so

$$\int \prod_{i=1}^{n} F_N(x_j) dP \approx [N \cdots (N - n + 1)/N^n] P[(x \leq x_1) \cap \cdots \cap (x \leq x_n)].$$

Recalling (*), we see that, as $N \to \infty$,

$$\int \prod_{j=1}^{n} F(x_j)\, dQ(F) \approx P(x_1, \ldots, x_n)$$

where $Q(F) = \lim_{N \to \infty} P(F_N)$. $\triangleleft$

The general form of representation for real-valued exchangeable random quantities is therefore *as if* we have independent observations $x_1, \ldots, x_n$ conditional on F, an unknown (i.e., random) distribution function (which plays the role of an infinite-dimensional "parameter" in this case), with a belief distribution Q for F, having the operational interpretation of "what we believe the empirical distribution function would look like for a large sample".

The structure of the learning process for a general exchangeable sequence of real-valued random quantities, with the distribution function representation given in Proposition 4.3, cannot easily be described explicitly. In what follows, we shall therefore find it convenient to restrict attention to those cases where a corresponding representation holds in terms of density functions, labelled by a finite-dimensional parameter, $\boldsymbol{\theta}$, say, rather than the infinite-dimensional label, F. For ease of reference, we present this representation as a corollary to Proposition 4.3.

Corollary 1. *Assuming the required densities to exist, under the conditions of Proposition 4.3 the joint density of* $x_1, \ldots, x_n$ *has the form*

$$p(x_1, \ldots, x_n) = \int_\Theta \prod_{i=1}^n p(x_i \,|\, \boldsymbol{\theta})\, dQ(\boldsymbol{\theta}),$$

with $p(\cdot \,|\, \boldsymbol{\theta})$ *denoting the density function corresponding to the "unknown parameter"* $\boldsymbol{\theta} \in \Theta$.

The role of Bayes' theorem in the learning process is now easily identified.

Corollary 2. *If* $x_1, x_2, \ldots$ *is an infinitely exchangeable sequence of real-valued random quantities admitting a density representation as in Corollary 1, then*

$$p(x_{m+1}, \ldots, x_n \,|\, x_1, \ldots, x_m) = \int_\Theta \prod_{i=m+1}^n p(x_i \,|\, \boldsymbol{\theta})\, dQ(\boldsymbol{\theta} \,|\, x_1, \ldots, x_m)$$

where

$$dQ(\boldsymbol{\theta} \,|\, x_1, \ldots, x_m) = \frac{\prod_{i=1}^m p(x_i \,|\, \boldsymbol{\theta})\, dQ(\boldsymbol{\theta})}{\int_\Theta \prod_{i=1}^m p(x_i \,|\, \boldsymbol{\theta})\, dQ(\boldsymbol{\theta})}.$$

Proof. This follows immediately on writing

$$p(x_{m+1}, \ldots, x_n \,|\, x_1, \ldots, x_m) = \frac{p(x_1, \ldots, x_n)}{p(x_1, \ldots, x_m)},$$

applying the density representation form to both $p(x_1, \ldots, x_n)$ and $p(x_1, \ldots, x_m)$, and rearranging the resulting expression. $\triangleleft$

The technical discussion in this section has centred on exchangeable sequences, $x_1, x_2, \ldots$, of real-valued random quantities. In fact, everything carries over in an obviously analogous manner to the case of exchangeable sequences $\boldsymbol{x}_1, \boldsymbol{x}_2, \ldots$, with $\boldsymbol{x}_i \in \Re^k$. All that happens, in effect, is that the distribution functions and densities referred to in Proposition 4.3 and its corollaries become the joint distribution functions and densities for the k components of the $\boldsymbol{x}_i$. To avoid tedious distinctions between $x \in \Re$ and $\boldsymbol{x} \in \Re^k$, in subsequent developments we shall often just write $x \in X$. In cases where the distinction between $k = 1$ and $k > 1$ matters, it will be clear from the context what is intended.

In Section 4.8.1, we shall give detailed references to the literature on representation theorems for exchangeable sequences, including far-reaching generalisations of the 0–1 and real-valued cases. However, even the simple cases we have presented already provide, from the subjectivist perspective, a deeply satisfying clarification of such fundamental notions as *models, parameters, conditional independence* and the relationship between *beliefs* and *limiting frequencies*.

In terms of Definition 4.1, the assumption of exchangeability for the real-valued random quantities $x_1, x_2, \ldots$ again places (as in the 0–1 case) a limitation on the family of probability measures P which can serve as predictive probability models. In this case, however, in the context of the general form of representation given in Proposition 4.3, the "parameter", F, underlying the conditional independence structure within the mixture is a random distribution function, so that the "parameter" is, in effect, *infinite dimensional*, and the family of coherent predictive probability models is generated by ranging through all possible prior distributions $Q(F)$. The mathematical form of the required representation is well-defined, but the practical task of translating actual beliefs about real-valued random quantities into the required mathematical form of a measure over a function space seems, to say the least, a somewhat daunting prospect. It is interesting therefore to see whether there exist more complex formal structures of belief, imposing further symmetries or structure beyond simple exchangeability, which lead to more specific and "familiar" model representations. In particular, it is of interest to identify situations in which exchangeability leads to a mixture of conditional independence structures which are defined in terms of a *finite dimensional* parameter so that the more explicit forms given in the corollaries to Proposition 4.3 can be invoked. Given the interpretation of the components of such a parameter as strong law limits of simple sequences of functions of the observations, the specification of Q, and hence of the complete predictive probability model P, then becomes a much less daunting task.

4.4 MODELS VIA INVARIANCE

4.4.1 The Normal Model

Suppose that in addition to judging an infinite sequence of real-valued random quantities $x_1, x_2, \ldots$ to be exchangeable, we consider the possibility of further judgements of invariance, perhaps relating to the "geometry" of the space in which a finite subset of observations, $x_1, \ldots, x_n$, say, lie. The following definitions describe two such possible judgements of invariance. As with exchangeability, there is no claim that such judgements have any a priori special status. They are intended, simply, as possible forms of judgement that *might* be made, and whose consequences might be interesting to explore.

Definition 4.4. (*Spherical symmetry*). *A sequence of random quantities $x_1, \ldots, x_n$ is said to have spherical symmetry under a predictive probability model P if the latter defines the distributions of $\boldsymbol{x} = (x_1, \ldots, x_n)$ and $\boldsymbol{A}\boldsymbol{x}$ to be identical, for any (orthogonal) $n \times n$ matrix $\boldsymbol{A}$ such that $\boldsymbol{A}^t\boldsymbol{A} = \boldsymbol{I}$.*

This definition encapsulates a judgement of rotational symmetry, in the sense that, although measurements happened to have been expressed in terms of a particular coordinate system (yielding $x_1, \ldots, x_n$), our quantitative beliefs would not change if they had been expressed in a rotated coordinate system. Since rotational invariance fixes "distances" from the origin, this is equivalent to a judgement of identical beliefs for all outcomes of $x_1, \ldots, x_n$ leading to the same value of $x_1^2 + \cdots + x_n^2$.

The next result states that if we make the judgement of spherical symmetry (which in turn implies a judgement of exchangeability, since permutation is a special case of orthogonal transformation), the general mixture representation given in Proposition 4.3 assumes a much more concrete and familiar form.

Proposition 4.4. ***(Representation theorem under spherical symmetry)***. *If $x_1, x_2, \ldots$ is an infinite sequence of real-valued random quantities with probability measure P, and if, for any n, $\{x_1, \ldots, x_n\}$ have spherical symmetry, there exists a distribution function Q on $\Re^+$ such that the joint distribution function of $x_1, \ldots, x_n$ has the form*

$$P(x_1, \ldots, x_n) = \int_{\Re^+} \prod_{i=1}^{n} \Phi(\lambda^{1/2} x_i) dQ(\lambda),$$

where Φ is the standard normal distribution function and

$$Q(\lambda) = \lim_{n \to \infty} P(s_n^{-2} \leq \lambda),$$

with $s_n^2 = n^{-1}(x_1^2 + \cdots + x_n^2)$, and $\lambda^{-1} = \lim_{n \to \infty} s_n^2$.

Proof. See, for example, Freedman (1963a) and Kingman (1972); details are omitted here, since the proof of a generalisation of this result will be given in full in Proposition 4.5. ◁

The form of representation obtained in Proposition 4.4 tells us that the judgement of spherical symmetry restricts the set of coherent predictive probability models to those which are generated by acting *as if*:

(i) observations are conditionally independent *normal* random quantities, given the random quantity λ (which, as a "labelling parameter", corresponds to the precision; i.e., the reciprocal of the variance);
(ii) λ is itself assigned a distribution Q;
(iii) by the strong law of large numbers, $\lambda^{-1} = \lim_{n \to \infty} s_n^2$, so that Q may be interpreted as "beliefs about the reciprocal of the limiting mean sum of squares of the observations".

For related work see Dawid (1977, 1978). To obtain a justification for the usual normal specification, with "unknown mean and precision", we need to generalise the above discussion slightly.

We note first that the judgement of spherical symmetry implicitly attaches a special significance to the *origin* of the coordinate system, since it is equivalent to a judgement of invariance in terms of distance from the origin. In general, however, if we were to feel able to make a judgement of spherical symmetry, it would typically only be *relative* to an "origin" defined in terms of the "centre" of the random quantities under consideration. This motivates the following definition.

Definition 4.5. (***Centred spherical symmetry***). *A sequence of random quantities $x_1, \ldots, x_n$ is said to have centred spherical symmetry if the random quantities $x_1 - \bar{x}_n, \ldots, x_n - \bar{x}_n$ have spherical symmetry, where $\bar{x}_n = n^{-1}\sum x_i$. This is equivalent to a judgement of identical beliefs for all outcomes of $x_1, \ldots, x_n$ leading to the same value of $(x_1 - \bar{x}_n)^2 + \cdots + (x_n - \bar{x}_n)^2$.*

Proposition 4.5. (***Representation under centred spherical symmetry***). *If $x_1, x_2, \ldots$ is an infinitely exchangeable sequence of real-valued random quantities with probability measure P, and if, for any n, $\{x_1, \ldots x_n\}$ have centred spherical symmetry, then there exists a distribution function Q on $\Re \times \Re^+$ such that the joint distribution of $x_1, \ldots, x_n$ has the form*

$$P(x_1, \ldots, x_n) = \int_{\Re \times \Re^+} \prod_{i=1}^{n} \Phi[\lambda^{1/2}(x_i - \mu)] dQ(\mu, \lambda),$$

where Φ is the standard normal distribution function and

$$Q(\mu, \lambda) = \lim_{n \to \infty} P\big[(\bar{x}_n \le \mu) \cap (s_n^{-2} \le \lambda)\big],$$

with $\bar{x}_n = n^{-1}(x_1 + \cdots + x_n)$, $s_n^2 = n^{-1}\big[(x_1 - \bar{x}_n)^2 + \cdots + (x_n - \bar{x}_n)^2\big]$, $\mu = \lim_{n\to\infty} \bar{x}_n$, and $\lambda^{-1} = \lim_{n\to\infty} s_n^2$.

Proof. (Smith, 1981). Since the sequence $x_1, x_2, \ldots$ is exchangeable, by Proposition 4.3 there exists a random distribution function F such that, conditional on F, the random quantities $x_1, \ldots, x_n$, for any n, are independent. There is therefore a random characteristic function, ϕ, corresponding to F, such that

$$E\left[\exp\left(i\sum_{j=1}^{n} t_j x_j\right) \middle| F\right] = \prod_{j=1}^{n} \phi(t_j)$$

and hence

$$E\left[\exp\left(i\sum_{j=1}^{n} t_j x_j\right)\right] = E\left[\prod_{j=1}^{n} \phi(t_j)\right].$$

If we now define $y_j = x_j - \bar{x}_n, j = 1, \ldots, n$, it follows that

$$E\left[\exp\left(i\sum_{j=1}^{n} s_j y_j\right)\right] = E\left[\prod_{j=1}^{n}\phi(s_j)\right] \tag{*}$$

for all real $s_1, \ldots, s_n$ such that $s_1 + \cdots + s_n = 0$. Since $y_1, \ldots, y_n$ are spherically symmetric, both sides of this latter equality depend only on $s_1^2 + \cdots + s_n^2$.

Recalling that $\phi(-t) = \bar{\phi}(t)$, the complex conjugate, and that $\phi(0) = 1$, it follows that, for any real u and v,

$$\begin{aligned}
E\{|\phi(u+v)\phi(u-v) - \phi^2(u)\phi(v)\phi(-v)|^2\} \\
&= E\{\phi(u+v)\phi(u-v)\phi(-u-v)\phi(v-u)\} \\
&\quad - E\{\phi(u+v)\phi(u-v)\phi^2(-u)\phi(-v)\phi(v)\} \\
&\quad - E\{\phi(-u-v)\phi(v-u)\phi^2(u)\phi(v)\phi(-v)\} \\
&\quad + E\{\phi^2(u)\phi^2(v)\phi^2(-v)\phi^2(-u)\},
\end{aligned}$$

where all four terms in this expression are of the form of the right-hand side of (*) with $n = 8$, $s_1 + \cdots + s_8 = 0$ and $s_1^2 + \cdots + s_8^2 = 4(u^2 + v^2)$. All the four terms are therefore equal, so that the overall expression is zero. This implies that, almost surely with respect to the probability measure P, ϕ satisfies the functional equation

$$\phi(u+v)\phi(u-v) = \phi^2(u)\phi(v)\phi(-v)$$

for all real u and v. This can be rewritten in the form

$$\Psi_1(u+v) + \Psi_2(u-v) = A(u) + B(v),$$

where $\Psi_1(t) = \Psi_2(t) = \log\phi(t)$, and where $A(u) = 2\log\phi(u)$ and $B(v) = \log[\phi(v)\phi(-v)]$; it follows that $\log\phi(t)$ is a quadratic in t (see, for example, Kagan, Linnik and Rao, 1973, Lemma 1.5.1). Again using $\phi(-t) = \bar{\phi}(t), \phi(0) = 1$, we see that, for this quadratic, the constant coefficient must be zero, the linear coefficient purely imaginary and the quadratic coefficient real and non-positive. This establishes that the random characteristic function ϕ takes the form

$$\phi(t) = \exp\left\{i\mu t - \frac{1}{2}\frac{t^2}{\lambda}\right\}$$

for some random quantities $\mu \in \Re$, $\lambda \in \Re^+$.

If we now define a random quantity z by

$$z = \exp\left(i\sum_{j=1}^{n} t_j x_j\right),$$

then, by iterated expectation, we have

$$E(z \mid \mu, \lambda) = E[E(z \mid F) \mid \mu, \lambda] = E\left[\prod_{j=1}^{n} \phi(t_j) \middle| \mu, \lambda\right]$$

so that

$$E\left[\exp\left(i\sum_{j=1}^{n} t_j x_j\right) \middle| \mu, \lambda\right] = \prod_{j=1}^{n} \exp\left(i\mu t_j - \frac{1}{2}\frac{t_j^2}{\lambda}\right).$$

This establishes that, conditional on μ and λ, $x_1, \ldots, x_n$ are independent normally distributed random quantities, each with mean μ and precision λ. The mixing distribution in the general representation theorem reduces therefore to a joint distribution over μ and λ. But, by the strong law of large numbers,

$$\lim_{n\to\infty} \frac{x_1 + \cdots + x_n}{n} = \mu,$$

$$\lim_{n\to\infty} \frac{(x_1 - \bar{x}_n)^2 + \cdots + (x_n - \bar{x}_n)^2}{n} = \frac{1}{\lambda},$$

and the result follows. $\triangleleft$

We see, therefore, that the combined judgements of exchangeability and centred spherical symmetry restrict the set of coherent predictive probability models to those which, expressed in conventional terminology, correspond to acting *as if*:

(i) we have a *random sample* from a *normal distribution* with *unknown mean and precision* parameters, μ and λ, generating a *likelihood*

$$p(x_1, \ldots, x_n \mid \mu, \lambda) = \prod_{i=1}^{n} N(x_i \mid \mu, \lambda);$$

(ii) we have a joint *prior distribution* $Q(\mu, \lambda)$ for the unknown parameters, μ and λ, which can be given an operational interpretation as "beliefs about the sample mean and reciprocal sample variance which would result from a large number of observations".

4.4.2 The Multivariate Normal Model

Suppose now that we have an infinitely exchangeable sequence of random vectors $\boldsymbol{x}_1, \boldsymbol{x}_2, \ldots$ taking values in $\Re^k, k \geq 2$, and that, in addition, we judge, for all n and for all $\boldsymbol{c} \in \Re^k$, that the random quantities $\boldsymbol{c}^t\boldsymbol{x}_1, \ldots, \boldsymbol{c}^t\boldsymbol{x}_n$ have centred spherical symmetry. The next result then provides a multivariate generalisation of Proposition 4.5.

Proposition 4.6. (***Multivariate representation theorem under centred spherical symmetry***). *If $\boldsymbol{x}_1, \boldsymbol{x}_2, \ldots$ is an infinitely exchangeable sequence of random vectors taking values in $\Re^k$, with probability measure P, such that, for any n and $\boldsymbol{c} \in \Re^k$, the random quantities $\boldsymbol{c}^t\boldsymbol{x}_1, \ldots, \boldsymbol{c}^t\boldsymbol{x}_n$ have centred spherical symmetry, the structure of evaluations under P of probabilities of events defined by $\boldsymbol{x}_1, \ldots, \boldsymbol{x}_n$ is as if the latter were independent, multivariate normally distributed random vectors, conditional on a random mean vector $\boldsymbol{\mu}$ and a random precision matrix $\boldsymbol{\lambda}$, with a distribution over $\boldsymbol{\mu}$ and $\boldsymbol{\lambda}$ induced by P, where*

$$\boldsymbol{\mu} = \lim_{n\to\infty} \frac{1}{n}\sum_{i=1}^{n} \boldsymbol{x}_i, \quad \boldsymbol{\lambda}^{-1} = \lim_{n\to\infty} \frac{1}{n}\sum_{j=1}^{n} (\boldsymbol{x}_j - \bar{\boldsymbol{x}}_n)(\boldsymbol{x}_j - \bar{\boldsymbol{x}}_n)^t.$$

Proof. Defining $y_j = \boldsymbol{c}^t\boldsymbol{x}_j, j = 1, \ldots, n$, we see that the random quantities $y_1, \ldots, y_n$ have centred spherical symmetry and so, by Proposition 4.5, there exist $\mu = \mu(\boldsymbol{c})$ and $\lambda = \lambda(\boldsymbol{c})$ such that, for all $t_j \in \Re, j = 1, \ldots, n$,

$$E\left[\exp\left(i\sum_{j=1}^{n} t_j y_j\right) \middle| \mu, \lambda\right] = \prod_{j=1}^{n} \exp\left(i\mu t_j - \frac{1}{2}\frac{t_j^2}{\lambda}\right),$$

where

$$\mu = \mu(\boldsymbol{c}) = \lim_{n\to\infty} \frac{1}{n}\sum_{j=1}^{n} y_j, \quad \lambda^{-1} = \lambda^{-1}(\boldsymbol{c}) = \lim_{n\to\infty} \sum_{j=1}^{n} (y_j - \bar{y}_n)^2.$$

But

$$\mu = \mu(\boldsymbol{c}) = \boldsymbol{c}^t \lim_{n\to\infty} \left(\frac{1}{n}\sum_{j=1}^{n} \boldsymbol{x}_j\right) = \boldsymbol{c}^t\boldsymbol{\mu}$$

and

$$\lambda^{-1} = \lambda^{-1}(\boldsymbol{c}) = \boldsymbol{c}^t \lim_{n\to\infty} \left[\frac{1}{n}\sum_{j=1}^{n} (\boldsymbol{x}_j - \bar{\boldsymbol{x}}_n)(\boldsymbol{x}_j - \bar{\boldsymbol{x}}_n)^t\right] \boldsymbol{c} = \boldsymbol{c}^t\boldsymbol{\lambda}^{-1}\boldsymbol{c},$$

so that

$$E\left[\exp\left(i\boldsymbol{c}^t\sum_{j=1}^{n} t_j\boldsymbol{x}_j\right) \middle| \boldsymbol{\mu}, \boldsymbol{\lambda}\right] = \prod_{j=1}^{n} \exp[i\boldsymbol{c}^t\boldsymbol{\mu}t_j - \tfrac{1}{2}(\boldsymbol{c}^t\boldsymbol{\lambda}^{-1}\boldsymbol{c}\,t_j^2)],$$

for all $\boldsymbol{c} \in \Re^k, t_j \in \Re, j = 1, \ldots, n$. It follows that, for all $\boldsymbol{t}_j \in \Re^k, j = 1, \ldots, n$,

$$E\left[\exp\left(i\sum_{j=1}^{n} \boldsymbol{t}_j^t\boldsymbol{x}_j\right) \middle| \boldsymbol{\mu}, \boldsymbol{\lambda}\right] = \prod_{j=1}^{n} \exp[i\boldsymbol{\mu}^t\boldsymbol{t}_j - \tfrac{1}{2}\,(\boldsymbol{t}_j^t\boldsymbol{\lambda}^{-1}\boldsymbol{t}_j)]$$

so that, conditional on $\boldsymbol{\mu}$ and $\boldsymbol{\lambda}, \boldsymbol{x}_1, \ldots, \boldsymbol{x}_n$ are independent multivariate normal random quantities each with mean $\boldsymbol{\mu}$ and precision matrix $\boldsymbol{\lambda}$. ◁

4.4.3 The Exponential Model

Suppose $x_1, x_2, \ldots$ is judged to be an infinitely exchangeable sequence of positive real-valued random quantities. In particular, we note that this implies, for any pair x_i, x_j, an identity of beliefs for any events in the positive quadrant which are symmetrically placed with respect to the 45° line through the origin.

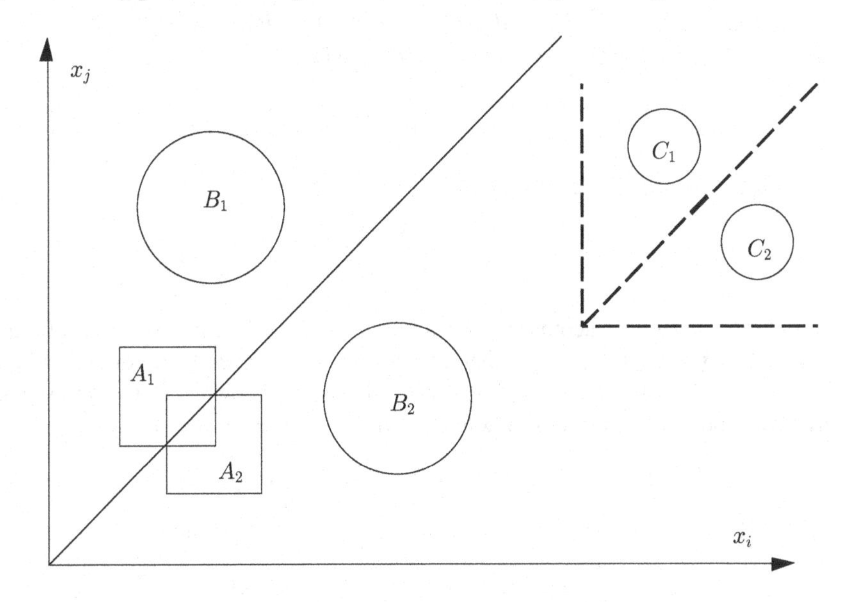

Figure 4.1 *A_1, A_2, B_1, B_2 reflections in 45° line. C_1, C_2 reflections in (dashed) 45° line*

Thus, for example, in Figure 4.1, the probabilities assigned to A_1 and A_2, B_1 and B_2, respectively, must be equal, for any $i \neq j$. In general, however, the assumption of exchangeability would not imply that events such as C_1 and C_2 have equal probabilities, even though they are symmetrically placed with respect to a 45° line (but not the one through the origin).

It is interesting to ask under what circumstances an individual *might* judge events such as C_1, C_2 to have equal probabilities. The answer is suggested by the additional (dashed) lines in the figure. *If* we added to the assumption of exchangeability the judgement that the "origins" of the x_i and x_j axes are "irrelevant", so far as probability judgements are concerned, then the probabilities of events such as C_1 and C_2 *would* be judged equal. In perhaps more familiar terms, this would be as though, when making judgements about events in the positive quadrant, an individual's judgement exhibited a form of "lack of memory" property with respect to the origin. If such a judgement is assumed to hold for all subsets of n (rather than just two) random quantities, the resulting representation is as follows.

Proposition 4.7. ***(Continuous representation under origin invariance).*** *If $x_1, x_2, \ldots$ is an infinitely exchangeable sequence of positive real-valued random quantities with probability measure P, such that, for all n, and any event A in $\Re^+ \times \ldots \times \Re^+$,*

$$P[(x_1, \ldots, x_n) \in A] = P[(x_1, \ldots, x_n) \in A + \boldsymbol{a}]$$

for all $\boldsymbol{a} \in \Re \times \ldots \times \Re$ such that $\boldsymbol{a}^t \boldsymbol{1} = 0$ and $A + \boldsymbol{a}$ is an event in $\Re^+ \times \ldots \times \Re^+$, then the joint density for $x_1, \ldots, x_n$ has the form

$$p(x_1, \ldots, x_n) = \int_0^\infty \prod_{i=1}^n \theta \exp(-\theta x_i) dQ(\theta),$$

where $\theta = \lim_{n\to\infty} \bar{x}_n^{-1}$, and

$$Q(\theta) = \lim_{n\to\infty} P[(\bar{x}_n^{-1}) \le \theta], \quad \bar{x}_n = n^{-1}(x_1 + \cdots + x_n).$$

Outline proof. (Diaconis and Ylvisaker, 1985). By the general representation theorem, there exists a random distribution function F, such that, conditional on $F, x_1, \ldots, x_n$ are independent, for any n. It can be shown that the additional invariance property continues to hold conditional on F, so that, for any $i \neq j$,

$$P[(x_i, x_j) \in A \,|\, F] = P[(x_i, x_j) \in A + \boldsymbol{a} \,|\, F]$$

for A and $\boldsymbol{a}$ as described above. If we now take $\boldsymbol{a}^t = (a_1, a_2)$ and

$$A = \{(x_i, x_j); \quad x_i > a_1 + a_2, x_j > 0\}$$

we have

$$\begin{aligned} P\big[(x_i > a_1 + a_2) \cap (x_j > 0) \,|\, F\big] &= P\big[(x_i > a_1) \cap (x_j > a_2) \,|\, F\big] \\ &= P\big[(x_i > a_1) \,|\, F\big] P\big[(x_j > a_2) \,|\, F\big]. \end{aligned}$$

By exchangeability, and recalling that x_j is certainly positive for all j, this implies that

$$P(x_i > a_1 + a_2 \,|\, F) = P(x_i > a_1 \,|\, F) P(x_i > a_2 \,|\, F).$$

But this functional relationship implies, for positive real-valued x_i, that

$$p(x_i > x \,|\, F) = e^{-\theta x}$$

for some θ, so that the density, $p(x_i \,|\, F) = p(x_i \,|\, \theta)$, is the derivative of

$$1 - \exp(-\theta x_i),$$

and hence given by $\theta \exp(-\theta x_i)$. The rest of the result follows on noting that, by the strong law of large numbers, $\theta^{-1} = \lim_{n\to\infty} [n^{-1}(x_1 + \cdots + x_n)]$. ◁

Thus, we see that judgements of exchangeability and "lack of memory" for sequences of positive real-valued random quantities constrain the possible predictive probability models for the sequence to be those which are generated by acting *as if* we have a *random sample* from an *exponential distribution* with *unknown parameter* θ, with a *prior distribution* Q for the latter. In fact, if Q^* denotes the corresponding distribution for $\phi = \theta^{-1} = \lim_{n\to\infty} \bar{x}_n$, it may be easier to use the "reparametrised" representation

$$p(x_1, \ldots, x_n) = \int_0^\infty \prod_{i=1}^n \phi^{-1} \exp(-\phi^{-1} x_i) dQ^*(\phi),$$

since Q^* is then more directly accessible as "beliefs about the sample mean from a large number of observations".

Recalling the possible motivation given above for the additional invariance assumption on the sequence $x_1, x_2, \ldots$, it is interesting to note the very specific and well-known "lack of memory" property of the exponential distribution; namely,

$$P(x_i > a_1 + a_2 \,|\, \theta, x_i > a_1) = P(x_i > a_2 \,|\, \theta),$$

which appears implicitly in the above proof.

4.4.4 The Geometric Model

Suppose $x_1, x_2, \ldots$ is judged to be an infinitely exchangeable sequence of strictly positive integer-valued random quantities. It is easy to see that we could repeat the entire introductory discussion of Section 4.4.3, except that events would now be defined in terms of sets of points on the lattice $\mathcal{Z}^+ \times \ldots \times \mathcal{Z}^+$, rather than as regions in $\Re^+ \times \ldots \times \Re^+$. This enables us to state the following representation result.

Proposition 4.8. ***(Discrete representation under origin invariance).***
If $x_1, x_2, \ldots$ is an infinitely exchangeable sequence of positive integer-valued random quantities with probability measure P, such that, for all n and any event A in $\mathcal{Z}^+ \times \ldots \times \mathcal{Z}^+$,

$$P[(x_1, \ldots, x_n) \in A] = P[(x_1, \ldots, x_n) \in A + \boldsymbol{a}]$$

for all $\boldsymbol{a} \in \mathcal{Z} \times \ldots \times \mathcal{Z}$ such that $\boldsymbol{a}^t\mathbf{1} = 0$ and $A + \boldsymbol{a}$ is an event in $\mathcal{Z}^+ \times \ldots \times \mathcal{Z}^+$, then the joint density for $x_1, \ldots, x_n$ has the form

$$p(x_1, \ldots, x_n) = \int_0^1 \prod_{i=1}^n \theta(1-\theta)^{x_i - 1} dQ(\theta),$$

where $\theta = \lim_{n\to\infty} \bar{x}_n^{-1}$, $\bar{x}_n = n^{-1}(x_1 + \cdots + x_n)$, and

$$Q(\theta) = \lim_{n\to\infty} P[(\bar{x}_n^{-1}) \leq \theta].$$

Outline proof. This follows precisely the steps in the proof of Proposition 4.7, except that, for positive integer-valued x_i, the functional equation

$$P(x_i > a_1 + a_2 \mid F) = P(x_i > a_1 \mid F)P(x_i > a_2 \mid F)$$

implies that

$$P(x_i > x \mid F) = \theta^x,$$

so that the probability function, $p(x_i \mid F) = p(x_i \mid \theta)$ is easily seen to be $\theta(1-\theta)^{x_i-1}$. Again, by the strong law of large numbers, $\theta^{-1} = \lim_{n\to\infty} \bar{x}_n$, where, since $x_i \geq 1$ for all $i, 0 < \theta \leq 1$. $\triangleleft$

In this case, the coherent predictive probability models must be those which are generated by acting *as if* we have a *random sample* from a *geometric distribution* with *unknown parameter* θ, with a *prior distribution* Q for the latter, where $\theta^{-1} = \lim_{n\to\infty} \bar{x}_n$.

Again, recalling the possible motivation for the additional invariance property, it is interesting to note the familiar "lack of memory" property of the geometric distribution;

$$P(x_i > a_1 + a_2 \mid \theta, x_i > a_1) = P(x_i > a_2 \mid \theta).$$

4.5 MODELS VIA SUFFICIENT STATISTICS

4.5.1 Summary Statistics

We begin with a formal definition, which enables us to discuss the process of *summarising* a sequence, or *sample*, of random quantities, $x_1, \ldots, x_m$. (In general, our discussion carries over to the case of random vectors, but for notational simplicity we shall usually talk in terms of random quantities.)

Definition 4.6. **(*Statistic*).** *Given random quantities (vectors) $x_1, \ldots, x_m$, with specified sets of possible values $X_1, \ldots, X_m$, respectively, a random vector $\boldsymbol{t}_m : X_1 \times \cdots \times X_m \to \Re^{k(m)} (k(m) \leq m)$ is called a $k(m)$-dimensional statistic.*

A trivial case of such a statistic would be $\boldsymbol{t}_m(x_1, \ldots, x_m) = (x_1, \ldots, x_m)$, but this clearly does not achieve much by way of summarisation, since $k(m) = m$. Familiar examples of summary statistics are:

$\boldsymbol{t}_m = m^{-1}(x_1 + \ldots + x_m)$, *the sample mean* $(k(m) = 1)$;

$\boldsymbol{t}_m = [m, (x_1 + \ldots + x_m), (x_1^2 + \cdots + x_m^2)]$, *the sample size, total and sum of squares* $(k(m) = 3)$;

$\boldsymbol{t}_m = [m, \text{med}\{x_1, \ldots, x_m\}]$, *the sample size and median* $(k(m) = 2)$;

$\boldsymbol{t}_m = \max\{x_1, \ldots, x_m\} - \min\{x_1, \ldots, x_m\}$, *the sample range* $(k(m) = 1)$.

To achieve *data reduction*, we clearly need $k(m) < m$: moreover, as with the above examples, further clarity of interpretation is achieved if $k(m) = k$, a fixed dimension independent of m.

In the next section, we shall examine the formal acceptability and implications of seeking to act *as if* particular summary statistics have a special status in the context of representing beliefs about a sequence of random vectors. We shall not concern ourselves at this stage with the origin of or motivation for any such choice of particular summary statistics. Instead, we shall focus attention on the general questions of whether, and under what circumstances, it is coherent to invoke such a form of data reduction and, if so, what forms of representation for predictive probability models might result. Throughout, we shall assume that beliefs can be represented in terms of density functions.

4.5.2 Predictive Sufficiency and Parametric Sufficiency

As an example of the way in which a summary statistic might be assumed to play a special role in the evolution of beliefs, let us consider the following general situation. Past observations $x_1, \ldots, x_m$ are available and an individual is contemplating, conditional on this given information, beliefs about future observations $x_{m+1}, \ldots, x_n$, to be described by $p(x_{m+1}, \ldots, x_n \mid x_1, \ldots, x_m)$. The following definition describes one possible way in which assumptions of systematic data reduction might be incorporated into the structure of such conditional beliefs.

Definition 4.7. (***Predictive sufficiency***).
Given a sequence of random quantities $x_1, x_2, \ldots$, with probability measure P, where x_i takes values in $X_i, i = 1, 2, \ldots$ the sequence of statistics $\boldsymbol{t}_1, \boldsymbol{t}_2, \ldots$, with $\boldsymbol{t}_j$ defined on $X_1 \times \cdots \times X_j$, is said to be predictive sufficient for the sequence $x_1, x_2, \ldots$ if, for all $m \geq 1, r \geq 1$ and $\{i_1, \ldots, i_r\} \cap \{1, \ldots, m\} = \emptyset$,

$$p(x_{i_1}, \ldots, x_{i_r} \mid x_1, \ldots, x_m) = p(x_{i_1}, \ldots, x_{i_r} \mid \boldsymbol{t}_m),$$

where $p(.\mid.)$ is the conditional density induced by P.

The above definition captures the idea that, given $\boldsymbol{t}_m = \boldsymbol{t}_m(x_1, \ldots, x_m)$, the individual values of $x_1, \ldots, x_m$ contribute nothing further to one's evaluation of probabilities of future events defined in terms of as yet unobserved random quantities. Another way of expressing this, as is easily verified from Definition 4.7, is that future observations $(x_{i_1}, \ldots, x_{i_r})$ and past observations $(x_1, \ldots, x_m)$ are conditionally independent given $\boldsymbol{t}_m$. Clearly, from a pragmatic point of view the assumption of a specified sequence of predictive sufficient statistics will, in general, greatly simplify the process of assessing probabilities of future events conditional on past observations. From a formal point of view, however, we shall need additional structure if we are to succeed in using this idea to identify specific forms of the general representation of the joint distribution of $x_1, \ldots, x_n$.

As a *particular illustration* of what might be achieved, we shall assume in what follows that *the probability measure P describing our beliefs implies both predictive sufficiency and exchangeability for the infinite sequence* $x_1, x_2, \ldots$. As with our earlier discussion in Section 4.4, a mathematically rigorous treatment is beyond the intended level of this book and so we shall confine ourselves to an informal presentation of the main ideas.

In particular, throughout this section we shall assume that the exchangeability assumption leads to a finitely parametrised mixture representation, as in Corollary 1 to Proposition 4.3, so that, as shown in Corollary 2 to that proposition, the conditional density function of $x_{m+1}, \ldots, x_n$, given $x_1, \ldots, x_m$, has the form

$$p(x_{m+1}, \ldots, x_n \mid x_1, \ldots, x_m) = \int \prod_{i=m+1}^{n} p(x_i \mid \boldsymbol{\theta}) dQ(\boldsymbol{\theta} \mid x_1, \ldots, x_m)$$

where

$$dQ(\boldsymbol{\theta} \mid x_1, \ldots, x_m) = \frac{\prod_{i=1}^{m} p(x_i \mid \boldsymbol{\theta}) dQ(\boldsymbol{\theta})}{\int \prod_{i=1}^{m} p(x_i \mid \boldsymbol{\theta}) dQ(\boldsymbol{\theta})}$$

and all integrals, here and in what follows, are assumed to be over the set of possible values of $\boldsymbol{\theta}$.

This latter form makes clear that, for such exchangeable beliefs, the learning process is "transmitted" within the mixture representation by the updating of beliefs about the "unknown parameter" $\boldsymbol{\theta}$. This suggests another possible way of defining a statistic $\boldsymbol{t}_m = \boldsymbol{t}_m(x_1, \ldots, x_m)$ to be a "sufficient summary" of $x_1, \ldots, x_m$.

Definition 4.8. (*Parametric sufficiency*). *If $x_1, x_2, \ldots$ is an infinitely exchangeable sequence of random quantities, where x_i takes values in $X_i = X, i = 1, 2, \ldots$, the sequence of statistics $\boldsymbol{t}_1, \boldsymbol{t}_2, \ldots$, with $\boldsymbol{t}_j$ defined on $X_1 \times \cdots \times X_j$, is said to be parametric sufficient for $x_1, x_2, \ldots$ if, for any $n \geq 1$,*

$$dQ(\boldsymbol{\theta} \mid x_1, \ldots, x_n) = dQ(\boldsymbol{\theta} \mid \boldsymbol{t}_n),$$

for any $dQ(\boldsymbol{\theta})$ defining an exchangeable predictive probability model via the representation

$$p(x_1, \ldots, x_n) = \int \prod_{i=1}^{n} p(x_i \mid \boldsymbol{\theta}) \, dQ(\boldsymbol{\theta}).$$

Definitions 4.7 and 4.8 both seem intuitively compelling as encapsulations of the notion of a statistic being a "sufficient summary". It is perhaps reassuring therefore that, within our assumed framework, we can establish the following.

Proposition 4.9. (***Equivalence of predictive and parametric sufficiencies***). *Given an infinitely exchangeable sequence of random quantities $x_1, x_2, \ldots$, where x_i takes values in $X_i = X, i = 1, 2, \ldots$, the sequence of statistics $\boldsymbol{t}_1, \boldsymbol{t}_2, \ldots$ with $\boldsymbol{t}_j$ defined on $X_1 \times \cdots \times X_j$ is predictive sufficient if, and only if, it is parametric sufficient.*

Heuristic proof. For any $x_1, \ldots, x_m, x_{m+1}, \ldots, x_n$ and any sequence of statistics $\boldsymbol{t}_m$, where $\boldsymbol{t}_m = \boldsymbol{t}_m(x_1, \ldots, x_m), m = 1, \ldots, n-1$, the representation theorem implies that

$$p(x_{m+1}, \ldots, x_n \,|\, \boldsymbol{t}_m) = \frac{1}{p(\boldsymbol{t}_m)} p(x_{m+1}, \ldots, x_n, \boldsymbol{t}_m),$$

$$= \frac{1}{p(\boldsymbol{t}_m)} \int_A p(x_1, \ldots, x_m, x_{m+1}, \ldots, x_n)\, dx_1, \ldots, dx_m$$

where $A = \{(x_1 \ldots, x_m);\ \boldsymbol{t}_m(x_1, \ldots, x_m) = \boldsymbol{t}_m\}$, which, in turn, can be easily shown to be expressible as

$$\frac{1}{p(\boldsymbol{t}_m)} \int_\Theta \left[\int_A \prod_{i=1}^n p(x_i \,|\, \boldsymbol{\theta})\, dx_1, \ldots, dx_m \right] dQ(\boldsymbol{\theta})$$

$$= \frac{1}{p(\boldsymbol{t}_m)} \int_\Theta \prod_{i=m+1}^n p(x_i \,|\, \boldsymbol{\theta})\, p(\boldsymbol{t}_m \,|\, \boldsymbol{\theta})\, dQ(\boldsymbol{\theta}) = \int_\Theta \prod_{i=m+1}^n p(x_i \,|\, \boldsymbol{\theta})\, dQ(\boldsymbol{\theta} \,|\, \boldsymbol{t}_m),$$

where

$$dQ(\boldsymbol{\theta} \,|\, \boldsymbol{t}_m) = \frac{1}{p(\boldsymbol{t}_m)} p(\boldsymbol{t}_m \,|\, \boldsymbol{\theta})\, dQ(\boldsymbol{\theta}) = \frac{p(\boldsymbol{t}_m \,|\, \boldsymbol{\theta})\, dQ(\boldsymbol{\theta})}{\int p(\boldsymbol{t}_m \,|\, \boldsymbol{\theta})\, dQ(\boldsymbol{\theta})}\,.$$

It follows that

$$p(x_{m+1}, \ldots, x_n \,|\, \boldsymbol{t}_m) = p(x_{m+1}, \ldots, x_n \,|\, x_1, \ldots, x_m)$$

$$= \int \prod_{i=m+1}^n p(x_i \,|\, \boldsymbol{\theta})\, dQ(\boldsymbol{\theta} \,|\, x_1, \ldots, x_m),$$

if, and only if, $dQ(\boldsymbol{\theta} \,|\, x_1, \ldots, x_m) = dQ(\boldsymbol{\theta} \,|\, \boldsymbol{t}_m)$ for all $dQ(\boldsymbol{\theta})$. ◁

To make further progress, we now establish that parametric sufficiency is itself equivalent to certain further conditions on the probability structure.

Proposition 4.10. (***Neyman factorisation criterion***). *The sequence $\boldsymbol{t}_1, \boldsymbol{t}_2, \ldots$ is parametric sufficient for infinitely exchangeable $x_1, x_2, \ldots$ admitting a finitely parametrised mixture representation if and only if, for any $m \geq 1$, the joint density for $x_1, \ldots, x_m$ given $\boldsymbol{\theta}$ has the form*

$$p(x_1, \ldots, x_m \,|\, \boldsymbol{\theta}) = h_m(\boldsymbol{t}_m, \boldsymbol{\theta}) g(x_1, \ldots, x_m),$$

for some functions $h_m \geq 0, g > 0$.

Outline proof. Given such a factorisation, for any $dQ(\boldsymbol{\theta})$ we have

$$dQ(\boldsymbol{\theta} \mid x_1, \ldots, x_m) = \frac{p(x_1, \ldots, x_m \mid \boldsymbol{\theta})dQ(\boldsymbol{\theta})}{\int_\Theta p(x_1, \ldots, x_m \mid \boldsymbol{\theta})dQ(\boldsymbol{\theta})} = \frac{h_m(\boldsymbol{t}_m, \boldsymbol{\theta})dQ(\boldsymbol{\theta})}{\int_\Theta h_m(\boldsymbol{t}_m, \boldsymbol{\theta})dQ(\boldsymbol{\theta})},$$

for some $h_m > 0$. The right-hand side depends on $x_1, \ldots, x_m$ only through $\boldsymbol{t}_m$ and, hence, $dQ(\boldsymbol{\theta} \mid x_1, \ldots, x_m) = dQ(\boldsymbol{\theta} \mid \boldsymbol{t}_m)$. Conversely, given parametric sufficiency, we have, for any $dQ(\boldsymbol{\theta})$ with support Θ,

$$\begin{aligned}\frac{p(x_1, \ldots, x_m \mid \boldsymbol{\theta})dQ(\boldsymbol{\theta})}{p(x_1, \ldots, x_m)} &= dQ(\boldsymbol{\theta} \mid x_1, \ldots, x_m) \\ &= dQ(\boldsymbol{\theta} \mid \boldsymbol{t}_m) = \frac{p(\boldsymbol{t}_m \mid \boldsymbol{\theta})dQ(\boldsymbol{\theta})}{p(\boldsymbol{t}_m)}\end{aligned}$$

so that

$$p(x_1, \ldots, x_m \mid \boldsymbol{\theta}) = h_m(\boldsymbol{t}_m, \boldsymbol{\theta})g(x_1, \ldots, x_m)$$

for some $h_m \geq 0, g > 0$ as required. ◁

Proposition 4.11. (***Sufficiency and conditional independence***).
The sequence $\boldsymbol{t}_1, \boldsymbol{t}_2, \ldots$ is parametric sufficient for infinitely exchangeable $x_1, x_2, \ldots$ if, and only if, for any $m \geq 1$, the density $p(x_1, \ldots, x_m \mid \boldsymbol{\theta}, \boldsymbol{t}_m)$ is independent of $\boldsymbol{\theta}$.

Outline proof. For any $\boldsymbol{t}_m = \boldsymbol{t}_m(x_1, \ldots, x_m)$ we have

$$p(x_1, \ldots, x_m \mid \boldsymbol{\theta}) = p(x_1, \ldots, x_m \mid \boldsymbol{\theta}, \boldsymbol{t}_m)p(\boldsymbol{t}_m \mid \boldsymbol{\theta}).$$

If $p(x_1, \ldots, x_m \mid \boldsymbol{\theta}, \boldsymbol{t}_m)$ is independent of $\boldsymbol{\theta}$, the parametric sufficiency of $\boldsymbol{t}_1, \boldsymbol{t}_2, \ldots$ follows immediately from Proposition 4.10.

Conversely, suppose that $\boldsymbol{t}_1, \boldsymbol{t}_2, \ldots$ is parametric sufficient, so that, by Proposition 4.10,

$$p(x_1, \ldots, x_m \mid \boldsymbol{\theta}) = h_m(\boldsymbol{t}_m, \boldsymbol{\theta})g(x_1, \ldots, x_m)$$

for some $h_m \geq 0, g > 0$. Integrating over all values $\{x_1, \ldots, x_m\}$ such that $\boldsymbol{t}_m(x_1, \ldots, x_m) = \boldsymbol{t}_m$, we obtain

$$p(\boldsymbol{t}_m \mid \boldsymbol{\theta}) = h_m(\boldsymbol{t}_m, \boldsymbol{\theta})G(\boldsymbol{t}_m)$$

for some $G > 0$. Substituting for $h_m(\boldsymbol{t}_m, \boldsymbol{\theta})$ in the expression for $p(x_1, \ldots, x_m \mid \boldsymbol{\theta})$, we obtain

$$p(x_1, \ldots, x_m \mid \boldsymbol{\theta}) = p(\boldsymbol{t}_m \mid \boldsymbol{\theta})\frac{g(x_1, \ldots, x_m)}{G(\boldsymbol{t}_m)}$$

so that

$$p(x_1, \ldots, x_m \mid \boldsymbol{\theta}, \boldsymbol{t}_m) = \frac{g(x_1, \ldots, x_m)}{G(\boldsymbol{t}_m)},$$

which is independent of $\boldsymbol{\theta}$. ◁

In the approach we have adopted, the definitions and consequences of predictive and parametric sufficiency have been motivated and examined within the general framework of seeking to find coherent representations of subjective beliefs about sequences of observables. Thus, for example, the notion of parametric sufficiency has so far only been put forward within the context of exchangeable beliefs, where the operational significance of "parameter" typically becomes clear from the relevant representation theorem.

In fact, however, as the reader familiar with more "conventional" approaches will have already realised, related concepts of "sufficiency" are also central to non-subjectivist theories. In particular, we note that the non-dependence of the density $p(x_1, \ldots, x_m \mid \boldsymbol{\theta}, \boldsymbol{t}_m)$ on $\boldsymbol{\theta}$, established here in Proposition 4.11 as a *consequence* of our definitions, was itself put forward as *the definition* of a "sufficient statistic" by Fisher (1922), and the factorisation given in Proposition 4.10 was established by Neyman (1935) as equivalent to the Fisher definition.

From an operational, subjectivist point of view, it seems to us rather mysterious to launch into fundamental definitions about learning processes expressed in terms of conditioning on "parameters" having no status other than as "labels". However, from a technical point of view, since our representation for exchangeable sequences provides, for us, a justification for regarding the usual (Fisher) definition as equivalent to predictive and parametric sufficiency, we can exploit many of the important mathematical results which have been established using that definition as a starting point.

In the context of our subjectivist discussion of beliefs and models, we shall mainly be interested in asking the following questions.

When is it coherent to act as if there is a sequence of predictive sufficient statistics associated with an exchangeable sequence of random quantities?

What forms of predictive probability model are implied in cases where we can assume a sequence of predictive sufficient statistics?

Aside from these foundational and modelling questions, however, the results given above also enable us to check the form of the predictive sufficient statistics for any given exchangeable representation. We shall illustrate this possibility with some simple examples before continuing with the general development.

Example 4.5. ***(Bernoulli model).*** We recall from Proposition 4.1 that if $x_1, x_2, \ldots$ is an infinitely exchangeable sequence of 0-1 random quantities, then we have the general representation

$$\begin{aligned} p(x_1, \ldots, x_n) &= \int_0^1 p(x_1, \ldots, x_n \mid \theta) dQ(\theta) \\ &= \int_0^1 \prod_{i=1}^n \mathrm{Br}(x_i \mid \theta) dQ(\theta) \\ &= \int_0^1 \theta^{s_n} (1-\theta)^{n-s_n} dQ(\theta), \end{aligned}$$

where $s_n = x_1 + \cdots + x_n$. Defining $\boldsymbol{t}_n = [n, s_n]$ and noting that we can write

$$p(x_1, \ldots, x_n \mid \theta) = h_n(\boldsymbol{t}_n, \theta) g(x_1, \ldots, x_n),$$

with

$$h_n(\boldsymbol{t}_n, \theta) = \theta^{s_n}(1-\theta)^{n-s_n}, g(x_1, \ldots, x_n) = 1,$$

it follows from Propositions 4.9 and 4.10 that the sequence $\boldsymbol{t}_1, \boldsymbol{t}_2, \ldots$ is predictive and parametric sufficient for $x_1, x_2, \ldots$. This corresponds precisely to the intuitive idea that the sequence length and total number of 1's summarises all the interesting information in any sequence of observed exchangeable 0-1 random quantities.

Example 4.6. ***(Normal model).*** We recall from Proposition 4.5 that if $x_1, x_2, \ldots$ is an exchangeable sequence of real-valued random quantities with the additional property of centred spherical symmetry then we have the general representation

$$\begin{aligned}
p(x_1, \ldots, x_n) &= \int_{-\infty}^{\infty} \int_0^{\infty} p(x_1, \ldots, x_n \mid \mu, \lambda) dQ(\mu, \lambda) \\
&= \int_{-\infty}^{\infty} \int_0^{\infty} \prod_{i=1}^{n} \mathrm{N}(x_i \mid \mu, \lambda) dQ(\mu, \lambda) \\
&= \int_{-\infty}^{\infty} \int_0^{\infty} \prod_{i=1}^{n} \left(\frac{\lambda}{2\pi}\right)^{1/2} \exp\left\{-\frac{\lambda}{2}(x_i - \mu)^2\right\} dQ(\mu, \lambda) \\
&= \int_{-\infty}^{\infty} \int_0^{\infty} \left(\frac{\lambda}{2\pi}\right)^{n/2} \exp\left\{-\frac{\lambda}{2}[n(\bar{x}_n - \mu)^2 + ns_n^2]\right\} dQ(\mu, \lambda)
\end{aligned}$$

where

$$\bar{x}_n = \frac{1}{n}\sum_{i=1}^{n} x_i, \quad s_n^2 = \frac{1}{n}\sum_{i=1}^{n}(x_i - \bar{x}_n)^2.$$

In the light of Propositions 4.10 and Proposition 4.11, inspection of $p(x_1, \ldots, x_n \mid \mu, \lambda)$ reveals that

$$\boldsymbol{t}_n = (n, \bar{x}_n, s_n^2)$$

defines a sequence of predictive and parametric sufficient statistics for $x_1, x_2, \ldots$. In view of the centring and spherical symmetry conditions, it is perhaps not surprising that the sample size, mean and sample mean sum of squares about the mean turn out to be sufficient summaries. Of course, $\boldsymbol{t}_n$ is not unique; for example, since

$$ns_n^2 = x_1^2 + \cdots + x_n^2 - n(\bar{x}_n)^2$$

we could equally well define $\boldsymbol{t}_n = [n, \bar{x}_n, n^{-1}(x_1^2 + \cdots + x_n^2)]$ as the sequence of sufficient statistics.

Example 4.7. ***(Exponential model).*** We recall from Proposition 4.7 that if $x_1, x_2, \ldots$ is an exchangeable sequence of positive real-valued random quantities with an additional "origin invariance" property, then we have the general representation

$$\begin{aligned} p(x_1, \ldots, x_n) &= \int_0^\infty p(x_1, \ldots, x_n \mid \theta) dQ(\theta) \\ &= \int_0^\infty \prod_{i=1}^n \mathrm{Ex}(x_i \mid \theta) dQ(\theta) \\ &= \int_0^\infty \theta^n \exp(-\theta s_n) dQ(\theta) \end{aligned}$$

where $s_n = x_1 + \cdots + x_n$. Again, it is immediate from Propositions 4.10 and 4.11 that $\boldsymbol{t}_n = [n, s_n]$ defines a sequence of predictive and parametric sufficient statistics, although, in this example, there is not such an obvious link between the form of invariance assumed and the form of the sufficient statistic.

It is clear from the general definition of a sufficient statistic (parametric or predictive) that $\boldsymbol{t}_n(x_1, \ldots, x_n) = [n, (x_1, \ldots, x_n)]$ is *always* a sufficient statistic. However, given our interest in achieving simplification through data reduction, it is equally clear that we should like to focus on sufficient statistics which are, in some sense, minimal. This motivates the following definition.

Definition 4.9. ***(Minimal sufficient statistic).*** *If $x_1, x_2, \ldots,$ is an infinitely exchangeable sequence of random quantities, where x_i takes values in $X_i = X$, the sequence of statistics $\boldsymbol{t}_1, \boldsymbol{t}_2, \ldots$, with $\boldsymbol{t}_j$ defined on $X_1 \times \ldots \times X_j$, is minimal sufficient for $x_1, x_2, \ldots$ if given any other sequence of sufficient statistics, $\boldsymbol{s}_1, \boldsymbol{s}_2, \ldots,$ there exist functions $\boldsymbol{g}_1(\cdot), \boldsymbol{g}_2(\cdot), \ldots$ such that $\boldsymbol{t}_i = \boldsymbol{g}_i(\boldsymbol{s}_i)$, $i = 1, 2, \ldots$*

It is easily seen that the forms of $\boldsymbol{t}(\boldsymbol{x})$ identified in Examples 4.5 to 4.7 are minimal sufficient statistics. From now on, references to sufficient statistics should be interpreted as intending minimal sufficient statistics.

Finally, since n very often appears as part of the sufficient statistic, we shall sometimes, to avoid tedious repetition, omit explicit mention of n and refer to the "interesting function(s) of $x_1, \ldots, x_n$" as the sufficient statistic.

4.5.3 Sufficiency and the Exponential Family

In the previous section, we identified some further potential structure in the general representation of joint densities for exchangeable random quantities when predictive sufficiency is assumed. We shall now take this process a stage further by examining in detail representations relating to sufficient statistics of fixed dimension.

Since we have established, in the finite parameter framework, the equivalence of predictive and parametric sufficiency for the case of exchangeable random quantities, and their equivalence with the factorisation criterion of Proposition 4.11, we shall from now on simply use the term *sufficient statistic*, without risk of confusion.

We begin by considering exchangeable beliefs constructed by mixing, with respect to some $dQ(\theta)$, over a specified parametric form

$$p(x_1, \ldots, x_n \mid \theta) = \prod_{i=1}^{k} p(x_i \mid \theta), \quad (x_1, \ldots, x_n) \in X^n \subseteq \Re^n$$

where θ is a one-dimensional parameter. By Proposition 4.10, if the form of $p(x \mid \theta)$ is such that $p(x_1, \ldots, x_n \mid \theta)$ factors into $h_n(\boldsymbol{t}_n, \theta) g(x_1, \ldots, x_n)$, for some h_n, g, the statistic $\boldsymbol{t}_n = \boldsymbol{t}_n(x_1, \ldots, x_n)$ would be sufficient. An important class of such $p(x \mid \theta)$ is identified in the following definition.

Definition 4.10. (***One-parameter exponential family***). *A probability density (or mass function) $p(x \mid \theta)$, labelled by $\theta \in \Theta \subseteq \Re$, is said to belong to the one-parameter exponential family if it is of the form*

$$p(x \mid \theta) = \mathrm{Ef}(x \mid f, g, h, \phi, \theta, c) = f(x) g(\theta) \exp\{c\phi(\theta) h(x)\}, \quad x \in X,$$

where, given f, h, ϕ, and c, $[g(\theta)]^{-1} = \int_X f(x) \exp\{c\phi(\theta) h(x)\} dx < \infty$. The family is called **regular** *if X does not depend on θ; otherwise it is called* ***non-regular***.

Proposition 4.12. (***Sufficient statistics for the one-parameter exponential family***). *If $x_1, x_2, \ldots, x_n \in X$, is an exchangeable sequence such that, given regular* $\mathrm{Ef}(\cdot \mid \cdot)$,

$$p(x_1, \ldots, x_n) = \int_\Theta \prod_{i=1}^{n} \mathrm{Ef}(x_i \mid f, g, h, \phi, \theta, c) dQ(\theta),$$

for some $dQ(\theta)$, then $\boldsymbol{t}_n = \boldsymbol{t}_n(x_1, \ldots, x_n) = [n, h(x_1) + \cdots + h(x_n)]$, for $n = 1, 2, \ldots$, is a sequence of sufficient statistics.

Proof. This follows immediately from Proposition 4.10 on noting that

$$\prod_{i=1}^{n} \mathrm{Ef}(x_i \mid f, g, h, \phi, \theta, c) = \prod_{i=1}^{n} f(x_i) \cdot [g(\theta)]^n \exp\left\{ c\phi(\theta) \sum_{i=1}^{n} h(x_i) \right\}$$

◁

The following standard univariate probability distributions are particular cases of the (regular) one-parameter exponential family with the appropriate choices of $f, g,$ etc. as indicated.

Bernoulli

$$p(x \mid \theta) = \mathrm{Br}(x \mid \theta) = \theta^x (1-\theta)^{1-x}, \quad x \in \{0,1\}, \quad \theta \in [0,1].$$

$$f(x) = 1, \quad g(\theta) = 1-\theta, \quad h(x) = x, \quad \phi(\theta) = \log \frac{\theta}{1-\theta}, \quad c = 1.$$

Poisson

$$p(x \mid \theta) = \mathrm{Po}(x \mid \theta) = \frac{\theta^x e^{-\theta}}{x!}, \quad x \in \{0,1,2,\ldots\}, \quad \theta \in \Re^+.$$

$$f(x) = (x!)^{-1}, \quad g(\theta) = e^{-\theta}, \quad h(x) = x, \quad \phi(\theta) = \log\theta, \quad c = 1.$$

Exponential

$$p(x \mid \theta) = \mathrm{Ex}(x \mid \theta) = \theta e^{-\theta x}, \quad x \in \Re^+, \quad \theta \in \Re^+.$$

$$f(x) = 1, \quad g(\theta) = \theta, \quad h(x) = x, \quad \phi(\theta) = \theta, \quad c = -1.$$

Normal (variance unknown)

$$p(x \mid \theta) = \mathrm{N}(x \mid 0, \theta) = \big(\theta/(2\pi)\big)^{1/2} \exp\big[-\tfrac{1}{2}\theta x^2\big], \quad x \in \Re, \quad \theta \in \Re^+.$$

$$f(x) = (2\pi)^{-1/2}, \quad g(\theta) = \theta^{1/2}, \quad h(x) = x^2, \quad \phi(\theta) = \theta, \quad c = -1/2.$$

We note that the term $c\phi(\theta)$ appearing in the general $\mathrm{Ef}(\cdot \mid \cdot)$ form could always be simply written as $\phi^*(\theta)$ with ϕ^* suitably defined (see, also, Definition 4.11). However, it is often convenient to be able to separate the "interesting" function of θ, $\phi(\theta)$, from the constant which happens to multiply it.

In Definition 4.10, we allowed for the possibility (the non-regular case) that the range, X, of possible values of x might itself depend on the labelling parameter θ. Although we have not yet made a connection between this case and forms of representation arising in the modelling of exchangeable sequences, it will be useful at this stage to note examples of the well-known forms of distribution which are covered by this definition. We shall indicate later how the use of such forms in the modelling process might be given a subjectivist justification.

Uniform

$$p(x \mid \theta) = \mathrm{U}(x \mid 0, \theta) = \theta^{-1}, \quad x \in (0, \theta), \quad \theta \in \Re^+.$$

$$f(x) = 1, \quad g(\theta) = \theta^{-1}, \quad h(x) = 0, \quad \phi(\theta) = \theta, \quad c = 1.$$

Shifted exponential

$$p(x \mid \theta) = \text{Shex}(x \mid \theta) = \exp[-(x-\theta)], \quad x-\theta \in \Re^{+}, \quad \theta \in \Re^{+}.$$

$$f(x) = e^{-x}, \quad g(\theta) = e^{\theta}, \quad h(x) = 0, \quad \phi(\theta) = \theta, \quad c = 1.$$

In order to identify sequences of sufficient statistics in these and similar cases, we make use of the factorisation criterion given in Proposition 4.10.

For the uniform, we rewrite the density in the form

$$p(x \mid \theta) = \theta^{-1} I_{(0,\theta)}(x), \qquad x \in \Re,$$

so that, for any sequence $x_1, \ldots, x_n$ which is conditionally independent given θ,

$$\begin{aligned} p(x_1, \ldots, x_n \mid \theta) &= \prod_{i=1}^{n} p(x_i \mid \theta) \\ &= \theta^{-n}\, I_{(0,\theta)}\Big(\max_{i=1,\ldots,n} \{x_i\}\Big), \qquad (x_1, \ldots, x_n) \in \Re^n. \end{aligned}$$

It then follows immediately from Proposition 4.10 that

$$t_n = t_n(x_1, \ldots, x_n) = \left[n, \ \max_{i=1,\ldots,n} \{x_i\}\right], \qquad n = 1, 2, \ldots$$

is a sequence of sufficient statistics in this case.

For the shifted exponential, if we rewrite the density in the form

$$p(x \mid \theta) = \exp[-(x-\theta)]\, I_{(\theta,\infty)}(x), \qquad x \in \Re,$$

a similar argument shows that, for $(x_1, \ldots, x_n) \in \Re^{+}$,

$$p(x_1, \ldots, x_n \mid \theta) = \exp[-n\bar{x}_n] \exp[n\theta]\, I_{(\theta,\infty)}\Big(\min_{i=1,\ldots,n} \{x_i\}\Big),$$

so that, for $n = 1, 2, \ldots$,

$$t_n = t_n(x_1, \ldots, x_n) = \left[n, \ \min_{i=1,\ldots,n} \{x_i\}\right]$$

provides a sequence of sufficient statistics.

The above discussion readily generalises to the case of exchangeable sequences generated by mixing over specified parametric forms involving a k-dimensional parameter θ.

Definition 4.11. ***(k-parameter exponential family).*** *A probability density (or mass function) $p(x \mid \boldsymbol{\theta})$, $x \in X$, which is labelled by $\boldsymbol{\theta} \in \Theta \subseteq \Re^k$, is said to belong to the k-parameter exponential family if it is of the form*

$$p(x \mid \boldsymbol{\theta}) = \mathrm{Ef}_k(x \mid f, g, \boldsymbol{h}, \boldsymbol{\phi}, \boldsymbol{\theta}, \boldsymbol{c}) = f(x) g(\boldsymbol{\theta}) \exp\left\{ \sum_{i=1}^{k} c_i \phi_i(\boldsymbol{\theta}) h_i(x) \right\},$$

where $\boldsymbol{h} = (h_1, \ldots, h_k)$, $\boldsymbol{\phi}(\boldsymbol{\theta}) = (\phi_1, \ldots, \phi_k)$ and, given the functions $f, \boldsymbol{h}, \boldsymbol{\phi}$, and the constants c_i,

$$\frac{1}{g(\boldsymbol{\theta})} = \int_X f(x) \exp\left\{ \sum_{i=1}^{k} c_i \phi_i(\boldsymbol{\theta}) h_i(x) \right\} dx < \infty.$$

The family is called ***regular*** *if X does not depend on $\boldsymbol{\theta}$; otherwise it is called* ***non-regular.***

Proposition 4.13. ***(Sufficient statistics for the k-parameter exponential family).*** *If $x_1, x_2, \ldots, x_i \in X$, is an exchangeable sequence such that, given regular k-parameter $\mathrm{Ef}_k(\cdot \mid \cdot)$,*

$$p(x_1, \ldots, x_n) = \int_{\Theta} \prod_{i=1}^{n} \mathrm{Ef}_k(x_i \mid f, g, \boldsymbol{h}, \boldsymbol{\phi}, \boldsymbol{\theta}, \boldsymbol{c}) dQ(\boldsymbol{\theta}),$$

for some $dQ(\boldsymbol{\theta})$, then

$$\boldsymbol{t}_n = \boldsymbol{t}_n(x_1, \ldots, x_n) = \left[n, \sum_{i=1}^{n} h_1(x_i), \ldots, \sum_{i=1}^{n} h_k(x_i) \right], n = 1, 2, \ldots$$

is a sequence of sufficient statistics.

Proof. This is analogous to Proposition 4.12 and is a straightforward consequence of Proposition 4.10. ◁

The following standard probability distributions are particular cases (the first regular, the second non-regular) of the k-parameter exponential family with the appropriate choices of f, g etc. as indicated.

Normal (unknown mean and variance)

$$p(x \mid \boldsymbol{\theta}) = p(x \mid \mu, \tau) = \mathrm{N}(x \mid \mu, \tau)$$
$$= \left(\frac{\tau}{2\pi}\right)^{1/2} \exp\left[-\frac{\tau}{2}(x-\mu)^2\right], \quad x \in \Re, \quad \mu \in \Re, \quad \tau \in \Re^+.$$

In this case, $k = 2$ and

$$f(x) = (2\pi)^{-1/2}, \quad g(\boldsymbol{\theta}) = \tau^{1/2} \exp[-\tfrac{1}{2}\tau\mu^2], \quad \boldsymbol{h}(x) = (x, x^2),$$

$$\boldsymbol{\phi}(\boldsymbol{\theta}) = (\tau\mu, \tau), \quad c_1 = 1, \quad c_2 = -1/2,$$

so that $\boldsymbol{t}_n = \left[n, \sum_{i=1}^n x_i, \sum_{i=1}^n x_i^2\right], n = 1, 2, \ldots$ is a sequence of sufficient statistics.

Uniform (over the interval $[\theta_1, \theta_2]$*)*

$$p(x \mid \boldsymbol{\theta}) = p(x \mid \theta_1, \theta_2) = \mathrm{U}(x \mid \theta_1, \theta_2) = (\theta_2 - \theta_1)^{-1},$$
$$x \in (\theta_1, \theta_2), \quad \theta_1 \in \Re, \quad \theta_2 - \theta_1 \in \Re^+.$$

In this case,

$$f(x) = 1, \quad g(\boldsymbol{\theta}) = (\theta_2 - \theta_1)^{-1}, \quad \boldsymbol{h}(x) = 0, \quad \boldsymbol{\phi}(\boldsymbol{\theta}) = (\theta_1, \theta_2), \quad c_1 = c_2 = 0,$$

and

$$\boldsymbol{t}_n = [n, \min\{x_1, \ldots, x_n\}, \ \max\{x_1, \ldots, x_n\}], n = 1, 2, \ldots$$

is easily seen to give a sequence of sufficient statistics.

The description of the exponential family forms given in Definitions 4.10 and 4.11, is convenient for some purposes (relating straightforwardly to familiar versions of parametric families), but somewhat cumbersome for others. This motivates the following definition, which we give for the general k-parameter case.

Definition 4.12. (***Canonical exponential family***).
The probability density (or mass function)

$$p(\boldsymbol{y} \mid \boldsymbol{\psi}) = \mathrm{Cef}(\boldsymbol{y} \mid a, b, \boldsymbol{\psi}) = a(\boldsymbol{y}) \exp\{\boldsymbol{y}^t \boldsymbol{\psi} - b(\boldsymbol{\psi})\}, \qquad \boldsymbol{y} \in Y,$$

derived from $\mathrm{Ef}_k(\cdot \mid \cdot)$ *in Definition 4.11, via the transformations*

$$\boldsymbol{y} = (y_1, \ldots, y_k), \qquad \boldsymbol{\psi} = (\psi_1, \ldots, \psi_k),$$

$$y_i = h_i(x), \qquad \psi_i = c_i \phi_i(\boldsymbol{\theta}), \qquad i = 1, \ldots, k,$$

is called the ***canonical form*** *of representation of the exponential family.*

Systematic use of this canonical form to clarify the nature of the Bayesian learning process will be presented in Section 5.2.2. Here, we shall use it to examine briefly the nature and interpretation of the function $b(\psi)$, and to identify the distribution of sums of independent Cef random quantities.

Proposition 4.14. (***First two moments of the canonical exponential family***). *For y in Definition 4.12,*

$$E(\boldsymbol{y} \mid \boldsymbol{\psi}) = \nabla b(\boldsymbol{\psi}), \qquad V(\boldsymbol{y} \mid \boldsymbol{\psi}) = \nabla^2 b(\boldsymbol{\psi}).$$

Proof. It is easy to verify that the characteristic function of $\boldsymbol{y}$ conditional on $\boldsymbol{\psi}$ is given by

$$E(\exp\{i\boldsymbol{u}^t\boldsymbol{y}\} \mid \boldsymbol{\psi}) = \exp\{b(i\boldsymbol{u} + \boldsymbol{\psi}) - b(\boldsymbol{\psi})\},$$

from which the result follows straightforwardly. $\triangleleft$

Proposition 4.15. (***Sufficiency in the canonical exponential family***). *If $\boldsymbol{y}_1, \cdots, \boldsymbol{y}_n$ are independent* Cef$(\boldsymbol{y} \mid a, b, \boldsymbol{\psi})$ *random quantities, then*

$$\boldsymbol{s} = \sum_{i=1}^{n} \boldsymbol{y}_i$$

is a sufficient statistic and has a distribution Cef$(\boldsymbol{s} \mid a^{(n)}, nb, \boldsymbol{\psi})$, *where $a^{(n)}$ is the n-fold convolution of a.*

Proof. Sufficiency is immediate from Proposition 4.12. We see immediately that the characteristic function of $\boldsymbol{s}$ is $\exp\{nb(i\boldsymbol{u} + \boldsymbol{\psi}) - nb(\boldsymbol{\psi})\}$, so that the distribution of $\boldsymbol{s}$ is as claimed, where $a^{(n)}$ satisfies

$$nb(\boldsymbol{\psi}) = \log \int a^{(n)}(\boldsymbol{s}) \exp\{\boldsymbol{\psi}^t \boldsymbol{s}\} d\boldsymbol{s}.$$

Examination of the density convolution form for $n = 1$, plus induction, establishes the form of $a^{(n)}$. $\triangleleft$

Our discussion thus far has considered the situation where exchangeable belief distributions are constructed by assuming a mixing over finite-parameter exponential family forms. A consequence is that sufficient statistics of fixed dimension exist. Moreover, classical results of Darmois (1936), Koopman (1936), Pitman (1936), Hipp (1974) and Huzurbazar (1976) establish, under various regularity conditions, that the exponential family is the only family of distributions for which such sufficient statistics exist.

In the second part of this subsection, we shall consider the question of whether there are structural assumptions about an exchangeable sequence $x_1, x_2, \ldots$, which *imply* that the mixing *must* be over exponential family forms.

Previously, in Section 4.4, we considered particular *invariance* assumptions, which, together with exchangeability, identified the parametric forms that had to appear in the mixture representation. Here, we shall consider, instead, whether characterisations can be established via assumptions about *conditional distributions*, motivated by *sufficiency* ideas.

As a preliminary, suppose for a moment that an exchangeable sequence, $\{\boldsymbol{y}_i\}$, is modelled by

$$p(\boldsymbol{y}_1, \ldots, \boldsymbol{y}_n) = \int \prod_{i=1}^{n} \mathrm{Cef}(\boldsymbol{y}_i \mid a, b, \boldsymbol{\psi}) dQ(\boldsymbol{\psi}).$$

Now consider the form of $p(\boldsymbol{y}_1, \ldots, \boldsymbol{y}_k \mid \boldsymbol{y}_1 + \cdots + \boldsymbol{y}_n = \boldsymbol{s})$, $k < n$. Because of exchangeability, this has a representation as a mixture over

$$p(\boldsymbol{y}_1, \ldots, \boldsymbol{y}_k \mid \boldsymbol{y}_1 + \cdots + \boldsymbol{y}_n = \boldsymbol{s}, \boldsymbol{\psi}).$$

But the latter does not involve $\boldsymbol{\psi}$ because of the sufficiency of $\boldsymbol{y}_1 + \cdots + \boldsymbol{y}_n$ (Propositions 4.11 and 4.15), so that

$$\begin{aligned} p(\boldsymbol{y}_1, \ldots, \boldsymbol{y}_k \mid \textstyle\sum_{i=1}^n \boldsymbol{y}_i = \boldsymbol{s}) &= p(\boldsymbol{y}_1, \ldots, \boldsymbol{y}_k \mid \textstyle\sum_{i=1}^n \boldsymbol{y}_i = \boldsymbol{s}, \boldsymbol{\psi}) \\ &= \prod_{i=1}^{k} a(\boldsymbol{y}_i) \exp\{\boldsymbol{\psi}^t \boldsymbol{y}_i - b(\boldsymbol{\psi})\} \frac{a^{(n-k)}(\boldsymbol{s} - \boldsymbol{s}_k) \exp\{\boldsymbol{\psi}^t(\boldsymbol{s} - \boldsymbol{s}_k) - (n-k)b(\boldsymbol{\psi})\}}{a^{(n)}(\boldsymbol{s}) \exp\{\boldsymbol{\psi}^t \boldsymbol{s} - nb(\boldsymbol{\psi})\}} \end{aligned}$$

where, in the numerator, $\boldsymbol{s}_k = \boldsymbol{y}_1 + \cdots + \boldsymbol{y}_k \leq \boldsymbol{s}$. The exponential family mixture representation thus *implies* that,

$$p(\textstyle\sum_{i=1}^k \boldsymbol{y}_i \mid \sum_{i=1}^n \boldsymbol{y}_i = \boldsymbol{s}) = \frac{\prod_{i=1}^k a(\boldsymbol{y}_i) a^{(n-k)}(\boldsymbol{s} - \boldsymbol{s}_k)}{a^{(n)}(\boldsymbol{s})}.$$

Now suppose we consider the converse. If we assume $\boldsymbol{y}_1, \boldsymbol{y}_2, \ldots$ to be exchangeable and also assume that, for all n and $k < n$, the conditional distributions have the above form (for some a defining a $\mathrm{Cef}(\boldsymbol{y} \mid a, b, \boldsymbol{\psi})$ form), does this imply that $p(\boldsymbol{y}_1, \cdots, \boldsymbol{y}_n)$ has the corresponding exponential family mixture form? A rigorous mathematical discussion of this question is beyond the scope of this volume (see Diaconis and Freedman, 1990). However, with considerable licence in ignoring regularity conditions, the result and the "flavour" of a proof are given by the following.

Proposition 4.16. (***Representation theorem under sufficiency***).
If $y_1, y_2, \ldots$ is any exchangeable sequence such that, for all $n \geq 2$ and $k < n$,

$$p(y_1, \ldots, y_k \mid y_1 + \cdots + y_n = s) = \prod_{i=1}^{k} a(y_i) a^{(n-k)}(s - s_k)/a^{(n)}(s),$$

where $s_k = y_1 + \cdots + y_k$ and $a(\cdot)$ defines $\text{Cef}(y \mid a, b, \psi)$, *then*

$$p(y_1, \ldots, y_n) = \int \prod_{i=1}^{n} \text{Cef}(y_i \mid a, b, \psi) dQ(\psi),$$

for some $dQ(\psi)$.

Outline proof. We first note that exchangeability implies a mixture representation, mixing over distributions which make the y_i independent. But each of the latter distributions, with densities denoted generically by f, themselves imply an exchangeable sequence, so that, for $n \geq 2, k < n, f(y_1, \ldots, y_k \mid y_1 + \cdots + y_n = s)$ also has the specified form in terms of $a(\cdot)$.

Now consider $n = 2, k = 1$. Independence implies that

$$f(y_1 \mid y_1 + y_2 = s) = \frac{f(y_1) f(s - y_1)}{f^{(2)}(s)},$$

where $f(\cdot)$ denotes the marginal density and $f^{(2)}(\cdot)$ its twofold convolution, so that $f(\cdot)$ must satisfy

$$\frac{f(y_1) f(s - y_1)}{f^{(2)}(s)} = \frac{a(y_1) a(s - y_1)}{a^{(2)}(s)}.$$

If we now define

$$u(y_1) = \log \frac{f(y_1)}{a(y_1)} - \log \frac{f(0)}{a(0)}$$

and

$$v(s) = \log \frac{f^{(2)}(s)}{a^{(2)}(s)} - 2 \log \frac{f(0)}{a(0)},$$

it follows that

$$u(y_1) + u(s - y_1) = v(s).$$

Setting $y_1 = s$, and noting that $u(0) = 0$, we obtain $u(s) = v(s)$, and hence

$$u(y_1) + u(y_2) = u(y_1 + y_2).$$

This implies that $u(y) = \psi^t y$, for some ψ, so that

$$f(y) = a(y) \exp\{\psi^t y - b(\psi)\}. \qquad \triangleleft$$

The following example provides a concrete illustration of the general result.

Example 4.8. ***(Characterisation of the Poisson model).*** Suppose that the sequence of non-negative integer valued random quantities $y_1, y_2, \ldots$ is judged exchangeable, with the conditional distribution of $\boldsymbol{y} = (y_1, \ldots, y_k)$ given $y_1 + \cdots + y_n = s$, $n \geq 2, k < n$, specified to be the multinomial $\text{Mu}_k(\boldsymbol{y} \mid s, \boldsymbol{\theta})$, where $\boldsymbol{\theta} = (1/n, \cdots, 1/n)$, so that

$$p(y_1, \ldots, y_k \mid y_1 + \cdots + y_n = s) = \frac{s!}{\prod_{i=1}^{k} y_i!(s - s_k)!} \prod_{i=1}^{k} \left(\frac{1}{n}\right)^{y_i} \left(1 - \frac{k}{n}\right)^{s - s_k},$$

where $s_k = y_1 + \cdots + y_k$. Noting that the Poisson distribution, $\text{Pn}(y \mid \psi)$, can be written in $\text{Cef}(y \mid a, b, \psi)$ form as

$$\text{Pn}(y \mid \psi) = \frac{1}{y!} \exp\left\{y\psi - e^{\psi}\right\} = a(y) \exp\{y\psi - b(\psi)\},$$

from which it easily follows that $a^{(n)}(s) = n^s/s!$, it is straightforward to check that, in terms of $a(\cdot)$ and $a^{(n)}(\cdot)$,

$$M_k(\boldsymbol{y} \mid s, \boldsymbol{\theta}) = \prod_{i=1}^{k} a(y_i) a^{(n-k)}(s - s_k)/a^{(n)}(s).$$

By Proposition 4.16, it follows that the belief specification for $y_1, y_2, \ldots$ is coherent and implies that

$$p(y_1, \ldots, y_n) = \int_0^{\infty} \prod_{i=1}^{n} \text{Pn}(y_i \mid \psi) dQ(\psi),$$

for some $dQ(\psi)$, $\psi \in \Re^+$. $\triangleleft$

As we remarked earlier, the above heuristic analysis and discussion for the k-parameter regular exponential family has been given without any attempt at rigour. For the full story the reader is referred to Diaconis and Freedman (1990). Other relevant references for the mathematics of exponential families include Barndorff-Nielsen (1978), Morris (1982) and Brown (1985).

We conclude this subsection by considering, briefly and informally, what can be said about characterisations of exchangeable sequences as mixtures of non-regular exponential families. For concreteness, we shall focus on the uniform, $U(x \mid 0, \theta)$, distribution, which has density $\theta^{-1} I_{(0,\theta)}(x)$, $x \in \Re$, and sufficient statistic $\max\{x_1, \ldots, x_n\}$, given a sample $x_1, \ldots, x_n$. This sufficient statistic is clearly not a summation, as is the case for regular families (and plays a key role in Proposition 4.16). However, conditional on $m_n = \max\{x_1, \ldots, x_n\}, x_1, \ldots, x_k, k << n$, are approximately independent $U(x_i \mid 0, m_n)$ and this will therefore be true for all exchangeable $x_1, x_2, \ldots$ constructed by mixing over independent $U(x_i \mid 0, \theta)$. Conversely, we might wonder whether positive exchangeable sequences having this conditional property are necessarily mixtures of independent $U(x_i \mid 0, \theta)$. Intuitively, if m_n tends to a finite θ from below, as $n \to \infty$, one might expect the result to be true. This is indeed the case, but a general account of the required mathematical results is beyond our intended scope in this volume. The interested reader is referred to Diaconis and Freedman (1984), and the further references discussed in Section 4.8.1.

4.5.4 Information Measures and the Exponential Family

Our approach to the exponential family has been through the concept of predictive or, equivalently, parametric sufficient statistics. It is interesting to note, however, that exponential family distributions can also be motivated through the concept of the utility of a distribution (c.f. Section 3.4), using the derived notions of approximation and discrepancy.

Consider the following problem. We seek to obtain a mathematical representation of a probability density $p(x)$, which satisfies the k (independent) constraints

$$\int_X h_i(x)p(x)dx = m_i < \infty, \quad i = 1, \ldots, k,$$

where $m_1, \ldots, m_k$ are specified constants, together with the normalizing constraint $\int_X p(x)dx = 1$, and, in addition, is to be approximated as closely as possible by a specified density $f(x)$.

We recall from Definition 3.20 (with a convenient change of notation) that the *discrepancy* from a probability density $p(x)$ *assumed to be true* of an *approximation* $f(x)$ is given by

$$\delta(f \mid p) = \int_X p(x) \log \frac{p(x)}{f(x)} dx,$$

where f and p are both assumed to be strictly positive densities over the same range, X, of possible values. Note that we are interested in deriving a mathematical representation of the *true* probability density $p(x)$, not of the (specified) approximation $f(x)$. Thus, we minimise $\delta(f \mid p)$ over p subject to the required constraints on p, rather than $\delta(f \mid p)$ over f subject to constraints on f. Hence, we seek p to minimise

$$\begin{aligned} F(p) = & \int_X p(x) \log \frac{p(x)}{f(x)} dx \\ & + \sum_{i=1}^{k} \theta_i \left[\int_X h_i(x)p(x)dx - m_i \right] + c \left[\int_X p(x)dx - 1 \right], \end{aligned}$$

where $\theta_1, \ldots, \theta_k$ and c are arbitrary constant multipliers.

Proposition 4.17. (***The exponential family as an approximation***).
The functional $F(p)$ defined above is minimised by

$$p(x) = \mathrm{Ef}_k(x \mid f, g, \boldsymbol{h}, \boldsymbol{\phi}, \boldsymbol{\theta}, \mathbf{c}), x \in X$$

where f and h are given in $F(p)$, $c_i = 1, \boldsymbol{\phi} = \boldsymbol{\theta} = (\theta_1, \ldots, \theta_k)$ and

$$\frac{1}{g(\boldsymbol{\theta})} = \int_X f(x) \exp\left\{\sum_{i=1}^{k} \theta_i h_i(x)\right\} dx.$$

Proof. By a standard variational argument (see, for example, Jeffreys and Jeffreys, 1946, Chapter 10), a necessary condition for p to give a stationary value of $F(p)$ is that

$$(\partial/\partial\alpha)F(p(x) + \alpha\tau(x)) \mid_{\alpha=0} = 0$$

for any function $\tau : x \to \Re$ of sufficiently small norm. This condition reduces to the equation

$$\int \left[\log(p(x)/f(x)) + \sum_{i=1}^{k} \theta_i h_i(x) + (c+1)\right] \tau(x) dx = 0,$$

from which it follows that

$$p(x) \propto f(x) \exp\left\{\sum_{i=1}^{k} \theta_i h_i(x)\right\},$$

as required. (For an alternative proof, see Kullback, 1959/1968, Chapter 3.) $\triangleleft$

The resulting exponential family form for $p(x)$ was derived on the basis of a given approximation $f(x)$ and a collection of "constant" functions $\boldsymbol{h}(x) = [h_i(x), \ldots, h_k(x)]$. If we wish to emphasise this derivation of the family, we shall refer to $\mathrm{Ef}(x \mid f, g, \boldsymbol{h}, \phi, \theta, \mathbf{c})$ as *the exponential family generated by f and $\boldsymbol{h}$.*

In general, specification of the sufficient statistic

$$\boldsymbol{t}_m = \left[m, \sum_{i=1}^{m} h_i(x_i), \ldots, \sum_{i=1}^{m} h_k(x_i)\right]$$

does not uniquely identify the form of $f(x)$ within the exponential family framework. Consider, for example, the $\mathrm{Ga}(x \mid \alpha, \theta)$ family with α known. Each distinct α defines a distinct exponential family with density

$$(x^{\alpha-1}/\Gamma(\alpha))\theta^{\alpha} \exp\{-\theta x\},$$

so that, in addition to $h(x) = x$, we need to specify $f(x) = x^{\alpha-1}/\Gamma(\alpha)$ in order to identify the family.

Returning to the general problem of choosing p to be "as close as possible" to an "approximation" f, subject to the k constraints defined by $\boldsymbol{h}(x)$, it is interesting to ask what happens if the approximation f is very "vague", in the sense that f is extremely diffusely spread over X. A limiting form of this would be to consider $f(x) =$ constant, which leads us to seek the p minimising $\int_X p(x) \log p(x) dx$ subject to the given constraints. The solution is then

$$p(x) = \frac{\exp\left\{\sum_{i=1}^{k} \theta_i h_i(x)\right\}}{\int_X \exp\left\{\sum_{i=1}^{k} \theta_i h_i(x)\right\} dx}$$

which, since minimising $\int_X p(x) \log p(x)\, dx$ is equivalent to maximising $H(p) = -\int_X p(x) \log p(x)\, dx$, is the so-called *maximum entropy* choice of p.

Thus, for example, if $X = \Re^+$ and $h(x) = x$, the maximum entropy choice for $p(x)$ is $\text{Ex}(x \mid \phi)$, the exponential distribution with $\phi^{-1} = \text{E}(x \mid \phi)$. If $X = \Re$ and $\boldsymbol{h}(x) = (x, x^2)$, the maximum entropy choice for $p(x)$ turns out to be $\text{N}(x \mid \mu, \lambda)$, the normal distribution with $\mu = E(x \mid \mu, \lambda), \lambda^{-1} = V(x \mid \mu, \lambda)$ (c.f. Example 3.4, following Definition 3.20).

Our discussion of modelling has so far concentrated on the case of beliefs about a single sequence of observations $x_1, x_2, \ldots$, judged to have various kinds of invariance or sufficiency properties. In the next section, we shall extend our discussion in order to relate these ideas to the more complex situations, which arise when several such sequences of observations are involved, or when there are several possible ways of making exchangeable or related judgements about sequences.

4.6 MODELS VIA PARTIAL EXCHANGEABILITY

4.6.1 Models for Extended Data Structures

In Section 4.5, we discussed various kinds of justification for modelling a sequence of random quantities $x_1, x_2, \ldots$ as a random sample from a parametric family with density $p(x \mid \theta)$, together with a prior distribution $dQ(\theta)$ for θ. We also briefly mentioned further possible kinds of judgements, involving assumptions about conditional moments or information considerations, which further help to pinpoint the appropriate specification of a parametric family.

However, in order to concentrate on the basic conceptual issues, we have thus far restricted attention to the case of a *single sequence* of random quantities, $x_1, x_2, \ldots$, labelled by a *single index*, $i = 1, 2, \ldots$, and *unrelated to other random quantities*. Clearly, in many (if not most) areas of application of statistical modelling the situation will be more complicated than this, and we shall need to extend and adapt the basic form of representation to deal with the perceived complexities of the situation. Among the typical (but by no means exhaustive) kinds of situation we shall wish to consider are the following.

(i) Sequences $x_{i1}, x_{i2}, \ldots$ of random quantities are to be observed in each of $i \in I$ contexts. For example: we may have sequences of clinical responses to each of I different drugs; or responses to the same drug used on I different subgroups of a population. A modelling framework is required which enables us to learn, in some sense, about differences between some aspect of the responses in the different sequences.

(ii) In each of $i \in I$ contexts, $j \in J$ different treatments are each replicated $k \in K$ times, and the random quantities x_{ijk} denote observable responses for each context/treatment/replicate combination. For example: we may have I different irrigation systems for fruit trees, J different tree pruning regimes and K trees exposed to each irrigation/pruning combination, with x_{ijk} denoting the total yield of fruit in a given year; or we may have I different geographical areas, J different age-groups and K individuals in each of the IJ combinations, with x_{ijk} denoting the presence or absence of a specific type of disease, or a coding of voting intention, or whatever. A modelling framework is required which enables us to investigate differences between either contexts, or treatments, or context/treatment combinations.

(iii) Sequences of random quantities $x_{i1}, x_{i2}, \ldots, i \in I$, are to be observed, where some form of qualitative assumption has been made about a form of relationship between the x_{ij} and other specified (controlled or observed) quantities $\boldsymbol{z}_i = (z_{i1}, \ldots, z_{ik}), k \geq 1$. For example: x_{ij} might denote the status (dead or alive) of the jth rat exposed to a toxic substance administered at dose level z_i, with an assumed form of relationship between z_i and the corresponding "death rate"; or x_{ij} might denote the height or weight at time z_i from the jth replicate measurement of a plant or animal following some assumed form of "growth curve"; or x_{ij} might denote the output yield on the jth run of a chemical process when k inputs are set at the levels $\boldsymbol{z}_i = (z_{1i}, \ldots, z_{ki})$ and the general form of relationship between process output and inputs is either assumed known or well-approximated by a specified mathematical form. In each case a modelling framework is required which enables us to learn about the quantitative form of the relationship, and to quantify beliefs (predictions) about the observable x^* corresponding to a specified input or control quantity $\boldsymbol{z}^*$.

(iv) Exchangeable sequences, $x_{i1}, x_{i2}, \ldots$, of random quantities are to be observed in each of $i \in I$ contexts, where I is itself a selection from a potentially larger index set I^*. Suppose that for each sequence,

$$t_m^{(i)} = \left[m, \; \sum_{j=1}^{m} s^{(i)}(x_j) \right], \quad i \in I,$$

is judged to be a sufficient statistic, that the strong law limits

$$\theta_i = \lim_{i \to \infty} \frac{1}{m} t_m^{(i)}, \quad i \in I,$$

exist and that the sequence $\theta_1, \theta_2, \ldots$ is itself judged exchangeable. For example: sequence i may consist of 0 – 1 (success-failure) outcomes on repeated trials with the ith of I similar electronic components; or sequence i may consist of quality measurements of known precision on replicate samples of the ith of I chemically similar dyestuffs. In the first case, the sequence of long-run frequencies of failures for each of the components might, a priori, be judged to be exchangeable; in the second case, the sequence of large-sample averages of quality for each of the dyestuffs might, a priori, be judged to be exchangeable. A modelling framework is required which enables us to exploit such further judgements of exchangeability in order to be able to use information from *all* the sequences to strengthen, in some sense, the learning process within an *individual* sequence.

4.6.2 Several Samples

We shall begin our discussion of possible forms of partial exchangeability judgements for several sequences of observables, $x_{i1}, x_{i2}, \ldots, i = 1, \ldots, m$, by considering the simple case of 0 – 1 random quantities.

In many situations, including that of a comparative clinical trial, joint beliefs about several sequences of 0 – 1 observables would typically have the property encapsulated in the following definition, where, here and throughout this section, $\boldsymbol{x}_i(n_i)$ denotes the vector of random quantities $(x_{i1}, \ldots, x_{in_i})$.

Definition 4.13. (***Unrestricted exchangeability for 0 – 1 sequences***).
Sequences of 0 – 1 random quantities, $x_{i1}, x_{i2}, \ldots, i = 1, \ldots, m$, are said to be unrestrictedly exchangeable if each sequence is infinitely exchangeable and, in addition, for all $n_i \leq N_i, i = 1, \ldots, m$,

$$p(\boldsymbol{x}_1(n_1), \ldots, \boldsymbol{x}_m(n_m) \,|\, y_1(N_1), \ldots, y_m(N_m)) = \prod_{i=1}^{m} p(\boldsymbol{x}_i(n_i) \,|\, y_i(N_i)),$$

where $y_i(N_i) = x_{i1} + \cdots + x_{iN_i}, i = 1, \ldots, m$.

In addition to the exchangeability of the individual sequences, this definition encapsulates the judgement that, given the total number of successes in the first N_i observations from the ith sequence, $i = 1, \ldots m$, *only the total for the ith sequence is relevant* when it comes to beliefs about the outcomes of any subset of n_i of the N_i observations from that sequence. Thus, for example, given 15 deaths in the first 100 patients receiving Drug 1 ($N_1 = 100$, $y_1(N_1) = 15$) and 20 deaths in the first 80 patients receiving Drug 2 ($N_2 = 80$, $y_2(N_2) = 20$), we would typically judge the latter information to be irrelevant to any assessment of the probability that the first three patients receiving Drug 1 survived and the fourth one died ($x_{11} = 0, x_{12} = 0, x_{13} = 0, x_{14} = 1$). Of course, the information might well be judged relevant if we were *not* informed of the value of $y_1(N_1)$. The definition thus encapsulates a kind of "conditional irrelevance" judgement.

As an example of a situation where this condition does *not* apply, suppose that $x_{11}, x_{12}, \ldots$ is an infinitely exchangeable 0–1 sequence and that another sequence $x_{21}, x_{22}, \ldots$ is defined by $x_{2j} = x_{1j}$ (or by $x_{2j} = 1 - x_{1j}$). Then $x_{21}, x_{22}, \ldots$ is certainly an exchangeable sequence (since $x_{11}, x_{12}, \ldots$ is), but, taking $x_{2j} = x_{1j}$ and $n_1 = n_2 = N_1 = N_2 = 2$,

$$p(x_{11} = 0, x_{12} = 1, x_{21} = 1, x_{22} = 0 \mid y_{12} = 1, y_{22} = 1) = 0,$$

whereas

$$p(x_{11} = 0, x_{12} = 1 \mid y_{12} = 1)\, p(x_{21} = 1, x_{22} = 0 \mid y_{22} = 1) = 1/2 \times 1/2 = 1/4.$$

Further insight is obtained by noting (from Definition 4.13) that unrestricted exchangeability implies that

$$\begin{aligned} &p(x_{11}, \ldots, x_{1n_1}, \ldots, x_{m1}, \ldots, x_{mn_m}) \\ &\qquad = p(x_{1\pi_1(1)}, \ldots, x_{1\pi_1(n_1)}, \ldots, x_{m\pi_m(1)}, \ldots, x_{m\pi_m(n_m)}) \end{aligned}$$

for any *unrestricted* choice of permutations π_i of $\{1, \ldots, n_i\}, i = 1, \ldots m$, whereas, in the case of the above counter-example, we only have invariance of the joint distribution when $\pi_1 = \pi_2$. For a development starting from this latter condition see de Finetti (1938).

We can now establish the following generalisation of Proposition 4.1.

Proposition 4.18. (***Representation theorem for several sequences of 0–1 random quantities***). *If $x_{i1}, x_{i2}, \ldots, i = 1, \ldots, m$ are unrestrictedly infinitely exchangeable sequences of 0–1 random quantities with joint probability measure P, there exists a distribution function Q such that*

$$p(\boldsymbol{x}_1(n_1), \ldots, \boldsymbol{x}_m(n_m)) = \int_{[0,1]^m} \prod_{i=1}^{m} \prod_{j=1}^{n_i} \theta_i^{x_{ij}} (1 - \theta_i)^{1 - x_{ij}} \, dQ(\boldsymbol{\theta})$$

where, with $y_i(n_i) = x_{i1} + \cdots + x_{in_i}, i = 1, \ldots, m$,

$$Q(\boldsymbol{\theta}) = \lim_{all\ n_i \to \infty} P\left[\left(\frac{y_1(n_1)}{n_1} \le \theta_1 \right) \cap \cdots \cap \left(\frac{y_m(n_m)}{n_m} \le \theta_m \right) \right].$$

Corollary. *Under the conditions of Proposition 4.18,*

$$p(y_1(n_1), \ldots, y_m(n_m)) = \int_{[0,1]^m} \prod_{i=1}^{m} \binom{n_i}{y_i(n_i)} \theta_i^{y_i(n_i)} (1-\theta_i)^{n_i - y_i(n_i)} dQ(\theta_1, \ldots, \theta_m)$$

Proof. We first note that

$$p(y_1(n_1), \ldots, y_m(n_m)) = \binom{n_1}{y_1(n_1)} \cdots \binom{n_m}{y_m(n_m)} p(\boldsymbol{x}_1(n_1), \ldots, \boldsymbol{x}_m(n_m))$$

so that, to prove the proposition, it suffices to establish the corollary. Moreover, for any $N_i \geq n_i, i = 1, \ldots, m$, we may express $p(y_1(n_1), \ldots, y_m(n_m))$ as

$$\sum p(y_1(n_1), \ldots, y_m(n_m) \mid y_1(N_1), \ldots, y_m(N_m))\; p(y_1(N_1), \ldots, y_m(N_m)),$$

where the ith of the m summations ranges from $y_i(N_i) = y_i(n_i)$ to $y_i(N_i) = N_i$, and where, by Definition 4.4 and a straightforward generalisation of the argument given in Proposition 4.1,

$$p(y_1(n_1), \ldots, y_m(n_m) \mid y_1(N_1), \ldots, y_m(N_m)) = \prod_{i=1}^{m} p(y_i(n_i) \mid y_i(N_i)) = \prod_{i=1}^{m} \binom{n_i}{y_i(n_i)} \binom{N_i - n_i}{n_i - y_i(n_i)} \Big/ \binom{N_i}{n_i}.$$

Writing $(y_N)_{y_n} = y_N(y_N - 1) \cdots (y_N - (y_n - 1))$, etc., and defining the function $Q_{N_1, \ldots, N_m}(\theta_1, \ldots, \theta_m)$ on $\Re^m$ to be the m-dimensional "step" function with "jumps" of $p(y_1(N_1), \ldots, y_m(N_m))$ at

$$(\theta_1, \ldots, \theta_m) = \left(\frac{y_1(N_1)}{N_1}, \ldots, \frac{y_m(N_m)}{N_m} \right),$$

where $y_i(N_i) = 0, \ldots, N_i, i = 1, \ldots, m$. We see that $p(y_1(n_1), \ldots, y_m(n_m))$ is equal to

$$\int_{[0,1]^m} \prod_{i=1}^{m} \left\{ \binom{n_i}{y_i(n_i)} \frac{(\theta_i N_i)_{y_i(n_i)} [(1-\theta_i)N_i]_{n_i - y_i(n_i)}}{(N_i)_{n_i}} \right\} dQ_{N_1, \ldots, N_m}(\boldsymbol{\theta}).$$

As $N_1, \ldots, N_m \to \infty$,

$$\prod_{i=1}^{m} \left\{ \frac{(\theta_i N_i)_{y_i(n_i)} [(1-\theta_i)N_i]_{n_i - y_i(n_i)}}{(N_i)_{n_i}} \right\} \to \prod_{i=1}^{m} \theta_i^{y_i(n_i)} (1-\theta_i)^{n_i - y_i(n_i)},$$

uniformly in $\theta_1, \ldots, \theta_m$, and, by the multidimensional version of Helly's theorem (see Section 3.2.3), there exists a subsequence $Q_{N_1(j), \ldots, N_m(j)}, j = 1, 2, \ldots$ having a limit Q, which is a distribution function on $\Re^m$. The result follows. ◁

Considering, for simplicity, $m = 2$, Proposition 4.18 (or its corollary) asserts that if we judge two sequences of 0 – 1 random quantities to be unrestrictedly exchangeable, we can proceed *as if*:

(i) the x_{ij} are judged to be independent Bernoulli random quantities (or the $y_i(n_i)$ to be independent binomial random quantities) conditional on random quantities $\theta_i, i = 1, 2$;

(ii) (θ_1, θ_2) are assigned a joint probability distribution Q;

(iii) by the strong law of large numbers, $\theta_i = \lim_{n_i \to \infty}(y_i(n_i)/n_i)$, so that Q may be interpreted as "joint beliefs about the limiting relative frequencies of 1's in the two sequences".

The model is completed by the specification of $dQ(\theta_1, \theta_2)$, whose detailed form will, of course, depend on the particular beliefs appropriate to the actual practical application of the model. At a qualitative level, we note the following possibilities:

(a) knowledge of the limiting relative frequency for one of the sequences would not change beliefs about outcomes in the other sequence, so that we have the independent form of prior specification, $dQ(\theta_1, \theta_2) = dQ(\theta_1)dQ(\theta_2)$;

(b) the limiting relative frequency for the second sequence will necessarily be greater than that for the first sequence (due, for example, to a known improvement in a drug or an electronic component under test), so that $dQ(\theta_1, \theta_2)$ is zero outside the range $0 \leq \theta_1 < \theta_2 \leq 1$;

(c) there is a real possibility, to which an individual assigns probability π, say, that, in fact, the limiting frequencies could turn out to be equal, so that, writing $\theta = \theta_1 = \theta_2$, in this case $dQ(\theta_1, \theta_2)$ has the form

$$\pi dQ^*(\theta) + (1 - \pi)dQ^+(\theta_1, \theta_2)$$

and the representation, for (y_{1n1}, y_{2n2}), say, has the form

$$p(y_1(n_1), y_2(n_2)) = \pi \int_0^1 \text{Bi}(y_1(n_1) \,|\, n_1, \theta) \, \text{Bi}(y_2(n_2) \,|\, n_2, \theta) \, dQ^*(\theta)$$
$$+ (1 - \pi) \int_0^1 \int_0^1 \text{Bi}(y_1(n_1) \,|\, n_1, \theta_1) \, \text{Bi}(y_2(n_2) \,|\, n_2, \theta_2) \, dQ^+(\theta_1, \theta_2),$$

where $dQ^+(\theta_1, \theta_2)$ assigns probability over the range of values of (θ_1, θ_2) such that $\theta_1 \neq \theta_2$.

As we shall see later, in Chapter 5, the general form of representation of beliefs for observables defined in terms of the two sequences, together with detailed specifications of $dQ(\theta_1, \theta_2)$, enables us to explore coherently any desired aspect of the learning process. For example, we may have observed that out of the first n_1, n_2

patients receiving drug treatments 1, 2, respectively, $y_1(n_1)$ and $y_2(n_2)$ survived, and, on the basis of this information, wish to make judgements about the relative performance of the drugs were they to be used on a large future sequence of patients. This might be done by calculating, for example,

$$p(\lim_{N\to\infty}(y_1(N)/N) - \lim_{N\to\infty}(y_2(N)/N) \,|\, y_1(n_1), y_2(n_2)),$$

which, in the language of the conventional paradigm, is the "posterior density for $\theta_1 - \theta_2$, given $y_1(n_1)$, $y_2(n_2)$".

Clearly, the discussion and resulting forms of representation which we have given for the case of unrestrictedly exchangeable sequences of 0–1 random quantities can be extended to more general cases. One possible generalisation of Definition 4.13 is the following.

Definition 4.14. (***Unrestricted exchangeability for sequences with predictive sufficient statistics***). *Sequences of random quantities $x_{i1}, x_{i2}, \ldots$ taking values in $X_i, i = 1, \ldots, m$, are said to be unrestrictedly infinitely exchangeable if each sequence is infinitely exchangeable and, in addition, for all $n_i \leq N_i$, $i = 1, \ldots, m$,*

$$p(\boldsymbol{x}_1(n_1), \ldots, \boldsymbol{x}_m(n_m) \,|\, \boldsymbol{t}_{N_1}, \ldots, \boldsymbol{t}_{N_m}) = \prod_{i=1}^{m} p(\boldsymbol{x}_i(n_i) \,|\, \boldsymbol{t}_{N_i})$$

where $\boldsymbol{t}_{N_i} = \boldsymbol{t}_{N_i}(\boldsymbol{x}_i(N_i))$, $i = 1, \ldots, m$, are separately predictive sufficient statistics for the individual sequences.

In general, given m unrestrictedly exchangeable sequences of random quantities, $x_{i1}, x_{i2}, \ldots$, with x_{ij} taking values in X_i, we typically arrive at a representation of the form

$$p(\boldsymbol{x}_1(n_1), \ldots, \boldsymbol{x}_m(n_m)) = \int_{\Theta^*} \prod_{i=1}^{m} \prod_{j=1}^{n_i} p_i(x_{ij} \,|\, \theta_i)\, dQ(\theta_1, \ldots, \theta_m),$$

where $\Theta^* = \prod_{i=1}^{m} \Theta_i$ and the parametric families

$$p_i(x \,|\, \theta_i), \quad x \in X_i, \quad \theta_i \in \Theta_i, \quad i = 1, \ldots, m,$$

have been identified through consideration of sufficient statistics of fixed dimension, or whatever, as discussed in previous sections. Most often, the fact that the k sequences are being considered together will mean that the random quantities $x_{i1}, x_{i2}, \ldots$ relate to the same form of measurement or counting procedure for all $i = 1, \ldots, m$, so that typically we will have $p_i(x \,|\, \theta_i) = p(x \,|\, \theta_i)$, $i = 1, \ldots, m$, where the parameters correspond to strong law limits of functions of the sufficient statistics. The following forms are frequently assumed in applications.

Example 4.9. ***(Binomial).*** If $y_i(n_i)$ denotes the number of 1's in the first n_i outcomes of the ith of m unrestrictedly exchangeable sequences of 0 – 1 random quantities, then

$$p(y_1(n_1), \ldots, y_m(n_m)) = \int_{[0,1]^m} \prod_{i=1}^{m} \mathrm{Bi}(y_i(n_i) \mid \theta_i, n_i)\, dQ(\theta_1, \ldots, \theta_m).$$

Example 4.10. ***(Multinomial).*** If $\boldsymbol{y}_i(n_i)$ denotes the category membership count (into the first k of $k+1$ exclusive categories) from the first n_i outcomes of the ith of m unrestrictedly exchangeable sequences of "0 – 1 random vectors" (see Section 4.3), then

$$p(\boldsymbol{y}_1(n_1), \ldots, \boldsymbol{y}_m(n_m)) = \int_{\Theta^m} \prod_{i=1}^{m} \mathrm{Mu}_k(\boldsymbol{y}_i(n_i) \mid \theta_i, n_i)\, dQ(\theta_1, \ldots, \theta_m),$$

where $\theta_i = \lim_{n\to\infty}(\boldsymbol{y}_i(n)/n)$ and $\Theta = \{\theta = (\theta_1, \ldots, \theta_k)$ such that $0 \le \theta_i \le 1, 1 \le i \le k$, and $\theta_1 + \cdots + \theta_k \le 1\}$. This model describes beliefs about an $m \times (k+1)$ *contingency table* of count data, with row totals $n_1, \ldots, n_m$. It generalises the case of the $m \times 2$ contingency table described in Example 4.9. ◁

Example 4.11. ***(Normal).*** If $x_{ij}, j = 1, \ldots, n_i, i = 1, \ldots, m$, denote real-valued observations from m unrestrictedly exchangeable sequences of real-valued random quantities, the assumed sufficiency of the sample sum and sum of squares within each sequence might lead to the representation

$$p(\boldsymbol{x}_1(n_1), \ldots, \boldsymbol{x}_m(n_m)) = \int_{\Theta^m} \prod_{i=1}^{m} \prod_{j=1}^{n_i} \mathrm{N}(x_{ij} \mid \mu_i, \lambda_i)\, dQ(\theta),$$

where, with $\bar{x}_n(i) = n^{-1}(x_{i1} + \cdots + x_{in})$ and $s_n^2(i) = n^{-1}\sum_{j=1}^{n}(x_{ij} - \bar{x}_n(i))^2$, we have $\mu_i = \lim_{n\to\infty} \bar{x}_n(i)$, $\lambda_i^{-1} = \lim_{n\to\infty} s_n^2(i)$, $\theta = (\mu_1, \ldots, \mu_m, \lambda_1, \ldots, \lambda_m)$ and $\Theta = \Re^m \times (\Re^+)^m$.

In many applications, the further judgement is made that $\lambda_1 = \cdots = \lambda_m = \lambda$, say, so that the representation then takes the form

$$p(\boldsymbol{x}_1(n_1), \ldots, \boldsymbol{x}_m(n_m)) = \int_{\Re^m \times \Re^+} \prod_{i=1}^{m} \prod_{j=1}^{n_i} \mathrm{N}(x_{ij} \mid \mu_i, \lambda)\, dQ(\mu_1, \ldots, \mu_m, \lambda).$$

This is the model most often used to describe beliefs about a *one-way layout* of measurement data. ◁

As in the case of 0 – 1 random quantities with $m = 2$, discussed earlier in this section, we could make analogous remarks concerning the various qualitative forms of specification of the prior distribution Q that might be made in these cases. We shall not pursue this further here, but will comment further in Section 4.7.5.

4.6.3 Structured Layouts

Let us now consider the situation described in (ii) of Section 4.6.1, where the random quantity x_{ijk} is triple-subscripted to indicate that it is the kth of K "replicates" of an observable in "context" $i \in I$, subject to "treatment" $j \in J$. In general terms, we have a *two-way layout*, having I rows and J columns, with K replicates in each of the IJ cells.

In such contexts, most individuals would find it unacceptable to make a judgement of complete exchangeability for the random quantities x_{ijk}. For example, if rows represent age-groups, columns correspond to different drug treatments, replicates refer to sequences of patients within each age-group/treatment combination and the x_{ijk} measure death-rates, say, it is typically not the case that beliefs about the x_{ijk} would be invariant under permutations of the subscript i. On the other hand, for the kinds of mechanisms routinely used to allocate patients to treatment groups in clinical trials, many individuals would have exchangeable beliefs about the sequence $x_{ij1}, x_{ij2}, \ldots$ for any fixed i, j.

Technically, such a situation corresponds to the invariance of joint beliefs for the collection of random quantities, x_{ijk}, under some restricted set of permutations of the subscripts, rather than under the unrestricted set of all possible permutations (which would correspond to complete exchangeability). The precise nature of the appropriate set of invariances encapsulating beliefs in a particular application will, of course, depend on the actual perceived partial exchangeabilities in that application. In what follows, we shall simply motivate, using very minimal exchangeability assumptions, a model which is widely used in the context of the two-way layout. There is no suggestion that the particular form discussed has any special status, or *ought* to be routinely adopted, or whatever.

Suppose that, for any fixed i, j, we think of $x_{ij1}, x_{ij2}, \ldots$ as a (potentially) infinite sequence of real-valued random quantities ($x \in \Re$), such that the IJ sequences of this kind, with I and J fixed, are judged to be unrestrictedly exchangeable. If further assumptions of centred spherical symmetry or sufficiency for each sequence then lead to the normal form of representation, we have

$$p(x_{11}(n_{11}), \ldots, x_{IJ}(n_{IJ})) = \int_{\Theta^{IJ}} \prod_{i=1}^{I} \prod_{j=1}^{J} \prod_{k=1}^{n_{ij}} \mathrm{N}(x_{ijk} \mid \mu_{ij}, \lambda_{ij})\, dQ(\theta),$$

where $\theta = (\mu_{11}, \ldots, \mu_{IJ}, \lambda_{11}, \ldots, \lambda_{IJ})$ and $\Theta = \Re^{IJ} \times (\Re^{+})^{IJ}$, so that conditional, for each (i, j), on the strong law limits

$$\mu_{ij} = \lim_{K \to \infty} K^{-1}(x_{ij1} + \cdots + x_{ijK}) = \lim_{K \to \infty} K^{-1} \bar{x}_{ij}(K),$$

$$(\lambda_{ij})^{-1} = \lim_{K \to \infty} K^{-1} \sum_{k=1}^{K} (x_{ijk} - \bar{x}_{ij}(K))^2 = \lim_{K \to \infty} s_{ij}^2(K),$$

the x_{ijk} are assumed independently and normally distributed with means μ_{ij} and variances $(\lambda_{ij})^{-1}$.

In many cases, the nature of the observational process leads to the judgement that $\lim_{K\to\infty} s_{ij}^2(K)$ may be assumed to be the same for all (i,j), so that $\lambda_{ij} = \lambda$, say, for all i,j. Letting

$$\mu_{i\bullet} = \lim_{K\to\infty} K^{-1}J^{-1}\sum_{i=1}^{J} \bar{x}_{ij}(K) = J^{-1}\sum_{j=1}^{J}\mu_{ij}$$

$$\mu_{\bullet j} = \lim_{K\to\infty} K^{-1}I^{-1}\sum_{i=1}^{I} \bar{x}_{ij}(K) = I^{-1}\sum_{i=1}^{I}\mu_{ij}$$

$$\mu_{\bullet\bullet} = \lim_{K\to\infty} K^{-1}I^{-1}J^{-1}\sum_{i=1}^{I}\sum_{j=1}^{J} \bar{x}_{ij}(K) = I^{-1}\sum_{i=1}^{I}\mu_{i\bullet} = J^{-1}\sum_{j=1}^{J}\mu_{\bullet j}$$

denote the strong law limits of the row averages, column averages and overall average, respectively, from the two-way layout with I and J fixed, we can always write

$$\mu_{ij} = \mu + \alpha_i + \beta_j + \gamma_{ij},$$

where

$$\alpha_i = (\mu_{i\bullet} - \mu), \quad \beta_j = (\mu_{\bullet j} - \mu), \quad \gamma_{ij} = (\mu_{ij} - \mu_{i\bullet} - \mu_{\bullet j}),$$

so that the random quantities x_{ijk} are conditionally independently distributed with

$$p(x_{ijk} \mid \mu, \alpha_i, \beta_j, \gamma_{ij}, \lambda) = \mathrm{N}(x_{ijk} \mid \mu + \alpha_i + \beta_j + \gamma_{ij}, \lambda).$$

The full model representation is then completed by the specification of a prior distribution Q for λ and any IJ linearly independent combinations of the μ_{ij}. In conventional terminology, μ is referred to as the *overall mean*, α_i as the ith *row effect*, β_j as the jth *column effect* and γ_{ij} as the (ij)th *interaction effect*. Collectively, the $\{\alpha_i\}$ and $\{\beta_j\}$ are referred to as the *main effects* and $\{\gamma_{ij}\}$ as the *interactions*. Interest in applications often centres on whether or not interactions or main effects are close to zero and, if not, on making inferences about the magnitudes of differences between different row or column effects.

In the above discussion, our exchangeability assumptions were restricted to the sequence $x_{ij1}, x_{ij2}, \ldots$ for fixed i,j. It is possible, of course, that further forms of symmetric beliefs might be judged reasonable for certain permutations of the i,j subscripts. We shall return to this possibility in Section 4.6.5, where we shall see that certain further assumptions of invariance lead naturally to the idea of hierarchical representations of beliefs.

4.6.4 Covariates

In (iii) of Section 4.6, we gave examples of situations where beliefs about sequences of observables $x_{i1}, x_{i2}, \ldots, i = 1, \ldots, m$ are functionally dependent, in some sense, on the observed values, $\boldsymbol{z}_i, i = 1, \ldots, m$, of a related sequence of (random) quantities. We shall refer to the latter as *covariates* and, in recognition of this dependency, we shall denote the joint density of $x_{ij}, j = 1, \ldots, n_i, i = 1, \ldots, m$, by

$$p(\boldsymbol{x}_1(n_1), \ldots, \boldsymbol{x}_m(n_m) \mid \boldsymbol{z}_1, \ldots, \boldsymbol{z}_m).$$

The examples which follow illustrate some of the typical forms assumed in applications. Again, there is no suggestion that these particular forms have any special status; they simply illustrate some of the kinds of models which are commonly used.

Example 4.12. ***(Bioassay).*** Suppose that at each of m specified dose levels, $z_1, \ldots, z_m$, of a toxic substance, typically measured on a logarithmic scale, sequences of 0 – 1 random quantities, $x_{i1}, x_{i2}, \ldots, i = 1, \ldots, m$, are to be observed, where $x_{ij} = 1$ if the jth animal receiving dose z_i survives, $x_{ij} = 0$ otherwise. If, for each $i = 1, \ldots, m$, the sequences $x_{i1}, x_{i2}, \ldots$ are judged exchangeable, and if we denote the number of survivors out of n_i animals observed in the ith sequence by $y_i(n_i) = x_{i1} + \cdots + x_{in_i}$, a straightforward generalisation of the corollary to Proposition 4.18 implies a representation of the form

$$p(y_1(n_1), \ldots, y_m(n_m) \mid \boldsymbol{z}) = \int_{[0,1]^m} \prod_{i=1}^{m} \mathrm{Bi}\Big(y_i(n_i) \mid \theta_i(\boldsymbol{z}), n_i\Big)\, dQ(\theta(\boldsymbol{z})),$$

where $\boldsymbol{z} = (z_1, \ldots, z_m)$, $\theta(\boldsymbol{z}) = (\theta_1(\boldsymbol{z}), \ldots, \theta_m(\boldsymbol{z}))$ and $\theta_i\,(\boldsymbol{z}) = \lim_{n\to\infty} n^{-1} y_i(n)$.

In many situations, investigators often find it reasonable to assume that

$$\theta_i(\boldsymbol{z}) = \theta(z_i) = G(\phi; z_i),$$

where the functional form G (usually monotone increasing from 0 to 1) is specified, but ϕ is a random quantity. Functions having the form $G(\phi; z_i) = G(\phi_1 + \phi_2 z_i)$, with $\phi_1 \in \Re, \phi_2 \in \Re^+$, are widely used (see, for example, Hewlett and Plackett, 1979), with

$$G(\phi_1 + \phi_2 z_i) = \int_{-\infty}^{\phi_1 + \phi_2 z_i} \mathrm{N}(\mu \mid 0, 1) d\mu \quad \text{(the \textit{probit} model)}$$

and

$$G(\phi_1 + \phi_2 z_i) = \exp(\phi_1 + \phi_2 z_i)/\{1 + \exp(\phi_1 + \phi_2 z_i)\} \quad \text{(the \textit{logit} model)}$$

being the most common. For any specified $G(.; z_i)$, the required representation has the form

$$p(y_1(n_1), \ldots, y_m(n_m) \mid \boldsymbol{z}) = \int_{\Phi} \prod_{i=1}^{m} \mathrm{Bi}\Big(y_i(n_i) \mid G(\phi; z_i), n_i\Big) dQ^*(\phi),$$

with $dQ^*(\phi)$ specifying a prior distribution for $\phi \in \Phi$. In practice, the specification of Q might be facilitated by reparametrising from ϕ to a more suitable (1-1) transformation $\Psi = \Psi(\phi)$. In the probit and logit cases, for example, $\psi_1 = -\phi_1/\phi_2$ corresponds to the (log) dose, z_i, at which $G(\phi_1 + \phi_2 z_i) = 1/2$. Beliefs about ψ_1 then correspond to beliefs about the (log-) dose level for which the survival frequency in a large series of animals would equal 1/2, the so-called LD50 dose. Experimenters might typically be more accustomed to thinking in terms of $(-\phi_1/\phi_2, \phi_2)$, say, than in terms of (ϕ_1, ϕ_2). ◁

Example 4.13. ***(Growth-curves).*** Suppose that at each of m specified time points, say $z_1, \ldots, z_m$, sequences of real-valued random quantities, $x_{i1}, x_{i2}, \ldots, i = 1, \ldots, m$, are to be observed, where x_{ij} is the jth replicate measurement (perhaps on a logarithmic scale) of the size or weight of the subject or object of interest at time z_i. Suppose further that the kinds of judgements outlined in Example 4.11 are made about the sequences $x_{i1}, x_{i2}, \ldots$, with $i = 1, \ldots, m$, so that we have the representation

$$p(\boldsymbol{x}_1(n_1), \ldots, \boldsymbol{x}_m(n_m)) = \int_{\Theta_z} \prod_{i=1}^{m} \prod_{j=1}^{n_j} \mathrm{N}(x_{ij} \mid \mu_i(\boldsymbol{z}), \lambda_i(\boldsymbol{z}))\, dQ(\boldsymbol{\theta}(z)),$$

where $\boldsymbol{\theta}(\boldsymbol{z}) = (\mu_1(\boldsymbol{z}), \ldots, \mu_m(\boldsymbol{z}), \lambda_1(\boldsymbol{z}), \ldots, \lambda_m(\boldsymbol{z}))$ and $\Theta_z = \Re^m \times (\Re^+)^m$.

In many such situations, the judgement is made that $\lambda_1(\boldsymbol{z}) = \cdots = \lambda_m(\boldsymbol{z}) = \lambda$ (particularly if measurements are made on a logarithmic scale) and that

$$\mu_i(\boldsymbol{z}) = \mu_i(z_i) = g(\boldsymbol{\phi}; z_i),$$

where the functional form g (usually monotone increasing) is specified, but ϕ is a random quantity. Commonly assumed forms include

$$g(\boldsymbol{\phi}; z_i) = (\phi_1 + \phi_2 \phi_3^{-z_i})^{-1}, \quad \text{(the } \textit{logistic} \text{ model)}$$

and

$$g(\boldsymbol{\phi}; z_i) = \phi_1 + \phi_2 z_i \quad \text{(the } \textit{straight-line} \text{ model)}.$$

For any specified $g(.; z_i)$, the joint predictive density representation has the form

$$p(\boldsymbol{x}_1(n_1), \ldots, \boldsymbol{x}_m(n_m)) = \int_{\Theta_z} \prod_{i=1}^{m} \prod_{j=1}^{n_j} \mathrm{N}\Big(x_{ij} \mid g(\boldsymbol{\phi}, z_i), \lambda\Big) dQ(\boldsymbol{\phi}, \lambda),$$

where $dQ(\boldsymbol{\phi}, \lambda)$ specifying a prior distribution for $\boldsymbol{\phi} \in \Phi$ and $\lambda \in \Re^+$.

As with Example 4.12, specification of Q might be facilitated if we reparametrise from ϕ to a more suitable (1-1) transformation, $\psi = \psi(\phi)$. In the logistic case, for example, we might take $\psi_1 = \phi_1^{-1}$, corresponding to the "saturation" growth level reached as $z_i \to \infty$, and $\psi_2 = (\phi_1 + \phi_2)^{-1}$, corresponding to the growth level at the "time origin", $z_i = 0$. Beliefs about ψ_1, ψ_2 then acquire an operational meaning as beliefs about the average growth-levels, at times "∞" and "0", respectively, that would be observed from a large number of replicate measurements. A third possible parameter to which investigators could easily relate in some applications might be $\psi_3 = \log[\phi_1\phi_2/(2\phi_1 + \phi_2)]/\log(\phi_3)$, the time at which growth is half-way from the initial to the final level. ◁

Example 4.14. ***(Multiple regression).*** Suppose that, for each $i = 1, \ldots, m$, sequences of real-valued random quantities $x_{i1}, x_{i2}, \ldots$ are to be observed, where each x_{ij} is related to certain specified observed quantities $\boldsymbol{z}_i = (z_{i1}, \ldots, z_{ik})$ and judgements are made which lead to the belief representation

$$p(\boldsymbol{x}_1(n_1), \ldots, \boldsymbol{x}_m(n_m)) = \int_{\Theta_z} \prod_{i=1}^{m} \prod_{j=1}^{n_j} N(x_{ij} \mid \mu_i(\boldsymbol{z}_i), \lambda_i(\boldsymbol{z}_i))\, dQ(\boldsymbol{\theta}(\boldsymbol{z})),$$

where

$$\mu_i(\boldsymbol{z}_i) = \lim_{n \to \infty} \bar{x}_n(i), \quad \lambda_i^{-1}(\boldsymbol{z}_i) = \lim_{n \to \infty} s_n^2(i),$$

$$\boldsymbol{\theta}(\boldsymbol{z}) = (\mu_1(\boldsymbol{z}_1), \ldots, \mu_m(\boldsymbol{z}_m), \quad \lambda_1(\boldsymbol{z}_1), \ldots, \lambda_m(\boldsymbol{z}_m))$$

and $\Theta = \Re^m \times (\Re^+)^m$, with $\boldsymbol{z} = (\boldsymbol{z}_1, \ldots, \boldsymbol{z}_m)$.

In many situations, the further judgements are made that $\lambda_i(\boldsymbol{z}_i) = \lambda_i = \lambda$ and $\mu_i(\boldsymbol{z}_i) = \mu(\boldsymbol{z}_i), i = 1, \ldots, m$, where λ and $\mu(.)$ are unknown, but the latter is assumed to be a "smooth" function, adequately approximated by a first-order Taylor expansion, so that, for some (unspecified) $\boldsymbol{z}^*$,

$$\mu(\boldsymbol{z}_i) = \mu(\boldsymbol{z}^*) + (\boldsymbol{z}_i - \boldsymbol{z}^*) \bigtriangledown \mu(\boldsymbol{z}^*) = \boldsymbol{a}_i \boldsymbol{\theta},$$

where we define

$$\boldsymbol{a}_i = (1, z_{i1}, \ldots, z_{ik}) \text{ (row vector)}$$

and

$$\boldsymbol{\theta} = (\theta_0, \theta_1, \ldots, \theta_k)^t \text{ (column vector)}$$

with

$$\theta_0 = \mu(\boldsymbol{z}^*) - \boldsymbol{z}^* \bigtriangledown \mu(\boldsymbol{z}^*), \theta_i = [\bigtriangledown \mu(\boldsymbol{z}^*)]_i, i = 1, \ldots, k.$$

Conditional on $\phi = (\boldsymbol{\theta}, \lambda)$, the joint distribution of

$$\boldsymbol{x} = (\boldsymbol{x}_1(n_1), \ldots, \boldsymbol{x}_m(n_m))$$

is thus seen to be multivariate normal, $N_n(\boldsymbol{x} \mid \boldsymbol{A}\boldsymbol{\theta}, \boldsymbol{\lambda})$, where $\boldsymbol{A}$ is an $n \times k$ matrix ($n = n_1 + \cdots + n_m$), whose rows consist of $\boldsymbol{a}_1$ replicated n_1 times, followed by $\boldsymbol{a}_2$ replicated n_2 times, and so on, and $\boldsymbol{\lambda} = \lambda \boldsymbol{I}_n$, with $\boldsymbol{I}_n$ denoting the $n \times n$ identity matrix. The unconditional representation can therefore be written as

$$p(\boldsymbol{x}) = \int_{\Re^{k+1} \times \Re^+} N_n(\boldsymbol{x} \mid \boldsymbol{A}\boldsymbol{\theta}, \boldsymbol{\lambda})\, dQ(\boldsymbol{\theta}, \boldsymbol{\lambda}).$$

It is conventional to refer to $z_{1j}, z_{2j}, \ldots$ as values of the *regressor variables* $z^{(j)}, j = 1, \ldots, k$, to $\boldsymbol{\theta}$ as the vector of *regression coefficients* and $\boldsymbol{A}$ as the *design matrix*. The form $\mu(\boldsymbol{z}) = \boldsymbol{A}\boldsymbol{\theta}$ is called a *regression equation* and the structure

$$E(\boldsymbol{x} \mid \boldsymbol{A}, \boldsymbol{\theta}, \boldsymbol{\lambda}) = \boldsymbol{A}\boldsymbol{\theta}$$

is said to define a *linear model*. If $k = 1$, we have the *simple regression* (*straight-line*) model, $E(x_{ij}) = \theta_0 + \theta_1 z_{i1}$; for $k \geq 2$, we have a *multiple regression* model.

From an operational point of view, beliefs about $\boldsymbol{\theta}$ in the general case relate to beliefs about the intercept (θ_0) of the regression equation and the marginal rates of change $(\theta_1, \ldots, \theta_k)$ of the x_{ij} with respect to the regressor variables $(z_1, \ldots, z_k)$. However, within this general structure we can represent various special cases such as $z^{(j)} = z^j$ (*polynomial regression*) or $z^{(j)} = \sin(jH/N)$, for some N (a version of *trigonometric regression*); in these cases, beliefs about $\boldsymbol{\theta}$ will stem from rather different considerations. ◁

Specification of the kinds of structures which we have illustrated in Examples 4.12 to 4.14 essentially reduces to the same process as we have seen in earlier representations of joint predictive densities as integral mixtures. We proceed *as if*:

(i) the random quantities are *conditionally independent*, given the values of the relevant *covariates*, $\boldsymbol{z}$, and given the *unknown parameters*, $\boldsymbol{\phi}$;
(ii) the latter are assigned a prior distribution, $dQ(\boldsymbol{\phi})$.

In many cases, the likelihood, defined through conditional independence, involves familiar probability models, often of exponential family form (as with the binomial, normal and multivariate examples seen above), but with at least some of the usual "labelling" parameters replaced by more complex functional forms involving the covariates. From a conceptual point of view, this is all that really needs to be said for the time being. However, when we consider the applications of such models, together with the problems of computation, approximation, etc., which arise in implementing the Bayesian learning process, it is often useful to have a more structured taxonomy in mind: for example, linear versus non-linear functional forms; normal versus non-normal distributions, and so on.

4.6.5 Hierarchical Models

In Section 4.6.2, we considered the general situation where several sequences of random quantities, $x_{i1}, x_{i2}, \ldots, i = 1, \ldots, m$ are judged unrestrictedly infinitely exchangeable, leading typically to a joint density representation of the form

$$p(\boldsymbol{x}_1(n_1), \ldots, \boldsymbol{x}_m(n_m)) = \int_{\Theta^m} \prod_{i=1}^{m} \prod_{j=1}^{n_i} p(x_{ij} \mid \theta_i) \, dQ(\theta_1, \ldots, \theta_m).$$

We remarked at that time that nothing can be said, *in general*, about the prior specification $Q(\theta_1, \ldots, \theta_m)$, since this must reflect whatever beliefs are appropriate for the specific application being modelled. However, it is often the case that additional judgements about relationships among the m sequences lead to interestingly structured forms of $Q(\theta_1, \ldots, \theta_m)$.

In Section 4.6.1, we noted some of the possible contexts in which judgements of exchangeability might be appropriate not only for the random quantities *within* each of m separate sequence of observables, but also *between* the m strong law limits of appropriately defined statistics for each of the sequences. The following examples illustrate this kind of structured judgement and the forms of *hierarchical model* which result.

Example 4.15. ***(Exchangeable binomial parameters)***. Suppose that we have unrestrictedly infinitely exchangeable sequences of 0–1 random quantities, $x_{i1}, x_{i2}, \ldots$, with $i = 1, \ldots, m$. Then, for $i = 1, 2, \ldots$, $[n_i, y_i(n_i) = x_{i1} + \cdots + x_{in_i}]$, is a sufficient statistic for the ith sequence and

$$p(y_1(n_1), \ldots, y_m(n_m)) = \int_{[0,1]^m} p(y_1(n_1), \ldots, y_m(n_m) \,|\, \theta_1, \ldots, \theta_m) dQ(Q_1, \ldots, \theta_n)$$
$$= \int_{[0,1]^m} \prod_{i=1}^{m} \textbf{Bi}(y_i(n_i) \,|\, \theta_i, n_i)\; dQ(\theta_1, \ldots, \theta_m),$$

where

$$\theta_i = \lim_{n \to \infty} (y_i(n)/n).$$

As we remarked in Section 4.6.1, if the sequences consists of success-failure outcomes on repeated trials with m different (but, to all intents and purposes, "similar") types of component, it might be reasonable to judge the m "long-run success frequencies" to be themselves exchangeable. This corresponds to specifying *an exchangeable form of prior distribution for the parameters* $\theta_1, \ldots, \theta_m$. If the m types of component can be thought of as a selection from a potentially infinite sequence of similar components, we then have (see Section 4.3.3) the general representation

$$Q(\theta_1, \ldots, \theta_m) = \int_{\Im} Q(\theta_1, \ldots, \theta_m \,|\, G)\; d\Pi(G)$$
$$= \int_{\Im} \prod_{i=1}^{m} G(\theta_i)\; d\Pi(G).$$

The complete model structure is then seen to have the *hierarchical form*

$$p(y_1(n_1), \ldots, y_m(n_m) \,|\, \theta_1, \ldots, \theta_m) = \prod_{i=1}^{m} \textbf{Bi}(y_i(n_i) \,|\, \theta_i, n_i)$$

$$Q(\theta_1, \ldots, \theta_m \,|\, G) = \prod_{i=1}^{m} G(\theta_i)$$

$$\Pi(G)$$

In conventional terminology, the first stage of the hierarchy relates data to parameters via binomial distributions; the second stage models the binomial parameters *as if* they were a random sample from a distribution G; the third, and final, stage specifies beliefs about G. $\triangleleft$

The above example is readily generalised to the case of exchangeable parameters for any one-parameter exponential family. In practice, beliefs about G might concentrate on a particular parametric family, so that, assuming the existence of the appropriate densities, the prior specification takes the form

$$g(\theta_1, \ldots, \theta_m) = \int_\Phi g(\theta_1, \ldots, \theta_m \mid \phi)\, d\Pi(\phi) = \int_\Phi \prod_{i=1}^m g(\theta_i \mid \phi)\, d\Pi(\phi)$$

and, for appropriate sufficient statistics $y_i(n_i), i = 1, \ldots, m$, defines the hierarchical structure

$$p(y_1(n_1), \ldots, y_m(n_m) \mid \theta_1, \ldots, \theta_m) = \prod_{i=1}^m p(y_i(n_i) \mid \theta_i)$$

$$g(\theta_1, \ldots, \theta_m \mid \phi) = \prod_{i=1}^m g(\theta_i \mid \phi)$$

$$\Pi(\phi).$$

As before, the first stage of the hierarchy relates data to parameters in a form assumed to be independent of G; the second stage now models the parameters *as if* they were a random sample from a parametric family labelled by the *hyperparameter* $\phi \in \Phi$; the third, and final, stage specifies beliefs about the hyperparameter. Such beliefs acquire operational significance by identifying the hyperparameter with appropriate strong law limits of observables, as we shall indicate in the following example.

Example 4.16. ***(Exchangeable normal mean parameters).*** Suppose that we have m unrestrictedly infinitely exchangeable sequences $x_{i1}, x_{i2}, \ldots, i = 1, \ldots, m$, of real valued random quantities, for which (see Example 4.11) the joint density has the representation

$$p(\boldsymbol{x}_1(n_1), \ldots, \boldsymbol{x}_m(n_m)) = \int_{\Re^m \times \Re^+} \prod_{i=1}^m \prod_{j=1}^{n_i} \mathrm{N}(x_{ij} \mid \mu_i, \lambda)\, dQ(\mu_1, \ldots, \mu_m, \lambda),$$

where we recall that $\lambda^{-1} = \lim_{n\to\infty} s_n^2(i)$ and $\mu_i = \lim_{n\to\infty} \bar{x}_n(i)$, where

$$n\bar{x}_n(i) = (x_{i1} + \cdots + x_{in}), \quad ns_n^2(i) = \sum_{j=1}^n (x_{ij} - \bar{x}_n(i))^2, i = 1, \ldots, m.$$

So far as the specification of $Q(\mu_1, \ldots, \mu_m, \lambda)$ is concerned, we first note that in many applications it is helpful to think in terms of

$$Q(\mu_1, \ldots, \mu_m, \lambda) = Q_\mu(\mu_1, \ldots, \mu_m \mid \lambda) Q_\lambda(\lambda),$$

for some Q_μ, Q_λ. In some cases, knowledge of the strong law limits of sums of squares about the mean may be judged irrelevant to the assessment of beliefs for strong law limits of the sample averages: in such cases, $Q_\mu(\mu_1, \ldots, \mu_m \mid \lambda)$ will not depend on λ. In other cases, we might believe, for example, that variation among the limiting sample averages is certainly bigger (or certainly smaller) than within-sequence variation of observations about the sample mean: in such cases, $Q_\mu(\mu_1, \ldots, \mu_m \mid \lambda)$ will involve λ. In either case, it is useful to think in terms of the product form of Q.

Now suppose that, conditional on λ, the limiting sample means are judged exchangeable. If the m sequences can be thought of as a selection from a potentially infinite collection of similar sequences, we have (see Section 4.3) a further representation of Q_μ in the form

$$Q_\mu(\mu_1, \ldots, \mu_m \mid \lambda) = \int_{\mathfrak{F}} Q_\mu(\mu_1, \ldots, \mu_m \mid \lambda, G)\, d\Pi(G \mid \lambda)$$

$$= \int_{\mathfrak{F}} \prod_{i=1}^{m} G(\mu_i \mid \lambda)\, d\Pi(G \mid \lambda).$$

The complete model then has the hierarchical structure

$$p(\boldsymbol{x}_1(n_1), \ldots, \boldsymbol{x}_m(n_m) \mid \mu_1, \ldots, \mu_m, \lambda) = \prod_{i=1}^{m} p(\boldsymbol{x}_i(n_i) \mid \mu_i, \lambda)$$

$$Q_\mu(\mu_1, \ldots, \mu_m \mid \lambda, G) = \prod_{i=1}^{m} G(\mu_i \mid \lambda)$$

$$\Pi(G \mid \lambda) Q_\lambda(\lambda).$$

In practice, beliefs about G, given λ, might concentrate on a particular parametric family, so that, assuming the existence of the appropriate densities, the hierarchical structure would take the form

$$p(\boldsymbol{x}_1(n_1), \ldots, \boldsymbol{x}_m(n_m) \mid \mu_1, \ldots, \mu_m, \lambda) = \prod_{i=1}^{m} p(\boldsymbol{x}_i(n_i) \mid \mu_i, \lambda)$$

$$g_\mu(\mu_1, \ldots, \mu_m \mid \lambda, \boldsymbol{\phi}) = \prod_{i=1}^{m} g_\mu(\mu_i \mid \lambda, \boldsymbol{\phi})$$

$$\Pi x(\boldsymbol{\phi} \mid \lambda)\; Q_\lambda(\lambda).$$

For an explicit example of this, suppose that, given a potentially infinite sequence $\mu_1, \mu_2, \ldots$ (or, more concretely, $\bar{x}_{n1}(1), \bar{x}_{n2}(2), \ldots$, for very large $n_1, n_2, \ldots$) the quantities $m, \bar{\mu}(m) = m^{-1}(\mu_1 + \cdots + \mu_m)$ and $s^2(m) = m^{-1} \sum_{j=1}^{m} (\mu_j - \bar{\mu}(m))^2$ (or the large sample analogues of $\bar{\mu}(m)$ and $s^2(m)$) were judged sufficient for the sequence. It would then be natural (see Section 4.5) to take $g_\mu(\mu_i \mid \lambda, \boldsymbol{\phi}) = N(\mu_i \mid \phi_1, \phi_2)$, where

$$\phi_1 = \lim_{m\to\infty} \bar{\mu}(m), \quad \phi_2 = \lim_{m\to\infty} s^2(m).$$

From an operational standpoint, the final stage specification of the joint prior distribution for ϕ_1, ϕ_2 and λ then reduces to a specification of beliefs about the following limits of observable quantities (for large m and $n_i, i = 1, \ldots, m$):

(i) the mean of all the observations from all the sequences (ϕ_1);

(ii) the mean sum of squares of the individual sequence means about the overall mean (ϕ_2);
(iii) the mean (over sequences) of the mean sum of squares of observations within a sequence about the sequence mean (λ).

The precise form of specification at this stage will, of course, depend on the particular situation in which the model is being applied. $\triangleleft$

Hierarchical modelling provides a powerful and flexible approach to the representation of beliefs about observables in extended data structures, and is being increasingly used in statistical modelling and analysis. This section has merely provided a brief introduction to the basic ideas and the way such structures arise naturally within a subjectivist, modelling framework. In the context of the Bayesian learning process, further brief discussion will be given in Section 5.6.4, where links will be made with *empirical Bayes* ideas.

An extensive discussion of hierarchical modelling will be given in the volumes *Bayesian Computation* and *Bayesian Methods.* A selection of references to the literature on inference for hierarchical models will be given in Section 5.6.4.

4.7 PRAGMATIC ASPECTS

4.7.1 Finite and Infinite Exchangeability

The de Finetti representation theorem for 0-1 random quantities, and the various extensions we have been considering in this chapter, characterise forms of $p(x_1, \ldots, x_n)$ for observables $x_1, \ldots, x_n$ assumed to be part of an *infinite* exchangeable sequence. However, in general, mathematical representations which correspond to probabilistic mixing over conditionally independent parametric forms do not hold for *finite* exchangeable sequences.

To see this, consider $n = 2$ and finitely exchangeable 0-1 x_1, x_2, such that

$$p(x_1 = 0, x_2 = 0) = p(x_1 = 1, x_2 = 1) = 0$$
$$p(x_1 = 1, x_2 = 0) = p(x_1 = 0, x_2 = 1) = \tfrac{1}{2}\,.$$

If the de Finetti representation held, we would have

$$\int_0^1 \theta^2 dQ(\theta) = \int_0^1 (1 - \theta)^2 dQ(\theta) = 0,$$

for some $Q(\theta)$, an impossibility since the latter would have to assign probability one to both $\theta = 0$ and $\theta = 1$ (Diaconis and Freedman, 1980a).

It appears, therefore, that there is a potential conflict between realistic modelling (acknowledging the necessarily finite nature of actual exchangeability judgements) and the use of conventional mathematical representations (derived on the basis of assumed infinite exchangeability).

To discuss this problem, let us call an exchangeable sequence, $x_1, \ldots, x_n$, with $x_i \in X$, *N-extendible* if it is part of the longer exchangeable sequence $x_1, \ldots, x_N$. Practical judgements of exchangeability for specific observables $x_1, \ldots, x_n$ are typically of this kind: the $x_1, \ldots, x_n$ can be considered as part of a larger, *but finite*, potential sequence of exchangeable observables. Infinite exchangeability corresponds to the possibly unrealistic assumption of N-extendibility for all $N > n$.

In general, the assumption of infinite exchangeability implies that the probability assigned to an event $(x_1, \ldots, x_n) \in E \subseteq X^n$ is of the form

$$P_Q(E) = \int F^n(E)dQ(F),$$

for some Q. If we denote by $P(E)$ the corresponding probability assigned under N-extendibility for a specific N, a possible measure of the "distortion" introduced by assuming infinite exchangeability is given by

$$\sup_E |P(E) - P_Q(E)|,$$

where the supremum is taken over all events in the appropriate σ-field on X^n. Intuitively, one might feel that if $x_1, \ldots, x_n$ is N-extendible for some $N >> n$, the "distortion" should be somewhat negligible. This is made precise by the following.

Proposition 4.19. (***Finite approximation of infinite exchangeability***). *With the preceding notation, there exists Q such that*

$$\sup_E |P(E) - P_Q(E)| \leq \frac{f(n)n}{N},$$

where $f(n)$ is the number of elements in X, if the latter is finite, and $f(n) = (n-1)$ otherwise.

Proof. See Diaconis and Freedman (1980a) for a rigorous statement and technical details. $\triangleleft$

The message is clear and somewhat comforting. If a realistic judgement of N-extendibility for large, but finite, N is replaced by the mathematically convenient assumption of infinite exchangeability, no important distortion will occur in quantifying uncertainties.

For further discussion, see Diaconis (1977), Jaynes (1986) and Hill (1992). For extensions of Proposition 4.19 to multivariate and linear model structures, see Diaconis *et al.* (1992).

4.7.2 Parametric and Nonparametric Models

In Section 4.3, we saw that the assumption of exchangeability for a sequence $x_1, x_2, \ldots$ of real-valued random quantities implied a general representation of the joint distribution function of $x_1, \ldots, x_n$ of the form

$$P(x_1, \ldots, x_n) = \int_{\Im} \prod_{i=1}^{n} F(x_i) dQ(F),$$

where

$$Q(F) = \lim_{n \to \infty} P(F_n)$$

and F_n is the empirical distribution function defined by $x_1, \ldots, x_n$. This implies that we should proceed *as if* we have a random sample from an unknown distribution function F, with Q representing our beliefs about "what the empirical distribution would look like for large n".

As we remarked at the end of Section 4.3.3, the task of assessing and representing such a belief distribution Q over the set $\Im$ of all possible distribution functions is by no means straightforward, since F is, effectively, an infinite-dimensional parameter. Most of this chapter has therefore been devoted to exploring additional features of beliefs which justify the restriction of $\Im$ to families of distributions having explicit mathematical forms involving only a finite-dimensional labelling parameter.

Conventionally, albeit somewhat paradoxically, representations in the finite-dimensional case are referred to as *parametric models*, whereas those involving the infinite-dimensional parameter are referred to as *nonparametric models*! The technical key to Bayesian nonparametric modelling is thus seen to be the specification of appropriate probability measures over function spaces, rather than over finite-dimensional real spaces, as in the parametric case. For this reason, the Bayesian analysis of nonparametric models requires considerably more mathematical machinery than the corresponding analysis of parametric models. In the rest of this volume we will deal exclusively with the parametric case, postponing a treatment of nonparametric problems to the volumes *Bayesian Computation* and *Bayesian Methods*.

Among important references on this topic, we note Whittle (1958), Hill (1968, 1988, 1992), Dickey (1969), Kimeldorf and Wahba (1970), Good and Gaskins (1971, 1980), Ferguson (1973, 1974), Leonard (1973), Antoniak (1974), Doksum (1974), Susarla and van Ryzin (1976), Ferguson and Phadia (1979), Dalal and Hall (1980), Dykstra and Laud (1981), Padgett and Wei (1981), Rolin (1983), Lo (1984), Thorburn (1986), Kestemont (1987), Berliner and Hill (1988), Wahba (1988), Hjort (1990), Lenk (1991) and Lavine (1992a).

As we have seen, the use of specific parametric forms can often be given a *formal* motivation or justification as the coherent representation of certain forms of belief characterised by invariance or sufficiency properties. In practice, of course, there are often less formal, more *pragmatic*, reasons for choosing to work with a particular parametric model (as there often are for acting, formally, *as if* particular forms of summary statistic were sufficient!). In particular, specific parametric models are often suggested by *exploratory data analysis* (typically involving graphical techniques to identify plausible distributional shapes and forms of relationship with covariates), or by *experience* (i.e., historical reference to "similar" situations, where a given model seemed "to work") or by *scientific theory* (which determines that a specific mathematical relationship "must" hold, in accordance with an assumed "law"). In each case, of course, the choice involves subjective judgements; for example, regarding such things as the "straightness" of a graphical normal plot, the "similarity" between a current and a previous trial, and the "applicability of a theory to the situation under study. From the standpoint of the general representation theorem, such judgements correspond to acting *as if* one has a Q which concentrates on a subset of $\Im$ defined in terms of a finite-dimensional labelling parameter.

4.7.3 Model Elaboration

However, in arriving at a particular parametric model specification, by means of whatever combination of formal and pragmatic judgements have been deemed appropriate, a number of simplifying assumptions will necessarily have been made (either consciously or unconsciously). It would always be prudent, therefore, to "expand one's consciousness" a little in relation to an intended model in order to review the judgements that have been made. Depending on the context, the following kinds of critical questions might be appropriate:

(i) is it reasonable to assume that all the observables form a "homogeneous sample", or might a few of them be "aberrant" in some sense?

(ii) is it reasonable to apply the modelling assumptions to the observables on their original scale of measurement, or should the scale be transformed to logarithms, reciprocals, or whatever?

(iii) when considering temporally or spatially related observables, is it reasonable to have made a particular conditional independence assumption, or should some form of correlation be taken into account?

(iv) if some, but not all, potential covariates have been included in the model, is it reasonable to have excluded the others, or might some of them be important, either individually or in conjunction with covariates already included?

We shall consider each of these possibilities in turn, indicating briefly the kinds of elaboration of the "first thought of" model that might be considered.

Outlier elaboration. Suppose that judgements about a sequence $x_1, x_2, \ldots$ of real-valued random quantities have led to serious consideration of the model

$$p(x_1, \ldots, x_n) = \int_{\Re \times \Re^+} \prod_{i=1}^{n} \mathrm{N}(x_i \,|\, \mu, \lambda)\, dQ(\mu, \lambda),$$

but, on reflection, it is thought wise to allow for the fact that (an unknown) one of $x_1, \ldots, x_n$ *might* be aberrant. If aberrant observations are assumed to be such that a sequence of them would have a limiting mean equal to μ, but a limiting mean square about the mean equal to $(\gamma\lambda)^{-1}, 0 < \gamma < 1$, where μ and λ^{-1} denote the corresponding limits for non-aberrant observations, a suitable form of elaborated model might be

$$p(x_1, \ldots, x_n) = \pi \int_{\Re \times \Re^+} \prod_{i=1}^{n} \mathrm{N}(x_i \,|\, \mu, \lambda)\, dQ(\mu, \lambda)$$

$$+ (1 - \pi) \int_{\Re \times \Re^+ \times [0,1]} \sum_{j=1}^{n} \frac{1}{n} \mathrm{N}(x_j \,|\, \mu, \gamma\lambda) \left\{ \prod_{i \neq j} \mathrm{N}(x_i \,|\, \mu, \lambda) \right\} dQ(\mu, \lambda)\, dQ(\gamma).$$

This model corresponds to an initial assumption that, with specified probability π, there are no aberrant observations, but, with probability $1 - \pi$, there is precisely one aberrant observation, which is equally likely to be any one of $x_1, \ldots, x_n$. Generalisations to cover more than one possible aberrant observation can be constructed in an obviously analogous manner. Such models are usually referred to as "outlier" models, since $\gamma < 1$ implies an increased probability that, in the observed sample $x_1, \ldots, x_n$, the aberrant observation will "outlie". Since for an aberrant observation x, $\mathrm{E}[(x - \mu)^2 \,|\, \mu, \lambda, \gamma] = (\gamma\lambda)^{-1}$, prior belief in the relative inaccuracy of an aberrant observation as a "predictor" of μ is reflected in the weight attached by the prior distribution $Q(\gamma)$ to values of γ much smaller than 1.

De Finetti (1961) and Box and Tiao (1968) are pioneering Bayesian papers on this topic. More recent literature includes; Dawid (1973), O'Hagan (1979, 1988b, 1990), Freeman (1980), Smith (1983), West (1984, 1985), Pettit and Smith (1985), Arnaiz and Ruíz-Rivas (1986), Muirhead (1986), Pettit (1986, 1992), Guttman and Peña (1988) and Peña and Guttman (1993).

Transformation elaboration. Suppose now that judgements about a sequence $x_1, x_2, \ldots$ of real-valued random quantities are such that it seems reasonable to suppose that, *if a suitable γ were identified*, beliefs about the sequence $x_1^{(\gamma)}, x_2^{(\gamma)}, \ldots,$ defined by

$$\begin{aligned} x_i^{(\gamma)} &= (x_i^{\gamma} - 1)/\gamma \qquad (\gamma \neq 0, \gamma \in \Gamma) \\ &= \log(x_i) \qquad (\gamma = 0), \end{aligned}$$

would plausibly have the representation

$$p(x_1^{(\gamma)}, \ldots, x_n^{(\gamma)}) = \int_{\Re \times \Re^+} \prod_{i=1}^{n} \mathbf{N}(x_i^{(\gamma)} \mid \mu, \lambda) dQ^*(\mu, \lambda \mid \gamma).$$

It then follows that

$$p(x_1, \ldots, x_n) = \int_{\Re \times \Re^+ \times \Gamma} \prod_{i=1}^{n} \mathbf{N}(x_1^{(\gamma)} \mid \mu, \lambda) J(\boldsymbol{x}, \gamma)\, dQ^*(\mu, \lambda \mid \gamma)\, dQ^+(\gamma)$$

where

$$J(\boldsymbol{x}, \gamma) = \prod_{i=1}^{n} [dx_i^{(\gamma)} / dx_i].$$

The case $\gamma = 1$ corresponds to assuming a normal parametric model for the observations on their original scale of measurement. If Γ includes values such as $\gamma = -1, \gamma = 1/2, \gamma = 0$, the elaborated model admits the possibility that transformations such as reciprocal, square root, or logarithm, might provide a better scale on which to assume a normal parametric model. Judgements about the relative plausibilities of these and other possible transformations are then incorporated in Q^+. For detailed developments see Box and Cox (1964), Pericchi (1981) and Sweeting (1984, 1985).

Correlation elaboration. Suppose that judgements about $x_1, x_2, \ldots$ again lead to a "first thought of model in which

$$p(x_1, \ldots, x_n \mid \mu, \lambda) = \prod_{i=1}^{n} \mathbf{N}(x_i \mid \mu, \lambda),$$

but that it is then recognised that there may be a serial correlation structure among $x_1, \ldots, x_n$ (since, for example, the observations correspond to successive time-points, $t = 1, t = 2$, etc.) A possible extension of the representation to incorporate such correlation might be to assume that, for a given $\gamma \in [-1, 1)$, and conditional on μ and λ, the correlation between x_i and x_{i+h} is given by $R(x_i, x_{i+h} \mid \mu, \lambda, \gamma) = \gamma^h$, so that

$$p(\boldsymbol{x} \mid \mu, \lambda, \gamma) = p(x_1, \ldots, x_n \mid \mu, \lambda, \gamma) = \mathbf{N}_n(\boldsymbol{x} \mid \mu \mathbf{1}, \lambda \Gamma^{-1}),$$

where

$$\Gamma = \begin{bmatrix} 1 & \gamma & \gamma^2 & \ldots & \gamma^{n-1} \\ \gamma & 1 & \gamma & & \gamma^{n-2} \\ \vdots & & \ldots & & \vdots \\ \gamma^{n-1} & \gamma^{n-2} & \gamma^{n-3} & \ldots & 1 \end{bmatrix}$$

The elaborated model then becomes, for some Q^*, Q^+,

$$p(x_1, \ldots, x_n) = \int_{\Re \times \Re^+ \times (-1,1)} \mathrm{N}_n(\boldsymbol{x} \mid \mu \mathbf{1}, \lambda \Gamma^{-1})\, dQ^*(\mu, \lambda \mid \gamma)\, dQ^+(\gamma)$$

The "first thought of" model corresponds to $\gamma = 0$ and beliefs about the relative plausibility of this value compared with other possible values of positive or negative correlation are reflected in the specification of Q^+.

Covariate elaboration. Suppose that the "first thought of" model for the observables $\boldsymbol{x} = (\boldsymbol{x}_1(n_1), \ldots, \boldsymbol{x}_n(n_m))$, where $\boldsymbol{x}_i(n_i) = (x_{i1}, \ldots, x_{in_i})$ denotes replicate observations corresponding to the observed value $\boldsymbol{z}_i = (z_{i1}, \ldots, z_{ik})$ of the covariates $z_1, \ldots, z_k$, is the multiple regression model with representation

$$p(\boldsymbol{x}) = \int_{\Re^{k+1} \times \Re^+} \mathrm{N}_n(\boldsymbol{x} \mid \boldsymbol{A\theta}, \lambda)\, dQ(\boldsymbol{\theta}, \lambda)$$

as described in Example 4.16 of Section 4.6. If it is subsequently thought that covariates $z^{k+1}, \ldots, z^l$ should also have been taken into account, a suitable elaboration might take the form of an extended regression model

$$p(\boldsymbol{x}) = \int_{\Re^{l+1} \times \Re^+} \mathrm{N}_n(\boldsymbol{x} \mid \boldsymbol{A\theta} + \boldsymbol{B\gamma}, \lambda)\, dQ^*(\boldsymbol{\theta}, \boldsymbol{\gamma}, \lambda),$$

where B consists of rows containing $\boldsymbol{b}_i = (z_{ik+1}, \ldots, z_{il})$ replicated n_i times, $i = 1, \ldots, m$ and $\boldsymbol{\gamma} = (\theta_{k+1}, \ldots, \theta_l)$ denotes the regression coefficients of the additional regressor variables $z_{k+1}, \ldots, z_l$. The value $\boldsymbol{\gamma} = \mathbf{0}$ corresponds to the "first thought of" model.

In all these cases, an initially considered representation of the form

$$p(\boldsymbol{x}) = \int p(\boldsymbol{x} \mid \boldsymbol{\phi})\, dQ(\boldsymbol{\phi})$$

is replaced by an elaborated representation

$$p(\boldsymbol{x}) = \int p(\boldsymbol{x} \mid \boldsymbol{\phi}, \boldsymbol{\gamma})\, dQ^*(\boldsymbol{\phi}, \boldsymbol{\gamma}),$$

the latter reducing to the original representation on setting the elaboration parameter $\boldsymbol{\gamma}$ equal to $\mathbf{0}$. Inference about such a $\boldsymbol{\gamma}$, imaginatively chosen to reflect interesting possible forms of departure from the original model, often provides a natural basis for checking on the adequacy of an initially proposed model, as well as learning about the directions in which the model needs extending.

Other Bayesian approaches to the problem of covariate selection include Bernardo and Bermúdez (1985), Mitchell and Beauchamp (1988) and George and McCulloch (1993a).

4.7.4 Model Simplification

The process of model elaboration, outlined in the previous section, consists in expanding a "first thought of" model to include additional parameters (and possibly covariates), reflecting features of the situation whose omission from the original model formulation is, on reflection, thought to be possibly injudicious.

The process of model simplification is, in a sense, the converse. In reviewing a currently proposed model, we might wonder whether some parameters (or covariates) have been unnecessary included, in the sense that a simpler form of model might be perfectly adequate. As it stands, of course, this latter consideration is somewhat ill-defined: the "adequacy", or otherwise, of a particular form of belief representation can only be judged in relation to the consequence arising from actions taken on the basis of such beliefs. These and other questions relating to the fundamentally important area of model comparison and model choice will be considered at length in Chapter 6. For the present, it will suffice just to give an indication of some particular forms of model simplification that are routinely considered.

Equality of parameters. In Section 4.6, we analysed the situation where several sequences of observables are judged unrestrictedly infinitely exchangeable, leading to a general representation of the form

$$p(\boldsymbol{x}_1(n_1), \ldots, \boldsymbol{x}_n(n_m)) = \int_{\Theta^*} \prod_{i=1}^{m} \prod_{j=1}^{n_i} p(x_{ij} \mid \theta_i) \, dQ(\theta_1, \ldots, \theta_m),$$

where $\theta_i \in \Theta_i, \Theta^* = \prod_{i=1}^{m} \Theta_i$ and the parameter θ_i relating to the ith sequence can typically be interpreted as the limit of a suitable summary statistic for the ith sequence. If, on the other hand, the simplifying judgement were made that, in fact, the labelling of the sequences is irrelevant and that any combined collection of observables from any or all of the sequences would be completely exchangeable, we would have the representation

$$p(\boldsymbol{x}_1(n_1), \ldots, \boldsymbol{x}_n(n_m)) = \int_{\Theta^*} \prod_{i=1}^{m} \prod_{j=1}^{n} p(x_{ij} \mid \theta) \, dQ(\theta)$$

where the same parameter $\theta \in \Theta$ now suffices to label the parametric model for each of the sequences. In conventional terminology, the simplified representation is sometimes referred to as the *null-hypothesis* $(\theta_1 = \cdots = \theta_m)$ and the original representation as the *alternative hypothesis* $(\theta_1 \neq \cdots \neq \theta_m)$. As we saw in Section 4.6 (for the case of two 0-1 sequences), rather than opt for sure for one or other of these representations, we could take a *mixture* of the two (with weight π, say, on the null representation and $1 - \pi$ on the alternative, general, representation). This form of representation will be considered in more detail in Chapter 6, where it will be shown to provide a possible basis for evaluating the relative plausibility of the "null and alternative hypotheses" in the light of data.

Absence of effects. In Section 4.6, we considered the situation of a structured layout with replicate sequences of observations in each of IJ cells, and a possible parametric model representation involving *row effects* $(\alpha_1, \ldots, \alpha_I)$, *column effects* $(\beta_1, \ldots, \beta_J)$ and *interaction effects* $(\gamma_{11}, \ldots, \gamma_{IJ})$. A commonly considered simplifying assumption is that there are no interaction effects $(\gamma_{11} = \cdots = \gamma_{IJ} = 0)$, so that large sample means in individual cells are just the additive combination of the corresponding large sample row and column means.

Further possible simplifying judgements would be that the row (or column) labelling is irrelevant, so that $\alpha_1 = \cdots = \alpha_I = 0$ (or $\beta_1 = \cdots = \beta_J = 0$) and large sample cell means coincide with column (or row) means. Again, conventional terminology would refer to these simplifying judgements as "null hypotheses".

Omission of covariates. Considering, for example, the multiple regression case, described in Example 4.14 of Section 4.6 and reconsidered in the previous section on model elaboration, we see that here the simplification process is very clearly just the converse of the elaboration process. If γ denotes the regression coefficients of the covariates we are considering omitting, then the model corresponding to $\gamma = 0$ provides the required simplification.

In fact, in all the cases of elaboration which we considered in the previous section, setting the "elaboration parameter" γ to 0 provides a natural form of simplification of potential interest. Whether the process of model comparison and choice is seen as one of elaboration or of simplification is then very much a pragmatic issue of whether we begin with a "smaller" model and consider making it "bigger", or vice versa. In any case, issues of model comparison and choice require a separate detailed and extensive treatment, which we defer until Chapter 6.

4.7.5 Prior Distributions

The operational subjectivist approach to modelling views predictive models as representations of beliefs about observables (including limiting, large-sample functions of observables, conventionally referred to as parameters). Invariance and sufficiency considerations have then been shown to justify a structured approach to predictive models in terms of integral mixtures of parametric models with respect to distributions for the labelling parameters. In familiar terminology, we specify a distribution for the observables conditional on unknown parameters (a *sampling distribution*, defining a *likelihood*), together with a distribution for the unknown parameters (a *prior distribution*). *It is the combination of prior and likelihood which defines the overall model.* In terms of the mixture representation, the specification of a prior distribution for unknown parameters is therefore an essential and unavoidable part of the process of representing beliefs about observables and hence of learning from experience.

> From the operational, subjectivist perspective, it is meaningless to approach modelling solely in terms of the parametric component and ignoring the prior distribution. We are, therefore, in fundamental disagreement with approaches to statistical modelling and analysis which proceed only on the basis of the sampling distribution or likelihood and treat the prior distribution as something optional, irrelevant, or even subversive (see Appendix B).

That said, it should be readily acknowledged that the process of representing prior beliefs itself involves a number of both conceptual and practical difficulties, and certainly cannot be summarily dealt with in a superficial or glib manner.

From a conceptual point of view, as we have repeatedly stressed throughout this chapter, prior beliefs about parameters typically acquire an operational significance and interpretation as beliefs about limiting (large-sample) functions of observables. Care must therefore obviously be taken to ensure that prior specifications respect logical or other constraints pertaining to such limits. Often, the specification process will be facilitated by suitable "reparametrisation".

From a practical point of view, detailed treatment of specific cases is very much a matter of "methods" rather than "theory" and will be dealt with in the third volume of this series. However, a general overview of representation strategies, together with a number of illustrative examples, will be given in the inference context in Chapter 5. In particular, we shall see that the range of creative possibilities opened up by the consideration of mixtures, asymptotics, robustness and sensitivity analysis, as well as novel and flexible forms of inference reporting, provides a rich and illuminating perspective and framework for inference, within which many of the apparent difficulties associated with the precise specification of prior distributions are seen to be of far less significance than is commonly asserted by critics of the Bayesian approach.

4.8 DISCUSSION AND FURTHER REFERENCES

4.8.1 Representation Theorems

The original representation theorem for exchangeable $0-1$ random quantities appears in de Finetti (1930), the concept of exchangeability having been considered earlier by Haag (1924) and also in the early 1930's by Khintchine (1932). Extensions to the case of general exchangeable random quantities appear in de Finetti (1937/1964) and Dynkin (1953), with an abstract analytical version appearing in Hewitt and Savage (1955). Seminal extensions to more complex forms of symmetry (partial exchangeability) can be found in de Finetti (1938) and Freedman

(1962). See Diaconis and Freedman (1980b) and Wechsler (1993) for overviews and generalisations of the concept of exchangeability.

Recent and current developments have generated an extensive catalogue of characterisations of distributions via both invariance and sufficiency conditions. Important progress is made in Diaconis and Freedman (1984, 1987, 1990) and Küchler and Lauritzen (1989). See, also, Ressel (1985). Useful reviews are given by Aldous (1985), Diaconis (1988a) and, from a rather different perspective, Lauritzen (1982, 1988). The conference proceedings edited by Koch and Spizzichino (1982) also provides a wealth of related material and references. For related developments from a reliability perspective, see Barlow and Mendel (1992, 1994) and Mendel (1992).

4.8.2 Subjectivity and Objectivity

Our approach to modelling has been dictated by a subjectivist, operational concern with individual beliefs about (potential) observables. Through judgements of symmetry, partial symmetry, more complex invariance or sufficiency, we have seen how mixtures over conditionally independent "parameter-labelled" forms arise as typical representations of such beliefs. We have noted how this illuminates, and puts into perspective, linguistic separation into "likelihood" (or "sampling model") and "prior" components. But we have also stressed that, from our standpoint, the two are actually inseparable in defining a belief model.

In contrast, traditional discussion of a statistical model typically refers to the parametric form as "the model". The latter then defines "objective" probabilities for outcomes defined in terms of observables, these probabilities being determined by the values of the "unknown parameters". It is often implicit in such discussion that if the "true" parameter were known, the corresponding parametric form would be the "true" model for the observables. Clearly, such an approach seeks to make a very clear distinction between the nature of observables and parameters. It is as if, given the "true" parameter, the corresponding parametric distribution is seen as part of "objective reality", providing the mechanism whereby the observables are generated. The "prior", on the other hand, is seen as a "subjective" optional extra, a potential contaminant of the objective statements provided by the parametric model.

Clearly, this view has little in common with the approach we have systematically followed in this volume. However, there is an interesting sense, even from our standpoint, in which the parametric model and the prior can be seen as having different roles.

Instead of viewing these roles as corresponding to an objective/subjective dichotomy, we view them in terms of an intersubjective/subjective dichotomy (following Dawid, 1982b, 1986b). To this end, consider a *group* of Bayesians, all concerned with their belief distributions for the same sequence of observables. In the absence of any general agreement over assumptions of symmetry, invariance or sufficiency, the individuals are each simply left with their own subjective assessments. However, given some set of common assumptions, the results of this chapter imply that the entire group will structure their beliefs using some common form of mixture representation. Within the mixture, the parametric forms adopted will be the same (the *intersubjective* component), while the priors for the parameter will differ from individual to individual (the *subjective* component). Such intersubjective agreement clearly facilitates communication within the group and reduces areas of potential disagreement to just that of different prior judgements for the parameter. As we shall see in Chapter 5, judgements about the parameter will tend more towards a consensus as more data are acquired, so that such a group of Bayesians may eventually come to share very similar beliefs, even if their initial judgements about the parameter were markedly different. We emphasise again, however, that the key element here is intersubjective agreement or consensus. We can find no real role for the idea of objectivity except, perhaps, as a possibly convenient, but potentially dangerously misleading, "shorthand" for intersubjective communality of beliefs.

4.8.3 Critical Issues

We conclude this chapter on modelling with some further comments concerning (i) *The Role and Nature of Models*, (ii) *Structural and Stochastic Assumptions*, (iii) *Identifiability* and (iv) *Robustness Considerations.*

The Role and Nature of Models

In the approach we have adopted, the fundamental notion of a model is that of a predictive probability specification for observables. However, the forms of representation theorems we have been discussing provide, in typical cases, a basis for separating out, if required, two components; the parametric model, and the belief model for the parameters. Indeed, we have drawn attention in Section 4.8.2 to the fact that shared structural belief assumptions among a group of individuals can imply the adoption of a common form of parametric model, while allowing the belief models for the parameters to vary from individual to individual. One might go further and argue that without some element of agreement of this kind there would be great difficulty in obtaining any meaningful form of scientific discussion or possible consensus.

Non-subjectivist discussions of the role and nature of models in statistical analysis tend to have a rather different emphasis (see, for example, Cox, 1990, and Lehmann, 1990). However, such discussions often end up with a similar message, implicit or explicit, about the importance of models in providing a focused framework to serve as a basis for subsequent identification of areas of agreement and disagreement. In order to think about complex phenomena, one must necessarily work with simplified representations. In any given context, there are typically a number of different choices of degrees of simplification and idealisation that might be adopted and these different choices correspond to what Lehmann calls "a reservoir of models", where

> ... particular emphasis is placed on transparent characterisations or descriptions of the models that would facilitate the understanding of when a given model is appropriate. (Lehmann, 1990)

But appropriate for what? Many authors—including Cox and Lehmann—highlight a distinction between what one might call *scientific* and *technological* approaches to models. The essence of the dichotomy is that scientists are assumed to seek *explanatory* models, which aim at providing insight into and understanding of the "true" mechanisms of the phenomenon under study; whereas technologists are content with *empirical* models, which are not concerned with the "truth", but simply with providing a reliable basis for practical action in predicting and controlling phenomena of interest.

Put very crudely, in terms of our generic notation, explanatory modellers take the form of $p(\boldsymbol{x} \mid \boldsymbol{\theta})$ very seriously, whereas empirical modellers are simply concerned that $p(\boldsymbol{x})$ "works". For an elaboration of the latter view, see Leonard (1980).

The approach we have adopted is compatible with either emphasis. As we have stressed many times, it is observables which provide the touchstone of experience. When comparing rival belief specifications, all other things being equal we are intuitively more impressed with the one which consistently assigns higher probabilities to the things that actually happen. If, in fact, a phenomenon is governed by the specific mechanism $p(\boldsymbol{x} \mid \boldsymbol{\theta})$ with $\boldsymbol{\theta} = \boldsymbol{\theta}_0$, a scientist who discovers this and sets $p(\boldsymbol{x}) = p(\boldsymbol{x} \mid \boldsymbol{\theta}_0)$ will certainly have a $p(\boldsymbol{x})$ that "works".

However, we are personally rather sceptical about taking the science versus technology distinction too seriously. Whilst we would not dispute that there are typically real differences in motivation and rhetoric between scientists and technologists, it seems to us that theories are always ultimately judged by the predictive power they provide. Is there really a meaningful concept of "truth" in this context other than a pragmatic one predicated on $p(\boldsymbol{x})$? We shall return to this issue in Chapter 6, but our prejudices are well-captured in the adage: "*all models are false, but some are useful*".

Structural and Stochastic Assumptions

In Section 4.6, we considered several illustrative examples where, separate from considerations about the complete form of probability specification to be adopted, the key role of the parametric model component $p(\boldsymbol{x} \mid \boldsymbol{\theta})$ was to specify structured forms of expectations for the observables conditional on the parameters. We recall two examples.

In the case of observables x_{ijk} in a two-way layout with replications (Section 4.6.3), with parameters corresponding to overall mean, main effects and interactions, we encountered the form

$$E(x_{ijk}) = \mu + \alpha_i + \beta_j + \gamma_{ij};$$

in the case of a vector of observables $\boldsymbol{x}$ in a multiple regression context with design matrix $\boldsymbol{A}$ (Section 4.6.4, Example 4.14), we encountered the form

$$E(\boldsymbol{x}) = \boldsymbol{A\theta}.$$

In both of these cases, fundamental explanatory or predictive structure is captured by the specification of the conditional expectation, and this aspect can in many cases be thought through separately from the choice of a particular specification of full probability distribution.

Identifiability

A parametric model for which an element of the parametrisation is redundant is said to be non-identified. Such models are often introduced at an early stage of model building (particularly in econometrics) in order to include all parameters which may originally be thought to be relevant. Identifiability is a property of the parametric model, but a Bayesian analysis of a non-identified model is always possible if a proper prior on all the parameters is specified. For detailed discussion of this issue, see Morales (1971), Drèze (1974), Kadane (1974), Florens and Mouchart (1986), Hills (1987) and Florens *et al.* (1990, Section 4.5).

Robustness Considerations

For concreteness, in our earlier discussion of these examples we assumed that the $p(\boldsymbol{x} \mid \boldsymbol{\theta})$ terms were specified in terms of normal distributions. As we demonstrated earlier in this chapter, under the a priori assumption of appropriate invariances, or on the basis of experience with particular applications, such a specification may well be natural and acceptable. However, in many situations the choice of a specific probability distribution may feel a much less "secure" component of the overall modelling process than the choice of conditional expectation structure.

For example, past experience might suggest that departures of observables from assumed expectations resemble a symmetric bell-shaped distribution centred around zero. But a number of families of distributions match these general characteristics, including the normal, Student and logistic families. Faced with a seemingly arbitrary choice, what can be done in a situation like this to obtain further insight and guidance? Does the choice matter? Or are subsequent inferences or predictions robust against such choices?

An exactly analogous problem arises with the choice of mathematical specifications for the prior model component.

In robustness considerations, theoretical analysis—sometimes referred to as "what if?" analysis—has an interesting role to play. Using the inference machinery which we shall develop in Chapter 5, the desired insight and guidance can often be obtained by studying mathematically the ways in which the various "arbitrary" choices affect subsequent forms of inferences and predictions. For example, a "what if?" analysis might consider the effect of a single, aberrant, outlying observation on inferences for main effects in a multiway layout under the alternative assumptions of a normal or Student parametric model distribution. It can be shown that the influence of the aberrant observation is large under the normal assumption, but negligible under the Student assumption, thus providing a potential basis for preferring one or other of the otherwise seemingly arbitrary choices.

More detailed analysis of such robustness issues will be given in Section 5.6.3.

Chapter 5

Inference

Summary

The role of Bayes' theorem in the updating of beliefs about observables in the light of new information is identified and related to conventional mechanisms of predictive and parametric inference. The roles of sufficiency, ancillarity and stopping rules in such inference processes are also examined. Forms of common statistical decisions and inference summaries are introduced and the problems of implementing Bayesian procedures are discussed at length. In particular, conjugate, asymptotic and reference forms of analysis and numerical approximation approaches are detailed.

5.1 THE BAYESIAN PARADIGM

5.1.1 Observables, Beliefs and Models

Our development has focused on the foundational issues which arise when we aspire to formal quantitative coherence in the context of decision making in situations of uncertainty. This development, in combination with an operational approach to the basic concepts, has led us to view the problem of statistical modelling as that of identifying or selecting particular forms of representation of beliefs about observables.

For example, in the case of a sequence $x_1, x_2, \ldots$, of $0-1$ random quantities for which beliefs correspond to a judgement of infinite exchangeability, Proposition 4.1, (de Finetti's theorem) identifies the representation of the joint mass function for $x_1, \ldots, x_n$ as having the form

$$p(x_1, \ldots, x_n) = \int_0^1 \prod_{i=1}^n \theta^{x_i}(1-\theta)^{1-x_i}\, dQ(\theta),$$

for some choice of distribution Q over the interval $[0, 1]$.

More generally, for sequences of real-valued or integer-valued random quantities, $x_1, x_2, \ldots$, we have seen, in Sections 4.3 – 4.5, that beliefs which combine judgements of exchangeability with some form of further structure (either in terms of invariance or sufficient statistics), often lead us to work with representations of the form

$$p(x_1, \ldots, x_n) = \int_{\Re^k} \prod_{i=1}^n p(x_i \mid \boldsymbol{\theta})\, dQ(\boldsymbol{\theta}),$$

where $p(x \mid \boldsymbol{\theta})$ denotes a specified form of labelled family of probability distributions and Q is some choice of distribution over $\Re^k$.

Such representations, and the more complicated forms considered in Section 4.6, exhibit the various ways in which the element of primary significance from the subjectivist, operationalist standpoint, namely the *predictive model* of beliefs about observables, can be thought of *as if* constructed from a *parametric model* together with a *prior distribution* for the labelling parameter.

Our primary concern in this chapter will be with the way in which the updating of beliefs in the light of new information takes place within the framework of such representations.

5.1.2 The Role of Bayes' Theorem

In its simplest form, within the formal framework of predictive model belief distributions derived from quantitative coherence considerations, the problem corresponds to identifying the joint conditional density of

$$p(x_{n+1}, \ldots, x_{n+m} \mid x_1, \ldots, x_n)$$

for any $m \geq 1$, given, for any $n \geq 1$, the form of representation of the joint density $p(x_1, \ldots, x_n)$.

In general, of course, this simply reduces to calculating

$$p(x_{n+1}, \ldots, x_{n+m} \mid x_1, \ldots, x_n) = \frac{p(x_1, \ldots, x_{n+m})}{p(x_1, \ldots, x_n)}$$

and, in the absence of further structure, there is little more that can be said. However, when the predictive model admits a representation in terms of parametric models and prior distributions, the learning process can be essentially identified, in conventional terminology, with the standard parametric form of Bayes' theorem.

Thus, for example, if we consider the general parametric form of representation for an exchangeable sequence, with $dQ(\boldsymbol{\theta})$ having density representation, $p(\boldsymbol{\theta})d\boldsymbol{\theta}$, we have

$$p(x_1, \ldots, x_n) = \int \prod_{i=1}^{n} p(x_i \,|\, \boldsymbol{\theta}) p(\boldsymbol{\theta})\, d\boldsymbol{\theta},$$

from which it follows that

$$\begin{aligned} p(x_{n+1}, \ldots, x_{n+m} \,|\, x_1, \ldots, x_n) &= \frac{\int \prod_{i=1}^{n+m} p(x_i \,|\, \boldsymbol{\theta}) p(\boldsymbol{\theta})\, d\boldsymbol{\theta}}{\int \prod_{i=1}^{n} p(x_i \,|\, \boldsymbol{\theta}) p(\boldsymbol{\theta})\, d\boldsymbol{\theta}} \\ &= \int \prod_{i=n+1}^{n+m} p(x_i \,|\, \boldsymbol{\theta}) p(\boldsymbol{\theta} \,|\, x_1, \ldots, x_n)\, d\boldsymbol{\theta}, \end{aligned}$$

where

$$p(\boldsymbol{\theta} \,|\, x_1, \ldots, x_n) = \frac{\prod_{i=1}^{n} p(x_i \,|\, \boldsymbol{\theta}) p(\boldsymbol{\theta})}{\int \prod_{i=1}^{n} p(x_i \,|\, \boldsymbol{\theta}) p(\boldsymbol{\theta})\, d\boldsymbol{\theta}}.$$

This latter relationship is just *Bayes' theorem*, expressing the *posterior density* for $\boldsymbol{\theta}$, given $x_1, \ldots, x_n$, in terms of the *parametric model* for $x_1, \ldots, x_n$ given $\boldsymbol{\theta}$, and the *prior density* for $\boldsymbol{\theta}$. The (conditional, or posterior) predictive model for $x_{n+1}, \ldots, x_{n+m}$, given $x_1, \ldots, x_n$ is seen to have precisely the same general form of representation as the initial predictive model, except that the corresponding parametric model component is now integrated with respect to the posterior distribution of the parameter, rather than with respect to the prior distribution.

We recall from Chapter 4 that, considered as a function of $\boldsymbol{\theta}$,

$$\text{lik}(\boldsymbol{\theta} \,|\, x_1, \ldots, x_n) = p(x_1, \ldots, x_n \,|\, \boldsymbol{\theta})$$

is usually referred to as the *likelihood function*. A formal definition of such a concept is, however, problematic; for details, see Bayarri *et al.* (1988) and Bayarri and DeGroot (1992b).

5.1.3 Predictive and Parametric Inference

Given our operationalist concern with modelling and reporting uncertainty in terms of *observables*, it is not surprising that Bayes' theorem, in its role as the key to a coherent learning process for *parameters*, simply appears as a step within the predictive process of passing from

$$p(x_1, \ldots, x_n) = \int p(x_1, \ldots, x_n \,|\, \boldsymbol{\theta}) p(\boldsymbol{\theta})\, d\boldsymbol{\theta}$$

to

$$p(x_{n+1},\ldots,x_{n+m}\,|\,x_1,\ldots,x_n) = \int p(x_{n+1},\ldots,x_{n+m}\,|\,\boldsymbol{\theta})p(\boldsymbol{\theta}\,|\,x_1,\ldots,x_n)\,d\boldsymbol{\theta},$$

by means of

$$p(\boldsymbol{\theta}\,|\,x_1,\ldots,x_n) = \frac{p(x_1,\ldots,x_n\,|\,\boldsymbol{\theta})p(\boldsymbol{\theta})}{\int p(x_1,\ldots,x_n\,|\,\boldsymbol{\theta})p(\boldsymbol{\theta})\,d\boldsymbol{\theta}}.$$

Writing $\boldsymbol{y} = \{y_1,\ldots,y_m\} = \{x_{n+1},\ldots,x_{n+m}\}$ to denote future (or, as yet unobserved) quantities and $\boldsymbol{x} = \{x_1,\ldots,x_n\}$ to denote the already observed quantities, these relations may be re-expressed more simply as

$$p(\boldsymbol{x}) = \int p(\boldsymbol{x}\,|\,\boldsymbol{\theta})p(\boldsymbol{\theta})\,d\boldsymbol{\theta},$$

$$p(\boldsymbol{y}\,|\,\boldsymbol{x}) = \int p(\boldsymbol{y}\,|\,\boldsymbol{\theta})p(\boldsymbol{\theta}\,|\,\boldsymbol{x})\,d\boldsymbol{\theta}$$

and

$$p(\boldsymbol{\theta}\,|\,\boldsymbol{x}) = p(\boldsymbol{x}\,|\,\boldsymbol{\theta})p(\boldsymbol{\theta})/p(\boldsymbol{x}).$$

However, as we noted on many occasions in Chapter 4, if we proceed purely formally, from an operationalist standpoint it is not at all clear, at first sight, how we should interpret "beliefs about parameters", as represented by $p(\boldsymbol{\theta})$ and $p(\boldsymbol{\theta}\,|\,\boldsymbol{x})$, or even whether such "beliefs" have any intrinsic interest. We also answered these questions on many occasions in Chapter 4, by noting that, in all the forms of predictive model representations we considered, the parameters had interpretations as strong law limits of (appropriate functions of) observables. Thus, for example, in the case of the infinitely exchangeable $0-1$ sequence (Section 4.3.1) beliefs about θ correspond to beliefs about what the long-run frequency of 1's would be in a future sample; in the context of a real-valued exchangeable sequence with centred spherical symmetry (Section 4.4.1), beliefs about μ and σ^2, respectively, correspond to beliefs about what the large sample mean, and the large sample mean sum of squares about the sample mean would be, in a future sample.

Inference about parameters is thus seen to be a limiting form of predictive inference about observables. This means that, although the predictive form is primary, and the role of parametric inference is typically that of an intermediate structural step, parametric inference will often itself be the legitimate end-product of a statistical analysis in situations where interest focuses on quantities which could be viewed as large-sample functions of observables. Either way, parametric inference is of considerable importance for statistical analysis in the context of the models we are mainly concerned with in this volume.

When a parametric form is involved simply as an intermediate step in the predictive process, we have seen that $p(\boldsymbol{\theta} \mid x_1, \ldots, x_n)$, the full joint posterior density for the parameter vector $\boldsymbol{\theta}$ is all that is required. However, if we are concerned with parametric inference *per se*, we may be interested in only some subset, $\boldsymbol{\phi}$, of the components of $\boldsymbol{\theta}$, or in some transformed subvector of parameters, $\boldsymbol{g}(\boldsymbol{\theta})$. For example, in the case of a real-valued sequence we may only be interested in the large-sample mean and not in the variance; or in the case of two $0-1$ sequences we may only be interested in the difference in the long-run frequencies.

In the case of interest in a subvector of $\boldsymbol{\theta}$, let us suppose that the full parameter vector can be partitioned into $\boldsymbol{\theta} = \{\boldsymbol{\phi}, \boldsymbol{\lambda}\}$, where $\boldsymbol{\phi}$ is the subvector of interest, and $\boldsymbol{\lambda}$ is the complementary subvector of $\boldsymbol{\theta}$, often referred to, in this context, as the vector of *nuisance parameters*. Since

$$p(\boldsymbol{\theta} \mid \boldsymbol{x}) = \frac{p(\boldsymbol{x} \mid \boldsymbol{\theta})p(\boldsymbol{\theta})}{p(\boldsymbol{x})},$$

the (marginal) posterior density for $\boldsymbol{\phi}$ is given by

$$p(\boldsymbol{\phi} \mid \boldsymbol{x}) = \int p(\boldsymbol{\theta} \mid \boldsymbol{x})\, d\boldsymbol{\lambda} = \int p(\boldsymbol{\phi}, \boldsymbol{\lambda} \mid \boldsymbol{x})\, d\boldsymbol{\lambda},$$

where

$$p(\boldsymbol{x}) = \int p(\boldsymbol{x} \mid \boldsymbol{\theta})p(\boldsymbol{\theta})\, d\boldsymbol{\theta} = \int p(\boldsymbol{x} \mid \boldsymbol{\phi}, \boldsymbol{\lambda})p(\boldsymbol{\phi}, \boldsymbol{\lambda}) d\boldsymbol{\phi}\, d\boldsymbol{\lambda},$$

with all integrals taken over the full range of possible values of the relevant quantities.

Expressed in terms of the notation introduced in Section 3.2.4, we have

$$p(\boldsymbol{x} \mid \boldsymbol{\phi}, \boldsymbol{\lambda}) \otimes p(\boldsymbol{\phi}, \boldsymbol{\lambda}) \equiv p(\boldsymbol{\phi}, \boldsymbol{\lambda} \mid \boldsymbol{x}),$$

$$p(\boldsymbol{\phi}, \boldsymbol{\lambda} \mid \boldsymbol{x}) \underset{\boldsymbol{\phi}}{\longrightarrow} p(\boldsymbol{\phi} \mid \boldsymbol{x}).$$

In some situations, the prior specification $p(\boldsymbol{\phi}, \boldsymbol{\lambda})$ may be most easily arrived at through the specification of $p(\boldsymbol{\lambda} \mid \boldsymbol{\phi})p(\boldsymbol{\phi})$. In such cases, we note that we could first calculate the *integrated likelihood* for $\boldsymbol{\phi}$,

$$p(\boldsymbol{x} \mid \boldsymbol{\phi}) = \int p(\boldsymbol{x} \mid \boldsymbol{\phi}, \boldsymbol{\lambda})p(\boldsymbol{\lambda} \mid \boldsymbol{\phi})\, d\boldsymbol{\lambda},$$

and subsequently proceed without any further need to consider the nuisance parameters, since

$$p(\boldsymbol{\phi} \mid \boldsymbol{x}) = \frac{p(\boldsymbol{x} \mid \boldsymbol{\phi})p(\boldsymbol{\phi})}{p(\boldsymbol{x})}.$$

In the case where interest is focused on a transformed parameter vector, $g(\boldsymbol{\theta})$, we proceed using standard change-of-variable probability techniques as described in Section 3.2.4. Suppose first that $\boldsymbol{\psi} = g(\boldsymbol{\theta})$ is a one-to-one differentiable transformation of $\boldsymbol{\theta}$. It then follows that

$$p_{\psi}(\boldsymbol{\psi} \mid \boldsymbol{x}) = p_{\theta}(\boldsymbol{g}^{-1}(\boldsymbol{\psi}) \mid \boldsymbol{x}) \mid \boldsymbol{J}_{g^{-1}}(\boldsymbol{\psi}) \mid ,$$

where

$$\boldsymbol{J}_{g^{-1}}(\boldsymbol{\psi}) = \frac{\partial \boldsymbol{g}^{-1}(\boldsymbol{\psi})}{\partial \boldsymbol{\psi}}$$

is the Jacobian of the inverse transformation $\boldsymbol{\theta} = \boldsymbol{g}^{-1}(\boldsymbol{\psi})$. Alternatively, by substituting $\boldsymbol{\theta} = \boldsymbol{g}^{-1}(\boldsymbol{\psi})$, we could write $p(\boldsymbol{x} \mid \boldsymbol{\theta})$ as $p(\boldsymbol{x} \mid \boldsymbol{\psi})$, and replace $p(\boldsymbol{\theta})$ by $p_{\theta}(\boldsymbol{g}^{-1}(\boldsymbol{\psi})) \mid \boldsymbol{J}_{g^{-1}}(\boldsymbol{\psi}) \mid$, to obtain $p(\boldsymbol{\psi} \mid \boldsymbol{x}) = p(\boldsymbol{x} \mid \boldsymbol{\psi})p(\boldsymbol{\psi})/p(\boldsymbol{x})$ directly.

If $\boldsymbol{\psi} = g(\boldsymbol{\theta})$ has dimension less than $\boldsymbol{\theta}$, we can typically define $\boldsymbol{\gamma} = (\boldsymbol{\psi}, \boldsymbol{\omega}) = \boldsymbol{h}(\boldsymbol{\theta})$, for some $\boldsymbol{\omega}$ such that $\boldsymbol{\gamma} = \boldsymbol{h}(\boldsymbol{\theta})$ is a one-to-one differentiable transformation, and then proceed in two steps. We first obtain

$$p(\boldsymbol{\psi}, \boldsymbol{\omega} \mid \boldsymbol{x}) = p_{\theta}(\boldsymbol{h}^{-1}(\boldsymbol{\gamma}) \mid \boldsymbol{x}) \mid \boldsymbol{J}_{h^{-1}}(\boldsymbol{\gamma}) \mid ,$$

where

$$\boldsymbol{J}_{h^{-1}}(\boldsymbol{\gamma}) = \frac{\partial \boldsymbol{h}^{-1}(\boldsymbol{\gamma})}{\partial \boldsymbol{\gamma}} ,$$

and then marginalise to

$$p(\boldsymbol{\psi} \mid \boldsymbol{x}) = \int p(\boldsymbol{\psi}, \boldsymbol{\omega} \mid \boldsymbol{x}) \, d\boldsymbol{\omega}.$$

These techniques will be used extensively in later sections of this chapter.

In order to keep the presentation of these basic manipulative techniques as simple as possible, we have avoided introducing additional notation for the ranges of possible values of the various parameters. In particular, all integrals have been assumed to be over the full ranges of the possible parameter values.

In general, this notational economy will cause no confusion and the parameter ranges will be clear from the context. However, there are situations where specific constraints on parameters are introduced and need to be made explicit in the analysis. In such cases, notation for ranges of parameter values will typically also need to be made explicit.

Consider, for example, a parametric model, $p(\boldsymbol{x} \mid \boldsymbol{\theta})$, together with a prior specification $p(\boldsymbol{\theta})$, $\boldsymbol{\theta} \in \Theta$, for which the posterior density, suppressing explicit use of Θ, is given by

$$p(\boldsymbol{\theta} \mid \boldsymbol{x}) = \frac{p(\boldsymbol{x} \mid \boldsymbol{\theta})p(\boldsymbol{\theta})}{\int p(\boldsymbol{x} \mid \boldsymbol{\theta})p(\boldsymbol{\theta}) \, d\boldsymbol{\theta}} .$$

Now suppose that it is required to specify the posterior subject to the constraint $\boldsymbol{\theta} \in \Theta_0 \subset \Theta$, where $\int_{\Theta_0} p(\boldsymbol{\theta}) d\boldsymbol{\theta} > 0$.

Defining the constrained prior density by

$$p_0(\boldsymbol{\theta}) = \frac{p(\boldsymbol{\theta})}{\int_{\Theta_0} p(\boldsymbol{\theta}) d(\boldsymbol{\theta})}, \quad \boldsymbol{\theta} \in \Theta_0,$$

we obtain, using Bayes' theorem,

$$p(\boldsymbol{\theta} \mid \boldsymbol{x}, \boldsymbol{\theta} \in \Theta_0) = \frac{p(\boldsymbol{x} \mid \boldsymbol{\theta}) p_0(\boldsymbol{\theta})}{\int_{\Theta_0} p(\boldsymbol{x} \mid \boldsymbol{\theta}) p_0(\boldsymbol{\theta}) d\boldsymbol{\theta}}, \quad \boldsymbol{\theta} \in \Theta_0.$$

From this, substituting for $p_0(\boldsymbol{\theta})$ in terms of $p(\boldsymbol{\theta})$ and dividing both numerator and denominator by

$$p(\boldsymbol{x}) = \int_{\Theta} p(\boldsymbol{x} \mid \boldsymbol{\theta}) p(\boldsymbol{\theta}) d\boldsymbol{\theta},$$

we obtain

$$p(\boldsymbol{\theta} \mid \boldsymbol{x}, \boldsymbol{\theta} \in \Theta_0) = \frac{p(\boldsymbol{\theta} \mid \boldsymbol{x})}{\int_{\Theta_0} p(\boldsymbol{\theta} \mid \boldsymbol{x}) \, d\boldsymbol{\theta}}, \quad \boldsymbol{\theta} \in \Theta_0,$$

expressing the constraint in terms of the unconstrained posterior (a result which could, of course, have been obtained by direct, straightforward conditioning).

Numerical methods are often necessary to analyze models with constrained parameters; see Gelfand *et al.* (1992) for the use of Gibbs sampling in this context.

5.1.4 Sufficiency, Ancillarity and Stopping Rules

The concepts of predictive and parametric sufficient statistics were introduced in Section 4.5.2, and shown to be equivalent, within the framework of the kinds of models we are considering in this volume. In particular, it was established that a (minimal) sufficient statistic, $\boldsymbol{t}(\boldsymbol{x})$, for $\boldsymbol{\theta}$, in the context of a parametric model $p(\boldsymbol{x} \mid \boldsymbol{\theta})$, can be characterised by either of the conditions

$$p(\boldsymbol{\theta} \mid \boldsymbol{x}) = p\big(\boldsymbol{\theta} \mid \boldsymbol{t}(\boldsymbol{x})\big), \qquad \text{for all } p(\boldsymbol{\theta}),$$

or

$$p\big(\boldsymbol{x} \mid \boldsymbol{t}(\boldsymbol{x}), \boldsymbol{\theta}\big) = p\big(\boldsymbol{x} \mid \boldsymbol{t}(\boldsymbol{x})\big).$$

The important implication of the concept is that $\boldsymbol{t}(\boldsymbol{x})$ serves as a sufficient summary of the complete data $\boldsymbol{x}$ in forming any required revision of beliefs. The resulting data reduction often implies considerable simplification in modelling and analysis. In many cases, the sufficient statistic $\boldsymbol{t}(\boldsymbol{x})$ can itself be partitioned into two component statistics, $\boldsymbol{t}(\boldsymbol{x}) = [\boldsymbol{a}(\boldsymbol{x}), \boldsymbol{s}(\boldsymbol{x})]$ such that, for all $\boldsymbol{\theta}$,

$$\begin{aligned} p(\boldsymbol{t}(\boldsymbol{x}) \mid \boldsymbol{\theta}) &= p(\boldsymbol{s}(\boldsymbol{x}) \mid \boldsymbol{a}(\boldsymbol{x}), \boldsymbol{\theta}) p(\boldsymbol{a}(\boldsymbol{x}) \mid \boldsymbol{\theta}) \\ &= p(\boldsymbol{s}(\boldsymbol{x}) \mid \boldsymbol{a}(\boldsymbol{x}), \boldsymbol{\theta}) p(\boldsymbol{a}(\boldsymbol{x})). \end{aligned}$$

It then follows that, for any choice of $p(\boldsymbol{\theta})$,

$$\begin{aligned} p(\boldsymbol{\theta} \mid \boldsymbol{x}) = p(\boldsymbol{\theta} \mid \boldsymbol{t}(\boldsymbol{x})) &\propto p(\boldsymbol{t}(\boldsymbol{x}) \mid \boldsymbol{\theta}) p(\boldsymbol{\theta}) \\ &\propto p(\boldsymbol{s}(\boldsymbol{x}) \mid \boldsymbol{a}(\boldsymbol{x}), \boldsymbol{\theta}) p(\boldsymbol{\theta}), \end{aligned}$$

so that, in the prior to posterior inference process defined by Bayes' theorem, it suffices to use $p(\boldsymbol{s}(\boldsymbol{x}) \mid \boldsymbol{a}(\boldsymbol{x}), \boldsymbol{\theta})$, rather than $p(\boldsymbol{t}(\boldsymbol{x}) \mid \boldsymbol{\theta})$ as the likelihood function. This further simplification motivates the following definition.

Definition 5.1. ***(Ancillary statistic)****. A statistic, $\boldsymbol{a}(\boldsymbol{x})$, is said to be ancillary, with respect to $\boldsymbol{\theta}$ in a parametric model $p(\boldsymbol{x} \mid \boldsymbol{\theta})$, if $p(\boldsymbol{a}(\boldsymbol{x}) \mid \boldsymbol{\theta}) = p(\boldsymbol{a}(\boldsymbol{x}))$ for all values of $\boldsymbol{\theta}$.*

Example 5.1. ***(Bernoulli model).*** In Example 4.5, we saw that for the Bernoulli parametric model

$$p(x_1, \ldots, x_n \mid \theta) = \prod_{i=1}^{n} p(x_i \mid \theta) = \theta^{r_n}(1-\theta)^{n-r_n},$$

which only depends on n and $r_n = x_1 + \cdots + x_n$. Thus, $\boldsymbol{t}_n = [n, r_n]$ provides a minimal sufficient statistic, and one may work in terms of the joint probability function $p(n, r_n \mid \theta)$.

If we now write

$$p(n, r_n \mid \theta) = p(r_n \mid n, \theta) p(n \mid \theta),$$

and make the assumption that, for all $n \geq 1$, the mechanism by which the sample size, n, is arrived at does not depend on θ, so that $p(n \mid \theta) = p(n)$, $n \geq 1$, we see that n *is ancillary for* θ, in the sense of Definition 5.1. It follows that prior to posterior inference for θ can therefore proceed on the basis of

$$p(\theta \mid \boldsymbol{x}) = p(\theta \mid n, r_n) \propto p(r_n \mid n, \theta) p(\theta),$$

for any choice of $p(\theta)$, $0 \leq \theta \leq 1$. From Corollary 4.1, we see that

$$\begin{aligned} p(r_n \mid n, \theta) &= \binom{n}{r_n} \theta^{r_n}(1-\theta)^{n-r_n}, \qquad 0 \leq r_n \leq n, \\ &= \mathrm{Bi}(r_n \mid \theta, n), \end{aligned}$$

so that inferences in this case can be made as if we had adopted a *binomial parametric model.* However, if we write

$$p(n, r_n \mid \theta) = p(n \mid r_n, \theta)p(r_n \mid \theta)$$

and make the assumption that, for all $r_n \geq 1$, termination of sampling is governed by a mechanism for selecting r_n, which does not depend on θ, so that $p(r_n \mid \theta) = p(r_n), r_n \geq 1$, we see that r_n is ancillary for θ, in the sense of Definition 5.1. It follows that prior to posterior inference for θ can therefore proceed on the basis of

$$p(\theta \mid \boldsymbol{x}) = p(\theta \mid n, r_n) \propto p(n \mid r_n, \theta)p(\theta),$$

for any choice of $p(\theta), 0 < \theta \leq 1$. It is easily verified that

$$\begin{aligned} p(n \mid r_n, \theta) &= \binom{n-1}{r_n - 1}\theta^{r_n}(1-\theta)^{n-r_n}, \qquad n \geq r_n, \\ &= \text{Nb}(n \mid \theta, r_n) \end{aligned}$$

(see Section 3.2.2), so that inferences in this case can be made as if we had adopted a *negative-binomial parametric model.*

We note, incidentally, that whereas in the binomial case it makes sense to consider $p(\theta)$ as specified over $0 \leq \theta \leq 1$, in the negative-binomial case it may only make sense to think of $p(\theta)$ as specified over $0 < \theta \leq 1$, since $p(r_n \mid \theta = 0) = 0$, for all $r_n \geq 1$.

So far as prior to posterior inference for θ is concerned, we note that, for any specified $p(\theta)$, and assuming that either $p(n \mid \theta) = p(n)$ or $p(r_n \mid \theta) = p(r_n)$, we obtain

$$p(\theta \mid x_1, \ldots, x_n) = p(\theta \mid n, r_n) \propto \theta^{r_n}(1-\theta)^{n-r_n}p(\theta)$$

since, considered as functions of θ,

$$p(r_n \mid n, \theta) \propto p(n \mid r_n, \theta) \propto \theta^{r_n}(1-\theta)^{n-r_n}.$$

The last part of the above example illustrates a general fact about the mechanism of parametric Bayesian inference which is trivially obvious; namely, *for any specified $p(\boldsymbol{\theta})$, if the likelihood functions $p_1(\boldsymbol{x}_1 \mid \boldsymbol{\theta}), p_2(\boldsymbol{x}_2 \mid \boldsymbol{\theta})$ are proportional as functions of $\boldsymbol{\theta}$, the resulting posterior densities for $\boldsymbol{\theta}$ are identical.* It turns out, as we shall see in Appendix B, that many non-Bayesian inference procedures do not lead to identical inferences when applied to such proportional likelihoods. The assertion that they *should*, the so-called *Likelihood Principle*, is therefore a controversial issue among statisticians . In contrast, in the Bayesian inference context described above, this is a straightforward consequence of Bayes' theorem, rather than an imposed "principle". Note, however, that the above remarks are predicated on a specified $p(\boldsymbol{\theta})$. It may be, of course, that knowledge of the particular sampling mechanism employed has implications for the specification of $p(\boldsymbol{\theta})$, as illustrated, for example, by the comment above concerning negative-binomial sampling and the restriction to $0 < \theta \leq 1$.

Although the likelihood principle is implicit in Bayesian statistics, it was developed as a separate principle by Barnard (1949), and became a focus of interest when Birnbaum (1962) showed that it followed from the widely accepted sufficiency and conditionality principles. Berger and Wolpert (1984/1988) provide an extensive discussion of the likelihood principle and related issues. Other relevant references are Barnard *et al.* (1962), Fraser (1963), Pratt (1965), Barnard (1967), Hartigan (1967), Birnbaum (1968, 1978), Durbin (1970), Basu (1975), Dawid (1983a), Joshi (1983), Berger (1985b), Hill (1987) and Bayarri *et al.* (1988).

Example 5.1 illustrates the way in which ancillary statistics often arise naturally as a consequence of the way in which data are collected. In general, it is very often the case that the sample size, n, is fixed in advance and that inferences are automatically made conditional on n, without further reflection. It is, however, perhaps not obvious that inferences can be made conditional on n if the latter has arisen as a result of such familiar imperatives as "stop collecting data when you feel tired", or "when the research budget runs out". The kind of analysis given above makes it intuitively clear that such conditioning is, in fact, valid, provided that the mechanism which has led to n "does not depend on $\boldsymbol{\theta}$". This latter condition may, however, not always be immediately obviously transparent, and the following definition provides one version of a more formal framework for considering sampling mechanisms and their dependence on model parameters.

Definition 5.2. (***Stopping rule***). *A* ***stopping rule***, *h, for (sequential) sampling from a sequence of observables $x_1 \in X_1, x_2 \in X_2, \ldots$, is a sequence of functions $h_n : X_1 \times \cdots \times X_n \to [0, 1]$, such that, if $\boldsymbol{x}_{(n)} = (x_1, \ldots, x_n)$ is observed, then sampling is terminated with probability $h_n(\boldsymbol{x}_{(n)})$; otherwise, the $(n + 1)$th observation is made. A stopping rule is* **proper** *if the induced probability distribution $p_h(n), n = 1, 2, \ldots$, for final sample size guarantees that the latter is finite. The rule is* **deterministic** *if $h_n(\boldsymbol{x}_{(n)}) \in \{0, 1\}$ for all $(n, \boldsymbol{x}_{(n)})$; otherwise, it is a* **randomised** *stopping rule.*

In general, we must regard the data resulting from a sampling mechanism defined by a stopping rule h as consisting of $(n, \boldsymbol{x}_{(n)})$, the sample size, together with the observed quantities $x_1, \ldots, x_n$. A parametric model for these data thus involves a probability density of the form $p(n, \boldsymbol{x}_{(n)} \mid \boldsymbol{h}, \boldsymbol{\theta})$, conditioning both on the stopping rule (i.e., sampling mechanism) and on an underlying labelling parameter $\boldsymbol{\theta}$. But, either through unawareness or misapprehension, this is typically ignored and, instead, we act as if the actual observed sample size n had been fixed in advance, in effect assuming that

$$p(n, \boldsymbol{x}_{(n)} \mid \boldsymbol{h}, \boldsymbol{\theta}) = p(\boldsymbol{x}_{(n)} \mid n, \boldsymbol{\theta}) = p(\boldsymbol{x}_{(n)} \mid \boldsymbol{\theta}),$$

using the standard notation we have hitherto adopted for fixed n. The important question that now arises is the following: under what circumstances, if any, can we proceed to make inferences about $\boldsymbol{\theta}$ on the basis of this (generally erroneous!) assumption, without considering explicit conditioning on the actual form of $\boldsymbol{h}$? Let us first consider a simple example.

Example 5.2. ***("Biased" stopping rule for a Bernoulli sequence).*** Suppose, given θ, that $x_1, x_2, \ldots$ may be regarded as a sequence of independent Bernoulli random quantities with $p(x_i \mid \theta) = \text{Bi}(x_i \mid \theta, 1)$, $x_i = 0, 1$, and that a sequential sample is to be obtained using the deterministic stopping rule $\boldsymbol{h}$, defined by: $h_1(1) = 1$, $h_1(0) = 0$, $h_2(x_1, x_2) = 1$ for all x_1, x_2. In other words, if there is a success on the first trial, sampling is terminated (resulting in $n = 1$, $x_1 = 1$); otherwise, two observations are obtained (resulting in either $n = 2$, $x_1 = 0$, $x_2 = 0$ or $n = 2$, $x_1 = 0$, $x_2 = 1$).

At first sight, it might appear essential to take explicit account of $\boldsymbol{h}$ in making inferences about θ, since the sampling procedure seems designed to bias us towards believing in large values of θ. Consider, however, the following detailed analysis:

$$\begin{aligned}p(n = 1, x_1 = 1 \mid \boldsymbol{h}, \theta) &= p(x_1 = 1 \mid n = 1, \boldsymbol{h}, \theta)p(n = 1 \mid \boldsymbol{h}, \theta)\\ &= 1 \cdot p(x_1 = 1 \mid \theta) = p(x_1 = 1 \mid \theta)\end{aligned}$$

and, for $x = 0, 1$,

$$\begin{aligned}p(n = 2, x_1 = 0, x_2 = x \mid \boldsymbol{h}, \theta) &= p(x_1 = 0, x_2 = x \mid n = 2, \boldsymbol{h}, \theta)p(n = 2 \mid \boldsymbol{h}, \theta)\\ = p(x_1 = 0|n = 2, \boldsymbol{h}, \theta)&p(x_2 = x \mid x_1 = 0, n = 2, \boldsymbol{h}, \theta)p(n = 2 \mid \boldsymbol{h}, \theta)\\ &= 1 \cdot p(x_2 = x \mid x_1 = 0, \theta)p(x_1 = 0 \mid \theta)\\ &= p(x_2 = x, x_1 = 0 \mid \theta).\end{aligned}$$

Thus, for all $(n, \boldsymbol{x}_{(n)})$ having non-zero probability, we obtain in this case

$$p(n, \boldsymbol{x}_{(n)} \mid \boldsymbol{h}, \theta) = p(\boldsymbol{x}_{(n)} \mid \theta),$$

the latter considered pointwise as functions of θ (i.e., likelihoods). It then follows trivially from Bayes' theorem that, *for any specified* $p(\theta)$, inferences for θ based on assuming n to have been fixed at its observed value will be identical to those based on a likelihood derived from explicit consideration of $\boldsymbol{h}$.

Consider now a randomised version of this stopping rule which is defined by $h_1(1) = \pi$, $h_1(0) = 0$, $h_2(x_1, x_2) = 1$ for all x_1, x_2. In this case, we have

$$\begin{aligned}p(n = 1, x_1 = 1 \mid \boldsymbol{h}, \theta) &= p(x_1 = 1 \mid n = 1, \boldsymbol{h}, \theta)p(n = 1 \mid \boldsymbol{h}, \theta)\\ &= 1 \cdot \pi \cdot p(x_1 = 1 \mid \theta),\end{aligned}$$

with, for $x = 0, 1$,

$$\begin{aligned}p(n =2, x_1 = 0, x_2 = x \mid \boldsymbol{h}, \theta)&\\ = p(n = 2 \mid x_1 = 0, \boldsymbol{h}, \theta)&\\ \times p(x_1 = 0 \mid \boldsymbol{h}, \theta)&p(x_2 = x \mid x_1 = 0, n = 2, \boldsymbol{h}, \theta)\\ = 1 \cdot p(x_1 = 0 \mid \theta)&p(x_2 = x \mid \theta)\end{aligned}$$

and

$$p(n = 2, x_1 = 1, x_2 = x \mid \boldsymbol{h}, \theta) = p(n = 2 \mid x_1 = 1, \boldsymbol{h}, \theta)p(x_1 = 1 \mid \boldsymbol{h}, \theta)$$
$$\times\, p(x_2 = x \mid x_1 = 1, n = 2, \boldsymbol{h}, \theta)$$
$$= (1 - \pi)p(x_1 = 1 \mid \theta)p(x_2 = x \mid \theta).$$

Thus, for all $(n, \boldsymbol{x}_{(n)})$ having non-zero probability, we again find that

$$p(n, \boldsymbol{x}_{(n)} \mid \boldsymbol{h}, \theta) \propto p(\boldsymbol{x}_{(n)} \mid \theta)$$

as functions of θ, so that the proportionality of the likelihoods once more implies identical inferences from Bayes' theorem, for any given $p(\theta)$.

The analysis of the preceding example showed, perhaps contrary to intuition, that, although seemingly biasing the analysis towards beliefs in larger values of θ, the stopping rule does not in fact lead to a different likelihood from that of the a priori fixed sample size. The following, rather trivial, proposition makes clear that this is true for all stopping rules as defined in Definition 5.2, which we might therefore describe as "likelihood non-informative stopping rules".

Proposition 5.1. ***(Stopping rules are likelihood non-informative***).
For any stopping rule $\boldsymbol{h}$, *for (sequential) sampling from a sequence of observables* $x_1, x_2, \ldots$, *having fixed sample size parametric model* $p(\boldsymbol{x}_{(n)} \mid n, \boldsymbol{\theta}) = p(\boldsymbol{x}_{(n)} \mid \boldsymbol{\theta})$,

$$p(n, \boldsymbol{x}_{(n)} \mid \boldsymbol{h}, \boldsymbol{\theta}) \propto p(\boldsymbol{x}_{(n)} \mid \boldsymbol{\theta}), \quad \boldsymbol{\theta} \in \Theta,$$

for all $(n, \boldsymbol{x}_{(n)})$ *such that* $p(n, \boldsymbol{x}_{(n)} \mid \boldsymbol{h}, \boldsymbol{\theta}) \neq 0$.

Proof. This follows straightforwardly on noting that

$$p(n, \boldsymbol{x}_{(n)} \mid \boldsymbol{h}, \boldsymbol{\theta}) = \left[h_n(\boldsymbol{x}_{(n)}) \prod_{i=1}^{n-1} \left(1 - h_i(\boldsymbol{x}_{(i)})\right)\right] p(\boldsymbol{x}_{(n)} \mid \boldsymbol{\theta}),$$

and that the term in square brackets does not depend on $\boldsymbol{\theta}$. ◁

Again, it is a trivial consequence of Bayes' theorem that, *for any specified prior density*, prior to posterior inference for $\boldsymbol{\theta}$ given data $(n, \boldsymbol{x}_{(n)})$ obtained using a likelihood non-informative stopping rule h can proceed by acting as if $\boldsymbol{x}_{(n)}$ were obtained using a fixed sample size n. However, a notationally precise rendering of Bayes' theorem,

$$p(\boldsymbol{\theta} \mid n, \boldsymbol{x}_{(n)}, \boldsymbol{h}) \propto p(n, \boldsymbol{x}_{(n)} \mid \boldsymbol{h}, \boldsymbol{\theta})p(\boldsymbol{\theta} \mid \boldsymbol{h})$$
$$\propto p(\boldsymbol{x}_{(n)} \mid \theta)p(\boldsymbol{\theta} \mid \boldsymbol{h}),$$

reveals that *knowledge of h might well affect the specification of the prior density*! It is for this reason that we use the term "likelihood non-informative" rather than just "non-informative" stopping rules. It cannot be emphasised too often that, although it is often convenient for expository reasons to focus at a given juncture on one or other of the "likelihood" and "prior" components of the model, our discussion in Chapter 4 makes clear their basic inseparability in coherent modelling and analysis of beliefs. This issue is highlighted in the following example.

Example 5.3. ***("Biased" stopping rule for a normal mean)***. Suppose, given θ, that $x_1, x_2, \ldots,$ may be regarded as a sequence of independent normal random quantities with $p(x_i \mid \theta) = \mathrm{N}(x_i \mid \theta, 1)$, $x_i \in \Re$. Suppose further that an investigator has a particular concern with the parameter value $\theta = 0$ and wants to stop sampling if $\overline{x}_n = \sum_i x_i/n$ ever takes on a value that is "unlikely", assuming $\theta = 0$ to be true.

For any fixed sample size n, if "unlikely" is interpreted as "an event having probability less than or equal to α", for small α, a possible stopping rule, using the fact that $p(\overline{x}_n \mid n, \theta) = \mathrm{N}(\overline{x}_n \mid \theta, n)$, might be

$$h_n(\boldsymbol{x}_{(n)}) = \begin{cases} 1, & \text{if } |\,\overline{x}_n\,| > k(\alpha)/\sqrt{n} \\ 0, & \text{if } |\,\overline{x}_n\,| \le k(\alpha)/\sqrt{n} \end{cases}$$

for suitable $k(\alpha)$ (for example, $k = 1.96$ for $\alpha = 0.05$, $k = 2.57$ for $\alpha = 0.01$, or $k = 3.31$ for $\alpha = 0.001$). It can be shown, using the law of the iterated logarithm (see, for example, Section 3.2.3), that this is a proper stopping rule, so that termination will certainly occur for some finite n, yielding data $(n, \boldsymbol{x}_{(n)})$. Moreover, defining

$$S_n = \left\{ \boldsymbol{x}_{(n)};\ |\overline{x}_1| \le k(\alpha),\ |\overline{x}_2| \le \frac{k(\alpha)}{\sqrt{2}}, \ldots, \right.$$
$$\left. |\overline{x}_{n-1}| \le \frac{k(\alpha)}{\sqrt{n-1}},\ |\overline{x}_n| > \frac{k(\alpha)}{\sqrt{n}} \right\},$$

we have

$$\begin{aligned} p(n, \boldsymbol{x}_{(n)} \mid \boldsymbol{h}, \theta) &= p(\boldsymbol{x}_{(n)} \mid n, \boldsymbol{h}, \theta) p(n \mid \boldsymbol{h}, \theta) \\ &= p(\boldsymbol{x}_{(n)} \mid S_n, \theta) p(S_n \mid \theta) \\ &= p(\boldsymbol{x}_{(n)} \mid \theta), \end{aligned}$$

as a function of θ, for all $(n, \boldsymbol{x}_{(n)})$ for which the left-hand side is non-zero. It follows that h is a likelihood non-informative stopping rule.

Now consider prior to posterior inference for θ, where, for illustration, we assume the prior specification $p(\theta) = \mathrm{N}(\theta \mid \mu, \lambda)$, with precision $\lambda \simeq 0$, to be interpreted as indicating extremely vague prior beliefs about θ, which take no explicit account of the stopping rule h. Since the latter is likelihood non-informative, we have

$$\begin{aligned} p(\theta \mid \boldsymbol{x}_{(n)}, n) &\propto p(\boldsymbol{x}_{(n)} \mid n, \theta) p(\theta) \\ &\propto p(\overline{x}_n \mid n, \theta) p(\theta) \\ &\propto \mathrm{N}(\overline{x}_n \mid \theta, n) \mathrm{N}(\theta \mid \mu, \lambda) \end{aligned}$$

by virtue of the sufficiency of $(n, \overline{x}_n)$ for the normal parametric model. The right-hand side is easily seen to be proportional to $\exp\{-\frac{1}{2}Q(\theta)\}$, where

$$Q(\theta) = (n + h)\left[\theta - \frac{n\overline{x}_n + \lambda\mu}{n + \lambda}\right]^2,$$

which implies that

$$p(\theta \,|\, \boldsymbol{x}_{(n)}, n) = \mathrm{N}\left(\theta \,\middle|\, \frac{n\overline{x}_n + \lambda\mu}{n + \lambda}, (n + \lambda)\right)$$
$$\simeq \mathrm{N}(\theta \,|\, \overline{x}_n, n)$$

for $\lambda \simeq 0$.

One consequence of this vague prior specification is that, having observed $(n, \boldsymbol{x}_{(n)})$, we are led to the posterior probability statement

$$P\left[\theta \in \left(\overline{x}_n \pm \frac{k(\alpha)}{\sqrt{n}}\right) \middle| n, \overline{x}_n\right] = 1 - \alpha.$$

But the stopping rule $\boldsymbol{h}$ ensures that $|\,\overline{x}_n\,| > k(\alpha)/\sqrt{n}$. This means that the value $\theta = 0$ certainly does not lie in the posterior interval to which someone with initially very vague beliefs would attach a high probability. An investigator *knowing* $\theta = 0$ to be the true value can therefore, by using this stopping rule, mislead someone who, unaware of the stopping rule, acts as if initially very vague.

However, let us now consider an analysis which takes into account the stopping rule. The nature of $\boldsymbol{h}$ might suggest a prior specification $p(\theta \,|\, \boldsymbol{h})$ that recognises $\theta = 0$ as a possibly "special" parameter value, which should be assigned non-zero prior probability (rather than the zero probability resulting from any continuous prior density specification). As an illustration, suppose that we specify

$$p(\theta \,|\, \boldsymbol{h}) = \pi\, 1_{(\theta=0)}(\theta) + (1 - \pi)1_{(\theta\neq 0)}(\theta)\mathrm{N}(\theta \,|\, 0, \lambda_0),$$

which assigns a "spike" of probability, π, to the special value, $\theta = 0$, and assigns $1 - \pi$ times a $\mathrm{N}(\theta \,|\, 0, \lambda_0)$ density to the range $\theta \neq 0$.

Since $\boldsymbol{h}$ is a likelihood non-informative stopping rule and $(n, \overline{x}_n)$ are sufficient statistics for the normal parametric model, we have

$$p(\theta \,|\, n, \boldsymbol{x}_{(n)}, \boldsymbol{h}) \propto \mathrm{N}(\overline{x}_n \,|\, \theta, n)p(\theta \,|\, \boldsymbol{h}).$$

The complete posterior $p(\theta \,|\, n, \boldsymbol{x}_{(n)}, \boldsymbol{h})$ is thus given by

$$\frac{\pi\, 1_{(\theta=0)}(\theta)\mathrm{N}(\overline{x}_n \,|\, 0, n) + (1 - \pi)1_{(\theta\neq 0)}(\theta)\mathrm{N}(\overline{x}_n \,|\, \theta, n)\mathrm{N}(\theta \,|\, 0, \lambda_0)}{\pi\, \mathrm{N}(\overline{x}_n \,|\, 0, n) + (1 - \pi)\int_{-\infty}^{\infty} \mathrm{N}(\overline{x}_n \,|\, \theta, n)\mathrm{N}(\theta \,|\, 0, \lambda_0)d\theta}$$

$$= \pi^* 1_{(\theta=0)}(\theta) + (1 - \pi^*)1_{(\theta\neq 0)}\mathrm{N}\left(\theta \,\middle|\, \frac{n\overline{x}_n}{n + \lambda_0}, n + \lambda_0\right),$$

where, since

$$\int_{-\infty}^{\infty} \mathrm{N}(\overline{x}_n \,|\, \theta, n)\mathrm{N}(\theta \,|\, 0, \lambda_0)d\theta = \mathrm{N}\left(\overline{x}_n \,|\, 0, n\frac{\lambda_0}{n+\lambda_0}\right),$$

it is easily verified that

$$\begin{aligned}\pi^* &= \left\{1 + \frac{1-\pi}{\pi} \cdot \frac{\mathrm{N}(\overline{x}_n \,|\, 0, n\lambda_0(n+\lambda_0)^{-1})}{\mathrm{N}(\overline{x}_n \,|\, 0, n)}\right\}^{-1} \\ &= \left\{1 + \frac{1-\pi}{\pi}\left(1+\frac{n}{\lambda_0}\right)^{-1/2} \exp\left[\tfrac{1}{2}(\sqrt{n}\overline{x}_n)^2\left(1+\frac{\lambda_0}{n}\right)^{-1}\right]\right\}^{-1}.\end{aligned}$$

The posterior distribution thus assigns a "spike" π^* to $\theta = 0$ and assigns $1 - \pi^*$ times a $\mathrm{N}(\theta \,|\, (n+\lambda_0)^{-1}n\overline{x}_n,\, n+\lambda_0)$ density to the range $\theta \neq 0$.

The behaviour of this posterior density, derived from a prior taking account of h, is clearly very different from that of the posterior density based on a vague prior taking no account of the stopping rule. For qualitative insight, consider the case where actually $\theta = 0$ and α has been chosen to be very small, so that $k(\alpha)$ is quite large. In such a case, n is likely to be very large and at the stopping point we shall have $\overline{x}_n \simeq k(\alpha)/\sqrt{n}$. This means that

$$\pi^* \simeq \left[1 + \frac{1-\pi}{\pi}\left(1+\frac{n}{\lambda_0}\right)^{-1/2} \exp\left(\tfrac{1}{2}k^2(\alpha)\right)\right]^{-1} \simeq 1,$$

for large n, so that knowing the stopping rule and then observing that it results in a large sample size leads to an increasing conviction that $\theta = 0$. On the other hand, if θ is appreciably different from 0, the resulting n, and hence π^*, will tend to be small and the posterior will be dominated by the $\mathrm{N}(\theta \,|\, (n+\lambda_0)^{-1}n\overline{x}_n, n+\lambda_0)$ component.

5.1.5 Decisions and Inference Summaries

In Chapter 2, we made clear that our central concern is the representation and revision of beliefs as the basis for decisions. Either beliefs are to be used directly in the choice of an action, or are to be recorded or reported in some selected form, with the possibility or intention of subsequently guiding the choice of a future action.

With slightly revised notation and terminology, we recall from Chapters 2 and 3 the elements and procedures required for coherent, quantitative decision-making. The elements of a decision problem in the inference context are:

(i) $\boldsymbol{a} \in \mathcal{A}$, available "answers" to the inference problem;

(ii) $\boldsymbol{\omega} \in \Omega$, unknown states of the world;

(iii) $u : \mathcal{A} \times \Omega \to \Re$, a function attaching utilities to each consequence $(\boldsymbol{a}, \boldsymbol{\omega})$ of a decision to summarise inference in the form of an "answer", $\boldsymbol{a}$, and an ensuing state of the world, $\boldsymbol{\omega}$;

(iv) $p(\omega)$, a specification, in the form of a probability distribution, of current beliefs about the possible states of the world.

The optimal choice of answer to an inference problem is an $a \in \mathcal{A}$ which *maximises the expected utility,*

$$\int_{\Omega} u(a, \omega) p(\omega)\, d\omega.$$

Alternatively, if instead of working with $u(a, \omega)$ we work with a so-called *loss function,*

$$l(a, \omega) = f(\omega) - u(a, \omega),$$

where f is an arbitrary, fixed function, the optimal choice of answer is an $a \in \mathcal{A}$ which *minimises the expected loss,*

$$\int_{\Omega} l(a, \omega) p(\omega)\, d\omega.$$

It is clear from the forms of the expected utilities or losses which have to be calculated in order to choose an optimal answer, that, if beliefs about unknown states of the world are to provide an appropriate basis for future decision making, where, as yet, $\mathcal{A}$ and u (or l) may be unspecified, we need to report the complete belief distribution $p(\omega)$.

However, if an immediate application to a particular decision problem, with specified $\mathcal{A}$ and u (or l), is all that is required, the optimal answer—maximising the expected utility or minimising the expected loss—may turn out to involve only limited, specific features of the belief distribution, so that these "summaries" of the full distribution suffice for decision-making purposes.

In the following headed subsections, we shall illustrate and discuss some of these commonly used forms of summary. Throughout, we shall have in mind the context of parametric and predictive inference, where the unknown states of the world are parameters or future data values (observables), and current beliefs, $p(\omega)$, typically reduce to one or other of the familiar forms:

$p(\theta)$	initial beliefs about a parameter vector, θ;
$p(\theta \mid x)$	beliefs about θ, given data x;
$p(\psi \mid x)$	beliefs about $\psi = g(\theta)$, given data x;
$p(y \mid x)$	beliefs about future data y, given data x.

Point Estimates

In cases where $\omega \in \Omega$ corresponds to an unknown quantity, so that Ω is $\Re$, or $\Re^k$, or $\Re^+$, or $\Re \times \Re^+$, etc., and the required answer, $a \in \mathcal{A}$, is an estimate of the true value of ω (so that $\mathcal{A} = \Omega$), the corresponding decision problem is typically referred to as one of *point estimation.*

If $\omega = \boldsymbol{\theta}$ or $\omega = \boldsymbol{\psi}$, we refer to *parametric* point estimation; if $\omega = \boldsymbol{y}$, we refer to *predictive* point estimation. Moreover, since one is almost certain not to get the answer exactly right in an estimation problem, statisticians typically work directly with the loss function concept, rather than with the utility function. A point estimation problem is thus completely defined once $\mathcal{A} = \Omega$ and $l(\boldsymbol{a}, \boldsymbol{\omega})$ are specified. Direct intuition suggests that in the one-dimensional case, distributional summaries such as the mean, median or mode of $p(\omega)$ could be reasonable point estimates of a random quantity ω. Clearly, however, these could differ considerably, and more formal guidance may be required as to when and why particular functionals of the belief distribution are justified as point estimates. This is provided by the following definition and result.

Definition 5.3. (***Bayes estimate***). *A Bayes estimate of $\boldsymbol{\omega}$ with respect to the loss function $l(\boldsymbol{a}, \boldsymbol{\omega})$ and the belief distribution $p(\boldsymbol{\omega})$ is an $\boldsymbol{a} \in \mathcal{A} = \Omega$ which minimises $\int_\Omega l(\boldsymbol{a}, \boldsymbol{\omega}) p(\boldsymbol{\omega})\, d\boldsymbol{\omega}$.*

Proposition 5.2. (***Forms of Bayes estimates***).

(i) *If $\mathcal{A} = \Omega = \Re^k$, $l(\boldsymbol{a}, \boldsymbol{\omega}) = (\boldsymbol{a} - \boldsymbol{\omega})^t \boldsymbol{H} (\boldsymbol{a} - \boldsymbol{\omega})$, and $\boldsymbol{H}$ is symmetric definite positive, the Bayes estimate satisfies*

$$\boldsymbol{H}\boldsymbol{a} = \boldsymbol{H} E(\boldsymbol{\omega}).$$

If $\boldsymbol{H}^{-1}$ exists, $\boldsymbol{a} = E(\boldsymbol{\omega})$, and so ***the Bayes estimate with respect to quadratic form loss is the mean*** *of $p(\boldsymbol{\omega})$, assuming the mean to exist.*

(ii) *If $\mathcal{A} = \Omega = \Re$ and $l(a, \omega) = c_1(a - \omega) 1_{(\omega \le a)}(a) + c_2(\omega - a) 1_{(\omega > a)}(a)$, the* ***Bayes estimate with respect to linear loss is the quantile*** *such that*

$$P(\omega \le a) = c_2/(c_1 + c_2).$$

If $c_1 = c_2$, the right-hand side equals $1/2$ and so ***the Bayes estimate with respect to absolute value loss is a median*** *of $p(\omega)$.*

(iii) *If $\mathcal{A} = \Omega \subseteq \Re^k$ and $l(\boldsymbol{a}, \boldsymbol{\omega}) = 1 - 1_{(B_\varepsilon(\boldsymbol{a}))}(\boldsymbol{\omega})$, where $B_\varepsilon(\boldsymbol{a})$ is a ball of radius ε in Ω centred at $\boldsymbol{a}$, the Bayes estimate maximises*

$$\int_{B_\varepsilon(\boldsymbol{a})} p(\boldsymbol{\omega})\, d\boldsymbol{\omega}.$$

As $\varepsilon \to 0$, the function to be maximised tends to $p(\boldsymbol{a})$ and so ***the Bayes estimate with respect to zero-one loss is a mode*** *of $p(\boldsymbol{\omega})$, assuming a mode to exist.*

Proof. Differentiating $\int (a - \omega)^t H(a - \omega)p(\omega)\, d\omega$ with respect to a and equating to zero yields

$$2H \int (a - \omega)p(\omega)\, d\omega = 0.$$

This establishes (i). Since

$$\int l(a, \omega)p(\omega)\, d\omega = c_1 \int_{\{\omega \le a\}} (a - \omega)p(\omega)\, d\omega + c_2 \int_{\{\omega > a\}} (\omega - a)p(\omega)\, d\omega,$$

differentiating with respect to a and equating to zero yields

$$c_1 \int_{\{\omega \le a\}} p(\omega)\, d\omega = c_2 \int_{\{\omega > a\}} p(\omega)\, d\omega,$$

whence, adding $c_2 \int_{\omega \le a} p(\omega)\, d\omega$ to each side, we obtain (ii). Finally, since

$$\int l(a, \omega)p(\omega)\, d\omega = 1 - \int 1_{B_\varepsilon(a)}(\omega)p(\omega)\, d\omega,$$

and this is minimised when $\int_{B_\varepsilon(a)} p(\omega)\, d\omega$ is maximised, we have (iii). $\triangleleft$

Further insight into the nature of case (iii) can be obtained by thinking of a unimodal, continuous $p(\omega)$ in one dimension. It is then immediate by a continuity argument that a should be chosen such that

$$p(a - \varepsilon) = p(a + \varepsilon).$$

In the case of a unimodal, symmetric belief distribution, $p(\omega)$, for a single random quantity ω, the mean, median and mode coincide. In general, for unimodal, positively skewed, densities we have the relation

$$\text{mean} > \text{median} > \text{mode}$$

and the difference can be substantial if $p(\omega)$ is markedly skew. Unless, therefore, there is a very clear need for a point estimate, and a strong rationale for a specific one of the loss functions considered in Proposition 5.2, the provision of a single number to summarise $p(\omega)$ may be extremely misleading as a summary of the information available about ω. Of course, such a comment acquires even greater force if $p(\omega)$ is multimodal or otherwise "irregular".

For further discussion of Bayes estimators, see, for example, DeGroot and Rao (1963, 1966), Sacks (1963), Farrell (1964), Brown (1973), Tiao and Box (1974), Berger and Srinivasan (1978), Berger (1979, 1986), Hwang (1985, 1988), de la Horra (1987, 1988, 1992), Ghosh (1992a, 1992b), Irony (1992) and Spall and Maryak (1992).

Credible regions

We have emphasised that, from a theoretical perspective, uncertainty about an unknown quantity of interest, ω, needs to be communicated in the form of the full (prior, posterior or predictive) density, $p(\omega)$, if formal calculation of expected loss or utility is to be possible for any arbitrary future decision problem. In practice, however, $p(\omega)$ may be a somewhat complicated entity and it may be both more convenient, and also sufficient for general orientation regarding the uncertainty about ω, simply to describe regions $C \subseteq \Omega$ of given probability under $p(\omega)$. Thus, for example, in the case where $\Omega \subseteq \Re$, the identification of intervals containing 50%, 90%, 95% or 99% of the probability under $p(\omega)$ might suffice to give a good idea of the general quantitative messages implicit in $p(\omega)$. This is the intuitive basis of popular graphical representations of univariate distributions such as *box plots*.

Definition 5.4. (***Credible Region***). *A region $C \subseteq \Omega$ such that*

$$\int_C p(\omega)\, d\omega = 1 - \alpha$$

is said to be a $100(1-\alpha)\%$ credible region for ω, with respect to $p(\omega)$.

If $\Omega \subseteq \Re$, connected credible regions will be referred to as ***credible intervals***.

If $p(\omega)$ is a (prior-posterior-predictive) density, we refer to (prior-posterior-predictive) credible regions.

Clearly, for any given α there is not a unique credible region—even if we restrict attention to connected regions, as we should normally wish to do for obvious ease of interpretation (at least in cases where $p(\omega)$ is unimodal). For given Ω, $p(\omega)$ and fixed α, the problem of choosing among the subsets $C \subseteq \Omega$ such that $\int_C p(\omega)\, d\omega = 1 - \alpha$ could be viewed as a decision problem, provided that we are willing to specify a loss function, $l(C, \omega)$, reflecting the possible consequences of quoting the $100(1-\alpha)\%$ credible region C. We now describe the resulting form of credible region when a loss function is used which encapsulates the intuitive idea that, for given α, we would prefer to report a credible region C whose size $||C||$ (volume, area, length) is minimised.

Proposition 5.3. (***Minimal size credible regions***). *Let $p(\omega)$ be a probability density for $\omega \in \Omega$ almost everywhere continuous; given α, $0 < \alpha < 1$, if $\mathcal{A} = \{C;\ P(\omega \in C) = 1 - \alpha\} \neq \emptyset$ and*

$$l(C, \omega) = k||C|| - 1_C(\omega), \quad C \in \mathcal{A}, \quad \omega \in \Omega, \quad k > 0,$$

then C is optimal if and only if it has the property that $p(\omega_1) \geq p(\omega_2)$ for all $\omega_1 \in C$, $\omega_2 \notin C$ (except possibly for a subset of Ω of zero probability).

Proof. It follows straightforwardly that, for any $C \in \mathcal{A}$,

$$\int_{\Omega} l(C,\omega)p(\omega)\,d\omega = k||C|| + 1 - \alpha,$$

so that an optimal C must have minimal size.

If C has the stated property and D is any other region belonging to $\mathcal{A}$, then since $C = (C \cap D) \cup (C \cap D^c)$, $D = (C \cap D) \cup (C^c \cap D)$ and $P(\omega \in C) = P(\omega \in D)$, we have

$$\begin{aligned}\inf_{\omega \in C \cap D^c} p(\omega)||C \cap D^c|| &\le \int_{C \cap D^c} p(\omega)\,d\omega \\ &= \int_{C^c \cap D} p(\omega)\,d\omega \le \sup_{\omega \in C^c \cap D} p(\omega)||C^c \cap D||\end{aligned}$$

with

$$\sup_{\omega \in C^c \cap D} p(\omega) \le \inf_{\omega \in C \cap D^c} p(\omega)$$

so that $||C \cap D^c|| \le ||C^c \cap D||$, and hence $||C|| \le ||D||$.

If C does not have the stated property, there exists $A \subseteq C$ such that for all $\omega_1 \in A$, there exists $\omega_2 \notin C$ such that $p(\omega_2) > p(\omega_1)$. Let $B \subseteq C^c$ be such that $P(\omega \in A) = P(\omega \in B)$ and $p(\omega_2) > p(\omega_1)$ for all $\omega_2 \in B$ and $\omega_1 \in A$. Define $D = (C \cap A^c) \cup B$. Then $D \in \mathcal{A}$ and by a similar argument to that given above the result follows by showing that $||D|| < ||C||$. $\triangleleft$

The property of Proposition 5.3 is worth emphasising in the form of a definition (Box and Tiao, 1965).

Definition 5.5. (***Highest probability density (HPD) regions***).
A region $C \subseteq \Omega$ is said to be a $100(1-\alpha)\%$ highest probability density region for ω with respect to $p(\omega)$ if

(i) $P(\omega \in C) = 1 - \alpha$

(ii) $p(\omega_1) \ge p(\omega_2)$ *for all $\omega_1 \in C$ and $\omega_2 \notin C$, except possibly for a subset of Ω having probability zero.*

If $p(\omega)$ is a (prior-posterior-predictive) density, we refer to highest (prior-posterior-predictive) density regions.

Clearly, the credible region approach to summarising $p(\omega)$ is not particularly useful in the case of discrete Ω, since such regions will only exist for limited choices of α. The above development should therefore be understood as intended for the case of continuous Ω.

For a number of commonly occurring univariate forms of $p(\omega)$, there exist tables which facilitate the identification of HPD intervals for a range of values of α

$p(\omega)$

ω

ω_0 C

Figure 5.1a ω_0 *almost as "plausible" as all* $\omega \in C$

$p(\omega)$

ω

ω_0 C

Figure 5.1b ω_0 *much less "plausible" than most* $\omega \in C$

(see, for example, Isaacs *et al.*, 1974, Ferrándiz and Sendra,1982, and Lindley and Scott, 1985).

In general, however, the derivation of an HPD region requires numerical calculation and, particularly if $p(\boldsymbol{\omega})$ does not exhibit markedly skewed behaviour, it may be satisfactory in practice to quote some more simply calculated credible region. For example, in the univariate case, conventional statistical tables facilitate the identification of intervals which exclude equi-probable tails of $p(\omega)$ for many standard distributions.

Although an appropriately chosen selection of credible regions can serve to give a useful summary of $p(\omega)$ when we focus just on the quantity ω, there is a fundamental difficulty which prevents such regions serving, in general, as a proxy for the actual density $p(\omega)$. The problem is that of lack of invariance under parameter transformation. Even if $v = g(\omega)$ is a one-to-one transformation, it is easy to see that there is no general relation between HPD regions for ω and v. In addition, there is no way of identifying a marginal HPD region for a (possibly transformed) subset of components of ω from knowledge of the joint HPD region.

In cases where an HPD credible region C is pragmatically acceptable as a crude summary of the density $p(\omega)$, then, particularly for small values of α (for example, 0.05, 0.01), a specific value $\omega_0 \in \Omega$ will tend to be regarded as somewhat "implausible" if $\omega_0 \notin C$. This, of course, provides no justification for actions such as "rejecting the hypothesis that $\omega = \omega_0$". If we wish to consider such actions, we must formulate a proper decision problem, specifying alternative actions and the losses consequent on correct and incorrect actions. Inferences about a specific hypothesised value ω_0 of a random quantity ω in the absence of alternative hypothesised values are often considered in the general statistical literature under the heading of "significance testing". We shall discuss this further in Chapter 6.

For the present, it will suffice to note—as illustrated in Figure 5.1—that even the intuitive notion of "implausibility if $\omega_0 \notin C$" depends much more on the complete characterisation of $p(\omega)$ than on an either-or assessment based on an HPD region.

For further discussion of credible regions see, for example, Pratt (1961), Aitchison (1964, 1966), Wright (1986) and DasGupta (1991).

Hypothesis Testing

The basic hypothesis testing problem usually considered by statisticians may be described as a decision problem with elements

$$\Omega = \{\omega_0 = [H_0 : \boldsymbol{\theta} \in \Theta_0], \quad \omega_1 = [H_1 : \boldsymbol{\theta} \in \Theta_1]\},$$

together with $p(\omega)$, where $\boldsymbol{\theta} \in \Theta = \Theta_0 \cup \Theta_1$, is the parameter labelling a parametric model, $p(\boldsymbol{x} \mid \boldsymbol{\theta})$, $\mathcal{A} = \{a_0, a_1\}$, with $a_1(a_0)$ corresponding to rejecting hypothesis $H_0(H_1)$, and loss function $l(a_i, \omega_j) = l_{ij}$, $i, j \in \{0, 1\}$, with the l_{ij} reflecting the relative seriousness of the four possible consequences and, typically, $l_{00} = l_{11} = 0$.

Clearly, the main motivation and the principal use of the hypothesis testing framework is in model choice and comparison, an activity which has a somewhat different flavour from decision-making and inference within the context of an accepted model. For this reason, we shall postpone a detailed consideration of the topic until Chapter 6, where we shall provide a much more general perspective on model choice and criticism.

General discussions of Bayesian hypothesis testing are included in Jeffreys (1939/1961), Good (1950, 1965, 1983), Lindley (1957, 1961b, 1965, 1977), Edwards *et al.* (1963), Pratt (1965), Smith (1965), Farrell (1968), Dickey (1971, 1974, 1977), Lempers (1971), Rubin (1971), Zellner (1971), DeGroot (1973), Leamer (1978), Box (1980), Shafer (1982b), Gilio and Scozzafava (1985), Smith, (1986), Berger and Delampady (1987), Berger and Sellke (1987) and Hodges (1990, 1992).

5.1.6 Implementation Issues

Given a likelihood $p(\boldsymbol{x} \mid \boldsymbol{\theta})$ and prior density $p(\boldsymbol{\theta})$, the starting point for any form of parametric inference summary or decision about $\boldsymbol{\theta}$ is the joint posterior density

$$p(\boldsymbol{\theta} \mid \boldsymbol{x}) = \frac{p(\boldsymbol{x} \mid \boldsymbol{\theta})p(\boldsymbol{\theta})}{\int p(\boldsymbol{x} \mid \boldsymbol{\theta})p(\boldsymbol{\theta})d\boldsymbol{\theta}},$$

and the starting point for any predictive inference summary or decision about future observables $\boldsymbol{y}$ is the predictive density

$$p(\boldsymbol{y} \mid \boldsymbol{x}) = \int p(\boldsymbol{y} \mid \boldsymbol{\theta})p(\boldsymbol{\theta} \mid \boldsymbol{x})\, d\boldsymbol{\theta}.$$

It is clear that to form these posterior and predictive densities there is a technical requirement to perform integrations over the range of $\boldsymbol{\theta}$. Moreover, further summarisation, in order to obtain marginal densities, or marginal moments, or expected utilities or losses in explicitly defined decision problems, will necessitate further integrations with respect to components of $\boldsymbol{\theta}$ or $\boldsymbol{y}$, or transformations thereof.

The key problem in implementing the formal Bayes solution to inference reporting or decision problems is therefore seen to be that of evaluating the required integrals. In cases where the likelihood just involves a single parameter, implementation just involves integration in one dimension and is essentially trivial. However, in problems involving a multiparameter likelihood the task of implementation is anything but trivial, since, if $\boldsymbol{\theta}$ has k components, two k-dimensional integrals are required just to form $p(\boldsymbol{\theta} \mid \boldsymbol{x})$ and $p(\boldsymbol{y} \mid \boldsymbol{x})$. Moreover, in the case of $p(\boldsymbol{\theta} \mid \boldsymbol{x})$, for example, k $(k-1)$-dimensional integrals are required to obtain univariate marginal density values or moments, $\binom{k}{2}$ $(k-2)$-dimensional integrals are required to obtain bivariate marginal densities, and so on. Clearly, if k is at all large, the problem of implementation will, in general, lead to challenging technical problems, requiring simultaneous analytic or numerical approximation of a number of multidimensional integrals.

The above discussion has assumed a given specification of a likelihood and prior density function. However, as we have seen in Chapter 4, although a specific mathematical form for the likelihood in a given context is very often implied or suggested by consideration of symmetry, sufficiency or experience, the mathematical specification of prior densities is typically more problematic. Some of the problems involved—such as the pragmatic strategies to be adopted in translating actual beliefs into mathematical form—relate more to practical methodology than to conceptual and theoretical issues and will be not be discussed in detail in this volume. However, many of the other problems of specifying prior densities are closely related to the general problems of implementation described above, as exemplified by the following questions:

(i) given that, for any specific beliefs, there is some arbitrariness in the precise choice of the mathematical representation of a prior density, are there choices which enable the integrations required to be carried out straightforwardly and hence permit the tractable implementation of a range of analyses, thus facilitating the kind of interpersonal analysis and scientific reporting referred to in Section 4.8.2 and again later in 6.3.3?

(ii) if the information to be provided by the data is known to be far greater than that implicit in an individual's prior beliefs, is there any necessity for a precise mathematical representation of the latter, or can a Bayesian implementation proceed purely on the basis of this qualitative understanding?

(iii) either in the context of interpersonal analysis, or as a special form of actual individual analysis, is there a formal way of representing the beliefs of an individual whose prior information is to be regarded as minimal, relative to the information provided by the data?

(iv) for general forms of likelihood and prior density, are there analytic/numerical techniques available for approximating the integrals required for implementing Bayesian methods?

Question (i) will be answered in Section 5.2, where the concept of a *conjugate* prior density will be introduced.

Question (ii) will be answered in part at the end of Section 5.2 and in more detail in Section 5.3, where an approximate "large sample" Bayesian theory involving *asymptotic posterior normality* will be presented.

Question (iii) will be answered in Section 5.4, where the information-based concept of a *reference* prior density will be introduced. An extended historical discussion of this celebrated philosophical problem of how to represent "ignorance" will be given in Section 5.6.2.

Question (iv) will be answered in Section 5.5, where classical applied analysis techniques such as *Laplace's approximation* for integrals will be briefly reviewed in the context of implementing Bayesian inference and decision summaries, together with classical numerical analytical techniques such as *Gauss-Hermite quadrature* and stochastic simulation techniques such as *importance sampling*, *sampling-importance-resampling* and *Markov chain Monte Carlo*.

5.2 CONJUGATE ANALYSIS

5.2.1 Conjugate Families

The first issue raised at the end of Section 5.1.6 is that of tractability. Given a likelihood function $p(\boldsymbol{x} \mid \boldsymbol{\theta})$, for what choices of $p(\boldsymbol{\theta})$ are integrals such as

$$p(\boldsymbol{x}) = \int p(\boldsymbol{x} \mid \boldsymbol{\theta}) p(\boldsymbol{\theta}) d\boldsymbol{\theta} \qquad \text{and} \qquad p(\boldsymbol{y} \mid \boldsymbol{x}) = \int p(\boldsymbol{y} \mid \boldsymbol{\theta}) p(\boldsymbol{\theta} \mid \boldsymbol{x}) d\boldsymbol{\theta}$$

easily evaluated analytically? However, since any particular mathematical form of $p(\boldsymbol{\theta})$ is acting as a representation of beliefs—either of an actual individual, or as part of a stylised sensitivity study involving a range of prior to posterior analyses—we require, in addition to tractability, that the class of mathematical functions from which $p(\boldsymbol{\theta})$ is to be chosen be both rich in the forms of beliefs it can represent and also facilitate the matching of beliefs to particular members of the class. Tractability can be achieved by noting that, since Bayes' theorem may be expressed in the form

$$p(\boldsymbol{\theta} \mid \boldsymbol{x}) \propto p(\boldsymbol{x} \mid \boldsymbol{\theta}) p(\boldsymbol{\theta}),$$

both $p(\boldsymbol{\theta} \mid \boldsymbol{x})$ and $p(\boldsymbol{\theta})$ can be guaranteed to belong to the same general family of mathematical functions by choosing $p(\boldsymbol{\theta})$ to have the same "structure" as $p(\boldsymbol{x} \mid \boldsymbol{\theta})$, when the latter is viewed as a function of $\boldsymbol{\theta}$. However, as stated, this is a rather vacuous idea, since $p(\boldsymbol{\theta} \mid \boldsymbol{x})$ and $p(\boldsymbol{\theta})$ would always belong to the same "general family" of functions if the latter were suitably defined. To achieve a more meaningful version of the underlying idea, let us first recall (from Section 4.5) that if $\boldsymbol{t} = \boldsymbol{t}(\boldsymbol{x})$ is a sufficient statistic we have

$$p(\boldsymbol{\theta} \mid \boldsymbol{x}) = p(\boldsymbol{\theta} \mid \boldsymbol{t}) \propto p(\boldsymbol{t} \mid \boldsymbol{\theta}) p(\boldsymbol{\theta}),$$

so that we can restate our requirement for tractability in terms of $p(\boldsymbol{\theta})$ having the same structure as $p(\boldsymbol{t} \mid \boldsymbol{\theta})$, when the latter is viewed as a function of $\boldsymbol{\theta}$. Again, however, without further constraint on the nature of the sequence of sufficient statistics the class of possible functions $p(\boldsymbol{\theta})$ is too large to permit easily interpreted matching of beliefs to particular members of the class. This suggests that it is only in the case of likelihoods admitting sufficient statistics of fixed dimension that we shall be able to identify a family of prior densities which ensures both tractability and ease of interpretation. This motivates the following definition.

Definition 5.6. **(*Conjugate prior family*).** *The conjugate family of prior densities for $\boldsymbol{\theta} \in \Theta$, with respect to a likelihood $p(\boldsymbol{x} \mid \boldsymbol{\theta})$ with sufficient statistic $\boldsymbol{t} = \boldsymbol{t}(\boldsymbol{x}) = \{n, \boldsymbol{s}(\boldsymbol{x})\}$ of a fixed dimension k independent of that of $\boldsymbol{x}$, is*

$$\{p(\boldsymbol{\theta} \mid \boldsymbol{\tau}), \boldsymbol{\tau} = (\tau_0, \tau_1, \ldots, \tau_k) \in \mathcal{T}\},$$

where

$$\mathcal{T} = \left\{ \boldsymbol{\tau};\ \int_{\Theta} p(\boldsymbol{s} = (\tau_1, \ldots, \tau_k) \,|\, \boldsymbol{\theta}, n = \tau_0) d\boldsymbol{\theta} < \infty \right\}$$

and

$$p(\boldsymbol{\theta} \,|\, \boldsymbol{\tau}) = \frac{p(\boldsymbol{s} = (\tau_1, \ldots, \tau_k) \,|\, \boldsymbol{\theta}, n = \tau_0)}{\int_{\Theta} p(\boldsymbol{s} = (\tau_1, \ldots, \tau_k) \,|\, \boldsymbol{\theta}, n = \tau_0) d\boldsymbol{\theta}} \,.$$

From Section 4.5 and Definition 5.6, it follows that the likelihoods for which conjugate prior families exist are those corresponding to general exponential family parametric models (Definitions 4.10 and 4.11), for which, given f, $\boldsymbol{h}$, $\boldsymbol{\phi}$ and $\boldsymbol{c}$,

$$p(x \,|\, \boldsymbol{\theta}) = f(x) g(\boldsymbol{\theta}) \exp\left\{ \sum_{i=1}^{k} c_i \phi_i(\boldsymbol{\theta}) h_i(x) \right\}, \quad x \in X,$$

$$\left(g(\boldsymbol{\theta})\right)^{-1} = \int_X f(x) \exp\left\{ \sum_{i=1}^{k} c_i \phi_i(\boldsymbol{\theta}) h_i(x) \right\} dx.$$

The exponential family model is referred to as regular or non-regular, respectively, according as X does not or does depend on $\boldsymbol{\theta}$.

Proposition 5.4. **(*Conjugate families for regular exponential families*).** *If $\boldsymbol{x} = (x_1, \ldots, x_n)$ is a random sample from a regular exponential family distribution such that*

$$p(\boldsymbol{x} \,|\, \boldsymbol{\theta}) = \prod_{j=1}^{n} f(x_j) \left[g(\boldsymbol{\theta})\right]^n \exp\left\{ \sum_{i=1}^{k} c_i \phi_i(\boldsymbol{\theta}) \left(\sum_{j=1}^{n} h_i(x_j) \right) \right\},$$

then the conjugate family for $\boldsymbol{\theta}$ has the form

$$p(\boldsymbol{\theta} \,|\, \boldsymbol{\tau}) = \left[K(\boldsymbol{\tau})\right]^{-1} \left[g(\boldsymbol{\theta})\right]^{\tau_0} \exp\left\{ \sum_{i=1}^{k} c_i \phi_i(\boldsymbol{\theta}) \tau_i \right\}, \quad \boldsymbol{\theta} \in \Theta,$$

where $\boldsymbol{\tau}$ is such that $K(\boldsymbol{\tau}) = \int_{\Theta} [g(\boldsymbol{\theta})]^{\tau_0} \exp\left\{ \sum_{i=1}^{k} c_i \phi_i(\boldsymbol{\theta}) \tau_i \right\} d\boldsymbol{\theta} < \infty$.

Proof. By Proposition 4.10 (the Neyman factorisation criterion), the sufficient statistics for $\boldsymbol{\phi}$ have the form

$$\boldsymbol{t}_n(x_1, \ldots, x_n) = \left[n, \sum_{j=1}^{n} h_1(x_j), \ldots, \sum_{j=1}^{n} h_k(x_j) \right] = [n, \boldsymbol{s}(\boldsymbol{x})],$$

so that, for any $\boldsymbol{\tau} = (\tau_0, \tau_1, \ldots, \tau_n)$ such that $\int_\Theta p(\boldsymbol{\theta} \mid \boldsymbol{\tau}) d\boldsymbol{\theta} < \infty$, a conjugate prior density has the form

$$\begin{aligned} p(\boldsymbol{\theta} \mid \boldsymbol{\tau}) &\propto p(s_1(\boldsymbol{x}) = \tau_1, \ldots, s_k(\boldsymbol{x}) = \tau_k \mid \boldsymbol{\theta}, n = \tau_0) \\ &\propto [g(\boldsymbol{\theta})]^{\tau_0} \exp\left\{ \sum_{i=1}^{k} c_i \phi_i(\boldsymbol{\theta}) \tau_i \right\} \end{aligned}$$

by Proposition 4.2. ◁

Example 5.4. ***(Bernoulli likelihood; beta prior).*** The Bernoulli likelihood has the form

$$\begin{aligned} p(x_1, \ldots x_n \mid \theta) &= \prod_{i=1}^{n} \theta^{x_i} (1-\theta)^{1-x_i} \qquad (0 \le \theta \le 1) \\ &= (1-\theta)^n \exp\left\{ \log\left(\frac{\theta}{1-\theta}\right) \sum_{i=1}^{n} x_i \right\}, \end{aligned}$$

so that, by Proposition 5.4, the conjugate prior density for θ is given by

$$\begin{aligned} p(\theta \mid \tau_0, \tau_1) &\propto (1-\theta)^{\tau_0} \exp\left\{ \log\left(\frac{\theta}{1-\theta}\right) \tau_1 \right\} \\ &= \frac{1}{K(\tau_0, \tau_1)} \theta^{\tau_1} (1-\theta)^{\tau_0 - \tau_1}, \end{aligned}$$

assuming the existence of

$$K(\tau_0, \tau_1) = \int_0^1 \theta^{\tau_1} (1-\theta)^{\tau_0 - \tau_1} \, d\theta.$$

Writing $\alpha = \tau_1 + 1$, and $\beta = \tau_0 - \tau_1 + 1$, we have $p(\theta \mid \alpha, \beta) \propto \theta^{\alpha-1}(1-\theta)^{\beta-1}$, and hence, comparing with the definition of a beta density,

$$p(\theta \mid \tau_0, \tau_1) = p(\theta \mid \alpha, \beta) = \text{Be}(\theta \mid \alpha, \beta), \qquad \alpha > 0, \quad \beta > 0.$$

Example 5.5. ***(Poisson likelihood; gamma prior).*** The Poisson likelihood has the form

$$\begin{aligned} p(x_1, \ldots, x_n \mid \theta) &= \prod_{i=1}^{n} \frac{\theta^{x_i} \exp(-\theta)}{x_i!} \qquad (\theta > 0) \\ &= \left(\prod_{i=1}^{n} x_i! \right)^{-1} \exp(-n\theta) \exp\left(\log\theta \sum_{i=1}^{n} x_i \right), \end{aligned}$$

so that, by Proposition 5.4, the conjugate prior density for θ is given by

$$\begin{aligned} p(\theta \mid \tau_0, \tau_1) &\propto \exp(-\tau_0 \theta) \exp(\tau_1 \log\theta) \\ &= \frac{1}{K(\tau_0, \tau_1)} \theta^{\tau_1} \exp(-\tau_0 \theta), \end{aligned}$$

assuming the existence of

$$K(\tau_0, \tau_1) = \int_0^\infty \theta^{\tau_1} \exp(-\tau_0\theta)\, d\theta.$$

Writing $\alpha = \tau_1 + 1$ and $\beta = \tau_0$ we have $p(\theta \mid \alpha, \beta) \propto \theta^{\alpha-1}\exp(-\beta\theta)$ and hence, comparing with the definition of a gamma density,

$$p(\theta \mid \tau_0, \tau_1) = p(\theta \mid \alpha, \beta) = \mathrm{Ga}(\theta \mid \alpha, \beta), \qquad \alpha > 0, \quad \beta > 0.$$

Example 5.6. ***(Normal likelihood; normal-gamma prior).*** The normal likelihood, with unknown mean and precision, has the form

$$\begin{aligned} p(x_1, \ldots, x_n \mid \mu, \lambda) &= \prod_{i=1}^{n} \left(\frac{\lambda}{2\pi}\right)^{\frac{1}{2}} \exp\left\{-\frac{\lambda}{2}(x_i - \mu)^2\right\} \\ &= (2\pi)^{-n/2}\left[\lambda^{1/2}\exp\left(-\frac{\lambda}{2}\mu^2\right)\right]^n \exp\left\{\mu\lambda\sum_{i=1}^{n} x_i - \frac{\lambda}{2}\sum_{i=1}^{n} x_i^2\right\}, \end{aligned}$$

so that, by Proposition 5.4, the conjugate prior density for $\theta = (\mu, \lambda)$ is given by

$$\begin{aligned} p(\mu, \lambda \mid \tau_0, \tau_1, \tau_2) &\propto \left[\lambda^{1/2}\exp\left(-\frac{1}{2}\lambda\mu^2\right)\right]^{\tau_0} \exp\left\{\mu\lambda\tau_1 - \frac{1}{2}\lambda\tau_2\right\} \\ &= \frac{1}{K(\tau_0, \tau_1, \tau_2)}\lambda^{(\tau_0-1)/2}\exp\left(-\frac{\lambda}{2}\left(\tau_2 - \frac{\tau_1^2}{\tau_0}\right)\right)\lambda^{\frac{1}{2}}\exp\left\{-\frac{\lambda\tau_0}{2}\left(\mu - \frac{\tau_1}{\tau_0}\right)^2\right\}, \end{aligned}$$

assuming the existence of $K(\tau_0, \tau_1, \tau_2)$, given by

$$\int_0^\infty \lambda^{\frac{\tau_0-1}{2}}\exp\left(-\frac{\lambda}{2}\left(\tau_2 - \frac{\tau_1^2}{\tau_0}\right)\right)\left\{\int_{-\infty}^{\infty}\lambda^{\frac{1}{2}}\exp\left[-\frac{\lambda\tau_0}{2}\left(\mu - \frac{\tau_1}{\tau_0}\right)^2\right] d\mu\right\} d\lambda.$$

Writing $\alpha = \frac{1}{2}(\tau_0 + 1)$, $\beta = \frac{1}{2}(\tau_2 - \frac{\tau_1^2}{\tau_0})$, $\gamma = \tau_1/\tau_0$, and comparing with the definition of a normal-gamma density, we have

$$\begin{aligned} p(\mu, \lambda \mid \tau_0, \tau_1, \tau_2) &= p(\mu, \lambda \mid \alpha, \beta, \gamma) \\ &= \mathrm{Ng}(\mu, \lambda \mid \alpha, \beta, \gamma) \\ &= \mathrm{N}\Big(\mu \mid \gamma, \lambda(2\alpha - 1)\Big)\mathrm{Ga}(\lambda \mid \alpha, \beta), \end{aligned}$$

with $\alpha > \frac{1}{2}$, $\beta > 0$, $\gamma \in \Re$.

5.2.2 Canonical Conjugate Analysis

Conjugate prior density families were motivated by considerations of tractability in implementing the Bayesian paradigm. The following proposition demonstrates that, in the case of regular exponential family likelihoods and conjugate prior densities, the analytic forms of the joint posterior and predictive densities which underlie any form of inference summary or decision making are easily identified.

Proposition 5.5. (***Conjugate analysis for regular exponential families***). *For the exponential family likelihood and conjugate prior density of Proposition 5.4:*

(i) *the posterior density for θ is*

$$p(\boldsymbol{\theta} \mid \boldsymbol{x}, \boldsymbol{\tau}) \underset{n}{=} p(\boldsymbol{\theta} \mid \boldsymbol{\tau} + \boldsymbol{t}_n(\boldsymbol{x}))$$

where

$$\boldsymbol{\tau} + \boldsymbol{t}_n(\boldsymbol{x}) = \left(\tau_0 + n,\ \tau_1 + \sum_{j=1}^{n} h_1(x_j), \ldots, \tau_k + \sum_{j=1}^{n} h_k(x_j) \right);$$

(ii) *the predictive density for future observables $\boldsymbol{y} = (y_1, \ldots, y_m)$ is*

$$\begin{aligned} p(\boldsymbol{y} \mid \boldsymbol{x}, \boldsymbol{\tau}) &= p(\boldsymbol{y} \mid \boldsymbol{\tau} + \boldsymbol{t}_n(\boldsymbol{x})) \\ &= \prod_{l=1}^{m} f(y_l) \frac{K(\boldsymbol{\tau} + \boldsymbol{t}_n(\boldsymbol{x}) + \boldsymbol{t}_m(\boldsymbol{y}))}{K(\boldsymbol{\tau} + \boldsymbol{t}_n(\boldsymbol{x}))}, \end{aligned}$$

where $\boldsymbol{t}_m(\boldsymbol{y}) = [m,\ \sum_{l=1}^{m} h_1(y_l), \ldots, \sum_{l=1}^{m} h_k(y_l)]$.

Proof. By Bayes' theorem,

$$\begin{aligned} p(\boldsymbol{\theta} \mid \boldsymbol{x}, \boldsymbol{\tau}) &\propto p(\boldsymbol{x} \mid \boldsymbol{\theta}) p(\boldsymbol{\theta} \mid \boldsymbol{\tau}) \\ &\propto [g(\boldsymbol{\theta})]^{\tau_0 + n} \exp\left\{ \sum_{i=1}^{k} c_i \phi_i(\boldsymbol{\theta}) \left(\tau_i + \sum_{j=1}^{n} h_i(x_j) \right) \right\} \\ &\propto p(\boldsymbol{\theta} \mid \boldsymbol{\tau} + \boldsymbol{t}_n(\boldsymbol{x})), \end{aligned}$$

which proves (i). Moreover,

$$\begin{aligned} p(\boldsymbol{y} \mid \boldsymbol{x}, \boldsymbol{\tau}) &= \int_{\Theta} p(\boldsymbol{y} \mid \boldsymbol{\theta}) p(\boldsymbol{\theta} \mid \boldsymbol{x}) d\boldsymbol{\theta} \\ &= \prod_{l=1}^{m} f(y_l) \cdot [K(\boldsymbol{\tau} + \boldsymbol{t}_n(\boldsymbol{x}))]^{-1} \int_{\Theta} [g(\boldsymbol{\theta})]^{\tau_0 + n + m} \\ &\qquad \times \exp\left\{ \sum_{i=1}^{k} c_i \phi_i(\boldsymbol{\theta}) \left(\tau_i + \sum_{j=1}^{n} h_i(x_j) + \sum_{l=1}^{m} h_i(y_l) \right) \right\} d\boldsymbol{\theta} \\ &= \prod_{l=1}^{m} f(y_l) \frac{K(\boldsymbol{\tau} + \boldsymbol{t}_n(\boldsymbol{x}) + \boldsymbol{t}_m(\boldsymbol{y}))}{K(\boldsymbol{\tau} + \boldsymbol{t}_n(\boldsymbol{x}))}, \end{aligned}$$

which proves (ii). ◁

Proposition 5.5(i) establishes that the conjugate family is *closed under sampling*, with respect to the corresponding exponential family likelihood, a concept which seems to be due to G. A. Barnard. This means that both the joint prior and posterior densities belong to the same, simply defined, family of distributions, the inference process being totally defined by the mapping $\boldsymbol{\tau} \to (\boldsymbol{\tau} + \boldsymbol{t}_n(\boldsymbol{x}))$, under which the labelling parameters of the prior density are simply modified by the addition of the values of the sufficient statistic to form the labelling parameter of the posterior distribution. The inference process defined by Bayes' theorem is therefore reduced from the essentially infinite-dimensional problem of the transformation of density functions, to a simple, additive finite-dimensional transformation. Proposition 5.5(ii) establishes that a similar, simplifying closure property holds for predictive densities.

The forms arising in the conjugate analysis of a number of standard exponential family forms are summarised in Appendix A. However, to provide some preliminary insights into the prior $\to$ posterior $\to$ predictive process described by Proposition 5.5, we shall illustrate the general results by reconsidering Example 5.4.

Example 5.4. ***(continued)***. With the Bernoulli likelihood written in its explicit exponential family form, and writing $r_n = x_1 + \cdots + x_n$, the posterior density corresponding to the conjugate prior density, $p(\theta \mid \tau_0, \tau_1)$, is given by

$$\begin{aligned} p(\theta \mid \boldsymbol{x}, \tau_0, \tau_1) &\propto p(\boldsymbol{x} \mid \theta) p(\theta \mid \tau_0, \tau_1) \\ &\propto (1-\theta)^n \exp\left\{ \log\left(\frac{\theta}{1-\theta}\right) r_n \right\} (1-\theta)^{\tau_0} \\ &\quad \times \exp\left\{ \log\left(\frac{\theta}{1-\theta}\right) \tau_1 \right\} \\ &= \frac{\Gamma(\tau_0(n)+2)}{\Gamma(\tau_1(n)+1)\Gamma(\tau_0(n)-\tau_1(n)+1)} (1-\theta)^{\tau_0(n)} \\ &\quad \times \exp\left\{ \log\left(\frac{\theta}{1-\theta}\right) \tau_1(n) \right\}, \end{aligned}$$

where $\tau_0(n) = \tau_0 + n$, $\tau_1(n) = \tau_1 + r_n$, showing explicitly how the inference process reduces to the updating of the prior to posterior hyperparameters by the addition of the sufficient statistics, n and r_n.

Alternatively, we could proceed on the basis of the original representation of the Bernoulli likelihood, combining it directly with the familiar beta prior density, $\text{Be}(\theta \mid \alpha, \beta)$, so that

$$\begin{aligned} p(\theta \mid \boldsymbol{x}, \alpha, \beta) &\propto p(\boldsymbol{x} \mid \theta) p(\theta \mid \alpha, \beta) \\ &\propto \theta^{r_n} (1-\theta)^{n-r_n} \theta^{\alpha-1} (1-\theta)^{\beta-1} \\ &= \frac{\Gamma(\alpha_n + \beta_n)}{\Gamma(\alpha_n)\Gamma(\beta_n)} \theta^{\alpha_n - 1} (1-\theta)^{\beta_n - 1}, \end{aligned}$$

where $\alpha_n = \alpha + r_n$, $\beta_n = \beta + n - r_n$ and, again, the process reduces to the updating of the prior to posterior hyperparameters.

Clearly, the two notational forms and procedures used in the example are equivalent. Using the standard exponential family form has the advantage of displaying the simple hyperparameter updating by the addition of the sufficient statistics. However, the second form seems much less cumbersome notationally and is more transparently interpretable and memorable in terms of the beta density.

In general, when analysing particular models we shall work in terms of whatever functional representation seems best suited to the task in hand.

Example 5.4. ***(continued)***. Instead of working with the original Bernoulli likelihood, $p(x_1, \ldots, x_n|\theta)$, we could, of course, work with a likelihood defined in terms of the sufficient statistic (n, r_n). In particular, if either n or r_n were ancillary, we would use one or other of $p(r_n \mid n, \theta)$ or $p(n \mid r_n, \theta)$ and, in either case,

$$p(\theta \mid n, r_n, \alpha, \beta) \propto \theta^{r_n}(1-\theta)^{n-r_n}\theta^{\alpha-1}(1-\theta)^{\beta-1}.$$

Taking the binomial form, $p(r_n \mid n, \theta)$, the prior to posterior operation defined by Bayes' theorem can be simply expressed, in terms of the notation introduced in Section 3.2.4, as

$$\text{Bi}(r_n \mid \theta, n) \otimes \text{Be}(\theta \mid \alpha, \beta) \equiv \text{Be}(\theta \mid \alpha + r_n, \beta + n - r_n).$$

The predictive density for future Bernoulli observables, which we denote by

$$\boldsymbol{y} = (y_1, \ldots, y_m) = (x_{n+1}, \ldots, x_{n+m}),$$

is also easily derived. Writing $r'_m = y_1 + \cdots + y_m$, we see that

$$\begin{aligned}
p(\boldsymbol{y} \mid \boldsymbol{x}, \alpha, \beta) &= p(\boldsymbol{y} \mid \alpha_n, \beta_n) \\
&= \int_0^1 p(\boldsymbol{y} \mid \theta) p(\theta \mid \alpha_n, \beta_n)\, d\theta \\
&= \frac{\Gamma(\alpha_n + \beta_n)}{\Gamma(\alpha_n)\Gamma(\beta_n)} \int_0^1 \theta^{\alpha_n + r'_m - 1}(1-\theta)^{\beta_n + m - r'_m - 1}\, d\theta \\
&= \frac{\Gamma(\alpha_n + \beta_n)}{\Gamma(\alpha_n)\Gamma(\beta_n)} \frac{\Gamma(\alpha_{n+m})\Gamma(\beta_{n+m})}{\Gamma(\alpha_{n+m} + \beta_{n+m})},
\end{aligned}$$

where

$$\begin{aligned}
\alpha_{n+m} &= \alpha_n + r'_m = \alpha + r_n + r'_m, \\
\beta_{n+m} &= \beta_n + m - r'_m = \beta + (n+m) - (r_n + r'_m),
\end{aligned}$$

a result which also could be obtained directly from Proposition 5.5(ii).

If, instead, we were interested in the predictive density for r'_m, it easily follows that

$$\begin{aligned}
p(r'_m \mid \alpha_n, \beta_n, m) &= \int_0^1 p(r'_m \mid m, \theta) p(\theta \mid \alpha_n, \beta_n)\, d\theta \\
&= \int_0^1 \binom{m}{r'_m} p(\boldsymbol{y} \mid \theta) p(\theta \mid \alpha_n, \beta_n)\, d\theta \\
&= \binom{m}{r'_m} p(\boldsymbol{y} \mid \alpha_n, \beta_n).
\end{aligned}$$

Comparison with Section 3.2.2 reveals this predictive density to have the binomial-beta form, $\text{Bb}(r'_m \mid \alpha_n, \beta_n, m)$.

The particular case $m = 1$ is of some interest, since $p(r'_m = 1 \mid \alpha_n, \beta_n, m = 1)$ is then the predictive probability assigned to a success on the $(n+1)$th trial, given r_n observed successes in the first n trials and an initial $\text{Be}(\theta \mid \alpha, \beta)$ belief about the limiting relative frequency of successes, θ.

We see immediately, on substituting into the above, that

$$p(r'_m = 1 \mid \alpha_n, \beta_n, m = 1) = \frac{\alpha_n}{\alpha_n + \beta_n} = E(\theta \mid \alpha_n, \beta_n),$$

using the fact that $\Gamma(t+1) = t\Gamma(t)$ and recalling, from Section 3.2.2, the form of the mean of a beta distribution.

With respect to quadratic loss, $E(\theta \mid \alpha_n, \beta_n) = (\alpha + r_n)/(\alpha + \beta + n)$ is the optimal estimate of θ given current information, and the above result demonstrates that this should serve as the evaluation of the probability of a success on the next trial. In the case $\alpha = \beta = 1$ this evaluation becomes $(r_n + 1)/(n + 2)$, which is the celebrated *Laplace's rule of succession* (Laplace, 1812), which has served historically to stimulate considerable philosophical debate about the nature of inductive inference. We shall consider this problem further in Example 5.16 of Section 5.4.4. For an elementary, but insightful, account of Bayesian inference for the Bernoulli case, see Lindley and Phillips (1976).

In presenting the basic ideas of conjugate analysis, we used the following notation for the k-parameter exponential family and corresponding prior form:

$$p(x \mid \boldsymbol{\theta}) = f(x) g(\boldsymbol{\theta}) \exp\left\{\sum_{i=1}^{k} c_i \phi_i(\boldsymbol{\theta}) h_i(x)\right\}, \qquad x \in X,$$

and

$$p(\boldsymbol{\theta} \mid \boldsymbol{\tau}) = [K(\boldsymbol{\tau})]^{-1} [g(\boldsymbol{\theta})]^{\tau_0} \exp\left\{\sum_{i=1}^{k} \phi_i(\boldsymbol{\theta}) \tau_i\right\}, \qquad \boldsymbol{\theta} \in \Theta,$$

the latter being defined for $\boldsymbol{\tau}$ such that $K(\boldsymbol{\tau}) < \infty$.

From a notational perspective (cf. Definition 4.12), we can obtain considerable simplification by defining $\boldsymbol{\psi} = (\psi_1, \ldots, \psi_k)$, $\boldsymbol{y} = (y_1, \ldots, y_k)$, where $\psi_i = c_i \phi_i(\boldsymbol{\theta})$ and $y_i = h(x_i)$, $i = 1, \ldots, k$, together with prior hyperparameters n_0, $\boldsymbol{y}_0$, so that these forms become

$$p(\boldsymbol{y} \mid \boldsymbol{\psi}) = a(\boldsymbol{y}) \exp\left\{\boldsymbol{y}^t \boldsymbol{\psi} - b(\boldsymbol{\psi})\right\}, \quad \boldsymbol{y} \in Y,$$

$$p(\boldsymbol{\psi} \mid n_0, \boldsymbol{y}_0) = c(n_0, \boldsymbol{y}_0) \exp\left\{n_0 \boldsymbol{y}_0^t \boldsymbol{\psi} - n_0 b(\boldsymbol{\psi})\right\}, \quad \boldsymbol{\psi} \in \Psi,$$

for appropriately defined Y, Ψ and real-valued functions a, b and c. We shall refer to these (Definition 4.12) as the *canonical (or natural) forms* of the exponential family and its conjugate prior family. If $\Psi = \Re^k$, we require $n_0 > 0$, $\boldsymbol{y}_0 \in Y$

in order for $p(\boldsymbol{\psi} \mid n_0, \boldsymbol{y}_0)$ to be a proper density; for $\Psi \neq \Re^k$, the situation is somewhat more complicated (see Diaconis and Ylvisaker, 1979, for details). We shall typically assume that Ψ consists of all $\boldsymbol{\psi}$ such that $\int_Y p(\boldsymbol{y} \mid \boldsymbol{\psi}) d\boldsymbol{y} = 1$ and that $b(\boldsymbol{\psi})$ is continuously differentiable and strictly convex throughout the interior of Ψ.

The motivation for choosing $n_0, \boldsymbol{y}_0$ as notation for the prior hyperparameter is partly clarified by the following proposition and becomes even clearer in the context of Proposition 5.7.

Proposition 5.6. ***(Canonical conjugate analysis).*** *If $\boldsymbol{y}_1, \ldots, \boldsymbol{y}_n$ are the values of $\boldsymbol{y}$ resulting from a random sample of size n from the canonical exponential family parametric model, $p(\boldsymbol{y} \mid \boldsymbol{\psi})$, then the posterior density corresponding to the canonical conjugate form, $p(\boldsymbol{\psi} \mid n_0, \boldsymbol{y}_0)$, is given by*

$$p(\boldsymbol{\psi} \mid n_0, \boldsymbol{y}_0, \boldsymbol{y}_1, \ldots, \boldsymbol{y}_n) = p\left(\boldsymbol{\psi} \;\middle|\; n_0 + n, \frac{n_0 \boldsymbol{y}_0 + n\overline{\boldsymbol{y}}_n}{(n + n_0)}\right),$$

where $\overline{\boldsymbol{y}}_n = \sum_{i=1}^n \boldsymbol{y}_i / n$.

Proof.

$$\begin{aligned} p(\boldsymbol{\psi} \mid n_0, \boldsymbol{y}_0, \boldsymbol{y}_1, \ldots, \boldsymbol{y}_n) &\propto \prod_{i=1}^n p(\boldsymbol{y}_i \mid \boldsymbol{\psi}) p(\boldsymbol{\psi} \mid n_0, \boldsymbol{y}_0) \\ &\propto \exp\left\{n\overline{\boldsymbol{y}}_n^t \boldsymbol{\psi} - nb(\boldsymbol{\psi})\right\} \\ &\quad \times \exp\left\{n_0 \boldsymbol{y}_0^t \boldsymbol{\psi} - n_0 b(\boldsymbol{\psi})\right\} \\ &\propto \exp\left\{(n_0 \boldsymbol{y}_0 + n\overline{\boldsymbol{y}}_n)^t \boldsymbol{\psi} - (n_0 + n) b(\boldsymbol{\psi})\right\}, \end{aligned}$$

and the result follows. ◁

Example 5.4. ***(continued).*** In the case of the Bernoulli parametric model, we have seen earlier that the pairing of the parametric model and conjugate prior can be expressed as

$$p(x \mid \theta) = (1 - \theta) \exp\left\{x \log\left(\frac{\theta}{1 - \theta}\right)\right\}$$

$$p(\theta \mid \tau_0, \tau_1) = [K(\boldsymbol{\tau})]^{-1} (1 - \theta)^{\tau_0} \exp\left\{\tau_1 \log\left(\frac{\theta}{1 - \theta}\right)\right\},$$

The canonical forms in this case are obtained by setting

$$y = x, \quad \psi = \log\left(\frac{\theta}{1 - \theta}\right), \quad a(y) = 1, \quad b(\psi) = \log(1 + e^{\psi}),$$

$$c(n_0, y_0) = \frac{\Gamma(n_0 + 2)}{\Gamma(n_0 y_0 + 1)\Gamma(n_0 - n_0 y_0 + 1)},$$

and, hence, the posterior distribution of the canonical parameter ψ is given by

$$p(\psi \mid n_0, y_0, y_1, \ldots, y_n) \propto \exp\left[(n_0 + n)\left\{\frac{n_0 y_0 + n\overline{y}_n}{n + n_0} \psi - b(\psi)\right\}\right].$$

Example 5.5. ***(continued).*** In the case of the Poisson parametric model, we have seen earlier that the pairings of the parametric model and conjugate form can be expressed as

$$p(x \mid \theta) = \frac{1}{x!} \exp(-\theta) \exp(x \log \theta)$$

$$p(\theta \mid \tau_0, \tau_1) = [K(\boldsymbol{\tau})]^{-1} \exp(-\tau_0 \theta) \exp(\tau_1 \log \theta),$$

The canonical forms in this case are obtained by setting

$$y = x, \quad \psi = \log \theta, \quad a(y) = \frac{1}{y!}, \quad b(\psi) = e^{\psi}, \quad c(n_0, y_0) = \frac{n_0^{y_0+1}}{\Gamma(y_0+1)}.$$

The posterior distribution of the canonical parameter ψ is now immediately given by Proposition 5.6.

Example 5.6. ***(continued).*** In the case of the normal parametric model, we have seen earlier that the pairings of the parametric model and conjugate form can be expressed as

$$p(x \mid \mu, \lambda) = (2\pi)^{-1/2} \left[\lambda^{1/2} \exp\left(-\frac{1}{2}\lambda\mu^2\right)\right] \exp\left\{x(\lambda\mu) - \frac{1}{2}x^2\lambda\right\}$$

$$p(\mu, \lambda \mid \tau_0, \tau_1, \tau_2) = [K(\tau)]^{-1} \left[\lambda^{1/2} \exp\left(-\frac{1}{2}\lambda\mu^2\right)\right]^{\tau_0} \exp\left\{\tau_1(\lambda\mu) - \frac{1}{2}\tau_2\lambda\right\}.$$

The canonical forms in this case are obtained by setting

$$\boldsymbol{y} = (y_1, y_2) = (x, x^2), \quad \boldsymbol{\psi} = (\psi_1, \psi_2) = \left(\lambda\mu, -\frac{1}{2}\lambda\right),$$

$$a(\boldsymbol{y}) = (2\pi)^{-1/2}, \quad b(\boldsymbol{\psi}) = \log(-2\psi_2)^{-1/2} - \frac{\psi_1^2}{4\psi_2},$$

$$c(n_0, \boldsymbol{y}_0) = \left(\frac{2\pi}{n_0}\right)^{1/2} \frac{\left(\frac{1}{2}(n_0 y_{02})\right)^{(n_0 y_{01}+1)/2}}{\Gamma\left(\frac{1}{2}(n_0+1)\right)}.$$

Again, the posterior distribution of the canonical parameters $\boldsymbol{\psi} = (\psi_1, \psi_2)$ is now immediately given by Proposition 5.6.

For specific applications, the choice of the representation of the parametric model and conjugate prior forms is typically guided by the ease of interpretation of the parametrisations adopted. Example 5.6 above suffices to demonstrate that the canonical forms may be very unappealing. From a *theoretical* perspective, however, the canonical representation often provides valuable unifying insight, as in Proposition 5.6, where the economy of notation makes it straightforward to demonstrate that the learning process just involves a simple weighted average,

$$\frac{n_0 \boldsymbol{y}_0 + n\bar{\boldsymbol{y}}_n}{n_0 + n},$$

of prior and sample information. Again using the canonical forms, we can give a more precise characterisation of this weighted average.

Proposition 5.7. (***Weighted average form of posterior expectation***).
If $y_1, \ldots, y_n$ are the values of y resulting from a random sample of size n from the canonical exponential family parametric model,

$$p(\boldsymbol{y} \mid \boldsymbol{\psi}) = a(\boldsymbol{y}) \exp\left\{\boldsymbol{y}^t \boldsymbol{\psi} - b(\boldsymbol{\psi})\right\},$$

with canonical conjugate prior $p(\boldsymbol{\psi} \mid n_0, \boldsymbol{y}_0)$, then

$$E\left[\nabla b(\boldsymbol{\psi}) \mid n_0, \boldsymbol{y}_0, \boldsymbol{y}\right] = \pi \overline{\boldsymbol{y}}_n + (1 - \pi)\boldsymbol{y}_0,$$

where

$$\pi = \frac{n}{n_0 + n}, \quad \left[\nabla b(\boldsymbol{\psi})\right]_i = \frac{\partial}{\partial \psi_i} b(\boldsymbol{\psi}).$$

Proof. By Proposition 5.6, it suffices to prove that $E(\nabla b(\boldsymbol{\psi}) \mid n_0, \boldsymbol{y}_0) = \boldsymbol{y}_0$. But

$$\begin{aligned} n_0\left[\boldsymbol{y}_0 - E(\nabla b(\boldsymbol{\psi}) \mid n_0, \boldsymbol{y}_0)\right] &= \int_\Psi n_0\left(\boldsymbol{y}_0 - \nabla b(\boldsymbol{\psi})\right) p(\boldsymbol{\psi} \mid n_0, \boldsymbol{y}_0)\, d\boldsymbol{\psi} \\ &= \int_\Psi \nabla p(\boldsymbol{\psi} \mid n_0, \boldsymbol{y}_0)\, d\boldsymbol{\psi}. \end{aligned}$$

This establishes the result. $\triangleleft$

Proposition 5.7 reveals, in this natural conjugate setting, that the posterior expectation of $\nabla b(\boldsymbol{\psi})$, that is its Bayes estimate with respect to quadratic loss (see Proposition 5.2), is a weighted average of $\boldsymbol{y}_0$ and $\overline{\boldsymbol{y}}_n$. The former is the prior estimate of $\nabla b(\boldsymbol{\psi})$; the latter can be viewed as an intuitively "natural" sample-based estimate of $\nabla b(\boldsymbol{\psi})$, since

$$\begin{aligned} E(\boldsymbol{y} \mid \boldsymbol{\psi}) - \nabla b(\boldsymbol{\psi}) &= \int (\boldsymbol{y} - \nabla b(\boldsymbol{\psi})) p(\boldsymbol{y} \mid \boldsymbol{\psi}) d\boldsymbol{y} \\ &= \int \nabla p(\boldsymbol{y} \mid \boldsymbol{\psi})\, d\boldsymbol{y} = \nabla \int p(\boldsymbol{y} \mid \boldsymbol{\psi}) d\boldsymbol{y} = 0 \end{aligned}$$

and hence $E(\boldsymbol{y} \mid \boldsymbol{\psi}) = E(\overline{\boldsymbol{y}}_n \mid \boldsymbol{\psi}) = \nabla b(\boldsymbol{\psi})$.

For any given prior hyperparameters, $(n_0, \boldsymbol{y}_0)$, as the sample size n becomes large, the weight, π, tends to one and the sample-based information dominates the posterior. In this context, we make an important point alluded to in our discussion of "objectivity and subjectivity", in Section 4.8.2. Namely, that in the stylised setting of a group of individuals agreeing on an exponential family parametric form, but assigning different conjugate priors, a sufficiently large sample will lead to more or less identical posterior beliefs. Statements based on the latter might well, in common parlance, be claimed to be "objective". One should always be aware, however, that this is no more than a conventional way of indicating a subjective consensus, resulting from a large amount of data processed in the light of a central core of shared assumptions.

Proposition 5.7 shows that conjugate priors for exponential family parameters imply that posterior expectations are linear functions of the sufficient statistics. It is interesting to ask whether other forms of prior specification can also lead to linear posterior expectations. Or, more generally, whether knowing or constraining posterior moments to be of some simple algebraic form suffices to characterise possible families of prior distributions. These kinds of questions are considered in detail in, for example, Diaconis and Ylvisaker (1979) and Goel and DeGroot (1980). In particular, it can be shown, under some regularity conditions, that, for continuous exponential families, linearity of the posterior expectation does imply that the prior must be conjugate.

The weighted average form of posterior mean,

$$E[\nabla b(\boldsymbol{\psi}) \,|\, n_0, \boldsymbol{y}_0, \boldsymbol{y}] = \frac{n_0 \boldsymbol{y}_0 + n\overline{\boldsymbol{y}}_n}{n_0 + n},$$

obtained in Proposition 5.7, and also appearing explicitly in the prior to posterior updating process given in Proposition 5.6 makes clear that the prior parameter, n_0, attached to the prior mean, $\boldsymbol{y}_0$ for $\nabla b(\boldsymbol{\psi})$, plays an analogous role to the sample size, n, attached to the data mean $\overline{\boldsymbol{y}}_n$. The choice of an n_0 which is large relative to n thus implies that the prior will dominate the data in determining the posterior (see, however, Section 5.6.3 for illustration of why a weighted-average form might not be desirable). Conversely, the choice of an n_0 which is small relative to n ensures that the form of the posterior is essentially determined by the data. In particular, this suggests that a tractable analysis which "lets the data speak for themselves" can be obtained by letting $n_0 \to 0$. Clearly, however, this has to be regarded as simply a convenient approximation to the posterior that would have been obtained from the choice of a prior with small, but positive n_0. The choice $n_0 = 0$ typically implies a form of $p(\boldsymbol{\psi} \,|\, n_0, \boldsymbol{y}_0)$ which does not integrate to unity (a so-called *improper* density) and thus cannot be interpreted as representing an actual prior belief. The following example illustrates this use of limiting, improper conjugate priors in the context of the Bernoulli parametric model with beta conjugate prior, using standard rather than canonical forms for the parametric models and prior densities.

Example 5.4. ***(continued)***. We have seen that if $r_n = x_1 + \cdots + x_n$ denotes the number of successes in n Bernoulli trials, the conjugate beta prior density, $\text{Be}(\theta \,|\, \alpha, \beta)$, for the limiting relative frequency of successes, θ, leads to a $\text{Be}(\theta \,|\, \alpha + r_n, \beta + n - r_n)$ posterior for θ, which has expectation

$$\frac{\alpha + r_n}{\alpha + \beta + n} = \pi \left(\frac{r_n}{n}\right) + (1 - \pi)\left(\frac{\alpha}{\alpha + \beta}\right),$$

where $\pi = (\alpha + \beta + n)^{-1} n$, providing a weighted average between the prior mean for θ and the frequency estimate provided by the data. In this notation, $n_0 \to 0$ corresponds to $\alpha \to 0$,

$\beta \to 0$, which implies a $\text{Be}(\theta \mid r_n, n-r_n)$ approximation to the posterior distribution, having expectation r_n/n. The limiting prior form, however, would be

$$p(\theta \mid \alpha = 0, \beta = 0) \propto \theta^{-1}(1-\theta)^{-1},$$

which is not a proper density. As a technique for arriving at the approximate posterior distribution, it is certainly convenient to make formal use of Bayes' theorem with this improper form playing the role of a prior, since

$$\begin{aligned} p(\theta \mid \alpha = 0, \beta = 0, n, r_n) &\propto p(r_n \mid n\theta) p(\theta \mid \alpha = 0, \beta = 0) \\ &\propto \theta^{r_n}(1-\theta)^{n-r_n}\theta^{-1}(1-\theta)^{-1} \\ &\propto \text{Be}(\theta \mid r_n, n-r_n). \end{aligned}$$

It is important to recognise, however, that this is merely an approximation device and in no way justifies regarding $p(\theta \mid \alpha = 0, \beta = 0)$ as having any special significance as a representation of "prior ignorance". Clearly, *any* choice of α, β small compared with r_n, $n - r_n$ (for example, $\alpha = \beta = \frac{1}{2}$ or $\alpha = \beta = 1$ for typical values of $r_n, n - r_n$) will lead to an almost identical posterior distribution for θ.

A further problem of interpretation arises if we consider inferences for functions of θ. Consider, for example, the choice $\alpha = \beta = 1$, which implies a uniform prior density for θ. At an intuitive level, it might be argued that this represents "complete ignorance" about θ, which should, presumably, entail "complete ignorance" about any function, $g(\theta)$, of θ. However, $p(\theta)$ uniform implies that $p(g(\theta))$ is not uniform. This makes it clear that *ad hoc* intuitive notions of "ignorance, or of what constitutes a "non-informative" prior distribution (in some sense), cannot be relied upon. There is a need for a more formal analysis of the concept and this will be given in Section 5.4, with further discussion in Section 5.6.2.

Proposition 5.2 established the general forms of Bayes estimates for some commonly used loss functions. Proposition 5.7 provided further insight into the (posterior mean) form arising from quadratic loss in the case of an exponential family parametric model with conjugate prior. Within this latter framework, the following development, based closely on Gutiérrez-Peña (1992), provides further insight into how the posterior mode can be justified as a Bayes estimate.

We recall, from the discussion preceding Proposition 5.6, the canonical forms of the k-parameter exponential family and its corresponding conjugate prior:

$$p(\boldsymbol{y}|\boldsymbol{\psi}) = a(\boldsymbol{y}) \exp\left\{\boldsymbol{y}^t\boldsymbol{\psi} - b(\boldsymbol{\psi})\right\}, \quad \boldsymbol{y} \in Y$$

and

$$p(\boldsymbol{\psi}|n_0, \boldsymbol{y}_0) = c(n_0, \boldsymbol{y}_0) \exp\left\{n_0\boldsymbol{y}_0^t\boldsymbol{\psi} - n_0 b(\boldsymbol{\psi})\right\}, \quad \boldsymbol{\psi} \in \Psi,$$

for appropriately defined Y, Ψ and real-valued functions a, b and c.

Consider $p(\boldsymbol{\psi}|n_0, \boldsymbol{y}_0)$ and define $d(s, \boldsymbol{t}) = -\log c(s, s^{-1}\boldsymbol{t})$, with $s > 0$ and $\boldsymbol{t} \in Y$. Further define

$$\nabla d(s, \boldsymbol{t}) = \left[\frac{\partial d(s, \boldsymbol{t})}{\partial t_1}, \ldots, \frac{\partial d(s, \boldsymbol{t})}{\partial t_k}\right]^t = [d_1(s, \boldsymbol{t}), \ldots, d_k(s, \boldsymbol{t})]^t$$

and $d_0(s, \boldsymbol{t}) = \partial d(s, \boldsymbol{t})/\partial s$. As a final preliminary, recall the logarithmic divergence measure

$$\delta(\boldsymbol{\theta} \mid \boldsymbol{\theta}_0) = \int p(\boldsymbol{x} \mid \boldsymbol{\theta}) \log \frac{p(\boldsymbol{x} \mid \boldsymbol{\theta})}{p(\boldsymbol{x} \mid \boldsymbol{\theta}_0)}\, d\boldsymbol{x}$$

between two distributions $p(\boldsymbol{x}|\boldsymbol{\theta})$ and $p(\boldsymbol{x}|\boldsymbol{\theta}_0)$. We can now establish the following technical results.

Proposition 5.8. (***Logarithmic divergence between conjugate distributions***). *With respect to the canonical form of the k-parameter exponential family and its corresponding conjugate prior:*

(i) $\delta(\boldsymbol{\psi}|\boldsymbol{\psi}_0) = b(\boldsymbol{\psi}_0) - b(\boldsymbol{\psi}) + (\boldsymbol{\psi} - \boldsymbol{\psi}_0)^t \nabla b(\boldsymbol{\psi})$;

(ii) $E[\delta(\boldsymbol{\psi}|\boldsymbol{\psi}_0)] = d_0(n_0, n_0\boldsymbol{y}_0) + b(\boldsymbol{\psi}_0) + n_0{}^{-1}\{k + [\nabla d(n_0, n_0\boldsymbol{y}_0) - \boldsymbol{\psi}_0]^t n_0\boldsymbol{y}_0\}$.

Proof. From the definition of logarithmic divergence we see that

$$\delta(\boldsymbol{\psi}|\boldsymbol{\psi}_0) = b(\boldsymbol{\psi}_0) - b(\boldsymbol{\psi}) + (\boldsymbol{\psi} - \boldsymbol{\psi}_0)^t E_{\boldsymbol{y}|\boldsymbol{\psi}}[\boldsymbol{y}],$$

and (i) follows. Moreover,

$$E[\delta(\boldsymbol{\psi}|\boldsymbol{\psi}_0)] = b(\boldsymbol{\psi}_0) - E[b(\boldsymbol{\psi})] + E[\boldsymbol{\psi}^t \nabla b(\boldsymbol{\psi})] - \boldsymbol{\psi}_0^t E[\nabla b(\boldsymbol{\psi})].$$

Differentiation of the identity

$$\log \int \exp\{\boldsymbol{t}^t \boldsymbol{\psi} - s b(\boldsymbol{\psi})\} d\boldsymbol{\psi} = d(s, \boldsymbol{t}),$$

with respect to s, establishes straightforwardly that

$$E[b(\boldsymbol{\psi})] = -d_0(n_0, n_0\boldsymbol{y}_0).$$

Recalling that $E[\nabla b(\boldsymbol{\psi})] = \boldsymbol{y}_0$, we can write, for $i = 1, \ldots, k$,

$$\log \int b_i(\boldsymbol{\psi}) \exp\{\boldsymbol{t}^t \boldsymbol{\psi} - s b(\boldsymbol{\psi})\} d\boldsymbol{\psi} = \log t_i - \log c(s, s^{-1}\boldsymbol{t}) - \log s.$$

Differentiating this identity with respect to t_i, and interchanging the order of differentiation and integration, we see that

$$\int \psi_i b_i(\boldsymbol{\psi}) c(s, s^{-1}\boldsymbol{t}) \exp\{\boldsymbol{t}^t \boldsymbol{\psi} - s b(\boldsymbol{\psi})\} d\boldsymbol{\psi} = s^{-1}[1 + d_i(s, \boldsymbol{t}) t_i],$$

for $i = 1, \ldots, k$, so that

$$E[\boldsymbol{\psi}^t \nabla b(\boldsymbol{\psi})] = n_0^{-1}[k + \nabla d(n_0, n_0\boldsymbol{y}_0)^t (n_0\boldsymbol{y}_0)] - n_0^{-1} \boldsymbol{\psi}_0^t (n_0, \boldsymbol{y}_0)$$

and (ii) follows. $\triangleleft$

This result now enables us to establish easily the main result of interest.

Proposition 5.9. ***(Conjugate posterior modes as Bayes estimates).*** *With respect to the loss function* $l(a, \psi) = \delta(\psi|a)$*, the Bayes estimate for* ψ*, derived from independent observations* $y_1, \ldots, y_n$ *from the canonical* k*-parameter exponential family* $p(y|\psi)$ *and corresponding conjugate prior* $p(\psi|n_0, y_0)$*, is the posterior mode,* ψ^**, which satisfies*

$$\nabla b(\psi^*) = (n_0 + n)^{-1}(n_0 \boldsymbol{y}_0 + n\bar{\boldsymbol{y}}_n),$$

with $\bar{\boldsymbol{y}}_n = n^{-1}(\boldsymbol{y}_1 + \cdots + \boldsymbol{y}_n)$.

Proof. We note first (see the proof of Proposition 5.6) that the logarithm of the posterior density is given by

$$\text{constant } + (n_0 \boldsymbol{y}_0 + n\bar{\boldsymbol{y}}_n)^t \psi - (n_0 + n) b(\psi),$$

from which the claimed estimating equation for the posterior mode, ψ^*, is immediately obtained. The result now follows by noting that the same equation arises in the minimisation of (ii) of Proposition 5.8, with $n_0 + n$ replacing n_0, and $n_0 \boldsymbol{y}_0 + n\bar{\boldsymbol{y}}_n$ replacing $n_0 \boldsymbol{y}_0$. $\triangleleft$

For a recent discussion of conjugate priors for exponential families, see Consonni and Veronese (1992b). In complex problems, conjugate priors may have strong, unsuspected implications; for an example, see Dawid (1988a).

5.2.3 Approximations with Conjugate Families

Our main motivation in considering conjugate priors for exponential families has been to provide tractable prior to posterior (or predictive) analysis. At the same time, we might hope that the conjugate family for a particular parametric model would contain a sufficiently rich range of prior density "shapes" to enable one to approximate reasonably closely any particular actual prior belief function of interest. The next example shows that might well not be the case. However, it also indicates how, with a suitable extension of the conjugate family idea, we can achieve both tractability and the ability to approximate closely any actual beliefs.

Example 5.7. ***(The spun coin).*** Diaconis and Ylvisaker (1979) highlight the fact that, whereas a tossed coin typically generates equal long-run frequencies of heads and tails, this is not at all the case if a coin is spun on its edge. Experience suggests that these long-run frequencies often turn out for some coins to be in the ratio 2:1 or 1:2, and for other coins even as extreme as 1:4. In addition, some coins do appear to behave symmetrically.

Let us consider the repeated spinning under perceived "identical conditions" of a given coin, about which we have no specific information beyond the general background set out above. Under the circumstances specified, suppose we judge the sequence of outcomes to be exchangeable, so that a Bernoulli parametric model, together with a prior density for the long-run frequency of heads, completely specifies our belief model. How might we represent this prior density mathematically?

We are immediately struck by two things: first, in the light of the information given, any realistic prior shape will be at least bimodal, and possibly trimodal; secondly, the conjugate family for the Bernoulli parametric model is the beta family (see Example 5.4), which does not contain bimodal densities. It appears, therefore, that an insistence on tractability, in the sense of restricting ourselves to conjugate priors, would preclude an honest prior specification.

However, we can easily generate multimodal shapes by considering *mixtures* of beta densities,

$$p(\theta \mid \boldsymbol{\pi}, \boldsymbol{\alpha}, \boldsymbol{\beta}) = \sum_{i=1}^{m} \pi_i \mathrm{Be}(\theta \mid \alpha_i, \beta_i),$$

with mixing weights $\pi_i > 0, \pi_1 + \cdots + \pi_m = 1$, attached to a selection of conjugate densities, $\mathrm{Be}(\theta \mid \alpha_i, \beta_i), i = 1, \ldots, m$. Figure 5.2 displays the prior density resulting from the mixture

$$0.5\, \mathrm{Be}(\theta \mid 10, 20) + 0.2\, \mathrm{Be}(\theta \mid 15, 15) + 0.3\, \mathrm{Be}(\theta \mid 20, 10),$$

which, among other things, reflects a judgement that about 20% of coins seem to behave symmetrically and most of the rest tend to lead to 2:1 or 1:2 ratios, with somewhat more of the latter than the former.

Suppose now that we observe n outcomes $\boldsymbol{x} = (x_1, \ldots, x_n)$ and that these result in $r_n = x_1 + \cdots + x_n$ heads, so that

$$p(x_1, \ldots, x_n \mid \theta) = \prod_{i=1}^{n} \theta^{x_i} (1 - \theta)^{1 - x_i} = \theta^{r_n} (1 - \theta)^{n - r_n}.$$

Considering the general mixture prior form

$$p(\theta \mid \boldsymbol{\pi}, \boldsymbol{\alpha}, \boldsymbol{\beta}) = \sum_{i=1}^{m} \pi_i\, \mathrm{Be}(\theta \mid \alpha_i, \beta_i),$$

we easily see from Bayes' theorem that

$$p(\theta \mid \boldsymbol{\pi}, \boldsymbol{\alpha}, \boldsymbol{\beta}, \boldsymbol{x}) = p(\theta \mid \boldsymbol{\pi}^*, \boldsymbol{\alpha}^*, \boldsymbol{\beta}^*),$$

where

$$\alpha_i^* = \alpha_i + r_n, \quad \beta_i^* = \beta_i + n - r_n$$

and

$$\begin{aligned} \pi_i^* &\propto \pi_i \int_0^1 \theta^{r_n} (1 - \theta)^{n - r_n} \mathrm{Be}(\theta \mid \alpha_i, \beta_i)\, d\theta \\ &\propto \pi_i \frac{\Gamma(\alpha_i + \beta_i)}{\Gamma(\alpha_i)\Gamma(\beta_i)} \cdot \frac{\Gamma(\alpha_i^*)\Gamma(\beta_i^*)}{\Gamma(\alpha_i^* + \beta_i^*)}, \end{aligned}$$

so that the resulting posterior density,

$$p(\theta \mid \boldsymbol{\pi}, \boldsymbol{\alpha}, \boldsymbol{\beta}, \boldsymbol{x}) = \sum_{i=1}^{m} \pi_i^* \, \mathrm{Be}(\theta \mid \alpha_i^*, \beta_i^*),$$

is itself a mixture of m beta components. This establishes that the general mixture class of beta densities is *closed under sampling* with respect to the Bernoulli model.

In the case considered above, suppose that the spun coin results in 3 heads after 10 spins and 14 heads after 50 spins. The suggested prior density corresponds to $m = 3$,

$$\boldsymbol{\pi} = (0.5, 0.2, 0.3), \quad \boldsymbol{\alpha} = (10, 15, 20), \quad \boldsymbol{\beta} = (20, 15, 10).$$

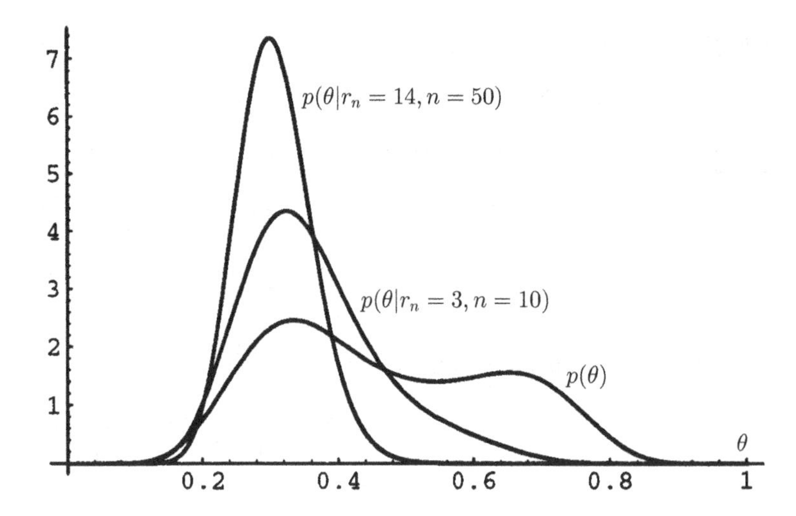

Figure 5.2 *Prior and posteriors from a three-component beta mixture prior density*

Detailed calculation yields:

$$\begin{aligned} \text{for } n = 10,\ r_n = 3;\ \boldsymbol{\pi}^* &= (0.77, 0.16, 0.07), \\ \boldsymbol{\alpha}^* &= (13, 18, 23),\ \boldsymbol{\beta}^* = (27, 22, 17) \\ \text{for } n = 50,\ r_n = 14;\ \boldsymbol{\pi}^* &= (0.90, 0.09, 0.006), \\ \boldsymbol{\alpha}^* &= (24, 29, 34),\ \boldsymbol{\beta}^* = (56, 51, 46), \end{aligned}$$

and the resulting posterior densities are shown in Figure 5.2.

This example demonstrates that, at least in the case of the Bernoulli parametric model and the beta conjugate family, the use of mixtures of conjugate densities both maintains the tractability of the analysis and provides a great deal of flexibility in approximating actual forms of prior belief. In fact, the same is true for any exponential family model and corresponding conjugate family, as we show in the following.

Proposition 5.10. ***(Mixtures of conjugate priors).*** *Let $\boldsymbol{x} = (x_1, \ldots, x_n)$ be a random sample from a regular exponential family distribution such that*

$$p(\boldsymbol{x} \mid \boldsymbol{\theta}) = \prod_{j=1}^{n} f(x_j)[g(\boldsymbol{\theta})]^n \exp\left\{\sum_{i=1}^{m} c_i\phi_i(\boldsymbol{\theta})\left(\sum_{j=1}^{n} h_i(x_j)\right)\right\}$$

and let

$$p(\theta \mid \boldsymbol{\pi}, \boldsymbol{\tau}_1, \ldots, \boldsymbol{\tau}_m) = \sum_{l=1}^{m} \pi_l p(\boldsymbol{\theta} \mid \boldsymbol{\tau}_l),$$

where, for $l = 1, \ldots, m$,

$$p(\theta \mid \boldsymbol{\tau}_l) = [K(\boldsymbol{\tau}_l)]^{-1}[g(\boldsymbol{\theta})]^{\tau_{l0}} \exp\left\{\sum_{i=1}^{k} c_i\phi_i(\boldsymbol{\theta})\tau_{li}\right\}$$

are elements of the conjugate family. Then

$$p(\theta \mid \boldsymbol{\pi}, \boldsymbol{\tau}_1, \ldots, \boldsymbol{\tau}_m, \boldsymbol{x}) = p(\theta \mid \boldsymbol{\pi}^*, \boldsymbol{\tau}_1^*, \ldots, \boldsymbol{\tau}_m^*) = \sum_{l=1}^{m} \pi_l^* p(\theta \mid \boldsymbol{\tau}_l^*),$$

where, with $\boldsymbol{t}_n(\boldsymbol{x}) = \left\{n, \sum_{j=1}^{n} h_1(x_j), \ldots, \sum_{j=1}^{n} h_k(j)\right\}$,

$$\boldsymbol{\tau}_l^* = \boldsymbol{\tau}_l + \boldsymbol{t}_n(\boldsymbol{x}),$$

and

$$\pi_l^* \propto \pi_l \prod_{j=1}^{n} f(x_j) \frac{K(\boldsymbol{\tau}_l^*)}{K(\boldsymbol{\tau}_l)}\,.$$

Proof. The results follows straightforwardly from Bayes' theorem and Proposition 5.5. ◁

It is interesting to ask just how flexible mixtures of conjugate prior are. The answer is that *any* prior density for an exponential family parameter can be approximated arbitrarily closely by such a mixture, as shown by Dalal and Hall (1983), and Diaconis and Ylvisaker (1985). However, their analyses do not provide a constructive mechanism for building up such a mixture. In practice, we are left with having to judge when a particular tractable choice, typically a conjugate form, a limiting conjugate form, or a mixture of conjugate forms, is "good enough, in the sense that probability statements based on the resulting posterior will not differ radically from the statements that would have resulted from using a more honest, but difficult to specify or intractable, prior.

The following result provides some guidance, in a much more general setting than that of conjugate mixtures, as to when an "approximate" (possibly improper) prior may be safely used in place of an "honest" prior.

Proposition 5.11. (***Prior approximation***). *Suppose that a belief model is defined by* $p(\boldsymbol{x} \mid \boldsymbol{\theta})$ *and* $p(\boldsymbol{\theta})$, $\boldsymbol{\theta} \in \Theta$ *and that* $q(\boldsymbol{\theta})$ *is a non-negative function such that* $q(\boldsymbol{x}) = \int_\Theta p(\boldsymbol{x} \mid \boldsymbol{\theta}) q(\boldsymbol{\theta}) d\boldsymbol{\theta} < \infty$, *where, for some* $\Theta_0 \subseteq \Theta$ *and* $\alpha, \beta \in \Re^*$,

(a) $1 \leq p(\boldsymbol{\theta})/q(\boldsymbol{\theta}) \leq 1 + \alpha$, *for all* $\boldsymbol{\theta} \in \Theta_0$,

(b) $p(\boldsymbol{\theta})/q(\boldsymbol{\theta}) \leq \beta$, *for all* $\boldsymbol{\theta} \in \Theta$.

Let $p = \int_{\Theta_0} p(\boldsymbol{\theta} \mid \boldsymbol{x}) d\boldsymbol{\theta}$, $q = \int_{\Theta_0} q(\boldsymbol{\theta} \mid \boldsymbol{x}) d\boldsymbol{\theta}$, *and*
$q(\boldsymbol{\theta} \mid \boldsymbol{x}) = p(\boldsymbol{x} \mid \boldsymbol{\theta}) q(\boldsymbol{\theta}) / \int p(\boldsymbol{x} \mid \boldsymbol{\theta}) q(\boldsymbol{\theta}) d\boldsymbol{\theta}$. *Then,*

(i) $(1 - p)/p \leq \beta(1 - q)/q$

(ii) $q \leq p(\boldsymbol{x})/q(\boldsymbol{x}) \leq (1 + \alpha)/p$

(iii) *for all* $\boldsymbol{\theta} \in \Theta$, $p(\boldsymbol{\theta} \mid \boldsymbol{x})/q(\boldsymbol{\theta} \mid \boldsymbol{x}) \leq [p(\boldsymbol{\theta})/q(\boldsymbol{\theta})]/q \leq \beta/q$

(iv) *for all* $\boldsymbol{\theta} \in \Theta_0$, $p/(1 + \alpha) \leq p(\boldsymbol{\theta} \mid \boldsymbol{x})/q(\boldsymbol{\theta} \mid \boldsymbol{x}) \leq (1 + \alpha)/q$

(v) *for* $\varepsilon = \max\{(1 - p), (1 - q)\}$ *and* $f : \Theta \to \Re$ *such that* $|\, f(\boldsymbol{\theta}) \,| \leq m$,

$$m^{-1} \left| \int_\Theta f(\boldsymbol{\theta}) p(\boldsymbol{\theta} \mid \boldsymbol{x}) d\boldsymbol{\theta} - \int_\Theta f(\boldsymbol{\theta}) q(\boldsymbol{\theta} \mid \boldsymbol{x}) d\boldsymbol{\theta} \right| \leq \alpha + 3\varepsilon$$

Proof. (Dickey, 1976). Part (i) clearly follows from

$$\frac{1 - p}{p} = \frac{\int_{\Theta_0^c} p(\boldsymbol{x} \mid \boldsymbol{\theta}) p(\boldsymbol{\theta}) d\boldsymbol{\theta}}{\int_{\Theta_0} p(\boldsymbol{x} \mid \boldsymbol{\theta}) p(\boldsymbol{\theta}) d\boldsymbol{\theta}} \leq \beta \frac{1 - q}{q} .$$

Clearly,

$$p(\boldsymbol{x}) \geq \int_{\Theta_0} p(\boldsymbol{x} \mid \boldsymbol{\theta}) p(\boldsymbol{\theta}) d\boldsymbol{\theta} \geq \int_{\Theta_0} p(\boldsymbol{x} \mid \boldsymbol{\theta}) q(\boldsymbol{\theta}) d\boldsymbol{\theta} = q \cdot q(\boldsymbol{x}),$$

$$q(\boldsymbol{x}) \geq \int_{\Theta_0} q(\boldsymbol{x} \mid \boldsymbol{\theta}) q(\boldsymbol{\theta}) d\boldsymbol{\theta} \geq \frac{1}{1+\alpha} \int_{\Theta_0} p(\boldsymbol{x} \mid \boldsymbol{\theta}) p(\boldsymbol{\theta}) d\boldsymbol{\theta} = \frac{p}{1+\alpha} p(\boldsymbol{x}),$$

which establishes (ii). Part (iii) follows from (b) and (ii), and part (iv) follows from (a) and (ii). Finally,

$$m^{-1} \left| \int_{\Theta} f(\boldsymbol{\theta}) p(\boldsymbol{\theta} \mid \boldsymbol{x}) d\boldsymbol{\theta} - \int_{\Theta} f(\boldsymbol{\theta}) q(\boldsymbol{\theta} \mid \boldsymbol{x}) d\boldsymbol{\theta} \right| \leq \int \left| p(\boldsymbol{\theta} \mid \boldsymbol{x}) - q(\boldsymbol{\theta} \mid \boldsymbol{x}) \right| d\boldsymbol{\theta}$$

$$\begin{aligned}
&\leq \int_{\Theta_0} \left| p(\boldsymbol{\theta} \mid \boldsymbol{x}) - q(\boldsymbol{\theta} \mid \boldsymbol{x}) \right| d\boldsymbol{\theta} + \int_{\Theta_0^c} \left| p(\boldsymbol{\theta} \mid \boldsymbol{x}) - q(\boldsymbol{\theta} \mid \boldsymbol{x}) \right| d\boldsymbol{\theta} \\
&\leq \int_{\Theta_0} \left| q(\boldsymbol{\theta} \mid \boldsymbol{x}) \left(\frac{p(\boldsymbol{\theta} \mid \boldsymbol{x})}{q(\boldsymbol{\theta} \mid \boldsymbol{x})} - 1 \right) \right| d\boldsymbol{\theta} + \int_{\Theta_0^c} \left| p(\boldsymbol{\theta} \mid \boldsymbol{x}) \right| d\boldsymbol{\theta} + \int_{\Theta_0^c} \left| q(\boldsymbol{\theta} \mid \boldsymbol{x}) \right| d\boldsymbol{\theta} \\
&\leq \int_{\Theta_0} \left| q(\boldsymbol{\theta} \mid \boldsymbol{x}) \left(\frac{1+\alpha}{q} - 1 \right) \right| d\boldsymbol{\theta} + (1-p) + (1-q) \qquad \text{(by iv)} \\
&= (1+\alpha-q) + (1-p) + (1-q) \leq \alpha + 3\varepsilon,
\end{aligned}$$

which proves (v). ◁

If, in the above, Θ_0 is a subset of Θ with high probability under $q(\boldsymbol{\theta} \mid \boldsymbol{x})$ and α is chosen to be small and β not too large, so that $q(\boldsymbol{\theta})$ provides a good approximation to $p(\boldsymbol{\theta})$ within Θ_0 and $p(\boldsymbol{\theta})$ is nowhere much greater than $q(\boldsymbol{\theta})$, then (i) implies that Θ_0 has high probability under $p(\boldsymbol{\theta} \mid \boldsymbol{x})$ and (ii), (iv) and (v) establish that both the respective predictive and posterior distributions, within Θ_0, and also the posterior expectations of bounded functions are very close. More specifically, if f is taken to be the indicator function of any subset $\Theta^* \subseteq \Theta$, (v) implies that

$$\left| \int_{\Theta^*} p(\boldsymbol{\theta} \mid \boldsymbol{x}) d\boldsymbol{\theta} - \int_{\Theta^*} q(\boldsymbol{\theta} \mid \boldsymbol{x}) d\boldsymbol{\theta} \right| \leq \alpha + 3\varepsilon,$$

providing a bound on the inaccuracy of the posterior probability statement made using $q(\boldsymbol{\theta} \mid \boldsymbol{x})$ rather than $p(\boldsymbol{\theta} \mid \boldsymbol{x})$.

Proposition 5.11 therefore asserts that if a mathematically convenient alternative, $q(\boldsymbol{\theta})$, to the would-be honest prior, $p(\boldsymbol{\theta})$, can be found, giving high posterior probability to a set $\Theta_0 \subseteq \Theta$ within which it provides a good approximation to $p(\boldsymbol{\theta})$ and such that it is nowhere orders of magnitude smaller than $p(\boldsymbol{\theta})$ outside Θ_0, then $q(\boldsymbol{\theta})$ may reasonably be used in place of $p(\boldsymbol{\theta})$.

In the case of $\Theta = \Re$, Figure 5.3 illustrates, in stylised form, a frequently occurring situation, where the choice $q(\boldsymbol{\theta}) = c$, for some constant c, provides

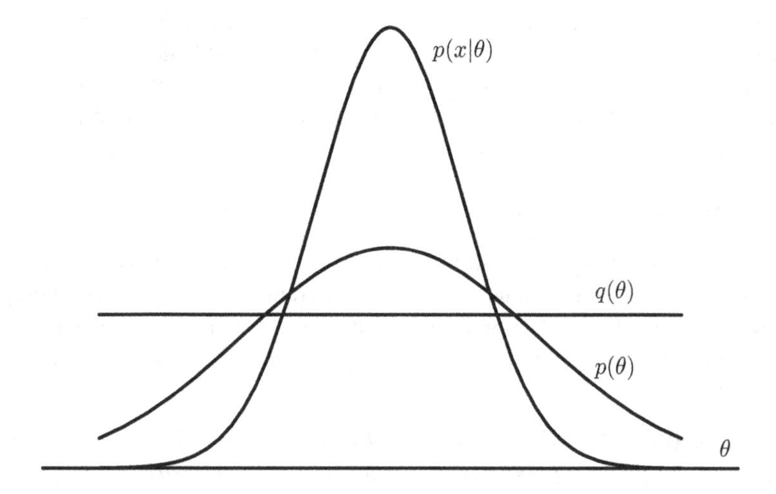

Figure 5.3 *Typical conditions for precise measurement*

a convenient approximation. In qualitative terms, the likelihood is highly peaked relative to $p(\boldsymbol{\theta})$, which has little curvature in the region of non-negligible likelihood.

In this situation of "precise measurement" (Savage, 1962), the choice of the function $q(\boldsymbol{\theta}) = c$, for an appropriate constant c, clearly satisfies the conditions of Proposition 5.10 and we obtain

$$p(\theta \mid \boldsymbol{x}) \simeq q(\theta \mid \boldsymbol{x}) = \frac{p(\boldsymbol{x} \mid \theta)c}{\int_{\Re} p(\boldsymbol{x} \mid \theta)c \, d\theta} = \frac{p(\boldsymbol{x} \mid \theta)}{\int_{\Re} p(\boldsymbol{x} \mid \theta) \, d\theta},$$

the normalised likelihood function.

The second of the implementation questions posed at the end of Section 5.1.6 concerned the possibility of avoiding the need for precise mathematical representation of the prior density in situations where the information provided by the data is far greater than that implicit in the prior. The above analysis goes some way to answering that question; the following section provides a more detailed analysis.

5.3 ASYMPTOTIC ANALYSIS

In Chapter 4, we saw that in representations of belief models for observables involving a parametric model $p(x \mid \boldsymbol{\theta})$ and a prior specification $p(\boldsymbol{\theta})$, the parameter $\boldsymbol{\theta}$ acquired an operational meaning as some form of strong law limit of observables. Given observations $\boldsymbol{x} = (x_1, \ldots, x_n)$, the posterior distribution, $p(\boldsymbol{\theta} \mid \boldsymbol{x})$, then describes beliefs about that strong law limit in the light of the information provided by $x_1, \ldots, x_n$. To answer the second question posed at the end of Section 5.1.6, we now wish to examine various properties of $p(\boldsymbol{\theta} \mid x)$ as the number of observations increases; i.e., as $n \to \infty$. Intuitively, we would hope that beliefs about $\boldsymbol{\theta}$ would become more and more concentrated around the "true" parameter value; i.e., the corresponding strong law limit. Under appropriate conditions, we shall see that this is, indeed, the case.

5.3.1 Discrete Asymptotics

We begin by considering the situation where $\Theta = \{\boldsymbol{\theta}_1, \boldsymbol{\theta}_2, \ldots,\}$ consists of a countable (possibly finite) set of values, such that the parametric model corresponding to the true parameter, $\boldsymbol{\theta}_t$, is "distinguishable" from the others, in the sense that the logarithmic divergences, $\int p(x \mid \boldsymbol{\theta}_t) \log[p(x \mid \boldsymbol{\theta}_t)/p(x \mid \boldsymbol{\theta}_i)]\, dx$ are strictly larger than zero, for all $i \neq t$.

Proposition 5.12. **(*Discrete asymptotics*).** *Let $\boldsymbol{x} = (x_1, \ldots, x_n)$ be observations for which a belief model is defined by the parametric model $p(x \mid \boldsymbol{\theta})$, where $\boldsymbol{\theta} \in \Theta = \{\boldsymbol{\theta}_1, \boldsymbol{\theta}_2, \ldots\}$, and the prior $p(\boldsymbol{\theta}) = \{p_1, p_2, \ldots\}$, $p_i > 0$, $\sum_i p_i = 1$. Suppose that $\boldsymbol{\theta}_t \in \Theta$ is the true value of $\boldsymbol{\theta}$ and that, for all $i \neq t$,*

$$\int p(x \mid \boldsymbol{\theta}_t) \log \left[\frac{p(x \mid \boldsymbol{\theta}_t)}{p(x \mid \boldsymbol{\theta}_i)} \right] dx > 0;$$

then

$$\lim_{n\to\infty} p(\boldsymbol{\theta}_t \mid \boldsymbol{x}) = 1, \quad \lim_{n\to\infty} p(\boldsymbol{\theta}_i \mid \boldsymbol{x}) = 0, \; i \neq t.$$

Proof. By Bayes' theorem, and assuming that $p(\boldsymbol{x}|\boldsymbol{\theta}) = \prod_{i=1}^n p(x_i|\boldsymbol{\theta})$,

$$\begin{aligned} p(\boldsymbol{\theta}_i \mid \boldsymbol{x}) &= p_i \frac{p(\boldsymbol{x} \mid \boldsymbol{\theta}_i)}{p(\boldsymbol{x})} \\ &= \frac{p_i \{p(\boldsymbol{x} \mid \boldsymbol{\theta}_i)/p(\boldsymbol{x} \mid \boldsymbol{\theta}_t)\}}{\sum_i p_i \{p(\boldsymbol{x} \mid \boldsymbol{\theta}_i)/p(\boldsymbol{x} \mid \boldsymbol{\theta}_t)\}} \\ &= \frac{\exp\{\log p_i + S_i\}}{\sum_i \exp\{\log p_i + S_i\}}, \end{aligned}$$

where

$$S_i = \sum_{j=1}^n \log \frac{p(x_j \mid \boldsymbol{\theta}_i)}{p(x_j \mid \boldsymbol{\theta}_t)}\,.$$

Conditional on $\boldsymbol{\theta}_t$, the latter is the sum of n independent identically distributed random quantities and hence, by the strong law of large numbers (see Section 3.2.3),

$$\lim_{n\to\infty} \frac{1}{n} S_i = \int p(x \mid \boldsymbol{\theta}_t) \log \left[\frac{p(x \mid \boldsymbol{\theta}_i)}{p(x \mid \boldsymbol{\theta}_t)} \right] dx.$$

The right-hand side is negative for all $i \neq t$, and equals zero for $i = t$, so that, as $n \to \infty$, $S_t \to 0$ and $S_i \to -\infty$ for $i \neq t$, which establishes the result. ◁

An alternative way of expressing the result of Proposition 5.12, established for countable Θ, is to say that the posterior distribution function for θ ultimately degenerates to a step function with a single (unit) step at $\theta = \theta_t$. In fact, this result can be shown to hold, under suitable regularity conditions, for much more general forms of Θ. However, the proofs require considerable measure-theoretic machinery and the reader is referred to Berk (1966, 1970) for details.

A particularly interesting result is that if the true θ is *not* in Θ, the posterior degenerates onto the value in Θ which gives the parametric model closest in logarithmic divergence to the true model.

5.3.2 Continuous Asymptotics

Let us now consider what can be said in the case of general Θ about the forms of probability statements implied by $p(\boldsymbol{\theta} \mid \boldsymbol{x})$ for large n. Proceeding heuristically for the moment, without concern for precise regularity conditions, we note that, in the case of a parametric representation for an exchangeable sequence of observables,

$$\begin{aligned} p(\boldsymbol{\theta} \mid \boldsymbol{x}) &\propto p(\boldsymbol{\theta}) \prod_{i=1}^{n} p(x_i \mid \boldsymbol{\theta}) \\ &\propto \exp\{\log p(\boldsymbol{\theta}) + \log p(\boldsymbol{x} \mid \boldsymbol{\theta})\}. \end{aligned}$$

If we now expand the two logarithmic terms about their respective maxima, $\boldsymbol{m}_0$ and $\hat{\boldsymbol{\theta}}_n$, assumed to be determined by setting $\nabla \log p(\boldsymbol{\theta}) = 0$, $\nabla \log p(\boldsymbol{x} \mid \boldsymbol{\theta}) = 0$, respectively, we obtain

$$\begin{aligned} \log p(\boldsymbol{\theta}) &= \log p(\boldsymbol{m}_0) - \frac{1}{2}(\boldsymbol{\theta} - \boldsymbol{m}_0)^t \boldsymbol{H}_0 (\boldsymbol{\theta} - \boldsymbol{m}_0) + R_0 \\ \log p(\boldsymbol{x} \mid \boldsymbol{\theta}) &= \log p(\boldsymbol{x} \mid \hat{\boldsymbol{\theta}}_n) - \frac{1}{2}(\boldsymbol{\theta} - \hat{\boldsymbol{\theta}}_n)^t \boldsymbol{H}(\hat{\boldsymbol{\theta}}_n)(\boldsymbol{\theta} - \hat{\boldsymbol{\theta}}_n) + R_n, \end{aligned}$$

where R_0, R_n denote remainder terms and

$$\boldsymbol{H}_0 = \left(-\frac{\partial^2 \log p(\boldsymbol{\theta})}{\partial\theta_i \partial\theta_j} \right)\Bigg|_{\boldsymbol{\theta}=\boldsymbol{m}_0} \qquad \boldsymbol{H}(\hat{\boldsymbol{\theta}}_n) = \left(-\frac{\partial^2 \log p(\boldsymbol{x} \mid \boldsymbol{\theta})}{\partial\theta_i \partial\theta_j} \right)\Bigg|_{\boldsymbol{\theta}=\hat{\boldsymbol{\theta}}_n}.$$

Assuming regularity conditions which ensure that R_0, R_n are small for large n, and ignoring constants of proportionality, we see that

$$\begin{aligned} p(\boldsymbol{\theta} \mid \boldsymbol{x}) &\propto \exp\left\{ -\frac{1}{2}(\boldsymbol{\theta} - \boldsymbol{m}_0)^t \boldsymbol{H}_0 (\boldsymbol{\theta} - \boldsymbol{m}_0) - \frac{1}{2}(\boldsymbol{\theta} - \hat{\boldsymbol{\theta}}_n)^t H(\hat{\boldsymbol{\theta}}_n)(\boldsymbol{\theta} - \hat{\boldsymbol{\theta}}_n) \right\} \\ &\propto \exp\left\{ -\frac{1}{2}(\boldsymbol{\theta} - \boldsymbol{m}_n)^t \boldsymbol{H}_n (\boldsymbol{\theta} - \boldsymbol{m}_n) \right\}, \end{aligned}$$

with

$$\begin{aligned}\boldsymbol{H}_n &= \boldsymbol{H}_0 + \boldsymbol{H}(\hat{\boldsymbol{\theta}}_n)\\ \boldsymbol{m}_n &= \boldsymbol{H}_n^{-1}\left(\boldsymbol{H}_0\boldsymbol{m}_0 + \boldsymbol{H}(\hat{\boldsymbol{\theta}}_n)\hat{\boldsymbol{\theta}}_n\right),\end{aligned}$$

where $\boldsymbol{m}_0$ (the *prior mode*) maximises $p(\boldsymbol{\theta})$ and $\hat{\boldsymbol{\theta}}_n$ (the *maximum likelihood estimate*) maximises $p(\boldsymbol{x} \mid \boldsymbol{\theta})$. The Hessian matrix, $\boldsymbol{H}(\hat{\boldsymbol{\theta}}_n)$, measures the local curvature of the log-likelihood function at its maximum, $\hat{\boldsymbol{\theta}}_n$, and is often called the *observed information matrix*.

This heuristic development thus suggests that $p(\boldsymbol{\theta} \mid \boldsymbol{x})$ will, for large n, tend to resemble a multivariate normal distribution, $\mathrm{N}_k(\boldsymbol{\theta} \mid \boldsymbol{m}_n, \boldsymbol{H}_n)$ (see Section 3.2.5) whose mean is a matrix weighted average of a prior (modal) estimate and an observation-based (maximum likelihood) estimate, and whose precision matrix is the sum of the prior precision matrix and the observed information matrix.

Other approximations suggest themselves: for example, for large n the prior precision will tend to be small compared with the precision provided by the data and could be ignored. Also, since, by the strong law of large numbers, for all i, j,

$$\begin{aligned}\lim_{n\to\infty}\left\{\frac{1}{n}\left(-\frac{\partial^2 \log p(\boldsymbol{x} \mid \boldsymbol{\theta})}{\partial\theta_i\partial\theta_j}\right)\right\} &= \lim_{n\to\infty}\left\{\frac{1}{n}\sum_{l=1}^{n}\left(-\frac{\partial^2 \log p(x_l \mid \boldsymbol{\theta})}{\partial\theta_i\partial\theta_j}\right)\right\}\\ &= \int p(x \mid \boldsymbol{\theta})\left(-\frac{\partial^2 \log p(x \mid \boldsymbol{\theta})}{\partial\theta_i\partial\theta_j}\right)dx\end{aligned}$$

we see that $H(\hat{\boldsymbol{\theta}}_n) \to n\boldsymbol{I}(\hat{\boldsymbol{\theta}}_n)$, where $\boldsymbol{I}(\boldsymbol{\theta})$, defined by

$$(\boldsymbol{I}(\boldsymbol{\theta}))_{ij} = \int p(x \mid \boldsymbol{\theta})\left(-\frac{\partial^2 \log p(x \mid \boldsymbol{\theta})}{\partial\theta_i\partial\theta_j}\right)dx,$$

is the so-called *Fisher (or expected) information matrix*. We might approximate $p(\boldsymbol{\theta} \mid \boldsymbol{x})$, therefore, by either $\mathrm{N}_k(\boldsymbol{\theta} \mid \hat{\boldsymbol{\theta}}_n, H(\hat{\boldsymbol{\theta}}_n))$ or $\mathrm{N}_k(\boldsymbol{\theta} \mid \hat{\boldsymbol{\theta}}_n, nI(\hat{\boldsymbol{\theta}}_n))$, where k is the dimension of $\boldsymbol{\theta}$.

In the case of $\theta \in \Theta \subseteq \Re$,

$$H(\hat{\theta}) = -\frac{\partial^2}{\partial\theta^2}\log p(\boldsymbol{x} \mid \theta),$$

so that the approximate posterior variance is the negative reciprocal of the rate of change of the first derivative of $\log p(\boldsymbol{x} \mid \theta)$ in the neighbourhood of its maximum. Sharply peaked log-likelihoods imply small posterior uncertainty and vice-versa.

There is a large literature on the regularity conditions required to justify mathematically the heuristics presented above. Those who have contributed to the field include: Laplace (1812), Jeffreys (1939/1961, Chapter 4), LeCam (1953, 1956, 1958,

1966, 1970, 1986), Lindley (1961b), Freedman (1963b, 1965), Walker (1969), Chao (1970), Dawid (1970), DeGroot (1970, Chapter 10), Ibragimov and Hasminski (1973), Heyde and Johnstone (1979), Hartigan (1983, Chapter 4), Bermúdez (1985), Chen (1985), Sweeting and Adekola (1987), Fu and Kass (1988), Fraser and McDunnough (1989), Sweeting (1992) and Ghosh *et al.* (1994). Related work on higher-order expansion approximations in which the normal appears as a leading term includes that of Hartigan (1965), Johnson (1967, 1970), Johnson and Ladalla (1979) and Crowder (1988). The account given below is based on Chen (1985).

In what follows, we assume that $\boldsymbol{\theta} \in \Theta \subseteq \Re^k$ and that $\{p_n(\boldsymbol{\theta}), n = 1, 2, \ldots\}$ is a sequence of posterior densities for $\boldsymbol{\theta}$, typically of the form $p_n(\boldsymbol{\theta}) = p(\boldsymbol{\theta} \mid x_1, \ldots, x_n)$, derived from an exchangeable sequence with parametric model $p(x \mid \boldsymbol{\theta})$ and prior $p(\boldsymbol{\theta})$, although the mathematical development to be given does not require this. We define $L_n(\boldsymbol{\theta}) = \log p_n(\boldsymbol{\theta})$, and assume throughout that, for every n, there is a strict local maximum, $\boldsymbol{m}_n$, of p_n (or, equivalently, L_n) satisfying:

$$\boldsymbol{L}'_n(\boldsymbol{m}_n) = \nabla L_n(\boldsymbol{\theta}) \mid_{\boldsymbol{\theta}=\boldsymbol{m}_n} = 0$$

and implying the existence and positive-definiteness of

$$\Sigma_n = \left(-\boldsymbol{L}''_n(\boldsymbol{m}_n)\right)^{-1},$$

where $[\boldsymbol{L}''_n(\boldsymbol{m}_n)]_{ij} = \left(\partial^2 L_n(\boldsymbol{\theta})/\partial\theta_i\partial\theta_j\right) \mid_{\boldsymbol{\theta}=\boldsymbol{m}_n}$.

Defining $\mid \boldsymbol{\theta} \mid = (\boldsymbol{\theta}^t\boldsymbol{\theta})^{1/2}$ and $B_\delta(\boldsymbol{\theta}^*) = \{\boldsymbol{\theta} \in \Theta;\ \mid \boldsymbol{\theta} - \boldsymbol{\theta}^* \mid < \delta\}$, we shall show that the following three basic conditions are sufficient to ensure a valid normal approximation for $p_n(\boldsymbol{\theta})$ in a small neighbourhood of $\boldsymbol{m}_n$ as n becomes large.

(c1) "*Steepness*". $\overline{\sigma}_n^2 \to 0$ as $n \to \infty$, where $\overline{\sigma}_n^2$ is the largest eigenvalue of Σ_n.

(c2) "*Smoothness*". For any $\varepsilon > 0$, there exists N and $\delta > 0$ such that, for any $n > N$ and $\boldsymbol{\theta} \in B_\delta(\boldsymbol{m}_n)$, $\boldsymbol{L}''_n(\boldsymbol{\theta})$ exists and satisfies

$$\boldsymbol{I} - \boldsymbol{A}(\varepsilon) \leq \boldsymbol{L}''_n(\boldsymbol{\theta})\{\boldsymbol{L}''(\boldsymbol{m}_n)\}^{-1} \leq \boldsymbol{I} + \boldsymbol{A}(\varepsilon),$$

where $\boldsymbol{I}$ is the $k \times k$ identity matrix and $\boldsymbol{A}(\varepsilon)$ is a $k \times k$ symmetric positive-semidefinite matrix whose largest eigenvalue tends to zero as $\varepsilon \to 0$.

(c3) "*Concentration*". For any $\delta > 0$, $\int_{B_\delta(\boldsymbol{m}_n)} p_n(\boldsymbol{\theta})d\boldsymbol{\theta} \to 1$ as $n \to \infty$.

Essentially, we shall see that (c1), (c2) together ensure that, for large n, inside a small neighbourhood of $\boldsymbol{m}_n$ the function p_n becomes highly peaked and behaves like the multivariate normal density kernel $\exp\{-\frac{1}{2}(\boldsymbol{\theta}-\boldsymbol{m}_n)^t\,\Sigma_n^{-1}\,(\boldsymbol{\theta}-\boldsymbol{m}_n)\}$. The final condition (c3) ensures that the probability outside any neighbourhood of $\boldsymbol{m}_n$ becomes negligible. We do not require any assumption that the $\boldsymbol{m}_n$ themselves converge, nor do we need to insist that $\boldsymbol{m}_n$ be a global maximum of p_n. We implicitly assume, however, that the limit of $p_n(\boldsymbol{m}_n) \mid \Sigma_n \mid^{1/2}$ exists as $n \to \infty$, and we shall now establish a bound for that limit.

Proposition 5.13. (***Bounded concentration***).
The conditions (c1), (c2) imply that

$$\lim_{n\to\infty} p_n(\boldsymbol{m}_n)\,|\Sigma_n|^{1/2} \leq (2\pi)^{-k/2},$$

with equality if and only if (c3) holds.

Proof. Given $\varepsilon > 0$, consider $n > N$ and $\delta > 0$ as given in (c2). Then, for any $\boldsymbol{\theta} \in B_\delta(\boldsymbol{m}_n)$, a simple Taylor expansion establishes that

$$\begin{aligned} p_n(\boldsymbol{\theta}) &= p_n(\boldsymbol{m}_n)\exp\{L_n(\boldsymbol{\theta}) - L_n(\boldsymbol{m}_n)\} \\ &= p_n(\boldsymbol{m}_n)\exp\left\{-\frac{1}{2}(\boldsymbol{\theta}-\boldsymbol{m}_n)^t(\boldsymbol{I}+\boldsymbol{R}_n)\Sigma_n^{-1}(\boldsymbol{\theta}-\boldsymbol{m}_n)\right\}, \end{aligned}$$

where

$$\boldsymbol{R}_n = \boldsymbol{L}_n''(\boldsymbol{\theta}^+)\{\boldsymbol{L}_n''(\boldsymbol{m}_n)\}^{-1}(\boldsymbol{m}_n) - \boldsymbol{I},$$

for some $\boldsymbol{\theta}^+$ lying between $\boldsymbol{\theta}$ and $\boldsymbol{m}_n$. It follows that

$$P_n(\delta) = \int_{B_\delta(\boldsymbol{m}_n)} p_n(\boldsymbol{\theta})d\boldsymbol{\theta}$$

is bounded above by

$$P_n^+(\delta) = p_n(\boldsymbol{m}_n)\,|\,\Sigma_n\,|^{1/2}\,|\,\boldsymbol{I}-\boldsymbol{A}(\varepsilon)\,|^{-1/2}\int_{|\,\boldsymbol{z}\,|<s_n} \exp\left\{-\tfrac{1}{2}\boldsymbol{z}^t\boldsymbol{z}\right\}d\boldsymbol{z}$$

and below by

$$P_n^-(\delta) = p_n(\boldsymbol{m}_n)\,|\,\Sigma_n\,|^{1/2}\,|\,\boldsymbol{I}+\boldsymbol{A}(\varepsilon)\,|^{-1/2}\int_{|\,\boldsymbol{z}\,|<t_n} \exp\left\{-\tfrac{1}{2}\boldsymbol{z}^t\boldsymbol{z}\right\}d\boldsymbol{z},$$

where $s_n = \delta(1-\underline{\alpha}(\varepsilon))^{1/2}/\underline{\sigma}_n$ and $t_n = \delta(1+\underline{\alpha}(\varepsilon))^{1/2}/\overline{\sigma}_n$, with $\overline{\sigma}_n^2(\underline{\sigma}_n^2)$ and $\overline{\alpha}(\varepsilon)(\underline{\alpha}(\varepsilon))$ the largest (smallest) eigenvalues of Σ_n and $\boldsymbol{A}(\varepsilon)$, respectively, since, for any $k \times k$ matrix $\boldsymbol{V}$,

$$B_{\delta/\overline{V}}(0) \subseteq \left\{\boldsymbol{z};(\boldsymbol{z}^t\boldsymbol{V}\boldsymbol{z})^{1/2} < \delta\right\} \subseteq B_{\delta/\underline{V}}(0),$$

where $\overline{V}^2(\underline{V}^2)$ are the largest (smallest) eigenvalues of $\boldsymbol{V}$.

Since (c1) implies that both s_n and t_n tend to infinity as $n \to \infty$, we have

$$\begin{aligned} |\boldsymbol{I}-\boldsymbol{A}(\varepsilon)|^{1/2}\lim_{n\to\infty} P_n(\delta) &\leq \lim_{n\to\infty} p_n(\boldsymbol{m}_n)|\Sigma_n|^{1/2}(2\pi)^{k/2} \\ &\leq |\boldsymbol{I}+\boldsymbol{A}(\varepsilon)|^{1/2}\lim_{n\to\infty} P_n(\delta), \end{aligned}$$

and the required inequality follows from the fact that $|\boldsymbol{I}\pm\boldsymbol{A}(\varepsilon)| \to 1$ as $\varepsilon \to 0$ and $P_n(\delta) \leq 1$ for all n. Clearly, we have equality if and only if $\lim_{n\to\infty} P_n(\delta) = 1$, which is condition (c3). ◁

We can now establish the main result, which may colloquially be stated as "$\boldsymbol{\theta}$ has an asymptotic posterior $\mathrm{N}_k(\boldsymbol{\theta}|\boldsymbol{m}_n, \Sigma_n^{-1})$ distribution, where $\boldsymbol{L}_n'(\boldsymbol{m}_n) = 0$ and $\Sigma_n^{-1} = -\boldsymbol{L}_n''(\boldsymbol{m}_n)$."

Proposition 5.14. (***Asymptotic posterior normality***). *For each n, consider $p_n(\cdot)$ as the density function of a random quantity $\boldsymbol{\theta}_n$, and define, using the notation above, $\boldsymbol{\phi}_n = \Sigma_n^{-1/2}(\boldsymbol{\theta}_n - \boldsymbol{m}_n)$. Then, given (c1) and (c2), (c3) is a necessary and sufficient condition for $\boldsymbol{\phi}_n$ to converge in distribution to $\boldsymbol{\phi}$, where $p(\boldsymbol{\phi}) = (2\pi)^{-k/2} \exp\left\{-\frac{1}{2}\boldsymbol{\phi}^t\boldsymbol{\phi}\right\}$.*

Proof. Given (c1) and (c2), and writing $\boldsymbol{b} \geq \boldsymbol{a}$, for $\boldsymbol{a}, \boldsymbol{b} \in \Re^k$, to denote that all components of $\boldsymbol{b} - \boldsymbol{a}$ are non-negative, it suffices to show that, as $n \to \infty$, $P_n(\boldsymbol{a} \leq \boldsymbol{\phi}_n \leq \boldsymbol{b}) \to P(\boldsymbol{a} \leq \boldsymbol{\phi} \leq \boldsymbol{b})$ if and only if (c3) holds.

We first note that

$$P_n(\boldsymbol{a} \leq \boldsymbol{\phi}_n \leq \boldsymbol{b}) = \int_{\Theta_n} p_n(\boldsymbol{\theta}) d\boldsymbol{\theta},$$

where, by (c1), for any $\delta > 0$ and sufficiently large n,

$$\Theta_n = \left\{\boldsymbol{\theta}; \Sigma_n^{1/2}\boldsymbol{a} \leq (\boldsymbol{\theta} - \boldsymbol{m}_n) \leq \Sigma_n^{1/2}\boldsymbol{b}\right\} \subset B_\delta(\boldsymbol{m}_n).$$

It then follows, by a similar argument to that used in Proposition 5.13, that, for any $\varepsilon > 0$, $P_n(\boldsymbol{a} \leq \boldsymbol{\phi}_n \leq \boldsymbol{b})$ is bounded above by

$$P_n(\boldsymbol{m}_n) \left|\boldsymbol{I} - \boldsymbol{A}(\varepsilon)\right|^{-1/2} \left|\Sigma_n\right|^{1/2} \int_{Z(\varepsilon)} \exp\left\{-\tfrac{1}{2}\boldsymbol{z}^t\boldsymbol{z}\right\} d\boldsymbol{z},$$

where

$$Z(\varepsilon) = \left\{\boldsymbol{z}; \left[\boldsymbol{I} - \boldsymbol{A}(\varepsilon)\right]^{1/2} \boldsymbol{a} \leq \boldsymbol{z} \leq \left[\boldsymbol{I} - \boldsymbol{A}(\varepsilon)\right]^{1/2} \boldsymbol{b}\right\},$$

and is bounded below by a similar quantity with $+\boldsymbol{A}(\varepsilon)$ in place of $-\boldsymbol{A}(\varepsilon)$.

Given (c1), (c2), as $\varepsilon \to 0$ we have

$$\lim_{n\to\infty} P_n(\boldsymbol{a} \leq \boldsymbol{\phi}_n \leq \boldsymbol{b}) = \lim_{n\to\infty} p_n(\boldsymbol{m}_n) \left|\, \Sigma_n \,\right|^{1/2} \int_{Z(0)} \exp\left\{-\tfrac{1}{2}\boldsymbol{z}^t\boldsymbol{z}\right\} d\boldsymbol{z},$$

where $Z(0) = \{\boldsymbol{z}; \boldsymbol{a} \leq \boldsymbol{z} \leq \boldsymbol{b}\}$. The result follows from Proposition 5.13. $\triangleleft$

Conditions (c1) and (c2) are often relatively easy to check in specific applications, but (c3) may not be so directly accessible. It is useful therefore to have available alternative conditions which, given (c1), (c2), imply (c3). Two such are provided by the following:

(c4) For any $\delta > 0$, there exists an integer N and $c, d \in \Re^+$ such that, for any $n > N$ and $\boldsymbol{\theta} \notin B_\delta(\boldsymbol{m}_n)$,

$$L_n(\boldsymbol{\theta}) - L_n(\boldsymbol{m}_n) < -c\left\{(\boldsymbol{\theta} - \boldsymbol{m}_n)^t \Sigma_n^{-1} (\boldsymbol{\theta} - \boldsymbol{m}_n)\right\}^d.$$

(c5) As (c4), but, with $G(\boldsymbol{\theta}) = \log g(\boldsymbol{\theta})$ for some density (or normalisable positive function) $g(\boldsymbol{\theta})$ over Θ,

$$L_n(\boldsymbol{\theta}) - L_n(\boldsymbol{m}_n) < -c\,|\Sigma_n|^{-d} + G(\boldsymbol{\theta}).$$

Proposition 5.15. (***Alternative conditions***). *Given (c1), (c2), either (c4) or (c5) implies (c3).*

Proof. It is straightforward to verify that

$$\int_{\Theta - B_\delta(\boldsymbol{m}_n)} p_n(\boldsymbol{\theta}) d\boldsymbol{\theta} \leq p_n(\boldsymbol{m}_n)\,|\Sigma_n|^{1/2} \int_{|\,\boldsymbol{z}\,| > \delta/\overline{\sigma}_n} \exp\left\{-c(\boldsymbol{z}^t \boldsymbol{z})^d\right\} d\boldsymbol{z},$$

given (c4), and similarly, that

$$\int_{\Theta - B_\delta(\boldsymbol{m}_n)} p_n(\boldsymbol{\theta}) d\boldsymbol{\theta} \leq p_n(\boldsymbol{m}_n)\,|\Sigma_n|^{1/2}\,|\Sigma_n|^{-1/2} \exp\left\{-c\,|\Sigma_n|^{-d}\right\},$$

given (c4).

Since $p_n(\boldsymbol{m}_n)\,|\,\Sigma_n\,|^{1/2}$ is bounded (Proposition 5.11) and the remaining terms or the right-hand side clearly tend to zero, it follows that the left-hand side tends to zero as $n \to \infty$. ◁

To understand better the relative ease of checking (c4) or (c5) in applications, we note that, if $p_n(\boldsymbol{\theta})$ is based on data $\boldsymbol{x}$,

$$L_n(\boldsymbol{\theta}) = \log p(\boldsymbol{\theta}) + \log p(\boldsymbol{x}\,|\,\boldsymbol{\theta}) - \log p(\boldsymbol{x}),$$

so that $L_n(\boldsymbol{\theta}) - L_n(\boldsymbol{m}_n)$ does not involve the, often intractable, normalising constant $p(\boldsymbol{x})$. Moreover, (c4) does not even require the use of a proper prior for the vector $\boldsymbol{\theta}$.

We shall illustrate the use of (c4) for the general case of canonical conjugate analysis for exponential families.

Proposition 5.16. ***(Asymptotic normality under conjugate analysis).*** *Suppose that $\boldsymbol{y}_1, \ldots, \boldsymbol{y}_n$ are data resulting from a random sample of size n from the canonical exponential family form*

$$p(\boldsymbol{y} \mid \boldsymbol{\psi}) = a(\boldsymbol{y}) \exp\left\{\boldsymbol{y}^t \boldsymbol{\psi} - b(\boldsymbol{\psi})\right\}$$

with canonical conjugate prior density

$$p(\boldsymbol{\psi} \mid n_0, \boldsymbol{y}_0) = c(n_0, \boldsymbol{y}_0) \exp\left\{n_0 \boldsymbol{y}_0^t \boldsymbol{\psi} - n_0 b(\boldsymbol{\psi})\right\}.$$

For each n, consider the posterior density

$$p_n(\boldsymbol{\psi}) = p(\boldsymbol{\psi} \mid n_0 + n, n_0 \boldsymbol{y}_0 + n\overline{\boldsymbol{y}}_n),$$

with $\overline{\boldsymbol{y}}_n = \sum_{i=1}^n \boldsymbol{y}_i / n$, to be the density function for a random quantity $\boldsymbol{\psi}_n$, and define $\boldsymbol{\phi}_n = \Sigma_n^{-1/2}(\boldsymbol{\psi}_n - \boldsymbol{b}'(\boldsymbol{m}_n))$, where

$$\boldsymbol{b}'(\boldsymbol{m}_n) = \nabla b(\boldsymbol{\psi})\Big|_{\boldsymbol{\psi}=\boldsymbol{m}_n} = \frac{n_0 \boldsymbol{y}_0 + n\overline{\boldsymbol{y}}_n}{n_0 + n}$$

$$\left(\boldsymbol{b}''(\boldsymbol{m}_n)\right)_{ij} = \left(\frac{\partial^2 b(\boldsymbol{\psi})}{\partial \psi_i \partial \psi_j}\right)\Bigg|_{\boldsymbol{\psi}=\boldsymbol{m}_n} = (n_0 + n)\,(\Sigma_n)_{ij}^{-1}.$$

Then $\boldsymbol{\phi}_n$ converges in distribution to $\boldsymbol{\phi}$, where

$$p(\boldsymbol{\phi}) = (2\pi)^{-k/2} \exp\left\{-\tfrac{1}{2}\boldsymbol{\phi}^t \boldsymbol{\phi}\right\}.$$

Proof. Colloquially, we have to prove that $\boldsymbol{\psi}$ has an asymptotic posterior $\mathrm{N}_k(\boldsymbol{\psi} \mid \boldsymbol{b}'(\boldsymbol{m}_n), \Sigma_n^{-1})$ distribution, where $\boldsymbol{b}'(\boldsymbol{m}_n') = (n_0 + n)^{-1}(n_0 \boldsymbol{y}_0 + n\overline{\boldsymbol{y}}_n)$ and $\Sigma_n^{-1} = (n_0 + n)^{-1} \boldsymbol{b}''(\boldsymbol{m}_n)$. From a mathematical perspective,

$$p_n(\boldsymbol{\psi}) \propto \exp\left\{(n_0 + n) h(\boldsymbol{\psi})\right\},$$

where $h(\boldsymbol{\psi}) = [\boldsymbol{b}'(\boldsymbol{m}_n)]^t \boldsymbol{\psi} - b(\boldsymbol{\psi})$, with $b(\boldsymbol{\psi})$ a continuously differentiable and strictly convex function (see Section 5.2.2). It follows that, for each n, $p_n(\boldsymbol{\psi})$ is unimodal with a maximum at $\boldsymbol{\psi} = \boldsymbol{m}_n$ satisfying $\nabla h(\boldsymbol{m}_n) = 0$. By the strict concavity of $h(\cdot)$, for any $\delta > 0$ and $\theta \notin B_\delta(\boldsymbol{m}_n)$, we have, for some $\boldsymbol{\psi}^+$ between $\boldsymbol{\psi}$ and $\boldsymbol{m}_n$, with angle θ between $\boldsymbol{\psi} - \boldsymbol{m}_n$ and $\nabla h(\boldsymbol{\psi}^+)$,

$$\begin{aligned} h(\boldsymbol{\psi}) - h(\boldsymbol{m}_n) &= (\boldsymbol{\psi} - \boldsymbol{m}_n)^t \nabla h(\boldsymbol{\psi}^+) \\ &= |\,\boldsymbol{\psi} - \boldsymbol{m}_n\,|\;\;|\,\nabla h(\boldsymbol{\psi}^+)\,|\,\cos\theta \\ &< -c\,|\,\boldsymbol{\psi} - \boldsymbol{m}_n\,|, \end{aligned}$$

for $c = \inf\left\{\,|\,\nabla h(\psi^+)\,|\,; \psi \notin B_\delta(m_n)\right\} > 0$. It follows that

$$L_n(\psi) - L_n(m_n) < -(n_0 + n)\,|\,\psi - m_n\,|$$
$$< -c_1\left\{(\psi - m_n)^t \Sigma_n^{-1}(\psi - m_n)\right\}^{1/2},$$

where $c_1 = c\lambda^{-1}$, with λ^2 the largest eigenvalue of $b''(m_n)$, and hence that (c4) is satisfied. Conditions (c1), (c2) follows straightforwardly from the fact that

$$(n_0 + n)\Sigma_n^{-1} = b''(m_n),$$
$$L_n''(\psi)\{L_n''(m_n)\}^{-1} = b''(\psi)\{b''(m_n)\}^{-1},$$

the latter not depending on $n_0 + n$, and so the result follows by Propositions 5.12 and 5.13. ◁

Example 5.4. ***(continued)***. Suppose that $\text{Be}(\theta\,|\,\alpha_n, \beta_n)$, where $\alpha_n = \alpha + r_n$, and $\beta_n = \beta + n - r_n$, is the posterior derived from n Bernoulli trials with r_n successes and a $\text{Be}(\theta\,|\,\alpha, \beta)$ prior. Proceeding directly,

$$L_n(\theta) = \log p_n(\theta) = \log p(\boldsymbol{x}\,|\,\theta) + \log p(\boldsymbol{\theta}) - \log p(\boldsymbol{x})$$
$$= (\alpha_n - 1)\log\theta + (\beta_n - 1)\log(1-\theta) - \log p(\boldsymbol{x})$$

so that

$$L_n'(\theta) = \frac{(\alpha_n - 1)}{\theta} - \frac{(\beta_n - 1)}{1 - \theta}$$

and

$$L_n''(\theta) = -\frac{(\alpha_n - 1)}{\theta^2} - \frac{(\beta_n - 1)}{(1-\theta)^2}\,.$$

It follows that

$$m_n = \frac{\alpha_n - 1}{(\alpha_n + \beta_n - 2)}, \quad (-L_n''(m_n))^{-1} = \frac{(\alpha_n - 1)(\beta_n - 1)}{(\alpha_n + \beta_n - 2)^3}\,.$$

Condition (c1) is clearly satisfied since $(-L_n''(m_n))^{-1} \to 0$ as $n \to \infty$; condition (c2) follows from the fact that $L_n''(\theta)$ is a continuous function of θ. Finally, (c4) may be verified with an argument similar to the one used in the proof of Proposition 5.16.

Taking $\alpha = \beta = 1$ for illustration, we see that

$$m_n = \frac{r_n}{n}, \quad (-L_n''(m_n))^{-1} = \frac{1}{n}\cdot\frac{r_n}{n}\left(1 - \frac{r_n}{n}\right),$$

and hence that the asymptotic posterior for θ is

$$\text{N}\left(\theta\,\middle|\,\frac{r_n}{n}, \left\{\frac{1}{n}\cdot\frac{r_n}{n}\left(1 - \frac{r_n}{n}\right)\right\}^{-1}\right).$$

(As an aside, we note the interesting "duality" between this asymptotic form for θ given n, r_n, and the asymptotic distribution for r_n/n given θ, which, by the central limit theorem, has the form

$$\text{N}\left(\frac{r_n}{n}\,\middle|\,\theta, \left\{\frac{1}{n}\theta(1-\theta)\right\}^{-1}\right).$$

Further reference to this kind of "duality" will be given in Appendix B.)

5.3.3 Asymptotics under Transformations

The result of Proposition 5.16 is given in terms of the canonical parametrisation of the exponential family underlying the conjugate analysis. This prompts the obvious question as to whether the asymptotic posterior normality "carries over, with appropriate transformations of the mean and covariance, to an arbitrary (one-to-one) reparametrisation of the model. More generally, we could ask the same question in relation to Proposition 5.14. A partial answer is provided by the following.

Proposition 5.17. (***Asymptotic normality under transformation***).
With the notation and background of Proposition 5.14, suppose that $\boldsymbol{\theta}$ *has an asymptotic* $\mathrm{N}_k(\boldsymbol{\theta}|\boldsymbol{m}_n, \Sigma_n^{-1})$ *distribution, with the additional assumptions that, with respect to a parametric model* $p(\boldsymbol{x}|\boldsymbol{\theta}_0)$, $\bar{\sigma}_n^2 \to 0$ *and* $\boldsymbol{m}_n \to \boldsymbol{\theta}_0$ *in probability, and that given any* $\delta > 0$, *there is a constraint* $c(\delta)$ *such that* $P(\bar{\sigma}_n^2 \underline{\sigma}_n^{-2} \leq c(\delta)) \geq 1 - \delta$ *for all sufficiently large* n, *where* $\bar{\sigma}_n^2$ $(\underline{\sigma}_n^2)$ *is the largest (smallest) eigenvalue of* Σ_n^2. *Then, if* $\boldsymbol{\nu} = \boldsymbol{g}(\boldsymbol{\theta})$ *is a transformation such that, at* $\boldsymbol{\theta} = \boldsymbol{\theta}_0$,

$$\boldsymbol{J}_{\boldsymbol{g}}(\boldsymbol{\theta}) = \frac{\partial \boldsymbol{g}(\boldsymbol{\theta})}{\partial \boldsymbol{\theta}}$$

is non-singular with continuous entries, $\boldsymbol{\nu}$ *has an asymptotic distribution*

$$\mathrm{N}_k\left(\boldsymbol{\nu} \,\Big|\, \boldsymbol{g}(\boldsymbol{m}_n), [\boldsymbol{J}_{\boldsymbol{g}}(\boldsymbol{m}_n)\Sigma_n \boldsymbol{J}_{\boldsymbol{g}}^t(\boldsymbol{m}_n)]^{-1}\right).$$

Proof. This is a generalization and Bayesian reformulation of classical results presented in Serfling (1980, Section 3.3). For details, see Mendoza (1994). ◁

For any finite n, the adequancy of the normal approximation provided by Proposition 5.17 may be highly dependent on the particular transformation used. Anscombe (1964a, 1964b) analyses the choice of transformations which improve asymptotic normality. A related issue is that of selecting appropriate parametrisations for various numerical approximation methods (Hills and Smith, 1992, 1993).

The expression for the asymptotic posterior precision matrix (inverse covariance matrix) given in Proposition 5.17 is often rather cumbersome to work with. A simpler, alternative form is given by the following.

Corollary 1. (***Asymptotic precision after transformation***).
In Proposition 5.10, if $\boldsymbol{H}_n = \Sigma_n^{-1}$ *denotes the asymptotic precision matrix for* $\boldsymbol{\theta}$, *then the asymptotic precision matrix for* $\boldsymbol{\nu} = \boldsymbol{g}(\boldsymbol{\theta})$ *has the form*

$$\boldsymbol{J}_{\boldsymbol{g}^{-1}}^t(\boldsymbol{g}(\boldsymbol{m}_n))\boldsymbol{H}_n \boldsymbol{J}_{\boldsymbol{g}^{-1}}(\boldsymbol{g}(\boldsymbol{m}_n)),$$

where

$$\boldsymbol{J}_{\boldsymbol{g}^{-1}}(\boldsymbol{\nu}) = \frac{\partial \boldsymbol{g}^{-1}(\boldsymbol{\nu})}{\partial \boldsymbol{\nu}}$$

is the Jacobian of the inverse transformation.

Proof. This follows immediately by reversing of the roles of $\boldsymbol{\theta}$ and $\boldsymbol{\nu}$. ◁

In many applications, we simply wish to consider one-to-one transformations of a single parameter. The next result provides a convenient summary of the required transformation result.

Corollary 2. (***Asymptotic normality after scalar transformation***). *Suppose that given the conditions of Propositions 5.14, 5.17 with scalar θ, the sequence m_n tends in probability to θ_0 under $p(\boldsymbol{x}|\theta_0)$, and that $L''_n(m_n) \to 0$ in probability as $n \to \infty$. Then, if $\nu = g(\theta)$ is such that $g'(\theta) = dg(\theta)/\,d\theta$ is continuous and non-zero at $\theta = \theta_0$, the asymptotic posterior distribution for ν is*

$$\mathrm{N}(\nu|g(m_n), -L''_n(m_n)[g'(m_n)]^{-2}).$$

Proof. The conditions ensure, by Proposition 5.14, that θ has an asymptotic posterior distribution of the form $\mathrm{N}(\theta|m_n, -L''_n(m_n))$, so that the result follows from Proposition 5.17. $\triangleleft$

Example 5.4. ***(continued)***. Suppose, again, that $\mathrm{Be}(\theta\,|\,\alpha_n, \beta_n)$, where $\alpha_n = \alpha + r_n$, and $\beta_n = \beta + n - r_n$, is the posterior distribution of the parameter of a Bernoulli distribution afte n trials, and suppose now that we are interested in the asymptotic posterior distribution of the variance stabilising transformation (recall Example 3.3)

$$\nu = g(\theta) = 2\sin^{-1}\sqrt{\theta}\,.$$

Straightforward application of Corollary 2 to Proposition 5.17, leads to the asymptotic distribution

$$\mathrm{N}(\nu|2\sin^{-1}(\sqrt{r_n/n}), n),$$

whose mean and variance can be compared with the forms given in Example 3.3.

It is clear from the presence of the term $[g'(m_n)]^{-2}$ in the form of the asymptotic precision given in Corollary 2 to Proposition 5.17 that things will go wrong if $g'(m_n) \to 0$ as $n \to \infty$. This is dealt with in the result presented by the requirement that $g'(\theta_0) \neq 0$, where $m_n \to \theta_0$ in probability. A concrete illustration of the problems that arise when such a condition is not met is given by the following.

Example 5.8. ***(Non-normal asymptotic posterior)***. Suppose that the asymptotic posterior for a parameter $\theta \in \Re$ is given by $\mathrm{N}(\theta|\bar{x}_n, n)$, $n\bar{x}_n = x_1 + \cdots + x_n$, perhaps derived from $\mathrm{N}(x_i|\theta, 1)$, $i = 1, \ldots, n$, with $\mathrm{N}(\theta|0, h)$, having $h \approx 0$. Now consider the transformation $\nu = g(\theta) = \theta^2$, and suppose that the actual value of θ generating the x_i through $\mathrm{N}(x_i|\theta, 1)$ is $\theta = 0$.

Intuitively, it is clear that ν cannot have an asymptotic normal distribution since the sequence $\bar{x}_n^2$ is converging in probability to 0 through *strictly positive* values. Technically, $g'(0) = 0$ and the condition of the corollary is not satisfied. In fact, it can be shown that the asymptotic posterior distribution of $n\nu$ is χ^2 in this case.

One attraction of the availability of the results given in Proposition 5.17 and Corollary 1 is that verification of the conditions for asymptotic posterior normality (as in, for example, Proposition 5.14) may be much more straightforward under one choice of parametrisation of the likelihood than under another. The result given enables us to identify the posterior normal form for any convenient choice of parameters, subsequently deriving the form for the parameters of interest by straightforward transformation. An indication of the usefulness of this result is given in the following example (and further applications can be found in Section 5.4).

Example 5.9. ***(Asymptotic posterior normality for a ratio).*** Suppose that we have a random sample $x_1, \ldots, x_n$ from the model $\{\prod_{i=1}^n \mathrm{N}(x_i|\theta_1, 1), \mathrm{N}(\theta_1|0, \lambda_1)\}$ and, independently, another random sample $y_1, \cdots, y_n$ from the model $\{\prod_{i=1}^n \mathrm{N}(y_i|\theta_2, 1), \mathrm{N}(\theta_2|0, \lambda_2)\}$, where $\lambda_1 \approx 0$, $\lambda_2 \approx 0$ and $\theta_2 \neq 0$. We are interested in the posterior distribution of $\phi_1 = \theta_1/\theta_2$ as $n \to \infty$.

First, we note that, for large n, it is very easily verified that the joint posterior distribution for $\boldsymbol{\theta} = (\theta_1, \theta_2)$ is given by

$$\mathrm{N}_2\left\{\begin{pmatrix}\theta_1\\ \theta_2\end{pmatrix} \middle| \begin{pmatrix}\bar{x}_n\\ \bar{y}_n\end{pmatrix}, \begin{pmatrix}n & 0\\ 0 & n\end{pmatrix}\right\},$$

where $n\bar{x}_n = x_1 + \cdots + x_n$, $n\bar{y}_n = y_1 + \cdots + y_n$. Secondly, we note that the marginal asymptotic posterior for ϕ_1 can be obtained by defining an appropriate ϕ_2 such that $(\theta_1, \theta_2) \to (\phi_1, \phi_2)$ is a one-to-one transformation, obtaining the distribution of $\boldsymbol{\phi} = (\phi_1, \phi_2)$ using Proposition 5.17, and subsequently marginalising to ϕ_1.

An obvious choice for ϕ_2 is $\phi_2 = \theta_2$, so that, in the notation of Proposition 5.17, $\boldsymbol{g}(\theta_1, \theta_2) = (\phi_1, \phi_2)$ and

$$\boldsymbol{J_g}(\boldsymbol{\theta}) = \begin{pmatrix}\partial\phi_1/\partial\theta_1 & \partial\phi_1/\partial\theta_2\\ \partial\phi_2/\partial\theta_1 & \partial\phi_2/\partial\theta_2\end{pmatrix} = \begin{pmatrix}\theta_2^{-1} & -\theta_1\theta_2^{-2}\\ 0 & 1\end{pmatrix}.$$

The determinant of this, θ_2^{-1}, is non-zero for $\theta_2 \neq 0$, and the conditions of Proposition 5.17 are clearly satisfied. It follows that the asymptotic posterior of $\boldsymbol{\phi}$ is

$$\mathrm{N}_2\left(\begin{pmatrix}\phi_1\\ \phi_2\end{pmatrix} \middle| \begin{pmatrix}\bar{x}_n/\bar{y}_n\\ \bar{y}_n\end{pmatrix}, \ n\bar{y}_n^2\begin{pmatrix}1 + (\bar{x}_n/\bar{y}_n)^2 & -\bar{x}_n\\ -\bar{x}_n & \bar{y}_n^2\end{pmatrix}^{-1}\right),$$

so that the required asymptotic posterior for $\phi_1 = \theta_1/\theta_2$ is

$$\mathrm{N}\left(\phi_1 \middle| \frac{\bar{x}_n}{\bar{y}_n}, \ n\bar{y}_n^2\left(\frac{\bar{y}_n^2}{\bar{x}_n^2 + \bar{y}_n^2}\right)\right).$$

Any reader remaining unappreciative of the simplicity of the above analysis may care to examine the form of the likelihood function, etc., corresponding to an initial parametrisation directly in terms of ϕ_1, ϕ_2, and to contemplate verifying directly the conditions of Proposition 5.14 using the ϕ_1, ϕ_2 parametrisation.

5.4 REFERENCE ANALYSIS

In the previous section, we have examined situations where data corresponding to large sample sizes come to dominate prior information, leading to inferences which are negligibly dependent on the initial state of information. The third of the questions posed at the end of Section 5.1.6 relates to specifying prior distributions in situations where it is felt that, *even for moderate sample sizes*, the data should be expected to dominate prior information because of the "vague" nature of the latter.

However, the problem of characterising a "*non-informative*" or "*objective*" prior distribution, representing "*prior ignorance*", "*vague prior knowledge*" and "*letting the data speak for themselves*" is far more complex than the apparent intuitive immediacy of these words and phrases would suggest.

In Section 5.6.2, we shall provide a brief review of the fascinating history of the quest for this "baseline", limiting prior form. However, it is as well to make clear straightaway our own view—very much in the operationalist spirit with which we began our discussion of uncertainty in Chapter 2—that "mere words" are an inadequate basis for clarifying such a slippery concept. Put bluntly: data cannot ever speak entirely for themselves; every prior specification has *some* informative posterior or predictive implications; and "vague" is itself much too vague an idea to be useful. There is no "objective" prior that represents ignorance.

On the other hand, we recognise that there *is* often a pragmatically important need for a form of prior to posterior analysis capturing, *in some well-defined sense*, the notion of the prior having a minimal effect, relative to the data, on the final inference. Such a *reference analysis* might be required as an approximation to actual individual beliefs; more typically, it might be required as a limiting "what if?" baseline in considering a range of prior to posterior analyses, or as a *default* option when there are insufficient resources for detailed elicitation of actual prior knowledge.

In line with the unified perspective we have tried to adopt throughout this volume, the setting for our development of such a reference analysis will be the general decision-theoretic framework, together with the specific information-theoretic tools that have emerged in earlier chapters as key measures of the discrepancies (or "distances") between belief distributions. From the approach we adopt, it will be clear that the *reference prior* component of the analysis is simply a mathematical tool. It has considerable pragmatic importance in implementing a *reference analysis*, whose role and character will be precisely defined, but it is not a privileged, "uniquely non-informative" or "objective" prior. Its main use will be to provide a "conventional" prior, to be used when a default specification having a claim to being *non-influential* in the sense described above is required. We seek to move away, therefore, from the rather philosophically muddled debates about "prior ignorance" that have all too often confused these issues, and towards well-defined decision-theoretic and information-theoretic procedures.

5.4.1 Reference Decisions

Consider a specific form of decision problem with possible decisions $d \in \mathcal{D}$ providing possible answers, $a \in \mathcal{A}$, to an inference problem, with unknown state of the world $\omega = (\omega_1, \omega_2)$, utilities for consequences (a, ω) given by $u(d(\omega_1)) = u(a, \omega_1)$ and the availability of an experiment e which consists of obtaining an observation x having parametric model $p(x \mid \omega_2)$ and a prior probability density $p(\omega) = p(\omega_1 \mid \omega_2)p(\omega_2)$ for the unknown state of the world, ω. This general structure describes a situation where practical consequences depend directly on the ω_1 component of ω, whereas inference from data $x \in X$ provided by experiment e takes place indirectly, through the ω_2 component of ω as described by $p(\omega_1 \mid \omega_2)$. If ω_1 is a function of ω_2, the prior density is, of course, simply $p(\omega_2)$.

To avoid subscript proliferation, let us now, without any risk of confusion, indulge in a harmless abuse of notation by writing $\omega_1 = \omega, \omega_2 = \boldsymbol{\theta}$. This both simplifies the exposition and has the mnemonic value of suggesting that ω is the state of the world of ultimate interest (since it occurs in the utility function), whereas $\boldsymbol{\theta}$ is a parameter in the usual sense (since it occurs in the probability model). Often ω is just some function $\omega = \phi(\boldsymbol{\theta})$ of $\boldsymbol{\theta}$; if ω is not a function of $\boldsymbol{\theta}$, the relationship between ω and $\boldsymbol{\theta}$ is that described in their joint distribution $p(\omega, \boldsymbol{\theta}) = p(\omega \mid \boldsymbol{\theta})p(\boldsymbol{\theta})$.

Now, for given conditional prior $p(\omega \mid \boldsymbol{\theta})$ and utility function $u(a, \omega)$, let us examine, *in utility terms*, the influence of the prior $p(\boldsymbol{\theta})$, relative to the observational information provided by e. We note that if a_0^* denotes the optimal answer under $p(\omega)$ and $a_{\boldsymbol{x}}^*$ denotes the optimal answer under $p(\omega \mid x)$, then, using Definition 3.13 (ii), with appropriate notational changes, and noting that

$$\int p(\boldsymbol{x}) \int p(\omega \mid \boldsymbol{x}) u(a_0^*, \omega)\, d\omega d\boldsymbol{x} = \int p(\omega) u(a_0^*, \omega)\, d\omega,$$

the expected (utility) value of the experiment e, given the prior $p(\boldsymbol{\theta})$, is

$$v_u\{e, p(\boldsymbol{\theta})\} = \int p(\boldsymbol{x}) \int p(\omega \mid \boldsymbol{x}) u(a_{\boldsymbol{x}}^*, \omega)\, d\omega d\boldsymbol{x} - \int p(\omega) u(a_0^*, \omega)\, d\omega,$$

where, assuming ω is independent of x, given $\boldsymbol{\theta}$,

$$p(\omega) = \int p(\omega \mid \boldsymbol{\theta}) p(\boldsymbol{\theta})\, d\boldsymbol{\theta}, \qquad p(\omega \mid \boldsymbol{x}) = \int \frac{p(\boldsymbol{x} \mid \boldsymbol{\theta}) p(\omega \mid \boldsymbol{\theta})}{p(\boldsymbol{x})} p(\boldsymbol{\theta})\, d\boldsymbol{\theta}$$

and

$$p(\boldsymbol{x}) = \int p(\boldsymbol{x} \mid \boldsymbol{\theta}) p(\boldsymbol{\theta})\, d\boldsymbol{\theta}.$$

If $e(k)$ denotes the experiment consisting of k independent replications of e, that is yielding observations $\{\boldsymbol{x}_1, \ldots, \boldsymbol{x}_k\}$ with joint parametric model $\prod_{i=1}^k p(\boldsymbol{x}_i \mid \boldsymbol{\theta})$, then $v_u\{e(k), p(\boldsymbol{\theta})\}$, the expected utility value of the experiment $e(k)$, has the same mathematical form as $v_u\{e, p(\boldsymbol{\theta})\}$, but with $\boldsymbol{x} = (\boldsymbol{x}_1, \ldots, \boldsymbol{x}_k)$ and $p(\boldsymbol{x} \mid \boldsymbol{\theta}) = \prod_{i=1}^k p(\boldsymbol{x}_i \mid \boldsymbol{\theta})$. Intuitively, at least in suitably regular cases, as $k \to \infty$ we obtain, from $e(\infty)$, perfect (i.e., complete) information about $\boldsymbol{\theta}$, so that, assuming the limit to exist,

$$v_u\{e(\infty), p(\boldsymbol{\theta})\} = \lim_{k\to\infty} v_u\{e(k), p(\boldsymbol{\theta})\}$$

is the expected (utility) *value of perfect information*, about $\boldsymbol{\theta}$, given $p(\boldsymbol{\theta})$.

Clearly, the more valuable the information contained in $p(\boldsymbol{\theta})$, the less will be the expected value of perfect information about $\boldsymbol{\theta}$; conversely, the less valuable the information contained in the prior, the more we would expect to gain from exhaustive experimentation. This, then, suggests a well-defined "thought experiment" procedure for characterising a "minimally valuable prior": choose, from the class of priors which has been identified as compatible with other assumptions about $(\boldsymbol{\omega}, \boldsymbol{\theta})$, that prior, $\pi(\boldsymbol{\theta})$, say, which *maximises the expected value of perfect information about* $\boldsymbol{\theta}$. Such a prior will be called a *u-reference prior*; the posterior distributions,

$$\pi(\boldsymbol{\omega} \mid \boldsymbol{x}) = \int p(\boldsymbol{\omega} \mid \boldsymbol{\theta})\pi(\boldsymbol{\theta} \mid \boldsymbol{x}) d\boldsymbol{\theta}$$

$$\pi(\boldsymbol{\theta} \mid \boldsymbol{x}) \propto p(\boldsymbol{x} \mid \boldsymbol{\theta})\pi(\boldsymbol{\theta})$$

derived from combining $\pi(\boldsymbol{\theta})$ with actual data $\boldsymbol{x}$, will be called *u-reference posteriors*; and the optimal decision derived from $\pi(\boldsymbol{\omega} \mid \boldsymbol{x})$ and $u(\boldsymbol{a}, \boldsymbol{\omega})$ will be called a *u-reference decision.*

It is important to note that the limit above is *not* taken in order to obtain some form of asymptotic "approximation" to reference distributions; the "exact" reference prior is *defined* as that which maximises the value of *perfect* information about θ, *not* as that which maximises the expected value of the experiment.

Example 5.10. ***(Prediction with quadratic loss).*** Suppose that beliefs about a sequence of observables, $\boldsymbol{x} = (x_1, \ldots, x_n)$, correspond to assuming the latter to be a random sample from an $N(x \mid \mu, \lambda)$ parametric model, with known precision λ, together with a prior for μ to be selected from the class $\{N(\mu \mid \mu_0, \lambda_0), \mu_0 \in \Re, \lambda_0 \geq 0\}$. Assuming a quadratic loss function, the decision problem is to provide a point estimate for x_{n+1}, given $x_1, \ldots, x_n$. We shall derive a reference analysis of this problem, for which $\mathcal{A} = \Re$, $\omega = x_{n+1}$, and $\theta = \mu$. Moreover,

$$u(a, \omega) = -(a - x_{n+1})^2, \quad p(\boldsymbol{x} \mid \theta) = \prod_{i=1}^{n} N(x_i \mid \mu, \lambda)$$

and, for given μ_0, λ_0, we have

$$p(\omega, \theta) = p(x_{n+1}, \mu) = p(x_{n+1} \mid \mu)p(\mu) = N(x_{n+1} \mid \mu, \lambda)N(\mu \mid \mu_0, \lambda_0).$$

For the purposes of the "thought experiment", let $\boldsymbol{z}_k = (\boldsymbol{x}_1, \ldots, \boldsymbol{x}_k)$ denote the (imagined) outcomes of k replications of the experiment yielding the observables $(x_1, \ldots, x_{kn})$, say, and let us denote the future observation to be predicted (x_{kn+1}) simply by x. Then

$$v_u\{e(k), N(\mu \mid \mu_0, \lambda_0)\} = -\int p(\boldsymbol{z}_k) \inf_a \int p(x \mid \boldsymbol{z}_k)(a-x)^2 dx d\boldsymbol{z}_k$$
$$+ \inf_a \int p(x)(a-x)^2 dx.$$

However, we know from Proposition 5.3 that optimal estimates with respect to quadratic loss functions are given by the appropriate means, so that

$$v_u\{e(k), N(\mu \mid \mu_0, \lambda_0)\} = -\int p(\boldsymbol{z}_k) V[x \mid \boldsymbol{z}_k] d\boldsymbol{z}_k + V[x]$$
$$= -V[x \mid \boldsymbol{z}_k] + V[x],$$

since, by virtue of the normal distributional assumptions, the predictive variance of x given $\boldsymbol{z}_k$ does not depend explicitly on $\boldsymbol{z}_k$. In fact, straightforward manipulations reveal that

$$v_u\{e(\infty), N(\mu \mid \mu_0, \lambda_0)\} = \lim_{k \to \infty} v_u\{e(k), N(\mu \mid \mu_0, \lambda_0)\}$$
$$= \lim_{k \to \infty} \{-\left[\lambda^{-1} + (\lambda_0 + kn\lambda)^{-1}\right] + (\lambda^{-1} + \lambda_0^{-1})\} = \lambda_0^{-1},$$

so that the *u-reference prior* corresponds to the choice $\lambda_0 = 0$, with μ_0 arbitrary.

Example 5.11. ***(Variance estimation)***. Suppose that beliefs about $\boldsymbol{x} = \{x_1, \ldots, x_n\}$ correspond to assuming $\boldsymbol{x}$ to be a random sample from $N(x \mid 0, \lambda)$ together with a gamma prior for λ centred on λ_0, so that $p(\lambda) = \text{Ga}(\lambda \mid \alpha, \alpha\lambda_0^{-1})$, $\alpha > 0$. The decision problem is to provide a point estimate for $\sigma^2 = \lambda^{-1}$, assuming a standardised quadratic loss function, so that

$$u(a, \sigma^2) = -\left[\frac{(a - \sigma^2)}{\sigma^2}\right]^2 = -(a\lambda - 1)^2.$$

Thus, we have $\mathcal{A} = \Re^+$, $\theta = \lambda$, $w = \sigma^2$, and

$$p(\boldsymbol{x}, \lambda) = \prod_{i=1}^n N(x_i \mid 0, \lambda)\, \text{Ga}(\lambda \mid \alpha, \alpha\lambda_0^{-1}).$$

Let $\boldsymbol{z}_k = \{\boldsymbol{x}_1, \ldots, \boldsymbol{x}_k\}$ denote the outcome of k replications of the experiment. Then

$$v_u\{e(k), p(\lambda)\} = -\int p(\boldsymbol{z}_k) \inf_a \int p(\lambda \mid \boldsymbol{z}_k)\, (a\lambda - 1)^2\, d\lambda\, d\boldsymbol{z}_k$$
$$+ \inf_a \int p(\lambda)\, (a\lambda - 1)^2\, d\lambda,$$

where

$$p(\lambda) = \text{Ga}(\lambda \mid \alpha, \alpha\lambda_0^{-1}), \quad p(\lambda \mid z_k) = \text{Ga}\left(\lambda \mid \alpha + \frac{kn}{2}, \alpha\lambda_0^{-1} + \frac{kns^2}{2}\right),$$

and $kns^2 = \sum_i \sum_j x_{ij}^2$. Since

$$\inf_a \int \text{Ga}(\lambda \mid \alpha, \beta)\, (a\lambda - 1)^2\, d\lambda = \frac{1}{\alpha + 1},$$

and this is attained when $a = \beta/(\alpha + 1)$, one has

$$\begin{aligned} v_u\{e(\infty), p(\lambda)\} &= \lim_{k\to\infty} v_u\{e(k), p(\lambda)\} \\ &= \lim_{k\to\infty} \left\{ -\frac{1}{1+\alpha+(kn)/2} + \frac{1}{1+\alpha} \right\} = \frac{1}{1+\alpha}. \end{aligned}$$

This is maximised when $\alpha = 0$ and, hence, the *u-reference prior* corresponds to the choice $\alpha = 0$, with λ_0 arbitrary. Given *actual* data, $x = (x_1, \ldots, x_n)$, the *u-reference posterior* for λ is $\text{Ga}(\lambda \mid n/2, ns^2/2)$, where $ns^2 = \sum_i x_i^2$ and, thus, the *u-reference decision* is to give the estimate

$$\hat{\sigma}^2 = \frac{ns^2/2}{(n/2)+1} = \frac{\Sigma x_i^2}{n+2}.$$

Hence, the reference estimator of σ^2 with respect to *standardised* quadratic loss is *not* the usual s^2, but a slightly smaller multiple of s^2.

It is of interest to note that, from a frequentist perspective, $\hat{\sigma}^2$ is the best invariant estimator of σ^2 and is admissible. Indeed, $\hat{\sigma}^2$ dominates s^2 or any smaller multiple of s^2 in terms of frequentist risk (cf. Example 45 in Berger, 1985a, Chapter 4). Thus, the u-reference approach has led to the "correct" multiple of s^2 as seen from a frequentist perspective.

Explicit reference decision analysis is possible when the parameter space $\Theta = \{\theta_1, \ldots, \theta_M\}$ is finite. In this case, the expected value of perfect information (cf. Definition 2.19) may be written as

$$v_u\{e(\infty), p(\theta)\} = \sum_{i=1}^{M} p(\theta_i) \sup_{\mathcal{D}} u(d(\theta_i)) - \sup_{\mathcal{D}} \sum_{i=1}^{M} p(\theta_i)\, u(d(\theta_i)),$$

and the u-reference prior, which is that $\pi(\theta)$ which maximises $v_u\{e(\infty), p(\theta)\}$, may be explicitly obtained by standard algebraic manipulations. For further information, see Bernardo (1981a) and Rabena (1998).

5.4.2 One-dimensional Reference Distributions

In Sections 2.7 and 3.4, we noted that reporting beliefs is itself a decision problem, where the "inference answer" space consists of the class of possible belief distributions that could be reported about the quantity of interest, and the utility function is a proper scoring rule which—in pure inference problems—may be identified with the logarithmic scoring rule.

Our development of reference analysis from now on will concentrate on this case, for which we simply denote $v_u\{\cdot\}$ by $v\{\cdot\}$, and replace the term "u-reference" by "reference".

In discussing reference decisions, we have considered a rather general utility structure where practical interest centred on a quantity $\boldsymbol{\omega}$ related to the $\boldsymbol{\theta}$ of an experiment by a conditional probability specification, $p(\boldsymbol{\omega} \,|\, \boldsymbol{\theta})$. Here, we shall consider the case where the quantity of interest is θ itself, with $\theta \in \Theta \subset \Re$. More general cases will be considered later.

If an experiment e consists of an observation $\boldsymbol{x} \in X$ having parametric model $p(\boldsymbol{x} \,|\, \theta)$, with $\boldsymbol{\omega} = \theta, \mathcal{A} = \{q(\cdot); q(\theta) > 0, \int_\Theta q(\theta)d\theta = 1\}$ and the utility function is the logarithmic scoring rule

$$u\{q(\cdot), \theta\} = A \log q(\theta) + B(\theta),$$

the expected utility value of the experiment e, given the prior density $p(\theta)$, is

$$v\{e, p(\theta)\} = \int p(\boldsymbol{x}) \int u\{q_x(\cdot), \theta\} p(\theta \,|\, \boldsymbol{x})\, d\theta d\boldsymbol{x} - \int u\{q_0(\cdot), \theta\}\, p(\theta)\, d\theta,$$

where $q_0(\cdot), q_x(\cdot)$ denote the optimal choices of $q(\cdot)$ with respect to $p(\theta)$ and $p(\theta \,|\, x)$, respectively. Noting that u is a proper scoring rule, so that, for any $p(\theta)$,

$$\sup_q \int u\{q(\cdot), \theta\} p(\theta)\, d\theta = \int u\{p(\cdot), \theta\} p(\theta)\, d\theta,$$

it is easily seen that

$$v\{e, p(\theta)\} \propto \int p(\boldsymbol{x}) \int p(\theta \,|\, \boldsymbol{x}) \log \frac{p(\theta \,|\, \boldsymbol{x})}{p(\theta)} d\theta d\boldsymbol{x} = I\{e, p(\theta)\}$$

the *amount of information* about θ which e may be expected to provide.

The corresponding expected information from the (hypothetical) experiment $e(k)$ yielding the (imagined) observation $\boldsymbol{z}_k = (\boldsymbol{x}_1, \ldots, \boldsymbol{x}_k)$ with parametric model

$$p(\boldsymbol{z}_k \,|\, \theta) = \prod_{i=1}^{k} p(\boldsymbol{x}_i \,|\, \theta)$$

is given by

$$I\{e(k), p(\theta)\} = \int p(\boldsymbol{z}_k) \int p(\theta \,|\, \boldsymbol{z}_k) \log \frac{p(\theta \,|\, \boldsymbol{z}_k)}{p(\theta)}\, d\theta d\boldsymbol{z}_k,$$

and so the expected (utility) value of perfect information about θ is

$$I\{e(\infty), p(\theta)\} = \lim_{k \to \infty} I\{e(k), p(\theta)\},$$

provided that this limit exists. This quantity measures the *missing information* about θ as a function of the prior $p(\theta)$.

The *reference prior* for θ, denoted by $\pi(\theta)$, is thus defined to be that prior which maximises the missing information functional. Given actual data $\boldsymbol{x}$, the *reference posterior* $\pi(\theta \mid \boldsymbol{x})$ to be reported is simply derived from Bayes' theorem, as $\pi(\theta \mid \boldsymbol{x}) \propto p(\boldsymbol{x} \mid \theta)\pi(\theta)$.

Unfortunately, $\lim_{k\to\infty} I\{e(k), p(\theta)\}$ is typically infinite (unless θ can only take a finite range of values) and a direct approach to deriving $\pi(\theta)$ along these lines cannot be implemented. However, a natural way of overcoming this technical difficulty is available: we derive the sequence of priors $\pi_k(\theta)$ which maximise $I\{e(k), p(\theta)\}, k = 1, 2, \ldots$, and subsequently take $\pi(\theta)$ to be a suitable limit. This approach will now be developed in detail.

Let e be the experiment which consists of one observation $\boldsymbol{x}$ from $p(\boldsymbol{x} \mid \theta)$, $\theta \in \Theta \subseteq \Re$. Suppose that we are interested in reporting inferences about θ and that no restrictions are imposed on the form of the prior distribution $p(\theta)$. It is easily verified that the amount of information about θ which k independent replications of e may be expected to provide may be rewritten as

$$I^{\theta}\{e(k), p(\theta)\} = \int p(\theta) \log \frac{f_k(\theta)}{p(\theta)}\, d\theta,$$

where

$$f_k(\theta) = \exp\left\{\int p(\boldsymbol{z}_k \mid \theta) \log p(\theta \mid \boldsymbol{z}_k) d\boldsymbol{z}_k\right\}$$

and $\boldsymbol{z}_k = \{\boldsymbol{x}_1, \ldots, \boldsymbol{x}_k\}$ is a possible outcome from $e(k)$, so that

$$p(\theta \mid \boldsymbol{z}_k) \propto \prod_{i=1}^{k} p(\boldsymbol{x}_i \mid \theta) p(\theta)$$

is the posterior distribution for θ after $\boldsymbol{z}_k$ has been observed. Moreover, for any prior $p(\theta)$ one must have the constraint $\int p(\theta)\, d\theta = 1$ and, therefore, the prior $\pi_k(\theta)$ which maximises $I^{\theta}\{e(k), p(\theta)\}$ must be an extremal of the functional

$$F\{p(\cdot)\} = \int p(\theta) \log \frac{f_k(\theta)}{p(\theta)} d\theta + \lambda \left\{\int p(\theta)\, d\theta - 1\right\}.$$

Since this is of the form $F\{p(\cdot)\} = \int g\{p(\cdot)\}\, d\theta$, where, as a functional of $p(\cdot)$, g is twice continuously differentiable, any function $p(\cdot)$ which maximises F must satisfy the condition

$$\frac{\partial}{\partial \varepsilon} F\{p(\cdot) + \varepsilon \tau(\cdot)\}\Big|_{\varepsilon=0} = 0, \quad \text{for all } \tau.$$

It follows that, for any function τ,

$$\int \left\{ \tau(\theta) \log f_k(\theta) + \frac{p(\theta)}{f_k(\theta)} f_k'(\theta) - \tau(\theta)\,(1 + \log p(\theta)) + \tau(\theta)\lambda \right\} d\theta = 0,$$

where, after some algebra,

$$\begin{aligned} f_k'(\theta) &= \frac{\partial}{\partial \varepsilon} \left\{ \exp \left[\int p(\boldsymbol{z}_k \,|\, \theta) \log \frac{p(\boldsymbol{z} \,|\, \theta)\{p(\theta) + \varepsilon\tau(\theta)\}}{\int p(\boldsymbol{z}_k \,|\, \theta)\{p(\theta) + \varepsilon\tau(\theta)\}\, d\theta} d\boldsymbol{z}_k \right] \right\} \Bigg|_{\varepsilon=0} \\ &= f_k(\theta) \frac{\tau(\theta)}{p(\theta)}\,. \end{aligned}$$

Thus, the required condition becomes

$$\int \tau(\theta)\, \{\log f_k(\theta) - \log p(\theta) + \lambda\}\, d\theta = 0, \quad \text{for all } \tau(\theta),$$

which implies that the desired extremal should satisfy, for all $\theta \in \Theta$,

$$\log f_k(\theta) - \log p(\theta) + \lambda = 0$$

and hence that $p(\theta) \propto f_k(\theta)$.

Note that, for each k, this only provides an *implicit* solution for the prior which maximises $I^\theta\{e(k), p(\theta)\}$, since $f_k(\theta)$ depends on the prior through the posterior distribution $p(\theta \,|\, \boldsymbol{z}_k) = p(\theta \,|\, \boldsymbol{x}_1, \ldots, \boldsymbol{x}_k)$. However, for large values of k, an approximation, $p^*(\theta \,|\, \boldsymbol{z}_k)$, say, may be found to the posterior distribution of θ, which *is* independent of the prior $p(\theta)$. It follows that, under suitable regularity conditions, the sequence of positive functions

$$p_k^*(\theta) = \exp\left\{ \int p(\boldsymbol{z}_k \,|\, \theta) \log p^*(\theta \,|\, \boldsymbol{z}_k) d\boldsymbol{z}_k \right\}$$

will induce, by formal use of Bayes' theorem, a sequence of posterior distributions

$$\pi_k(\theta \,|\, \boldsymbol{x}) \propto p(\boldsymbol{x} \,|\, \theta) p_k^*(\theta)$$

with the same limiting distributions that would have been obtained from the sequence of posteriors derived from the sequence of priors $\pi_k(\theta)$ which maximise $I^\theta\{e(k), p(\theta)\}$. This completes our motivation for Definition 5.7. For further information see Bernardo (1979b) and ensuing discussion.

Definition 5.7. (***One-dimensional reference distributions***).
Let x be the result of an experiment e which consists of one observation from $p(x \mid \theta), x \in X, \theta \in \Theta \subseteq \Re$, let $z_k = \{x_1, \ldots, x_k\}$ be the result of k independent replications of e, and define

$$f_k^*(\theta) = \exp\left\{ \int p(z_k \mid \theta) \log p^*(\theta \mid z_k) dz_k \right\},$$

where

$$p^*(\theta \mid z_k) = \frac{\prod_{i=1}^k p(x_i \mid \theta)}{\int \prod_{i=1}^k p(x_i \mid \theta)\, d\theta}.$$

The ***reference posterior*** *density of θ after x has been observed is defined to be the log-divergence limit, $\pi(\theta \mid x)$, of $\pi_k(\theta \mid x)$, assuming this limit to exist, where*

$$\pi_k(\theta \mid x) = c_k(x)\, p(x \mid \theta)\, f_k^*(\theta),$$

the $c_k(x)$'s are the required normalising constants and, for almost all x,

$$\lim_{k \to \infty} \int \pi_k(\theta \mid x) \log \frac{\pi_k(\theta \mid x)}{\pi(\theta \mid x)}\, d\theta = 0.$$

Any positive function $\pi(\theta)$ such that, for some $c(x) > 0$ and for all $\theta \in \Theta$,

$$\pi(\theta \mid x) = c(x)\, p(x \mid \theta)\, \pi(\theta)$$

will be called a ***reference prior*** *for θ relative to the experiment e.*

It should be clear from the argument which motivates the definition that any asymptotic approximation to the posterior distribution may be used in place of the asymptotic approximation $p^*(\theta \mid z_k)$ defined above. The use of convergence in the information sense, the natural convergence in this context, rather than just pointwise convergence, is necessary to avoid possibly pathological behaviour; for details, see Berger and Bernardo (1992c).

Although most of the following discussion refers to reference priors, it must be stressed that *only reference posterior* distributions are directly interpretable in probabilistic terms. The positive functions $\pi(\theta)$ are merely pragmatically convenient *tools* for the derivation of reference posterior distributions via Bayes' theorem. An explicit form for the reference prior is immediately available from Definition 5.7, and it will be clear from later illustrative examples that the forms which arise may have no direct probabilistic interpretation.

We should stress that the definitions and "propositions" in this section are by and large heuristic in the sense that they are lacking statements of the technical conditions which would make the theory rigorous. Making the statements and proofs precise, however, would require a different level of mathematics from that used in this book and, at the time of writing, is still an active area of research. The reader interested in the technicalities involved is referred to Berger and Bernardo (1989, 1992a, 1992b, 1992c) and Berger *et al.* (1989). So far as the contents of this section are concerned, the reader would be best advised to view the procedure as an "algorithm, which compared with other proposals—discussed in Section 5.6.2—appears to produce appealing solutions in all situations thus far examined.

Proposition 5.18. (*Explicit form of the reference prior*).
A reference prior for θ relative to the experiment which consists of one observation from $p(\boldsymbol{x} \mid \theta)$, $\boldsymbol{x} \in X$, $\theta \in \Theta \subseteq \Re$, is given, provided the limit exists, and convergence in the information sense is verified, by

$$\pi(\theta) = c \lim_{k\to\infty} \frac{f_k^*(\theta)}{f_k^*(\theta_0)}, \qquad \theta \in \Theta$$

where $c > 0$, $\theta_0 \in \Theta$,

$$f_k^*(\theta) = \exp\left\{ \int p(\boldsymbol{z}_k \mid \theta) \log p^*(\theta \mid \boldsymbol{z}_k) d\boldsymbol{z}_k \right\},$$

with $\boldsymbol{z}_k = \{\boldsymbol{x}_1, \ldots, \boldsymbol{x}_k\}$ a random sample from $p(\boldsymbol{x} \mid \theta)$, and $p^*(\theta \mid \boldsymbol{z}_k)$ is an asymptotic approximation to the posterior distribution of θ.

Proof. Using $\pi(\theta)$ as a formal prior,

$$\pi(\theta \mid \boldsymbol{x}) \propto p(\boldsymbol{x} \mid \theta)\pi(\theta) \propto p(\boldsymbol{x} \mid \theta) \lim_{k\to\infty} \frac{f_k^*(\theta)}{f_k^*(\theta_0)} \propto \lim_{k\to\infty} \frac{p(\boldsymbol{x} \mid \theta) f_k^*(\theta)}{\int p(\boldsymbol{x} \mid \theta) f_k^*(\theta)\, d\theta},$$

and hence

$$\pi(\theta \mid \boldsymbol{x}) = \lim_{k\to\infty} \pi_k(\theta \mid \boldsymbol{x}), \quad \pi_k(\theta \mid \boldsymbol{x}) \propto p(\boldsymbol{x} \mid \theta) f_k^*(\theta)$$

as required. Note that, under suitable regularity conditions, the limits above will not depend on the particular asymptotic approximation to the posterior distribution used to derive $f_k^*(\theta)$. ◁

If the parameter space is finite, it turns out that the reference prior is uniform, independently of the experiment performed.

Proposition 5.19. (*Reference prior in the finite case*). *Let $\boldsymbol{x}$ be the result of one observation from $p(\boldsymbol{x} \mid \theta)$, where $\theta \in \Theta = \{\theta_1, \ldots, \theta_M\}$. Then, any function of the form $\pi(\theta_i) = a$, $a > 0$, $i = 1, \ldots, M$, is a reference prior and the reference posterior is*

$$\pi(\theta_i \mid \boldsymbol{x}) = c(\boldsymbol{x}) p(\boldsymbol{x} \mid \theta_i), \quad i = 1, \ldots, M$$

where $c(\boldsymbol{x})$ is the required normalising constant.

Proof. We have already established (Proposition 5.12) that if Θ is finite then, for any strictly positive prior, $p(\theta_i \mid x_1, \ldots, x_k)$ will converge to 1 if θ_i is the true value of θ. It follows that the integral in the exponent of

$$f_k(\theta_i) = \exp\left\{\int p(\boldsymbol{z}_k \mid \theta_i) \log p(\theta_i \mid \boldsymbol{z}_k) d\boldsymbol{z}_k\right\}, \quad i = 1, \ldots, M,$$

will converge to zero as $k \to \infty$. Hence, a reference prior is given by

$$\pi(\theta_i) = \lim_{k\to\infty} \frac{f_k(\theta_i)}{f_k(\theta_j)} = 1.$$

The general form of reference prior follows immediately. ◁

The preceding result for the case of a finite parameter space is easily derived from first principles. Indeed, in this case the expected missing information is finite and equals the entropy

$$H\{p(\theta)\} = -\sum_{i=1}^{M} p(\theta_i) \log p(\theta_i)$$

of the prior. This is maximised if and only if the prior is uniform.

The technique encapsulated in Definition 5.7 for identifying the reference prior depends on the asymptotic behaviour of the posterior for the parameter of interest under (imagined) replications of the experiment to be actually performed. Thus far, our derivations have proceeded on the basis of an assumed single observation from a parametric model, $p(\boldsymbol{x} \mid \theta)$. The next proposition establishes that for experiments involving a sequence of $n \geq 1$ observations, which are to be modelled as if they are a random sample, conditional on a parametric model, the reference prior does not depend on the size of the experiment and can thus be derived on the basis of a single observation experiment. Note, however, that for experiments involving more structured designs (for example, in linear models) the situation is much more complicated.

Proposition 5.20. ***(Independence of sample size).***
Let e_n, $n \geq 1$, be the experiment which consists of the observation of a random sample $\boldsymbol{x}_1, \ldots, \boldsymbol{x}_n$ from $p(\boldsymbol{x} \mid \theta), \boldsymbol{x} \in X$, $\theta \in \Theta$, and let $\mathcal{P}_n$ denote the class of reference priors for θ with respect to e_n, derived in accordance with Definition 5.7, by considering the sample to be a single observation from $\prod_{i=1}^{n} p(\boldsymbol{x}_i \mid \theta)$. Then $\mathcal{P}_1 = \mathcal{P}_n$, for all n.

Proof. If $z_k = \{x_1, \ldots, x_k\}$ is the result of a k-fold independent replicate of e_1, then, by Proposition 5.18, $\mathcal{P}_1$ consists of $\pi(\theta)$ of the form

$$\pi(\theta) = c \lim_{k\to\infty} \frac{f_k^*(\theta)}{f_k^*(\theta_0)},$$

with $c > 0$, $\theta, \theta_0 \in \Theta$ and

$$f_k^*(\theta) = \exp\left\{\int p(z_k \mid \theta) \log p^*(\theta \mid z_k)\, dz_k\right\},$$

where $p^*(\theta \mid z_k)$ is an asymptotic approximation (as $k \to \infty$) to the posterior distribution of θ given z_k.

Now consider $z_{nk} = \{x_1, \ldots, x_n, x_{n+1}, \ldots, x_{2n}, \ldots, x_{kn}\}$ which can be considered as the result of a k-fold independent replicate of e_n, so that $\mathcal{P}_n$ consists of $\pi(\theta)$ of the form

$$\pi(\theta) = c \lim_{k\to\infty} \frac{f_{nk}^*(\theta)}{f_{nk}^*(\theta_0)}.$$

But z_{nk} can equally be considered as a nk-fold independent replicate of e_1 and so the limiting ratios are clearly identical. $\triangleleft$

In considering experiments involving random samples from distributions admitting a sufficient statistic of fixed dimension, it is natural to wonder whether the reference priors derived from the distribution of the sufficient statistic are identical to those derived from the joint distribution for the sample. The next proposition guarantees us that this is indeed the case.

Proposition 5.21. (***Compatibility with sufficient statistics***).
Let $e_n, n \geq 1$, be the experiment which consists of the observation of a random sample $x_1, \ldots, x_n$ from $p(x \mid \theta), x \in X, \theta \in \Theta$, where, for all n, the latter admits a sufficient statistic $t_n = t(x_1, \ldots, x_n)$. Then, for any n, the classes of reference priors derived by considering replications of $(x_1, \ldots, x_n)$ and t_n respectively, coincide, and are identical to the class obtained by considering replications of e_1.

Proof. If z_k denotes a k-fold replicate of $(x_1, \ldots, x_n)$ and y_k denotes the corresponding k-fold replicate of t_n, then, by the definition of a sufficient statistic, $p(\theta \mid z_k) = p(\theta \mid y_k)$, for any prior $p(\theta)$. It follows that the corresponding asymptotic distributions are identical, so that $p^*(\theta \mid z_k) = p^*(\theta \mid y_k)$. We thus have

$$\begin{aligned} f_k^*(\theta) &= \exp\left\{\int p(z_k \mid \theta) \log p^*(\theta \mid z_k) dz_k\right\} \\ &= \exp\left\{\int p(z_k \mid \theta) \log p^*(\theta \mid y_k) dz_k\right\} \\ &= \exp\left\{\int p(y_k \mid \theta) \log p^*(\theta \mid y_k) dy_k\right\} \end{aligned}$$

so that, by Definition 5.7, the reference priors are identical. Identity with those derived from e_1 follows from Proposition 5.20. $\triangleleft$

Given a parametric model, $p(\boldsymbol{x} \mid \theta)$, $x \in X$, $\theta \in \Theta$, we could, of course, reparametrise and work instead with $p(\boldsymbol{x} \mid \phi)$, $x \in X$, $\phi = \phi(\theta)$, for any monotone one-to-one mapping $g : \Theta \to \Phi$. The question now arises as to whether reference priors for θ and ϕ, derived from the parametric models $p(\boldsymbol{x} \mid \theta)$ and $p(\boldsymbol{x} \mid \phi)$, respectively, are consistent, in the sense that their ratio is the required Jacobian element. The next proposition establishes this form of consistency and can clearly be extended to mappings which are piecewise monotone.

Proposition 5.22. (***Invariance under one-to-one transformations***). *Suppose that $\pi_\theta(\theta)$, $\pi_\phi(\phi)$ are reference priors derived by considering replications of experiments consisting of a single observation from $p(\boldsymbol{x} \mid \theta)$, with $\boldsymbol{x} \in X$, $\theta \in \Theta$ and from $p(\boldsymbol{x} \mid \phi)$, with $x \in X, \phi \in \Phi$, respectively, where $\phi = g(\theta)$ and $g : \Theta \to \Phi$ is a one-to-one monotone mapping. Then, for some $c > 0$ and for all $\phi \in \Phi$:*

(i) $\pi_\phi(\phi) = c\,\pi_\theta\left(g^{-1}(\phi)\right)$, *if Θ is discrete;*

(ii) $\pi_\phi(\phi) = c\,\pi_\theta\left(g^{-1}(\phi)\right) \,|\, J_\phi \,|$, *if* $J_\phi = \dfrac{\partial g^{-1}(\phi)}{\partial \phi}$ *exists.*

Proof. If Θ is discrete, so is Φ and the result follows from Proposition 5.19. Otherwise, if $\boldsymbol{z}_k$ denotes a k-fold replicate of a single observation from $p(\boldsymbol{x} \mid \theta)$, then, for any proper prior $p(\theta)$, the corresponding prior for ϕ is given by $p_\phi(\phi) = p_\theta\left(g^{-1}(\phi)\right) |J_\phi|$ and hence, for all $\phi \in \Phi$,

$$p_\phi(\phi \mid \boldsymbol{z}_k) = p_\theta\left(g^{-1}(\phi) \mid \boldsymbol{z}_k\right) |J_\phi|.$$

It follows that, as $k \to \infty$, the asymptotic posterior approximations are related by the same Jacobian element and hence

$$\begin{aligned} f_k^*(\theta) &= \exp\left\{\int p(\boldsymbol{z}_k \mid \theta) \log p^*(\theta \mid \boldsymbol{z}_k) d\boldsymbol{z}_k\right\} \\ &= |J_\phi|^{-1} \exp\left\{\int p(\boldsymbol{z}_k \mid \phi) \log p^*(\phi \mid \boldsymbol{z}_k) d\boldsymbol{z}_k\right\} \\ &= |J_\phi|^{-1}\, f_k^*(\phi). \end{aligned}$$

The second result now follows from Proposition 5.18. $\triangleleft$

The assumed existence of the asymptotic posterior distributions that would result from an imagined k-fold replicate of the experiment under consideration clearly plays a key role in the derivation of the reference prior. However, it is important to note that no assumption has thus far been required concerning the form of this asymptotic posterior distribution. As we shall see later, we shall typically consider the case of asymptotic posterior normality, but the following example shows that the technique is by no means restricted to this case.

Example 5.12. ***(Uniform model)*.** Let e be the experiment which consists of observing the sequence $x_1, \ldots, x_n, n \geq 1$, whose belief distribution is represented as that of a random sample from a uniform distribution on $[\theta - \frac{1}{2}, \theta + \frac{1}{2}], \theta \in \Re$, together with a prior distribution $p(\theta)$ for θ. If

$$\boldsymbol{t}_n = \left[x_{\min}^{(n)}, x_{\max}^{(n)}\right], \quad x_{\min}^{(n)} = \min\{x_1, \ldots, x_n\}, \quad x_{\max}^{(n)} = \max\{x_1, \ldots, x_n\},$$

then $\boldsymbol{t}_n$ is a sufficient statistic for θ, and

$$p(\theta \mid \boldsymbol{x}) = p(\theta \mid \boldsymbol{t}_n) \propto p(\theta), \qquad x_{\max}^{(n)} - \tfrac{1}{2} \leq \theta \leq x_{\min}^{(n)} + \tfrac{1}{2} \cdot$$

It follows that, as $k \to \infty$, a k-fold replicate of e with a uniform prior will result in the posterior uniform distribution

$$p^*(\theta \mid \boldsymbol{t}_{kn}) \propto c, \qquad x_{\max}^{(kn)} - \tfrac{1}{2} \leq \theta \leq x_{\min}^{(kn)} + \tfrac{1}{2} \cdot$$

It is easily verified that

$$\int p(\boldsymbol{t}_{kn} \mid \theta) \log p^*(\theta \mid \boldsymbol{t}_{kn}) d\boldsymbol{t}_{kn} = E\left[-\log\left\{1 - (x_{\max}^{(kn)} - x_{\min}^{(kn)})\right\} \middle| \theta\right],$$

the expectation being with respect to the distribution of $\boldsymbol{t}_{kn}$. For large k, the right-hand side is well-approximated by

$$-\log\left\{1 - \left(E\left[x_{\max}^{(kn)}\right] - E\left[x_{\min}^{(kn)}\right]\right)\right\},$$

and, noting that the distributions of

$$u = x_{\max}^{(kn)} - \theta - \tfrac{1}{2}, \quad v = x_{\min}^{(kn)} - \theta + \tfrac{1}{2}$$

are $\text{Be}(u \mid kn, 1)$ and $\text{Be}(v \mid 1, kn)$, respectively, we see that the above reduces to

$$-\log\left[1 - \frac{kn}{kn+1} + \frac{1}{kn+1}\right] = \log\left(\frac{kn+1}{2}\right).$$

It follows that $f_{kn}^*(\theta) = (kn+1)/2$, and hence that

$$\pi(\theta) = c \lim_{k \to \infty} \frac{(kn+1)/2}{(kn+1)/2} = c\,.$$

Any reference prior for this problem is therefore a constant and, therefore, given a set of actual data $\boldsymbol{x} = (x_1, \ldots, x_n)$, the reference posterior distribution is

$$\pi(\theta \mid \boldsymbol{x}) \propto c\,, \qquad x_{\max}^{(n)} - \tfrac{1}{2} \leq \theta \leq x_{\min}^{(n)} + \tfrac{1}{2},$$

a uniform distribution over the set of θ values which remain possible after x has been observed.

Typically, under suitable regularity conditions, the asymptotic posterior distribution $p^*(\theta \mid z_{kn})$, corresponding to an imagined k-fold replication of an experiment e_n involving a random sample of n from $p(x \mid \theta)$, will only depend on z_{kn} through an *asymptotically sufficient, consistent estimate of* θ, a concept which is made precise in the next proposition. In such cases, the reference prior can easily be identified from the form of the asymptotic posterior distribution.

Proposition 5.23. (***Explicit form of the reference prior when there is a consistent, asymptotically sufficient, estimator***). *Let e_n be the experiment which consists of the observation of a random sample $x = \{x_1, \ldots, x_n\}$ from $p(x \mid \theta), x \in X, \theta \in \Theta \subseteq \Re$, and let z_{kn} be the result of a k-fold replicate of e_n. If there exists $\hat{\theta}_{kn} = \hat{\theta}_{kn}(z_{kn})$ such that, with probability one*

$$\lim_{k \to \infty} \hat{\theta}_{kn} = \theta$$

and, as $k \to \infty$,

$$\int p(z_{kn} \mid \theta) \log \frac{p^*(\theta \mid z_{kn})}{p^*(\theta \mid \hat{\theta}_{kn})} dz_{kn} \to 0,$$

then, for any $c > 0, \theta_0 \in \Theta$, reference priors are defined by

$$\pi(\theta) = c \lim_{k \to \infty} \frac{f^*_{kn}(\theta)}{f^*_{kn}(\theta_0)},$$

where

$$f^*_{kn}(\theta) = p^*(\theta \mid \hat{\theta}_{kn})\Big|_{\hat{\theta}_{kn} = \theta}.$$

Proof. As $k \to \infty$, it follows from the assumptions that

$$\begin{aligned} f^*_{kn}(\theta) &= \exp\left\{\int p(z_{kn} \mid \theta) \log p^*(\theta \mid z_{kn}) dz_{kn}\right\} \\ &= \exp\left\{\int p(z_{kn} \mid \theta) \log p^*(\theta \mid \hat{\theta}_{kn}) dz_{kn}\right\} \\ &= \exp\left\{\int p(\hat{\theta}_{kn} \mid \theta) \log p^*(\theta \mid \hat{\theta}_{kn}) d\hat{\theta}_{kn}\right\} \\ &= \exp\left\{\log p^*(\theta \mid \hat{\theta}_{kn})\Big|_{\hat{\theta}_{kn} = \theta}\right\} = p^*(\theta \mid \hat{\theta}_{kn})\Big|_{\hat{\theta}_{kn} = \theta}. \end{aligned}$$

The result now follows from Proposition 5.18. $\triangleleft$

Example 5.13. ***(Deviation from uniformity model).*** Let e_n be the experiment which consists of obtaining a random sample from $p(x \mid \theta), 0 \leq x \leq 1, \theta > 0$, where

$$p(x \mid \theta) = \begin{cases} \theta\{2x\}^{\theta-1} & \text{for } \; 0 \leq x \leq \frac{1}{2} \\ \\ \theta\{2(1-x)\}^{\theta-1} & \text{for } \; \frac{1}{2} \leq x \leq 1 \end{cases}$$

defines a one-parameter probability model on $[0,1]$, which finds application (see Bernardo and Bayarri, 1985) in exploring deviations from the standard uniform model on $[0,1]$ (given by $\theta = 1$).

It is easily verified that if $\boldsymbol{z}_{kn} = \{x_1, \ldots, x_{kn}\}$ results from a k-fold replicate of e_n, the sufficient statistic t_{kn} is given by

$$t_{kn} = -\frac{1}{nk} \sum_{i=1}^{kn} \{\log\{2x_i\} 1_{[0,1/2]}(x_i) + \log\{2(1-x_i)\} 1_{]1/2,1]}(x_i)\}$$

and, for any prior $p(\theta)$,

$$\begin{aligned} p(\theta \mid \boldsymbol{z}_{kn}) &= p(\theta \mid t_{kn}) \\ &\propto p(\theta)\theta^{kn} \exp\{-kn(\theta - 1)t_{kn}\}. \end{aligned}$$

It is also easily shown that $p(t_{kn} \mid \theta) = \text{Ga}(t_{kn} \mid kn, kn\theta)$, so that

$$E[t_{kn} \mid \theta] = \frac{1}{\theta}, \quad V[t_{kn} \mid \theta] = \frac{1}{kn\theta^2},$$

from which we can establish that $\hat{\theta}_{kn} = t_{kn}^{-1}$ is a sufficient, consistent estimate of θ. It follows that

$$p^*(\theta \mid \hat{\theta}_{kn}) \propto \theta^{kn} \exp\left\{-\frac{kn(\theta - 1)}{\hat{\theta}_{kn}}\right\}$$

provides, for large k, an asymptotic posterior approximation which satisfies the conditions required in Proposition 5.23. From the form of the right-hand side, we see that

$$\begin{aligned} p^*(\theta \mid \hat{\theta}_{kn}) &= \text{Ga}(\theta \mid kn + 1, kn/\hat{\theta}_{kn}) \\ &= \frac{(kn/\hat{\theta}_{kn})^{kn+1}}{\Gamma(kn+1)} \theta^{kn} \exp\left\{\frac{-kn\theta}{\hat{\theta}_{kn}}\right\}, \end{aligned}$$

so that

$$f_{kn}^*(\theta) = p^*(\theta \mid \hat{\theta}_{kn}) \Big|_{\hat{\theta}_{kn} = \theta} = \frac{(kn)^{kn+1} e^{-nk}}{\Gamma(kn+1)\theta},$$

and, from Proposition 5.18, for some $c > 0$, $\theta_0 > 0$,

$$\pi(\theta) = c \lim_{k \to \infty} \frac{f_{kn}^*(\theta)}{f_{kn}^*(\theta_0)} = \frac{c\theta_0}{\theta} \propto \frac{1}{\theta}.$$

The reference posterior for θ having observed actual data $\boldsymbol{x} = (x_1, \ldots, x_n)$, producing the sufficient statistic $t = t(\boldsymbol{x})$, is therefore

$$\begin{aligned} \pi(\theta \mid \boldsymbol{x}) = \pi(\theta \mid t) &\propto p(\boldsymbol{x} \mid \theta) \frac{1}{\theta} \\ &\propto \theta^{n-1} \exp\{-n(\theta - 1)t\}, \end{aligned}$$

which is a $\text{Ga}(\theta \mid n, nt)$ distribution.

Under regularity conditions similar to those described in Section 5.2.3, the asymptotic posterior distribution of θ tends to normality. In such cases, we can obtain a characterisation of the reference prior directly in terms of the parametric model in which θ appears.

Proposition 5.24. (***Reference priors under asymptotic normality***).
Let e_n be the experiment which consists of the observation of a random sample $\boldsymbol{x}_1, \ldots, \boldsymbol{x}_n$ from $p(\boldsymbol{x} \mid \theta)$, $\boldsymbol{x} \in X$, $\theta \in \Theta \subset \Re$. Then, if the asymptotic posterior distribution of θ, given a k-fold replicate of e_n, is normal with precision $knh(\hat{\theta}_{kn})$, where $\hat{\theta}_{kn}$ is a consistent estimate of θ, reference priors have the form

$$\pi(\theta) \propto \{h(\theta)\}^{1/2}.$$

Proof. Under regularity conditions such as those detailed in Section 5.2.3, it follows that an asymptotic approximation to the posterior distribution of θ, given a k-fold replicate of e_n, is

$$p^*(\theta \mid \hat{\theta}_{kn}) = N\left(\theta \mid \hat{\theta}_{kn}, knh(\hat{\theta}_{kn})\right),$$

where $\hat{\theta}_{kn}$ is some consistent estimator of θ. Thus, by Proposition 5.23,

$$\begin{aligned} f_{kn}^*(\theta) &= p^*(\theta \mid \hat{\theta}_{kn})\Big|_{\hat{\theta}_{kn}=\theta} \\ &= \left(\frac{kn}{2\pi}\right)^{1/2} \{h(\theta)\}^{1/2}, \end{aligned}$$

and therefore, for some $c > 0$, $\theta_0 \in \Theta$,

$$\pi(\theta) = c \lim_{k\to\infty} \frac{f_{kn}^*(\theta)}{f_{kn}^*(\theta_0)} = \frac{\{h(\theta)\}^{1/2}}{\{h(\theta_0)\}^{1/2}} \propto \{h(\theta)\}^{1/2},$$

as required. $\triangleleft$

The result of Proposition 5.24 is closely related to the "rules" proposed by Jeffreys (1946, 1939/1961) and by Perks (1947) to derive "non-informative" priors. Typically, under the conditions where asymptotic posterior normality obtains we find that

$$h(\theta) = \int p(\boldsymbol{x} \mid \theta) \left(-\frac{\partial^2}{\partial\theta^2} \log p(\boldsymbol{x} \mid \theta)\right) d\boldsymbol{x},$$

i.e., *Fisher's information* (Fisher, 1925), and hence the reference prior,

$$\pi(\theta) \propto h(\theta)^{1/2},$$

becomes Jeffreys' (or Perks') prior. See Polson (1992) for a related derivation.

It should be noted however that, even under conditions which guarantee asymptotic normality, Jeffreys' formula is not necessarily the easiest way of deriving a reference prior. As illustrated in Examples 5.12 and 5.13 above, it is often simpler to apply Proposition 5.18 using an asymptotic approximation to the posterior distribution.

It is important to stress that reference distributions are, by definition, a function of the *entire* probability model $p(\boldsymbol{x} \mid \theta)$, $\boldsymbol{x} \in X$, $\boldsymbol{\theta} \in \Theta$, not only of the observed likelihood. Technically, this is a consequence of the fact that the amount of information which an experiment may be *expected* to provide is the value of an integral over the entire sample space X, which, therefore, has to be specified. We have, of course, already encountered in Section 5.1.4 the idea that knowledge of the data generating mechanism may influence the prior specification.

Example 5.14. ***(Binomial and negative binomial models).*** Consider an experiment which consists of the observation of n Bernoulli trials, with n fixed in advance, so that $\boldsymbol{x} = \{x_1, \ldots, x_n\}$,

$$p(x \mid \theta) = \theta^x (1-\theta)^{1-x}, \quad x \in \{0, 1\}, \quad 0 \le \theta \le 1,$$

$$h(\theta) = -\sum_{x=0}^{1} p(x \mid \theta) \frac{\partial^2}{\partial \theta^2} \log p(x \mid \theta) = \theta^{-1}(1-\theta)^{-1},$$

and hence, by Proposition 5.24, the reference prior is

$$\pi(\theta) \propto \theta^{-1/2}(1-\theta)^{-1/2}.$$

If $r = \sum_{i=1}^{n} x_i$, the reference posterior,

$$\pi(\theta \mid \boldsymbol{x}) \propto p(\boldsymbol{x} \mid \theta)\pi(\theta) \propto \theta^{r-1/2}(1-\theta)^{n-r-1/2},$$

is the beta distribution $\text{Be}(\theta \mid r + \frac{1}{2}, n - r + \frac{1}{2})$. Note that $\pi(\theta \mid \boldsymbol{x})$ is proper, whatever the number of successes r. In particular, if $r = 0$, $\pi(\theta \mid \boldsymbol{x}) = \text{Be}(\theta \mid \frac{1}{2}, n + \frac{1}{2})$, from which sensible inference summaries can be made, *even though there are no observed successes.* (Compare this with the Haldane (1948) prior, $\pi(\theta) \propto \theta^{-1}(1-\theta)^{-1}$, which produces an improper posterior until at least one success is observed.)

Consider now, however, an experiment which consists of counting the number x of Bernoulli trials which it is necessary to perform in order to observe a prespecified number of successes, $r \ge 1$. The probability model for this situation is the negative binomial

$$p(x \mid \theta) = \binom{x-1}{r-1} \theta^r (1-\theta)^{x-r}, \quad x = r, r+1, \ldots$$

from which we obtain

$$h(\theta) = -\sum_{x=r}^{\infty} p(x \mid \theta) \frac{\partial^2}{\partial \theta^2} \log p(x \mid \theta) = r\theta^{-2}(1-\theta)^{-1}$$

and hence, by Proposition 5.24, the reference prior is $\pi(\theta) \propto \theta^{-1}(1-\theta)^{-1/2}$. The reference posterior is given by

$$\pi(\theta \mid x) \propto p(x \mid \theta)\pi(\theta) \propto \theta^{r-1}(1-\theta)^{x-r-1/2}, \quad x = r, r+1, \ldots,$$

which is the beta distribution $\text{Be}(\theta \mid r, x - r + \frac{1}{2})$. Again, we note that this distribution is proper, whatever the number of observations x required to obtain r successes. Note that $r = 0$ is *not* possible under this model: the use of an inverse binomial sampling design implicitly assumes that r successes *will* eventually occur *for sure*, which is not true in direct binomial sampling. This difference in the underlying assumption about θ is duly reflected in the slight difference which occurs between the respective reference prior distributions.

See Geisser (1984) and ensuing discussion for further analysis and discussion of this canonical example.

In reporting results, scientists are typically required to specify not only the data but *also* the conditions under which the data were obtained (the *design* of the experiment), so that the data analyst has available the *full* specification of the probability model $p(x \mid \theta)$, $x \in X$, $\theta \in \Theta$. In order to carry out the reference analysis described in this section, such a full specification is clearly required.

We want to stress, however, that the preceding argument is totally compatible with a full personalistic view of probability. A reference prior is nothing but a (limiting) form of rather *specific* beliefs; namely, those which maximise the missing information which a *particular* experiment could possibly be expected to provide. Consequently, different experiments generally define different types of limiting beliefs. To report the corresponding reference posteriors (possibly for a range of possible alternative models) is only part of the general prior-to-posterior mapping which interpersonal or sensitivity considerations would suggest should always be carried out. Reference analysis provides an answer to an important "what if?" question: namely, what can be said about the parameter of interest *if* prior information were minimal *relative* to the maximum information which a well-defined, specific experiment could be expected to provide?

5.4.3 Restricted Reference Distributions

When analysing the inferential implications of the result of an experiment for a quantity of interest, θ, where, for simplicity, we continue to assume that $\theta \in \Theta \subseteq \Re$, it is often interesting, either *per se*, or on a "what if?" basis, to *condition* on some assumed features of the prior distribution $p(\theta)$, thus defining a restricted class, Q, say, of priors which consists of those distributions compatible with such conditioning. The concept of a reference posterior may easily be extended to this situation by maximising the missing information which the experiment may possibly be expected to provide *within* this restricted class of priors.

Repeating the argument which motivated the definition of (unrestricted) reference distributions, we are led to seek the limit of the sequence of posterior distributions, $\pi_k(\theta \mid x)$, which correspond to the sequence of priors, $\pi_k(\theta)$, which are obtained by maximising, *within* Q, the amount of information

$$I\{e(k), p(\theta)\} = \int p(\theta) \log \frac{f_k(\theta)}{p(\theta)}\, d\theta,$$

where

$$f_k(\theta) = \exp\left\{ \int p(\boldsymbol{z}_k \mid \theta) \log p(\theta \mid \boldsymbol{z}_k) d\boldsymbol{z}_k \right\},$$

which could be expected from k independent replications $z = \{\boldsymbol{x}_1, \ldots, \boldsymbol{x}_k\}$ of the single observation experiment.

Definition 5.8. (***Restricted reference distributions***).
Let x be the result of an experiment e *which consists of one observation from $p(\boldsymbol{x} \mid \theta)$, $x \in X$, with $\theta \in \Theta \subseteq \Re$, let Q be a subclass of the class of all prior distributions for θ, let $\boldsymbol{z}_k = \{\boldsymbol{x}_1, \ldots, \boldsymbol{x}_k\}$ be the result of k independent replications of* e *and define*

$$f_k^*(\theta) = \exp\left\{ \int p(\boldsymbol{z}_k \mid \theta) \log p^*(\theta \mid \boldsymbol{z}_k) d\boldsymbol{z}_k \right\},$$

where

$$p^*(\theta \mid \boldsymbol{z}_k) = \frac{\prod_{i=1}^k p(\boldsymbol{x}_i \mid \theta)}{\int \prod_{i=1}^k p(\boldsymbol{x}_i \mid \theta)\, d\theta}$$

Provided it exists, the Q-reference posterior distribution of θ, after x has been observed, is defined to be $\pi^Q(\theta \mid \boldsymbol{x})$, such that

$$E[\delta\{\pi_k^Q(\theta \mid \boldsymbol{x}), \pi^Q(\theta \mid \boldsymbol{x})\}] \to 0, \quad \textit{as} \quad k \to \infty,$$

$$\pi_k^Q(\theta \mid \boldsymbol{x}) \propto p(\boldsymbol{x} \mid \theta) \pi_k^Q(\theta),$$

where δ is the logarithmic divergence specified in Definition 5.7, and $\pi_k^Q(\theta)$ is a prior which minimises, within Q

$$\int p(\theta) \log \frac{p(\theta)}{f_k^*(\theta)} d\theta.$$

A positive function $\pi^Q(\theta)$ in Q such that

$$\pi^Q(\theta \mid \boldsymbol{x}) \propto p(\boldsymbol{x} \mid \theta) \pi^Q(\theta), \quad \textit{for all } \theta \in \Theta,$$

is then called a Q-reference prior for θ relative to the experiment e.

The intuitive content of Definition 5.8 is illuminated by the following result, which essentially establishes that the Q-reference prior is the closest prior in Q to the unrestricted reference prior $\pi(\theta)$, in the sense of minimising its logarithmic divergence from $\pi(\theta)$.

Proposition 5.25. (***The restricted reference prior as an approximation***). *Suppose that an unrestricted reference prior $\pi(\theta)$ relative to a given experiment is proper; then, if it exists, a Q-reference prior $\pi_Q(\theta)$ satisfies*

$$\int \pi^Q(\theta) \log \frac{\pi^Q(\theta)}{\pi(\theta)} \, d\theta = \inf_{p \in \mathcal{Q}} \int p(\theta) \log \frac{p(\theta)}{\pi(\theta)} \, d\theta.$$

Proof. It follows from Proposition 5.18 that $\pi(\theta)$ is proper if and only if

$$\int f_k^*(\theta) \, d\theta = c_k < \infty,$$

in which case,

$$\pi(\theta) = \lim_{k \to \infty} \pi_k(\theta) = \lim_{k \to \infty} c_k^{-1} f_k^*(\theta).$$

Moreover,

$$\begin{aligned}\int p(\theta) \log \frac{f_k^*(\theta)}{p(\theta)} \, d\theta &= - \int p(\theta) \log \frac{c_k^{-1} p(\theta)}{c_k^{-1} f_k^*(\theta)} d\theta \\ &= \log c_k - \int p(\theta) \log \frac{p(\theta)}{\pi_k(\theta)} d\theta,\end{aligned}$$

which is maximised if the integral is minimised. Let $\pi_k^Q(\theta)$ be the prior which minimises the integral within Q. Then, by Definition 5.8,

$$\pi^Q(\theta \mid x) \propto p(x \mid \theta) \lim_{k \to \infty} \pi_k^Q(\theta) = p(x \mid \theta) \pi^Q(\theta),$$

where, by the continuity of the divergence functional, $\pi^Q(\theta)$ is the prior which minimises, within Q,

$$\int p(\theta) \log \left\{ \frac{p(\theta)}{\lim\limits_{k \to \infty} \pi_k(\theta)} \right\} d\theta = \int p(\theta) \log \left\{ \frac{p(\theta)}{\pi(\theta)} \right\} d\theta.$$

◁

If $\pi(\theta)$ is not proper, it is necessary to apply Definition 5.8 directly in order to characterise $\pi^Q(\theta)$. The following result provides an explicit solution for the rather large class of problems where the conditions which define Q may be expressed as a collection of expected value restrictions.

Proposition 5.26. (***Explicit form of restricted reference priors***).
Let e be an experiment which provides information about θ, and, for given $\{(g_i(\cdot), \beta_i),\, i = 1, \ldots, m\}$, let Q be the class of prior distributions $p(\theta)$ of θ which satisfy

$$\int g_i(\theta) p(\theta) d\theta = \beta_i, \quad i = 1, \ldots, m.$$

Let $\pi(\theta)$ be an unrestricted reference prior for θ relative to e; then, a Q-reference prior of θ relative to e, if it exists, is of the form

$$\pi^Q(\theta) \propto \pi(\theta) \exp\left\{\sum_{i=1}^{m} \lambda_i g_i(\theta)\right\},$$

where the λ_i's are constants determined by the conditions which define Q.

Proof. The calculus of variations argument which underlay the derivation of reference priors may be extended to include the additional restrictions imposed by the definition of Q, thus leading us to seek an extremal of the functional

$$\int p(\theta) \log \frac{f_k^*(\theta)}{p(\theta)}\, d\theta + \lambda \left\{ \int p(\theta)\, d\theta - 1 \right\} + \sum_{i=1}^{m} \lambda_i \left\{ \int g_i(\theta)\, p(\theta)\, d\theta - \beta_i \right\},$$

corresponding to the assumption of a k-fold replicate of e. A standard argument now shows that the solution must satisfy

$$\log f_k^*(\theta) - \log p(\theta) + \lambda + \sum_{i=1}^{m} \lambda_i g_i(\theta) \equiv 0$$

and hence that

$$p(\theta) \propto f_k^*(\theta) \exp\left\{\sum_{i=1}^{m} \lambda_i g_i(\theta)\right\}.$$

Taking $k \to \infty$, the result follows from Proposition 5.18. $\triangleleft$

Example 5.15. ***(Location models).*** Let $x = \{x_1, \ldots, x_n\}$ be a random sample from a location model $p(x \mid \theta) = h(x - \theta)$, $x \in \Re$, $\theta \in \Re$, and suppose that the prior mean and variance of θ are restricted to be $E[\theta] = \mu_0$, $V[\theta] = \sigma_0^2$. Under suitable regularity conditions, the asymptotic posterior distribution of θ will be of the form $p^*(\theta \mid x_1, \ldots, x_n) \propto f(\hat{\theta}_n - \theta)$, where $\hat{\theta}_n$ is an asymptotically sufficient, consistent estimator of θ. Thus, by Proposition 5.23,

$$\pi(\theta) \propto p^*(\theta \mid \hat{\theta}_n)\Big|_{\hat{\theta}_n = \theta} \propto f(0),$$

which is constant, so that the unrestricted reference prior will be *uniform*. It now follows from Proposition 5.26 that the restricted reference prior will be

$$\pi^Q(\theta) \propto \exp\left\{\lambda_1\theta + \lambda_2(\theta - \mu_0)^2\right\},$$

with $\int \theta\pi^Q(\theta)\, d\theta = \mu_0$ and $\int (\theta - \mu_0)^2\pi^Q(\theta)\, d\theta = \sigma_0^2$. Thus, the restricted reference prior is the *normal* distribution with the specified mean and variance.

5.4.4 Nuisance Parameters

The development given thus far has assumed that θ was one-dimensional and that interest was centred on θ or on a one-to-one transformation of θ. We shall next consider the case where $\boldsymbol{\theta}$ is two-dimensional and interest centres on reporting inferences for a one-dimensional function, $\phi = \phi(\boldsymbol{\theta})$. Without loss of generality, we may rewrite the vector parameter in the form $\boldsymbol{\theta} = (\phi, \lambda)$, $\phi \in \Phi$, $\lambda \in \Lambda$, where ϕ is the parameter of interest and λ is a nuisance parameter. The problem is to *identify a reference prior for $\boldsymbol{\theta}$, when the decision problem is that of reporting marginal inferences for ϕ*, assuming a logarithmic score (utility) function.

To motivate our approach to this problem, consider $\boldsymbol{z}_k$ to be the result of a k-fold replicate of the experiment which consists in obtaining a single observation, $\boldsymbol{x}$, from $p(\boldsymbol{x} \mid \boldsymbol{\theta}) = p(\boldsymbol{x} \mid \phi, \lambda)$. Recalling that $p(\boldsymbol{\theta})$ can be thought of in terms of the decomposition

$$p(\boldsymbol{\theta}) = p(\phi, \lambda) = p(\phi)p(\lambda \mid \phi),$$

suppose, for the moment, that a *suitable reference form*, $\pi(\lambda \mid \phi)$, for $p(\lambda \mid \phi)$ has been specified and that only $\pi(\phi)$ remains to be identified. Proposition 5.18 then implies that the "marginal reference prior" for ϕ is given by

$$\pi(\phi) \propto \lim_{k \to \infty} \left[f_k^*(\phi)/f_k^*(\phi_0)\right], \quad \phi, \phi_0 \in \Phi,$$

where

$$f_k^*(\phi) = \exp\left\{\int p(\boldsymbol{z}_k \mid \phi) \log p^*(\phi \mid \boldsymbol{z}_k) d\boldsymbol{z}_k\right\},$$

$p^*(\phi \mid \boldsymbol{z}_k)$ is an asymptotic approximation to the marginal posterior for ϕ, and

$$\begin{aligned} p(\boldsymbol{z}_k \mid \phi) &= \int p(\boldsymbol{z}_k \mid \phi, \lambda)\pi(\lambda \mid \phi)\, d\lambda \\ &= \int \prod_{i=1}^{k} p(\boldsymbol{x}_i \mid \phi, \lambda)\pi(\lambda \mid \phi)\, d\lambda. \end{aligned}$$

By conditioning throughout on ϕ, we see from Proposition 5.18 that the "conditional reference prior" for λ given ϕ has the form

$$\pi(\lambda \mid \phi) \propto \lim_{k \to \infty} \left[\frac{f_k^*(\lambda \mid \phi)}{f_k^*(\lambda_0 \mid \phi)} \right], \qquad \lambda, \lambda_0 \in \Lambda, \phi \in \Phi,$$

where

$$f_k^*(\lambda \mid \phi) = \exp \left\{ \int p(\boldsymbol{z}_k \mid \phi, \lambda) \log p^*(\lambda \mid \phi, \boldsymbol{z}_k) d\boldsymbol{z}_k \right\},$$

$p^*(\lambda \mid \phi, \boldsymbol{z}_k)$ is an asymptotic approximation to the conditional posterior for λ given ϕ, and

$$p(\boldsymbol{z}_k \mid \phi, \lambda) = \prod_{i=1}^{k} p(\boldsymbol{x}_i \mid \phi, \lambda).$$

Given actual data $\boldsymbol{x}$, the marginal reference posterior for ϕ, corresponding to the reference prior

$$\pi(\boldsymbol{\theta}) = \pi(\phi, \lambda) = \pi(\phi)\pi(\lambda \mid \phi)$$

derived from the above procedure, would then be

$$\begin{aligned} \pi(\phi \mid \boldsymbol{x}) &\propto \int \pi(\phi, \lambda \mid \boldsymbol{x})\, d\lambda \\ &\propto \pi(\phi) \int p(\boldsymbol{x} \mid \phi, \lambda)\pi(\lambda \mid \phi) d\lambda. \end{aligned}$$

This would appear, then, to provide a straightforward approach to deriving reference analysis procedures in the presence of nuisance parameters. *However, there is a major difficulty.*

In general, as we have already seen, reference priors are typically *not* proper probability densities. This means that the integrated form derived from $\pi(\lambda \mid \phi)$,

$$p(\boldsymbol{z}_k \mid \phi) = \int p(\boldsymbol{z}_k \mid \phi, \lambda)\pi(\lambda \mid \phi)\, d\lambda,$$

which plays a key role in the above derivation of $\pi(\phi)$, will typically not be a proper probability model. The above approach will fail in such cases.

Clearly, a more subtle approach is required to overcome this technical problem. However, before turning to the details of such an approach, we present an example, involving *finite* parameter ranges, where the approach outlined above does produce an interesting solution.

Example 5.16. ***(Induction)*****.** Consider a large, finite dichotomised population, all of whose elements individually may or may not have a specified property. A random sample is taken without replacement from the population, the sample being large in absolute size, but still relatively small compared with the population size. *All* the elements sampled turn out to have the specified property. Many commentators have argued that, in view of the large absolute size of the sample, one should be led to believe quite strongly that all elements of the *population* have the property, irrespective of the fact that the population size is greater still, an argument related to Laplace's rule of succession. (See, for example, Wrinch and Jeffreys, 1921, Jeffreys, 1939/1961, pp. 128–132 and Geisser, 1980a.)

Let us denote the population size by N, the sample size by n, the observed number of elements having the property by x, and the actual number of elements in the population having the property by θ. The probability model for the sampling mechanism is then the hypergeometric, which, for possible values of x, has the form

$$p(x \mid \theta) = \frac{\binom{\theta}{x}\binom{N-\theta}{n-x}}{\binom{N}{n}}.$$

If $p(\theta = r)$, $r = 0, \ldots, N$ defines a prior distribution for θ, the posterior probability that $\theta = N$, having observed $x = n$, is given by

$$p(\theta = N \mid x = n) = \frac{p(x = n \mid \theta = N)p(\theta = N)}{\sum_{r=n}^{N} p(x = n \mid \theta = r)p(\theta = r)}.$$

Suppose we considered θ to be the parameter of interest, and wished to provide a reference analysis. Then, since the set of possible values for θ is finite, Proposition 5.19 implies that

$$p(\theta = r) = \frac{1}{N+1}, \quad r = 0, 1, \ldots, N,$$

is a reference prior. Straightforward calculation then establishes that

$$p(\theta = N \mid x = n) = \frac{n+1}{N+1},$$

which is *not* close to unity when n is large but n/N is small.

However, careful consideration of the problem suggests that it is *not* θ which is the parameter of interest: rather it is the parameter

$$\phi = \begin{cases} 1 & \text{if } \theta = N \\ 0 & \text{if } \theta \neq N. \end{cases}$$

To obtain a representation of θ in the form (ϕ, λ), let us define

$$\lambda = \begin{cases} 1 & \text{if } \theta = N \\ \theta & \text{if } \theta \neq N. \end{cases}$$

By Proposition 5.19, the reference priors $\pi(\phi)$ and $\pi(\lambda \mid \phi)$ are both uniform over the appropriate ranges, and are given by

$$\pi(\phi = 0) = \pi(\phi = 1) = \tfrac{1}{2}\,,$$

$$\pi(\lambda = 1 \mid \phi = 1) = 1, \quad \pi(\lambda = r \mid \phi = 0) = \frac{1}{N}\,, \quad r = 0, 1, \ldots, N-1.$$

These imply a reference prior for θ of the form

$$p(\theta) = \begin{cases} \dfrac{1}{2} & \text{if } \theta = N \\ \dfrac{1}{2N} & \text{if } \theta \neq N \end{cases}$$

and straightforward calculation establishes that

$$p(\theta = N \mid x = n) = \left[1 + \frac{1}{(n+1)}\left(1 - \frac{n}{N}\right)\right]^{-1} \approx \frac{n+1}{n+2}\,,$$

which clearly displays the irrelevance of the sampling fraction and the approach to unity for large n (see Bernardo, 1985b, for further discussion).

We return now to the general problem of defining a reference prior for $\boldsymbol{\theta} = (\phi, \lambda)$, $\phi \in \Phi$, $\lambda \in \Lambda$, where ϕ is the parameter vector of interest and λ is a nuisance parameter. We shall refer to the pair (ϕ, λ) as an *ordered parametrisation* of the model. We recall that the problem arises because in order to obtain the marginal reference prior $\pi(\phi)$ for the first parameter we need to work with the integrated model

$$p(\boldsymbol{z}_k \mid \phi) = \int p(\boldsymbol{z}_k \mid \phi, \lambda)\pi(\lambda \mid \phi)\, d\lambda.$$

However, this will only be a proper model if the conditional prior $\pi(\lambda \mid \phi)$ for the second parameter is a proper probability density and, typically, this will not be the case.

This suggests the following strategy: identify an increasing sequence $\{\Lambda_i\}$ of subsets of Λ, $\bigcup_i \Lambda_i = \Lambda$, which may depend on ϕ, such that, on each Λ_i, the conditional reference prior, $\pi(\lambda \mid \phi)$ restricted to Λ_i can be normalised to give a reference prior, $\pi_i(\lambda \mid \phi)$, which is proper. For each i, a proper integrated model can then be obtained and a marginal reference prior $\pi_i(\phi)$ identified. The required reference prior $\pi(\phi, \lambda)$ is then obtained by taking the limit as $i \to \infty$. The strategy clearly requires a choice of the Λ_i's to be made, but in any specific problem a "natural" sequence usually suggests itself. We formalise this procedure in the next definition.

Definition 5.9. (***Reference distributions given a nuisance parameter***).
Let x be the result of an experiment e *which consists of one observation from the probability model $p(x \mid \phi, \lambda)$, $x \in X$, $(\phi, \lambda) \in \Phi \times \Lambda \subset \Re \times \Re$. The reference posterior, $\pi(\phi \mid x)$, for the parameter of interest ϕ, relative to the experiment* e *and to the increasing sequences of subsets of Λ, $\{\Lambda_i(\phi)\}$, $\phi \in \Phi$, $\bigcup_i \Lambda_i(\phi) = \Lambda$, is defined to be the result of the following procedure:*

(i) *applying Definition 5.7 to the model $p(x \mid \phi, \lambda)$, for fixed ϕ, obtain the conditional reference prior, $\pi(\lambda \mid \phi)$, for Λ;*

(ii) *for each ϕ, normalise $\pi(\lambda \mid \phi)$ within each $\Lambda_i(\phi)$ to obtain a sequence of proper priors, $\pi_i(\lambda \mid \phi)$;*

(iii) *use these to obtain a sequence of integrated models*

$$p_i(x \mid \phi) = \int_{\Lambda_i(\phi)} p(x \mid \phi, \lambda)\pi_i(\lambda \mid \phi)\, d\lambda;$$

(iv) *use those to derive the sequence of reference priors*

$$\pi_i(\phi) = c \lim_{k \to \infty} \frac{f_k^*(\phi)}{f_k^*(\phi_0)},$$

$$f_k^*(\phi) = \exp\left\{ \int p_i(z_k \mid \phi) \log p^*(\phi \mid z_k) dz_k \right\},$$

and, for data x, obtain the corresponding reference posteriors

$$\pi_i(\phi \mid x) \propto \pi_i(\phi) \int_{\Lambda_i(\phi)} p(x \mid \phi, \lambda)\pi_i(\lambda \mid \phi)\, d\lambda;$$

(v) *define $\pi(\phi \mid x)$ such that, for almost all x,*

$$\lim_{i \to \infty} \int \pi_i(\phi \mid x) \log \frac{\pi_i(\phi \mid x)}{\pi(\phi \mid x)} = 0.$$

The reference prior, relative to the ordered parametrisation (ϕ, λ), is any positive function $\pi(\phi, \lambda)$, such that

$$\pi(\phi \mid x) \propto \int p(x \mid \phi, \lambda)\pi(\phi, \lambda)\, d\lambda.$$

This will typically be simply obtained as

$$\pi(\phi, \lambda) = \lim_{i \to \infty} \frac{\pi_i(\phi)\pi_i(\lambda \mid \phi)}{\pi_i(\phi_0)\pi_i(\lambda_0 \mid \phi_0)}.$$

Ghosh and Mukerjee (1992) showed that, in effect, the reference prior thus defined maximises the missing information about the parameter of interest, ϕ, subject to the condition that, given ϕ, the missing information about the nuisance parameter, λ, is maximised.

In a model involving a parameter of interest and a nuisance parameter, the form chosen for the latter is, of course, arbitrary. Thus, $p(\boldsymbol{x} \mid \phi, \lambda)$ can be written alternatively as $p(\boldsymbol{x} \mid \phi, \psi)$, for any $\psi = \psi(\phi, \lambda)$ for which the transformation $(\phi, \lambda) \to (\phi, \psi)$ is one-to-one. Intuitively, we would hope that the reference posterior for ϕ derived according to Definition 5.9 would not depend on the particular form chosen for the nuisance parameters. The following proposition establishes that this is the case.

Proposition 5.27. ***(Invariance with respect to the choice of the nuisance parameter).*** *Let e be an experiment which consists in obtaining one observation from $p(\boldsymbol{x} \mid \phi, \lambda)$, $(\phi, \lambda) \in \Phi \times \Lambda \subset \Re \times \Re$, and let e' be an experiment which consists in obtaining one observation from $p(\boldsymbol{x} \mid \phi, \psi)$, $(\phi, \psi) \in \Phi \times \Psi \subseteq \Re \times \Re$, where $(\phi, \lambda) \to (\phi, \psi)$ is one-to-one transformation, with $\psi = g_\phi(\lambda)$. Then, the reference posteriors for ϕ, relative to $[e, \{\Lambda_i(\phi)\}]$ and $[e', \{\Psi_i(\phi)\}]$, where $\Psi_i(\phi) = g_\phi\{\Lambda_i(\phi)\}$, are identical.*

Proof. By Proposition 5.22, for given ϕ,

$$\pi_\psi(\psi \mid \phi) = \pi_\lambda(g_\phi^{-1}(\psi) \mid \phi) \mid J_{g_\phi^{-1}}(\psi) \mid,$$

where

$$\psi = g_\phi(\lambda), \qquad J_\psi(\phi) = \frac{\partial g_\phi^{-1}(\psi)}{\partial \psi}.$$

Hence, if we define

$$\Psi_i(\phi) = \{\psi;\ \psi = g_\phi(\lambda),\ \lambda \in \Lambda_i(\phi)\}$$

and normalise $\pi_\psi(\psi \mid \phi)$ over $\Psi_i(\phi)$ and $\pi_\lambda(g_\phi^{-1}(\psi) \mid \phi)$ over $\Lambda_i(\phi)$, we see that the normalised forms are consistently related by the appropriate Jacobian element. If we denote these normalised forms, for simplicity, by $\pi_i(\lambda \mid \phi)$, $\pi_i(\psi \mid \phi)$, we see that, for the integrated models used in steps (iii) and (iv) of Definition 5.9,

$$\begin{aligned} p_i(\boldsymbol{x} \mid \phi) &= \int_{\Lambda_i(\phi)} p(\boldsymbol{x} \mid \phi, \lambda) \pi_i(\lambda \mid \phi)\, d\lambda \\ &= \int_{\Psi_i(\phi)} p(\boldsymbol{x} \mid \phi, \psi) \pi_i(\psi \mid \phi)\, d\psi, \end{aligned}$$

and hence that the procedure will lead to identical forms of $\pi(\phi \mid \boldsymbol{x})$. $\triangleleft$

Alternatively, we may wish to consider retaining the same form of nuisance parameter, λ, but redefining the parameter of interest to be a one-to-one function of ϕ. Thus, $p(\boldsymbol{x} \mid \phi, \lambda)$ might be written as $p(\boldsymbol{x} \mid \gamma, \lambda)$, where $\gamma = g(\phi)$ is now the parameter vector of interest. Intuitively, we would hope that the reference posterior for γ would be consistently related to that of ϕ by means of the appropriate Jacobian element. The next proposition establishes that this is the case.

Proposition 5.28. ***(Invariance under one-to-one transformations)*.**
Let e *be an experiment which consists in obtaining one observation from* $p(\boldsymbol{x} \mid \phi, \lambda)$, $\phi \in \Phi$, $\lambda \in \Lambda$, *and let* e' *be an experiment which consists in obtaining one observation from* $p(\boldsymbol{x} \mid \gamma, \lambda)$, $\gamma \in \Gamma, \lambda \in \Lambda$, *where* $\gamma = g(\phi)$. *Then, given data* $\boldsymbol{x}$, *the reference posteriors for* ϕ *and* γ, *relative to* $[e, \{\Lambda_i(\phi)\}]$ *and* $[e', \{\Phi_i(\gamma)\}]$, $\Phi_i(\gamma) = \Lambda_i\{g(\phi)\}$ *are related by:*

(i) $\pi_\gamma(\gamma \mid \boldsymbol{x}) = \pi_\phi(g^{-1}(\gamma) \mid \boldsymbol{x})$, *if* Φ *is discrete;*

(ii) $\pi_\gamma(\gamma \mid \boldsymbol{x}) = \pi_\phi(g^{-1}(\gamma) \mid \boldsymbol{x}) \mid J_{g^{-1}}(\gamma) \mid$, *if* $J_{g^{-1}}(\gamma) = \dfrac{\partial g^{-1}(\gamma)}{\partial \gamma}$ *exists.*

Proof. In all cases, step (i) of Definition 5.9 clearly results in a conditional reference prior $\pi(\lambda \mid \phi) = \pi(\lambda \mid g^{-1}(\gamma))$. For discrete $\Phi, \Lambda, \pi_i(\phi)$ and $\pi_i(\gamma)$ defined by steps (ii)–(iv) of Definition 5.9 are both uniform distributions, by Proposition 5.18, and the result follows straightforwardly. If $J_{g^{-1}}(\gamma)$ exists, $\pi_i(\phi)$ and $\pi_i(\gamma)$ defined by steps (ii)–(iv) of Definition 5.9 are related by the claimed Jacobian element, $\mid J_{g^{-1}}(\gamma) \mid$, by Proposition 5.22, and the result follows immediately. ◁

In Proposition 5.23, we saw that the identification of explicit forms of reference prior can be greatly simplified if the approximate asymptotic posterior distribution is of the form

$$p^*(\theta \mid \boldsymbol{z}_k) = p^*(\theta \mid \hat{\theta}_k),$$

where $\hat{\theta}_k$ is an asymptotically sufficient, consistent estimate of θ. Proposition 5.24 establishes that even greater simplification results when the asymptotic distribution is normal. We shall now extend this to the nuisance parameter case.

Proposition 5.29. ***(Bivariate reference priors under asymptotic normality)*.**
Let e_n *be the experiment which consists of the observation of a random sample* $\boldsymbol{x}_1, \ldots, \boldsymbol{x}_n$ *from* $p(\boldsymbol{x} \mid \phi, \lambda)$, $(\phi, \lambda) \in \Phi \times \Lambda \subseteq \Re \times \Re$, *and let* $\{\Lambda_i(\phi)\}$ *be suitably defined sequences of subsets of* λ, *as required by Definition 5.9. Suppose that the joint asymptotic posterior distribution of* (ϕ, λ), *given a* k*-fold replicate of* e_n, *is multivariate normal with precision matrix* $kn\boldsymbol{H}(\hat{\phi}_{kn}, \hat{\lambda}_{kn})$, *where* $(\hat{\phi}_{kn}, \hat{\lambda}_{kn})$ *is a consistent estimate of* (ϕ, λ) *and suppose that* $\hat{h}_{ij} =$

$h_{ij}(\hat{\phi}_{kn}, \hat{\lambda}_{kn})$, *$i = 1, 2$, $j = 1, 2$, is the partition of $\boldsymbol{H}$ corresponding to ϕ, λ. Then*

$$\pi(\lambda \mid \phi) \propto \{h_{22}(\phi, \lambda)\}^{1/2};$$

$$\pi(\phi, \lambda) = \pi(\lambda \mid \phi) \lim_{i \to \infty} \left\{ \frac{\pi_i(\phi) c_i(\phi)}{\pi_i(\phi_0) c_i(\phi_0)} \right\}, \quad \phi_0 \in \Phi,$$

define a reference prior relative to the ordered parametrisation (ϕ, λ), where

$$\pi_i(\phi) \propto \exp\left\{ \int_{\Lambda_i(\phi)} \pi_i(\lambda \mid \phi) \log\left(\{h_\phi(\phi, \lambda)\}^{1/2} \right) d\lambda \right\},$$

with

$$\pi_i(\lambda \mid \phi) = c_i(\phi) \pi(\lambda \mid \phi) = \frac{\pi(\lambda \mid \phi)}{\int_{\Lambda_i(\phi)} \pi(\lambda \mid \phi)\, d\lambda},$$

and

$$h_\phi = (h_{11} - h_{12} h_{22}^{-1} h_{21}).$$

Proof. Given ϕ, the asymptotic conditional distribution of λ is normal with precision $knh_{22}(\hat{\phi}_{kn}, \hat{\lambda}_{kn})$. The first part of Proposition 5.29 then follows from Proposition 5.24.

Marginally, the asymptotic distribution of ϕ is univariate normal with precision $kn\hat{h}_\phi$, where $h_\phi = (h_{11} - h_{12} h_{22}^{-1} h_{21})$. To derive the form of $\pi_i(\phi)$, we note that if $\boldsymbol{z}_k \in Z$ denotes the result of a k-fold replication of e_n,

$$f_{kn}^*(\phi) = \exp\left\{ \int_Z \pi_i(\boldsymbol{z}_k \mid \phi) \log p^*(\phi \mid \boldsymbol{z}_k) d\boldsymbol{z}_k \right\},$$

where, with $\pi_i(\lambda \mid \phi)$ denoting the normalised version of $\pi(\lambda \mid \phi)$ over $\Lambda_i(\phi)$, the integrand has the form

$$\begin{aligned}
&\int_Z \left[\int_{\Lambda_i(\phi)} p(\boldsymbol{z}_k \mid \phi, \lambda) \pi_i(\lambda \mid \phi)\, d\lambda \right] \log N(\phi \mid \hat{\phi}_{kn}, kn\hat{h}_\phi) d\boldsymbol{z}_k \\
&\qquad = \int_{\Lambda_i(\phi)} \pi_i(\lambda \mid \phi) \left[\int_Z p(\boldsymbol{z}_k \mid \phi, \lambda) \log N(\phi \mid \hat{\phi}_{kn}, kn\hat{h}_\phi) d\boldsymbol{z}_k \right] d\lambda \\
&\qquad \approx \int_{\Lambda_i(\phi)} \pi_i(\lambda \mid \phi) \log \left[\frac{\{h_\phi(\phi, \lambda)\}}{2\pi} \right]^{1/2} d\lambda,
\end{aligned}$$

for large k, so that

$$\pi_i(\phi) = \lim_{k \to \infty} \frac{f_{kn}^*(\phi)}{f_{kn}^*(\phi_0)}$$

has the stated form. Since, for data x, the reference prior $\pi(\phi, \lambda)$ is defined by

$$\begin{aligned}\pi(\phi \mid \boldsymbol{x}) &= \lim_{i\to\infty} \pi_i(\phi \mid \boldsymbol{x}) \propto \lim_{i\to\infty} p_i(\boldsymbol{x} \mid \phi)\pi_i(\phi) \\ &\propto \lim_{i\to\infty} \pi_i(\phi) \int_{\Lambda_i} p(\boldsymbol{x} \mid \phi, \lambda) c_i(\phi) \pi(\lambda \mid \phi) d\lambda \\ &\propto \int p(\boldsymbol{x} \mid \phi, \lambda)\pi(\phi, \lambda) d\lambda,\end{aligned}$$

the result follows. ◁

In many cases, the forms of $\{h_{22}(\phi, \lambda)\}$ and $\{h_\phi(\phi, \lambda)\}$ factorise into products of separate functions of ϕ and λ, and the subsets $\{\Lambda_i\}$ do not depend on ϕ. In such cases, the reference prior takes on a very simple form.

Corollary. *Suppose that, under the conditions of Proposition 5.29, we choose a suitable increasing sequence of subsets $\{\Lambda_i\}$ of Λ, which do not depend on ϕ, and suppose also that*

$$\{h_\phi(\phi, \lambda)\}^{1/2} = f_1(\phi)g_1(\lambda), \quad \{h_{22}(\phi, \lambda)\}^{1/2} = f_2(\phi)g_2(\lambda).$$

Then a reference prior relative to the ordered parametrisation (ϕ, λ) is

$$\pi(\phi, \lambda) \propto f_1(\phi)g_2(\lambda)$$

Proof. By Proposition 5.29, $\pi(\lambda \mid \phi) \propto f_2(\phi)g_2(\lambda)$, and hence

$$\pi_i(\lambda \mid \phi) = a_i g_2(\lambda),$$

where $a_i^{-1} = \int_{\Lambda_i} g_2(\lambda)\, d\lambda$. It then follows that

$$\begin{aligned}\pi_i(\phi) &\propto \exp\left\{\int_{\Lambda_i} a_i g_2(\lambda) \log[f_1(\phi)g_1(\lambda)]\, d\lambda\right\} \\ &\propto b_i f_1(\phi),\end{aligned}$$

where $b_i = \int_{\Lambda_i} a_i g_2(\lambda) \log g_1(\lambda)\, d\lambda$, and the result easily follows. ◁

Example 5.17. ***(Normal mean and standard deviation).*** Let e_n be the experiment which consists in the observation of a random sample $x = \{x_1, \ldots, x_n\}$ from a normal distribution, with both mean, μ, and standard deviation, σ, unknown. We shall first obtain a reference analysis for μ, taking σ to be the nuisance parameter.

Since the distribution belongs to the exponential family, asymptotic normality obtains and the results of Proposition 5.29 can be applied. We therefore first obtain the Fisher (expected) information matrix, whose elements we recall are given by

$$h_{ij}(\mu,\sigma) = \int \mathrm{N}(x\,|\,\mu,\sigma^{-2})\left\{-\frac{\partial^2 \log N(x\,|\,\mu,\sigma^{-2})}{\partial\theta_i\partial\theta_j}\right\}dx,$$

from which it is easily verified that the asymptotic precision matrix as a function of $\boldsymbol{\theta} = (\mu, \sigma)$ is given by

$$\boldsymbol{H}_{\boldsymbol{\theta}}(\mu,\sigma) = \begin{pmatrix} \sigma^{-2} & 0 \\ 0 & 2\sigma^{-2} \end{pmatrix},$$

$$\{h_\mu(\mu,\sigma)\}^{1/2} = \sigma^{-1},$$

$$\{h_{22}(\mu,\sigma)\}^{1/2} = \sqrt{2}\sigma^{-1}.$$

This implies that

$$\pi(\sigma\,|\,\mu) \propto \{h_{22}(\mu,\sigma)\}^{1/2} \propto \sigma^{-1},$$

so that, for example, $\Lambda_i = \{\sigma; e^{-i} \le \sigma \le e^i\}, i = 1, 2, \ldots$, provides a suitable sequence of subsets of $\Lambda = \Re^+$ not depending on μ, over which $\pi(\sigma\,|\,\mu)$ can be normalised and the corollary to Proposition 5.29 can be applied. It follows that

$$\pi(\mu,\sigma) = \pi(\mu)\pi(\sigma\,|\,\mu) \propto 1 \times \sigma^{-1}$$

provides a reference prior relative to the ordered parametrisation (μ, σ). The corresponding reference posterior for μ, given $\boldsymbol{x}$, is

$$\begin{aligned}
\pi(\mu\,|\,\boldsymbol{x}) &\propto \int p(\boldsymbol{x}\,|\,\mu,\sigma)\pi(\mu,\sigma)\,d\sigma \\
&\propto \pi(\mu)\int \prod_{i=1}^n \mathrm{N}(x_i\,|\,\mu,\sigma)\pi(\sigma\,|\,\mu)\,d\sigma \\
&\propto \int \sigma^{-n}\exp\left\{-\frac{n}{2\sigma^2}\left[(\overline{x}-\mu)^2 + s^2\right]\right\}\sigma^{-1}\,d\sigma \\
&\propto \int \lambda^{n/2-1}\exp\left\{-\frac{n\lambda}{2}\left[(\overline{x}-\mu)^2 + s^2\right]\right\}d\lambda \\
&\propto \left[s^2 + (\mu-\overline{x})^2\right]^{-n/2} \\
&= \mathrm{St}(\mu\,|\,\overline{x}, (n-1)s^{-2}, n-1),
\end{aligned}$$

where $ns^2 = \Sigma(x_i - \overline{x})^2$.

If we now reverse the roles of μ and σ, so that the latter is now the parameter of interest and μ is the nuisance parameter, we obtain, writing $\boldsymbol{\phi} = (\sigma, \mu)$

$$\boldsymbol{H}_{\boldsymbol{\phi}}(\sigma,\mu) = \begin{pmatrix} 2\sigma^{-2} & 0 \\ 0 & \sigma^{-2} \end{pmatrix},$$

so that $\{h_\sigma(\sigma,\mu)\}^{1/2} = \sqrt{2}\sigma^{-1}, h_{22}(\sigma,\mu)\}^{1/2} = \sigma^{-1}$ and, by a similar analysis to the above,

$$\pi(\mu \,|\, \sigma) \propto \sigma^{-1}$$

so that, for example, $\Lambda_i = \{\mu; -e^i \leq \mu \leq e^i\}, i = 1, 2, \ldots$ provides a suitable sequence of subsets of $\Lambda = \Re$ not depending on σ, over which $\pi(\mu \,|\, \sigma)$ can be normalised and the corollary to Proposition 5.29 can be applied. It follows that

$$\pi(\mu, \sigma) = \pi(\sigma)\pi(\mu \,|\, \sigma) \propto 1 \times \sigma^{-1}$$

provides a reference prior relative to the ordered parametrisation (σ, μ). The corresponding reference posterior for σ, given $\boldsymbol{x}$, is

$$\begin{aligned}\pi(\sigma \,|\, \boldsymbol{x}) &\propto \int p(\boldsymbol{x} \,|\, \mu, \sigma)\; \pi(\mu, \sigma)\; d\mu \\ &\propto \pi(\sigma) \int \prod_{i=1}^{n} \mathrm{N}(x_i \,|\, \mu, \sigma)\; \pi(\mu \,|\, \sigma)\; d\mu,\end{aligned}$$

the right-hand side of which can be written in the form

$$\sigma^{-n} \exp\left\{-\frac{ns^2}{2\sigma^2}\right\} \int \sigma^{-1} \exp\left\{-\frac{n}{2\sigma^2}(\mu - \overline{x})^2\right\} d\mu.$$

Noting, by comparison with a $N(\mu \,|\, \overline{x}, n\lambda)$ density, that the integral is a constant, and changing the variable to $\lambda = \sigma^{-2}$, implies that

$$\begin{aligned}\pi(\lambda \,|\, \boldsymbol{x}) &\propto \lambda^{(n-1)/2-1} \exp\left\{\tfrac{1}{2}ns^2\lambda\right\} \\ &= \mathrm{Ga}\left(\lambda \,\middle|\, \tfrac{1}{2}(n-1),\; \tfrac{1}{2}ns^2\right),\end{aligned}$$

or, alternatively,

$$\begin{aligned}\pi(\lambda ns^2 \,|\, \boldsymbol{x}) &= \mathrm{Ga}\left(\lambda ns^2 \,\middle|\, \tfrac{1}{2}(n-1),\; \tfrac{1}{2}\right) \\ &= \chi^2(\lambda ns^2 \,|\, n-1).\end{aligned}$$

One feature of the above example is that the reference prior did not, in fact, depend on which of the parameters was taken to be the parameter of interest. In the following example the form does change when the parameter of interest changes.

Example 5.18. ***(Standardised normal mean).*** We consider the same situation as that of Example 5.17, but we now take $\phi = \mu/\sigma$ to be the parameter of interest. If σ is taken as the nuisance parameter (by Proposition 5.27 the choice is irrelevant), $\psi = (\phi, \sigma) = g(\mu, \sigma)$ is clearly a one-to-one transformation, with

$$\boldsymbol{J}_{\boldsymbol{g}^{-1}}(\boldsymbol{\psi}) = \begin{pmatrix} \dfrac{\partial\mu}{\partial\phi} & \dfrac{\partial\mu}{\partial\sigma} \\ \dfrac{\partial\sigma}{\partial\phi} & \dfrac{\partial\sigma}{\partial\sigma} \end{pmatrix} = \begin{pmatrix} \sigma & \phi \\ 0 & 1 \end{pmatrix}$$

and using Corollary 1 to Proposition 5.17.

$$\boldsymbol{H}_{\boldsymbol{\psi}}(\boldsymbol{\psi}) = \boldsymbol{J}^t_{\boldsymbol{g}^{-1}}(\boldsymbol{\psi})\boldsymbol{H}_{\boldsymbol{\theta}}(\boldsymbol{g}^{-1}(\boldsymbol{\psi}))\boldsymbol{J}_{\boldsymbol{g}^{-1}}(\boldsymbol{\psi}) = \begin{pmatrix} 1 & \phi\sigma^{-1} \\ \phi\sigma^{-1} & \sigma^{-2}(2+\phi^2) \end{pmatrix}.$$

Again, the sequence $\Lambda_i = \{\sigma; e^{-i} \leq \sigma \leq e^i\}, i = 1, 2, \ldots$, provides a reasonable basis for applying the corollary to Proposition 5.29. It is easily seen that

$$\begin{aligned} |\, h_\phi(\phi,\sigma)\,|^{1/2} &= \frac{|\, h(\phi,\sigma)\,|^{1/2}}{|\, h_{22}(\phi,\sigma)\,|^{1/2}} \propto (2+\phi^2)^{-1/2}, \\ |\, h_{22}(\phi,\sigma)\,|^{1/2} &\propto (2+\phi^2)^{1/2}\sigma^{-1}, \end{aligned}$$

so that the reference prior relative to the ordered parametrisation (ϕ, σ) is given by

$$\pi(\phi,\sigma) \propto (2+\phi^2)^{-1/2}\sigma^{-1}.$$

In the (μ, σ) parametrisation this corresponds to

$$\pi(\mu,\sigma) \propto \left(2 + \frac{\mu^2}{\sigma^2}\right)^{-1/2} \sigma^{-2},$$

which is clearly different from the form obtained in Example 5.17. Further discussion of this example will be provided in Example 5.26 of Section 5.6.2.

We conclude this subsection by considering a rather more involved example, where a natural choice of the required $\Lambda_i(\phi)$ subsequence *does* depend on ϕ. In this case, we use Proposition 5.29, since its corollary does not apply.

Example 5.19. ***(Product of normal means).*** Consider the case where independent random samples $\boldsymbol{x} = \{x_1, \ldots, x_n\}$ and $\boldsymbol{y} = \{y_1, \ldots, y_m\}$ are to be taken, respectively, from $N(x \mid \alpha, 1)$ and $N(y \mid \beta, 1)$, $\alpha > 0$, $\beta > 0$, so that the complete parametric model is

$$p(\boldsymbol{x}, \boldsymbol{y} \mid \alpha, \beta) = \prod_{i=1}^{n} N(x_i \mid \alpha, 1) \prod_{j=1}^{m} N(y_j \mid \beta, 1),$$

for which, writing $\boldsymbol{\theta} = (\alpha, \beta)$ the Fisher information matrix is easily seen to be

$$\boldsymbol{H}_{\boldsymbol{\theta}}(\boldsymbol{\theta}) = \boldsymbol{H}(\alpha,\beta) = \begin{pmatrix} n & 0 \\ 0 & m \end{pmatrix}.$$

Suppose now that we make the one-to-one transformation $\boldsymbol{\psi} = (\phi, \lambda) = (\alpha\beta, \alpha/\beta) = \boldsymbol{g}(\alpha, \beta) = \boldsymbol{g}(\boldsymbol{\theta})$, so that $\phi = \alpha\beta$ is taken to be the parameter of interest and $\lambda = \alpha/\beta$ is taken to be the nuisance parameter. Such a parameter of interest arises, for example, when inference about the area of a rectangle is required from data consisting of measurements of its sides.

The Jacobian of the inverse transformation is given by

$$J_{g^{-1}}(\psi) = \begin{pmatrix} \frac{\partial \alpha}{\partial \phi} & \frac{\partial \alpha}{\partial \lambda} \\ \frac{\partial \beta}{\partial \phi} & \frac{\partial \beta}{\partial \lambda} \end{pmatrix} = \frac{1}{2} \begin{pmatrix} \left(\frac{\lambda}{\phi}\right)^{1/2} & \left(\frac{\phi}{\lambda}\right)^{1/2} \\ \left(\frac{1}{\phi\lambda}\right)^{1/2} & -\frac{1}{\lambda}\left(\frac{\phi}{\lambda}\right)^{1/2} \end{pmatrix}$$

and hence, using Corollary 1 to Proposition 5.17

$$H_{\psi}(\psi) = J^t_{g^{-1}}(\psi) H_{\theta}(g^{-1}(\psi)) J_{g^{-1}}(\psi) = \frac{nm}{4\lambda^2} \begin{bmatrix} \frac{\lambda}{\phi}\left(\frac{\lambda^2}{m} + \frac{1}{n}\right) & \left(\frac{\lambda^2}{m} - \frac{1}{n}\right) \\ \left(\frac{\lambda^2}{m} - \frac{1}{n}\right) & \phi\left(\frac{\lambda}{m} + \frac{1}{n\lambda}\right) \end{bmatrix},$$

with $|\,H_{\psi}(\psi)\,| = \dfrac{nm}{4\lambda^2}$, so that

$$\pi(\lambda\,|\,\phi) \propto |h_{22}(\phi,\lambda)|^{1/2} \propto \frac{(nm\phi)^{1/2}}{\lambda}\left(\frac{\lambda}{m} + \frac{1}{n\lambda}\right)^{1/2}.$$

The question now arises as to what constitutes a "natural" sequence $\{\lambda_i(\phi)\}$, over which to define the normalised $\pi_i(\lambda\,|\,\phi)$ required by Definition 5.9. A natural increasing sequence of subsets of the original parameter space, $\Re^+ \times \Re^+$, for (α,β) would be the sets

$$S_i = \{(\alpha,\beta);\quad 0 < \alpha < i, 0 < \beta < i\},\quad i = 1,2,\ldots,$$

which transform, in the space of $\lambda \in \Lambda$, into the sequence

$$\Lambda_i(\phi) = \left\{\lambda;\quad \frac{\phi}{i^2} < \lambda < \frac{i^2}{\phi}\right\}.$$

We note that unlike in the previous cases we have considered, this does depend on ϕ.

To complete the analysis, it can be shown, after some manipulation, that, for large i,

$$\pi_i(\lambda\,|\,\phi) = \frac{\sqrt{nm}}{i(\sqrt{m}+\sqrt{n})}\phi^{1/2}\lambda^{-1}\left(\frac{1}{m} + \frac{1}{n\lambda}\right)^{1/2}$$

and

$$\pi_i(\phi) = \frac{\sqrt{nm}}{i\,(\sqrt{m}+\sqrt{n})}\int_{\Lambda_i(\phi)}\left(\frac{\lambda}{m} + \frac{1}{n\lambda}\right)^{1/2}\lambda^{-1}\log\left(\frac{\lambda}{m} + \frac{1}{n\lambda}\right)^{-1/2} d\lambda,$$

which leads to a reference prior relative to the ordered parametrisation (ϕ,λ) given by

$$\pi(\phi,\lambda) \propto \phi^{1/2}\lambda^{-1}\left(\frac{\lambda}{m} + \frac{1}{n\lambda}\right)^{1/2}$$

In the original parametrisation, this corresponds to

$$\pi(\alpha, \beta) \propto (n\alpha^2 + m\beta^2)^{1/2},$$

which depends on the sample sizes through the ratio m/n and reduces, in the case $n = m$, to $\pi(\alpha, \beta) \propto (\alpha^2 + \beta^2)^{1/2}$, a form originally proposed for this problem in an unpublished 1982 Stanford University technical report by Stein, who showed that it provides approximate agreement between Bayesian credible regions and classical confidence intervals for ϕ. For a detailed discussion of this example, and of the consequences of choosing a different sequence $\Lambda_i(\phi)$, see Berger and Bernardo (1989).

We note that the preceding example serves to illustrate the fact that reference priors may depend explicitly on the sample sizes defined by the experiment. There is, of course, nothing paradoxical in this, since the underlying notion of a reference analysis is a "minimally informative" prior *relative* to the actual experiment to be performed.

5.4.5 Multiparameter Problems

The approach to the nuisance parameter case considered above was based on the use of an ordered parametrisation whose first and second components were (ϕ, λ), referred to, respectively, as the *parameter of interest* and the nuisance parameter. The reference prior for the *ordered* parametrisation (ϕ, λ) was then constructed by conditioning to give the form $\pi(\lambda \mid \phi)\pi(\phi)$.

When the model parameter vector $\boldsymbol{\theta}$ has more than two components, this successive conditioning idea can obviously be extended by considering $\boldsymbol{\theta}$ as an ordered parametrisation, $(\theta_1, \ldots, \theta_m)$, say, and generating, by successive conditioning, a reference prior, *relative to this ordered parametrisation*, of the form

$$\pi(\boldsymbol{\theta}) = \pi(\theta_m \mid \theta_1, \ldots, \theta_{m-1}) \cdots \pi(\theta_2 \mid \theta_1)\pi(\theta_1).$$

In order to describe the algorithm for producing this successively conditioned form, in the standard, regular case we shall first need to introduce some notation.

Assuming the parametric model $p(\boldsymbol{x} \mid \boldsymbol{\theta})$, $\boldsymbol{\theta} \in \Theta$, to be such that the Fisher information matrix

$$\boldsymbol{H}(\boldsymbol{\theta}) = -E_{\boldsymbol{x} \mid \boldsymbol{\theta}} \left\{ \frac{\partial^2}{\partial\theta_i \partial\theta_j} \log p(\boldsymbol{x} \mid \boldsymbol{\theta}) \right\}$$

has full rank, we define $\boldsymbol{S}(\theta) = \boldsymbol{H}^{-1}(\theta)$, define the component vectors

$$\boldsymbol{\theta}^{[j]} = (\theta_1, \ldots, \theta_j), \quad \boldsymbol{\theta}_{[j]} = (\theta_{j+1}, \ldots, \theta_m),$$

and denote by $\boldsymbol{S}_j(\boldsymbol{\theta})$ the corresponding upper left $j \times j$ submatrix of $\boldsymbol{S}(\boldsymbol{\theta})$, and by $h_j(\boldsymbol{\theta})$ the lower right element of $\boldsymbol{S}_j^{-1}(\boldsymbol{\theta})$.

Finally, we assume that $\Theta = \Theta_1 \times \cdots \times \Theta_m$, with $\theta_i \in \Theta_i$, and, for $i = 1, 2, \ldots$, we denote by $\{\Theta_i^l\}$, $l = 1, 2, \ldots$, an increasing sequence of compact subsets of Θ_i, and define $\Theta_{[j]}^l = \Theta_{j+1}^l \times \cdots \times \Theta_m^l$.

Proposition 5.30. (***Ordered reference priors under asymptotic normality***). *With the above notation, and under regularity conditions extending those of Proposition 5.29 in an obvious way, the reference prior $\pi(\boldsymbol{\theta})$, relative to the ordered parametrisation $(\theta_1, \ldots, \theta_m)$, is given by*

$$\pi(\boldsymbol{\theta}) = \lim_{l \to \infty} \frac{\pi^l(\boldsymbol{\theta})}{\pi^l(\boldsymbol{\theta}^*)}, \quad \text{for some } \boldsymbol{\theta}^* \in \Theta,$$

where $\pi^l(\boldsymbol{\theta})$ is defined by the following recursion:

(i) For $j = m$, and $\theta_m \in \Theta_m^l$,

$$\pi_m^l\left(\boldsymbol{\theta}_{[m-1]} \mid \boldsymbol{\theta}^{[m-1]}\right) = \pi_m^l(\theta_m \mid \theta_1, \ldots, \theta_{m-1}) = \frac{\{h_m(\boldsymbol{\theta})\}^{1/2}}{\int_{\Theta_m^l} \{h_m(\boldsymbol{\theta})\}^{1/2} \, d\theta_m}.$$

(ii) For $j = m-1, m-2, \ldots, 2$, and $\theta_j \in \Theta_j^l$,

$$\pi_j^l\left(\boldsymbol{\theta}_{[j-1]} \mid \boldsymbol{\theta}^{[j-1]}\right) = \pi_{j+1}^l\left(\boldsymbol{\theta}_{[j]} \mid \boldsymbol{\theta}^{[j]}\right) \frac{\exp\left\{E_j^l\left[\log\{h_j(\boldsymbol{\theta})\}^{1/2}\right]\right\}}{\int_{\Theta_j^l} \exp\left\{E_j^l\left[\log\{h_j(\boldsymbol{\theta})\}^{1/2}\right]\right\} d\theta_j},$$

where

$$E_j^l\left[\log\{h_j(\boldsymbol{\theta})\}^{1/2}\right] = \int_{\Theta_{[j]}^l} \log\{h_j(\boldsymbol{\theta})\}^{1/2} \pi_{j+1}^l\left(\boldsymbol{\theta}_{[j]} \mid \boldsymbol{\theta}^{[j]}\right) d\boldsymbol{\theta}_{[j]}.$$

(iii) For $j = 1$, $\boldsymbol{\theta}_{[0]} = \boldsymbol{\theta}$, with $\boldsymbol{\theta}^{[0]}$ vacuous, and

$$\pi^l(\boldsymbol{\theta}) = \pi_1^l\left(\boldsymbol{\theta}_{[0]} | \boldsymbol{\theta}^{[0]}\right).$$

Proof. This follows closely the development given in Proposition 5.29. For details see Berger and Bernardo (1992a, 1992b, 1992c). ◁

The derivation of the ordered reference prior is greatly simplified if the $\{h_j(\boldsymbol{\theta})\}$ terms in the above depend only on $\theta^{[j]}$: even greater simplification obtains if $\boldsymbol{H}(\boldsymbol{\theta})$ is block diagonal, particularly, if, for $j = 1, \ldots, m$, the jth term can be factored into a product of a function of θ_j and a function not depending on θ_j.

Corollary. *If $h_j(\boldsymbol{\theta})$ depends only on $\boldsymbol{\theta}^{[j]}$, $j = 1, \ldots, m$, then*

$$\pi^l(\boldsymbol{\theta}) = \prod_{j=1}^m \frac{\{h_j(\boldsymbol{\theta})\}^{1/2}}{\int_{\Theta_j^l} \{h_j(\boldsymbol{\theta})\}^{1/2} \, d\theta_j}, \quad \boldsymbol{\theta} \in \Theta^l.$$

If $\boldsymbol{H}(\boldsymbol{\theta})$ is block diagonal (i.e., $\theta_1, \ldots, \theta_m$ are mutually orthogonal), with

$$\boldsymbol{H}(\boldsymbol{\theta}) = \begin{pmatrix} h_{11}(\boldsymbol{\theta}) & 0 & \cdots & 0 \\ 0 & h_{22}(\boldsymbol{\theta}) & \cdots & 0 \\ \cdots & \cdots & \cdots & \cdots \\ 0 & 0 & \cdots & h_{mm}(\boldsymbol{\theta}) \end{pmatrix},$$

then $h_j(\boldsymbol{\theta}) = h_{jj}(\boldsymbol{\theta}), j = 1, \ldots, m$. Furthermore, if, in this latter case,

$$\{h_{jj}(\boldsymbol{\theta})\}^{1/2} = f_j(\theta_j) g_j(\boldsymbol{\theta}),$$

where $g_j(\boldsymbol{\theta})$ does not depend on θ_j, and if the Θ_j^l's do not depend on $\boldsymbol{\theta}$, then

$$\pi(\boldsymbol{\theta}) \propto \prod_{j=1}^{m} f_j(\theta_j).$$

Proof. The results follow from the recursion of Proposition 5.29. ◁

The question obviously arises as to the appropriate ordering to be adopted in any specific problem. At present, no formal theory exists to guide such a choice, but experience with a wide range of examples suggests that—at least for non-hierarchical models (see Section 4.6.5), where the parameters may have special forms of interrelationship—the best procedure is to order the components of θ on the basis of their inferential interest.

Example 5.20. ***(Reference analysis for m normal means).*** Let e_n be an experiment which consists in obtaining $\{\boldsymbol{x}_1, \ldots, \boldsymbol{x}_n\}$, $n \geq 2$, a random sample from the multivariate normal model $N_m(\boldsymbol{x} \mid \boldsymbol{\mu}, \tau \boldsymbol{I}_m)$, $m \geq 1$, for which the Fisher information matrix is easily seen to be

$$\boldsymbol{H}(\boldsymbol{\mu}, \tau) = \begin{pmatrix} \tau \boldsymbol{I}_m & 0 \\ 0 & mn/(2\tau^2) \end{pmatrix}.$$

It follows from Proposition 5.30 that the reference prior relative to the natural parametrisation $(\mu_1, \ldots, \mu_m, \tau)$, is given by

$$\pi(\mu_1, \ldots, \mu_m, \tau) \propto \tau^{-1}.$$

Clearly, in this example the result does not, in fact, depend on the order in which the parametrisation is taken, since the parameters are all mutually orthogonal.

The reference prior $\pi(\mu_1, \ldots, \mu_m, \tau) \propto \tau^{-1}$ or $\pi(\mu_1, \ldots, \mu_m, \sigma) \propto \sigma^{-1}$ if we parametrise in terms of $\sigma = \tau^{-1/2}$, is thus the appropriate reference form if we are interested in any of the individual parameters. The reference posterior for any μ_j is easily shown to be the Student density

$$\pi(\mu_j \mid \boldsymbol{x}_1, \ldots, \boldsymbol{x}_n) = \mathrm{St}\left(\mu_j \mid \overline{x}_j, (n-1)s^{-2}, m(n-1)\right)$$

$$n\overline{x}_j = \sum_{i=1}^{n} x_{ij}, \qquad nms^2 = \sum_{i=1}^{n} \sum_{j=1}^{n} (x_{ij} - \overline{x}_j)^2$$

which agrees with the standard argument according to which one degree of freedom should be lost by each of the unknown means.

Example 5.21. ***(Multinomial model)*****.** Let $x = \{r_1, \ldots, r_m\}$ be an observation from a multinomial distribution (see Section 3.2), so that

$$p(r_1, \ldots, r_m \mid \theta_1, \ldots, \theta_m) = \frac{n!}{r_1! \cdots r_m!(n - \Sigma r_i)!} \theta_1^{r_1} \cdots \theta_m^{r_m} (1 - \Sigma\theta_i)^{n - \Sigma r_i},$$

from which the Fisher information matrix

$$\boldsymbol{H}(\theta_1, \ldots, \theta_m) = \frac{n}{1 - \Sigma\theta_i} \begin{bmatrix} \dfrac{1 + \theta_1 - \Sigma\theta_i}{\theta_1} & 1 & \cdots & 1 \\ 1 & \dfrac{1 + \theta_2 - \Sigma\theta_i}{\theta_2} & \cdots & 1 \\ \cdots & \cdots & \cdots & \cdots \\ 1 & 1 & \cdots & \dfrac{1 + \theta_m - \Sigma\theta_i}{\theta_m} \end{bmatrix}$$

is easily derived, with

$$|\boldsymbol{H}| = n^m \left[\left(1 - \sum_{i=1}^{m} \theta_1\right) \prod_{i=1}^{m} \theta_i \right]^{-1}.$$

In this case, the conditional reference priors derived using Proposition 5.28 turn out to be proper, and there is no need to consider subset sequences $\{\Theta_i^l\}$. In fact, noting that $H^{-1}(\theta_1, \ldots, \theta_m)$ is given by

$$\frac{1}{n} \begin{bmatrix} \theta_1(1 - \theta_1) & -\theta_1\theta_2 & \cdots & -\theta_1\theta_m \\ -\theta_1\theta_2 & \theta_2(1 - \theta_2) & \cdots & -\theta_2\theta_m \\ \cdots & \cdots & \cdots & \cdots \\ -\theta_1\theta_m & -\theta_2\theta_m & \cdots & \theta_m(1 - \theta_m) \end{bmatrix},$$

we see that the conditional asymptotic precisions used in Proposition 5.29 are easily identified, and hence that

$$\pi(\theta_j \mid \theta_1, \ldots, \theta_{j-1}) \propto \left(\frac{1 - \sum_{i=1}^{j-1} \theta_i}{\theta_j} \right)^{1/2} \left(\frac{1}{1 - \sum_{i=1}^{j} \theta_i} \right)^{1/2}, \quad \theta_j \leq 1 - \sum_{i=1}^{j-1} \theta_i.$$

The required reference prior relative to the ordered parametrisation $(\theta_1, \ldots, \theta_m)$, say, is then given by

$$\begin{aligned} \pi(\theta_1, \ldots, \theta_m) &\propto \pi(\theta_1)\pi(\theta_2 \mid \theta_1) \cdots \pi(\theta_m \mid \theta_1, \ldots, \theta_{m-1}) \\ &\propto \theta_1^{-1/2}(1 - \theta_1)^{-1/2}\theta_2^{-1/2}(1 - \theta_1 - \theta_2)^{-1/2} \cdots \theta_m^{-1/2}(1 - \theta_1 - \cdots - \theta_m)^{-1/2}, \end{aligned}$$

and corresponding reference posterior for θ_1 is

$$\pi(\theta_1 \mid r_1, \ldots, r_m) \propto \int p(r_1, \ldots, r_m \mid \theta_1, \ldots, \theta_m)\, \pi(\theta_1, \ldots, \theta_m)\, d\theta_2 \ldots d\theta_m,$$

which is proportional to

$$\int \theta_1^{r_1-1/2} \cdots \theta_m^{r_m-1/2}(1-\Sigma\theta_i)^{n-\Sigma r_i} \\ \times (1-\theta_1)^{-1/2}(1-\theta_1-\theta_2)^{-1/2}\cdots(1-\theta_1-\cdots-\theta_m)^{-1/2} d\theta_2 \cdots d\theta_m.$$

After some algebra, this implies that

$$\pi(\theta_1 \mid r_1, \ldots, r_m) = \text{Be}\left(\theta_1 \mid r_1 + \tfrac{1}{2}, n - r_1 + \tfrac{1}{2}\right),$$

which, as one could expect, coincides with the reference posterior which would have been obtained had we initially collapsed the multinomial analysis to a binomial model and then carried out a reference analysis for the latter. Clearly, by symmetry considerations, the above analysis applies to any θ_i, $i = 1, \ldots, m$, after appropriate changes in labelling and it is independent of the particular order in which the parameters are taken. For a detailed discussion of this example see Berger and Bernardo (1992a). Further comments on ordering of parameters are given in Section 5.6.2.

Example 5.22. ***(Normal correlation coefficient).*** Let $\{x_1, \ldots, x_n\}$ be a random sample from a bivariate normal distribution, $N_2(\boldsymbol{x} \mid \boldsymbol{\mu}, \boldsymbol{\tau})$, where

$$\boldsymbol{\mu} = \begin{pmatrix} \mu_1 \\ \mu_2 \end{pmatrix}, \quad \boldsymbol{\tau}^{-1} = \begin{pmatrix} \sigma_1^2 & \rho\sigma_1\sigma_2 \\ \rho\sigma_1\sigma_2 & \sigma_2^2 \end{pmatrix}.$$

Suppose that the correlation coefficient ρ is the parameter of interest, and consider the ordered parametrisation $\{\rho, \mu_1, \mu_2, \sigma_1, \sigma_2\}$. It is easily seen that

$$\boldsymbol{H}(\rho, \mu_1, \mu_2, \sigma_1, \sigma_2) = (1-\rho^2)^{-1} \begin{bmatrix} \dfrac{1+\rho^2}{1-\rho^2} & 0 & 0 & \dfrac{-\rho}{\sigma_1} & \dfrac{-\rho}{\sigma_2} \\ 0 & \dfrac{1}{\sigma_1^2} & \dfrac{-\rho}{\sigma_1\sigma_2} & 0 & 0 \\ 0 & \dfrac{-\rho}{\sigma_1\sigma_2} & \dfrac{1}{\sigma_2^2} & 0 & 0 \\ \dfrac{-\rho}{\sigma_1} & 0 & 0 & \dfrac{2-\rho^2}{\sigma_1^2} & \dfrac{-\rho^2}{\sigma_1\sigma_2} \\ \dfrac{-\rho}{\sigma_2} & 0 & 0 & \dfrac{-\rho^2}{\sigma_1\sigma_2} & \dfrac{2-\rho^2}{\sigma_2^2} \end{bmatrix},$$

so that

$$\boldsymbol{H}^{-1} = \begin{bmatrix} (1-\rho^2)^2 & 0 & 0 & \dfrac{\sigma_1}{2}\rho(1-\rho^2) & \dfrac{\sigma_2}{2}\rho(1-\rho^2) \\ 0 & \sigma_1^2 & \rho\sigma_1\sigma_2 & 0 & 0 \\ 0 & \rho\sigma_1\sigma_2 & \sigma_2^2 & 0 & 0 \\ \dfrac{\sigma_1}{2}\rho(1-\rho^2) & 0 & 0 & \dfrac{\sigma_1^2}{2} & \rho^2\dfrac{\sigma_1\sigma_2}{2} \\ \dfrac{\sigma_2}{2}\rho(1-\rho^2) & 0 & 0 & \rho^2\dfrac{\sigma_1\sigma_2}{2} & \dfrac{\sigma_2^2}{2} \end{bmatrix}.$$

After some algebra it can be shown that this leads to the reference prior

$$\pi(\rho, \mu_1, \mu_2, \sigma_1, \sigma_2) \propto (1 - \rho^2)^{-1}\sigma_1^{-1}\sigma_2^{-1},$$

whatever ordering of the nuisance parameters $\mu_1, \mu_2, \sigma_1, \sigma_2$ is taken. This agrees with Lindley's (1965, p. 219) analysis. Furthermore, as one could expect from Fisher's (1915) original analysis, the corresponding reference posterior distribution for ρ

$$\pi(\rho \mid \boldsymbol{x}_1, \ldots, \boldsymbol{x}_n) \propto \frac{(1 - \rho^2)^{(n-3)/2}}{(1 - \rho r)^{n-3/2}} F\left(\frac{1}{2}, \frac{1}{2}, n - \frac{1}{2}, \frac{1 + \rho r}{2}\right),$$

(where F is the hypergeometric function), only depends on the data through the sample correlation coefficient r, whose sampling distribution only depends on ρ. For a detailed analysis of this example, see Bayarri (1981); further discussion will be provided in Section 5.6.2.

See, also, Hills (1987), Ye and Berger (1991) and Berger and Bernardo (1992b) for derivations of the reference distributions for a variety of other interesting models.

Infinite discrete parameter spaces

The infinite discrete case presents special problems, due to the non-existence of an asymptotic theory comparable to that of the continuous case. It is, however, often possible to obtain an approximate reference posterior by embedding the discrete parameter space within a continuous one.

Example 5.23. ***(Infinite discrete case).*** In the context of capture-recapture problems, suppose it is of interest to make inferences about an integer $\theta \in \{1, 2, \ldots\}$ on the basis of a random sample $\boldsymbol{z} = \{x_1, \ldots, x_n\}$ from

$$p(x|\theta) = \frac{\theta(\theta + 1)}{(x + \theta)^2}, \quad 0 \leq x \leq 1$$

For several plausible "diffuse looking" prior distributions for θ one finds that the corresponding posterior virtually ignores the data. Intuitively, this has to be interpreted as suggesting that such priors actually contain a large amount of information about θ compared with that provided by the data. A more careful approach to providing a "non-informative" prior is clearly required. One possibility would be to embed the discrete space $\{1, 2, \ldots\}$ in the continuous space $]0, \infty[$ since, for each $\theta > 0$, $p(x|\theta)$ is still a probability density for x. Then, using Proposition 5.24, the appropriate refrence prior is

$$\pi(\theta) \propto h(\theta)^{1/2} \propto (\theta + 1)^{-1}\theta^{-1}$$

and it is easily verified that this prior leads to a posterior in which the data are no longer overwhelmed. If the physical conditions of the problem require the use of discrete θ values, one could always use, for example,

$$p(\theta = 1 \mid \boldsymbol{z}) = \int_0^{3/2} \pi(\theta \mid \boldsymbol{z}) d\theta, \qquad p(\theta = j \mid \boldsymbol{z}) = \int_{j-1/2}^{j+1/2} \pi(\theta|\boldsymbol{z}) d\theta, \quad j > 1$$

as an approximate discrete reference posterior.

Prediction and Hierarchical Models

Two classes of problems that are not covered by the methods so far discussed are hierarchical models and prediction problems. The difficulty with these problems is that there are unknowns (typically the unknowns of interest) that have specified distributions. For instance, if one wants to predict y based on z when (y, z) has density $p(y, z \mid \boldsymbol{\theta})$, the unknown of interest is y, but its distribution is conditionally specified. One needs a reference prior for $\boldsymbol{\theta}$, not y. Likewise, in a hierarchical model with, say, $\mu_1, \mu_2, \ldots, \mu_p$ being $\mathrm{N}(\mu_i \mid \mu_0, \lambda)$, the μ_i's may be the parameters of interest but a prior is only needed for the hyperparameters μ_0 and λ.

The obvious way to approach such problems is to integrate out the variables with conditionally known distributions (y in the predictive problem and the $\{\mu_i\}$ in the hierarchical model), and find the reference prior for the remaining parameters based on this marginal model. The difficulty that arises is how to then identify parameters of interest and nuisance parameters to construct the ordering necessary for applying the reference prior method, the real parameters of interest having been integrated out.

In future work, we propose to deal with this difficulty by defining the parameter of interest in the reduced model to be the conditional mean of the original parameter of interest. Thus, in the prediction problem, $E[y|\boldsymbol{\theta}]$ (which will be either $\boldsymbol{\theta}$ or some transformation thereof) will be the parameter of interest, and in the hierarchical model $E[\mu_i \mid \mu_0, \lambda] = \mu_0$ will be defined to be the parameter of interest. This technique has so far worked well in the examples to which it has been applied, but further study is clearly needed.

5.5 NUMERICAL APPROXIMATIONS

Section 5.3 considered forms of approximation appropriate as the sample size becomes large relative to the amount of information contained in the prior distribution. Section 5.4 considered the problem of approximating a prior specification maximising the expected information to be obtained from the data. In this section, we shall consider numerical techniques for implementing Bayesian methods for arbitrary forms of likelihood and prior specification, and arbitrary sample size.

We note that the technical problem of evaluating quantities required for Bayesian inference summaries typically reduces to the calculation of a ratio of two integrals. Specifically, given a likelihood $p(\boldsymbol{x} \mid \boldsymbol{\theta})$ and a prior density $p(\boldsymbol{\theta})$, the starting point for all subsequent inference summaries is the joint posterior density for $\boldsymbol{\theta}$ given by

$$p(\boldsymbol{\theta} \mid \boldsymbol{x}) = \frac{p(\boldsymbol{x} \mid \boldsymbol{\theta}) p(\boldsymbol{\theta})}{\int p(\boldsymbol{x} \mid \boldsymbol{\theta}) p(\boldsymbol{\theta})\, d\boldsymbol{\theta}} \,.$$

From this, we may be interested in obtaining univariate marginal posterior densities for the components of $\boldsymbol{\theta}$, bivariate joint marginal posterior densities for pairs of components of $\boldsymbol{\theta}$, and so on. Alternatively, we may be interested in marginal posterior densities for functions of components of $\boldsymbol{\theta}$ such as ratios or products.

In all these cases, the technical key to the implementation of the formal solution given by Bayes' theorem, for specified likelihood and prior, is the ability to perform a number of integrations. First, we need to evaluate the denominator in Bayes' theorem in order to obtain the normalising constant of the posterior density; then we need to integrate over complementary components of $\boldsymbol{\theta}$, or transformations of $\boldsymbol{\theta}$, in order to obtain marginal (univariate or bivariate) densities, together with summary moments, highest posterior density intervals and regions, or whatever. Except in certain rather stylised problems (e.g., exponential families together with conjugate priors), the required integrations will not be feasible analytically and, thus, efficient approximation strategies will be required.

In this section, we shall outline five possible numerical approximation strategies, which will be discussed under the subheadings: *Laplace Approximation; Iterative Quadrature; Importance Sampling; Sampling-importance-resampling; Markov Chain Monte Carlo.* An exhaustive account of these and other methods will be given in the second volume in this series, *Bayesian Computation.*

5.5.1 Laplace Approximation

We motivate the approximation by noting that the technical problem of evaluating quantities required for Bayesian inference summaries, is typically that of evaluating an integral of the form

$$E\left[g(\boldsymbol{\theta}) \mid \boldsymbol{x}\right] = \int g(\boldsymbol{\theta}) p(\boldsymbol{\theta} \mid \boldsymbol{x}) d\boldsymbol{\theta},$$

where $p(\boldsymbol{\theta} \mid \boldsymbol{x})$ is derived from a predictive model with an appropriate representation as a mixture of parametric models, and $g(\boldsymbol{\theta})$ is some real-valued function of interest. Often, $g(\boldsymbol{\theta})$ is a first or second moment, and since $p(\boldsymbol{\theta} \mid \boldsymbol{x})$ is given by

$$p(\boldsymbol{\theta} \mid \boldsymbol{x}) = \frac{p(\boldsymbol{x} \mid \boldsymbol{\theta}) p(\boldsymbol{\theta})}{\int p(\boldsymbol{x} \mid \boldsymbol{\theta}) p(\boldsymbol{\theta}) d\boldsymbol{\theta}},$$

we see that $E[g(\boldsymbol{\theta}) \mid \boldsymbol{x})]$ has the form of a ratio of two integrals.

Focusing initially on this situation of a required inference summary for $g(\boldsymbol{\theta})$, and assuming $g(\boldsymbol{\theta})$ almost everywhere positive, we note that the posterior expectation of interest can be written in the form

$$E\left[g(\boldsymbol{\theta}) \mid \boldsymbol{x}\right] = \frac{\int \exp\{-nh^*(\boldsymbol{\theta})\} d\boldsymbol{\theta}}{\int \exp\{-nh(\boldsymbol{\theta})\} d\boldsymbol{\theta}}$$

where, with the vector $\boldsymbol{x} = (x_1, \ldots, x_n)$ of observations fixed, the functions $h(\boldsymbol{\theta})$ and $h^*(\boldsymbol{\theta})$ are defined by

$$-nh(\boldsymbol{\theta}) = \log p(\boldsymbol{\theta}) + \log p(\boldsymbol{x} \mid \boldsymbol{\theta}),$$
$$-nh^*(\boldsymbol{\theta}) = \log g(\boldsymbol{\theta}) + \log p(\boldsymbol{\theta}) + \log p(\boldsymbol{x} \mid \boldsymbol{\theta}).$$

Let us consider first the case of a single unknown parameter, $\boldsymbol{\theta} = \theta \in \Re$, and define $\hat{\theta}, \theta^*$ and $\hat{\sigma}, \sigma^*$ such that

$$-h(\hat{\theta}) = \sup_{\theta} \{-h(\theta)\}, \qquad \hat{\sigma} = \left[h''(\theta)\right]^{-1/2}\Big|_{\theta=\hat{\theta}},$$
$$-h^*(\theta^*) = \sup_{\theta} \{-h^*(\theta)\}, \quad \sigma^* = \left[h^{*\prime\prime}(\theta)\right]^{-1/2}\Big|_{\theta=\theta^*}.$$

Assuming $h(\cdot)$, $h^*(\cdot)$ to be suitably smooth functions, the *Laplace approximations* for the two integrals defining the numerator and denominator of $E[g(\theta) \mid \boldsymbol{x}]$ are given (see, for example, Jeffreys and Jeffreys, 1946) by

$$\sqrt{2\pi}\sigma^* n^{-1/2} \exp\{-nh^*(\theta^*)\},$$

and

$$\sqrt{2\pi}\hat{\sigma} n^{-1/2} \exp\left\{-nh(\hat{\theta})\right\}.$$

Essentially, the approximations consist of retaining quadratic terms in Taylor expansions of $h(\cdot)$ and $h^*(\cdot)$, and are thus equivalent to normal-like approximations to the integrands. In the context we are considering, it then follows immediately that the resulting approximation for $E[g(\theta) \mid \boldsymbol{x}]$ has the form

$$\hat{E}[g(\theta) \mid \boldsymbol{x}] = \left(\frac{\sigma^*}{\hat{\sigma}}\right) \exp\left\{-n\left[h^*(\theta^*) - h(\hat{\theta})\right]\right\},$$

and Tierney and Kadane (1986) have shown that

$$E[g(\theta) \mid \boldsymbol{x}] = \hat{E}[g(\theta) \mid \boldsymbol{x}]\left(1 + O(n^{-2})\right).$$

The Laplace approximation approach, exploiting the fact that Bayesian inference summaries typically involve ratios of integrals, is thus seen to provide a potentially very powerful general approximation technique. See, also, Tierney, Kass and Kadane (1987, 1989a, 1989b), Kass, Tierney and Kadane (1988, 1989a, 1989b, 1991) and Wong and Li (1992) for further underpinning of, and extensions to, this methodology.

Considering now the general case of $\boldsymbol{\theta} \in \Re^k$, the Laplace approximation to the denominator of $E[g(\boldsymbol{\theta}) \mid \boldsymbol{x}]$ is given by

$$\int \exp\{-nh(\boldsymbol{\theta})\} d\boldsymbol{\theta} = (2\pi)^{k/2} \left|n\nabla^2 h(\hat{\boldsymbol{\theta}})\right|^{-1/2} \exp\left\{-nh(\hat{\boldsymbol{\theta}})\right\},$$

where $\hat{\theta}$ is defined by

$$-h(\hat{\boldsymbol{\theta}}) = \sup_{\boldsymbol{\theta}} h(\boldsymbol{\theta})$$

and

$$\left[\nabla^2 h(\hat{\boldsymbol{\theta}})\right]_{ij} = \left.\frac{\partial^2 h(\boldsymbol{\theta})}{\partial\theta_i \partial\theta_j}\right|_{\boldsymbol{\theta}=\hat{\boldsymbol{\theta}}},$$

the Hessian matrix of h evaluated at $\hat{\boldsymbol{\theta}}$, with an exactly analogous expression for the numerator, defined in terms of $h^*(\cdot)$ and $\boldsymbol{\theta}^*$. Writing

$$\hat{\sigma} = \left|n\nabla^2 h(\hat{\boldsymbol{\theta}})\right|^{-1/2}$$

$$\sigma^* = \left|n\nabla^2 h^*(\boldsymbol{\theta}^*)\right|^{-1/2},$$

the Laplace approximation to $E[g(\boldsymbol{\theta})\,|\,\boldsymbol{x}]$ is given by

$$\hat{E}[g(\boldsymbol{\theta})\,|\,\boldsymbol{x}] = \left(\frac{\sigma^*}{\sigma}\right) \exp\{-n[h^*(\boldsymbol{\theta}^*) - h(\hat{\boldsymbol{\theta}})]\},$$

completely analogous to the univariate case.

If $\boldsymbol{\theta} = (\boldsymbol{\phi}, \boldsymbol{\lambda})$ and the required inference summary is the marginal posterior density for $\boldsymbol{\phi}$, application of the Laplace approximation approach corresponds to obtaining $p(\boldsymbol{\phi}\,|\,\boldsymbol{x})$ pointwise by fixing $\boldsymbol{\phi}$ in the numerator and defining $g(\boldsymbol{\lambda}) = 1$. It is easily seen that this leads to

$$\hat{p}(\boldsymbol{\phi}\,|\,\boldsymbol{x}) \propto \int \exp\left\{-nh_{\boldsymbol{\phi}}(\boldsymbol{\lambda})\right\} d\boldsymbol{\lambda}$$

$$\propto \left|\nabla^2 h_{\phi}(\hat{\boldsymbol{\lambda}}_{\phi})\right|^{-1/2} \exp\left\{-nh_{\phi}(\hat{\boldsymbol{\lambda}}_{\phi})\right\},$$

where

$$-nh_{\phi}(\boldsymbol{\lambda}) = \log p(\boldsymbol{\phi}, \boldsymbol{\lambda}) + \log p(\boldsymbol{x}\,|\,\boldsymbol{\phi}, \boldsymbol{\lambda}),$$

considered as a function of $\boldsymbol{\lambda}$ for fixed $\boldsymbol{\phi}$, and

$$-h_{\phi}(\hat{\boldsymbol{\lambda}}_{\phi}) = -\sup_{\boldsymbol{\lambda}} h_{\phi}(\boldsymbol{\lambda}).$$

The form $\hat{p}(\boldsymbol{\phi}\,|\,\boldsymbol{x})$ thus provides (up to proportionality) a pointwise approximation to the ordinates of the marginal posterior density for $\boldsymbol{\phi}$. Considering this form in more detail, we see that, if $p(\boldsymbol{\phi}, \boldsymbol{\lambda})$ is constant,

$$\hat{p}(\boldsymbol{\phi}\,|\,\boldsymbol{x}) \propto \left|-\nabla^2 \log p(\boldsymbol{x}\,|\,\boldsymbol{\phi}, \hat{\boldsymbol{\lambda}}_{\phi})\right| p(\boldsymbol{x}\,|\,\boldsymbol{\phi}, \hat{\boldsymbol{\lambda}}_{\phi}).$$

The form $\nabla^2 \log p(\boldsymbol{x} \mid \boldsymbol{\phi}, \hat{\boldsymbol{\lambda}}_\phi)$ is the Hessian of the log-likelihood function, considered as a function of $\boldsymbol{\lambda}$ for fixed $\boldsymbol{\phi}$, and evaluated at the value $\hat{\boldsymbol{\lambda}}_\phi$ which maximises the log-likelihood over $\boldsymbol{\lambda}$ for fixed $\boldsymbol{\phi}$; the form $p(\boldsymbol{x} \mid \boldsymbol{\phi}, \hat{\boldsymbol{\lambda}}_\phi)$ is usually called the *profile likelihood* for $\boldsymbol{\phi}$, corresponding to the parametric model $p(\boldsymbol{x} \mid \boldsymbol{\phi}, \boldsymbol{\lambda})$. The approximation to the marginal density for $\boldsymbol{\phi}$ given by $\hat{p}(\boldsymbol{\phi} \mid \boldsymbol{x})$ has a form often referred to as the *modified profile likelihood* (see, for example, Cox and Reid, 1987, for a convenient discussion of this terminology). Approximation to Bayesian inference summaries through Laplace approximation is therefore seen to have links with forms of inference summary proposed and derived from a non-Bayesian perspective. For further references, see Appendix B, Section 4.2.

In relation to the above analysis, we note that the Laplace approximation is essentially derived by considering normal approximations to the integrands appearing in the numerator and denominator of the general form $E[g(\boldsymbol{\theta}) \mid \boldsymbol{x}]$. If the forms concerned are not well approximated by second-order Taylor expansions of the exponent terms of the integrands, which may be the case with small or moderate samples, particularly when components of $\boldsymbol{\theta}$ are constrained to ranges other than the real line, we may be able to improve substantially on this direct Laplace approximation approach.

One possible alternative, at least if $\boldsymbol{\theta} = \theta$ is a scalar parameter, is to attempt to approximate the integrands by forms other than normal, perhaps resembling more the actual posterior shapes, such as gammas or betas. Such an approach has been followed in the one-parameter case by Morris (1988), who develops a general approximation technique based around the Pearson family of densities. These are characterised by parameters m, μ_0 and a quadratic function Q, which specify a density for θ of the form

$$q_Q(\theta \mid m, \mu_0) = K_Q(m, \mu_0) \frac{p(\theta)}{Q(\theta)},$$

where

$$p(\theta) = \exp\left\{-m \int \left(\frac{\theta - \mu_0}{Q(\theta)}\right) d\theta\right\},$$

$$K_Q^{-1}(m, \mu_0) = \int \left(\frac{p(\theta)}{Q(\theta)}\right) d\theta,$$

$$Q(\theta) = q_0 + q_1\theta + q_2\theta^2$$

and the range of θ is such that $0 < Q(\theta) < \infty$.

It is shown by Morris (1988) that, for a given choice of quadratic function Q, an analogue to the Laplace-type approximation of an integral of a unimodal function $f(\theta)$ is given by

$$\int f(\theta)\, d\theta = \frac{f(\hat{\theta})}{q_Q(\hat{\theta} \mid \hat{m}, \hat{\theta})},$$

where $\hat{m} = r''(\hat{\theta})Q(\hat{\theta})$ and $\hat{\theta}$ maximises $r(\theta) = \log[f(\theta)Q(\theta)]$. Details of the forms of K^{-1}, Q and p for familiar forms of Pearson densities are given in Morris (1988), where it is also shown that the approximation can often be further simplified to the expression

$$\int f(\theta)d\theta = \frac{f(\hat{\theta})\sqrt{2\pi}}{[-r''(\hat{\theta})]^{1/2}}.$$

A second alternative is to note that the version of the Laplace approximation proposed by Tierney and Kadane (1986) is not invariant to changes in the (arbitrary) parametrisation chosen when specifying the likelihood and prior density functions. It may be, therefore, that by judicious reparametrisation (of the likelihood, together with the appropriate, Jacobian adjusted, prior density) the Laplace approximation can itself be made more accurate, even in contexts where the original parametrisation does not suggest the plausibility of a normal-type approximation to the integrands. We, note, incidentally, that such a strategy is also available in multiparameter contexts, whereas the Pearson family approach does not seem so readily generalisable.

To provide a concrete illustration of these alternative analytic approximation approaches consider the following.

Example 5.24. ***(Approximating the mean of a beta distribution).***
Suppose that a posterior beta distribution, $\text{Be}(\theta \mid r_n - \frac{1}{2}, n - r_n + \frac{1}{2})$, has arisen from a $\text{Bi}(r_n \mid n, \theta)$ likelihood, together with, $\text{Be}(\theta \mid \frac{1}{2}, \frac{1}{2})$ prior (the reference prior, derived in Example 5.14). Writing $r_n = x$, we can, in fact, identify the analytic form of the posterior mean in this case,

$$E[\theta \mid x] = \frac{x + \frac{1}{2}}{n + 1},$$

but we shall ignore this for the moment and examine approximations implied by the techniques discussed above.

First, defining $g(\theta) = \theta$, we see, after some algebra, that the Tierney-Kadane form of the Laplace approximation gives the estimated posterior mean

$$\hat{E}[\theta \mid x] = \frac{(n-1)^{n+1/2}(x + \frac{1}{2})^{x+1}}{n^{n+3/2}(x - \frac{1}{2})^{x}}.$$

If, instead, we reparametrise to $\phi = \sin^{-1}\sqrt{\theta}$, the required integrals are defined in terms of

$$g(\phi) = \sin^2\phi, \quad p(x \mid \phi) \propto (\sin^2\phi)^x(1 - \sin^2\phi)^{n-x}, \quad \pi(\phi) \propto 1,$$

and the Laplace approximation can be shown to be

$$\tilde{E}[\theta \mid x] = \frac{n^{n+1/2}(x+1)^{x+1}}{(n+1)^{n+3/2}x^{x}}.$$

Alternatively, if we work via the Pearson family, with $Q(\theta) = \theta(1-\theta)$ as the "natural" choice for a beta-like posterior, we obtain

$$E^*[\theta \mid x] = \frac{(n+1)^{n+1/2}\,(x+3/2)^{x+1}}{(n+2)^{n+\frac{3}{2}}\,(x+\frac{1}{2})^{x}}\,.$$

By considering the percentage errors of estimation, defined by

$$100 \times \left| \frac{\text{true} - \text{estimated}}{\text{true}} \right|,$$

we can study the performance of the three estimates for various values of n and x. Details are given in Achcar and Smith (1989); here, we simply summarise, in Table 5.1, the results for $n = 5$, $x = 3$, which typify the performance of the estimates for small n.

Table 5.1 *Approximation of* $E[\theta \mid x]$ *from* $\text{Be}(\theta \mid x+\frac{1}{2}, n-x+\frac{1}{2})$ *(percentage errors in parentheses)*

True value	*Laplace approximations*		*Pearson approximation*
	$\hat{E}[\theta \mid x]$	$\tilde{E}[\theta \mid x]$	$E^*[\theta \mid x]$
0.583	0.563	0.580	0.585
	(3.6%)	(0.6%)	(0.3%)

We see from Table 5.1 that the Pearson approximation, which is, in some sense, preselected to be best, does, in fact, outperform the others. However, it is striking that the performance of the Laplace approximation under reparametrisation leads to such a considerable improvement over that based on the original parametrisation, and is a very satisfactory alternative to the "optimal" Pearson form. Further examples are given in Achcar and Smith (1989).

In general, it would appear that, in cases involving a relatively small number of parameters, the Laplace approach, in combination with judicious reparametrisation, can provide excellent approximations to general Bayesian inference summaries, whether in the form of posterior moments or marginal posterior densities. However, in multiparameter contexts there may be numerical problems with the evaluation of local derivatives in cases where analytic forms are unobtainable or too tedious to identify explicitly. In addition, there are awkward complications if the integrands are multimodal. At the time of writing, this area of approximation theory is very much still an active research field and the full potential of this and related methods (see, also, Lindley, 1980b, Leonard *et al.*, 1989) has yet to be clarified.

5.5.2 Iterative Quadrature

It is well known that univariate integrals of the type

$$\int_{-\infty}^{\infty} e^{-t^2} f(t)dt$$

are often well approximated by Gauss-Hermite quadrature rules of the form

$$\sum_{i=1}^{n} w_i f(t_i),$$

where t_i is the ith zero of the Hermite polynomial $H_n(t)$. In particular, if $f(t)$ is a polynomial of degree at most $2n - 1$, then the quadrature rule approximates the integral without error. This implies, for example, that, if $h(t)$ is a suitably well behaved function and

$$g(t) = h(t)\,(2\pi\sigma^2)^{-1/2} \exp\left\{ -\frac{1}{2}\left(\frac{t-\mu}{\sigma}\right)^2 \right\},$$

then

$$\int_{-\infty}^{\infty} g(t)dt \approx \sum_{i=1}^{n} m_i g(z_i),$$

where

$$m_i = w_i \exp(t_i^2)\sqrt{2}\sigma, \quad z_i = \mu + \sqrt{2}\sigma\, t_i$$

(see, for example, Naylor and Smith, 1982).

It follows that Gauss-Hermite rules are likely to prove very efficient for functions which, expressed in informal terms, closely resemble "polynomial × normal" forms. In fact, this is a rather rich class which, even for moderate n (less than 12, say), covers many of the likelihood × prior shapes we typically encounter for parameters defined on $(-\infty, \infty)$. Moreover, the applicability of this approximation is vastly extended by working with suitable transformations of parameters defined on other ranges such as $(0, \infty)$ or (a, b), using, for example, $\log(t)$ or $\log(t - a) - \log(b - t)$, respectively. Of course, to use the above form we must specify μ and σ in the normal component. It turns out that, given reasonable starting values (from any convenient source, prior information, maximum likelihood estimates, etc.), we typically can successfully iterate the quadrature rule, substituting estimates of the posterior mean and variance obtained using previous values of m_i and z_i. Moreover, we note that if the posterior density is well-approximated by the product of a normal and a polynomial of degree at most $2n - 3$, then an n-point Gauss-Hermite rule will prove effective for *simultaneously evaluating the normalising constant and the first and second moments*, using the same (iterated) set of m_i and z_i. In practice, it is efficient to begin with a small grid size ($n = 4$ or $n = 5$) and then to gradually increase the grid size until stable answers are obtained both within and between the last two grid sizes used.

Our discussion so far has been for the one-dimensional case. Clearly, however, the need for an efficient strategy is most acute in higher dimensions. The "obvious" extension of the above ideas is to use a cartesian product rule giving the approximation

$$\int \cdots \int f(t_1, \ldots, t_k) dt_1 \ldots dt_k \approx \sum_{i_k} m_{i_k}^{(k)} g(z_{i_1}^{(k)}, \ldots, z_{i_k}^{(k)}),$$

where the grid points and the weights are found by substituting the appropriate iterated estimates of μ and σ^2 corresponding to the marginal component t_j.

The problem with this "obvious" strategy is that the product form is only efficient if we are able to make an (at least approximate) assumption of posterior independence among the individual components. In this case, the lattice of integration points formed from the product of the two one-dimensional grids will efficiently cover the bulk of the posterior density. However, if high posterior correlations exist, these will lead to many of the lattice points falling in areas of negligible posterior density, thus causing the cartesian product rule to provide poor estimates of the normalising constant and moments.

To overcome this problem, we could first apply individual parameter transformations of the type discussed above and then attempt to transform the resulting parameters, via an appropriate linear transformation, to a new, approximately orthogonal, set of parameters. At the first step, this linear transformation derives from an initial guess or estimate of the posterior covariance matrix (for example, based on the observed information matrix from a maximum likelihood analysis). Successive transformations are then based on the estimated covariance matrix from the previous iteration.

The following general strategy has proved highly effective for problems involving up to six parameters (see, for example, Naylor and Smith, 1982, Smith *et al.*, 1985, 1987, Naylor and Smith, 1988).

(1) Reparametrise individual parameters so that the resulting working parameters all take values on the real line.

(2) Using initial estimates of the joint posterior mean vector and covariance matrix for the working parameters, transform further to a centred, scaled, more "orthogonal" set of parameters.

(3) Using the derived initial location and scale estimates for these "orthogonal" parameters, carry out, on suitably dimensioned grids, cartesian product integration of functions of interest.

(4) Iterate, successively updating the mean and covariance estimates, until stable results are obtained both within and between grids of specified dimension.

For problems involving larger numbers of parameters, say between six and twenty, cartesian product approaches become computationally prohibitive and alternative approaches to numerical integration are required.

One possibility is the use of spherical quadrature rules (Stroud, 1971, Sections 2.6, and 2.7), derived by transforming from cartesian to spherical polar coordinates and constructing optimal integration formulae based on symmetric configurations over concentric spheres. Full details of this approach will be given in the volume *Bayesian Computation*. For a brief introduction, see Smith (1991). Other relevant references on numerical quadrature include Shaw (1988b), Flournoy and Tsutakawa (1991), O'Hagan (1991) and Dellaportas and Wright (1992).

The efficiency of numerical quadrature methods is often very dependent on the particular parametrisation used. For further information on this topic, see Marriott (1988), Hills and Smith (1992, 1993) and Marriott and Smith (1992). For related discussion, see Kass and Slate (1992).

The ideas outlined above relate to the use of numerical quadrature formulae to implement Bayesian statistical methods. It is amusing to note that the roles can be reversed and Bayesian statistical methods used to derive optimal numerical quadrature formulae! See, for example, Diaconis (1988b) and O'Hagan (1992).

5.5.3 Importance Sampling

The importance sampling approach to numerical integration is based on the observation that, if f is a function and g is a probability density function

$$\begin{aligned}\int f(x)dx &= \int \left[\frac{f(x)}{g(x)}\right] g(x)dx \\ &= \int \left[\frac{f(x)}{g(x)}\right] dG(x) \\ &= E_G\left[\frac{f(x)}{g(x)}\right],\end{aligned}$$

which suggest the "statistical" approach of generating a sample from the distribution function G—referred to in this context as the *importance sampling* distribution—and using the average of the values of the ratio f/g as an unbiased estimator of $\int f(x)dx$. However, the variance of such an estimator clearly depends critically on the choice of G, it being desirable to choose g to be "similar" to the shape of f.

In multiparameter Bayesian contexts, exploitation of this idea requires designing importance sampling distributions which are efficient for the kinds of integrands arising in typical Bayesian applications. A considerable amount of work has focused on the use of multivariate normal or Student forms, or modifications thereof, much of this work motivated by econometric applications. We note, in particular, the contributions of Kloek and van Dijk (1978), van Dijk and Kloek (1983, 1985), van Dijk *et al.* (1987) and Geweke (1988, 1989).

An alternative line of development (Shaw, 1988a) proceeds as follows. In the univariate case, if we choose g to be heavier-tailed than f, and if we work with $y = G(x)$, the required integral is the expected value of $f[G^{-1}(x)]/g[G^{-1}(x)]$ with respect to a uniform distribution on the interval $(0, 1)$. Owing to the periodic nature of the ratio function over this interval, we are likely to get a reasonable approximation to the integral by simply taking some equally spaced set of points on $(0, 1)$, rather than actually generating "uniformly distributed" random numbers. If f is a function of more than one argument (k, say), an exactly parallel argument suggess that the choice of a suitable g followed by the use of a suitably selected "uniform" configuration of points in the k-dimensional unit hypercube will provide an efficient multidimensional integration procedure.

However, the effectiveness of all this depends on choosing a suitable G, bearing in mind that we need to have available a flexible set of possible distributional shapes, for which G^{-1} is available explicitly. In the univariate case, one such family defined on $\Re$ is provided by considering the random variable

$$x_a = a\,h(u) - (1-a)\,h(1-u),$$

where u is uniformly distributed on $(0, 1)$, $h : (0, 1) \to \Re$ is a monotone increasing function such that

$$\lim_{u \to 0} h(u) = -\infty$$

and $0 \leq a \leq 1$ is a constant. The choice $a = 0.5$ leads to symmetric distributions; as $a \to 0$ or $a \to 1$ we obtain increasingly skew distributions (to the left or right). The tail-behaviour of the distributions is governed by the choice of the function h. Thus, for example, $h(u) = \log(u)$ leads to a family whose symmetric member is the logistic distribution; $h(u) = -\tan[\pi(1-u)/2]$ leads to a family whose symmetric member is the Cauchy distribution. Moreover, the moments of the distributions of the x_a are polynomials in a (of corresponding order), the median is linear in a, etc., so that sample information about such quantities provides (for any given choice of h) operational guidance on the appropriate choice of a. To use this family in the multiparameter case, we again employ individual parameter transformations, so that all parameters belong to $\Re$, together with "orthogonalising" transformations, so that parameters can be treated "independently". In the transformed setting, it is natural to consider an iterative importance sampling strategy which attempts to learn about an appropriate choice of G for each parameter.

As we remarked earlier, part of this strategy requires the specification of "uniform" configurations of points in the k-dimensional unit hypercube. This problem has been extensively studied by number theorists and systematic experimentation with various suggested forms of "quasi-random" sequences has identified effective forms of configuration for importance sampling purposes: for details, see Shaw (1988a). The general strategy is then the following.

(1) Reparametrise individual parameters so that resulting working parameters all take values on the real line.

(2) Using initial estimates of the posterior mean vector and covariance matrix for the working parameters, transform to a centred, scaled, more "orthogonal" set of parameters.

(3) In terms of these transformed parameters, set

$$g(\boldsymbol{x}) = \prod_{j=1}^{k} g_j(x_j),$$

for "suitable" choices of g_j, $j = 1, \ldots, k$.

(4) Use the inverse distribution function transformation to reduce the problem to that of calculating an average over a "suitable" uniform configuration in the k-dimensional hypercube.

(5) Use information from this "sample" to learn about skewness, tailweight, etc. for each g_j, and hence choose "better" g_j, $j = 1, \ldots, k$, and revise estimates of the mean vector and covariance matrix.

(6) Iterate until the sample variance of replicate estimates of the integral value is sufficiently small.

Teichroew (1965) provides a historical perspective on simulation techniques. For further advocacy and illustration of the use of (non-Markov-chain) Monte Carlo methods in Bayesian Statistics, see Stewart (1979, 1983, 1985, 1987), Stewart and Davis (1986), Shao (1989, 1990) and Wolpert (1991).

5.5.4 Sampling-importance-resampling

Instead of just using importance sampling to estimate integrals—and hence calculate posterior normalising constants and moments—we can also exploit the idea in order to produce simulated samples from posterior or predictive distributions. This technique is referred to by Rubin (1988) as *sampling-importance-resampling* (SIR).

We begin by taking a fresh look at Bayes' theorem from this sampling-importance-resampling perspective, shifting the focus in Bayes' theorem from densities to samples. Our account is based on Smith and Gelfand (1992).

As a first step, we note the essential duality between a sample and the distribution from which it is generated: clearly, the distribution can generate the sample; conversely, given a sample we can re-create, at least approximately, the distribution (as a histogram, an empirical distribution function, a kernel density estimate, or whatever). In terms of densities, Bayes' theorem defines the inference process as the modification of the prior density $p(\boldsymbol{\theta})$ to form the posterior density $p(\boldsymbol{\theta} \mid \boldsymbol{x})$, through the medium of the likelihood function $p(\boldsymbol{x} \mid \boldsymbol{\theta})$. Shifting to a sampling perspective, this corresponds to the modification of a sample from $p(\boldsymbol{\theta})$ to form a sample from $p(\boldsymbol{\theta} \mid \boldsymbol{x})$ through the medium of the likelihood function $p(\boldsymbol{x} \mid \boldsymbol{\theta})$.

To gain insight into the general problem of how a sample from one density may be modified to form a sample from a different density, consider the following. Suppose that a sample of random quantities has been generated from a density $g(\boldsymbol{\theta})$, but that what it is required is a sample from the density

$$h(\boldsymbol{\theta}) = \frac{f(\boldsymbol{\theta})}{\int f(\boldsymbol{\theta})\, d\boldsymbol{\theta}},$$

where only the functional form of $f(\boldsymbol{\theta})$ is specified. Given $f(\boldsymbol{\theta})$ and the sample from $g(\boldsymbol{\theta})$, how can we derive a sample from $h(\boldsymbol{\theta})$?

In cases where there exists an identifiable constant $M > 0$ such that

$$f(\boldsymbol{\theta})/g(\boldsymbol{\theta}) \leq M, \quad \text{for all } \boldsymbol{\theta},$$

an exact sampling procedure follows immediately from the well known rejection method for generating random quantities (see, for example, Ripley, 1987, p. 60):

(i) consider a $\boldsymbol{\theta}$ generated from $g(\boldsymbol{\theta})$;

(ii) generate u from $\text{Un}(u \mid 0, 1)$;

(iii) if $u \leq f(\boldsymbol{\theta})/Mg(\boldsymbol{\theta})$ accept $\boldsymbol{\theta}$; otherwise repeat (i)–(iii).

Any accepted $\boldsymbol{\theta}$ is then a random quantity from $h(\boldsymbol{\theta})$. Given a sample of size N for $g(\boldsymbol{\theta})$, it is immediately verified that the expected sample size from $h(\boldsymbol{\theta})$ is $M^{-1}N \int f(x)dx$.

In cases where the bound M in the above is not readily available, we can approximate samples from $h(\boldsymbol{\theta})$ as follows. Given $\boldsymbol{\theta}_1, \ldots, \boldsymbol{\theta}_N$ from $g(\boldsymbol{\theta})$, calculate

$$q_i = \frac{w_i}{\sum_{i=1}^N w_i}, \quad \text{where} \quad w_i = \frac{f(\boldsymbol{\theta}_i)}{g(\boldsymbol{\theta}_i)}.$$

If we now draw $\boldsymbol{\theta}^*$ from the discrete distribution $\{\boldsymbol{\theta}_1, \ldots, \boldsymbol{\theta}_N\}$ having mass q_i on $\boldsymbol{\theta}_i$, then $\boldsymbol{\theta}^*$ is approximately distributed as a random quantity from $h(\boldsymbol{\theta})$. To see this, consider, for mathematical convenience, the univariate case. Then, under appropriate regularity conditions, if P describes the actual distribution of θ^*,

$$\begin{aligned} P(\theta^* \leq a) &= \sum_{i=1}^{N} q_i \mathbf{1}_{(-\infty,a]}(\theta_i) \\ &= \frac{n^{-1} \sum_{i=1}^N w_i \mathbf{1}_{(-\infty,a]}(\theta_i)}{n^{-1} \sum_{i=1}^N w_i}, \end{aligned}$$

so that

$$\lim_{n\to\infty} P(\theta^* \le a) = \frac{E_g\left\{\dfrac{f(\theta)}{g(\theta)}\mathbf{1}_{(-\infty,a]}(\theta)\right\}}{E_g\left\{\dfrac{f(\theta)}{g(\theta)}\right\}}$$

$$= \frac{\displaystyle\int_{-\infty}^{a} f(\theta)\,d\theta}{\displaystyle\int_{-\infty}^{\infty} f(\theta)\,d\theta} = \int_{-\infty}^{a} h(\theta)\,d\theta\,.$$

Since sampling with replacement is not ruled out, the sample size generated in this case can be as large as desired. Clearly, however, the less $h(\theta)$ resembles $g(\theta)$ the larger N will need to be if the distribution of θ^* is to be a reasonable approximation to $h(\theta)$.

With this sampling-importance-resampling procedure in mind, let us return to the prior to posterior sample process defined by Bayes' theorem. For fixed x, define $f_{\boldsymbol{x}}(\boldsymbol{\theta}) = p(\boldsymbol{x}\,|\,\boldsymbol{\theta})p(\boldsymbol{\theta})$. Then, if $\hat{\boldsymbol{\theta}}$ maximising $p(\boldsymbol{x}\,|\,\boldsymbol{\theta})$ is available, the rejection procedure given above can be applied to a sample for $p(\boldsymbol{\theta})$ to obtain a sample from $p(\boldsymbol{\theta}\,|\,\boldsymbol{x})$ by taking $g(\boldsymbol{\theta}) = p(\boldsymbol{\theta})$, $f(\boldsymbol{\theta}) = f_{\boldsymbol{x}}(\boldsymbol{\theta})$ and $M = p(\boldsymbol{x}\,|\,\hat{\boldsymbol{\theta}})$. Bayes' theorem then takes the simple form:

For each $\boldsymbol{\theta}$ in the prior sample, accept $\boldsymbol{\theta}$ into the posterior sample with probability

$$\frac{f_{\boldsymbol{x}}(\boldsymbol{\theta})}{Mp(\boldsymbol{\theta})} = \frac{p(\boldsymbol{x}\,|\,\boldsymbol{\theta})}{p(\boldsymbol{x}\,|\,\hat{\boldsymbol{\theta}})}\,.$$

The likelihood therefore acts in an intuitive way to define the resampling probability: those $\boldsymbol{\theta}$ with high likelihoods are more likely to be represented in the posterior sample. Alternatively, if M is not readily available, we can use the approximate resampling method, which selects $\boldsymbol{\theta}_i$ into the posterior sample with probability

$$q_i = \frac{p(\boldsymbol{x}\,|\,\boldsymbol{\theta}_i)}{\sum_{j=1}^{N} p(\boldsymbol{x}\,|\,\boldsymbol{\theta}_j)}\,.$$

Again we note that this is proportional to the likelihood, so that the inference process via sampling proceeds in an intuitive way.

The sampling-resampling perspective outlined above opens up the possibility of novel applications of exploratory data analytic and computer graphical techniques in Bayesian statistics. We shall not pursue these ideas further here, since the topic is more properly dealt with in the subsequent volume *Bayesian Computation*. For an illustration of the method in the context of sensitivity analysis and intractable reference analysis, see Stephens and Smith (1992); for pedagogical illustration, see Albert (1993).

5.5.5 Markov Chain Monte Carlo

The key idea is very simple. Suppose that we wish to generate a sample from a posterior distribution $p(\boldsymbol{\theta}|\boldsymbol{x})$ for $\boldsymbol{\theta} \in \Theta \subset \Re^k$ but cannot do this directly. However, suppose that we can construct a Markov chain with state space Θ, which is straightforward to simulate from, and whose equilibrium distribution is $p(\boldsymbol{\theta}|\boldsymbol{x})$. If we then run the chain for a long time, simulated values of the chain can be used as a basis for summarising features of the posterior $p(\boldsymbol{\theta}|\boldsymbol{x})$ of interest. To implement this strategy, we simply need algorithms for constructing chains with specified equilibrium distributions. For recent accounts and discussion, see, for example, Gelfand and Smith (1990), Casella and George (1992), Gelman and Rubin (1992a, 1992b), Geyer (1992), Raftery and Lewis (1992), Ritter and Tanner (1992), Roberts (1992), Tierney (1992), Besag and Green (1993), Chan (1993), Gilks *et al.* (1993) and Smith and Roberts (1993); see, also, Tanner and Wong (1987) and Tanner (1991).

Under suitable regularity conditions, asymptotic results exist which clarify the sense in which the sample output from a chain with equilibrium distribution $p(\boldsymbol{\theta}|\boldsymbol{x})$ can be used to mimic a random sample from $p(\boldsymbol{\theta}|\boldsymbol{x})$ or to estimate the expected value, with respect to $p(\boldsymbol{\theta}|\boldsymbol{x})$, of a function $g(\boldsymbol{\theta})$ of interest.

If $\boldsymbol{\theta}^1, \boldsymbol{\theta}^2, \ldots, \boldsymbol{\theta}^t, \ldots$ is a realisation from an appropriate chain, typically available asymptotic results as $t \to \infty$ include

$$\boldsymbol{\theta}^t \to \boldsymbol{\theta} \sim p(\boldsymbol{\theta}|\boldsymbol{x}), \text{ in distribution}$$

and

$$\frac{1}{t}\sum_{i=1}^{t} g(\boldsymbol{\theta}^i) \to E_{\boldsymbol{\theta}|\boldsymbol{x}}\{g(\boldsymbol{\theta})\} \quad \text{almost surely.}$$

Clearly, successive $\boldsymbol{\theta}^t$ will be correlated, so that, if the first of these asymptotic results is to be exploited to mimic a random sample from $p(\boldsymbol{\theta}|\boldsymbol{x})$, suitable spacings will be required between realisations used to form the sample, or parallel independent runs of the chain might be considered. The second of the asymptotic results implies that *ergodic averaging* of a function of interest over realisations from a single run of the chain provides a consistent estimator of its expectation.

In what follows, we outline two particular forms of Markov chain scheme, which have proved particularly convenient for a range of applications in Bayesian statistics.

The Gibbs Sampling Algorithm

Suppose that $\boldsymbol{\theta}$, the vector of unknown quantities appearing in Bayes' theorem, has components $\theta_1, \ldots, \theta_k$, and that our objective is to obtain summary inferences from the joint posterior $p(\boldsymbol{\theta} \mid \boldsymbol{x}) = p(\theta_1, \ldots, \theta_k|\boldsymbol{x})$. As we have already observed in this section, except in simple, stylised cases, this will typically lead, unavoidably, to challenging problems of numerical integration.

In fact, this apparent need for sophisticated numerical integration technology can often be avoided by recasting the problem as one of iterative sampling of random quantities from appropriate distributions to produce an appropriate Markov chain. To this end, we note that

$$p(\theta_i \,|\, \boldsymbol{x}, \theta_j, j \neq i), \quad i = 1, \ldots, k\,,$$

the so-called *full conditional* densities for the individual components, given the data and specified values of all the other components of $\boldsymbol{\theta}$, are typically easily identified, as functions of θ_i, by inspection of the form of $p(\boldsymbol{\theta} \,|\, \boldsymbol{x}) \propto p(\boldsymbol{x} \,|\, \boldsymbol{\theta})p(\boldsymbol{\theta})$ in any given application. Suppose then, that given an arbitrary set of starting values,

$$\theta_1^{(0)}, \ldots, \theta_k^{(0)}$$

for the unknown quantities, we implement the following iterative procedure:

$$\begin{aligned}
&\text{draw } \theta_1^{(1)} \text{ from } p(\theta_1 \,|\, \boldsymbol{x}, \theta_2^{(0)}, \ldots, \theta_k^{(0)}),\\
&\text{draw } \theta_2^{(1)} \text{ from } p(\theta_2 \,|\, \boldsymbol{x}, \theta_1^{(1)}, \theta_3^{(0)}, \ldots, \theta_k^{(0)}),\\
&\text{draw } \theta_3^{(1)} \text{ from } p(\theta_3 \,|\, \boldsymbol{x}, \theta_1^{(1)}, \theta_2^{(1)}, \theta_4^{(0)}, \ldots, \theta_k^{(0)}),\\
&\quad\vdots\\
&\text{draw } \theta_k^{(1)} \text{ from } p(\theta_k \,|\, \boldsymbol{x}, \theta_1^{(1)}, \ldots, \theta_{k-1}^{(1)}),\\
&\text{draw } \theta_1^{(2)} \text{ from } p(\theta_1 \,|\, \boldsymbol{x}, \theta_2^{(1)}, \ldots, \theta_k^{(1)}),\\
&\quad\vdots
\end{aligned}$$

and so on.

Now suppose that the above procedure is continued through t iterations and is independently replicated m times so that from the current iteration we have m replicates of the sampled vector $\boldsymbol{\theta}^t = (\theta_1^{(t)}, \ldots, \theta_k^{(t)})$, where $\boldsymbol{\theta}^t$ is a realisation of a Markov chain with transition probabilities given by

$$\pi(\boldsymbol{\theta}^t, \boldsymbol{\theta}^{t+1}) = \prod_{l=1}^{k} p(\theta_l^{t+1} \,|\, \theta_j^t, j > l,\ \theta_j^{t+1}, j < l,\ \boldsymbol{x}).$$

Then (see, for example, Geman and Geman, 1984, Roberts and Smith, 1994), as $t \to \infty$, $(\theta_1^{(t)}, \ldots, \theta_k^{(t)})$ tends in distribution to a random vector whose joint density is $p(\boldsymbol{\theta} \,|\, \boldsymbol{x})$. In particular, $\theta_i^{(t)}$ tends in distribution to a random quantity whose density is $p(\theta_i \,|\, \boldsymbol{x})$. Thus, for large t, the replicates $(\theta_{i1}^{(t)}, \ldots, \theta_{im}^{(t)})$ are approximately a random sample from $p(\theta_i \,|\, \boldsymbol{x})$. It follows, by making m suitably large, that an estimate $\hat{p}(\theta_i \,|\, \boldsymbol{x})$ for $p(\theta_i \,|\, \boldsymbol{x})$ is easily obtained, either as a kernel density estimate derived from $(\theta_{i1}^{(t)}, \ldots, \theta_{im}^{(t)})$, or from

$$\hat{p}(\theta_i \,|\, \boldsymbol{x}) = \frac{1}{m} \sum_{l=1}^{m} p(\theta_i \,|\, \boldsymbol{x},\ \theta_{jl}^{(t)}, j \neq i).$$

So far as sampling from the $p(\theta_i \,|\, x,\ \theta_{jl}^{(t)}, j \neq i)$ is concerned, $i = 1, \ldots, k$, either the full conditionals assume familiar forms, in which case computer routines are typically already available, or they are simple arbitrary mathematical forms, in which case general stochastic simulation techniques are available—such as envelope rejection and ratio of uniforms—which can be adapted to the specific forms (see, for example, Devroye, 1986, Ripley, 1987, Wakefield *et al.,* 1991, Gilks, 1992, Gilks and Wild, 1992, and Dellaportas and Smith, 1993). See, also, Carlin and Gelfand (1991).

The potential of this iterative scheme for routine implementation of Bayesian analysis has been demonstrated in detail for a wide variety of problems: see, for example, Gelfand and Smith (1990), Gelfand *et al.* (1990) and Gilks *et al.* (1993). We shall not provide a more extensive discussion here, since illustration of the technique in complex situations more properly belongs to the second volume of this work. We note, however, that simulation approaches are ideally suited to providing *summary inferences* (we simply report an appropriate summary of the sample), *inferences for arbitrary functions* of $\theta_1, \ldots, \theta_k$ (we simply form a sample of the appropriate function from the samples of the θ_i's) or *predictions* (for example, in an obvious notation, $p(\boldsymbol{y} \,|\, \boldsymbol{x}) = m^{-1} \sum_{i=1}^{m} p(\boldsymbol{y} \,|\, \boldsymbol{\theta}_i^{(t)})$, the average being over the $\boldsymbol{\theta}_i^{(t)}$, which have an approximate $p(\boldsymbol{\theta} \,|\, \boldsymbol{x})$ distribution for large t).

The Metropolis-Hastings algorithm

This algorithm constructs a Markov chain $\boldsymbol{\theta}^1, \boldsymbol{\theta}^2, \ldots, \boldsymbol{\theta}^t, \ldots$ with state space Θ and equilibrium distribution $p(\boldsymbol{\theta}|\boldsymbol{x})$ by defining the transition probability from $\boldsymbol{\theta}^t = \boldsymbol{\theta}$ to the next realised state $\boldsymbol{\theta}^{t+1}$ as follows.

Let $q(\boldsymbol{\theta}, \boldsymbol{\theta}')$ denote a (for the moment arbitrary) transition probability function, such that, if $\boldsymbol{\theta}^t = \boldsymbol{\theta}$, the vector $\boldsymbol{\theta}'$ drawn from $q(\boldsymbol{\theta}, \boldsymbol{\theta}')$ is considered as a proposed possible value for $\boldsymbol{\theta}^{t+1}$. However, a further randomisation now takes place. With some probability $\alpha(\boldsymbol{\theta}, \boldsymbol{\theta}')$, we actually accept $\boldsymbol{\theta}^{t+1} = \boldsymbol{\theta}'$; otherwise, we reject the value generated from $q(\boldsymbol{\theta}, \boldsymbol{\theta}')$ and set $\boldsymbol{\theta}^{t+1} = \boldsymbol{\theta}$. This construction defines a Markov chain with transition probabilities given by

$$\begin{aligned} p(\boldsymbol{\theta}, \boldsymbol{\theta}') &= q(\boldsymbol{\theta}, \boldsymbol{\theta}')\, \alpha(\boldsymbol{\theta}, \boldsymbol{\theta}') \\ &\quad + I(\boldsymbol{\theta} = \boldsymbol{\theta}') \left[1 - \int q(\boldsymbol{\theta}, \boldsymbol{\theta}'')\, \alpha(\boldsymbol{\theta}, \boldsymbol{\theta}'')\, d\boldsymbol{\theta}'' \right], \end{aligned}$$

where $I(.)$ is the indicator function. If now we set

$$\alpha(\boldsymbol{\theta}, \boldsymbol{\theta}') = \min\left\{1, \quad \frac{p(\boldsymbol{\theta}' \mid \boldsymbol{x})q(\boldsymbol{\theta}', \boldsymbol{\theta})}{p(\boldsymbol{\theta} \mid \boldsymbol{x})q(\boldsymbol{\theta}, \boldsymbol{\theta}')}\right\}$$

it is easy to check that $p(\boldsymbol{\theta}|\boldsymbol{x})p(\boldsymbol{\theta}, \boldsymbol{\theta}') = p(\boldsymbol{\theta}|\boldsymbol{x})p(\boldsymbol{\theta}', \boldsymbol{\theta})$, which, provided that the thus far arbitrary $q(\boldsymbol{\theta}, \boldsymbol{\theta}')$ is chosen to be irreducible and aperiodic on a suitable state space, is a sufficient condition for $p(\boldsymbol{\theta}|\boldsymbol{x})$ to be the equilibrium distribution of the constructed chain.

This general algorithm is due to Hastings (1970); see, also, Metropolis *et al.* (1953), Peskun (1973), Tierney (1992), Besag and Green (1993), Roberts and Smith (1994) and Smith and Roberts (1993). It is important to note that the (equilibrium) distribution of interest, $p(\boldsymbol{\theta}|\boldsymbol{x})$, only enters $p(\boldsymbol{\theta}, \boldsymbol{\theta}')$ through the ratio $p(\boldsymbol{\theta}'|\boldsymbol{x})/p(\boldsymbol{\theta}|\boldsymbol{x})$. This is quite crucial since it means that knowledge of the distribution up to proportionality (given by the likelihood multiplied by the prior) is sufficient for implementation.

5.6 DISCUSSION AND FURTHER REFERENCES

5.6.1 An Historical Footnote

Blackwell (1988) gave a very elegant demonstration of the way in which a simple finite additivity argument can be used to give powerful insight into the relation between frequency and belief probability. The calculation involved has added interest in that—according to Stigler (1982)—it might very well have been made by Bayes himself.

The argument goes as follows. Suppose that 0–1 observables $x_1, \ldots, x_{n+1}$ are finitely exchangeable. We observe $\boldsymbol{x} = (x_1, \ldots, x_n)$ and wish to evaluate

$$\frac{P(x_{n+1} = 1|\boldsymbol{x})}{P(x_{n+1} = 0|\boldsymbol{x})}.$$

Writing $s = x_1 + \cdots + x_n$, $p(t) = P(x_1 + \cdots + x_{n+1} = t)$, this ratio, by virtue of exchangeability, is easily seen to be equal to

$$\frac{p(s+1)\Big/\binom{n+1}{s+1}}{p(s)\Big/\binom{n+1}{s}} = \frac{p(s+1)}{p(s)} \cdot \frac{s+1}{n-s+1} \approx \frac{s}{n-s},$$

if $p(s) \simeq p(s+1)$ and s and $n-s$ are not too small.

This can be interpreted as follows. If, before observing x, we considered s and $s+1$ to be about equally plausible as values for $x_1 + \cdots + x_{n+1}$, the resulting posterior odds for $x_{n+1} = 1$ will be essentially the frequency odds based on the first n trials.

Inverting the argument, we see that if one wants to have this "convergence" of beliefs and frequencies it is necessary that $p(s) \approx p(s+1)$. But what does this entail?

Reverting to an infinite exchangeability assumption, and hence the familiar binomial framework, suppose we require that $p(\theta)$ be chosen such that

$$p(s) = \int_0^1 \binom{n}{s} \theta^s (1-\theta)^{n-s} p(\theta) d\theta$$

does not depend on s. An easy calculation shows that this is satisfied if $p(\theta)$ is taken to be uniform on $(0,1)$—the so-called *Bayes (or Bayes-Laplace) Postulate.*

Stigler (1982) has argued that an argument like the above could have been Bayes' motivation for the adoption of this uniform prior.

5.6.2 Prior Ignorance

To many attracted to the formalism of the Bayesian inferential paradigm, the idea of a *non-informative* prior distribution, representing "ignorance" and "letting the data speak for themselves" has proved extremely seductive, often being regarded as synonymous with providing *objective* inferences. It will be clear from the general subjective perspective we have maintained throughout this volume, that we regard this search for "objectivity" to be misguided. However, it will also be clear from our detailed development in Section 5.4 that we recognise the rather special nature and role of the concept of a "minimally informative" prior specification —appropriately defined! In any case, the considerable body of conceptual and theoretical literature devoted to identifying "appropriate" procedures for formulating prior representations of "ignorance" constitutes a fascinating chapter in the history of Bayesian Statistics. In this section we shall provide an overview of some of the main directions followed in this search for a Bayesian "Holy Grail".

In the early works by Bayes (1763) and Laplace (1814/1952), the definition of a non-informative prior is based on what has now become known as the principle of *insufficient reason*, or the Bayes-Laplace postulate (see Section 5.6.1). According to this principle, in the absence of evidence to the contrary, all possibilities should have the same initial probability. This is closely related to the so-called Laplace-Bertrand paradox; see Jaynes (1971) for an interesting Bayesian resolution.

In particular, if an unknown quantity, ϕ, say, can only take a finite number of values, M, say, the non-informative prior suggested by the principle is the discrete uniform distribution $p(\phi) = \{1/M, \ldots, 1/M\}$. This may, at first sight, seem intuitively reasonable, but Example 5.16 showed that even in simple, finite, discrete cases care can be required in appropriately defining the unknown *quantity of interest*. Moreover, in countably infinite, discrete cases the uniform (now *improper*) prior is known to produce unappealing results. Jeffreys (1939/1961, p. 238) suggested, for the case of the integers, the prior

$$\pi(n) \propto n^{-1}, \quad n = 1, 2, \ldots.$$

More recently, Rissanen (1983) used a coding theory argument to motivate the prior

$$\pi(n) \propto \frac{1}{n} \times \frac{1}{\log n} \times \frac{1}{\log \log n} \times \ldots, \qquad n = 1, 2, \ldots.$$

However, as indicated in Example 5.23, embedding the discrete problem within a continuous framework and subsequently discretising the resulting reference prior for the continuous case may produce better results.

If the space, Φ, of ϕ values is a continuum (say, the real line) the principle of insufficient reason has been interpreted as requiring a uniform distribution over Φ. However, a uniform distribution for ϕ implies a non-uniform distribution for any non-linear monotone transformation of ϕ and thus the Bayes-Laplace postulate is inconsistent in the sense that, intuitively, "ignorance about ϕ" should surely imply "equal ignorance" about a one-to-one transformation of ϕ. Specifically, if some procedure yields $p(\phi)$ as a non-informative prior for ϕ and the same procedure yields $p(\zeta)$ as a non-informative prior for a one-to-one transformation $\zeta = \zeta(\phi)$ of ϕ, consistency would seem to demand that $p(\zeta)d\zeta = p(\phi)d\phi$; thus, a procedure for obtaining the "ignorance" prior should presumably be invariant under one-to-one reparametrisation.

Based on these invariance considerations, Jeffreys (1946) proposed as a non-informative prior, with respect to an experiment $e = \{X, \phi, p(x \mid \phi)\}$, involving a parametric model which depends on a single parameter ϕ, the (often improper) density

$$\pi(\phi) \propto h(\phi)^{1/2},$$

where

$$h(\phi) = -\int_X p(x \mid \phi) \frac{\partial^2}{\partial \phi^2} \log p(x \mid \phi)\, dx\ .$$

In effect, Jeffreys noted that the logarithmic divergence locally behaves like the square of a distance, determined by a Riemannian metric, whose natural length element is $h(\phi)^{1/2}$, and that natural length elements of Riemannian metrics are invariant to reparametrisation. In an illuminating paper, Kass (1989) elaborated on this *geometrical* interpretation by arguing that, more generally, natural volume elements generate "uniform" measures on manifolds, in the sense that equal mass is assigned to regions of equal volume, the essential property that makes Lebesgue measure intuitively appealing.

In his work, Jeffreys explored the implications of such a non-informative prior for a large number of inference problems. He found that his *rule* (by definition restricted to a continuous parameter) works well in the one-dimensional case, but can lead to unappealing results (Jeffreys, 1939/1961, p. 182) when one tries to extend it to multiparameter situations.

The procedure proposed by Jeffreys' preferred rule was rather *ad hoc*, in that there are many other procedures (some of which he described) which exhibit the required type of invariance. His intuition as to what is required, however, was rather good. Jeffreys' solution for the one-dimensional continuous case has been widely adopted, and a number of alternative justifications of the procedure have been provided.

Perks (1947) used an argument based on the asymptotic size of confidence regions to propose a non-informative prior of the form

$$\pi(\phi) \propto s(\phi)^{-1}$$

where $s(\phi)$ is the asymptotic standard deviation of the maximum likelihood estimate of ϕ. Under regularity conditions which imply asymptotic normality, this turns out to be equivalent to Jeffreys' rule.

Lindley (1961b) argued that, in practice, one can always replace a continuous range of ϕ by discrete values over a grid whose mesh size, $\delta(\phi)$, say, describes the precision of the measuring process, and that a possible operational interpretation of "ignorance" is a probability distribution which assigns equal probability to all points of this grid. In the continuous case, this implies a prior proportional to $\delta(\phi)^{-1}$. To determine $\delta(\phi)$ in the context of an experiment $e = \{X, \phi, p(x \mid \phi)\}$, Lindley showed that if the quantity can only take the values ϕ or $\phi + \delta(\phi)$, the amount of information that e may be expected to provide about ϕ, if $p(\phi) = p(\phi + \delta(\phi)) = \frac{1}{2}$, is $2\delta^2(\phi)h(\phi)$. This expected information will be independent of ϕ if $\delta(\phi) \propto h(\phi)^{-1/2}$, thus defining an appropriate mesh; arguing as before, this suggests Jeffreys' prior $\pi(\phi) \propto h(\theta)^{1/2}$. Akaike (1978a) used a related argument to justify Jeffreys' prior as "locally impartial".

Welch and Peers (1963) and Welch (1965) discussed conditions under which there is formal mathematical equivalence between one-dimensional Bayesian credible regions and corresponding frequentist confidence intervals. They showed that, under suitable regularity assumptions, one-sided intervals asymptotically coincide if the prior used for the Bayesian analysis is Jeffreys' prior. Peers (1965) later showed that the argument does not extend to several dimensions. Hartigan (1966b) and Peers (1968) discuss two-sided intervals. Tibshirani (1989), Mukerjee and Dey (1993) and Nicolau (1993) extend the analysis to the case where there are nuisance parameters.

Hartigan (1965) reported that the prior density which minimises the bias of the estimator d of ϕ associated with the loss function $l(d, \phi)$ is

$$\pi(\phi) = h(\phi) \left[\frac{\partial^2}{\partial d^2} l(d, \phi) \right]^{-1/2} \Bigg|_{d=\phi} .$$

If, in particular, one uses the discrepancy measure

$$l(d, \phi) = \int p(x \mid \phi) \log \frac{p(x \mid \phi)}{p(x \mid d)} \, dx$$

as a natural loss function (see Definition 3.15), this implies that $\pi(\phi) = h(\phi)^{1/2}$, which is, again, Jeffreys' prior.

Good (1969) derived Jeffreys' prior as the "least favourable" initial distribution with respect to a logarithmic scoring rule, in the sense that it minimises the expected score from reporting the true distribution. Since the logarithmic score is proper, and hence is maximised by reporting the true distribution, Jeffreys' prior may technically be described, under suitable regularity conditions, as a minimax solution to the problem of scientific reporting when the utility function is the logarithmic score function. Kashyap (1971) provided a similar, more detailed argument; an axiom system is used to justify the use of an information measure as a payoff function and Jeffreys' prior is shown to be a minimax solution in a —two person— zero sum game, where the statistician chooses the "non-informative" prior and nature chooses the "true" prior.

Hartigan (1971, 1983, Chapter 5) defines a similarity measure for events E, F to be $P(E \cap F)/P(E)P(F)$ and shows that Jeffreys' prior ensures, asymptotically, constant similarily for current and future observations.

Following Jeffreys (1955), Box and Tiao (1973, Section 1.3) argued for selecting a prior by convention to be used as a *standard of reference*. They suggested that the principle of insufficient reason may be sensible in location problems, and proposed as a conventional prior $\pi(\phi)$ for a model parameter ϕ that $\pi(\phi)$ which implies a uniform prior

$$\pi(\zeta) = \pi(\phi) \left| \frac{\partial \zeta}{d\phi} \right|^{-1} \propto c$$

for a function $\zeta = \zeta(\phi)$ such that $p(x \mid \zeta)$ is, at least approximately, a location parameter family; that is, such that, for some functions g and f,

$$p(x \mid \phi) \sim g\left[\zeta(\phi) - f(x)\right] .$$

Using standard asymptotic theory, they showed that, under suitable regularity conditions and for large samples, this will happen if

$$\zeta(\phi) = \int^{\phi} h(\phi)^{1/2} d\phi ,$$

i.e., if the non-informative prior is Jeffreys' prior. For a recent reconsideration and elaboration of these ideas, see Kass (1990), who extends the analysis by conditioning on an ancillary statistic.

Unfortunately, although many of the arguments summarised above generalise to the multiparameter continuous case, leading to the so-called multivariate Jeffreys' rule

$$\pi(\boldsymbol{\theta}) \propto |\, \boldsymbol{H}(\boldsymbol{\theta}) \,|^{1/2},$$

where

$$[\boldsymbol{H}(\boldsymbol{\theta})]_{ij} = -\int p(x \mid \boldsymbol{\theta}) \frac{\partial^2}{\partial\theta_i \partial\theta_j} \log p(x \mid \boldsymbol{\theta})\, dx$$

is Fisher's *information matrix*, the results thus obtained typically have intuitively unappealing implications. An example of this, pointed out by Jeffreys himself (Jeffreys, 1939/1961 p. 182) is provided by the simple location-scale problem, where the multivariate rule leads to $\pi(\theta, \sigma) \propto \sigma^{-2}$, where θ is the location and σ the scale parameter. See, also, Stein (1962).

Example 5.25. ***(Univariate normal model).*** Let $\{x_1, \ldots, x_n\}$ be a random sample from $N(x \mid \mu, \lambda)$, and consider $\sigma = \lambda^{-1/2}$, the (unknown) standard deviation. In the case of known mean, $\mu = 0$, say, the appropriate (univariate) Jeffreys' prior is $\pi(\sigma) \propto \sigma^{-1}$ and the posterior distribution of σ would be such that $[\Sigma_{i=1}^n x_i^2]/\sigma^2$ is χ_n^2. In the case of unknown mean, if we used the multivariate Jeffreys' prior $\pi(\mu, \sigma) \propto \sigma^{-2}$ the posterior distribution of σ would be such that $[\Sigma_{i=1}^n (x_i - \overline{x})^2]/\sigma^2$ is, again, χ_n^2. This is widely recognised as unacceptable, in that one does not lose any degrees of freedom even though one has lost the knowledge that $\mu = 0$, and conflicts with the use of the widely adopted reference prior $\pi(\mu, \sigma) = \sigma^{-1}$ (see Example 5.17 in Section 5.4), which implies that $[\Sigma_{i=1}^n (x_i - \overline{x})^2]/\sigma^2$ is χ_{n-1}^2.

The kind of problem exemplified above led Jeffreys to the *ad hoc* recommendation, widely adopted in the literature, of independent a priori treatment of location and scale parameters, applying his rule separately to each of the two subgroups of parameters, and then multiplying the resulting forms together to arrive at the overall prior specification. For an illustration of this, see Geisser and Cornfield (1963): for an elaboration of the idea, see Zellner (1986a).

At this point, one may wonder just what has become of the intuition motivating the arguments outlined above. Unfortunately, although the implied information limits are mathematically well-defined in one dimension, in higher dimensions the forms obtained may depend on the path followed to obtain the limit. Similar problems arise with other intuitively appealing desiderata. For example, the Box and Tiao suggestion of a uniform prior following transformation to a parametrisation ensuring data translation generalises, in the multiparameter setting, to the requirement of uniformity following a transformation which ensures that credible regions are of the same size. The problem, of course, is that, in several dimensions, such regions can be of the same size but very different in form.

Jeffreys' original requirement of invariance under reparametrisation remains perhaps the most intuitively convincing. If this is conceded, it follows that, whatever their apparent motivating intuition, approaches which do not have this property should be regarded as unsatisfactory. Such approaches include the use of limiting forms of conjugate priors, as in Haldane (1948), Novick and Hall (1965), Novick (1969), DeGroot (1970, Chapter 10) and Piccinato (1973, 1977), a predictivistic version of the principle of insufficient reason, Geisser (1984), and different forms of information-theoretical arguments, such as those put forward by Zellner (1977, 1991), Geisser (1979) and Torgesen (1981).

Maximising the expected information (as opposed to maximising the expected *missing* information) gives invariant, but unappealing results, producing priors that can have finite support (Berger *et al.*, 1989). Other information-based suggestions are those of Eaton (1982), Spall and Hill (1990) and Rodríguez (1991).

Partially satisfactory results have nevertheless been obtained in multiparameter problems where the parameter space can be considered as a group of transformations of the sample space. Invariance considerations within such a group suggest the use of *relatively invariant* (Hartigan, 1964) priors like the Haar measures. This idea was pioneered by Barnard (1952). Stone (1965) recognised that, in an appropriate sense, it should be possible to approximate the results obtained using a non-informative prior by those obtained using a convenient sequence of proper priors. He went on to show that, if a group structure is present, the corresponding *right* Haar measure is the only prior for which such a desirable convergence is obtained. It is reassuring that, in those one-dimensional problems for which a group of transformations does exist, the right Haar measures coincides with the relevant Jeffreys' prior. For some undesirable consequences of the *left* Haar measure see Bernardo (1978b). Further developments involving Haar measures are provided by Zidek (1969), Villegas (1969, 1971, 1977a, 1977b, 1981), Stone (1970), Florens (1978, 1982), Chang and Villegas (1986) and Chang and Eaves (1990). Dawid (1983b) provides an excellent review of work up to the early 1980's. However, a large group of interesting models do not have any group structure, so that these arguments cannot produce general solutions.

Even when the parameter space may be considered as a group of transformations there is no definitive answer. In such situations, the right Haar measures are the obvious choices and yet even these are open to criticism.

Example 5.26. ***(Standardised mean).*** Let $x = \{x_1, \ldots, x_n\}$ be a random sample from a normal distribution $N(x \mid \mu, \lambda)$. The standard prior recommended by group invariance arguments is $\pi(\mu, \sigma) = \sigma^{-1}$ where $\lambda = \sigma^{-2}$. Although this gives adequate results if one wants to make inferences about either μ or σ, it is quite unsatisfactory if inferences about the standardised mean $\phi = \mu/\sigma$ are required. Stone and Dawid (1972) show that the posterior distribution of ϕ obtained from such a prior depends on the data through the statistic

$$t = \frac{\sum_{i=1}^n x_i}{(\sum_{i=1}^n x_i^2)^{1/2}},$$

whose sampling distribution,

$$\begin{aligned} p(t \mid \mu, \sigma) &= p(t \mid \phi) \\ &= e^{-n\phi^2/2} \left\{ 1 - \frac{t^2}{n} \right\}^{(n-3)/2} \int_0^\infty \omega^{n-1} \exp\left\{ -\frac{\omega^2}{2} + t\phi\omega \right\} d\omega, \end{aligned}$$

only depends on ϕ. One would, therefore, expect to be able to "match" the original inferences about ϕ by the use of $p(t \mid \phi)$ together with some appropriate prior for ϕ. However, no such prior exists.

On the other hand, the reference prior relative to the ordered partition (ϕ, σ) is (see Example 5.18)

$$\pi(\phi, \sigma) = (2 + \phi^2)^{-1/2} \sigma^{-1}$$

and the corresponding posterior distribution for ϕ is

$$\pi(\phi \mid \boldsymbol{x}) \propto (2 + \phi^2)^{-1/2} \left[e^{-n\phi^2/2} \int_0^\infty \omega^{n-1} \exp\left\{ -\frac{\omega^2}{2} + t\phi\omega \right\} d\omega \right].$$

We observe that the factor in square brackets is proportional to $p(t \mid \phi)$ and thus the inconsistency disappears.

This type of *marginalisation paradox*, further explored by Dawid, Stone and Zidek (1973), appears in a large number of multivariate problems and makes it difficult to believe that, for any given model, a *single* prior may be usefully regarded as "universally" non-informative. Jaynes (1980) disagrees.

An acceptable general theory for non-informative priors should be able to provide consistent answers to the same inference problem whenever this is posed in different, but equivalent forms. Although this idea has failed to produce a constructive procedure for deriving priors, it may be used to discard those methods which fail to satisfy this rather intuitive requirement.

Example 5.27. ***(Correlation coefficient).*** Let $(\boldsymbol{x}, \boldsymbol{y}) = \{(x_1, y_1), \ldots, (x_n, y_n)\}$ be a random sample from a bivariate normal distribution, and suppose that inferences about the correlation coefficient ρ are required. It may be shown that if the prior is of the form

$$\pi(\mu_1, \mu_2, \sigma_1, \sigma_2, \rho) = \pi(\rho) \sigma_1^{-a} \sigma_2^{-a},$$

which includes all proposed "non-informative" priors for this model that we are aware of, then the posterior distribution of ρ is given by

$$\begin{aligned} \pi(\rho \mid \boldsymbol{x}, \boldsymbol{y}) &= \pi(\rho \mid r) \\ &= \frac{\pi(\rho)(1 - \rho^2)^{(n+2a-3)/2}}{(1 - \rho r)^{n+a-(5/2)}} F\left(\tfrac{1}{2}, \tfrac{1}{2}, n + a - \tfrac{3}{2}, \frac{1 + \rho r}{2} \right), \end{aligned}$$

where

$$r = \frac{\sum_i x_i y_i - n\overline{x}\,\overline{y}}{[\sum_i (x_i - \overline{x})^2]^{1/2}[\sum_i (y_i - \overline{y})^2]^{1/2}}$$

is the sample correlation coefficient, and F is the hypergeometric function. This posterior distribution only depends on the data through the sample correlation coefficient r; thus, with this form of prior, r is sufficient. On the other hand, the sampling distribution of r is

$$\begin{aligned} p(r \mid \mu_1, \mu_2, \sigma_1, \sigma_2, \rho) &= p(r \mid \rho) \\ &= \frac{(1-\rho^2)^{(n-1)/2}(1-r^2)^{(n-4)/2}}{(1-\rho r)^{n-3/2}} F\left(\tfrac{1}{2}, \tfrac{1}{2}, n - \tfrac{1}{2}, \frac{1+\rho r}{2}\right). \end{aligned}$$

Moreover, using the transformations $\delta = \tanh^{-1}\rho$ and $t = \tanh^{-1} r$, Jeffreys' prior for this univariate model is found to be $\pi(\rho) \propto (1-\rho^2)^{-1}$ (see Lindley, 1965, pp. 215–219).

Hence one would expect to be able to match, using this reduced model, the posterior distribution $\pi(\rho \mid r)$ given previously, so that

$$\pi(\rho \mid r) \propto p(r \mid \rho)(1-\rho^2)^{-1}.$$

Comparison between $\pi(\rho \mid r)$ and $p(r \mid \rho)$ shows that this is possible if and only if $a = 1$, and $\pi(\rho) = (1-\rho^2)^{-1}$. Hence, to avoid inconsistency the joint reference prior must be of the form

$$\pi(\mu_1, \mu_2, \sigma_1, \sigma_2, \rho) = (1-\rho^2)^{-1}\sigma_1^{-1}\sigma_2^{-1},$$

which is precisely (see Example 5.22, p. 337) the reference prior relative to the natural order, $\{\rho, \mu_1, \mu_2, \sigma_1, \sigma_2\}$.

However, it is easily checked that Jeffreys' multivariate prior is

$$\pi(\mu_1, \mu_2, \sigma_1, \sigma_2, \rho) = (1-\rho^2)^{-3/2}\sigma_1^{-2}\sigma_2^{-2}$$

and that the "two-step" Jeffreys' multivariate prior which separates the location and scale parameters is

$$\pi(\mu, \mu_2)\pi(\sigma_1, \sigma_2, \rho) = (1-\rho^2)^{-3/2}\sigma_1^{-1}\sigma_2^{-1}.$$

For further detailed discussion of this example, see Bayarri (1981).

Once again, this example suggests that different non-informative priors may be appropriate *depending on the particular function of interest* or, more generally, on the ordering of the parameters.

Although marginalisation paradoxes disappear when one uses proper priors, to use proper approximations to non-informative priors as an approximate description of "ignorance" does not solve the problem either.

Example 5.28. ***(Stein's paradox)***. Let $x = \{x_1, \ldots, x_n\}$ be a random sample from a multivariate normal distribution $N_k(x \mid \mu, I_k\}$. Let $\overline{x}_i$ be the mean of the n observations from coordinate i and let $t = \sum_i \overline{x}_i^2$. The universally recommended "non-informative" prior for this model is $\pi(\mu_1, \ldots, \mu_k) = 1$, which may be approximated by the proper density

$$\pi(\mu_1, \ldots, \mu_k) = \prod_{i=1}^{m} N(\mu_i \mid 0, \lambda),$$

where λ is very small. However, if inferences about $\phi = \sum_i \mu_i^2$ are desired, the use of this prior overwhelms, for large k, what the data have to say about ϕ. Indeed, with such a prior the posterior distribution of $n\phi$ is a non-central χ^2 distribution with k degrees of freedom and non-centrality parameter nt, so that

$$E[\phi \mid x] = t + \frac{k}{n}, \quad V[\phi \mid x] = \frac{2}{n}\left[2t + \frac{k}{n}\right],$$

while the sampling distribution of nt is a non-central χ^2 distribution with k degrees of freedom and parameter $n\theta$ so that $E[t \mid \phi] = \phi + k/n$. Thus, with, say, $k = 100$, $n = 1$ and $t = 200$, we have $E[\phi \mid x] \approx 300$, $V[\phi \mid x] \approx 32^2$, whereas the unbiased estimator based on the sampling distribution gives $\hat{\phi} = t - k \approx 100$.

However, the asymptotic posterior distribution of ϕ is $N(\phi \mid \hat{\phi}, (4\hat{\phi})^{-1})$ and hence, by Proposition 5.2, the reference posterior for ϕ relative to $p(t \mid \phi)$ is

$$\pi(\phi \mid x) \propto \pi(\phi)p(t \mid \phi) \propto \phi^{-1/2}\chi^2(nt \mid k, n\phi)$$

whose mode is close to $\hat{\phi}$. It may be shown that this is also the posterior distribution of ϕ derived from the reference prior relative to the ordered partition $\{\phi, \omega_1, \ldots, \omega_{k-1}\}$, obtained by reparametrising to polar coordinates in the full model. For further details, see Stein (1959), Efron (1973), Bernardo (1979b) and Ferrándiz (1982).

Naïve use of apparently "non-informative" prior distributions can lead to posterior distributions whose corresponding credible regions have untenable coverage probabilities, in the sense that, for some region C, the corresponding posterior probabilities $P(C \mid z)$ may be completely different from the conditional values $P(C \mid \theta)$ for almost all θ values.

Such a phenomenon is often referred to as *strong inconsistency* (see, for example, Stone, 1976). However, by carefully distinguishing between parameters of interest and nuisance parameters, reference analysis avoids this type of inconsistency. An illuminating example is provided by the reanalysis by Bernardo (1979b, reply to the discussion) of Stone's (1976) *Flatland* example. For further discussion of strong inconsistency and related topics, see Appendix B, Section 3.2.

Jaynes (1968) introduced a more general formulation of the problem. He allowed for the existence of a certain amount of initial "objective" information and then tried to determine a prior which reflected this initial information, but nothing else (see, also, Csiszár, 1985). Jaynes considered the entropy of a distribution to be the appropriate measure of uncertainty subject to any "objective" information one might have. If no such information exists and ϕ can only take a finite number of values, Jaynes' *maximum entropy* solution reduces to the Bayes-Laplace postulate. His arguments are quite convincing in the finite case; however, if ϕ is continuous, the non-invariant entropy functional, $H\{p(\phi)\} = -\int p(\phi)\log p(\phi)d\phi$, no longer has a sensible interpretation in terms of uncertainty. Jaynes' solution is to introduce a "reference" density $\pi(\phi)$ in order to define an "invariantised" entropy,

$$-\int p(\phi)\log\frac{p(\phi)}{\pi(\phi)}d\phi,$$

and to use the prior which maximises this expression, subject, again, to any initial "objective" information one might have. Unfortunately, $\pi(\phi)$ must itself be a representation of ignorance about ϕ so that no progress has been made. If a convenient group of transformations is present, Jaynes suggests invariance arguments to select the reference density. However, no general procedure is proposed.

Context-specific "non-informative" Bayesian analyses have been produced for specific classes of problems, with no attempt to provide a general theory. These include dynamic models (Pole and West, 1989) and finite population survey sampling (Meeden and Vardeman, 1991).

The quest for non-informative priors could be summarised as follows.

(i) In the finite case, Jaynes' principle of maximising the entropy is convincing, but cannot be extended to the continuous case.

(ii) In one-dimensional continuous regular problems, Jeffreys' prior is appropriate.

(iii) The infinite discrete case can often be handled by suitably embedding the problem within a continuous framework.

(iv) In continuous multiparameter situations there is no hope for a single, unique, "non-informative prior", appropriate for all the inference problems within a given model. To avoid having the prior dominating the posterior for *some* function ϕ of interest, the prior has to depend not only on the model but also on the parameter of interest or, more generally, on some notion of the order of importance of the parameters.

The reference prior theory introduced in Bernardo (1979b) and developed in detail in Section 5.4 avoids most of the problems encountered with other proposals. It reduces to Jaynes' form in the finite case and to Jeffreys' form in one-dimensional regular continuous problems, avoiding marginalisation paradoxes by insisting that the reference prior be tailored to the parameter of interest. However, subsequent work by Berger and Bernardo (1989) has shown that the heuristic arguments in Bernardo (1979b) can be misleading in complicated situations, thus necessitating more precise definitions. Moreover, Berger and Bernardo (1992a, 1992b, 1992c) showed that the partition into parameters of interest and nuisance parameter may not go far enough and that reference priors should be viewed relative to a given ordering—or, more generally, a given ordered grouping—of the parameters. This approach was described in detail in Section 5.4. Ye (1993) derives reference priors for sequential experiments.

A completely different objection to such approaches to non-informative priors lies in the fact that, for continuous parameters, they depend on the likelihood function. This is recognised to be potentially inconsistent with a personal interpretation of probability. For many subjectivists, the initial density $p(\phi)$ is a description of the opinions held about ϕ, independent of the experiment performed;

> why should one's knowledge, or ignorance, of a quantity depend on the experiment being used to determine it? Lindley (1972, p. 71).

In many situations, we would accept this argument. However, as we argued earlier, priors which reflect knowledge of the experiment can sometimes be genuinely appropriate in Bayesian inference, and may also have a useful role to play (see, for example, the discussion of stopping rules in Section 5.1.4) as technical devices to produce *reference* posteriors. Posteriors obtained from actual prior opinions could then be compared with those derived from a reference analysis in order to assess the relative importance of the initial opinions on the final inference.

> In general we feel that it is sensible to choose a non-informative prior which expresses ignorance *relative* to information which can be supplied by a particular experiment. If the experiment is changed, then the expression of relative ignorance can be expected to change correspondingly. (Box and Tiao, 1973, p. 46).

Finally, "non-informative" distributions have sometimes been criticised on the grounds that they are typically improper and may lead, for instance, to inadmissible estimates (see, e.g. Stein, 1956). However, sensible "non-informative" priors may be seen to be, in an appropriate sense, limits of proper priors (Stone, 1963, 1965, 1970; Stein, 1965; Akaike, 1980a). Regarded as a "baseline" for admissible inferences, posterior distributions derived from "non-informative" priors need not be themselves admissible, but only arbitrarily close to admissible posteriors.

However, there can be no final word on this topic! For example, recent work by Eaton (1992), Clarke and Wasserman (1993), George and McCulloch (1993b) and Ye (1993) seems to open up new perspectives and directions.

5.6.3 Robustness

In Section 4.8.3, we noted that some aspects of model specification, either for the parametric model or the prior distribution components, can seem arbitrary, and cited as an example the case of the choice between normal and Student-t distributions as a parametric model component to represent departures of observables from their conditional expected values. In this section, we shall provide some discussion of how insight and guidance into appropriate choices might be obtained.

We begin our discussion with a simple, direct approach to examining the ways in which a posterior density for a parameter depends on the choices of parametric model or prior distribution components. Consider, for simplicity, a single observable $x \in \Re$ having a parametric density $p(x|\theta)$, with $\theta \in \Re$ having prior density $p(\theta)$. The mechanism of Bayes' theorem,

$$p(\theta|x) = \frac{p(x|\theta)p(\theta)}{p(x)},$$

involves multiplication of the two model components, $p(x|\theta)$, $p(\theta)$, followed by normalisation, a somewhat "opaque" operation from the point of view of comparing specifications of $p(x|\theta)$ or $p(\theta)$ on a "what if?" basis.

However, suppose we take logarithms in Bayes' theorem and subsequently differentiate with respect to θ. This now results in a *linear* form

$$\frac{\partial}{\partial\theta}\log p(\theta|x) = \frac{\partial}{\partial\theta}\log p(x|\theta) + \frac{\partial}{\partial\theta}\log p(\theta).$$

The first term on the right-hand side is (apart from a sign change) a quantity known in classical statistics as the *efficient score function* (see, for example, Cox and Hinkley, 1974). On the linear scale, this is the quantity which transforms the prior into the posterior and hence opens the way, perhaps, to insight into the effect of a particular choice of $p(x|\theta)$ given the form of $p(\theta)$. See, for example, Ramsey and Novick (1980) and Smith (1983). Conversely, examination of the second term on the right-hand side for given $p(x|\theta)$ may provide insight into the implications of the mathematical specification of the prior.

For convenience of exposition—and perhaps because the prior component is often felt to be the less secure element in the model specification—we shall focus the following discussion on the sensitivity of characteristics of $p(\theta \mid x)$ to the choice of $p(\theta)$. Similar ideas apply to the choice of $p(x \mid \theta)$.

With x denoting the mean of n independent observables from a normal distribution with mean θ and precision λ, we shall illustrate these ideas by considering the form of the posterior mean for θ when $p(x|\theta) = \mathrm{N}(x|\theta, n\lambda)$ and $p(\theta)$ is of "arbitrary" form.

Defining

$$p(x) = \int p(x|\theta)p(\theta)d\theta,$$

$$s(x) = \frac{\partial \log p(x)}{\partial x},$$

it can be shown (see, for example, Pericchi and Smith, 1992) that

$$E(\theta|x) = x - n^{-1}\lambda^{-1}s(x).$$

Suppose we carry out a "what if?" analysis by asking how the behaviour of the posterior mean depends on the mathematical form adopted for $p(\theta)$.

What if we take $p(\theta)$ to be *normal*? With $p(\theta) = \mathrm{N}(\theta|\mu, \lambda_0)$, the reader can easily verify that in this case $p(x)$ will be normal, and hence $s(x)$ will be a linear combination of x and the prior mean. The formula given for $E(\theta|x)$ therefore reproduces the weighted average of sample and prior means that we obtained in Section 5.2, so that

$$E(\theta|x) = (n\lambda + \lambda_0)^{-1}(n\lambda x + \lambda_0\mu).$$

What if we take $p(\theta)$ to be *Student-t*? With $p(\theta) = \mathrm{St}(\theta|\mu, \lambda_0, \alpha)$ the exact treatment of $p(x)$ and $s(x)$ becomes intractable. However, detailed analysis (Pericchi and Smith, 1992) provides the approximation

$$E(\theta|x) = x - \frac{(\alpha+1)(x-\mu)}{n\lambda[\alpha\lambda_0^{-1} + (x-\mu)^2]}.$$

What if we take $p(\theta)$ to be *double-exponential*? In this case,

$$p(\theta) = \frac{1}{\nu\sqrt{2}} \exp\left(-\frac{\sqrt{2}}{\nu}|\theta - \mu|\right),$$

for some $\nu > 0$, $\mu \in \Re$ and the evaluation of $p(x)$ and $s(x)$ is possible, but tedious. After some algebra—see Pericchi and Smith (1992)—it can be shown that, if $b = n^{-1}\nu^{-1}\lambda^{-1}\sqrt{2}$,

$$E(\theta|x) = w(x)(x+b) + [1 - w(x)](x-b),$$

where $w(x)$ is a weight function, $0 \le w(x) \le 1$, so that

$$x - b \le E(\theta|x) \le x + b.$$

Examination of the three forms for $E(\theta|x)$ reveals striking qualitative differences. In the case of the normal, the posterior mean is unbounded in $x - \mu$, the departure of the observed mean from the prior mean. In the case of the Student-t, we see that for very small $x - \mu$ the posterior mean is approximately linear in $x - \mu$, like the normal, whereas for $x - \mu$ very large the posterior mean approaches x. In the case of the double-exponential, the posterior mean is bounded, with limits equal to x plus or minus a constant.

Consideration of these qualitative differences might provide guidance regarding an otherwise arbitrary choice if, for example, one knew how one would like the Bayesian learning mechanism to react to an "outlying" x, which was far from μ. See Smith (1983) and Pericchi *et al.* (1993) for further discussion and elaboration. See Jeffreys (1939/1961) for seminal ideas relating to the effect of the tail-weight of the distribution of the parametric model on posterior inferences. Other relevant references include Masreliez (1975), O'Hagan (1979, 1981, 1988b), West (1981), Maín (1988), Polson (1991), Gordon and Smith (1993) and O'Hagan and Le (1994).

The approach illustrated above is well-suited to probing qualitative differences in the posterior by considering, individually, the effects of a small number of potential alternative choices of model component (parametric model or prior distribution).

Suppose, instead, that someone has in mind a specific candidate component specification, p_0, say, but is all too aware that aspects of the specification have involved somewhat arbitrary choices. It is then natural to be concerned about whether posterior conclusions might be highly sensitive to the particular specification p_0, viewed in the context of alternative choices in an appropriately defined *neighbourhood* of p_0.

In the case of specifying a parametric component p_0—for example an "error" model for differences between observables and their (conditional) expected values—such concern might be motivated by definite knowledge of symmetry and unimodality, but an awareness of the arbitrariness of choosing a conventional distributional form such as normality. Here, a suitable neighbourhood might be formed by taking p_0 to be normal and forming a class of distributions whose tail-weights deviate (lighter and heavier) from normal: see, for example, the seminal papers of Box and Tiao (1962, 1964).

In the case of specifying a prior component p_0, such concern might be motivated by the fact that elicitation of prior opinion has only partly determined the specification (for example, by identifying a few quantiles), with considerable remaining arbitrariness in "filling out" the rest of the distribution. Here, a suitable neighbourhood of p_0 might consist of a class of priors all having the specified quantiles but with other characteristics varying: see, for example, O'Hagan and Berger (1988).

From a mathematical perspective, this formulation of the robustness problem presents some intriguing challenges. How to formulate interesting neighbourhood classes of distributions? How to calculate, with respect to such prior classes, bounds on posterior quantities of interest such as expectations or probabilities?

At the time of writing, this is an area of intensive research. For example, should neighbourhoods be defined parametrically or non-parametrically? And, if nonparametrically, what measures of distance should be used to define a neighbourhood "close" to p_0? Should the elements, p, of the neighbourhood be those such that the density ratio p/p_0 is bounded in some sense? Or such that the maximum difference in the probability assigned to any event under p and p_0 is bounded? Or such that p can be written as a "contamination" of p_0, $p = (1-\varepsilon)p_0 + \varepsilon q$, for small ε and q belonging to a suitable class?

As yet, few issues seem to be resolved and we shall not, therefore, attept a detailed overview. Relevant references include; Edwards *et al.* (1963), Dawid (1973), Dempster (1975), Hill (1975), Meeden and Isaacson (1977), Rubin (1977, 1988a, 1988b), Kadane and Chuang (1978), Berger (1980, 1982, 1985a), DeRobertis and Hartigan (1981), Hartigan (1983), Kadane (1984), Berger and Berliner (1986), Kempthorne (1986), Berger and O'Hagan (1988), Cuevas and Sanz (1988), Pericchi and Nazaret (1988), Polasek and Pötzelberger (1988, 1994), Carlin and Dempster (1989), Delampady (1989), Sivaganesan and Berger (1989, 1993), Wasserman (1989, 1992a, 1992b), Berliner and Goel (1990), Delampady and Berger (1990), Doksum and Lo (1990), Wasserman and Kadane (1990, 1992a, 1992b), Ríos (1990, 1992), Angers and Berger (1991), Berger and Fan (1991), Berger and Mortera (1991b, 1994), Lavine (1991a, 1991b, 1992a, 1992b, 1994), Lavine *et al.* (1991, 1993), Moreno and Cano (1991), Pericchi and Walley (1991), Pötzelberger and Polasek (1991), Sivaganesan (1991), Walley (1991), Berger and Jefferys (1992), Gilio (1992b), Gómez-Villegas and Maín (1992), Moreno and Pericchi (1992, 1993), Nau (1992), Sansó and Pericchi (1992), Liseo *et al.* (1993), Osiewalski and Steel (1993), Bayarri and Berger (1994), de la Horra and Fernández (1994), Delampady and Dey (1994), O'Hagan (1994b), Pericchi and Pérez (1994), Ríos and Martín (1994), Salinetti (1994). There are excellent reviews by Berger (1984a, 1985a, 1990, 1994) and Wasserman (1992a), which together provide a wealth of further references.

Finally, in the case of a large data sample, one might wonder whether the data themselves could be used to suggest a suitable form of parametric model component, thus removing the need for detailed specification and hence the arbitrariness of the choice. The so-called *Bayesian bootstrap* provides such a possible approach; see, for instance, Rubin (1981) and Lo (1987, 1993). However, since it is a heavily computationally based method we shall defer discussion to the volume *Bayesian Computation.*

The term *Bootstrap* is more familiar to most statisticians as a computationally intensive *frequentist* data-based simulation method for statistical inference; in particular, as a computer-based method for assigning frequentist measures of accuracy to point estimates. For an introduction to the method–and to the related technique of *jackknifing*—see Efron (1982). For a recent textbook treatment, see Efron and Tibshirani (1993). See, also, Hartigan (1969, 1975).

5.6.4 Hierarchical and Empirical Bayes

In Section 4.6.5, we motivated and discussed model structures which take the form of an hierarchy. Expressed in terms of generic densities, a simple version of such an hierarchical model has the form

$$p(\boldsymbol{x}|\boldsymbol{\theta}) = p(\boldsymbol{x}_1, \ldots, \boldsymbol{x}_k|\boldsymbol{\theta}_1, \ldots, \boldsymbol{\theta}_k) = \prod_{i=1}^{k} p(\boldsymbol{x}_i|\boldsymbol{\theta}_i),$$

$$p(\boldsymbol{\theta}|\boldsymbol{\phi}) = p(\boldsymbol{\theta}_1, \ldots, \boldsymbol{\theta}_k|\boldsymbol{\phi}) = \prod_{i=1}^{k} p(\boldsymbol{\theta}_i|\boldsymbol{\phi}),$$

$$p(\boldsymbol{\phi}).$$

The basic interpretation is as follows. Observables $\boldsymbol{x}_1, \ldots, \boldsymbol{x}_k$ are available from k different, but related, sources: for example, k individuals in a homogeneous population, or k clinical trial centres involved in the same study. The first stage of the hierarchy specifies parametric model components for each of the k observables. But because of the "relatedness" of the k observables, the parameters $\boldsymbol{\theta}_1, \ldots, \boldsymbol{\theta}_k$ are themselves judged to be exchangeable. The second and third stages of the hierarchy thus provide a prior for $\boldsymbol{\theta}$ of the familiar mixture representation form

$$p(\boldsymbol{\theta}) = p(\boldsymbol{\theta}_1, \ldots, \boldsymbol{\theta}_k) = \int \prod_{i=1}^{k} p(\boldsymbol{\theta}_i|\boldsymbol{\phi}) p(\boldsymbol{\phi}) d\boldsymbol{\phi}.$$

Here, the "hyperparameter" $\boldsymbol{\phi}$ typically has an interpretation in terms of characteristics—for example, mean and covariance—of the population (of individuals, trial centres) from which the k units are drawn.

In many applications, it may be of interest to make inferences both about the unit characteristics, the $\boldsymbol{\theta}_i$'s, and the population characteristics, $\boldsymbol{\phi}$. In either case, straightforward probability manipulations involving Bayes' theorem provide the required posterior inferences as follows:

$$p(\boldsymbol{\theta}_i|\boldsymbol{x}) = \int p(\boldsymbol{\theta}_i|\boldsymbol{\phi}, \boldsymbol{x}) p(\boldsymbol{\phi}|\boldsymbol{x}) d\boldsymbol{\phi},$$

where

$$p(\boldsymbol{\theta}_i|\boldsymbol{\phi}, \boldsymbol{x}) \propto p(\boldsymbol{x}|\boldsymbol{\theta}_i) p(\boldsymbol{\theta}_i|\boldsymbol{\phi})$$

$$p(\boldsymbol{\phi}|\boldsymbol{x}) \propto p(\boldsymbol{x}|\boldsymbol{\phi}) p(\boldsymbol{\phi}),$$

and

$$p(\boldsymbol{x}|\boldsymbol{\phi}) = \int p(\boldsymbol{x}|\boldsymbol{\theta}) p(\boldsymbol{\theta}|\boldsymbol{\phi}) d\boldsymbol{\theta}.$$

Of course, actual implementation requires the evaluation of the appropriate integrals and this may be non-trivial in many cases. However, as we shall see in the volumes *Bayesian Computation* and *Bayesian Methods*, such models can be implemented in a fully Bayesian way using appropriate computational techniques.

A detailed analysis of hierarchical models will be provided in those volumes; some key references are Good (1965, 1980b), Ericson (1969a, 1969b), Hill (1969, 1974), Lindley (1971), Lindley and Smith (1972), Smith (1973a, 1973b), Goldstein and Smith (1974), Leonard (1975), Mouchart and Simar (1980), Goel and DeGroot (1981), Goel (1983), Dawid (1988b), Berger and Robert (1990), Pérez and Pericchi (1992), Schervish *et al.* (1992), van der Merwe and van der Merwe (1992), Wolpert and Warren-Hicks (1992) and George *et al.* (1993, 1994).

A tempting approximation is suggested by the first line of the analysis above. We note that if $p(\phi|x)$ were fairly sharply peaked around its mode, ϕ^*, say, we would have

$$p(\boldsymbol{\theta}_i|\boldsymbol{x}) \approx p(\boldsymbol{\theta}_i \mid \boldsymbol{\phi}^*, \boldsymbol{x}).$$

The form that results can be thought of *as if* we first use the data to estimate ϕ and then, with ϕ^* as a "plug-in" value, use Bayes' theorem for the first two stages of the hierarchy. The analysis thus has the flavour of a Bayesian analysis, but with an "empirical" prior based on the data.

Such short-cut approximations to a fully Bayesian analysis of hierarchical models have become known as *Empirical Bayes* methods. This is actually slightly confusing, since the term was originally used to describe frequentist estimation of the second-stage distribution: see Robbins (1955, 1964, 1983). However, more recently, following the line of development of Efron and Morris (1972, 1975) and Morris (1983), the term has come to refer mainly to work aimed at approximating (aspects of) posterior distributions arising from hierarchical models.

The naïve approximation outlined above is clearly deficient in that it ignores uncertainty in ϕ. Much of the development following Morris (1983) has been directed to finding more defensible approximations. For more whole-hearted Bayesian approaches, see Deely and Lindley (1981), Gilliland *et al.* (1982), Kass and Steffey (1989) and Ghosh (1992a). An eclectic account of empirical Bayes methods is given by Maritz and Lwin (1989).

5.6.5 Further Methodological Developments

The distinction between *theory* and *methods* is not always clear-cut and the extensive Bayesian literature on specific methodological topics obviously includes a wealth of material relating to Bayesian concepts and theory. We shall review this material in the volume *Bayesian Methods* and confine ourselves here to simply providing a few references.

Among the areas which have stimulated the development of Bayesian theory, we note the following: *Actuarial Science and Insurance* (Jewell, 1974, 1988; Singpurwalla and Wilson, 1992), *Calibration* (Dunsmore, 1968; Hoadley, 1970; Brown and Mäkeläinen, 1992), *Classification and Discrimination* (Geisser, 1964, 1966; Binder, 1978; Bernardo, 1988, 1994; Bernardo and Girón, 1989; Dawid and Fang, 1992), *Contingency Tables* (Lindley, 1964; Good, 1965, 1967; Leonard, 1975; Leonard and Hsu, 1994), *Control Theory* (Aoki, 1967; Sawagari *et al.*, 1967), *Econometrics* (Mills, 1992; Steel, 1992), *Finite Population Sampling* (Basu, 1969, 1971; Ericson, 1969b, 1988; Godambe, 1969, 1970; Smouse, 1984; Lo, 1986), *Image Analysis* (Geman and Geman, 1984; Besag, 1986, 1989; Geman, 1988; Mardia *et al.*, 1992; Grenander and Miller, 1994), *Law* (Dawid, 1994), *Meta-Analysis* (DuMouchel and Harris, 1992; Wolpert and Warren-Hicks, 1992), *Missing Data* (Little and Rubin, 1987; Rubin, 1987; Meng and Rubin, 1992), *Mixtures* (Titterington *et al.*, 1985; Berliner, 1987; Bernardo and Girón, 1988; Florens *et al.*, 1992; West, 1992b; Diebolt and Robert, 1994; Robert and Soubiran, 1993; West *et al.*, 1994), *Multivariate Analysis* (Brown *et al.*, 1994), *Quality Assurance* (Wetherill and Campling, 1966; Hald, 1968; Booth and Smith, 1976; Irony *et al.*, 1992; Singpurwalla and Soyer, 1992), *Splines* (Wahba, 1978, 1983, 1988; Gu, 1992; Ansley *et al.*, 1993; Cox, 1993), *Stochastic Approximation* (Makov, 1988) and *Time Series and Forecasting* (Meinhold and Singpurwalla, 1983; West and Migon, 1985; Mortera, 1986; Smith and Gathercole, 1986; West and Harrison, 1986, 1989; Harrison and West, 1987; Ameen, 1992; Carlin and Polson, 1992; Gamerman, 1992; Smith, 1992; Gamerman and Migon, 1993; McCulloch and Tsay, 1993; Pole *et al.*, 1994).

5.6.6 Critical Issues

We conclude this chapter on inference by briefly discussing some further issues under the headings: (i) *Model Conditioned Inference*, (ii) *Prior Elicitation*, (iii) *Sequential Methods* and (iv) *Comparative Inference*.

Model Conditioned Inference

We have remarked on several occasions that the Bayesian learning process is predicated on a more or less formal framework. In this chapter, this has translated into model conditioned inference, in the sense that all prior to posterior or predictive inferences have taken place within the closed world of an assumed model structure.

It has therefore to be frankly acknowledged and recognised that all such inference is conditional. *If* we accept the model, *then* the mechanics of Bayesian learning—derived ultimately from the requirements of quantitative coherence—provide the appropriate uncertainty accounting and dynamics.

But what if, as individuals, we acknowledge some insecurity about the model? Or need to communicate with other individuals whose own models differ?

Clearly, issues of model criticism, model comparison, and, ultimately, model choice, are as much a part of the general world of confronting uncertainty as model conditioned thinking. We shall therefore devote Chapter 6 to a systematic exploration of these issues.

Prior Elicitation

We have emphasised, over and over, that our interpretation of a model requires—in conventional parametric representations—both a likelihood *and* a prior.

In accounts of Bayesian Statistics from a theoretical perspective—like that of this volume—discussions of the prior component inevitably focus on stylised forms, such as conjugate or reference specifications, which are amenable to a mathematical treatment, thus enabling general results and insights to be developed.

However, there is a danger of losing sight of the fact that, in real applications, prior specifications should be encapsulations of actual beliefs rather than stylised forms. This, of course, leads to the problem of how to elicit and encode such beliefs, i.e., how to structure questions to an individual, and how to process the answers, in order to arrive at a formal representation.

Much has been written on this topic, which clearly goes beyond the boundaries of statistical formalism and has proved of interest and importance to researchers from a number of other disciplines, including psychology and economics. However, despite its importance, the topic has a focus and flavour substantially different from the main technical concerns of this volume, and will be better discussed in the volume *Bayesian Methods*.

We shall therefore not attempt here any kind of systematic review of the very extensive literature. Very briefly, from the perspective of applications the best known protocol seems to be that described by Stäel von Holstein and Matheson (1979), the use of which in a large number of case studies has been reviewed by Merkhofer (1987). General discussion in a text-book setting is provided, for example, by Morgan and Henrion (1990), and Goodwin and Wright (1991). Warnings about the problems and difficulties are given in Kahneman *et al.* (1982). Some key references are de Finetti (1967), Winkler (1967a, 1967b), Edwards *et al.* (1968), Hogarth (1975, 1980) Dickey (1980), French (1980), Kadane (1980), Lindley (1982d), Jaynes (1985), Garthwaite and Dickey (1992), Leonard and Hsu (1992) and West and Crosse (1992).

Sequential Methods

In Section 2.6 we gave a brief overview of sequential decision problems but for most of our developments, we assumed that data were treated globally. It is obvious, however, that data are often available in sequential form and, moreover, there are often computational advantages in processing data sequentially, even if they are all immediately available.

There is a large Bayesian literature on sequential analysis and on sequential computation, which we will review in the volumes *Bayesian Computation* and *Bayesian Methods*. Key references include the seminal monograph of Wald (1947), Jackson (1960), who provides a bibliography of early work, Wetherill (1961), and the classic texts of Wetherill (1966) and DeGroot (1970). Berger and

Berry (1988) discuss the relevance of *stopping rules* in statistical inference. Some other references, primarily dealing with the analysis of stopping rules, are Amster (1963), Barnard (1967), Bartholomew (1967), Roberts (1967), Basu (1975) and Irony (1993). Witmer (1986) reviews multistage decision problems.

Comparative Inference

In this and in other chapters, our main concern has been to provide a self-contained systematic development of Bayesian ideas. However, both for completeness, and for the very obvious reason that there are still some statisticians who do not currently subscribe to the position adopted here, it seems necessary to make some attempt to compare and contrast Bayesian and non-Bayesian approaches.

We shall therefore provide, in Appendix B, a condensed critical overview of mainstream non-Bayesian ideas and developments. Any reader for whom our treatment is too condensed, should consult Thatcher (1964), Pratt (1965), Bartholomew (1971), Press (1972/1982), Barnett (1973/1982), Cox and Hinkley (1974), Box (1983), Anderson (1984), Casella and Berger (1987, 1990), DeGroot (1987), Piccinato (1992) and Poirier (1993).

Chapter 6

Remodelling

Summary

It is argued that whether viewed from the perspective of a sensitive individual modeller, or from that of a group of modellers, there are good reasons for systematically entertaining a range of possible belief models. A variety of decision problems are examined within this framework: some involving model choice only; some involving model choice followed by a terminal action, such as prediction; other involving only a terminal action. Throughout, a clear distinction is drawn between three rather different perspectives: first, the case where the range of models under consideration is assumed to include the "'true" belief model; secondly, the case where the range of models is being considered in order to provide a proxy for a specified, but intractable, actual belief model; finally, the case where the range of models is being considered in the absence of specification of an actual belief model. Links with hypothesis testing, significance testing and cross-validation are established.

6.1 MODEL COMPARISON

6.1.1 Ranges of Models

We recall from Chapter 4 that our ultimate modelling concern is with predictive beliefs for sequences of observables. More specifically, most of our detailed development has centred on belief models corresponding to judgements of exchangeability or, more generally, various forms of partial exchangeability.

In such cases, the predictive model typically has a mixture representation in terms of a random sample from a labelled model, together with a prior distribution for the label, the latter being interpretable in terms of a strong law limit of observables. For example, we saw that for an exchangeable real-valued sequence, a predictive belief distribution, P, has the general representation

$$P(x_1, \ldots, x_n) = \int_{\mathcal{F}} \prod_{i=1}^{n} F(x_i) dQ(F).$$

This corresponds to an (*as if*) assumption of a random sample from the unknown distribution function, F, together with a prior distribution, Q, for F, defined over the space, $\mathcal{F}$, of all distribution functions on $\Re$.

However, the very general nature of this representation precludes it —at least in terms of current limitations on intuition and technique— from providing a practical basis for routine concrete applications. This is why, in Chapter 4, much of our subsequent development was based on formal assumptions of further invariance or sufficiency structure, or pragmatic appeal to historical experience or scientific authority, in order to replace the general representation by mixtures involving finite-parameter families of densities.

Inescapably, however, this passage from the general, but intractable, form to a specific, but tractable, model involves judgements and assumptions going far beyond the simple initial judgement of exchangeability. These further judgements, and hence the models that result from them, are therefore typically much less securely based in terms of individual beliefs, and certainly much less likely to be mutually acceptable in an interpersonal context, than the straightforward symmetry judgement. Both from the perspective of a sensitive individual modeller and also from that of a group of modellers, there are therefore strong reasons for systematically entertaining *a range of possible models* (see, for example, Dickey, 1973, and Smith, 1986).

Given the assumption of exchangeability, a range of different belief models, $P_1, P_2, \ldots$ can each be represented in the general form

$$P_j(x_1, \ldots, x_n) = \int_{\mathcal{F}} \prod_{i=1}^{n} F(x_i) dQ_j(F),$$

for some $Q_1, Q_2, \ldots$, the latter encapsulating the particular alternative judgements that characterise the different models. The following stylised examples serve to illustrate some of the kinds of ranges of models that might be entertained in applications involving simple exchangeability judgements. In each case, the range of models can either be thought of as generated by a single, non-dogmatic individual (seeking to avoid commitment to one specific form); or generated as concrete suggestions by a group of individuals (each committed to one of the forms); or generated purely formally, as an imaginative proxy for models thought likely to correspond to the ranges of judgements which might be made by the eventual readership of inference reports based on the models. In general, our subsequent development will be expressed in terms of a possibly infinite sequence of models $P_1, P_2, \ldots$; in practice, we typically only work with a finite range, $P_1, \ldots, P_k$ for some $k \geq 2$.

Inference for a Location Parameter

Suppose that observations $x_1, \ldots, x_n, \ldots,$ can be thought of, conditional on

$$\mu = \lim_{n\to\infty} \frac{1}{n} \sum_{i=1}^{n} x_i\,,$$

as measurements of μ with errors $e_1, \ldots, e_n, \ldots,$, so that

$$x_i = \mu + e_i, \quad i = 1, \ldots, n,$$

with $e_1, e_2, \ldots,$ exchangeable. Various beliefs are then possible about the "error distribution. For example, appeal to the central limit theorem (Section 3.2.3) might suggest the assumption of normality; however, past experience might suggest a substantial proportion of "aberrant" or "outlying" measurements, thus requiring a distribution with heavier tails than normality; different past experience might suggest that the experimenter automatically suppresses any observations suspected of being "aberrant, thus requiring the assumption of a distribution with lighter tails than normality. With $k = 3$, and using density representations throughout, a choice of a range of models to cover these possibilities might be:

$$p_j(x_1, \ldots, x_n) = \int_{\Re} \int_{\Re^+} \prod_{i=1}^{n} p_j(x_i \mid \mu, \sigma) p_j(\mu, \sigma) d\mu d\sigma,$$

where

$$p_1(x \mid \mu, \sigma) = \frac{1}{\sqrt{2\pi}\sigma} \exp\left\{-\frac{1}{2\sigma^2}(x-\mu)^2\right\}, \qquad x \in \Re$$

$$p_2(x \mid \mu, \sigma) = \frac{1}{\sqrt{2}\sigma} \exp\left\{-\frac{\sqrt{2}}{\sigma}|x-\mu|\right\}, \qquad x \in \Re$$

and

$$p_3(x \mid \mu, \sigma) = \frac{1}{2\sqrt{3}\sigma}, \qquad x \in \left(\mu - \sqrt{3}\sigma, \mu + \sqrt{3}\sigma\right),$$

with $p_j(\mu, \sigma)$, $j = 1, 2, 3$, specifying prior beliefs for the location and scale parameters appearing in these normal, double-exponential and uniform parametric models. Thus $p_j(\mu, \sigma) = dQ_j(F)$ corresponds to a belief over $\mathcal{F}$ which assigns probability one to the family with parametric form $p_j(\cdot \,|\, \mu, \sigma)$, with density $p_j(\mu, \sigma)$ for the two parameters of this family. If these modelling possibilities emanate from a single individual, $p_j(\mu, \sigma)$ might not depend on j; in general, however, the $p_j(\mu, \sigma)$ could differ, even though, in this case, the interpretations of the parameter as strong law limits of observable measures of the location and spread of the measurements are the same.

Normality versus non-Normality

Suppose that $\mathcal{N} \subset \mathcal{F}$ is the set of all normal distributions on the real line, and hence that $\mathcal{N}^c = \mathcal{F} - \mathcal{N}$ is the set of all distributions other than normal. Then, given the assumption of exchangeability for a real-valued sequence, an individual dogmatically asserting normality is specifying, in the general representation, a $Q_1(F)$ which concentrates with probability one on $\mathcal{N}$. Conversely, an individual dogmatically asserting non-normality is specifying a $Q_2(F)$ which concentrates with probability one on $\mathcal{N}^c$. Our purpose here is mainly to point out how choices within the general exchangeable framework correspond to specification of Q. However, given the "size" of $\mathcal{F}$, one cannot but be struck by the monumental dogmatism implicit in Q_1!

Parametric Hypotheses

Suppose that $Q_j, j = 1, \ldots, k$, are even more dogmatic, in that they not only all focus on a single parametric family, $p(x \,|\, \boldsymbol{\theta})$, but, within the family, they specify $\boldsymbol{\theta}_1, \ldots, \boldsymbol{\theta}_k$, respectively, as the values of the parameter, so that

$$p_j(x_1, \ldots, x_n) = \prod_{i=1}^{n} p(x_i \,|\, \boldsymbol{\theta}_j).$$

If $k = 2$, this is often referred to as a situation of two *simple hypotheses*.

A somewhat different situation arises if $k = 2$, Q_1 again focuses on a specific parameter value, $\boldsymbol{\theta}$, but Q_2 simply assigns a prior density $p(\boldsymbol{\theta})$ over $\boldsymbol{\theta}$ in $p(x \,|\, \boldsymbol{\theta})$. The rival models then have the representations

$$p_1(x_1, \ldots, x_n) = \prod_{i=1}^{n} p(x_i \,|\, \boldsymbol{\theta}_1)$$

$$p_2(x_1, \ldots, x_n) = \int \prod_{i=1}^{n} p(x_i \,|\, \boldsymbol{\theta}) p(\boldsymbol{\theta}) d\boldsymbol{\theta},$$

corresponding to what are usually referred to, within the context of the parametric family $p(x \,|\, \boldsymbol{\theta})$, as a *simple hypothesis* and a *general alternative*.

In the contexts of judgements of partial rather than full exchangeability, the many versions of the former discussed in Chapter 4 clearly provide considerable scope for positing interesting ranges of models in any given application. The examples which follow illustrate just a few of these possibilities, expanding somewhat on the earlier discussion of model elaboration and simplification given in Sections 4.7.3 and 4.7.4.

Several Samples

Consider the situation of m unrestrictedly exchangeable sequences of zero-one random quantities, discussed in detail in Section 4.6.2. We recall that, if $\boldsymbol{x}(n_i) = (x_{i1}, \ldots, x_{in_i})$, $i = 1, \ldots, m$, the general representation of the joint predictive density for $\boldsymbol{x}(n_1), \ldots, \boldsymbol{x}(n_m)$ is given by

$$\int_{[0,1]^m} \prod_{i=1}^{m} \prod_{j=1}^{n_i} \theta_i^{x_{ij}} (1 - \theta_i)^{1-x_{ij}} \, dQ(\theta_1, \ldots, \theta_m),$$

so that, given a basic assumption of unrestricted exchangeability, alternative models are defined by different forms of Q.

As a stylised illustration of the possibilities, we might consider:

Q_1: assigning probability one to $\theta_1 = \cdots = \theta_m = \theta$, say, corresponding to the assumed equality of the limiting frequencies of ones in each of the m sequences, so that $dQ(\theta_1, \ldots, \theta_m)$ reduces to $dQ_1(\theta)$;

Q_2: assigning probability one to $\theta_1 = \phi_1, \theta_2 = \cdots = \theta_m = \phi_2$, say, so that $dQ(\theta_1, \ldots, \theta_m)$ reduces to $dQ_2(\phi_1, \phi_2)$;

Q_3: retaining a general, non-degenerate, form $dQ(\theta_1, \ldots, \theta_m)$ over the limiting frequencies.

For example, in the context of 0-1 responses in m clinical trial treatment groups, Q_1 corresponds, loosely speaking, to the hypothesis that all treatments have the same effect; Q_2 corresponds to the hypothesis that one of the treatments (possibly a "control") is different from all the other treatments, which themselves have the same effects; Q_3 corresponds to a general hypothesis that all treatments have different effects, any further (non-degenerate) relationships among them being defined by the specific form of Q_3.

Structured Layouts

In Section 4.6.3, we considered triply subscripted random quantities, x_{ijk}, representing the kth of a number of replicates of an observable in "context" $i \in I$, subject to "treatment" $j \in J$. In particular, we considered the situation where the predictive model might be thought of as generated via conditionally independent normal

$$N(x_{ijk} \mid \mu + \alpha_i + \beta_j + \gamma_{ij}, \tau)$$

distributions, together with a prior distribution Q for τ and any IJ linearly independent combinations of $\{\alpha_i\}, \{\beta_j\}, \{\gamma_{ij}\}, i \in I, j \in J$.

As a stylised illustration of alternative modelling possibilities, we might consider:

Q_1: specifying $\gamma_{ij} = 0$ for all i, j, together with a non-degenerate specification for $\{\alpha_i\}, \{\beta_j\}$ and μ;

Q_2: specifying $\gamma_{ij} = 0$ for all i, j and $\beta_j = 0$ for all j, together with a non-degenerate specification for $\{\alpha_i\}$ and μ;

Q_3: specifying $\gamma_{ij} = 0$, for all i, j and $\alpha_i = 0$ for all i, together with a non-degenerate specification for $\{\beta_j\}$ and μ;

Q_4: specifying $\gamma_{ij} = 0, \alpha_i = 0, \beta_j = 0$, for all i, j, together with a non-degenerate specification for μ.

The reader familiar with analysis of variance methods will readily identify these prior specifications with conventional forms of hypotheses regarding absences of interaction and main effects.

Covariates

In Section 4.6.4, we discussed a variety of models involving *covariates*, where beliefs about the sequence of observables (x's) were structurally dependent on another set of observables (z's). Given the enormous potential variety of such covariate dependent models, it does not seem appropriate to attempt a notationally precise illustration of all possibilities. Instead, we shall simply indicate in general terms, for each of the cases considered in Section 4.6.4, the kinds of alternative models that might be considered.

Example 6.1. ***(Bioassay).*** Alternative models for a single experiment might correspond to different assumptions about the functional dependence of the survival probabilities on the dose (for example, logit versus probit). In the case of several separate experiments, alternative models might assume the same functional form, but differ in whether or not they constrain model parameters—for example, the LD50's—to be equal.

Example 6.2. ***(Growth curves).*** Alternative models for an individual growth curve might correspond to different assumptions about the functional dependence of the response on time (for example, linear versus logistic). In the case of several growth curves for subjects from a relatively homogeneous population, alternative models might be concerned with whether some or all of the parameters defining the growth curves are identical or differ across subjects.

Example 6.3. ***(Multiple regression).*** Alternative models in the multiple regression context typically correspond to whether or not various regressor variables can be omitted from the linear regression form; equivalently, to whether or not various regression coefficients can be set equal to zero.

In the third volume of this work, *Bayesian Methods*, we shall discuss in detail a number of practical applications of this kind.

Hierarchical Models

Given the enormous variety of potential hierarchical models and alternative forms, we shall just content ourselves with some general comments for one of the specific cases considered in Section 4.6.5.

Example 6.4. ***(Exchangeable normal mean parameters).***
In Example 4.16 of Section 4.6.5, we considered a case where all the means, $\mu_1, \ldots, \mu_m$ of the m groups of observables with normal parametric models were judged exchangeable, and where this latter relationship was modelled as a mixture over a further parametric form, reflecting a symmetric judgement of "similarity" for $\mu_1, \ldots, \mu_m$. However, other symmetry judgements are possible: for example, that $m-1$ of the μ_i's are exchangeable, the other one is not, but all are equally likely, a priori, to be the odd one out. This would create a model allowing potential "outliers" among the m groups themselves. (See Section 4.7.3 for further development of this idea in a non-hierarchical setting.)

Confronted with a range of possible models, how should an individual or a group proceed?

From the perspective adopted throughout this book, clearly the answer depends on the perceived decision problem to which the modelling is a response. In the remainder of this chapter, we shall therefore illustrate various of the kinds of decision problems that might be considered. The emphasis will be on somewhat stylised, typically simple, versions of such problems, in order to highlight the conceptual issues. Detailed case-studies, involving the substantive complexities of context and the computational complexities of implementation will be more appropriately presented in the volumes *Bayesian Computation* and *Bayesian Methods*.

6.1.2 Perspectives on Model Comparison

To be concrete, let us assume that all the belief models $P_i, i \in I$, say, under consideration for observations x can be described in terms of finite parameter mixture representations. Given the specifications of the various densities forming the mixtures, the predictive distributions for the alternative models are described by

$$p_i(\boldsymbol{x}) = p(\boldsymbol{x} \mid M_i) = \int p_i(\boldsymbol{x} \mid \boldsymbol{\theta}_i) p_i(\boldsymbol{\theta}_i) d\boldsymbol{\theta}_i, \quad i \in I.$$

For mnemonic convenience, from now on we shall denote the alternative models by $\{M_i, i \in I\}$ (rather than $P_i, i \in I$, as in our previous discussions) and the set of these models by $\mathcal{M} = \{M_i, i \in I\}$.

Before we turn to a detailed discussion of decisions concerning model choice or comparison among $\{M_i, i \in I\}$, we need to draw attention to important distinctions among three alternative ways in which these possible models might be viewed.

The first alternative, which we shall call the $\mathcal{M}$-*closed* view, corresponds to believing that one of the models $\{M_i, i \in I\}$ is "true", without the explicit knowledge of which of them is the true model. From this perspective, which may reflect either the range of uncertainties within an undecided individual, or the range of different beliefs of a group of individuals, the overall model specifies beliefs for $\boldsymbol{x}$ of the form

$$p(\boldsymbol{x}) = \sum_{i \in I} P(M_i) p(\boldsymbol{x} \mid M_i),$$

with $P(M_i)$ denoting prior weights on the component models $\{M_i, i \in I\}$. There is, of course, some ambiguity as to what should be regarded as a component model (for example, the renormalised mixture of M_1 and M_2 could itself be regarded as a model), but this can be resolved pragmatically by taking $\{M_i, i \in I\}$ to be those individual models we are interested in comparing or choosing among.

But, continuing the discussion of Section 4.8.3 on the role and nature of models, when does it actually make sense to speak of a "true" model and hence to adopt the $\mathcal{M}$-closed perspective?

Clearly, this would be appropriate whenever one knew for sure that the real world mechanism involved was one of a specified finite set. One rather artificial situation where this would apply would be that of a computer simulation "inference game", where data are known to have been generated using one of a set of possible simulation programs, each a coded version of a different specified probability model, but it is not known which program was used.

Beyond such "controlled" situations, it seems to us to be difficult to accept the $\mathcal{M}$-closed perspective in a literal sense. However, there may be situations where one might not feel too uncomfortable in proceeding "as if" one meant it. For example, suppose that a parametric model with a specified parameter has been extensively adopted and found to be a successful predictive device in a range of applications. Now suppose that a new application context arises and that it is felt necessary to reconsider whether to continue with the previous specified parameter value or, in this new context, to incorporate uncertainty about the appropriate value. Provided we feel comfortable, in principle, with assigning prior weights to these two alternative formulations, we can exploit the $\mathcal{M}$-closed framework.

However, reality is typically not as relatively straightforward as this. Nature does *not* provide us with an exhaustive list of possible mechanisms and a guarantee that one of them is true. Instead, we ourselves choose the lists as part of the process of settling on a predictive specification that we hope will prove "fit for purpose" —in the jargon of modern quality assurance.

But if we abandon the $\mathcal{M}$-closed perspective, how else might we approach the very real and important problem of comparing or choosing among alternative models? It seems to us that the approach depends critically on whether one has oneself separately formulated a clear belief model or not.

In the former case, the alternative models are presumably being contemplated as a proxy because the actual belief model is too cumbersome to implement; however, they will still have to be evaluated and compared in the light of these actual beliefs. In the latter case, in the absence of an actual specified belief model, it would seem intuitively—and we shall see this more formally later—that the alternative models have to battle it out among themselves on some cross-validatory basis.

We now proceed to give these alternative perspectives a somewhat more formal description.

The second alternative, which we shall call the $\mathcal{M}$-*completed* view, corresponds to an individual acting as if $\{M_i, i \in I\}$ simply constitute a range of specified models currently available for comparison, to be evaluated in the light of the individuals separate actual belief model, which we shall denote by M_t. From this perspective, assigning the probabilities $\{P(M_i),\ i \in I\}$ does not make sense and the actual overall model specifies beliefs for x of the form $p(\boldsymbol{x}) = p_t(\boldsymbol{x}) = p(\boldsymbol{x} \mid M_t)$. $\mathcal{M}$-completed models, relative to a given proposed range of models $M_i, i \in I$ might be adopted for a variety of reasons. Typically, $\{M_i, i \in I\}$ will have been proposed largely because they are attractive from the point of view of tractability of analysis or communication of results compared with the actual belief model M_t.

The third alternative, which we shall call the $\mathcal{M}$-*open* view, also acknowledges that $\{M_i, i \in I\}$ are simply a range of specified models available for comparison, so that assigning probabilities $\{P(M_i), i \in I\}$ does not make sense. However, in this case, there is no separate overall actual belief specification, $p(x)$—perhaps because we lack the time or competence to provide it.

Examples of lists of "proxy models that are widely used include familiar ones based on parametric components, corresponding to: regression models with different choices of regressors; generalised linear models with different choices of covariates, link functions, etc.; contingency table structures with different patterns of independence and dependence assumptions.

The $\mathcal{M}$-open perspective requires comparison of such models in the absence of a separate belief model. The $\mathcal{M}$-completed perspective will typically have selected the particular proxy models in the light of an actual belief model. For example, if the actual belief model is based on non-linear functions of many covariates, together with Student probability distribution specifications, the proxy models to be evaluated might be various linear regression models with limited numbers of covariates and normal probability distribution specifications.

6.1.3 Model Comparison as a Decision Problem

We shall now discuss various possible decision problems where the answer to an inference problem involves model choice or comparison among the alternatives in $\mathcal{M} = \{M_i, i \in I\}$. Some of these only make sense from an $\mathcal{M}$-closed perspective; others can be approached from either an $\mathcal{M}$-closed, an $\mathcal{M}$-completed or an $\mathcal{M}$-open perspective. Throughout the following development, observed data on which decisions are to be based will be denoted by x, and the choice of model M_i, either as an end in itself, or as the basis for a subsequent answer to an inference problem, will be denoted by m_i, $i \in I$.

The first decision problem we shall consider involves only the choice of an M_i, without any subsequent action, so that the utility function has the form $u(m_i, \omega)$, where ω is some unknown of interest. This decision structure is shown schematically in Figure 6.1.

Provided we feel comfortable, in principle, with assigning prior weights to these two alternative formulations, we can exploit the $\mathcal{M}$-closed framework.

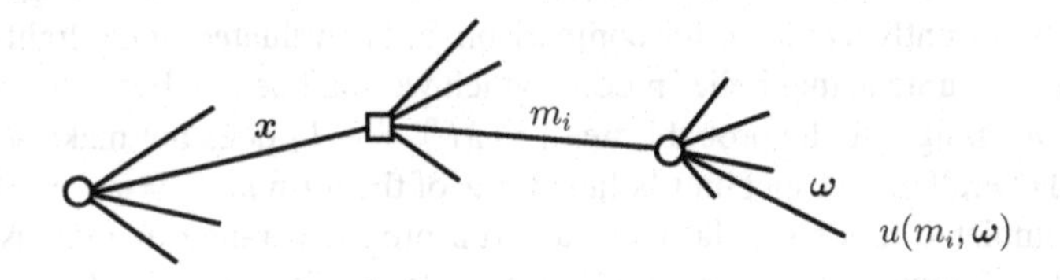

Figure 6.1 *A decision problem involving model choice only*

It is perhaps not obvious why such a problem would be of interest from an $\mathcal{M}$-open perspective. However, from an $\mathcal{M}$-closed perspective, an example of an obvious ω of interest might be the M_i for which, imagining a large future sample of observations, $y = (y_1, \ldots, y_s)$, $P(M_i \mid y) \to 1$ as $s \to \infty$. Recalling Proposition 5.9 of Section 5.2.3, ω in this case labels the "true model, and the utility of choosing a particular model then depends on whether a correct choice has been made.

Whatever the forms of ω and $u(m_i, \omega)$, in the general decision problem defined by Figure 6.1, maximising expected utility implies that the optimal model choice m^* is given by

$$\bar{u}(m^* \mid x) = \sup_{i \in I} \bar{u}(m_i \mid x) \; ,$$

where

$$\bar{u}(m_i \mid x) = \int u(m_i, \omega) p(\omega \mid x) d\omega, \quad i \in I,$$

with $p(\omega \mid x)$ representing *actual beliefs* about ω having observed x.

In the $\mathcal{M}$-closed case,

$$p(\omega \mid x) = \sum_{i \in I} p_i(\omega \mid x) P(M_i \mid x),$$

where

$$P(M_i \mid x) = \frac{P(M_i)p(x \mid M_i)}{\sum_{j \in I} P(M_j)p(x \mid M_j)},$$

and $p_i(\omega \mid x) = p(\omega \mid M_i, x)$ is given by standard (posterior or predictive) manipulations conditional on model $M_i, i \in I$. We note, in particular, the key role played by the quantities $\{P(M_i \mid x), i \in I\}$, which, within the purview of the $\mathcal{M}$-closed framework, are the posterior probabilities, given x, of model $M_i, i \in I$, being true.

From the $\mathcal{M}$-completed perspective, we can, at least in principle, obtain $p(\omega \mid x)$ and evaluate $\bar{u}(m_i \mid x), i \in I$, even though this may require extensive (Monte Carlo) numerical calculations in specific applications. In this way, one can compare the models in $\mathcal{M}$, even though none of them corresponds to one's own assumption regarding the true model.

From the $\mathcal{M}$-open perspective, nothing can be said in general about the explicit form of $p(\omega \mid x)$. It turns out, however, perhaps surprisingly, that, at least approximately, the same analysis can be carried out in the $\mathcal{M}$-open as in the $\mathcal{M}$-completed framework; in other words, one can compare the models in $\mathcal{M}$ on the basis of their expected utilities without actually having specified an alternative "true" model. We shall defer a detailed discussion of this until Section 6.1.6.

Let us now consider a rather different form of decision problem which first requires the choice of model M_i from $\mathcal{M}$, which we denote by m_i, and then, *assuming M_i to be the model*, requires an answer $a_j, j \in J_i$ relating to an unknown "state of the world" ω of interest. For example, we may wish to predict a future observation, or estimate a parameter common to all the models in $\mathcal{M}$. If $u(m_i, a_j, \omega)$ denotes the utility resulting from the successive choices m_i (i.e., model M_i) and $a_j, j \in J_i$ (answer to inference question, given M_i), when ω is the actual "state of the world", the resulting decision problem is shown schematically in Figure 6.2.

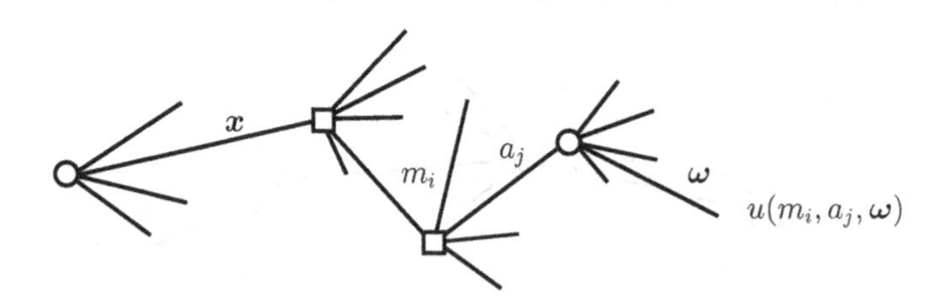

Figure 6.2 *A decision problem involving model choice and subsequent inference*

Systematic application of the criterion of maximising expected utility establishes that the optimal model choice is that $m_{\boldsymbol{x}}^*$ for which

$$\bar{u}(m_{\boldsymbol{x}}^* \,|\, \boldsymbol{x}) = \sup_{i \in I} \bar{u}(m_i \,|\, \boldsymbol{x}),$$

where

$$\bar{u}(m_i \,|\, \boldsymbol{x}) = \int u(m_i, a_{\boldsymbol{x}}^*, \omega) p(\omega \,|\, \boldsymbol{x}) d\omega$$

is the expected utility, given $\boldsymbol{x}$, of optimal behaviour given model M_i, so that $a_{\boldsymbol{x}}^*$ is obtained from maximising

$$\int u(m_i, a_j, \omega) p_i(\omega \,|\, \boldsymbol{x}) d\omega$$

The form $p_i(\omega \,|\, \boldsymbol{x})$ in the above is again given by standard (posterior or predictive) manipulation conditional on model $M_i, i \in I$, while the form $p(\omega \,|\, \boldsymbol{x})$ again represents actual beliefs about ω given $\boldsymbol{x}$.

The explicit form of $p(\omega \,|\, \boldsymbol{x})$ as a mixture of the $p_i(\omega \,|\, \boldsymbol{x})$, has been given above in the $\mathcal{M}$-closed case.

In the $\mathcal{M}$-completed case, we have also noted above that evaluation of $p(\omega \,|\, \boldsymbol{x})$ and $\{\bar{u}(m_i \,|\, \boldsymbol{x}), i \in I\}$ can in principle be carried out, numerically if necessary. Detailed analysis for the $\mathcal{M}$-open case will be given in Section 6.1.6.

From a conceptual perspective, it is important to recognise that different choices of ω and different forms of utility structure will naturally imply different forms of solution to the problem of model choice. In the next two subsections, we shall explore a number of specific cases, in order to underline the general message that coherent comparison of a finite or countable set of alternative models depends on the specification (at least implicitly) of a decision structure, including a utility function.

Before proceeding to further aspects of model choice and comparison, however, it is worth remarking that, in the above context, it is not necessary to choose among the elements of $\mathcal{M}$ in order to provide an answer a, to an inference problem. If we omit the explicit model choice step, the resulting, different form of decision problem is that shown schematically in Figure 6.3.

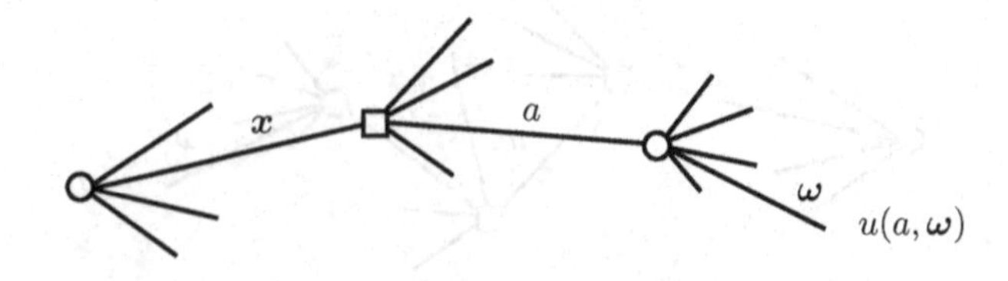

Figure 6.3 *A decision problem involving terminal decision only*

In this case, maximising expected utility leads immediately to the optimal answer a^*, given by

$$\bar{u}(a^* \mid \boldsymbol{x}) = \sup_a \bar{u}(a \mid \boldsymbol{x})$$

where

$$\bar{u}(a^* \mid \boldsymbol{x}) = \int u(a, \boldsymbol{\omega}) p(\boldsymbol{\omega} \mid \boldsymbol{x})\, d\boldsymbol{\omega},$$

with $p(\boldsymbol{\omega} \mid \boldsymbol{x})$ as discussed above. In the particular case of an $\mathcal{M}$-closed perspective, it follows from the posterior weighted mixture form of $p(\boldsymbol{\omega} \mid \boldsymbol{x})$ that, although we have omitted the model *choice* step, model *comparison* in the light of the data $\boldsymbol{x}$ is still being effected through the presence of $\{P(M_i \mid \boldsymbol{x}), i \in I\}$. In general, if we entertain a range of possible models for data $\boldsymbol{x}$, solutions to decision problems conditional on $\boldsymbol{x}$ will always implicitly depend on a comparison of the models in the light of the data, even if explicit choice among the models is not part of the decision problem.

6.1.4 Zero-one Utilities and Bayes Factors

In this section, we confine attention to the $\mathcal{M}$-closed perspective and consider first the problem of choosing a model from $\mathcal{M}$, without any subsequent decision, when the "state of the world" of interest is defined to be the "true" model, M_t, so that assuming a future sample $\boldsymbol{y} = (y_1, \ldots, y_s)$, $P(M_t \mid \boldsymbol{y}) \to 1$ as $s \to \infty$. From the $\mathcal{M}$-closed perspective, the problem, stated colloquially, is that of *choosing the true model.*

In this case, a natural form of utility function may be

$$u(m_i, \boldsymbol{\omega}) = \begin{cases} 1 & \text{if } \boldsymbol{\omega} = M_i \\ 0 & \text{if } \boldsymbol{\omega} \neq M_i. \end{cases}$$

It is then easily seen from the analysis relating to Figure 6.1 that

$$p_i(\boldsymbol{\omega} \mid \boldsymbol{x}) = \begin{cases} 1 & \text{if } \boldsymbol{\omega} = M_i \\ 0 & \text{if } \boldsymbol{\omega} \neq M_i, \end{cases}$$

and

$$p(\boldsymbol{\omega} \mid \boldsymbol{x}) = \begin{cases} P(M_i \mid \boldsymbol{x}), & \text{if } \boldsymbol{\omega} = M_i \\ 0, & \text{if } \boldsymbol{\omega} \neq M_i. \end{cases}$$

The expected utility of the decision m_i (choosing M_i), given $\boldsymbol{x}$, is hence

$$\begin{aligned} \bar{u}(m_i \mid \boldsymbol{x}) &= \int u(m_i, \boldsymbol{\omega}) p(\boldsymbol{\omega} \mid \boldsymbol{x}) d\boldsymbol{\omega} \\ &= P(M_i \mid \boldsymbol{x}), \quad i \in I. \end{aligned}$$

The optimal decision is therefore to "*choose the model which has the highest posterior probability*".

Bayes Factors
Less formally, suppose that some form of intuitive measure of pairwise comparison of plausibility is required between any two of the models $\{M_i, i \in I\}$. The above analysis suggests that M_i, M_j may be usefully compared using the *posterior odds ratio*,

$$\frac{P(M_i \mid \boldsymbol{x})}{P(M_j \mid \boldsymbol{x})} = \frac{p(\boldsymbol{x} \mid M_i)}{p(\boldsymbol{x} \mid M_j)} \times \frac{P(M_i)}{P(M_j)},$$

where, for example,

$$p(\boldsymbol{x} \mid M_i) = \int p_i(\boldsymbol{x} \mid \boldsymbol{\theta}_i) p_i(\boldsymbol{\theta}_i) d\boldsymbol{\theta}_i.$$

In words, the above comparison can be described as

"*posterior odds ratio* = *integrated likelihood ratio* × *prior odds ratio*",

making explicit the key role of the *ratio of integrated likelihoods* in providing the mechanism by which the data transform relative prior beliefs into relative posterior beliefs, in the context of parametric models.

The fundamental importance of this transformation warrants the following definition, apparently due to Turing (see, for example, Good, 1988b).

Definition 6.1. (***Bayes factor***). *Given two hypotheses H_i, H_j corresponding to assumptions of alternative models, M_i, M_j, for data $\boldsymbol{x}$, the Bayes factor in favour of H_i (and against H_j) is given by the* ***posterior to prior odds ratio.***

$$B_{ij}(\boldsymbol{x}) = \frac{p(\boldsymbol{x} \mid M_i)}{p(\boldsymbol{x} \mid M_j)} = \left\{\frac{P(M_i \mid \boldsymbol{x})}{P(M_j \mid \boldsymbol{x})}\right\} \Big/ \left\{\frac{P(M_i)}{P(M_j)}\right\}$$

Intuitively, the Bayes factor provides a measure of whether the data $\boldsymbol{x}$ have increased or decreased the odds on H_i relative to H_j. Thus, $B_{ij}(\boldsymbol{x}) > 1$ signifies that H_i is now more relatively plausible in the light of $\boldsymbol{x}$; $B_{ij}(\boldsymbol{x}) < 1$ signifies that the relative plausibility of H_j has increased.

Good (1950) has suggested that the logarithms of the various ratios in the above be called *weights of evidence* (a term apparently first used in a related context by Peirce, 1878), so that $\log B_{ij}(\boldsymbol{x})$ corresponds to the integrated likelihood weight of evidence in favour of M_i (and against M_j). On this logarithmic scale, the prior weight of evidence and $\log B_{ij}(\boldsymbol{x})$ combine *additively* to give the posterior weight of evidence.

In Section 6.1.1, we noted the extremely simple forms of predictive models which result when beliefs not only concentrate on a specific parametric family of distributions, but also identify the value of the parameter. An alternative set of such models, M_i, $i \in I$, then just corresponds to the specifications $\{p_i(\boldsymbol{x} \mid \boldsymbol{\theta}_i),\ i \in I\}$, and the integrated likelihood ratios reduce to simple ratios of likelihoods.

Hypothesis Testing

The problem of *hypothesis testing* has its own conventional terminology, which, within the framework we are adopting, can be described as follows. Two alternative models, M_1, M_2 are under consideration and both are special cases of the predictive model

$$p(\boldsymbol{x}) = \int p(\boldsymbol{x} \mid \boldsymbol{\theta}) dQ(\boldsymbol{\theta}),$$

with the same assumed parametric form $p(\boldsymbol{x} \mid \boldsymbol{\theta}), \boldsymbol{\theta} \in \Theta$, but with different choices of Q. If, for model M_i, Q_i assigns probability one to a specific value, $\boldsymbol{\theta}_i$, say, the model is said to reduce to a *simple hypothesis* for $\boldsymbol{\theta}$ (recalling that the form $p(\boldsymbol{x} \mid \boldsymbol{\theta})$ is assumed throughout). If, for model M_j, Q_j defines a non-degenerate density $p_j(\boldsymbol{\theta})$ over $\Theta_j \subseteq \Theta$, the model is said to reduce to a *composite hypothesis* for $\boldsymbol{\theta}$. If a simple hypothesis is being compared with a composite hypothesis, so that $\Theta_j = \Theta - \{\boldsymbol{\theta}_i\}$, the latter is called a *general alternative hypothesis.*

In the situation where the "state of the world" of interest, ω, is defined to be the true model M_t, we can generalise slightly the zero-one utility structure used earlier by assuming that

$$u(m_i, \omega) = -l_{ij}, \qquad \omega = M_j, \quad i = 1, 2, \quad j = 1, 2\ ,$$

with $l_{11} = l_{22} = 0$ and $l_{12}, l_{21} > 0$. Intuitively, there is a (possibly asymmetric) *loss* in choosing the wrong model, and there is no loss in choosing the correct model.

Given data x, and using, again $p_i(\omega|x) = 1$ if $\omega = M_i$ and 0 otherwise, the expected utility of m_i is then easily seen to be

$$\bar{u}(m_i \mid \boldsymbol{x}) = -\left[l_{i1} P(M_1 \mid \boldsymbol{x}) + l_{i2} P(M_2 \mid \boldsymbol{x})\right],$$

so that

$$\bar{u}(m_1 \mid \boldsymbol{x}) < \bar{u}(m_2 \mid \boldsymbol{x}) \quad \text{iff} \quad l_{12} p(M_2 \mid \boldsymbol{x}) > l_{21} p(M_1 \mid \boldsymbol{x}).$$

We thus prefer M_2 to M_1, if and only if

$$\frac{P(M_1 \mid \boldsymbol{x})}{P(M_2 \mid \boldsymbol{x})} < \frac{l_{12}}{l_{21}}\ ,$$

revealing a balancing of the posterior odds against the relative seriousness of the two possible ways of selecting the wrong model. In the symmetric case, $l_{12} = l_{21}$, the choice reduces to choosing the a posteriori most likely model, as shown earlier for the zero-one case.

The following describes the forms of so-called *Bayes tests* which arise in comparing models when the latter are defined by parametric hypotheses.

Proposition 6.1. ***(Forms of Bayes tests).*** *In comparing two models, M_1, M_2, defined by parametric hypotheses for $p(\boldsymbol{x} \mid \boldsymbol{\theta})$, with utility structure*

$$u(m_i, \boldsymbol{\omega}) = -l_{ij}, \qquad \boldsymbol{\omega} = M_j, \quad i = 1, 2, \quad j = 1, 2\,,$$

with $l_{11} = l_{22} = 0$ and $l_{12}, l_{21} > 0$, M_2 is preferred to M_1, if and only if

$$B_{12}(\boldsymbol{x}) < \frac{l_{12}}{l_{21}} \frac{P(M_2)}{P(M_1)}\,,$$

where:

$$B_{12}(\boldsymbol{x}) = \frac{p(\boldsymbol{x} \mid \boldsymbol{\theta}_1)}{p(\boldsymbol{x} \mid \boldsymbol{\theta}_2)} \qquad \textit{(simple versus simple test)},$$

$$B_{12}(\boldsymbol{x}) = \frac{p(\boldsymbol{x} \mid \boldsymbol{\theta}_1)}{\int_{\Theta_2} p(\boldsymbol{x} \mid \boldsymbol{\theta}) p_2(\boldsymbol{\theta}) d\boldsymbol{\theta}} \qquad \textit{(simple versus composite test)},$$

$$B_{12}(\boldsymbol{x}) = \frac{\int_{\Theta_1} p(\boldsymbol{x} \mid \boldsymbol{\theta}) p_1(\boldsymbol{\theta}) d\boldsymbol{\theta}}{\int_{\Theta_2} p(\boldsymbol{x} \mid \boldsymbol{\theta}) p_2(\boldsymbol{\theta}) d\boldsymbol{\theta}} \qquad \textit{(composite versus composite test)}.$$

Proof. The results follow directly from the preceding discussion. ◁

The following examples illustrate both general model comparison and a specific instance of hypothesis testing.

Example 6.5. ***(Geometric versus Poisson).*** Suppose we wish to compare the two completely specified parametric models, Negative-Binomial and Poisson, defined for conditionally independent $x_1, \ldots, x_n$, by

$$M_1 : \text{Nb}(x_i \mid \theta_1, 1), \; M_2 : \text{Pn}(x_i \mid \theta_2), \quad i = 1, \ldots, n.$$

The Bayes factor in this case is given by the simple likelihood ratio

$$B_{12}(\boldsymbol{x} \mid \theta_1, \theta_2) = \frac{\prod_{i=1}^n \text{Nb}(x_i \mid \theta_1, 1)}{\prod_{i=1}^n \text{Pn}(x_i \mid \theta_2)} = \frac{\theta_1^n (1 - \theta_1)^{n\overline{x}}}{\theta_2^{n\overline{x}} e^{-n\theta_2} \left\{\prod_{i=1}^n x_i!\right\}^{-1}}\,.$$

Suppose for illustration that $\theta_1 = \frac{1}{3}$, $\theta_2 = 2$ (implying equal mean values $E[x] = 2$ for both models); then, for example, with $n = 2$, $x_1 = x_2 = 0$, we have $B_{12}(\boldsymbol{x}) = e^4/9 \approx 6.07$, indicating an increase in plausibility for M_1; whereas with $n = 2$, $x_1 = x_2 = 2$, we have $B_{12}(\boldsymbol{x}) = 4e^4/729 \approx 0.30$, indicating a slight increase in plausibility for M_2.

Suppose now that θ_1, θ_2 are not known and are assigned the prior distributions

$$p_1(\theta_1) = \text{Be}(\theta_1 \mid \alpha_1, \beta_1), \quad p_2(\theta_2) = \text{Ga}(\theta_2 \mid \alpha_2, \beta_2),$$

whose forms are given in Section 3.2.2 (where details of $\text{Nb}(x_i \mid 1, \theta_1)$ and $\text{Pn}(x_i \mid \theta_2)$ can also be found). It follows straightforwardly that

$$p(x_1, \dots x_n \mid M_1) = \frac{\Gamma(\alpha_1 + \beta_1)}{\Gamma(\alpha_1)\Gamma(\beta_1)} \int_0^1 \theta_1^{n+\alpha_1-1}(1-\theta_1)^{n\overline{x}+\beta_1-1} d\theta$$
$$= \frac{\Gamma(\alpha_1 + \beta_1)}{\Gamma(\alpha_1)\Gamma(\beta_1)} \frac{\Gamma(n+\alpha_1)\Gamma(n\overline{x}+\beta_1)}{\Gamma(n+n\overline{x}+\alpha_1+\beta_1)},$$

and that

$$p(x_1, \dots, x_n \mid M_2) = \frac{\beta_2^{\alpha_2}}{\Gamma(\alpha_2)\prod_{i=1}^n x_i!} \int_0^\infty \theta_2^{n\overline{x}+\alpha_2-1} e^{-(n+\beta_2)\theta_2} d\theta_2$$
$$= \frac{\Gamma(n\overline{x}+\alpha_2)\beta_2^{\alpha_2}}{\Gamma(\alpha_2)(n+\beta_2)^{n\overline{x}+\alpha_2}} \frac{1}{\prod_{i=1}^n x_i!}.$$

so that

$$B_{12}(\boldsymbol{x}) = \frac{\Gamma(\alpha_1 + \beta_1)}{\Gamma(\alpha_1)\Gamma(\beta_1)} \frac{\Gamma(n+\alpha_1)\Gamma(n\overline{x}+\beta_1)}{\Gamma(n+n\overline{x}+\alpha_1+\beta_1)} \frac{\Gamma(\alpha_2)(n+\beta_2)^{n\overline{x}+\alpha_2}}{\Gamma(n\overline{x}+\alpha_2)\beta_2^{\alpha_2}} (\prod_{i=1}^n x_i!)$$

We further note that

$$E(x_i \mid M_1) = \int_0^1 E(x_i \mid M_1, \theta_1)\text{Be}(\theta_1 \mid \alpha_1, \beta_1) d\theta_1$$
$$= \int_0^1 \frac{(1-\theta_1)}{\theta_1} \text{Be}(\theta_1 \mid \alpha_1, \beta_1) d\theta_1$$
$$= \frac{\Gamma(\alpha_1 + \beta_1)}{\Gamma(\alpha_1)\Gamma(\beta_1)} \cdot \frac{\Gamma(\alpha_1 - 1)\Gamma(\beta_1 + 1)}{\Gamma(\alpha_1 + \beta_1)} = \frac{\beta_1}{\alpha_1 - 1},$$

and

$$E(x_i \mid M_2) = \int_0^\infty E(x_i \mid M_2, \theta_2)\, \text{Ga}(\theta_2 \mid \alpha_2, \beta_2) d\theta_2$$
$$= \int_0^\infty \theta_2\, \text{Ga}(\theta_2 \mid \alpha_2, \beta_2) = \frac{\alpha_2}{\beta_2},$$

so that prior specifications with $(\alpha_1 - 1)\alpha_2 = \beta_1\beta_2$ imply the same means for the two predictive models.

Table 6.1 *Dependence of $B_{12}(\boldsymbol{x})$ on prior-data combinations*

	$\alpha_1 = 2, \beta_1 = 2$ $\alpha_2 = 2, \beta_2 = 1$	$\alpha_1 = 31, \beta_1 = 60$ $\alpha_2 = 60, \beta_2 = 30$	$\alpha_1 = 2, \beta_1 = 3$ $\alpha_2 = 3, \beta_2 = 2$
$x_1 = x_2 = 0$	2.70	5.69	0.80
$x_1 = x_2 = 2$	0.29	0.30	0.49

As an illustration of the way in which the prior specification can affect the inferences, we present in Table 6.1 a selection of values of $B_{12}(\boldsymbol{x})$ resulting from particular prior-data combinations.

In the first two columns, the priors specify the same predictive means for the two models, namely $E[x \mid M_i] = 2$, but the priors in the second column are much more informative. In the final column, different predictive means are specified. Column 2 gives Bayes factors close to those obtained above assuming θ_1, θ_2 known, as might be expected from prior distributions concentrating sharply around the values $\theta_1 = \frac{1}{3}$, and $\theta_2 = 2$. However, comparison of the first and third columns for $x_1 = x_2 = 0$ makes clear that, with small data sets, seemingly minor changes in the priors for model parameters can lead to changes in direction in the Bayes factor.

The point made at the end of the above example is, of course, a general one. In any model comparison, the Bayes factor will depend on the prior distributions specified for the parameters of each model. That such dependence can be rather striking is well illustrated in the following example.

Example 6.6. ***(Lindley's paradox).*** Suppose that for $\boldsymbol{x} = (x_1, \ldots, x_n)$ two alternative models M_1, M_2, with $P(M_i) > 0, i = 1, 2$, correspond to simple and composite hypotheses about μ in $N(x_i \mid \mu, \lambda)$ defined by

$$M_1: \quad p_1(\boldsymbol{x}) = \prod_{i=1}^{n} N(x_i \mid \mu_0, \lambda), \qquad \mu_0, \lambda \text{ known},$$

$$M_2: \quad p_2(\boldsymbol{x}) = \int \prod_{i=1}^{n} N(x_i \mid \mu, \lambda) N(\mu \mid \mu_1, \lambda_1) d\mu, \qquad \mu_1, \lambda_1, \lambda \text{ known}.$$

In more conventional terminology, $x_1, \ldots, x_n$ are a random sample from $N(x \mid \mu, \lambda)$, with precision λ known; the null hypothesis is that $\mu = \mu_0$, and the alternative hypothesis is that $\mu \neq \mu_0$, with uncertainty about μ described by $N(\mu \mid \mu_1, \lambda_1)$.

Since $\overline{x} = n^{-1} \sum_{i=1}^{n} x_i$ is a sufficient statistic under both models, we easily see that

$$\begin{aligned} B_{12}(\boldsymbol{x}) &= \frac{N(\overline{x} \mid \mu_0, n\lambda)}{\int N(\overline{x} \mid \mu, n\lambda) N(\mu \mid \mu_1, \lambda_1) d\mu} \\ &= \left(\frac{\lambda_1 + n\lambda}{\lambda_1} \right)^{1/2} \frac{\exp\left\{ \frac{1}{2} \left(\lambda_1^{-1} + (n\lambda)^{-1} \right)^{-1} (\overline{x} - \mu_1)^2 \right\}}{\exp\left\{ \frac{1}{2} n\lambda (\overline{x} - \mu_0)^2 \right\}}. \end{aligned}$$

It is easily checked that, *for any fixed* $\overline{x}$, $B_{12}(\boldsymbol{x}) \to \infty$ as $\lambda_1 \to 0$, so that evidence in favour of M_1 becomes overwhelming as the prior precision in M_2 gets vanishingly small, and hence $P(M_1 \mid \boldsymbol{x}) \to 1$. In particular, this is true for $\boldsymbol{x}$ such that $\lambda^{1/2} \mid \overline{x} - \mu_0 \mid$ is large enough to cause the "null hypothesis" to be "rejected" at any arbitrary, prespecified level using a conventional significance test! This "paradox" was first discussed in detail by Lindley (1957) and has since occasioned considerable debate: see Smith (1965), Bernardo (1980), Shafer (1982b), Berger and Delampady (1987), Moreno and Cano (1989), Berger and Mortera (1991a) and Robert (1993) for further contributions and references.

A model comparison procedure which seems to be widely used implicitly in statistical practice, but rarely formalised, is the following. Given the assumption of a particular predictive model, $\{p(\boldsymbol{x} \mid \boldsymbol{\theta}), p(\boldsymbol{\theta})\}, \boldsymbol{\theta} \in \Theta$, a posterior density, $p(\boldsymbol{\theta} \mid \boldsymbol{x})$, is derived and, as we have seen in Section 5.1.5, may be at least partially summarised by identifying, for some $0 < p < 1$, a highest posterior density credible region $R_p(\boldsymbol{x})$, which is typically the smallest region such that

$$\int_{R_p(\boldsymbol{x})} p(\boldsymbol{\theta} \mid \boldsymbol{x}) d\boldsymbol{\theta} = p.$$

Intuitively, for large p, $R_p(\boldsymbol{x})$ contains those values of $\boldsymbol{\theta}$ which are most plausible given the model and the data. Conversely, $R_p^c(\boldsymbol{x})$ consists of those values of $\boldsymbol{\theta}$ which are rather implausible.

Now suppose that, given a specified p and derived $R_p(\boldsymbol{x})$, one is going to assert that the "true value" of $\boldsymbol{\theta}$ (i.e., the value onto which $p(\boldsymbol{\theta} \mid \boldsymbol{y})$ would concentrate as the size of a future sample tended to ∞) lies in $R_p(\boldsymbol{x})$. Defining the decision problem to be the choice of p, so that the possible answers to the inference problem are in $\mathcal{A} = [0, 1]$, with the state of the world ω defined to be the true $\boldsymbol{\theta}$, a value $a_p = p$ has to be chosen. An appropriate utility function may be

$$u(a_p, \boldsymbol{\theta}) = \begin{cases} f(p) & \text{for } \boldsymbol{\theta} \in R_p(\boldsymbol{x}) \\ g(1-p) & \text{for } \boldsymbol{\theta} \in R_p^c(\boldsymbol{x}) \end{cases}$$

where f and g are decreasing functions defined on $[0, 1]$. Essentially, such a utility function extends the idea of a zero-one function by reflecting the desire for a "correct" decision, but modified to allow for the fact that choosing p close to one leads to a rather vacuous assertion, whereas a correct assertion with p small is rather impressive.

The expected utility of choosing $a_p = p$ is easily seen to be

$$\bar{u}(a_p) = pf(p) + (1-p)g(1-p),$$

from which the optimal p may be derived for any specific choices of f and g. We note that if $f = g$, the unique maximum is at $p = 0.50$, so that it becomes optimal to quote a 50% highest posterior density credible region. If, for example, $f(p) = 1 - p$, $g(1-p) = [1 - (1-p)]^2 = p^2$, the resulting optimal value of p is $1/\sqrt{3} \approx 0.58$, so that a 58% credible region is appropriate. More exotically, if $f(p) = 1 - (2.7)p^2$, $g(1-p) = (1-p)^{-1}$, the reader might like to verify that a 95% credible region is optimal.

6.1.5 General Utilities

Continuing for the present with the ($\mathcal{M}$-closed) hypothesis testing framework, the consequences of incorrectly choosing a model may be less serious if the alternative models are "close" in some sense, in which case utilities of the zero-one type, which take no account of such "closeness", may be inappropriate.

One-sided Tests

We shall illustrate this idea, and forms of possibly more reasonable utility functions, by considering the special case of $\theta \in \Theta \subseteq \Re$, with parametric form $p(\boldsymbol{x} \mid \theta)$ and models M_1, M_2 defined by

$$p_1(x \mid \theta) = p(x \mid \theta), \quad \theta \in \Theta_1 = \{\theta;\ \theta \leq \theta_0\}$$
$$p_2(x \mid \theta) = p(x \mid \theta), \quad \theta \in \Theta_2 = \{\theta;\ \theta > \theta_0\}$$

for some $\theta_0 \in \Theta$. The models thus correspond to the hypotheses that the parameter is smaller or larger than some specified value θ_0.

It seems reasonable in such a situation to suppose that if one were to incorrectly choose M_2 $(\theta > \theta_0)$ rather than M_1 $(\theta \leq \theta_0)$, in many cases this would be much less serious if the true value of θ were actually $\theta_0 - \varepsilon$ than if it were $\theta_0 - 100\varepsilon$, say, for $\varepsilon > 0$. Such arguments suggest that, with the state of the world ω now representing the true parameter value θ, we might specify a utility function of the form

$$u(m_i, \omega) = u(m_i, \theta) = \begin{cases} 0 & \text{for } \theta \in \Theta_i, \\ -l_i(\theta) & \text{for } \theta \in \Theta_i^c, \end{cases}$$

for $i = 1, 2$ where l_1, l_2 are increasing positive functions of $(\theta - \theta_0)$ and $(\theta_0 - \theta)$, respectively. The expected utility of the decision corresponding to m_i (i.e., the choice of M_i) is therefore given by

$$\bar{u}(m_i \mid \boldsymbol{x}) = -\int_{\Theta_i^c} l_i(\theta) p(\theta \mid \boldsymbol{x}) d\theta,$$

where

$$p(\theta \mid \boldsymbol{x}) = \frac{p(\boldsymbol{x} \mid \theta) p(\theta)}{\int_\Theta p(\boldsymbol{x} \mid \theta) p(\theta) d\theta}.$$

The optimal answer to the inference problem is to prefer M_1 to M_2 if and only if

$$\bar{u}(m_1 \mid \boldsymbol{x}) > \bar{u}(m_2 \mid \boldsymbol{x}),$$

with explicit solutions depending, of course, on the choices of l_1, l_2, and the form of $p(\theta \mid \boldsymbol{x})$, as illustrated in the following example.

Example 6.7. ***(Normal posterior; linear losses).***

If $l_1(\theta) = \theta - \theta_0$, $l_2(\theta) = k(\theta_0 - \theta)$, with k reflecting the relative seriousness of "over-estimating" by choosing model M_1, and $p(\theta \mid \boldsymbol{x})$, given $\boldsymbol{x} = x_1, \ldots, x_n$, is $N(\theta \mid \mu_n, \lambda_n)$, say, then we have

$$\begin{aligned} \bar{u}(m_1 \mid \boldsymbol{x}) &= -\int_{\theta > \theta_0} (\theta - \theta_0) N(\theta \mid \mu_n, \lambda_n)\, d\theta \\ &= -\lambda_n^{-1/2} \Psi_1[\lambda_n^{-1/2}(\theta_0 - \mu_n)], \end{aligned}$$

where

$$\Psi_1(t) = N(t \mid 0, 1) - t \int_t^\infty N(s \mid 0, 1)\, ds\ ,$$

and

$$\begin{aligned}\bar{u}(m_2 \mid \boldsymbol{x}) &= -k \int_{\theta \le \theta_0} (\theta_0 - \theta) N(\theta \mid \mu_n, \lambda_n)\, d\theta \\ &= -k\lambda_n^{-1/2} \Psi_2 \left[\lambda_n^{-1/2}(\theta_0 - \mu_n)\right],\end{aligned}$$

where

$$\Psi_2(t) = N(t \mid 0, 1) + t \int_0^t N(s \mid 0, 1)\, ds.$$

It is therefore optimal to prefer M_1 to M_2 if and only if

$$k\Psi_2 \left[\lambda_n^{-1/2}(\theta_0 - \mu_n)\right] > \Psi_1 \left[\lambda_n^{-1/2}(\theta_0 - \mu_n)\right].$$

In the symmetric case, $k = 1$, it is easily seen that this reduces to preferring M_1 if and only if $\mu_n < \theta_0$, as one might intuitively have expected. For references and further discussion of related topics, see DeGroot (1970, Chapter 11), and Winkler (1972, Chapter 6).

Prediction

Moving away now from model comparisons which reduce to hypothesis tests in parametric models, let us consider the problem of model comparison or choice, given data x, in order to make a point prediction for a future observation y.

The general decision structure is that given schematically in Figure 6.2, where, assuming real-valued observables, m_i corresponds to acting in accordance with model M_i, a_j, $j \in J_i$ denotes the choice, based on M_i, of a prediction, $\hat{y}_i$, for a future observation y, and we shall assume a "quadratic loss" utility,

$$u(m_i, \hat{y}_i, y) = -(\hat{y}_i - y)^2, \quad i \in I.$$

We recall from the analysis given in Section 6.1.3 that the optimal model choice is m^*, given by

$$\bar{u}(m^* \mid \boldsymbol{x}) = \sup_{i \in I} \int u(m_i, \hat{y}_i^*, y) p(y \mid \boldsymbol{x}) dy,$$

where $\hat{y}_i^*$ is the optimal prediction of a future observation y, given data x and assuming model M_i; that is, the value $\hat{y}$ which minimises

$$\int (\hat{y} - y)^2 p_i(y \mid \boldsymbol{x}) dy,$$

where $p_i(y \mid \boldsymbol{x})$ is the predictive density for y given model M_i. It then follows immediately that

$$\hat{y}_i^* = \int y\, p_i(y \mid \boldsymbol{x}) dy = E[y \mid \boldsymbol{x}, M_i] \,,$$

the predictive mean, given model M_i, so that

$$\int u(m_i, \hat{y}_i^*, y) p(y \mid \boldsymbol{x}) dy = -\int (\hat{y}_i^* - y)^2 p(y \mid \boldsymbol{x}) dy, \quad i \in I.$$

Completion of the analysis now depends on the specification of the overall actual belief distribution $p(y \mid \boldsymbol{x})$ and the computation of the expectation of $(y_i^* - y)^2$, $i \in I$, with respect to $p(y \mid \boldsymbol{x})$.

Again, in the $\mathcal{M}$-completed case there is nothing further to be said explicitly; one simply carries out the necessary evaluations, using the appropriate form of $p(y \mid \boldsymbol{x})$, by numerical integration if necessary.

In the $\mathcal{M}$-open case, the detailed analysis of the problem of point prediction with quadratic loss will be given in Section 6.1.6.

In the $\mathcal{M}$-closed case, we have

$$p(y \mid \boldsymbol{x}) = \sum_{i \in I} p(M_i \mid \boldsymbol{x}) p_i(y \mid \boldsymbol{x}),$$

and, after some rearrangement, it is easily seen that

$$\int (\hat{y}_i^* - y)^2 p(y \mid \boldsymbol{x}) dy = \sum_{j \in I} p(M_j \mid \boldsymbol{x}) \int (\hat{y}_i^* - \hat{y}_j^* + \hat{y}_j^* - y)^2 p_j(y \mid \boldsymbol{x}) dy,$$

which reduces to

$$\sum_{j \in I} p(M_j \mid \boldsymbol{x}) V[y \mid M_j, \boldsymbol{x}] + \sum_{j \in I} (\hat{y}_i^* - \hat{y}_j^*)^2 p(M_j \mid \boldsymbol{x}).$$

The first term does not depend on i, and the second term can be rearranged in the form

$$(\hat{y}_i^* - \hat{y}^*)^2 + \sum_{j \in I} (\hat{y}_j^* - \hat{y}^*)^2 p(M_j \mid \boldsymbol{x}),$$

where $\hat{y}^*$ is the weighted prediction

$$\hat{y}^* = \sum_{j \in I} \hat{y}_j^*\, p(M_j \mid \boldsymbol{x}).$$

The preferred model M_i is therefore seen to be that for which the resulting prediction, $\hat{y}_i^*$, is closest to $\hat{y}^*$, the posterior weighted-average, over models, of the individual model predictions. If $k = 2$, it is easily checked that the preferred model is simply that with the highest posterior probability.

If we wish to make a prediction, but without first choosing a specific model, it is easily seen that the analysis of the problem in terms of the schematic decision problem given in Figure 6.3 of Section 6.1.2 leads directly to $\hat{y}^*$ as the optimal prediction.

Clearly, the above analyses go through in an obvious way, with very few modifications, if, instead of prediction, we were to consider point estimation, with quadratic loss, of a parameter common to all the models. More generally, the analysis can be carried out for loss functions other than the quadratic.

Reporting Inferences

Generalising beyond the specific problems of point prediction and estimation, let us consider the problem of model comparison or choice *in order to report inferences* about some unknown state of the world ω. For example, the latter might be a common model parameter, a function of future observables, an indicator function of the future realisation of a specified event, or whatever.

A major theme of our development in Chapters 2 and 3 has been that the problem of reporting beliefs about ω is itself a decision problem, where the possible answers to the inference problem are the consists of the class of probability distributions for ω which are compatible with given data. The appropriate utility functions in such problems were seen to be the *score functions* discussed in Sections 2.7 and 3.4. This general decision problem is thus a special case of that represented by Figure 6.2, where, given data x, m_i represents the choice of model M_i, the subsequent answer a_j, $j \in J_i$, to the inference problem is some report of beliefs about ω, assuming M_i, and the utility function is defined by

$$u(m_i, a_j, \omega) = u_i(q_i(\cdot \,|\, x), \omega),$$

for some score function u_i, and form of belief report, $q_i(\cdot \,|\, x)$, about ω, corresponding to d_j, $j \in J_i$.

If $p_i(\cdot \,|\, x)$ is the form of belief report for ω actually implied by m_i and if u_i is a *proper* scoring rule (see, for example, Definition 3.16) then it follows that the optimal a_j, $j \in J_i$ must be $a_i^* = p_i(\cdot \,|\, x)$ and that

$$u(m_i, a_i^*, \omega) = u_i(p_i(\cdot \,|\, x), \omega), \quad i \in I.$$

If, moreover, the score function is local (see, for example, Definition 3.18), we have the logarithmic form

$$u(m_i, a_i^*, \omega) = A \log p_i(\omega \,|\, x) + B(\omega), \quad i \in I, \quad A > 0,$$

for $a > 0$ and $B(\omega)$ arbitrary, in accordance with Proposition 3.13. The expected utility of m_i is therefore given by

$$\bar{u}(m_i \mid x) = \int \{a \log p_i(\omega \mid x) + B(\omega)\} \, p(\omega \mid x) d\omega,$$

and the preferred model is the M_i for which this is maximised over $i \in I$.

Comments about the detailed implementation of the analysis in the $\mathcal{M}$-open case are similar to those made in the previous problem.

For $\mathcal{M}$-closed models, we have the more explicit form

$$\bar{u}(m_i \mid x) = a \sum_{j \in I} p(M_j \mid x) \int p_j(\omega \mid x) \log p_i(\omega \mid x) d\omega + \int B(\omega) p(\omega \mid x) d\omega,$$

which, after straightforward rearrangement, shows that the preferred m_i is given by minimising, over $i \in I$,

$$\sum_{j \in I} p(M_j \mid x) \int p_j(\omega \mid x) \log \left[\frac{p_j(\omega \mid x)}{p_i(\omega \mid x)} \right] d\omega,$$

the posterior weighted-average, over models, of the logarithmic divergence (or discrepancy) between $p_i(\omega \mid x)$ and each of $p_j(\omega \mid x)$, $j \neq i \in I$.

If, instead, we were to adopt the (proper) quadratic scoring rule (see, for example, Definition 3.17), we obtain, ignoring irrelevant constants,

$$\bar{u}(m_i \mid x) \propto \int \left\{ 2p_i(\omega \mid x) - \int p_i^2(\omega \mid x) d\omega \right\} p(\omega \mid x) \, d\omega,$$

so that, after some algebraic rearrangement, in the case of $\mathcal{M}$-closed models the preferred M_i is seen to be that which minimises, over $i \in I$,

$$\sum_{j \in I} p(M_j \mid x) \int p_j(\omega \mid x) \left[f\{p_j(\omega \mid x)\} - f\{p_i(\omega \mid x)\} \right] d\omega,$$

where

$$f\{p_j(\omega \mid x)\} = 2p_j(\omega \mid x) - \int p_j^2(\omega \mid x) d\omega.$$

Comparison of the solutions for the logarithmic and quadratic cases reveals that if, for arbitrary f,

$$\delta\{q(\omega) \mid p(\omega)\} = \int p(\omega)[f\{p(\omega)\} - f\{q(\omega)\}] d\omega,$$

defines a discrepancy measure between p and q, both may be characterised as identifying the M_i for which

$$\sum_{j \in I} p(M_j \mid x) \delta\{p_j(\omega \mid x) \mid p_i(\omega \mid x)\}$$

is minimised over $i \in I$, the differences in the two cases corresponding to the form of f (logarithmic or quadratic, respectively).

Example 6.8. ***(Point prediction versus predictive beliefs).***
To illustrate the different potential implications of model comparison on the basis of quadratic loss for point prediction versus model comparison on the basis of logarithmic score for predictive belief distributions, consider the following simple ($\mathcal{M}$-closed) example.

Suppose that alternative models M_1, M_2 for $\boldsymbol{x} = (x_1, \ldots, x_n)$ are defined by:

$$M_j: \quad p(\boldsymbol{x}) = \int \prod_{i=1}^{n} N(x_i \mid \mu, \lambda_j) N(\mu \mid \mu_0, \lambda_0) d\mu, \quad j = 1, 2,$$

with $\lambda_1, \lambda_2, \mu_0, \lambda_0$ known: we are thus assuming normal data models with precisions λ_1, λ_2, respectively, and uncertainty about μ described by $N(\mu \mid \mu_0, \lambda_0)$ in both cases.

Now, given $\boldsymbol{x}$, consider two decision problems: the first problem consists in selecting a model and then providing a *point prediction*, with respect to quadratic loss, for the next observable, x_{n+1}; the second problem consists in selecting a model and then providing a *predictive distribution* for x_{n+1}, with respect to a logarithmic score function.

For the first problem, straightforward manipulation shows that the predictive distribution for x_{n+1} assuming model M_j is given by

$$p(x_{n+1} \mid M_j, \boldsymbol{x}) = p_j(x_{n+1} \mid \boldsymbol{x}) = N\left(x_{n+1} \,\middle|\, \mu_n(j), \lambda_j\left(\frac{n + \lambda_0/\lambda_j}{n + 1 + \lambda_0/\lambda_j}\right)\right),$$

where

$$\mu_n(j) = \frac{\lambda_0 \mu_0 + n\lambda_j \overline{x}}{\lambda_0 + n\lambda_j},$$

so that, corresponding to the analysis given earlier in this section, model M_j leads to the prediction $\mu_n(j)$, $j = 1, 2$, and the preferred model is M_1 if and only if

$$\frac{P(M_1 \mid \boldsymbol{x})}{P(M_2 \mid \boldsymbol{x})} > 1.$$

To identify these posterior probabilities, we note that, if $s^2 = n^{-1}\Sigma(x_i - \overline{x})^2$,

$$\begin{aligned} B_{12}(\boldsymbol{x}) = \frac{p(\boldsymbol{x} \mid M_1)}{p(\boldsymbol{x} \mid M_2)} &= \frac{p(\overline{x}, s^2 \mid M_1)}{p(\overline{x}, s^2 \mid M_2)} \\ &= \frac{\lambda_1 \chi^2_{n-1}(n\lambda_1 s^2) N(\overline{x} \mid \mu_0, \lambda_0 + n\lambda_1)}{\lambda_2 \chi^2_{n-1}(n\lambda_2 s^2) N(\overline{x} \mid \mu_0, \lambda_0 + n\lambda_2)}, \end{aligned}$$

which, for small λ_0, is well approximated by

$$\left(\frac{\lambda_1}{\lambda_2}\right)^{(n-1)/2} \exp\left\{-\frac{ns^2}{2}(\lambda_1 - \lambda_2)\right\}.$$

The posterior model probabilities are then given by

$$P(M_1 \mid \boldsymbol{x}) = \frac{B_{12}(\boldsymbol{x}) p_{12}}{1 + B_{12}(\boldsymbol{x}) p_{12}}, \quad P(M_2 \mid \boldsymbol{x}) = \frac{1}{1 + B_{12}(\boldsymbol{x}) p_{12}},$$

where $p_{12} = P(M_1)/P(M_2)$. Model M_1 is therefore preferred if and only if

$$\log[B_{12}(\boldsymbol{x})p_{12}] > 0.$$

In the case of equal prior weights, $p_{12} = 1$ and, assuming small λ_0, if we write the condition in terms of the model variances $\sigma_j^2 = \lambda_j^{-1}, j = 1, 2$, we prefer M_1 when

$$\frac{\Sigma(x_i - \overline{x})^2}{n-1} < \frac{\sigma_1^2\sigma_2^2[\log \sigma_2^2 - \log \sigma_1^2]}{\sigma_2^2 - \sigma_1^2}.$$

Noting that the left-hand side is an intuitively reasonable data-based estimate of the model variance, we see that model choice reduces to a simple cut-off rule in terms of this estimate.

For the second decision problem, the logarithmic divergence of $p(x_{n+1} \mid M_1, \boldsymbol{x})$ from $p(x_{n+1} \mid M_2, \boldsymbol{x})$ is given, for small λ_0, by

$$\begin{aligned}
\delta_{12} &= \int N\left(x_{n+1} \,\middle|\, \overline{x}, \lambda_2 \frac{n}{n+1}\right) \log \frac{N\left(x_{n+1} \,|\, \overline{x}, \lambda_2\left(n/(n+1)\right)\right)}{N\left(x_{n+1} \,|\, \overline{x}, \lambda_1\left(n/(n+1)\right)\right)} \, dx_{n+1} \\
&= \frac{1}{2}\log\frac{\lambda_2}{\lambda_1} + \frac{1}{2}\left[\frac{\lambda_2}{\lambda_1} - 1\right] \\
&= \frac{1}{2}\left\{\log\frac{\sigma_1^2}{\sigma_2^2} + \left(\frac{\sigma_1^2}{\sigma_2^2} - 1\right)\right\},
\end{aligned}$$

with a corresponding expression for δ_{21}. The general analysis given above thus implies that model M_1 is preferred if and only if

$$P(M_1 \mid \boldsymbol{x})\delta_{21} > P(M_2 \mid \boldsymbol{x})\delta_{12},$$

i.e., if and only if

$$\frac{P(M_1 \mid \boldsymbol{x})}{P(M_2 \mid \boldsymbol{x})} > \frac{\delta_{12}}{\delta_{21}},$$

(rather than > 1, as in the point prediction case). Note, incidentally, that should it happen that $P(M_1 \mid \boldsymbol{x}) = P(M_2 \mid \boldsymbol{x})$, model M_1, would be preferred if and only if $\delta_{12} < \delta_{21}$, which happens if and only if $\lambda_1 > \lambda_2$. Intuitively, all other things being equal, we prefer in this case the model with the smallest predictive variance.

To obtain some limited insight into the numerical implications of these results, consider the case where $\sigma_1^2 = 1$, $\sigma_2^2 = 25$, $n = 4$, $P(M_1) = P(M_2) = \frac{1}{2}$ and $s^2 = 3$, which gives $B_{12} = 0.394$, so that $P(M_1 \mid \boldsymbol{x}) = 0.28$, $P(M_2 \mid \boldsymbol{x}) = 0.72$. Using the point prediction with quadratic loss criterion, we therefore prefer M_2. However, $\delta_{12} = 1.129$ and $\delta_{21} = 10.31$, so that if we want to choose a predictive distribution in accordance with the logarithmic score criterion we prefer M_1, since $(0.28)/(0.72) > (1.129)/(10.31)$. However, if $s^2 = 4$, the reader might like to verify that M_2 is preferred under both criteria ($B_{12} = 0.058$, implying that $\Pr(M_1 \mid \boldsymbol{x}) = 0.055$).

6.1.6 Approximation by Cross-validation

For the general problem of model choice followed by a subsequent answer to an inference problem, the analysis based on Figure 6.2 implies that the optimal choice of model from $\mathcal{M}$ is the M_i for which

$$\int u(m_i, a_i^*, \boldsymbol{\omega}) p(\boldsymbol{\omega} \mid \boldsymbol{x}) d\boldsymbol{\omega}$$

is maximised over $i \in I$, where a_i^* denotes the optimal subsequent decision given M_i. In the $\mathcal{M}$-closed case, we have seen that the mixture form of $p(\boldsymbol{\omega} \mid \boldsymbol{x})$ enables an explicit form of general solution to be exhibited; in the $\mathcal{M}$-completed case, we have noted that the solution is in principle available, given appropriate computation.

We turn now to the case of model comparison within the $\mathcal{M}$-open framework. What can be done to compare the values of M_i, $i \in I$, as proxies for an actual belief model which itself has not been specified, so that $p(\boldsymbol{\omega} \mid \boldsymbol{x})$ is not available?

We shall illustrate a possible approach to this problem by detailed consideration of the special case where $\boldsymbol{\omega} = y$, a future observation, for which a point prediction with respect to quadratic loss, or a predictive distribution with respect to logarithmic or quadratic score, is required.

First, we note that, in all these cases, the expected utility of the choice M_i, $i \in I$, has the mathematical form

$$\int u(m_i, a_i^*, y) p(y \mid \boldsymbol{x}) dy = \int f_i(y, \boldsymbol{x}) p(y \mid \boldsymbol{x}) dy,$$

for some function f_i of y and x, depending on i, whose form can be explicitly identified. For example, for point prediction with quadratic loss, we have

$$f_i(y, \boldsymbol{x}) = -\{E[y \mid M_i, \boldsymbol{x}] - y\}^2;$$

for a predictive distribution with logarithmic score function we have, ignoring irrelevant terms,

$$f_i(y, \boldsymbol{x}) = \log p(y \mid M_i, \boldsymbol{x});$$

and with a quadratic score function we have

$$f_i(y, \boldsymbol{x}) = 2p(y \mid M_i, \boldsymbol{x}) - \int p^2(y \mid M_i, \boldsymbol{x}) dy.$$

Secondly, we note that there are n possible partitions of $\boldsymbol{x} = \boldsymbol{x}_n = (x_1, \ldots, x_n)$ into $\boldsymbol{x}_n = [\boldsymbol{x}_{n-1}(j), x_j]$, $j = 1, \ldots, n$, where $\boldsymbol{x}_{n-1}(j) = \boldsymbol{x}_n - \{x_j\}$ denotes $\boldsymbol{x}_n$ with x_j deleted, and that, if n is reasonably large, and the x's are exchangeable, each such partition effectively provides $\boldsymbol{x}_{n-1}(j)$ as a "proxy" for $\boldsymbol{x}$ and x_j as a "proxy" for y. If we now randomly select k from these n partitions, a standard law of large numbers argument suggests that, as $n, k \to \infty$,

$$\left| \int u(m_i, a_i^*, y) p(y \,|\, \boldsymbol{x}) dy - \frac{1}{k} \sum_{j=1}^{k} f_i(x_j, \boldsymbol{x}_{n-1}(j)) \right| \to 0,$$

so that the expected utilities of M_i, $i \in I$, can be compared on the basis of the quantities

$$\frac{1}{k} \sum_{j=1}^{k} f_i(x_j, \boldsymbol{x}_{n-1}(j)), \quad i \in I.$$

In the case of point prediction, if y is a future observation and $\hat{y}_i^*(j)$ denotes the value of $E[y \,|\, M_i, x]$ when x is replaced by $\boldsymbol{x}_{n-1}(j)$, this approximation implies that we minimise, over $i \in I$,

$$\frac{1}{k} \sum_{j=1}^{k} \left(\hat{y}_i^*(j) - x_j \right)^2,$$

which is an average measure, using squared distance, of how well M_i performs when, on a leave-one-out-at-a-time basis, it attempts to predict a missing part of the data from an available subset of the data.

In the case of a predictive distribution with a logarithmic score, we maximise, over $i \in I$,

$$\frac{1}{k} \sum_{j=1}^{k} \log p(x_j \,|\, M_i, \boldsymbol{x}_{n-1}(j)),$$

which can be regarded as an average measure based on the logarithm of the integrated likelihood under model M_i, and can be conveniently rewritten, for computational purposes, in the form

$$\log p(\boldsymbol{x} \,|\, M_i) - \frac{1}{k} \sum_{j=1}^{k} \log p(\boldsymbol{x}_{n-1}(j) \,|\, M_i).$$

In the case of comparing two models, M_1, M_2, this criterion can be given an interesting reformulation. Under the logarithmic prediction distribution utility, and writing $p_i(y \,|\, \boldsymbol{x}) = p(y \,|\, M_i, \boldsymbol{x})$, we can rearrange the criterion to see that we prefer model M_1 if

$$\int \log \frac{p_1(y \,|\, \boldsymbol{x})}{p_2(y \,|\, \boldsymbol{x})} p(y \,|\, \boldsymbol{x}) \, dy > 0,$$

where, however, in this $\mathcal{M}$-open perspective, $p(y \,|\, \boldsymbol{x})$ is ***not*** specified. But, as we saw above, we can form n partitions $\boldsymbol{x}_n = [\boldsymbol{x}_{n-1}(j), x_j]$ such that $\boldsymbol{x}_{n-1}(j) = \boldsymbol{x}_n - x_j$ is, for large n, a "proxy" for $\boldsymbol{x}$ and x_j is a "proxy" for y. It follows that, if we randomly select k of these partitions, the quantity

$$\frac{1}{k} \sum_{j=1}^{k} \log \frac{p_1(x_j \,|\, \boldsymbol{x}_{n-1}(j))}{p_2(x_j \,|\, \boldsymbol{x}_{n-1}(j))}$$

provides a (consistent, as $n \to \infty$) Monte Carlo estimate of the left-hand side of the model criterion above. But this, in turn, can be rewritten so that the criterion implies preferring model M_1 if

$$\prod_{j=1}^{k}\left[\frac{p_1(x_j \mid \boldsymbol{x}_{n-1}(j))}{p_2(x_j \mid \boldsymbol{x}_{n-1}(j))}\right]^{1/k} = \prod_{j=1}^{k}\left[B_{12}(x_j, \boldsymbol{x}_{n-1}(j))\right]^{1/k} > 1,$$

for $j = 1, \ldots, k$, where $B_{12}(x_j, \boldsymbol{x}_{n-1}(j))$ denotes the Bayes factor for M_1 against M_2, based on the versions

$$\{p_i(x_j \mid \boldsymbol{\theta}_i), p_i(\boldsymbol{\theta}_i \mid \boldsymbol{x}_{n-1}(j))\}$$

of M_1, M_2. We recall from Section 6.1.4 the role of Bayes factors based on the versions $\{p_i(\boldsymbol{x} \mid \boldsymbol{\theta}_i), p_i(\boldsymbol{\theta}_i)\}$ of M_1, M_2, in the context of zero-one loss functions and the $\mathcal{M}$-closed perspective. Although there are clear differences here in formulation ($\mathcal{M}$-open versus $\mathcal{M}$-closed, log-predictive utility versus 0-1 loss), it is interesting to note the role played again by the Bayes factor. One interesting difference is the following (Pericchi, 1993). In Section 6.1.4, the Bayes factor is evaluating M_1, M_2 on the basis of the models' ability to "predict" x given no data (beyond what has been used to specify $p_i(\boldsymbol{\theta}_i)$). In contrast, in the above we are taking a geometric average of Bayes factors which are evaluating M_1, M_2 on the basis of the models' ability to predict one further observable, given $n-1$ observations. The former situation puts the emphasis on "fidelity to the observed data"; the latter puts the emphasis on "future predictive power".

These kinds of approximate performance measurements for comparing models could obviously be generalised by considering random partitions of x involving leave-several-out-at-a-time techniques. We shall not develop such ideas further here—apart from giving one further interesting illustration in Section 6.3.3—but merely note that the above approximation to the optimal Bayesian procedure leads naturally to a *cross-validation* process, which results in a preference for models under which the data achieve the highest levels of "internal consistency". Thus, for example, in both the quadratic loss and logarithmic score cases, if under model M_i there are x_j which are "surprising" in the light of $\boldsymbol{x}_{n-1}(j)$, thus leading to large squared distance terms or small log-integrated-likelihood values, respectively, the performance measure will penalise M_i.

Model choice and estimation procedures involving cross-validation (sometimes called *predictive sample reuse*) have been proposed by several authors, from a mainly non-Bayesian perspective, as a pragmatic device for generating statistical methods without seeming to invoke a "true" sampling model: see, for example, Stone (1974) and Geisser (1975) for early accounts and Shao (1993) for a recent perspective. The above development clearly establishes that such cross-validatory techniques do indeed have an interesting role in a Bayesian decision-theoretic setting for approximating expected utilities in decision problems where a set of alternative models are to be compared without wishing to act as if one of the models were "true", and in the absence of a specified actual belief model.

Example 6.9. ***(Lindley's paradox revisited).***
In Example 6.6, we considered the case of two alternative models, M_1, M_2, for $\boldsymbol{x} = (x_1, \ldots, x_n)$, corresponding to simple and composite hypotheses about μ in $N(x_i \mid \mu, \lambda)$ and defined by:

$$M_1 : \; p_1(\boldsymbol{x}) = \prod_{i=1}^{n} N(x_i \mid \mu_0, \lambda), \quad \mu_0, \lambda \text{ known};$$

$$M_2 : \; p_2(\boldsymbol{x}) = \int \prod_{i=1}^{n} N(x_i \mid \mu, \lambda) N(\mu \mid \mu_1, \lambda_1) d\mu, \quad \mu_1, \lambda_1, \lambda \text{ known}.$$

The analysis given in Example 6.6 was within the $\mathcal{M}$-closed context with $P(M_i) > 0$, $i = 1, 2$, and it was shown that, as $\lambda_1 \to 0$, $P(M_1 \mid \boldsymbol{x}) \to 1$ for *any* fixed $\boldsymbol{x}$. It follows from results given in Sections 6.1.3 and 6.1.4 that, as $\lambda_1 \to 0$, M_1 would be the preferred model under either zero-one utility, or quadratic loss utility for point prediction (since in this latter case, the criterion reduces to the comparison of posterior probabilities when just two models are being compared).

We shall now reconsider the case of quadratic loss for point prediction in the $\mathcal{M}$-open context.

First, we note that, given $\boldsymbol{x}$, the optimal prediction of a future observation, y, under M_1 is just $\hat{y}_1^* = \mu_0$, whereas (making appropriate notational changes to the results given in Example 5.10) under M_2 the optimal prediction is

$$\hat{y}_2^* = \mu_n = \frac{(\lambda_1 \mu_1 + n\lambda\overline{x})}{\lambda_1 + n\lambda} = (1 - w_n)\mu_1 + w_n \overline{x}$$

where $w_n = n\lambda(\lambda_1 + n\lambda)^{-1}$. Secondly, from the cross-validation approximation analysis given above we see that M_1 is preferred to M_2 if and only if, based on k random partitions of $\boldsymbol{x}$ into x_j and $\boldsymbol{x}_{n-1}(j)$,

$$\sum_{j=1}^{k} (\mu_0 - x_j)^2 < \sum_{j=1}^{k} [\{(1 - w_{n-1})\mu_1 + w_{n-1}\overline{x}_{n-1}(j)\} - x_j]^2$$

where $\overline{x}_{n-1}(j) = \overline{x} + (n-1)^{-1}(\overline{x} - x_j)$ is the mean of the sample $\boldsymbol{x}$ with x_j omitted. Intuitively, M_2 will be preferred if the posterior mean on average does better as a predictor than μ_0. In particular, if $\lambda_1 \to 0$, and $k = n$, an approximate analysis shows that M_1 is preferred to M_2 if and only if

$$\sum_{j=1}^{k} (\mu_0 - x_j)^2 < \sum_{j=1}^{k} (\overline{x}_{n-1}(j) - x_j)^2$$

$$< \left(\frac{n}{n-1}\right)^2 \sum_{j=1}^{k} (\overline{x} - x_j)^2.$$

This is easily seen (Pericchi, 1993) to be equivalent to preferring M_1 if, and only if,

$$\frac{n(\mu_0 - \overline{x})^2}{\sum_{j=1}^{n}(x_j - \overline{x})^2/(n-1)} < \frac{2n-1}{n-1} \approx 2,$$

which, under M_1, is equivalent to rejecting M_1 if a Snedecor $F_{1,n-1}$ random quantity exceeds the value 2. See Leonard and Ord (1976) for a related argument.

This result provides a marked contrast to that obtained in Example 6.6 and makes clear that, even given the same data and utility criterion, preferences among models in $\mathcal{M}$ may differ radically, depending on whether one is approaching their comparison from an $\mathcal{M}$-closed or $\mathcal{M}$-open perspective.

6.1.7 Covariate Selection

We have already had occasion to remark several times that our emphasis in this volume is on concepts and theory, and that complex case-studies and associated computation will be more appropriately discussed in subsequent volumes. That said, it might be illuminating at this juncture to indicate briefly how the theory we have developed above can be applied in contexts which are much more complicated than those of the simple, stylised examples on which most of our discussion has been based. To this end, we shall consider the important problem of model comparison which arises when we try to identify appropriate covariates for use in practical prediction and classification problems.

To fix ideas, consider the following problem. Some kind of decision is to be made regarding an unknown state of the world ω relating to an individual unit of a population: for example, classifying the, as yet unknown, disease state of a specific patient, or predicting the, as yet unknown, quality level of the output from a particular run of an industrial production process. Possible predictive models are to be based, for various choices of m, on covariates $y_1(\boldsymbol{z}), \ldots, y_m(\boldsymbol{z})$, which are themselves selected functions of $\boldsymbol{z} = (z_1, \ldots, z_s)$, representing all possible observed relevant attributes (discrete or continuous) for the individual population unit: for example, the patient's complete recorded clinical history, or a record of all the input and control parameters of the production run. To aid the modelling process, a data bank (of "training data") is available consisting of

$$\boldsymbol{D} = \{(\omega^j, z_1^j, \ldots z_s^j), j = 1, \ldots, n\},$$

recording all the attributes and (eventually known) states of the world for n previously observed units of the same population: for example, n previous patients presenting at the same clinic, or n previous runs of the same production process.

We shall suppose that the ultimate objective is to provide, for the state of the world ω of the new individual unit, a predictive distribution $p(\omega \mid \boldsymbol{y}(\boldsymbol{z}), \boldsymbol{D})$, where y denotes a generic element of the set of all possible $\{y_i(\cdot), i = 1, \ldots, m\}$ under consideration for defining covariates. If ω is discrete, we typically refer to the problem as one of *classification*; if ω is continuous, we refer to it as one of *prediction*.

To simplify the exposition, we shall suppose that identification of the density $p(\cdot \mid \boldsymbol{y}(\boldsymbol{z}), \boldsymbol{D})$ is equivalent to the identification of $\boldsymbol{y} \in \mathcal{Y}$, where $\mathcal{Y}$ denotes the class of all y under consideration. The particular forms in $\mathcal{Y}$ will depend, of course, on the practical problem under consideration: typically, however, it will include functions mapping z to z_i, $i = 1, \ldots, s$, so that individual attributes themselves are also eligible to be chosen as covariates. Then, if $u\{p(\cdot \mid \boldsymbol{y}(\boldsymbol{z}), \boldsymbol{D}), \omega\}$ denotes a utility function for using the predictive form $p(\cdot \mid \boldsymbol{y}(\boldsymbol{z}), \boldsymbol{D})$ when ω turns out to be the true state of the world, the resulting decision problem is shown schematically in Figure 6.4.

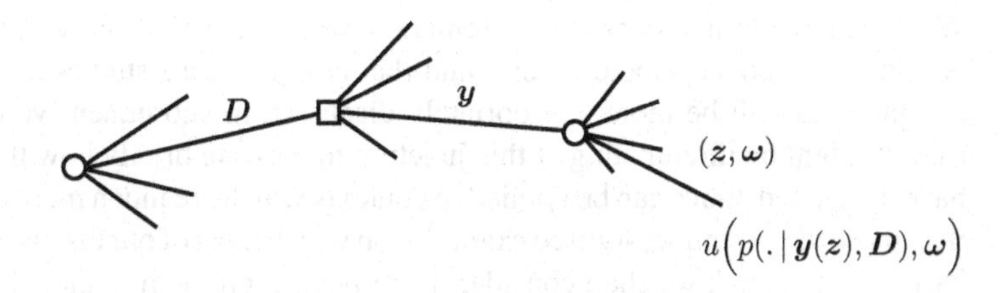

Figure 6.4 *Selection of covariates as a decision problem*

If $p(\boldsymbol{z}, \omega \mid \boldsymbol{D})$ represents the predictive distribution for $(\boldsymbol{z}, \omega)$, given the "training data" $\boldsymbol{D}$, the different possible models corresponding to the different possible choices of covariates, $\boldsymbol{y} \in Y$, are then compared on the basis of their expected utilities

$$\bar{u}(\boldsymbol{y} \mid \boldsymbol{D}) = \int u\{p(\cdot \mid \boldsymbol{y}(\boldsymbol{z}), \boldsymbol{D}), \omega\} p(\boldsymbol{z}, \omega \mid \boldsymbol{D}) d\boldsymbol{z} d\omega, \quad \boldsymbol{y} \in Y.$$

The resulting optimal choice will, of course, depend on the form of the utility function. Typically, the latter will not only incorporate a score function component for assessing $p(\cdot \mid \boldsymbol{y}(\boldsymbol{z}), \boldsymbol{D})$, but possibly also a cost component, reflecting the different costs associated with the use of different covariates $\boldsymbol{y}$. For example, in the case of disease classification the use of fewer covariates could well mean cheaper and quicker diagnoses; in the case of predicting production quality the use of fewer covariates could cut costs by requiring less on-line measurement and monitoring. If we suppose, for simplicity, that the utility function can be decomposed into additive score and cost components,

$$s\{p(\cdot \mid \boldsymbol{y}(\boldsymbol{z}), \boldsymbol{D}), \omega\} - c\{\boldsymbol{y}(\boldsymbol{x})\},$$

the expected utility of the choice y is given by

$$\bar{u}(\boldsymbol{y} \mid D) = \int s\{p(\cdot \mid \boldsymbol{y}(\boldsymbol{z}), \boldsymbol{D}), \boldsymbol{\omega}\} p(\boldsymbol{z}, \boldsymbol{\omega} \mid \boldsymbol{D}) d\boldsymbol{z} d\boldsymbol{\omega} - \int c\{\boldsymbol{y}(\boldsymbol{z})\} d\boldsymbol{z}.$$

In many cases, it will be natural to use proper score functions, for example, quadratic or logarithmic. If costs are omitted, the optimal model will typically involve a large number of covariates; if cost functions are used which increase with the number of covariates in the model, a small subset of the latter will typically be optimal. More pragmatically, one could ignore costs, identify the optimal $\boldsymbol{y}^*_{(i)}$, $i = 1, 2, \ldots$ over all possible choices of one covariate, two covariates, etc., observe that

$$\bar{u}(\boldsymbol{y}^*_{(1)} \mid D) \leq \bar{u}(\boldsymbol{y}^*_{(2)} \mid D) \leq \cdots \leq \bar{u}(\boldsymbol{y}^*_{(i)} \mid \boldsymbol{D}) \leq \cdots$$

is typically concave, reflecting marginal expected utility for the incorporation of further covariates, and hence select that $\boldsymbol{y}^*_{(i)}$ for which $\bar{u}(\boldsymbol{y}^*_{(i+1)} \mid \boldsymbol{D}) - \bar{u}(\boldsymbol{y}^*_{(i)} \mid \boldsymbol{D})$ is less than some appropriately predefined small constant.

Given the complexity of problems of this type, the set $\mathcal{Y} = \mathcal{M}$ of possible models is typically a rather pragmatically defined collection of stylised forms, and, recalling the discussion of Section 6.1.2, an $\mathcal{M}$-closed perspective would not usually be appropriate. In fact, in most applications $p(\boldsymbol{z}, \boldsymbol{\omega} \mid \boldsymbol{D})$ is likely to prove far too complicated for any honest representation, so that, in the terminology of Section 6.1.2, we need to perform a comparison of the models in $\mathcal{Y}$ from the $\mathcal{M}$-open perspective. There are interesting open problems in the development of the cross-validation techniques, that might be employed in particular cases, but discussion of these would take us far into the realm of methods and case-studies, and so will be deferred to the second and third volumes of this work.

6.2 MODEL REJECTION

6.2.1 Model Rejection through Model Comparison

In the previous section, we considered model comparison problems arising from the existence of a proposed range of well-defined possible models, $\mathcal{M} = \{M_i,\ i \in I\}$, for observations x, where the primary decision consisted in choosing m_i, $i \in I$, with the implication of subsequently acting as if the corresponding M_i, $i \in I$, were the predictive model.

In this section, we shall be concerned with the situation which arises when *just one specific well-defined model for* x, M_0 *say, has been proposed initially*, and the primary decision corresponds either to the choice m_0, which corresponds to subsequently acting as if M_0 were the predictive model, or to the choice m_0^c (thus, in a sense, *rejecting* M_0), with the implication of "doing something else. If, given

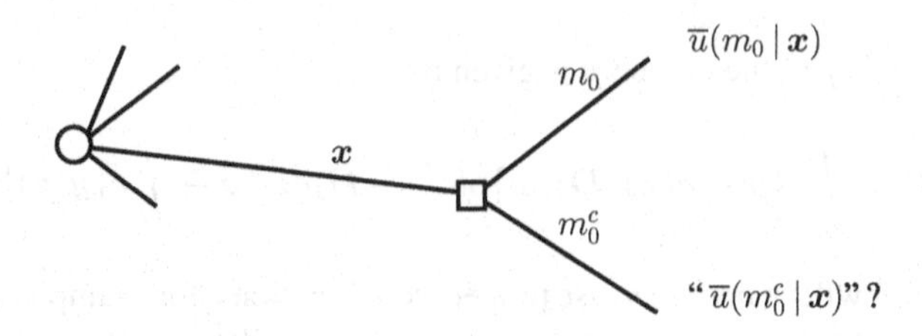

Figure 6.5 *Model rejection as a decision problem*

x, $\bar{u}(\cdot \mid \boldsymbol{x})$ denotes the ultimate expected utility of a primary action, this *model rejection* problem might be represented schematically by Figure 6.5.

Such a structure may arise, for example, as a consequence of M_0 being the only predictive model thus far put forward in a specific decision problem context; or, as a consequence of the application of some kind of principle of simplicity or parsimony, as an attempt to "get away with" using M_0, instead of using more complicated (but, in this context, unstated) alternatives.

What perspectives might one adopt in relation to this, thus far clearly ill-defined, problem of model rejection?

If we are concerned with coherent action in the context of a well-posed decision problem, we see from Figure 6.5 that we cannot proceed further unless we have some method for arriving at a value of $\bar{u}(m_0^c \mid \boldsymbol{x})$ to compare with $\bar{u}(m_0 \mid \boldsymbol{x})$. One way or another, we are forced to consider alternative models to M_0.

Let us suppose therefore that we have embedded M_0 in some larger class of models $\mathcal{M} = \{M_i, i \in I\}$. This might be done, particularly where M_0 has been put forward for reasons of simplicity or parsimony, by consideration of *actual alternatives* to M_0 thought (by someone) to be of practical interest. Otherwise, it might be done by consideration of *formal alternatives*, generated by selecting, in some way, a "mathematical neighbourhood" of M_0 (which might also, of course, contain alternatives of practical interest). For this redefined problem of *model rejection within* $\mathcal{M}$, shown schematically in Figure 6.6, the hitherto undefined value of $\bar{u}(m_0^c \mid \boldsymbol{x})$ becomes

$$\bar{u}(m_0^c \mid \boldsymbol{x}) = \max_{i \in I'} \bar{u}(m_i \mid \boldsymbol{x}).$$

where $I' = I - \{0\}$ indexes the models in $\mathcal{M}$ distinct from M_0.

For any specific decision problem, the calculation of $\bar{u}(m_i \mid \boldsymbol{x}), i \in I$ proceeds as indicated in Section 6.1. Thus, if we adopt the $\mathcal{M}$-closed perspective, evaluations are based on mixture forms involving prior and posterior probabilities of the M_i, $i \in I$; if we adopt the $\mathcal{M}$-completed perspective, the calculation is, in principle, well-defined, but may be numerically involved; if we adopt the $\mathcal{M}$-open perspective, we may use a cross-validation procedure to estimate the expected utilities.

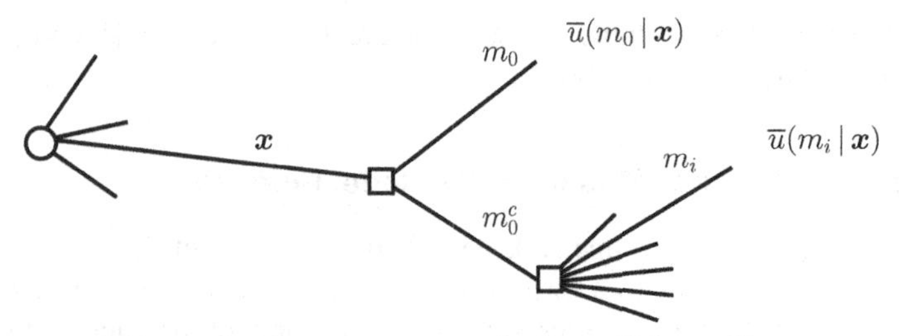

Figure 6.6 *Model rejection within* $\mathcal{M} = \{M_i, i \in I\}$

In the case where $\{M_i, i \in I'\}$ consists of actual alternatives to M_0, we might regard the redefined model rejection problem as essentially identical to the model comparison problem, so that rejecting m_0 corresponds to choosing the best of m_i, $i \in I'$. However, this would seem to ignore the fact that when M_0 has been put forward for reasons of simplicity or parsimony there is an implicit assumption that the latter has some "extra utility", over and above the expected utility $\bar{u}(m_0 \,|\, \boldsymbol{x})$. Thus, if $\bar{u}(m_0^c \,|\, \boldsymbol{x}) - \bar{u}(m_0 \,|\, \boldsymbol{x})$ were positive, but not "too large", we might still prefer M_0 because of the special "simple" status of M_0. The same argument applies even more forcibly in the case where $\{M_i, i \in I'\}$ consists of formal alternatives to M_0, since rejecting M_0 may not lead obviously to an actual alternative model, and the "extra utility" of choosing M_0 if at all possible may be greater. From this perspective, the redefinition of the problem of model rejection as one of model comparison corresponds to modifying slightly the representation given in Figure 6.6, by replacing $\bar{u}(m_0 \,|\, \boldsymbol{x})$ by $\bar{u}(m_0 \,|\, \boldsymbol{x}) + \varepsilon_0(\boldsymbol{x})$ where $\varepsilon_0(\boldsymbol{x})$ represents, given x, an implicit (but as yet undefined) *extra* utility relating to the special status of M_0. (See Dickey and Kadane, 1980, for related discussion.)

The formulation of the model rejection problem given above is rather too general to develop further in any detail. In order, therefore, to provide concrete illustrative analyses, we shall assume, for the remainder of this chapter that $\mathcal{M} = \{M_0, M_1\}$, where, for some parametric family $\{p(\cdot \,|\, \boldsymbol{\theta}), \boldsymbol{\theta} \in \Theta\}$, predictive models for x are defined, for some $\Theta_0 \subset \Theta$, by:

$$M_0: \quad p_0(x) = \int_{\Theta_0} p(\boldsymbol{x} \,|\, \boldsymbol{\theta}) p_0(\boldsymbol{\theta}) d\boldsymbol{\theta}$$

$$M_1: \quad p_1(x) = \int_{\Theta} p(\boldsymbol{x} \,|\, \boldsymbol{\theta}) p_1(\boldsymbol{\theta}) d\boldsymbol{\theta}.$$

Specially, M_0 will correspond to either:

(i) $p(\boldsymbol{x} \,|\, \boldsymbol{\theta}_0)$, a simple hypothesis that $\boldsymbol{\theta} = \boldsymbol{\theta}_0$ (specified by a degenerate prior $p_0(\boldsymbol{\theta})$ which concentrates on $\boldsymbol{\theta}_0$), or

(ii) $p(\boldsymbol{x} \,|\, \boldsymbol{\phi}_0, \boldsymbol{\lambda})$, a simple hypothesis on a parameter of interest $\boldsymbol{\phi}$ where $\boldsymbol{\lambda}$ is a nuisance parameter (specified by a prior $p_0(\boldsymbol{\theta}) = p(\boldsymbol{\lambda} \,|\, \boldsymbol{\phi}_0)$ which concentrates on the subspace defined by $\boldsymbol{\phi} = \boldsymbol{\phi}_0$).

The next three sections consider some detailed model rejection procedures within this parametric framework.

6.2.2 Discrepancy Measures for Model Rejection

Within the parametric framework described at the end of the previous section, model M_0 corresponds to a form of parametric restriction (or "null hypothesis") imposed on model M_1. In such situations, it is common practice to consider the decision problem of model rejection to be that of assessing the *compatibility* of model M_0 with the data $\boldsymbol{x}$, this being calibrated relative to the wider model M_1 within which M_0 is embedded. We shall focus on this version of the model rejection problem with utilities defined by $u(m_0, \boldsymbol{\theta})$, $u(m_1, \boldsymbol{\theta})$, $\boldsymbol{\theta} \in \Theta$, and overall beliefs, $p(\boldsymbol{\theta} \mid \boldsymbol{x}), \boldsymbol{\theta} \in \Theta$, defined either by an $\mathcal{M}$-closed form,

$$P(M_0 \mid \boldsymbol{x})p(\boldsymbol{\theta} \mid M_0, \boldsymbol{x}) + P(M_1 \mid \boldsymbol{x})p(\boldsymbol{\theta} \mid M_1, \boldsymbol{x}),$$

with $\mathcal{M} = \{M_0, M_1\}$, or by the $\{M_1\}$-closed form, $p(\boldsymbol{\theta} \mid M_1, \boldsymbol{x})$, the latter providing a kind of "adversarial" analysis, since it assigns M_0 no special status.

Noting that there are only two alternatives in this decision problem, it suffices to specify the (conditional) difference in utilities, say in favour of the larger model M_1,

$$\delta(\boldsymbol{\theta}) = u(m_1, \boldsymbol{\theta}) - u(m_0, \boldsymbol{\theta}),$$

since the optimal inference will clearly be to reject model M_0 if and only if

$$\int [u(m_1, \boldsymbol{\theta}) - u(m_0, \boldsymbol{\theta})]\, p(\boldsymbol{\theta} \mid \boldsymbol{x})\, d\boldsymbol{\theta} > \varepsilon_0(\boldsymbol{x}),$$

where $\varepsilon_0(\boldsymbol{x})$ represents, as before, the utility premium attached to keeping the simpler model M_0. We shall refer to $\delta(\boldsymbol{\theta})$ as a (utility-based) *discrepancy measure* between M_0 and M_1 when $\boldsymbol{\theta} \in \Theta$ is the true parameter. In terms of the discrepancy, the optimal action is to reject model M_0 if and only if

$$t(\boldsymbol{x}) > \varepsilon_0(\boldsymbol{x}),$$

where

$$t(\boldsymbol{x}) = \int_{\Theta} \delta(\boldsymbol{\theta}) p(\boldsymbol{\theta} \mid \boldsymbol{x}) d\boldsymbol{\theta}.$$

With a considerable reinterpretation of conventional statistical terminology, we might refer to $t(\cdot)$ as a *test statistic*, leading, for given data $\boldsymbol{x}$, to the rejection of model M_0 if the observed value of the test statistic exceeds a *critical value*, $c(\boldsymbol{x}) = \varepsilon_0(\boldsymbol{x})$.

How might $c(\boldsymbol{x})$ be chosen? One possible approach could be to consider, prior to observing $\boldsymbol{x}$ and assuming M_0 to be true, a choice of $c(\cdot)$ which would lead M_0 to only be rejected with low probability (α, say, for values of α of the order of 0.05 or 0.01). Under this approach, we would choose $c(\cdot)$ such that

$$P(t(\boldsymbol{x}) > c(\boldsymbol{x}) \mid M_0) = \alpha,$$

thus obtaining the $(1-\alpha)$th percentage point of the predictive distribution of $t(\boldsymbol{x})$, conditional on information available prior to observing $\boldsymbol{x}$ and assuming M_0 to be true. Of course, this is just one possible approach to selecting $c(\cdot)$ and has no special theoretical significance. It is interesting that this choice turns out to lead to commonly used procedures for model rejection which have typically not been derived or justified previously from the perspective of a decision problem (see, for example, Box, 1980, for a non-decision-theoretic approach). Examples will be given in the following two sections. However, for criticism of the practice of working with a fixed α, see Appendix B, Section 3.3.

6.2.3 Zero-one Discrepancies

Suppose that the discrepancy measure introduced in Section 6.2.2 is defined to be

$$\delta(\boldsymbol{\theta}) = \begin{cases} 0 & \text{if } \boldsymbol{\theta} = \boldsymbol{\theta}_0 \\ 1 & \text{if } \boldsymbol{\theta} \neq \boldsymbol{\theta}_0. \end{cases}$$

Assuming the decision problem of model rejection to be defined from the $\mathcal{M}$-closed perspective, we obtain

$$\begin{aligned} t(\boldsymbol{x}) &= \int \delta(\boldsymbol{\theta}) p(\boldsymbol{\theta} \mid \boldsymbol{x}) d\boldsymbol{\theta} \\ &= p(M_1 \mid \boldsymbol{x}) \\ &= \frac{p(M_1) p(\boldsymbol{x} \mid M_1)}{p(M_0) p(\boldsymbol{x} \mid M_0) + p(M_1) p(\boldsymbol{x} \mid M_1)}, \end{aligned}$$

where

$$p(\boldsymbol{x} \mid M_0) = \begin{cases} p(\boldsymbol{x} \mid \boldsymbol{\theta}_0) & \text{if } \boldsymbol{\theta}_0 \text{ specifies a simple hypothesis} \\ \int p(\boldsymbol{x} \mid \boldsymbol{\phi}_0, \boldsymbol{\lambda}) p(\boldsymbol{\lambda} \mid \boldsymbol{\phi}_0) d\boldsymbol{\lambda} & \text{if } \boldsymbol{\theta}_0 = (\boldsymbol{\phi}_0, \boldsymbol{\lambda}), \end{cases}$$

and

$$p(\boldsymbol{x} \mid M_1) = \int p(\boldsymbol{x} \mid \boldsymbol{\theta}) p_1(\boldsymbol{\theta}) d\boldsymbol{\theta}.$$

It follows from the analysis given in the previous section that M_0 should be rejected if and only if, for specified critical value $c(\boldsymbol{x})$,

$$P(M_1 \mid \boldsymbol{x}) > c(\boldsymbol{x}),$$

i.e., if the posterior odds on M_0 and against M_1 are smaller than $(1 - c(\boldsymbol{x}))/c(\boldsymbol{x})$. In the case where the prior odds are equal, the rejection criterion for M_0 in terms of the Bayes factor is given by

$$B_{01}(\boldsymbol{x}) < \frac{1 - c(\boldsymbol{x})}{c(\boldsymbol{x})}.$$

Example 6.10. ***(Null hypothesis for a binomial parameter).*** Suppose that x represents x successes in n binomial trials, with M_0, M_1 defined by

$$M_0: \quad p_0(\boldsymbol{x}) = p_0(x \mid n) = \mathrm{Bi}(x \mid n, \theta_0),$$

$$M_1: \quad p_1(\boldsymbol{x}) = p_1(x \mid n) = \int_0^1 \mathrm{Bi}(x \mid n, \theta)\mathrm{Un}(\theta \mid 0, 1)d\theta,$$

and $P(M_0) = P(M_1) = \frac{1}{2}$. Straightforward manipulation shows that

$$B_{01}(\boldsymbol{x}) = \frac{\Gamma(n+2)}{\Gamma(x+1)\Gamma(n-x+1)}\theta_0^x(1-\theta_0)^{n-x},$$

which, assuming, for purposes of illustration, large $x, n - x$, and applying Stirling's formula, $\log\Gamma(n+1) \approx (n+1)\log(n) - n + \frac{1}{2}\log(2\pi) + (12n)^{-1}$, can be approximated by

$$B_{01}(\boldsymbol{x}) = \left[\frac{n}{2\pi\theta_0(1-\theta_0)}\right]^{1/2} \exp\left\{-\tfrac{1}{2}\chi^2(\boldsymbol{x})\right\},$$

where

$$\chi^2(\boldsymbol{x}) = \frac{(x - n\theta_0)^2}{n\theta_0} + \frac{(n - x - n(1-\theta_0))^2}{n(1-\theta_0)}$$

is the usual chi-squared test statistic. By considering $-2\log B_{12}(\boldsymbol{x})$, for given n, θ_0, a value of $c(\boldsymbol{x}) = c(\theta_0, n)$ which calibrates the procedure to only reject M_0 with probability α when M_0 is true, is defined by the equation

$$\left[\frac{n}{2\pi\theta_0(1-\theta_0)}\right]\left(\frac{c(\theta_0, n)}{1 - c(\theta_0, n)}\right)^2 = \exp\left\{\chi^2_{1,\alpha}\right\},$$

where $\chi^2_{1,\alpha}$ denotes the upper $100\alpha\%$ point of a χ^2_1 distribution. Of course, having decided on this particular approach to the choice of $c(\boldsymbol{x})$, there is no real need to identify it! The rejection procedure is simply defined by comparing the test statistic value, $\chi^2(\boldsymbol{x})$, with its tail critical value, $\chi^2_{1,\alpha}$. The reader might like to verify (perhaps with the aid of Jeffreys, 1939/1961, Chapter 5) that similar results can be obtained for a variety of "null models" in more general contingency tables.

6.2.4 General Discrepancies

Given our systematic use throughout this volume of the logarithmic divergence between two distributions, an obviously appealing form of discrepancy measure is given by

$$\delta(\boldsymbol{\theta}) = \int p(\boldsymbol{x} \mid \boldsymbol{\theta}) \log \frac{p(\boldsymbol{x} \mid \boldsymbol{\theta})}{p(\boldsymbol{x} \mid \boldsymbol{\theta}_0)} d\boldsymbol{x},$$

where $p(\boldsymbol{x} \mid \boldsymbol{\theta})$ is the general parametric form of model M_1.

In the case of a location-scale model, $\boldsymbol{\theta}_0 = (\mu_0, \sigma)$, $\boldsymbol{\theta} = (\mu, \sigma)$, we might consider the standardised measure

$$\delta(\boldsymbol{\theta}) = \left(\frac{\mu - \mu_0}{\sigma}\right)^2.$$

In any case, the general prescription will be to reject M_0 if $t(\boldsymbol{x}) > c(\boldsymbol{x})$, for some appropriate $c(\boldsymbol{x})$, where

$$t(\boldsymbol{x}) = \int \delta(\boldsymbol{\theta}) p(\boldsymbol{\theta} \mid \boldsymbol{x}) d\boldsymbol{\theta},$$

with $p(\boldsymbol{\theta} \mid \boldsymbol{x})$ derived either from an $\mathcal{M}$-closed model, as illustrated in Section 6.2.3, or from the "adversarial" form corresponding to assuming model M_1.

Example 6.11. ***(Lindley's paradox revisited, again).*** In Examples 6.6 and 6.9 we considered the use of models M_1, M_2 defined, for $\boldsymbol{x} = (x_1, \ldots, x_n)$, by

$$M_1: \quad p(\boldsymbol{x}) = \prod_{i=1}^{n} N(x_i \mid \mu_0, \lambda), \quad \mu_0, \lambda \text{ known},$$

$$M_2: \quad p(\boldsymbol{x}) = \int \prod_{i=1}^{n} N(x_i \mid \mu, \lambda), N(\mu \mid \mu_1, \lambda_1) d\mu, \quad \mu_1, \lambda_1, \lambda \text{ known}.$$

Using the logarithmic divergence discrepancy, we obtain

$$\begin{aligned}
\delta(\mu) &= \int p(\boldsymbol{x} \mid \mu, \lambda) \log \frac{p(\boldsymbol{x} \mid \mu, \lambda)}{p(\boldsymbol{x} \mid \mu_0, \lambda)} d\boldsymbol{x} \\
&= n \int N(x \mid \mu, \lambda) \log \frac{N(x \mid \mu, \lambda)}{N(x \mid \mu_0, \lambda)} dx \\
&= \frac{n\lambda}{2} (\mu - \mu_0)^2,
\end{aligned}$$

which is just a multiple (by $n/2$) of a natural, standardised, measure (the non-centrality parameter) suggested by intuition as a discrepancy measure for a location scale family.

Assuming the reference prior for μ derived from an $\{M_1\}$-closed perspective, which, as a special case of Proposition 5.24, is easily seen to be uniform, we have the reference posterior

$$p(\mu \,|\, \boldsymbol{x}) = N(\mu \,|\, \overline{x}, n\lambda),$$

and hence

$$\begin{aligned} t(\boldsymbol{x}) &= \frac{n\lambda}{2} \int_{-\infty}^{\infty} (\mu - \mu_0)^2 N(\mu \,|\, \overline{x}, n\lambda) d\mu \\ &= \frac{n\lambda}{2} \left[(n\lambda)^{-1} + (\overline{x} - \mu_0)^2 \right] \\ &= \frac{1}{2}[1 + z^2(\boldsymbol{x})] \end{aligned}$$

where, with $\sigma^2 = \lambda^{-1}$, the statistic $z(\boldsymbol{x}) = \sqrt{n}(\overline{x} - \mu_0)/\sigma$ is seen to be a version of the standard significance test statistic for a normal location null hypothesis. With respect to $p(\boldsymbol{x} \,|\, \boldsymbol{\theta}_0)$, this has an $N(z \,|\, 0, 1)$ distribution and the appropriately calibrated value of $c(\boldsymbol{x})$ is thus implicitly defined, for example, by rejecting M_0 if $|\, z(\boldsymbol{x}) \,|$ exceeds the upper $100(\alpha/2)\%$ point of an $N(\cdot \,|\, 0, 1)$ distribution.

The above analysis is easily generalised to the case of unknown λ. Here, the reference posterior for (μ, λ) from an $\{M_1\}$-closed perspective (see Example 5.17) is given by

$$p(\mu, \lambda \,|\, \boldsymbol{x}) = N(\mu \,|\, \overline{x}, n\lambda) \, \text{Ga}\left(\lambda \,|\, \tfrac{1}{2}(n-1), \tfrac{1}{2}ns^2\right),$$

where $ns^2 = \Sigma(x_i - \overline{x})^2$. It follows that

$$\begin{aligned} t(\boldsymbol{x}) &= \frac{1}{2} \int_{-\infty}^{\infty} \int_{0}^{\infty} n\lambda(\mu - \mu_0)^2 N(\mu \,|\, \overline{x}, n\lambda) \, \text{Ga}\left(\lambda \,|\, \tfrac{1}{2}(n-1),\, \tfrac{1}{2}ns^2\right) d\mu d\lambda \\ &= \frac{1}{2} \int_{0}^{\infty} \left[1 + n\lambda(\overline{x} - \mu_0)^2\right] \text{Ga}\left(\lambda \,|\, \tfrac{1}{2}(n-1),\, \tfrac{1}{2}ns^2\right) d\mu d\lambda \\ &= \frac{1}{2}\left[1 + \frac{n}{s'^2}(\overline{x} - \mu_0)^2\right] \end{aligned}$$

where, with $s' = [ns^2/(n-1)]^{1/2}$, we see that $\sqrt{n}(\overline{x} - \mu_0)/s'$ is a version of the standard significance test statistic for a normal location null hypothesis in the presence of an unknown scale parameter. With respect to $p(\boldsymbol{x} \,|\, \boldsymbol{\theta}_0)$, this has a $\text{St}(t \,|\, 0, 1, n-1)$ distribution and the appropriately calibrated value of $c(\boldsymbol{x})$ is defined by the standard rejection procedure.

The reader can easily extend the above analyses to other stylised test situations: for example, testing the equality of means in two independent normal samples, with known or unknown (equal) precisions. Rueda (1992) provides the general expressions for one-dimensional regular exponential family models. Multivariate normal location cases are also easily dealt with, the logarithmic divergence discrepancy in this case being proportional to the Mahalanobis distance (see Ferrándiz, 1985). We shall not pursue such cases further here, since it seems to us that detailed discussion of model rejection and comparison procedures all too easily becomes artificial outside the disciplined context of real applications of the kind we shall introduce in the second and third volumes of this work. From the perspective of this volume, we have taken the analyses of this chapter sufficiently far to demonstrate the creative possibilities for model choice and comparison within the disciplined framework of Bayesian decision theory.

6.3 DISCUSSION AND FURTHER REFERENCES

6.3.1 Overview

We have argued that both from the perspective of a sensitive individual modeller, and also from that of a group of modellers, there are frequently strong reasons for considering a *range* of possible models.

This obviously leads to the problem of *model comparison*, or *model choice*, and our approach has been to consider formally a decision problem where the action space is the class of available models. In this setting, we have shown that "natural" Bayesian solutions, such as choosing the model with the highest posterior probability, are obtained as particular cases of the general structure for stylised, appropriately chosen, loss functions.

We have also considered the generally ill-posed problem of *model rejection*, where the primary decision consists in acting as if the proposed model were true —without having specific alternatives in mind—and have shown that useful results may be obtained by embedding the proposed model within a larger class, and then using discrepancy measures as loss functions in order to decide whether or not the original simpler model may be retained after all.

There is an extensive Bayesian literature directly related to the issues discussed in this chapter. Some authors adopt a purely inferential approach, by deriving either posterior probabilities, or Bayes factors for competing models; see, for example, Lindley (1965, 1972), Dickey and Lientz (1970), Dickey (1971, 1977), Leamer (1978), Bernardo (1980), Smith and Spiegelhalter (1980), Zellner and Siow (1980), Spiegelhalter and Smith (1982), Zellner (1984), Berger and Delampady (1987), Pettit and Young (1990), Aitkin (1991), Gómez-Villegas and Gómez (1992), Kass and Vaidyanathan (1992), McCulloch and Rossi (1992) and Lindley (1993). Others openly adopt a decision-theoretic approach; see, for example, Karlin and Rubin (1956), Raiffa and Schlaifer (1961), Schlaifer (1961), Box and Hill (1967), DeGroot (1970), Zellner (1971), Bernardo (1982, 1985a), San Martini and Spezzaferri (1984), Berger (1985a), Bernardo and Bayarri (1985), Ferrándiz (1985), Poskitt (1987), Felsenstein (1992) and Rueda (1992).

6.3.2 Modelling and remodelling

We have already argued that we see Bayesian statistics as a rather formalised procedure for inference and decision making within a well-defined probabilistic structure.

Fully specified belief models are an integral part of this structure, but it would be highly unrealistic to expect that in any particular application such a belief model will be general enough to pragmatically encompass a defensible description of reality from the very beginning.

In practice, we typically first consider simple models, which may have been informally suggested by a combination of exploratory data analysis, graphical analysis and prior experience with similar situations. And even with such a simple model, more formal investigation of its adequacy and the consequences of using it will often be necessary before one is prepared to seriously consider the model as a predictive specification. Such investigations will typically include *residual analysis*, *identification of outliers* and/or *influential data*, *cluster analysis*, and the behaviour of *diagnostic statistics* when compared with their predictive distributions. We shall not elaborate on this here. Some relevant references are Johnson and Geisser (1982, 1983, 1985), Pettit and Smith (1985), Pettit (1986), Geisser (1987, 1992, 1993), Chaloner and Brant (1988), McCulloch (1989), Verdinelli and Wasserman (1991), Gelfand, Dey and Chang (1992), Weiss and Cook (1992), Guttman and Peña (1993), Peña and Guttman (1993), and Chaloner (1994).

Bayarri and DeGroot (1987, 1990, 1992a) provide a Bayesian analysis of *selection models*, where data are randomly selected from a proper subset of the sample space rather than from the entire population.

As a consequence of this probing, mainly exploratory analysis, a *class* of alternative models will typically emerge. In this chapter we have discussed some of the procedures which may be useful in a *formal* comparison of such alternative models. The outcome of this strategy will typically be a more refined model for which a similar type of analysis may be repeated again.

Naturally, a pragmatic combination of time constraints, data limitations, and capacity of imagination, will force this sequence of informal exploration and formal analysis to eventually settle on the use of a particular belief model, which hopefully can be defended as a sensible and useful conceptual representation of the problem.

This remodelling process is never fully completed, however, in that either new data, more time, or an imaginative new idea, may force one to make yet another iteration towards the never attainable "perfect", all powerful predictive machine.

6.3.3 Critical issues

We shall comment further on six aspects of the general topic of remodelling under the following subheadings: (i) *Model Choice and Model Criticism*, (ii) *Inference and Decision*, (iii) *Overfitting and Cross-validation*, (iv) *Improper Priors*, (v) *Scientific Reporting*, and (vi) *Computer Software*.

Model Choice and Model Criticism

We have reviewed several procedures which, under different headings such as model comparison, model choice or model selection, may be used to choose among a class of alternative models, and we have argued that, from a decision-theoretical point of view, the problem of accepting that a particular model is suitable, is ill-defined unless an alternative is formally considered. See, also, Hodges (1990, 1992).

However, partly due to the classical heritage of significance testing, and given the obvious attraction of being able to check the adequacy of a given model without explicit consideration of any alternative, non-decision-theoretic Bayesians have often tried to produce procedures which evaluate the compatibility of the data with specific models.

As clearly stated by Box (1980), the posterior distribution of the parameters only permits the

> *estimation* of the parameters conditional on the adequacy of the entertained model, while the predictive distribution makes possible *criticism* of the entertained model in the light of current data.

Moreover, the predictive distributions which correspond to different models are comparable among themselves, while—in general—the posteriors are not. The use of predictive distributions to check model assumptions was pioneered by Jeffreys (1939/1961). Additional references include Geisser (1966, 1971, 1985, 1987, 1988, 1993), Box and Tiao (1973), Dempster (1975), Geisser and Eddy (1979), Rubin (1984), Bernardo and Bermúdez (1985), Clayton *et al.* (1986), Gelfand, Dey and Chang (1992) and Girón *et al.* (1992).

The basic idea consists of defining a set of appropriate *diagnostic functions* $\boldsymbol{t}_i = \boldsymbol{t}_i(\boldsymbol{x}_{n+1}, \ldots, \boldsymbol{x}_{n+k})$ of the data and comparing their actual values in a sample with their predictive distributions $p_{\boldsymbol{t}_i}(\cdot \mid \boldsymbol{x}_1, \ldots, \boldsymbol{x}_n)$ based on a different sample from the same population. Possible comparisons include checking whether or not the observed $\boldsymbol{t}_i$'s belong to appropriate predictive HPD intervals, or determining the predictive probability of observing $\boldsymbol{t}_i$'s more "outlying" than those observed. The reader will readily appreciate that common techniques such as residual analysis, identification of influential observations, segregation of homogeneous clusters, or outlier screening, can all be reformulated as particular implementations of this general framework.

As mentioned before, we see these very useful activities as part of the informal process that necessarily precedes the formulation of a model which we can then seriously entertain as an adequate predictive tool. However, it seems to us inescapable that if a formal *decision* is to be reached on whether or not to operate with a given model, then some form of alternative must be considered.

For further discussion of model choice, see Winkler (1980), Klein and Brown (1984), Krasker (1984), Florens and Mouchart (1985), Poirier (1985), Hill (1986, 1990), Skene *et al.* (1986) and West (1986).

Inference and Decision

Throughout this volume, we have emphasised the advantages of using a formal decision-oriented approach to the stylised statistical problems which represent a large proportion of the theoretical statistical literature. These advantages are specially obvious in model comparison since, by requiring the specification of an appropriate utility function, they make explicit the identification of those aspects of the model which really matter.

We have seen, moreover, that the more traditional Bayesian approaches to model comparison, such us determining the posterior probabilities of competing models or computing the relevant Bayes factors, can be obtained as particular cases of the general structure by using appropriately chosen, stylised utility functions.

Very often, the consequences of entertaining a particular model may usefully be examined in terms of the discrepancies between the prediction provided by the model for the value of a relevant observable vector, $\boldsymbol{t}$, say, and its actual, eventually observed, value. Scoring rules, of the general type $u(p_{\boldsymbol{t}}(\cdot \mid \boldsymbol{x}_1, \ldots, \boldsymbol{x}_n), \boldsymbol{t})$ provide natural utility functions to use in this context, by explicitly evaluating the degree of compatibility between the observed $\boldsymbol{t}$ and its predictive distribution.

Overfitting and Cross-Validation

If we are hoping for a positive evaluation of a prediction it is crucial that the predictive distribution is based on data which do not include the value to be predicted; otherwise, severe *overfitting* may occur. Pragmatically, however, although it is sometimes possible to check the predictions of the model under investigation by using a totally different set of data than that used to develop the model, it is far more common to be obliged to do both model construction and model checking with the same data set.

A natural solution consists of randomly partitioning the available sample $\boldsymbol{z} = \{\boldsymbol{x}_1, \ldots, \boldsymbol{x}_n\}$ say, into two subsamples $\{z_1, z_2\}$ one of which is used to produce the relevant predictive distributions, and the other to compute the diagnostic functions; the procedure then being repeated as many times as is necessary to reach stable conclusions. This technique is usually known as *cross-validation*. For recent work, see Peña and Tiao (1992), Gelfand, Dey and Chang (1992), and references therein. For a discussion of how cross validation may be seen as approximating formal Bayes procedures, see Section 6.1.6.

A possible systematic approach to cross-validation starting from a sample of size n, $\boldsymbol{z} = \{\boldsymbol{x}_1, \ldots, \boldsymbol{x}_n\}$, and a model $\{p(\boldsymbol{x} \mid \boldsymbol{\theta}), p(\boldsymbol{\theta})\}$, involves the following steps:

(i) Define a sample size k, where $k \leq n/2$ is large enough to evaluate the relevant observable function $\boldsymbol{t} = \boldsymbol{t}(\boldsymbol{x}_1, \ldots, \boldsymbol{x}_k)$. The observable function could either be that predictive aspect of the model which is of interest, as described by the utility function, or a diagnostic function, as described in the above approach to model criticism.

(ii) Determine the set of all predictive distributions of the form

$$p_j(\boldsymbol{t}(\boldsymbol{z}_j) \mid \boldsymbol{z}_{[j]}), \quad j = 1, \ldots, \binom{n}{k},$$

where $\boldsymbol{z}_j$ is a subsample of $\boldsymbol{z}$ of size k and $\boldsymbol{z}_{[j]}$ consists of all the $\boldsymbol{x}_i$'s in $\boldsymbol{z}$ which are *not* in $\boldsymbol{z}_j$.

(iii) Estimate the expected utility of the model by

$$\binom{n}{k}^{-1} \sum_j u(p_j(\cdot \mid \boldsymbol{z}_{[j]}), \boldsymbol{t}(\boldsymbol{z}_j)) \cdot$$

Note that the last expression is simply a Monte Carlo approximation to the exact value of the expected utility. We also note that this programme may be carried out with reference distributions since the corresponding reference (posterior) predictive distributions $\pi_j(\boldsymbol{t}(\boldsymbol{z}_j) \mid \boldsymbol{z}_{[j]})$ will be proper even if the reference prior $\pi(\boldsymbol{\theta})$ is not.

Improper Priors

In the context of analysis predicated on a fixed model, we have seen in Chapter 5 that perfectly proper posterior parametric and predictive inferences can be obtained for improper prior specifications.

When it comes to comparing models, however, in general the use of improper priors is much more problematic. We first note that for models

$$M_i\{p_i(\boldsymbol{x}|\boldsymbol{\theta}), p_i(\boldsymbol{\theta})\}, \quad i \in I,$$

the predictive quantities

$$p_i(\boldsymbol{x}) = \int p_i(\boldsymbol{x}|\boldsymbol{\theta}) p_i(\boldsymbol{\theta}) d\boldsymbol{\theta}, \quad i \in I,$$

typically play a key role in model comparisons for a range of specific decision problems and perspectives on model choice. But if one or more of the $p_i(\boldsymbol{\theta})$ is not a proper density, the corresponding $p_i(\boldsymbol{x})$'s will also be improper, thus precluding, for example, the calculation of posterior probabilities for models in an $\mathcal{M}$-closed analysis.

Essentially, with a formal improper specification for the prior component of a model an initial amount of data needs to be passed through the prior to posterior process before the model attains proper status as a predictive tool and can hence be compared with other alternative predictive tools.

An exception to this arises when two models M_i, M_j, say, have common $\boldsymbol{\theta}$ and improper $p(\boldsymbol{\theta})$, in which case it can be argued that the ratio $p_i(\boldsymbol{x})/p_j(\boldsymbol{x})$, the Bayes factor in favour of M_i, does provide a meaningful comparison between the two models, However, there is an inherent difficulty with these methods when the models compared have different dimensions.

Indeed, with the reference prior approach some models are implicitly disadvantaged relative to others; technically, this is due to the fact that the amount of information about the parameters of interest to be expected from the data crucially depends on the prior distribution of the nuisance parameters present in the model. Lindley's paradox (see, also, Bartlett, 1957), discussed earlier in this chapter, is a well known simple example of this behaviour.

A possible solution consists in specifying the improper prior probabilities of the models—or, equivalently, weighting the Bayes factors—in a way which may be expected to achieve neutral discrimination between the models. Some suggestions along these and similar lines include Bernardo (1980), Spiegelhalter and Smith (1982), Pericchi (1984), Eaves (1985) and Consonni and Veronese (1992a).

Another possible solution to the problem of comparing two models in the case of improper prior specification for the parameters is to exploit the use of cross-validation as a Monte Carlo approximation to a Bayes decision rule.

As we saw in Section 6.1.6, for the problem of predicting a future observation using a log-predictive utility function the (Monte Carlo approximated) criterion for model choice involves the geometric mean of Bayes factors of the form

$$B_{12}(x_j, \boldsymbol{x}_{n-1}(j)) = \frac{p_1(x_j \mid \boldsymbol{x}_{n-1}(j))}{p_2(x_j \mid \boldsymbol{x}_{n-1}(j))},$$

where $[\boldsymbol{x}_{n-1}(j), x_j]$ denotes a partition of $\boldsymbol{x} = (x_1, \ldots, x_n)$, and

$$p_i(x_j \mid \boldsymbol{x}_{n-1}(j)) = \int p_i(x_j \mid \boldsymbol{\theta}_i) p_i(\boldsymbol{\theta}_i \mid \boldsymbol{x}_{n-1}(j))\, d\boldsymbol{\theta}_i.$$

Since, for sufficiently large n, $p_i(\boldsymbol{\theta}_i \mid \boldsymbol{x}_{n-1}(j))$ will be proper, even for improper (non-pathological) priors $p_i(\boldsymbol{\theta}_i)$, no problem arises.

However, recalling from our discussion in Section 6.1.6 that the conventional Bayes factor is used to assess the models' ability to "predict x" from $\{p_i(\boldsymbol{x} \mid \boldsymbol{\theta}_i), p_i(\boldsymbol{\theta}_i)\}$, we see that the latter does run into trouble if $p_i(\boldsymbol{\theta}_i)$ is improper, since, then, $p_i(\boldsymbol{x})$ is not proper.

Proceeding formally, if we were to take $\log p_i(\boldsymbol{x})$ as the utility of choosing M_i, the $\mathcal{M}$-open perspective prefers M_1 if

$$\int \log \frac{p_1(\boldsymbol{x})}{p_2(\boldsymbol{x})} p(\boldsymbol{x})\, d\boldsymbol{x} > 0,$$

where $p(\boldsymbol{x})$ is not specified. Again, we can approach the evaluation of the left-hand side as a Monte Carlo calculation, based on partitioning $\boldsymbol{x}$ and averaging over random partitions. However, in this contect we want partitions where the proxy for the predictive part resembles data $\boldsymbol{x}$ and the proxy for the conditioning part resembles "no data". The closest we can come to this, and overcome the problem of the impropriety of $p_i(\boldsymbol{\theta}_i)$, is to take partitions of the form $\boldsymbol{x} = [\boldsymbol{x}_s(j), \boldsymbol{x}_{n-s}(j)]$, where $s\,(\geq 1)$ is the smallest integer such that both $p_1(\boldsymbol{\theta}_1 \mid \boldsymbol{x}_s(j))$ and $p_2(\boldsymbol{\theta}_2 \mid \boldsymbol{x}_s(j))$ are proper, and $j = 1, \ldots, \binom{n}{s}$.

The proposal is now to select randomly k such partitions, and approximate the left-hand side of the criterion inequality by

$$\frac{1}{k} \sum_{j=1}^{k} \log \left\{ \frac{p_1(\boldsymbol{x}_{n-s}(j) \mid \boldsymbol{x}_s(j))}{p_2(\boldsymbol{x}_{n-s}(j) \mid \boldsymbol{x}_s(j))} \right\}.$$

The (Monte Carlo approximated) model choice criterion then becomes, prefer M_1 if

$$\prod_{j=1}^{k} \left[\frac{p_1(\boldsymbol{x}_{n-s}(j) \mid \boldsymbol{x}_s(j))}{p_1(\boldsymbol{x}_{n-s}(j) \mid \boldsymbol{x}_s(j))} \right]^{1/k} = \prod_{j=1}^{k} \left[B_{12}(\boldsymbol{x}_{n-s}(j), \boldsymbol{x}_s(j)) \right]^{1/k} > 1,$$

where $B_{12}(\boldsymbol{x}_{n-s}(j), \boldsymbol{x}_s(j))$ denotes the Bayes factor for M_1 against M_2, based on the versions

$$\left\{ p_i(\boldsymbol{x}_{n-s}(j) \mid \boldsymbol{\theta}_i), p_i(\boldsymbol{\theta}_i \mid \boldsymbol{x}_s(j)) \right\}$$

of M_1, M_2. Again, we see the explicit role of the geometric average of Bayes factors, but with the latter "reversing", in a sense, the role of past and future data compared with the form obtained in Section 6.1.6.

At the time of writing we are aware of work in progress by several researchers who propose forms related to those discussed here. These include: J. O. Berger and L. R. Pericchi (*intrinsic* Bayes factors), A. O'Hagan (*fractional* Bayes factors) and A. F. de Vos (*fair* Bayes factors).

From a practical perspective, it might be desirable to trade-off, in utility terms, "fidelity to the observed data" and "future predictive power". This can be formalised by adopting a utility function of the form

$$\alpha \log[p_i(\boldsymbol{x})] + (1-\alpha)\log[p_i(y \mid \boldsymbol{x})], \quad 0 \le \alpha \le 1,$$

which, in turn, leads to a criterion of the form, prefer M_1 if

$$\left\{\prod_{j=1}^{k}\left[B_{12}(\boldsymbol{x}_{n-s}(j), \boldsymbol{x}_s(j))\right]^{1/k}\right\}^{\alpha}\left\{\prod_{j=1}^{k}\left[B_{12}(x_j, \boldsymbol{x}_{n-1}(j))\right]^{1/k}\right\}^{1-\alpha} > 1.$$

Work in progress (by L. R. Pericchi and A. F. M. Smith) suggests that such a criterion effectively encompasses and extends a number of current criteria.

We conclude by emphasising again that a predictivistic decision-theoretical approach to model comparison, where models are evaluated in terms of their predictive behaviour, bypasses the dimensionality issue, since posterior *predictive* distributions obtained from models with different dimensions are always directly comparable.

Scientific Reporting

Our whole development has been predicated on the central idea of normative standards for an individual wishing to act coherently in response to uncertainty. Beliefs as individual, personal probabilities are the key element in this process and, at any given moment in the learning cycle, are the encapsulation of the current response to the uncertainties of interest.

However, while many are willing to concede that, in narrowly focused decision-making, such beliefs are an essential element, there has also been a widespread view (see, for example, Fisher, 1956/1973) that it would be somehow subversive to sully the nobler, objective processes of science by allowing subjective beliefs to enter the picture. As Dickey (1973) has remarked, rhetorically:

> But is not personal knowledge, or opinion, like superstition, non-objective and unscientific, and therefore to be avoided in science? Who cares to read about a scientific reporter's opinion as described by his prior and posterior probabilities?

We have already made clear our own general view that objectivity has no meaning in this context apart from that pragmatically endowed by thinking of it as a shorthand for subjective consensus. However, there are clearly practical problems of communication between analysts and audiences which need addressing.

The solution to such problems lies in combining ideas from Chapters 4 and 6. On the one hand, we have seen that shared assumptions about structural aspects of beliefs (for example, exchangeability) can lead a group of individuals to have shared assumptions about the parametric model component, while perhaps differing over the prior component specification. On the other hand, we have seen, from several perspectives, that entertaining and comparing a range of models fit perfectly naturally within the formalism. There is nothing within the world of Bayesian Statistics that prohibits a scientist from performing and reporting a range of "what if?" analyses. To quote Dickey again:

> ... communicating a single opinion ought not to be the purpose of a scientific report; but, rather, to let the data speak for themselves by giving the effect of the data on the wide diversity of real prior opinions. ... an experimenter can summarise the application of Bayes' theorem to whole ranges of prior distributions, derived to include the opinions of his readers. Scientific reports should objectively exhibit as much as possible of the *inferential content of the data*, the data-specific prior-to-posterior *transformation* of the collection of all personal probability distributions on the parameters of a realistically rich statistical model.

We believe that this is the way forward—although, it has to be said, there is a great deal of work to be done in effecting such a cultural change. There are also some obvious technical challenges in making such a programme routinely implementable in practice. However, computational power grows apace, as does the sophistication of graphical displays. We shall return to this general problem in the second and third volumes of this work.

For early thoughts on these issues, see Edwards *et al.* (1963) and Hildreth (1963); for technical expositions, see Dickey (1973), Roberts (1974) and Dickey and Freeman (1975); for a discussion in the context of a public policy debate, see Smith (1978).

Computer Software

Despite the emphasis in Chapter 2 and 3 of this volume on foundational issues—necessary for a complete treatment of Bayesian theory—we are well aware that the majority of practising statisticians are more likely to be influenced by positive, preferably hands-on, experience with applications of methods to concrete problems than they ever will be by philosophical victories attained through the (empirically) bloodless means of axiomatics and stylised counter-examples. We are also well aware that the availability of suitable software is the key to the possibility of obtaining that hands-on experience.

But what are the appropriate software tools for Bayesian Statistics? What software? For whom? For what kinds of problems and purposes?

A number of such issues were reviewed in Smith (1988), but at the time of writing, many still remain unresolved. Goel (1988) provided a review of Bayesian software in the late 80's. Examples of creative use of modern software in Bayesian analysis include: Smith *et al.* (1985, 1987) and Racine-Poon *et al.* (1986), who describe the use of the *Bayes Four* package; Grieve (1987); Lauritzen and Spiegelhalter (1988), on which the commercial expert system builder *Ergo*™ is based; Albert (1990), Korsan (1992), and Ley and Steel (1992), who make use of the commercial package *Mathematica*™; Tierney (1990, 1992), who presents LISP-STAT, an object oriented environment for statistical computing and discusses possible uses of graphical animation; Cowell (1992) and Spiegelhalter and Cowell (1992), who, respectively, describe and apply the probabilistic expert system shell *BAIES*; Racine-Poon (1992), who discusses sample-assisted graphical analysis; Thomas *et al.* (1992), who describes *BUGS*, a program to perform Bayesian inference using Gibbs sampling; Wooff (1992), who describes *[B/D]*, an implementation of subjectivist analysis of beliefs, as described by Goldstein (1981, 1988, 1991, and references therein); and Marriot and Naylor (1993), who discuss the use of MINITAB to teach Bayesian statistics. Further review and detailed illustration will be provided in the volumes *Bayesian Computation* and *Bayesian Methods*.

Appendix A

Summary of Basic Formulae

Summary

Two sets of tables are provided for reference. The first records the definition, and the first two moments of the most common probability distributions used in this volume. The second records the basic elements of standard Bayesian inference processes for a number of special cases. In particular, it records the appropriate likelihood function, the sufficient statistics, the conjugate prior and corresponding posterior and predictive distributions, the reference prior and corresponding reference posterior and predictive distributions.

A.1 PROBABILITY DISTRIBUTIONS

The first section of this Appendix consists of a set of tables which record the notation, parameter range, variable range, definition, and first two moments of the probability distributions (discrete and continuous, univariate and multivariate) used in this volume.

Univariate Discrete Distributions

Br$(x \mid \theta)$ ***Bernoulli (p. 115)***

$0 < \theta < 1$ $\qquad x = 0, 1$

$p(x) = \theta^x (1-\theta)^{1-x}$

$E[x] = \theta$ $\qquad V[x] = \theta(1-\theta)$

Bi$(x \mid \theta, n)$ ***Binomial (p. 115)***

$0 < \theta < 1, n = 1, 2, \ldots$ $\qquad x = 0, 1, \ldots, n$

$$p(x) = \binom{n}{x} \theta^x (1-\theta)^{n-x}$$

$E[x] = n\theta$ $\qquad V[x] = n\theta(1-\theta)$

Bb$(x \mid \alpha, \beta, n)$ ***Binomial-Beta (p. 117)***

$\alpha > 0, \beta > 0, n = 1, 2, \ldots$ $\qquad x = 0, 1, \ldots, n$

$$p(x) = c \binom{n}{x} \Gamma(\alpha + x)\Gamma(\beta + n - x) \qquad c = \frac{\Gamma(\alpha+\beta)}{\Gamma(\alpha)\Gamma(\beta)\Gamma(\alpha+\beta+n)}$$

$$E[x] = n\frac{\alpha}{\alpha+\beta} \qquad V[x] = \frac{n\alpha\beta}{(\alpha+\beta)^2} \frac{(\alpha+\beta+n)}{(\alpha+\beta+1)}$$

Hy$(x \mid N, M, n)$ ***Hypergeometric (p. 115)***

$N = 1, 2, \ldots$ $\qquad x = a, a+1, \ldots, b$

$M = 1, 2, \ldots$ $\qquad a = \max(0, n-M)$

$n = 1, \ldots, N+M$ $\qquad b = \min(n, N)$

$$p(x) = c \binom{N}{x} \binom{M}{n-x} \qquad c = \binom{N+M}{n}^{-1}$$

$$E[x] = n\frac{N}{N+M} \qquad V[x] = \frac{nNM}{(N+M)^2} \frac{N+M-n}{N+M-1}$$

Univariate Discrete Distributions (continued)

Nb$(x \mid \theta, r)$ *Negative-Binomial (p. 116)*

$0 < \theta < 1, r = 1, 2, \ldots$ $\qquad x = 0, 1, 2, \ldots$

$$p(x) = c \binom{r+x-1}{r-1} (1-\theta)^x \qquad c = \theta^r$$

$$E[x] = r\theta \qquad V[x] = r\frac{1-\theta}{\theta^2}$$

Nbb$(x \mid \alpha, \beta, r)$ *Negative-Binomial-Beta (p. 118)*

$\alpha > 0,\ \beta > 0,\ r = 1, 2 \ldots$ $\qquad x = 0, 1, 2, \ldots$

$$p(x) = c \binom{r+x-1}{r-1} \frac{\Gamma(\beta+x)}{\Gamma(\alpha+\beta+r+x)} \qquad c = \frac{\Gamma(\alpha+\beta)\Gamma(\alpha+r)}{\Gamma(\alpha)\Gamma(\beta)}$$

$$E[x] = \frac{r\beta}{\alpha-1} \qquad V[x] = \frac{r\beta}{(\alpha-1)} \left[\frac{\alpha+\beta+r-1}{(\alpha-2)} + \frac{r\beta}{(\alpha-1)(\alpha-2)} \right]$$

Pn$(x \mid \lambda)$ *Poisson (p. 116)*

$\lambda > 0$ $\qquad x = 0, 1, 2, \ldots$

$$p(x) = c\ \frac{\lambda^x}{x!} \qquad c = e^{-\lambda}$$

$$E[x] = \lambda \qquad V[x] = \lambda$$

Pg$(x \mid \alpha, \beta, n)$ *Poisson-Gamma (p. 119)*

$\alpha > 0,\ \beta > 0,\ \nu > 0$ $\qquad x = 0, 1, 2, \ldots$

$$p(x) = c \frac{\Gamma(\alpha+x)}{x!} \frac{\nu^x}{(\beta+\nu)^{\alpha+x}} \qquad c = \frac{\beta^\alpha}{\Gamma(\alpha)}$$

$$E[x] = \nu\frac{\alpha}{\beta} \qquad V[x] = \frac{\nu\alpha}{\beta}\left[1 + \frac{\nu}{\beta}\right]$$

Univariate Continuous Distributions

Be$(x \mid \alpha, \beta)$ ***Beta (p. 116)***

$\alpha > 0, \beta > 0$	$0 < x < 1$
$p(x) = c\, x^{\alpha-1}(1-x)^{\beta-1}$	$c = \dfrac{\Gamma(\alpha+\beta)}{\Gamma(\alpha)\Gamma(\beta)}$
$E[x] = \dfrac{\alpha}{\alpha+\beta}$	$V[x] = \dfrac{\alpha\beta}{(\alpha+\beta)^2(\alpha+\beta+1)}$

Un$(x \mid a, b)$ ***Uniform (p. 117)***

$b > a$	$a < x < b$
$p(x) = c$	$c = (b-a)^{-1}$
$E[x] = \frac{1}{2}(a+b)$	$V[x] = \frac{1}{12}(b-a)^2$

Ga$(x \mid \alpha, \beta)$ ***Gamma (p. 118)***

$\alpha > 0, \beta > 0$	$x > 0$
$p(x) = c\, x^{\alpha-1}e^{-\beta x}$	$c = \dfrac{\beta^\alpha}{\Gamma(\alpha)}$
$E[x] = \alpha\beta^{-1}$	$V[x] = \alpha\beta^{-2}$

Ex$(x \mid \theta)$ ***Exponential (p. 118)***

$\theta > 0$	$x > 0$
$p(x) = c\, e^{-\theta x}$	$c = \theta$
$E[x] = 1/\theta$	$V[x] = 1/\theta^2$

Gg$(x \mid \alpha, \beta, n)$ ***Gamma-Gamma (p. 120)***

$\alpha > 0, \beta > 0, n > 0$	$x > 0$
$p(x) = c\, \dfrac{x^{n-1}}{(\beta+x)^{\alpha+n}}$	$c = \dfrac{\beta^\alpha}{\Gamma(\alpha)}\, \dfrac{\Gamma(\alpha+n)}{\Gamma(n)}$
$E[x] = n\dfrac{\beta}{\alpha-1}$	$V[x] = \dfrac{\beta^2(n^2+n(\alpha-1))}{(\alpha-1)^2(\alpha-2)}$

Univariate Continuous Distributions (continued)

$\chi^2(x \mid \nu) = \chi^2_\nu$ *Chi-squared (p. 120)*	
$\nu > 0$	$x > 0$
$p(x) = c\, x^{(\nu/2)-1} e^{-x/2}$	$c = \dfrac{(1/2)^{\nu/2}}{\Gamma(\nu/2)}$
$E[x] = \nu$	$V[x] = 2\nu$

$\chi^2(x \mid \nu, \lambda)$ *Non-central Chi-squared (p. 121)*	
$\nu > 0, \lambda > 0$	$x > 0$
$p(x) = \sum_{i=0}^{\infty} \mathrm{Pn}\left(i \,\middle|\, \frac{\lambda}{2}\right) \chi^2(x \mid \nu + 2i)$	
$E[x] = \nu + \lambda$	$V[x] = 2(\nu + 2\lambda)$

$\mathrm{Ig}(x \mid \alpha, \beta)$ *Inverted-Gamma (p. 119)*	
$\alpha > 0, \beta > 0$	$x > 0$
$p(x) = c\, x^{-(\alpha+1)} e^{-\beta/x}$	$c = \dfrac{\beta^\alpha}{\Gamma(\alpha)}$
$E[x] = \dfrac{\beta}{\alpha - 1}$	$V[x] = \dfrac{\beta^2}{(\alpha-1)^2(\alpha-2)}$

$\chi^{-1}(x \mid \nu)$ *Inverted-Chi-squared (p. 119)*	
$\nu > 0$	$x > 0$
$p(x) = c\, x^{-(\nu/2+1)} e^{-1/2x^2}$	$c = \dfrac{(1/2)^{\nu/2}}{\Gamma(\nu/2)}$
$E[x] = \dfrac{1}{\nu - 2}$	$V[x] = \dfrac{2}{(\nu-2)^2(\nu-4)}$

$\mathrm{Ga}^{-1/2}(x \mid \alpha, \beta)$ *Square-root Inverted-Gamma (p. 119)*	
$\alpha > 0, \beta > 0$	$x > 0$
$p(x) = c\, x^{-(2\alpha+1)} e^{-\beta/x^2}$	$c = \dfrac{2\beta^\alpha}{\Gamma(\alpha)}$
$E[x] = \dfrac{\sqrt{\beta}\,\Gamma(\alpha - 1/2)}{\Gamma(\alpha)}$	$V[x] = \dfrac{\beta}{\alpha - 1} - E[x]^2$

Univariate Continuous Distributions (continued)

$\text{Pa}(x \mid \alpha, \beta)$ *Pareto (p. 120)*

$\alpha > 0, \beta > 0$	$\beta \le x < +\infty$
$p(x) = c\, x^{-(\alpha+1)}$	$c = \alpha\beta^\alpha$
$E[x] = \frac{\beta\alpha}{\alpha-1}, \quad \alpha > 1$	$V[x] = \frac{\beta^2\alpha}{(\alpha-1)^2(\alpha-2)}, \quad \alpha > 2$

$\text{Ip}(x \mid \alpha, \beta)$ *Inverted-Pareto (p. 120)*

$\alpha > 0, \quad \beta > 0$	$0 < x < \beta^{-1}$
$p(x) = c\, x^{\alpha-1}$	$c = \alpha\beta^\alpha$
$E[x] = \beta^{-1}\alpha(\alpha+1)^{-1}$	$V[x] = \beta^{-2}\alpha(\alpha+1)^{-2}(\alpha+2)^{-1}$

$\text{N}(x \mid \mu, \lambda)$ *Normal (p. 121)*

$-\infty < \mu < +\infty, \lambda > 0$	$-\infty < x < +\infty$
$p(x) = c\, \exp\left\{-\frac{1}{2}\lambda(x-\mu)^2\right\}$	$c = \lambda^{1/2}(2\pi)^{-1/2}$
$E[x] = \mu$	$V[x] = \lambda^{-1}$

$\text{St}(x \mid \mu, \lambda, \alpha)$ *Student t (p. 122)*

$-\infty < \mu < +\infty, \lambda > 0, \alpha > 0$	$-\infty < x < +\infty$
$p(x) = c\, \left[1 + \alpha^{-1}\lambda(x-\mu)^2\right]^{-(\alpha+1)/2}$	$c = \dfrac{\Gamma\left(\frac{1}{2}(\alpha+1)\right)}{\Gamma(\frac{1}{2}\alpha)} \left(\dfrac{\lambda}{\alpha\pi}\right)^{1/2}$
$E[x] = \mu$	$V[x] = \lambda^{-1}\alpha(\alpha-2)^{-1}$

$\text{F}(x \mid \alpha, \beta) = \text{F}_{\alpha,\beta}$ *Snedecor F (p. 123)*

$\alpha > 0, \beta > 0$	$x > 0$
$p(x) = c\, \dfrac{x^{\alpha/2-1}}{(\beta+\alpha x)^{(\alpha+\beta)/2}}$	$c = \dfrac{\Gamma\left(\frac{1}{2}(\alpha+\beta)\right)\alpha^{\alpha/2}\beta^{\beta/2}}{\Gamma(\frac{1}{2}\alpha)\Gamma(\frac{1}{2}\beta)}$
$E[x] = \dfrac{\beta}{\beta-2}, \quad \beta > 2$	$V[x] = \dfrac{2\beta^2(\alpha+\beta-2)}{\alpha(\beta-2)^2(\beta-4)}, \beta > 4$

Univariate Continuous Distributions (continued)

Lo$(x \mid \alpha, \beta)$ *Logistic (p. 122)*

$-\infty < \alpha < +\infty, \beta > 0$ $\qquad -\infty < x < +\infty$

$$p(x) = \beta^{-1} \exp\left\{-\beta^{-1}(x-\alpha)\right\} \left[1 + \exp\left\{-\beta^{-1}(x-\alpha)\right\}\right]^{-2}$$

$E[x] = \alpha$ $\qquad V[x] = \beta^2\pi^2/3$

Multivariate Discrete Distributions

Mu$_k(\boldsymbol{x} \mid \boldsymbol{\theta}, n)$ *Multinomial (p. 133)*

$\boldsymbol{\theta} = (\theta_1, \ldots, \theta_k)$ $\qquad \boldsymbol{x} = (x_1, \ldots, x_k)$

$0 < \theta_i < 1, \quad \sum_{\ell=1}^k \theta_\ell \leq 1$ $\qquad \sum_{\ell=1}^k x_\ell \leq n$

$n = 1, 2, \ldots$ $\qquad x_i = 0, 1, 2, \ldots$

$$p(\boldsymbol{x}) = \frac{n!}{\prod_{\ell=1}^{k+1} x_\ell!} \prod_{\ell=1}^{k+1} \theta^{x_\ell}, \qquad \theta_{k+1} = 1 - \sum_{\ell=1}^k \theta_\ell, \quad x_{k+1} = n - \sum_{\ell=1}^k x_\ell$$

$E[x_i] = n\theta_i$ $\qquad V[x_i] = n\theta_i(1-\theta_i)$ $\qquad C[x_i, x_j] = -n\theta_i\theta_j$

Md$_k(\boldsymbol{x} \mid \boldsymbol{\theta}, n)$ *Multinomial-Dirichlet (p. 135)*

$\boldsymbol{\alpha} = (\alpha_1, \ldots, \alpha_{k+1})$ $\qquad \boldsymbol{x} = (x_1, \ldots, x_k)$

$\alpha_i > 0$ $\qquad x_i = 0, 1, 2, \ldots$

$n = 1, 2, \ldots$ $\qquad \sum_{\ell=1}^n x_l \leq n$

$$p(\boldsymbol{x}) = c \prod_{\ell=1}^{k+1} \frac{\alpha_\ell^{[x_\ell]}}{x_\ell!} \qquad c = \frac{n!}{\left(\sum_{\ell=1}^{k+1} \alpha_\ell\right)^{[n]}}$$

$\alpha^{[s]} = \prod_{\ell=1}^s (\alpha + \ell - 1)$ $\qquad x_{k+1} = n - \sum_{\ell=1}^k x_\ell$

$$E[x_i] = np_i \qquad V[x_i] = \frac{n + \sum_{\ell=1}^{k+1} \alpha_\ell}{1 + \sum_{\ell=1}^{k+1} \alpha_\ell} np_i(1-p_i)$$

$$p_i = \frac{\alpha_i}{\sum_{\ell=1}^{k+1} \alpha_\ell} \qquad C[x_i, x_j] = -\frac{n + \sum_{\ell=1}^{k+1} \alpha_\ell}{1 + \sum_{\ell=1}^{k+1} \alpha_\ell} np_ip_j$$

Multivariate Continuous Distributions

$\mathrm{Di}_k(\boldsymbol{x} \mid \boldsymbol{\alpha})$ *Dirichlet (p. 134)*

$\boldsymbol{\alpha} = (\alpha_1, \ldots, \alpha_{k+1})$ $\quad$ $\boldsymbol{x} = (x_1, \ldots, x_k)$

$\alpha_i > 0$ $\quad$ $0 < x_i < 1, \quad \sum_{\ell=1}^{k} x_\ell \le 1$

$$p(\boldsymbol{x}) = c \left(1 - \sum_{\ell=1}^{k} x_\ell\right)^{\alpha_{k+1}-1} \prod_{\ell=1}^{k} x_\ell^{\alpha_\ell - 1} \qquad c = \frac{\Gamma(\sum_{\ell=1}^{k+1} \alpha_\ell)}{\prod_{\ell=1}^{k+1} \Gamma(\alpha_\ell)}$$

$$E[x_i] = \frac{\alpha_i}{\sum_{\ell=1}^{k+1} \alpha_\ell} \qquad V[x_i] = \frac{E[x_i](1 - E[x_i])}{1 + \sum_{\ell=1}^{k+1} \alpha_\ell} \qquad C[x_i, x_j] = \frac{-E[x_i]E[x_j]}{1 + \sum_{\ell=1}^{k+1} \alpha_\ell}$$

$\mathrm{Ng}(x, y \mid \mu, \lambda, \alpha, \beta)$ *Normal-Gamma (p. 136)*

$\mu \in \Re, \ \lambda > 0, \ \alpha > 0, \ \beta > 0,$ $\quad$ $x \in \Re, \ y > 0$

$p(x, y) = \mathrm{N}(x \mid \mu, \lambda y) \, \mathrm{Ga}(y \mid \alpha, \beta)$

$E[x] = \mu \qquad E[y] = \alpha\beta^{-1} \qquad V[x] = \beta\lambda^{-1}(\alpha - 1)^{-1} \qquad V[y] = \alpha\beta^{-2}$

$p(x) = \mathrm{St}(x \mid \mu, \alpha\beta^{-1}\lambda, 2\alpha)$

$\mathrm{N}_k(\boldsymbol{x} \mid \boldsymbol{\mu}, \boldsymbol{\lambda})$ *Multivariate Normal (p. 136)*

$\boldsymbol{\mu} = (\mu_1, \ldots, \mu_k) \in \Re^k$ $\quad$ $\boldsymbol{x} = (x_1, \ldots, x_k) \in \Re^k$

$\boldsymbol{\lambda}$ symmetric positive-definite

$p(\boldsymbol{x}) = c \, \exp\left\{-\frac{1}{2}(\boldsymbol{x} - \boldsymbol{\mu})^t \boldsymbol{\lambda} (\boldsymbol{x} - \boldsymbol{\mu})\right\}$ $\quad$ $c = |\,\boldsymbol{\lambda}\,|^{1/2}(2\pi)^{-k/2}$

$E[\boldsymbol{x}] = \boldsymbol{\mu}$ $\quad$ $V[\boldsymbol{x}] = \boldsymbol{\lambda}^{-1}$

$\mathrm{Pa}_2(x, y \mid \alpha, \beta_0, \beta_1)$ *Bilateral Pareto (p. 141)*

$(\beta_0, \beta_1) \in \Re^2, \ \beta_0 < \beta_1, \ \alpha > 0$ $\quad$ $(x, y) \in \Re^2, \ x < \beta_0, \ y > \beta_1$

$p(x, y) = c \, (y - x)^{-(\alpha+2)}$ $\quad$ $c = \alpha(\alpha + 1)(\beta_1 - \beta_0)^\alpha$

$$E[x] = \frac{\alpha\beta_0 - \beta_1}{\alpha - 1} \qquad E[y] = \frac{\alpha\beta_1 - \beta_0}{\alpha - 1} \qquad V[x] = V[y] = \frac{\alpha(\beta_1 - \beta_0)^2}{(\alpha - 1)^2(\alpha - 2)}$$

Multivariate Continuous Distributions (continued)

$\text{Ng}_k(\boldsymbol{x}, y \mid \boldsymbol{\mu}, \boldsymbol{\lambda}, \alpha, \beta)$ *Multivariate Normal-Gamma (p. 140)*

$-\infty < \mu_i < +\infty,\ \alpha > 0,\ \beta > 0$ $\quad (\boldsymbol{x}, y) = (x_1, \ldots, x_k, y)$

$\boldsymbol{\lambda}$ symmetric positive-definite $\quad -\infty < x_i < \infty, \quad y > 0$

$p(\boldsymbol{x}, y) = \text{N}_k(\boldsymbol{x} \mid \boldsymbol{\mu}, \boldsymbol{\lambda} y)\ \text{Ga}(y \mid \alpha, \beta)$

$E[\boldsymbol{x}, y] = (\boldsymbol{\mu},\ \alpha\beta^{-1}), \qquad V[\boldsymbol{x}] = (\alpha - 1)^{-1}\beta\boldsymbol{\lambda}^{-1}, \qquad V[y] = \alpha\beta^{-2}$

$p(\boldsymbol{x}) = \text{St}_k(\boldsymbol{x} \mid \boldsymbol{\mu}, \boldsymbol{\lambda}\alpha\beta^{-1}, 2\alpha) \qquad p(y) = Ga(y \mid \alpha, \beta)$

$\text{Nw}_k(\boldsymbol{x}, \boldsymbol{y} \mid \boldsymbol{\mu}, \lambda, \alpha, \boldsymbol{\beta})$ *Multivariate Normal-Wishart (p. 140)*

$-\infty < \mu_i < +\infty,\ \lambda > 0$ $\quad \boldsymbol{x} = (x_1, \ldots, x_k)$

$2\alpha > k - 1$ $\quad -\infty < x_i < +\infty$

$\boldsymbol{\beta}$ symmetric non-singular $\quad \boldsymbol{y}$ symmetric positive-definite

$p(\boldsymbol{x}, \boldsymbol{y}) = \text{N}_k(\boldsymbol{x} \mid \boldsymbol{\mu}, \lambda\boldsymbol{y})\ \text{Wi}_k(\boldsymbol{y} \mid \alpha, \boldsymbol{\beta})$

$E[\boldsymbol{x}, \boldsymbol{y}] = \{\boldsymbol{\mu}, \alpha\boldsymbol{\beta}^{-1}\} \qquad V[\boldsymbol{x}] = (\alpha - 1)^{-1}\boldsymbol{\beta}\lambda^{-1}$

$p(\boldsymbol{x}) = \text{St}_k(\boldsymbol{x} \mid \boldsymbol{\mu}, \lambda\alpha\boldsymbol{\beta}^{-1}, 2\alpha) \qquad p(\boldsymbol{y}) = \text{Wi}_k(\boldsymbol{y} \mid \alpha, \boldsymbol{\beta})$

$\text{St}_k(\boldsymbol{x} \mid \boldsymbol{\mu}, \boldsymbol{\lambda}, \alpha)$ *Multivariate Student (p. 139)*

$-\infty < \mu_i < +\infty,\ \alpha > 0$ $\quad \boldsymbol{x} = (x_1, \ldots, x_k)$

$\boldsymbol{\lambda}$ symmetric positive-definite $\quad -\infty < x_i < +\infty$

$$p(\boldsymbol{x}) = c\left[1 + \frac{1}{\alpha}(\boldsymbol{x} - \boldsymbol{\mu})^t\boldsymbol{\lambda}(\boldsymbol{x} - \boldsymbol{\mu})\right]^{-(\alpha+k)/2} \qquad c = \frac{\Gamma\left(\frac{1}{2}(\alpha + k)\right)}{\Gamma(\frac{1}{2}\alpha)(\alpha\pi)^{k/2}}|\boldsymbol{\lambda}|^{1/2}$$

$E[\boldsymbol{x}] = \boldsymbol{\mu}, \quad V[\boldsymbol{x}] = \boldsymbol{\lambda}^{-1}(\alpha - 2)^{-1}\alpha$

$\text{Wi}_k(\boldsymbol{x} \mid \alpha, \boldsymbol{\beta})$ *Wishart (p. 138)*

$2\alpha > k - 1$ $\quad \boldsymbol{x}$ symmetric positive-definite

$\boldsymbol{\beta}$ symmetric non-singular

$$p(\boldsymbol{x}) = c\,|\boldsymbol{x}|^{\alpha-(k+1)/2}\exp\{-\operatorname{tr}(\boldsymbol{\beta}\boldsymbol{x})\} \qquad c = \frac{\pi^{-k(k-1)/4}|\boldsymbol{\beta}|^{\alpha}}{\prod_{\ell=1}^{k}\Gamma(\frac{1}{2}(2\alpha + 1 - \ell))}$$

$E[\boldsymbol{x}] = \alpha\boldsymbol{\beta}^{-1}, \quad E[\boldsymbol{x}^{-1}] = (\alpha - \frac{k+1}{2})^{-1}\boldsymbol{\beta}$

A.2 INFERENTIAL PROCESSES

The second section of this Appendix records the basic elements of the Bayesian learning processes for many commonly used statistical models.

For each of these models, we provide, in separate sections of the table, the following: the sufficient statistic and its sampling distribution; the conjugate family, the conjugate prior predictives for a single observable and for the sufficient statistic, the conjugate posterior and the conjugate posterior predictive for a single observable.

When clearly defined, we also provide, in a final section, the reference prior and the corresponding reference posterior and posterior predictive for a single observable. In the case of uniparameter models this can always be done. We recall, however, from Section 5.2.4 that, in multiparameter problems, the reference prior is only defined *relative to an ordered parametrisation.* In the univariate normal model (Example 5.17), the reference prior for (μ, λ) happens to be the same as that for (λ, μ), namely $\pi(\mu, \lambda) = \pi(\lambda, \mu) \propto \lambda^{-1}$, and we provide the corresponding reference posteriors for μ and λ, together with the reference predictive distribution for a future observation.

In the multinomial, multivariate normal and linear regression models, however, there are very many different reference priors, corresponding to different inference problems, and specified by different ordered parametrisations. These are not reproduced in this Appendix.

Bernoulli model

$\boldsymbol{z} = \{x_1, \ldots, x_n\}, \qquad x_i \in \{0, 1\}$
$p(x_i \mid \theta) = \text{Br}(x_i \mid \theta), \qquad 0 < \theta < 1$

$t(\boldsymbol{z}) = r = \sum_{i=1}^{n} x_i$
$p(r \mid \theta) = \text{Bi}(r \mid \theta, n)$

$p(\theta) = \text{Be}(\theta \mid \alpha, \beta)$
$p(x) = \text{Bb}(r \mid \alpha, \beta, 1)$
$p(r) = \text{Bb}(r \mid \alpha, \beta, n)$
$p(\theta \mid \boldsymbol{z}) = \text{Be}(\theta \mid \alpha + r, \beta + n - r)$
$p(x \mid \boldsymbol{z}) = \text{Bb}(x \mid \alpha + r, \beta + n - r, 1)$

$\pi(\theta) = \text{Be}(\theta \mid \tfrac{1}{2}, \tfrac{1}{2})$
$\pi(\theta \mid \boldsymbol{z}) = \text{Be}(\theta \mid \tfrac{1}{2} + r, \tfrac{1}{2} + n - r)$
$\pi(x \mid \boldsymbol{z}) = \text{Bb}(x \mid \tfrac{1}{2} + r, \tfrac{1}{2} + n - r, 1)$

Poisson Model

$$z = \{x_1, \ldots, x_n\}, \qquad x_i = 0, 1, 2, \ldots$$
$$p(x_i \mid \lambda) = \mathrm{Pn}(x_i \mid \lambda), \qquad \lambda \geq 0$$

$$t(z) = r = \sum_{i=1}^{n} x_i$$
$$p(r \mid \lambda) = \mathrm{Pn}(r \mid n\lambda)$$

$$p(\lambda) = \mathrm{Ga}(\lambda \mid \alpha, \beta)$$
$$p(x) = \mathrm{Pg}(x \mid \alpha, \beta, 1)$$
$$p(r) = \mathrm{Pg}(x \mid \alpha, \beta, n)$$
$$p(\lambda \mid z) = \mathrm{Ga}(\lambda \mid \alpha + r, \beta + n)$$
$$p(x \mid z) = \mathrm{Pg}(x \mid \alpha + r, \beta + n, 1)$$

$$\pi(\lambda) \propto \lambda^{-1/2}$$
$$\pi(\lambda \mid z) = \mathrm{Ga}(\lambda \mid r + \tfrac{1}{2}, n)$$
$$\pi(x \mid z) = \mathrm{Pg}(x \mid r + \tfrac{1}{2}, n, 1)$$

Negative-Binomial model

$$z = (x_1, \ldots, x_n), \qquad x_i = 0, 1, 2, \ldots$$
$$p(x_i \mid \theta) = \mathrm{Nb}(x_i \mid \theta, r), \qquad 0 < \theta < 1$$

$$t(z) = s = \sum_{i=1}^{n} x_i$$
$$p(s \mid \theta) = \mathrm{Nb}(s \mid \theta, nr)$$

$$p(\theta) = \mathrm{Be}(\theta \mid \alpha, \beta)$$
$$p(x) = \mathrm{Nbb}(x \mid \alpha, \beta, r)$$
$$p(s) = \mathrm{Nbb}(s \mid \alpha, \beta, nr)$$
$$p(\theta \mid z) = \mathrm{Be}(\theta \mid \alpha + nr, \beta + s)$$
$$p(x \mid z) = \mathrm{Nbb}(x \mid \alpha + nr, \beta + s, r)$$

$$\pi(\theta) \propto \theta^{-1}(1 - \theta)^{-1/2}$$
$$\pi(\theta \mid z) = \mathrm{Be}(\theta \mid nr, s + \tfrac{1}{2})$$
$$\pi(x \mid z) = \mathrm{Nbb}(x \mid nr, s + \tfrac{1}{2}, r)$$

Exponential Model

$$z = \{x_1, \ldots, x_n\}, \qquad 0 < x_i < \infty$$
$$p(x_i \mid \theta) = \mathrm{Ex}(x_i \mid \theta), \qquad \theta > 0$$

$$t(z) = t = \textstyle\sum_{i=1}^n x_i$$
$$p(t \mid \theta) = \mathrm{Ga}(t \mid n, \theta)$$

$$p(\theta) = \mathrm{Ga}(\theta \mid \alpha, \beta)$$
$$p(x) = \mathrm{Gg}(x \mid \alpha, \beta, 1)$$
$$p(t) = \mathrm{Gg}(t \mid \alpha, \beta, n)$$
$$p(\theta \mid z) = \mathrm{Ga}(\theta \mid \alpha + n, \beta + t)$$
$$p(x \mid z) = \mathrm{Gg}(x \mid \alpha + n, \beta + t, 1)$$

$$\pi(\theta) \propto \theta^{-1}$$
$$\pi(\theta \mid z) = \mathrm{Ga}(\theta \mid n, t)$$
$$\pi(x \mid z) = \mathrm{Gg}(x \mid n, t, 1)$$

Uniform Model

$$z = \{x_1, \ldots, x_n\}, \qquad 0 < x_i < \theta$$
$$p(x_i \mid \theta) = \mathrm{Un}(x_i \mid 0, \theta), \qquad \theta > 0$$

$$t(z) = t = \max\{x_1, \ldots, x_n\}$$
$$p(t \mid \theta) = \mathrm{Ip}(t \mid n, \theta^{-1})$$

$$p(\theta) = \mathrm{Pa}(\theta \mid \alpha, \beta)$$
$$p(x) = \tfrac{\alpha}{\alpha+1}\mathrm{Un}(x \mid 0, \beta), \text{ if } x \le \beta, \quad \tfrac{1}{\alpha+1}\mathrm{Pa}(x \mid \alpha, \beta), \text{ if } x > \beta$$
$$p(t) = \tfrac{\alpha}{\alpha+n}\mathrm{Ip}(t \mid n, \beta^{-1}), \text{ if } t \le \beta, \quad \tfrac{n}{\alpha+n}\mathrm{Pa}(t \mid \alpha, \beta), \text{ if } t > \beta$$
$$p(\theta \mid z) = \mathrm{Pa}(\theta \mid \alpha + n, \beta_n), \quad \beta_n = \max\{\beta, t\}$$
$$p(x \mid z) = \tfrac{\alpha+n}{\alpha+n+1}\mathrm{Un}(x \mid 0, \beta_n), \text{ if } x \le \beta_n, \quad \tfrac{1}{\alpha+n+1}\mathrm{Pa}(x \mid \alpha, \beta_n), \text{ if } x > \beta_n$$

$$\pi(\theta) \propto \theta^{-1}$$
$$\pi(\theta \mid z) = \mathrm{Pa}(\theta \mid n, t)$$
$$\pi(x \mid z) = \tfrac{n}{n+1}\mathrm{Un}(x \mid 0, t), \text{ if } x \le t, \quad \tfrac{1}{n+1}\mathrm{Pa}(x \mid n, t), \text{ if } x > t$$

Normal Model (known precision λ)

$z = \{x_1, \ldots, x_n\}, \qquad -\infty < x_i < \infty$
$p(x_i \mid \mu, \lambda) = \mathbf{N}(x_i \mid \mu, \lambda), \qquad -\infty < \mu < \infty$

$t(z) = \bar{x} = n^{-1} \sum_{i=1}^{n} x_i$
$p(\bar{x} \mid \mu, \lambda) = \mathbf{N}(x \mid \mu, n\lambda)$

$p(\mu) = \mathbf{N}(\mu \mid \mu_0, \lambda_0)$
$p(x) = \mathbf{N}\left(x \mid \mu_0, \lambda\,\lambda_0(\lambda_0 + \lambda)^{-1}\right)$
$p(\overline{x}) = \mathbf{N}\left(\overline{x} \mid \mu_0, n\lambda\,\lambda_0\lambda_n^{-1}\right), \qquad \lambda_n = \lambda_0 + n\lambda,$
$p(\mu \mid z) = \mathbf{N}(\mu \mid \mu_n, \lambda_n), \qquad \mu_n = \lambda_n^{-1}(\lambda_0\mu_0 + n\lambda\overline{x})$
$p(x \mid z) = \mathbf{N}\left(x \mid \mu_n, \lambda\,\lambda_n(\lambda_n + \lambda)^{-1}\right)$

$\pi(\mu) = \text{constant}$
$\pi(\mu \mid z) = \mathbf{N}(\mu \mid \bar{x}, n\lambda)$
$\pi(x \mid z) = \mathbf{N}(x \mid \bar{x}, \lambda\, n(n+1)^{-1})$

Normal Model (known mean μ)

$z = \{x_1, \ldots, x_n\}, \qquad -\infty < x_i < \infty$
$p(x_i \mid \mu, \lambda) = \mathbf{N}(x_i \mid \mu, \lambda), \qquad \lambda > 0$

$t(z) = t = \sum_{i=1}^{n} (x_i - \mu)^2$
$p(t \mid \mu, \lambda) = \mathbf{Ga}(t \mid \tfrac{1}{2}n, \tfrac{1}{2}\lambda), \qquad p(\lambda t) = \chi^2(\lambda t \mid n)$

$p(\lambda) = \mathbf{Ga}(\lambda \mid \alpha, \beta)$
$p(x) = \mathbf{St}(x \mid \mu, \alpha\beta^{-1}, 2\alpha)$
$p(t) = \mathbf{Gg}(x \mid \alpha, 2\beta, \tfrac{1}{2}n)$
$p(\lambda \mid z) = \mathbf{Ga}(\lambda \mid \alpha + \tfrac{1}{2}n, \beta + \tfrac{1}{2}t)$
$p(x \mid z) = \mathbf{St}(x \mid \mu, (\alpha + \tfrac{1}{2}n)(\beta + \tfrac{1}{2}t)^{-1}, 2\alpha + n)$

$\pi(\lambda) \propto \lambda^{-1}$
$\pi(\lambda \mid z) = \mathbf{Ga}(\lambda \mid \tfrac{1}{2}n, \tfrac{1}{2}t)$
$\pi(x \mid z) = \mathbf{St}(x \mid \mu, nt^{-1}, n)$

Normal Model (both parameters unknown)

$$z = \{x_1, \ldots, x_n\}, \qquad -\infty < x_i < \infty$$
$$p(x_i \mid \mu, \lambda) = \mathbf{N}(x_i \mid \mu, \lambda), \qquad -\infty < \mu < \infty, \quad \lambda > 0$$

$$t(z) = (\bar{x}, s), \qquad n\bar{x} = \textstyle\sum_{i=1}^n x_i, \qquad ns^2 = \sum_{i=1}^n (x_i - \bar{x})^2$$
$$p(\bar{x} \mid \mu, \lambda) = \mathbf{N}(\bar{x} \mid \mu, n\lambda)$$
$$p(ns^2 \mid \mu, \lambda) = \mathbf{Ga}(ns^2 \mid \tfrac{1}{2}(n-1), \tfrac{1}{2}\lambda), \qquad p(\lambda ns^2) = \chi^2(\lambda ns^2 \mid n-1)$$

$$p(\mu, \lambda) = \mathbf{Ng}(\mu, \lambda \mid \mu_0, n_0, \alpha, \beta) = \mathbf{N}(\mu \mid \mu_0, n_0\lambda)\ \mathbf{Ga}(\lambda \mid \alpha, \beta)$$
$$p(\mu) = \mathbf{St}(\mu \mid \mu_0, n_0\alpha\beta^{-1}, 2\alpha)$$
$$p(\lambda) = \mathbf{Ga}(\lambda \mid \alpha, \beta)$$
$$p(x) = \mathbf{St}(x \mid \mu_0, n_0(n_0+1)^{-1}\alpha\beta^{-1}, 2\alpha)$$
$$p(\bar{x}) = \mathbf{St}(\bar{x} \mid \mu_0, n_0 n(n_0+n)^{-1}\alpha\beta^{-1}, 2\alpha)$$
$$p(ns^2) = \mathbf{Gg}(ns^2 \mid \alpha, 2\beta, \tfrac{1}{2}(n-1))$$
$$p(\mu \mid z) = \mathbf{St}(\mu \mid \mu_n, (n+n_0)(\alpha+\tfrac{1}{2}n)\beta_n^{-1}, 2\alpha+n),$$
$$\mu_n = (n_0+n)^{-1}(n_0\mu_0 + n\bar{x}),$$
$$\beta_n = \beta + \tfrac{1}{2}ns^2 + \tfrac{1}{2}(n_0+n)^{-1}n_0 n(\mu_0 - \bar{x})^2$$
$$p(\lambda \mid z) = \mathbf{Ga}(\lambda \mid \alpha + \tfrac{1}{2}n, \beta_n)$$
$$p(x \mid z) = \mathbf{St}(x \mid \mu_n, (n+n_0)(n+n_0+1)^{-1}(\alpha+\tfrac{1}{2}n)\beta_n^{-1}, 2\alpha+n)$$

$$\pi(\mu, \lambda) = \pi(\lambda, \mu) \propto \lambda^{-1}, \qquad n > 1$$
$$\pi(\mu \mid z) = \mathbf{St}(\mu \mid \bar{x}, (n-1)s^{-2}, n-1)$$
$$\pi(\lambda \mid z) = \mathbf{Ga}\left(\lambda \mid \tfrac{1}{2}(n-1), \tfrac{1}{2}ns^2\right)$$
$$\pi(x \mid z) = \mathbf{St}\left(x \mid \bar{x}, (n-1)(n+1)^{-1}s^{-2}, n-1\right)$$

Multinomial Model

$$z = \{r_1, \ldots, r_k, n\}, \qquad r_i = 0, 1, 2, \ldots, \qquad \sum_{\ell=1}^{k} r_i \leq n$$
$$p(z \mid \theta) = \mathrm{Mu}_k(z \mid \theta, n), \qquad 0 < \theta_i < 1, \quad \sum_{i=1}^{k} \theta_i \leq 1$$

$$t(z) = (r, n), \quad r = (r_1, \ldots, r_k)$$
$$p(r \mid \theta) = \mathrm{Mu}_k(r \mid \theta, n)$$

$$p(\theta) = \mathrm{Di}_k(\theta \mid \alpha), \quad \alpha = \{\alpha_1, \ldots, \alpha_{k+1}\}$$
$$p(r) = \mathrm{Md}_k(r \mid \alpha, n)$$
$$p(\theta \mid z) = \mathrm{Di}_k\left(\theta \mid \alpha_1 + r_1, \ldots, \alpha_k + r_k, \alpha_{k+1} + n - \sum_{\ell=1}^{k} r_\ell\right)$$
$$p(x \mid z) = \mathrm{Md}_k\left(x \mid \alpha_1 + r_1, \ldots, \alpha_k + r_k, \alpha_{k+1} + n - \sum_{\ell=1}^{k} r_\ell, n\right)$$

Multivariate Normal Model

$$z = \{x_1, \ldots, x_n\}, \qquad x_i \in \Re^k$$
$$p(x_i \mid \mu, \lambda) = \mathrm{N}_k(x_i \mid \mu, \lambda), \qquad \mu \in \Re^k, \quad \lambda \ k \times k \text{ positive-definite}$$

$$t(z) = (\bar{x}, S), \qquad \bar{x} = n^{-1} \sum_{i=1}^{n} x_i, \qquad S = \sum_{i=1}^{n} (x_i - \bar{x})(x_i - \bar{x})^t$$
$$p(\bar{x} \mid \mu, \lambda) = \mathrm{N}_k(\bar{x} \mid \mu, n\lambda)$$
$$p(S \mid \lambda) = \mathrm{Wi}_k\left(S \mid \tfrac{1}{2}(n-1), \tfrac{1}{2}\lambda\right)$$

$$p(\mu, \lambda) = \mathrm{Nw}_k(\mu, \lambda \mid \mu_0, n_0, \alpha, \beta) = \mathrm{N}_k(\mu \mid \mu_0, n_0\lambda)\, \mathrm{Wi}_k(\lambda \mid \alpha, \beta)$$
$$p(x) = \mathrm{St}_k\left(x \mid \mu_0, (n_0+1)^{-1} n_0 (\alpha - \tfrac{1}{2}(k-1)) \beta^{-1}, 2\alpha - k + 1\right)$$
$$p(\mu \mid z) = \mathrm{St}_k\left(\mu \mid \mu_n, (n + n_0)\alpha_n \beta_n^{-1}, 2\alpha_n\right),$$
$$\mu_n = (n_0 + n)^{-1}(n_0 \mu_0 + n\bar{x}),$$
$$\beta_n = \beta + \tfrac{1}{2} S + \tfrac{1}{2}(n + n_0)^{-1} n n_0 (\mu_0 - \bar{x})(\mu_0 - \bar{x})^t$$
$$p(\lambda \mid z) = \mathrm{Wi}_k(\lambda \mid \alpha + \tfrac{1}{2}n, \beta_n)$$
$$p(x \mid z) = \mathrm{St}_k(x \mid \mu_n, (n_0 + n + 1)^{-1}(n_0 + n)\alpha_n \beta_n^{-1}, 2\alpha_n),$$
$$\alpha_n = \alpha + \tfrac{1}{2}n - \tfrac{1}{2}(k-1)$$

Linear Regression

$\boldsymbol{z} = (\boldsymbol{y}, \boldsymbol{X}), \quad \boldsymbol{y} = (y_1, \ldots, y_n)^t \in \Re^n, \; \boldsymbol{x}_i = (x_{i1}, \ldots, x_{ik}) \in \Re^k, \; \boldsymbol{X} = (x_{ij})$

$p(\boldsymbol{y} \mid \boldsymbol{X}, \boldsymbol{\theta}, \lambda) = \mathrm{N}_n(\boldsymbol{y} \mid \boldsymbol{X}\boldsymbol{\theta}, \lambda \boldsymbol{I}_n), \qquad \boldsymbol{\theta} \in \Re^k, \quad \lambda > 0$

$t(\boldsymbol{z}) = (\boldsymbol{X}^t\boldsymbol{X}, \boldsymbol{X}^t\boldsymbol{y})$

$p(\boldsymbol{\theta}, \lambda) = \mathrm{Ng}(\boldsymbol{\theta}, \lambda \mid \boldsymbol{\theta}_0, \boldsymbol{n}_0, \alpha, \beta) = \mathrm{N}_k(\boldsymbol{\theta} \mid \boldsymbol{\theta}_0, \boldsymbol{n}_0\lambda)\, \mathrm{Ga}(\lambda \mid \alpha, \beta)$

$p(\boldsymbol{\theta}) = \mathrm{St}_k(\boldsymbol{\theta} \mid \boldsymbol{\theta}_0, \; \boldsymbol{n}_0\alpha\beta^{-1}, \; 2\alpha)$

$p(\lambda) = \mathrm{Ga}(\lambda \mid \alpha, \beta)$

$p(y \mid \boldsymbol{x}) = \mathrm{St}\left(y \mid \boldsymbol{x}\boldsymbol{\theta}_0, \; f(\boldsymbol{x})\alpha\beta^{-1}, \; 2\alpha\right)$

$\qquad f(\boldsymbol{x}) = 1 - \boldsymbol{x}(\boldsymbol{x}^t\boldsymbol{x} + \boldsymbol{n}_0)^{-1}\boldsymbol{x}^t,$

$p(\boldsymbol{\theta} \mid \boldsymbol{z}) = \mathrm{St}_k\left(\boldsymbol{\theta} \mid \boldsymbol{\theta}_n, \; (\boldsymbol{n}_0 + \boldsymbol{X}^t\boldsymbol{X})(\alpha + \tfrac{1}{2}n)\beta_n^{-1}, \; 2\alpha + n\right),$

$\qquad \boldsymbol{\theta}_n = (\boldsymbol{n}_0 + \boldsymbol{X}^t\boldsymbol{X})^{-1}(\boldsymbol{n}_0\boldsymbol{\theta}_0 + \boldsymbol{X}^t\boldsymbol{y}),$

$\qquad \beta_n = \beta + \tfrac{1}{2}(\boldsymbol{y} - \boldsymbol{X}\boldsymbol{\theta}_n)^t\boldsymbol{y} + \tfrac{1}{2}(\boldsymbol{\theta}_0 - \boldsymbol{\theta}_n)^t\boldsymbol{n}_0\boldsymbol{\theta}_0$

$p(\lambda \mid \boldsymbol{z}) = \mathrm{Ga}(\lambda \mid \alpha + \tfrac{1}{2}n, \beta_n)$

$p(y \mid \boldsymbol{x}, \boldsymbol{z}) = \mathrm{St}(y \mid \boldsymbol{x}\boldsymbol{\theta}_n, \; f_n(\boldsymbol{x})(\alpha + \tfrac{1}{2}n)\beta_n^{-1}, \; 2\alpha + n),$

$\qquad f_n(\boldsymbol{x}) = 1 - \boldsymbol{x}(\boldsymbol{x}^t\boldsymbol{x} + \boldsymbol{n}_0 + \boldsymbol{X}^t\boldsymbol{X})^{-1}\boldsymbol{x}^t$

$\pi(\boldsymbol{\theta}, \lambda) = \pi(\lambda, \boldsymbol{\theta}) \propto \lambda^{-(k+1)/2}$ (for all reorderings of the θ_j)

$\pi(\boldsymbol{\theta} \mid \boldsymbol{z}) = \mathrm{St}_k\left(\boldsymbol{\theta} \mid \hat{\boldsymbol{\theta}}_n, \; \tfrac{1}{2}\boldsymbol{X}^t\boldsymbol{X}(n-k)\hat{\beta}_n^{-1}, \; n-k\right),$

$\qquad \hat{\boldsymbol{\theta}}_n = (\boldsymbol{X}^t\boldsymbol{X})^{-1}\boldsymbol{X}^t\boldsymbol{y},$

$\qquad \hat{\beta}_n = \tfrac{1}{2}(\boldsymbol{y} - \boldsymbol{X}\hat{\boldsymbol{\theta}}_n)^t\boldsymbol{y}$

$\pi(\lambda \mid \boldsymbol{z}) = \mathrm{Ga}(\lambda \mid \tfrac{1}{2}(n-k), \hat{\beta}_n)$

$\pi(y \mid \boldsymbol{x}, \boldsymbol{z}) = \mathrm{St}(y \mid \boldsymbol{x}\hat{\boldsymbol{\theta}}_n, \; \tfrac{1}{2}f_n(\boldsymbol{x})(n-k)\hat{\beta}_n^{-1}, \; n-k),$

$\qquad f_n(\boldsymbol{x}) = 1 - \boldsymbol{x}(\boldsymbol{x}^t\boldsymbol{x} + \boldsymbol{X}^t\boldsymbol{X})^{-1}\boldsymbol{x}^t$

Appendix B

Non-Bayesian Theories

Summary

A summary is given of a number of non-Bayesian statistical approaches and procedures. The main theories reviewed include classical decision theory, frequentism, likelihood, and fiducial inference. These are illustrated and compared and contrasted with Bayesian methods in the stylised contexts of point and interval estimation, and hypothesis and significance testing. Further issues discussed include: conditional and unconditional inferences; nuisance parameters and marginalisation; prediction; asymptotics and criteria for model choice.

B.1 OVERVIEW

Bayesian statistical theory as presented in this book is self-contained and can be understood and applied without reference to alternative statistical theories. There are, however, two broad reasons why we think it appropriate to give a summary overview of our attitude to other theories.

First, many, if not most, readers will have some previous exposure to "classical" statistics, and the material in this Appendix may help them to put the contents of this book into perspective.

Secondly, our own experience has been that some element of comparative analysis contributes significantly to an appreciation of the attractions of the Bayesian paradigm in statistics.

As a preliminary, we recall from Chapter 1 our acknowledgment that Bayesian analysis takes place in a rather *formal* framework, and that exploratory data analysis and graphical displays are often prerequisite, *informal* activities. It is important, therefore, to be clear that in this Appendix we are discussing non-Bayesian *formal* procedures.

We begin by making explicit some of the key differences between Bayesian and non-Bayesian theories:

(i) As we showed in detail in Chapter 2, Bayesian statistics has an *axiomatic foundation* which guarantees quantitative coherence. Non-Bayesian statistical theories typically lack foundational support of this kind and essentially consist of a set of recipes which are not necessarily internally consistent.

(ii) Non-Bayesian theories typically use *only* a parametric model family of the form $\{p(x \mid \boldsymbol{\theta}),\, x \in \boldsymbol{X},\, \boldsymbol{\theta} \in \Theta\}$, ignoring the prior distribution $p(\boldsymbol{\theta})$. The implications of this fact are so far reaching that sometimes Bayesian statistics is simplistically thought of as statistics with the "optional extra" of a prior distribution. In Chapters 2 and 3, we have argued that the *existence* of a prior distribution is a mathematical consequence of the foundational axioms. In Chapter 4, we stressed that predictive models, typically derived from combining $p(x \mid \boldsymbol{\theta})$ and $p(\boldsymbol{\theta})$, are primary.

(iii) The decision theoretical foundations of Bayesian statistics provide a natural framework within which specific problems can easily be structured, with solutions directly tailored to problems. In contrast, most non-Bayesian theories essentially consist of stylised *procedures*, such as those for point or interval estimation, or hypothesis testing, designed to satisfy or optimise an *ad hoc criterion*, and often lacking the necessary flexibility to be adaptable to specific problem situations.

(iv) We have argued that, from a Bayesian viewpoint, a *decision structure* is the natural framework for *any* formal statistical problem, and have described how a "pure" inference problem may be seen as a particular decision problem. Non-Bayesian theories depart radically from this viewpoint; classical decision theory is only partially relevant to inference, and non-Bayesian inference theories typically ignore the decision aspects of inference problems.

In Section B.2, we will revise the key ideas of a number of non-Bayesian statistical theories, specifically reviewing *Classical Decision Theory, Frequentist Procedures, Likelihood Inference* and *Fiducial and Related Theories.*

In Section B.3, we will follow the typical methodological partition of non-Bayesian textbooks into the topics of *Point Estimation*, *Interval Estimation*, *Hypothesis Testing* and *Significance Testing*. Within each of those subheadings we will comment on the internal logic, the relevance to actual statistical problems, and the performance of classical procedures relative to their Bayesian counterparts.

In Section B.4, we will discuss in detail some key comparative issues: *Conditional and Unconditional Inference*, *Nuisance Parameters and Marginalisation*, *Approaches to Prediction*, *Aspects of Asymptotics* and *Model Choice Criteria*.

For readers seeking further comparative discussion at textbook level, we recall, from our discussion at the end of Chapter 5, the books by Barnett (1971/1982), Press (1972/1982), Cox and Hinkley (1974), Anderson (1984), DeGroot (1987), Casella and Berger (1990) and Poirier (1993).

B.2 ALTERNATIVE APPROACHES

B.2.1 Classical Decision Theory

We recall from Section 3.3 the basic structure of a general decision problem, consisting of a set of a possible decisions D, a parameter space Ω, a prior distribution $p(\boldsymbol{\omega})$ over Ω, and a utility function $u\big(d(\boldsymbol{\omega})\big)$ which we shall denote by $u(d, \boldsymbol{\omega})$ to conform more closely to standard notation in classical decision theory. We established that the existence of *both* the prior distribution $p(\boldsymbol{\omega})$ and the utility function $u(d, \boldsymbol{\omega})$ is a mathematical consequence of the axioms of quantitative coherence and that the best decision d^* is that which maximises the expected utility

$$\overline{u}(d) = \int u(d, \boldsymbol{\omega}) p(\boldsymbol{\omega}) d\boldsymbol{\omega}.$$

We established furthermore that, if additional information x is obtained which is probabilistically related to $\boldsymbol{\omega}$ by $p(\boldsymbol{x} \mid \boldsymbol{\omega})$, then the best decision d_x^* is that which maximises the *posterior* expected utility

$$\overline{u}(d \mid \boldsymbol{x}) = \int u(d, \boldsymbol{\omega}) p(\boldsymbol{\omega} \mid \boldsymbol{x}) d\boldsymbol{\omega},$$

where

$$p(\boldsymbol{\omega} \mid \boldsymbol{x}) \propto p(\boldsymbol{x} \mid \boldsymbol{\omega}) p(\boldsymbol{\omega}).$$

Some authors prefer to use loss functions instead of utilities. A *regret function*, or *decision loss*, is easily defined from the utility function (at least in bounded cases) by

$$l(d, \boldsymbol{\omega}) = \sup_{d_i \in D} u(d_i, \boldsymbol{\omega}) - u(d, \boldsymbol{\omega}),$$

which quantifies the maximum loss that, for each ω, one may suffer as a consequence of a wrong decision. Since $\sup_D u(d, \omega)$ only depends on ω, the expected loss

$$\bar{l}(d) = \int l(d, \omega) p(\omega) d\omega$$

is minimised by the same decision d^* which maximises $\overline{u}(d)$ and hence, from a Bayesian point of view, the two formulations are essentially equivalent.

In contrast to this Bayesian formulation, the core framework of classical decision theory may be loosely described as decision theory without a prior distribution. A utility function (or a loss function) is accepted, perhaps justified by utility-only axiomatics of the type pioneered by von Neumann and Morgenstern (1944/1953), but a prior distribution for ω is not.

Although some of the basic ideas in classical statistical theory were present in the work of Neyman and Pearson (1933), it was Wald (1950) who introduced a systematic decision theory framework, excluding prior distributions as core elements, but including a formulation of standard statistical problems within a decision framework. This work was continued by Girshick and Savage (1951) and by Stein (1956). An excellent textbook introduction is that of Ferguson (1967).

Classical decision theory focuses on the way in which additional information x should be used to assist the decision process. Consequently, the basic space is not the class of decisions, D, but the class of *decision rules*, Δ, consisting of functions $\delta : X \to D$ which attach a decision $\delta(x)$ to each possible data set x. It is then suggested that decision rules should be evaluated in terms of their *average* loss with respect to the data which might arise. Thus, the *risk function* $r(\delta, \omega)$ of a decision rule δ is defined as

$$r(\delta, \omega) = \int l(\delta(x), \omega) p(x \mid \omega) dx$$

and subsequent comparison of decision rules is based on their risk functions.

The formulation includes, as a special case, the situation with no additional data (the no-data case), where the risk function reduces to the loss function.

Example B.1. ***(Estimation of the mean of a normal distribution).*** Let $x = \{x_1, \ldots, x_n\}$ be a random sample from a $N(x \mid \mu, 1)$ distribution, and suppose that we want to select an estimator for μ, so that $D = \Re$, under the assumption of a quadratic loss function $l(d, \mu) = (\mu - d)^2$. Some possible decision rules are

(i) $\delta_1(x) = \overline{x}$, the sample mean
(ii) $\delta_2(x) = \tilde{x}$, the sample median
(iii) $\delta_3(x) = \mu_0$, a fixed value
(iv) $\delta_4(x) = (n + n_0)^{-1}(n_0\mu_0 + n\overline{x})$, the posterior mean from an $N(\mu \mid \mu_0, n_0)$ prior, centred on μ_0 and with precision n_0.

Using the fact that the variance of the sample median is approximately $\pi/2n$, the corresponding risk functions are easily seen to be

(i) $r(\delta_1, \mu) = 1/n$
(ii) $r(\delta_2, \mu) \simeq \pi/2n$,
(iii) $r(\delta_3, \mu) = (\mu - \mu_0)^2$
(iv) $r(\delta_4, \mu) = (n + n_0)^{-2}\{n + n_0^2(\mu_0 - \mu)^2\}$

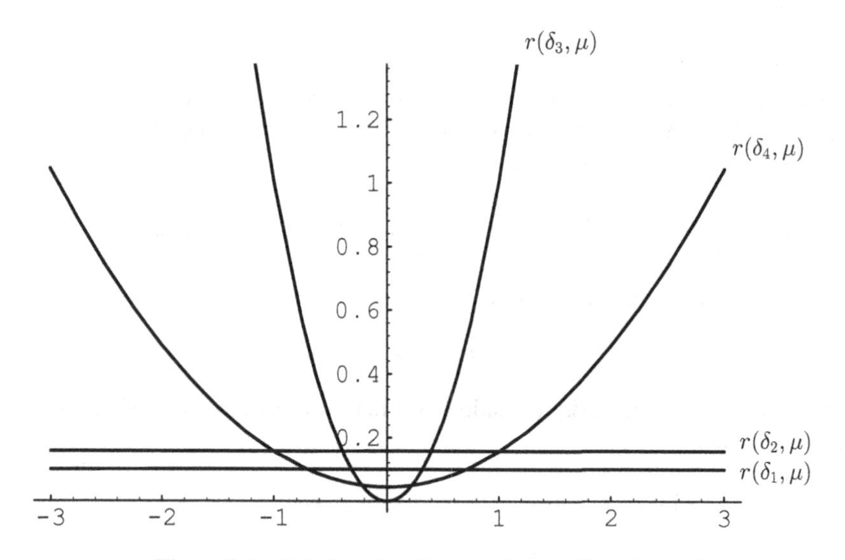

Figure B.1 *Risk functions for* $\mu_0 = 0$, $n = 10$ *and* $n_0 = 5$

Note that (iv) includes both (i) and (iii) as limiting cases when $n_0 \to 0$ and $n_0 \to \infty$ respectively. Figure B.1 provides a graphical comparison of the risk functions.

It is easily seen that, whatever the value of μ, δ_2 has larger risk than δ_1 but, otherwise, the best decision rule markedly depends on μ. The closer μ_0 is to the true value of μ, the more attractive δ_3 and δ_4 will obviously be.

Admissibility

The decision rule δ_2 in Example B.1 can hardly be considered a good decision rule since, for any value of the unknown parameter μ, the rule δ_1 has a smaller risk. This is formalised within classical decision theory by saying that a decision rule δ' is *dominated* by another decision rule δ if, for all ω,

$$r(\delta', \omega) \geq r(\delta, \omega)$$

with strict inequality for some ω, and that a decision rule is *admissible* if there is no other decision rule which dominates it. A class of decision rules is *complete* if for any δ' not in the class there is a δ in the class which dominates it, and a class is *minimal complete* if it does not contain a complete subclass. If one is to choose among decision rules in terms of their risk functions, classical decision theory establishes that one can limit attention to a minimal complete class. However, for guidance on how to choose among admissible decision rules, further concepts and criteria are required.

Bayes Rules

If the existence of a prior distribution $p(\omega)$ for the unknown parameter is accepted, classical decision theory focuses on the decision rule which minimises *expected* risk (or so-called *Bayes risk*)

$$\min_{\delta} \int_{\Omega} r(\delta, \omega)\, p(\omega)\, d\omega = \min_{\delta} \int_{\Omega} \int_{X} l(\delta(\boldsymbol{x}), \omega)\, p(\boldsymbol{x} \mid \omega)\, p(\omega)\, d\boldsymbol{x}\, d\omega,$$

which it calls a *Bayes decision rule.* Note that, since

$$p(\boldsymbol{x} \mid \omega)p(\omega) = p(\omega \mid \boldsymbol{x})p(\boldsymbol{x}),$$

under appropriate regularity conditions we may reverse the order of integration above to obtain

$$\begin{aligned} \min_{\delta} \int_{\Omega} \int_{X} & r(\delta(x), \omega) p(\omega \mid \boldsymbol{x}) p(\boldsymbol{x})\, d\boldsymbol{x} d\omega \\ &= \int_{X} p(\boldsymbol{x}) \min_{\delta} \int_{\Omega} l(d, \omega) p(\omega \mid \boldsymbol{x})\, d\omega d\boldsymbol{x}, \end{aligned}$$

so that the Bayes rule may be simply described as the decision rule which maps each data set $\boldsymbol{x}$ to the decision $\delta^*(\boldsymbol{x})$ which minimises the corresponding posterior expected loss. Note that this interpretation does *not* require the evaluation of any risk function.

It is easily shown that *any Bayes rule which corresponds to a proper prior distribution is admissible.* Indeed, if δ^* is the Bayes decision rule which corresponds to $p(\omega)$ and δ' were another decision rule such that $r(\delta', \omega) \leq r(\delta^*, \omega)$ with strict inequality on some subset of Ω with positive probability under $p(\omega)$, then one would have

$$\int r(\delta', \omega) p(\omega) d\omega < \int r(\delta^*, \omega) p(\omega) d\omega,$$

which would contradict the definition of a Bayes rule as one which minimises the expected risk. Wald (1950) proved the important converse result that, under rather general conditions, *any admissible decision rule is a Bayes decision rule with respect to some, possibly improper, prior distribution.*

There is, however, no guarantee that improper priors lead to admissible decision rules. A famous example is the inadmissibility of the sample mean of multivariate normal data as an estimator of the population mean, even though it is the Bayes estimator which corresponds to a uniform prior. For details, see James and Stein (1961).

Minimax rules

The combined facts that admissible rules must be Bayes, and that to derive the Bayes rule does not require computation of the risk function but simply the minimisation of the posterior expected loss, make it clear that, apart from purely mathematical interest, it is rather pointless to work in decision theory outside the Bayesian framework. Indeed, this has been the mainstream view since the early 1960's, with the authoritative monographs by DeGroot (1970) and Berger (1985a) becoming the most widely used decision theory texts.

Nevertheless, some textbooks continue to propose as a criterion for choosing among decisions (without using a prior distribution) the rather unappealing *minimax principle*. This asserts that one should choose that decision (or decision rule) for which the maximum possible loss (or risk) is minimal. It can be shown, under rather general conditions, and certainly in the finite spaces of real world applications, that the minimax rule is the Bayes decision rule which corresponds to the *least favourable prior distribution*, i.e., that which gives the highest expected risk.

The intuitive basis of the minimax principle is that one should guard against the largest possible loss. While this may have some value in the context of game theory, where a player may expect the opponent to try to put him or her in the worst possible situation, it has no obvious intuitive merit in standard decision problems. The idea that the minimax rule should be preferred to a rule which has better properties for nearly all plausible ω values, but has a slightly higher maximum risk for an extremely unlikely ω value seems absurd. Moreover, even as a formal decision criterion, minimax has very unattractive features; for instance, it gives different answers if applied to losses rather than to regret functions, and it can violate the transitivity of preferences (see e.g., Lindley, 1972).

Thus, although in specific instances—namely when prior beliefs happen to be close to the least favourable distribution—the minimax *solution* may be reasonable (essentially coinciding with the Bayes solution), the minimax *criterion* seems entirely unreasonable.

B.2.2 Frequentist Procedures

We recall from Section 5.1 the basic structure of a stylised inference problem, where inferences about $\boldsymbol{\theta} \in \Theta$ are to be drawn from data x, probabilistically related to $\boldsymbol{\theta}$ by the parametric model component $\{p(x \mid \boldsymbol{\theta}), \boldsymbol{\theta} \in \Theta\}$.

We established that the existence of a prior distribution $p(\boldsymbol{\theta})$ is a mathematical consequence of the axioms of quantitative coherence, and that the required inferential statement about $\boldsymbol{\theta}$ given $\boldsymbol{x}$ is simply provided by the full posterior distribution

$$p(\boldsymbol{\theta} \mid \boldsymbol{x}) = p(\boldsymbol{x} \mid \boldsymbol{\theta})p(\boldsymbol{\theta})/p(\boldsymbol{x}),$$

where

$$p(\boldsymbol{x}) = \int p(\boldsymbol{x} \mid \boldsymbol{\theta})p(\boldsymbol{\theta})d\boldsymbol{\theta}.$$

Frequentist statistical procedures are mainly distinguished by two related features; (i) they regard the information provided by the data $\boldsymbol{x}$ as the *sole* quantifiable form of relevant probabilistic information and (ii) they use, as a basis for both the construction and the assessment of statistical procedures, long-run frequency behaviour under hypothetical *repetition* of similar circumstances.

Although some of the ideas probably date back to the early 1800's, most of the basic concepts were brought together in the 1930's from two somewhat different perspectives, the work of Neyman and Pearson, being critically opposed by Fisher, as reflected in discussions at the time published in the Royal Statistical Society journals. Convenient references are Neyman and Pearson (1967) and Fisher (1990). See, also, Wald (1947) for specific methods for sequential problems.

Frequentist procedures make extensive use of the *likelihood function*

$$\text{lik}(\boldsymbol{\theta} \mid \boldsymbol{x}) = p(\boldsymbol{x} \mid \boldsymbol{\theta})$$

(or variants thereof), essentially taking the mathematical form of the sampling distribution of the observed data $\boldsymbol{x}$ and considering it as a function of the unknown parameter $\boldsymbol{\theta}$. If $\boldsymbol{z} = \boldsymbol{z}(\boldsymbol{x})$ is a one-to-one transformation of $\boldsymbol{x}$, the likelihood in terms of the sampling distribution of $\boldsymbol{z}$ becomes (in the above variant)

$$\text{lik}(\boldsymbol{\theta} \mid \boldsymbol{z}) = p(\boldsymbol{z} \mid \boldsymbol{\theta}) = p(\boldsymbol{x} \mid \boldsymbol{\theta}) \left| \frac{\partial \boldsymbol{x}}{\partial \boldsymbol{z}} \right| = \text{lik}(\boldsymbol{\theta} \mid \boldsymbol{x}) \left| \frac{\partial \boldsymbol{x}}{\partial \boldsymbol{z}} \right|,$$

which suggests that meaningful likelihood comparisons should be made in the form of ratios rather than, say, differences, in order for such comparisons not to depend on the use of $\boldsymbol{z}$ rather than $\boldsymbol{x}$.

The basic ideas behind frequentist statistics consist of (i) selecting a function of the data $\boldsymbol{t} = \boldsymbol{t}(\boldsymbol{x})$, called a *statistic*, which is related to the parameter $\boldsymbol{\theta}$ in a convenient way, (ii) deriving the sampling distribution of $\boldsymbol{t}$, i.e., the conditional distribution $p(\boldsymbol{t} \mid \boldsymbol{\theta})$, and (iii) measuring the "plausibility" of each possible $\boldsymbol{\theta}$ by calibrating the observed value of the statistic $\boldsymbol{t}$ against its expected long-run behaviour given $\boldsymbol{\theta}$, described by $p(\boldsymbol{t} \mid \boldsymbol{\theta})$. For a specific parameter value $\boldsymbol{\theta} = \boldsymbol{\theta}_0$, if the observed value of $\boldsymbol{t}$ is well within the area where most of the probability density of $p(\boldsymbol{t} \mid \boldsymbol{\theta}_0)$ lies, then $\boldsymbol{\theta}_0$ is claimed to be compatible with the data; otherwise it is said that either $\boldsymbol{\theta}_0$ is not the true value of $\boldsymbol{\theta}$, or a rare event has happened.

Such an approach is clearly far removed from the (to a Bayesian rather intuitively obvious) view that relevant inferences about $\boldsymbol{\theta}$ should be probability statements about $\boldsymbol{\theta}$ *given* the observed data, rather than probability statements about hypothetical repetitions of the data conditional on (the unknown) $\boldsymbol{\theta}$. This contrast is highlighted by the following example taken from Jaynes (1976).

Example B.2. ***(Cauchy observations).*** Let $x = \{x_1, x_2\}$ consist of two independent observations from a Cauchy distribution $p(x \mid \theta) = \mathrm{St}(x \mid \theta, 1, 1)$. Common sense (supported by translational and permutational symmetry arguments) suggests that $\tilde{\theta} = (x_1 + x_2)/2$ may be a sensible estimate of θ. Yet, the sampling distribution of $\tilde{\theta}$ is again $\mathrm{St}(x \mid \theta, 1, 1)$ so that, according to a naïve frequentist, it cannot make any difference whether one uses x_1, x_2 or $\tilde{\theta}$ to estimate θ. Clearly, there is more to inference than the choice of estimators and their assessment on the basis of sampling distributions.

Sufficiency

We recall from Section 4.5 that a statistic $\boldsymbol{t}$ is *sufficient* if $p(\boldsymbol{x} \mid \boldsymbol{t}, \boldsymbol{\theta}) = p(\boldsymbol{x} \mid \boldsymbol{t})$; i.e., if the conditional distribution of the data given $\boldsymbol{t}$ is independent of $\boldsymbol{\theta}$ (Proposition 4.11), and that a necessary and sufficient condition for $\boldsymbol{t}$ to be sufficient for $\boldsymbol{\theta}$ is that the likelihood function may be factorised as

$$\mathrm{lik}(\boldsymbol{\theta} \mid \boldsymbol{x}) = p(\boldsymbol{x} \mid \boldsymbol{\theta}) = h(\boldsymbol{t}, \boldsymbol{\theta}) g(\boldsymbol{x}),$$

in which case, for any prior $p(\boldsymbol{\theta})$, the posterior distribution of $\boldsymbol{\theta}$ only depends on $\boldsymbol{x}$ through $\boldsymbol{t}$, i.e., $p(\boldsymbol{\theta} \mid \boldsymbol{x}) = p(\boldsymbol{\theta} \mid \boldsymbol{t})$. The concept of sufficiency in the presence of nuisance parameters is controversial; see, for example, Cano *et al.* (1988) and references therein.

The *sufficiency principle* in classical statistics essentially states that, for any given model $p(\boldsymbol{x} \mid \boldsymbol{\theta})$ with sufficient statistic $\boldsymbol{t}$, identical conclusions should be drawn from data $\boldsymbol{x}_1$ and $\boldsymbol{x}_2$ with the same value of $\boldsymbol{t}$. The idea was introduced by Fisher (1922) and developed mathematically by Halmos and Savage (1949) and Bahadur (1954).

From a Bayesian viewpoint there is obviously nothing new in this "principle"; it is a simple mathematical consequence of Bayes' theorem.

However, from a "textbook" perspective, other frequentist developments of the sufficiency concept have little or no interest from a Bayesian perspective. For example: a sufficient statistic $\boldsymbol{t}$ is *complete* if for all $\boldsymbol{\theta}$ in Θ, $\int h(\boldsymbol{t}) p(\boldsymbol{t} \mid \boldsymbol{\theta}) d\boldsymbol{t} = 0$ implies that $h(\boldsymbol{t}) = 0$. The property of completeness guarantees the uniqueness of certain frequentist statistical procedures based on $\boldsymbol{t}$, but otherwise seems inconsequential.

Ancillarity

In Section 5.1 we demonstrated how a sufficient statistic $\boldsymbol{t} = \boldsymbol{t}(\boldsymbol{x})$ may often be partitioned into two component statistics $\boldsymbol{t}(\boldsymbol{x}) = [\boldsymbol{a}(\boldsymbol{x}), \boldsymbol{s}(\boldsymbol{x})]$ such that the sampling distribution of $\boldsymbol{a}(\boldsymbol{x})$ is independent of $\boldsymbol{\theta}$. We defined such an $a(\boldsymbol{x})$ to be an *ancillary* statistic and showed that, if $\boldsymbol{a}$ is ancillary, then

$$p(\boldsymbol{\theta} \mid \boldsymbol{x}) = p(\boldsymbol{\theta} \mid \boldsymbol{t}) \propto p(\boldsymbol{s} \mid \boldsymbol{a}, \boldsymbol{\theta}) p(\boldsymbol{\theta})$$

so that, in the inferential process described by Bayes' theorem, it suffices to work conditionally on the value of the ancillary statistic. For further information, see Basu (1959).

The *conditionality principle* in classical statistics states that, whenever there is an ancillary statistic a, the conclusions about the parameter should be drawn as if a were fixed at its observed value. The apparent need for such a principle in frequentist procedures is well illustrated in the following simple example.

Example B.3. ***(Conditional versus unconditional arguments).*** A 0-1 signal comes from one of two sources θ_1 or θ_2, and there are two receivers R_1 and R_2 such that

$$p(x = 0 \mid R_1, \theta_1) = p(x = 1 \mid R_1, \theta_2) = 0.9$$

$$p(x = 0 \mid R_2, \theta_1) = p(x = 1 \mid R_2, \theta_2) = 0.2$$

where the receiver is selected at random, with $p(R_1) = 0.99$. If R_2 were the receiver and $x = 1$ were obtained, the conditional likelihood would have been

$$\text{lik}(\theta_1 \mid R_2, x = 1) = 0.8, \qquad \text{lik}(\theta_2 \mid R_2, x = 1) = 0.2,$$

suggesting θ_1 as the true value of θ. On the other hand, the unconditional likelihood given $x = 1$ would have been

$$\text{lik}(\theta_1 \mid x = 1) = 0.107, \qquad \text{lik}(\theta_2 \mid x = 1) = 0.843,$$

suggesting θ_2 instead. The conflict arises because the latter (unconditional) argument takes undue account of what *might* have happened (i.e., R_1 might have been the receiver) but *did not*.

A further example regarding ancillarity is provided by reconsidering Example B.2. The difficulty in this case disappears if one works conditionally on the ancillary statistic $\mid x_1 - x_2 \mid /2$.

These examples serve to underline the obvious appeal of a trivial consequence of Bayes' theorem: namely, that one should always condition inferences on whatever information is available; the conditionality "principle" is just a small *ad hoc* step towards this rather obvious desideratum (which is, in any case, "automatic" in the Bayesian approach).

From a frequentist viewpoint, however, the conditionality "principle" is not necessarily easy to apply, since ancillary statistics are not readily identified, and are not necessarily unique. Moreover, applying the conditionality principle may leave the frequentist statistician in an impasse. For example, Basu (1964) noted that if x is uniform on $[\theta, 1+\theta[$, then the fractional part of x is uniformly distributed on $[0, 1[$ and hence ancillary, but the conditional distribution of x given its fractional part is a useless one-point distribution! See Basu (1992) for further elegant demonstration of the difficulties with ancillary statistics in the frequentist approach.

The Repeated Sampling Principle

A weak version of the repeated sampling principle states that one should not follow statistical procedures which, for some possible value of the parameter, would too frequently give misleading conclusions in hypothetical repetitions. Although this is too vague a formulation on which to base a formal critique, it can be used to criticise specific solutions to concrete problems.

A much stronger version of this "principle", whose essence is at the heart of frequentist statistics, states that *statistical procedures have to be assessed by their performance in hypothetical repetitions under identical conditions*. This implies that (i) measures of uncertainty have to be interpreted as long-run hypothetical frequencies, that (ii) optimality criteria have to be defined in terms of long-run behaviour under hypothetical repetitions and, that (iii) there are no means of assessing any finite-sample *realised* accuracy of the procedures.

Example B.4. ***(Confidence versus HPD intervals).*** Let $\boldsymbol{x} = \{x_1, \ldots, x_n\}$ be a random sample from $\mathrm{N}(x \mid \mu, 1)$. It is easily seen that $\overline{x}$ is a sufficient statistic, whose sampling distribution is $\mathrm{N}(\overline{x} \mid \mu, n)$, a normal distribution centred at the true value of the parameter, with precision n. Since the sampling distribution of $\overline{x}$ concentrates around μ, one might *expect* $\overline{x}$ to be close to μ on the basis of a large number of hypothetical repetitions of the sample, so that $\overline{x}$ suggests itself as an *estimator* of μ. Moreover, conditional on μ,

$$P\left[\bar{x} \in \mu \pm 1.96/\sqrt{n} \,\middle|\, \mu\right] = 0.95$$

so that, if we define a statistical procedure to consist of producing the interval $\overline{x} \pm 1.96/\sqrt{n}$ whenever a random sample of size n from $\mathrm{N}(x \mid \mu, 1)$ is obtained, we are producing an interval which will include the true value of the parameter 95% of the time, *in the long run.* Note that this says *nothing* about the probability that μ belongs to that interval for any *given* sample. In contrast, the superficially similar statement

$$P\left[\mu \in \bar{x} \pm 1.96/\sqrt{n} \,\middle|\, \overline{x}\right] = 0.95$$

which is derived from the reference posterior distribution of μ given $\overline{x}$, explicitly says that given $\overline{x}$, the *degree of belief* is 0.95 that μ belongs to $\overline{x} \pm 1.96\sqrt{n}$, and is not concerned at all with *hypothetical* repetitions of the experiment.

Invariance

If a parametric model $p(\boldsymbol{x} \mid \boldsymbol{\theta})$ is such that two different data sets, $\boldsymbol{x}_1$ and $\boldsymbol{x}_2$, have the same distribution for every $\boldsymbol{\theta}$, then both the likelihood principle and the mechanics of Bayes' theorem imply that one should derive the same conclusions about $\boldsymbol{\theta}$ from observing $\boldsymbol{x}_1$ as from observing $\boldsymbol{x}_2$.

A more elaborate form of invariance principle involves transformations of both the sample and the parameter spaces. Suppose that, with $\boldsymbol{X} = \Theta$, for all the elements of a group $\mathcal{G}$ of transformations there is a unique transformation g such that $g(\Theta) = \Theta$ and $p(\boldsymbol{x} \mid \boldsymbol{\theta}) = p(g(\boldsymbol{x}) \mid g(\boldsymbol{\theta}))$. Then the *invariance principle* would require the conclusions about $\boldsymbol{\theta}$ drawn from the statistic $t(g(\boldsymbol{x}))$ to be the same as those drawn from $g(t(\boldsymbol{x}))$. For example, in estimating $\theta \in \Re$ from a location model $p(x \mid \theta) = h(x - \theta)$ it may be natural to consider the group of translations. In this case, $g(x) = x + a$, $a \in \Re$, and $g(\theta) = \theta + a$. The invariance principle then requires that any estimate $t(\boldsymbol{x})$ of θ should satisfy $t(\boldsymbol{x} + a\mathbf{1}) = t(\boldsymbol{x}) + a$, where $\mathbf{1}$ is a vector of unities.

Note that the argument only works if there is no reason to believe that some θ values are more likely than others. From a Bayesian point of view, for invariance to be a relevant notion it must be true that the transformation involved also applies to the prior distribution (otherwise, one may have a uniform loss of expected utility from following the invariance principle). Another limitation to the practical usefulness of invariance ideas is the condition that $g(\Theta) = \Theta$. Thus, in the location/translation example, the invariance principle could not be applied if it were known that $\theta \geq 0$.

A final general comment. Frequentist procedures centre their attention on producing inference statements about *unobservable* parameters. As we shall see in Section B.4, such an approach typically fails to produce a sensible solution to the more fundamental problem of *predicting future observations*.

B.2.3 Likelihood Inference

We recall from Section 5.1 the following trivial consequence of Bayes' theorem. Consider two experiments yielding, respectively, data $\boldsymbol{x}$ and $\boldsymbol{z}$ and with model representation involving the same parameter $\boldsymbol{\theta} \in \Theta$, the same prior, and proportional likelihoods, so that

$$p(\boldsymbol{x} \mid \boldsymbol{\theta}) = h(\boldsymbol{x}, \boldsymbol{z}) p(\boldsymbol{z} \mid \boldsymbol{\theta}).$$

Then the experiments produce the same conclusions about θ, since they induce the same posterior distribution. The *likelihood principle* suggests that this should indeed be the case, for the relative support given by the two sets of data to the possible values of θ is precisely the same.

Frequentist procedures typically violate the likelihood principle, since long run behaviour under hypothetical repetitions depends on the entire distribution $\{p(\boldsymbol{x} \mid \boldsymbol{\theta}), \boldsymbol{x} \in X\}$ and not only on the likelihood.

As mentioned before, when common priors are used across models with proportional likelihoods, the Bayesian approach automatically obeys the likelihood principle and certainly accepts the likelihood function as a complete summary of the information provided by the data about the parameter of interest. With a uniform prior, the posterior distribution is, of course, proportional to the likelihood function. Proponents of the likelihood approach to inference go further, however, in their uses of the likelihood function, in that they regard it not only as the sole expression of the relevant information, but also as a meaningful relative numerical measure of support for different possible models, or for alternative parameter values within the same model. The basic ideas of this pure likelihood approach were established by Barnard (1949, 1963), Barnard *et al.* (1962), Birnbaum (1962, 1968, 1972) and Edwards (1972/1992). They essentially argue that (i) the likelihood function conveys all the information provided by a set of data about the relative plausibility of any two different possible values of θ and (ii) the ratio of the likelihood at two different θ values may be interpreted as a measure of the strength of evidence in favour of one value relative to the other.

Both claims make sense from a Bayesian point of view *when there are no nuisance parameters*. Indeed, (i) is just a restatement of the likelihood principle and, moreover, it follows from Bayes' theorem that

$$\frac{p(\boldsymbol{\theta}_1 \mid \boldsymbol{x})}{p(\boldsymbol{\theta}_2 \mid \boldsymbol{x})} = \frac{p(\boldsymbol{x} \mid \boldsymbol{\theta}_1)\, p(\boldsymbol{\theta}_1)}{p(\boldsymbol{x} \mid \boldsymbol{\theta}_2)\, p(\boldsymbol{\theta}_2)},$$

so that the likelihood ratio satisfies (ii), since it is the factor which modifies prior odds into posterior odds.

However, the pure likelihood approach, i.e., the attempt to produce inferences solely based on the likelihood function, breaks down immediately when there are nuisance parameters. The use of "marginal likelihoods" necessarily requires the elimination of nuisance parameters, but the suggested procedures for doing this seem hard to justify in terms of the likelihood approach. For early attempts, see Kalbfleish and Sprott (1970, 1973) and Andersen (1970, 1973). In recent years, work has focused on the properties of *profile likelihood* and its variants. Useful references include: Barnard and Sprott (1968), Barndorff-Nielsen (1980, 1983, 1991), Butler (1986), Davison (1986), Cox and Reid (1987, 1992), Cox (1988), Fraser and Reid (1989), Bjrnstad (1990) and Monahan and Boos (1992). Other references relevant to the interface between likelihood inference and Bayesian statistics include Hartigan (1967), Plante (1971), Akaike (1980b), Pereira and Lindley (1987), Bickel and Ghosh (1990), Goldstein and Howard (1991) and Royall (1992). See, also, Section 5.5.1 for a link with Laplace approximations of posterior densities.

For further information on the history of likelihood, see Edwards (1974).

The likelihood approach can also conflict with the weak repeated sampling principle, in that examples exist where, for some possible parameter values, hypothetical repetitions result in mostly misleading conclusions. Frequentist statistics solves the difficulty by comparing the observed likelihood function with the distribution of the likelihood functions which could have been obtained in hypothetical repetitions; Bayesian statistics solves the problem by working, not with the likelihood function, but with the posterior distribution defined as the weighted average of the likelihood function with respect to the prior. The following example is due to Birnbaum (1969).

Example B.5. ***(Naïve likelihood versus reference analysis).*** Consider the model $p(x \mid \theta)$, $x \in \{1, 2, \ldots, 100\}$, $\theta \in \{0, 1, 2, \ldots, 100\}$, where, for $x = 1, 2, \ldots, 100$,

$$p(x \mid \theta = 0) = 1/100,$$
$$p(x \mid \theta \neq 0) = I_{(x=\theta)}(x)$$

Then, whatever x is observed, if $\theta = 0$ the likelihood of the true value is always $1/100$th of the likelihood of the only other possible θ value, namely $\theta = x$.

From a Bayesian point of view, the answer obviously depends on the prior distribution. If all θ are judged a priori to have the same probability, then one certainly has

$$p(\theta = 0 \mid x) = 1/101$$
$$p(\theta = x \mid x) = 100/101, \quad x \neq 0.$$

However, if, say, $\theta = 0$ is considered to be special, as might well be the case in any real application of such a model, and is declared to be the parameter of interest, then the reference prior turns out to be

$$p(\theta = 0) = 1/2, \qquad p(\theta = r) = 1/200, \quad r = 1, \ldots, 100,$$

and a straightforward calculation reveals that this is *also the posterior*, given a single observation, x. Thus, with this prior, one observation from the model provides no information. Of course, for any prior, a second observation would, with high probability, reveal the true value of θ.

Finally, as we shall discuss further in Section B.4, we note that, like frequentist procedures, the likelihood approach has difficulties in producing an agreed solution to prediction problems.

B.2.4 Fiducial and Related Theories

We noted in Section B.2.3 that frequentist approaches are inherently unable to produce probability statements about the parameter of interest conditional on the data, a form of inference summary that seems most intuitively useful. This fact, coupled with the seeming aversion of most statisticians to the use of prior distributions, has led to a number of attempts to produce "posterior" distributions without using priors. We now review some of those proposals.

Fiducial Inference

In a series of papers published in the thirties, Fisher (1930, 1933, 1935, 1939) developed, through a series of examples and without any formal structure or theory, what he termed the *fiducial argument*. Essentially, he proposed using the distribution function $F(\boldsymbol{t} \mid \boldsymbol{\theta})$ of a sufficient estimator $\boldsymbol{t} \in T$ for $\boldsymbol{\theta} \in \Theta$ in order to make conditional probability statements about $\boldsymbol{\theta}$ given $\boldsymbol{t}$, thus somehow transferring the probability measure from T to Θ. However, no formal justification was offered for this controversial "transfer".

The basic characteristics of the argument may be described as follows. Let $p(\boldsymbol{x} \mid \theta), \theta \in (\theta_0, \theta_1) \subseteq \Re$ be a one-dimensional parametric model and let $\boldsymbol{t} = \boldsymbol{t}(\boldsymbol{x})$ be a sufficient statistic for θ. Suppose further that the distribution function of $\boldsymbol{t}$, $F(\boldsymbol{t} \mid \theta)$, is monotone decreasing in θ, with $F(\boldsymbol{t} \mid \theta_0) = 1$ and $F(\boldsymbol{t} \mid \theta_1) = 0$. Then, $G(\theta \mid \boldsymbol{t}) = 1 - F(\boldsymbol{t} \mid \theta)$ has the mathematical properties of a distribution function over (θ_0, θ_1) and, hence,

$$f(\theta \mid \boldsymbol{t}) = -\frac{\partial}{\partial \theta} F(\boldsymbol{t} \mid \theta)$$

has the mathematical structure of a "posterior density" for θ. This is the *fiducial distribution* of θ, as proposed by Fisher (1930, 1956/1973). The argument is trivially modified if $F(\boldsymbol{t} \mid \theta)$ is monotone *increasing* in θ, by using $G(\theta \mid \boldsymbol{t}) = F(\boldsymbol{t} \mid \theta)$.

Example B.6. ***(Fiducial and reference distributions).*** Let $x = \{x_1, \ldots, x_n\}$ be a random sample from an exponential distribution $p(x \mid \theta) = \text{Ex}(x \mid \theta) = \theta e^{-\theta x}$, with mean θ^{-1}. It is easily verified that $\bar{x}$ is a sufficient statistic for θ, and has a distribution function

$$F(\bar{x} \mid \theta) = \frac{1}{(n-1)!} \int_0^{n\bar{x}\theta} t^{n-1} e^{-t} dt,$$

which is monotone increasing in θ. Hence, $G(\theta \mid \bar{x}) = F(\bar{x} \mid \theta)$ is monotone increasing from 0 to 1 as θ ranges over $(0, \infty)$, and the fiducial distribution of θ is obtained as

$$f(\theta \mid \bar{x}) = \frac{\partial}{\partial \theta} F(\bar{x} \mid \theta) = \frac{1}{(n-1)!} \frac{1}{\theta} (n\bar{x}\theta)^n e^{-n\bar{x}\theta}.$$

Note that this has the form $f(\theta \mid \bar{x}) \propto p(\boldsymbol{x} \mid \theta)\pi(\theta)$, with $\pi(\theta) = \theta^{-1}$. Since $\pi(\theta) = \theta^{-1}$ is in this case the reference prior for θ, it follows that, in this example, the fiducial distribution coincides with the reference posterior distribution.

This last example suggests that the fiducial argument might simply be a re-expression of Bayesian inference with some appropriately chosen "non-informative" prior. However, Lindley (1958) established that this is true if, and only if, the probability model $p(x \mid \theta)$ is such that x and θ may separately be transformed so as to obtain a new parameter which is a *location* parameter for the transformed variable. See Seidenfeld (1992) for further discussion.

In one-dimensional problems, the fiducial argument, when applicable, is more or less well defined and often produces reasonable answers, which are nevertheless far better justified from a Bayesian reference analysis viewpoint. However, it is by no means clear—and, in fact, a matter of considerable controversy—how the argument might be extended to multiparameter problems. The Royal Statistical Society discussions following the papers by Fieller (1954) and Wilkinson (1977) serve to illustrate the difficulties. Other relevant references are Brillinger (1962) and Barnard (1963).

From a modern perspective, the fiducial argument seems now to have at most historical interest, and that mainly due to the perceived stature of its proponent. As Good (1971) puts it

> ... if we do not examine the fiducial argument carefully, it seems almost inconceivable that Fisher should have made the error which he did in fact make. It is because (i) it seemed so unlikely that a man of his stature should *persist* in the error, and (ii) because, as he modestly says, his 1930 'explanation left a good deal to be desired, that so many people assumed for so long that the argument was correct. They lacked the *daring* to question it.

See, however, Efron (1993) for a recent suggested modification of the fiducial distribution which may have better Bayesian properties.

Pivotal Inference

Suppose that, for a given model $p(\boldsymbol{x} \mid \theta)$, with sufficient statistic $\boldsymbol{t}$, it is possible to find some function $h(\theta, \boldsymbol{t})$ which is monotone increasing in θ for fixed $\boldsymbol{t}$, and in $\boldsymbol{t}$ for fixed θ, and which has a distribution which only depends on θ through $h(\theta, \boldsymbol{t})$. Then, $h(\theta, \boldsymbol{t})$ is called a *pivotal function* and the fiducial distribution of θ may simply be obtained by reinterpreting the probability distribution of h over T as a probability distribution over Θ. Fisher's original argument, as described above, is a special case of this formulation, since $G(\theta \mid \boldsymbol{t})$ is a pivotal function with a uniform distribution over $[0, 1]$, which is independent of θ.

Barnard (1980b) has tried to extend this idea into a general approach to inference. His basic idea is to produce statements derived from the distribution of an appropriately chosen pivotal function, possibly conditional on the observed values of an ancillary statistic $a(\boldsymbol{x})$.

Partitioning a pivotal function $h(\theta, \boldsymbol{x}) = [g(\theta, \boldsymbol{x}), a(\boldsymbol{x})]$ to identify a possibly uniquely defined ancillary statistic $a(\boldsymbol{x})$, and using the distribution of $g(\theta, \boldsymbol{x})$ conditional on the observed value of $a(\boldsymbol{x})$, does produce some interesting results in multiparameter problems where the standard fiducial argument fails. However, the mechanism by which the probability measure is transferred from the sample space to the parameter space remains without foundational justification, and the argument is limited to the availability—by no means obvious—of an appropriate pivotal function for the envisaged problem.

Structural Inference

Yet another attempt at justifying the transfer of the probability measure from the sample space into the parameter space is the *structural approach* proposed by Fraser (1968, 1972, 1979).

Fraser claimed that one often knows more about the relationship between data and parameters than that described by the standard parametric model $p(x \mid \boldsymbol{\theta})$. He proposes the specification of what he terms a *structural model*, having two parts: a *structural equation*, which relates data x and parameter $\boldsymbol{\theta}$ to some error variable e; and a *probability distribution for* e which is assumed known, and independent of $\boldsymbol{\theta}$. Thus, the observed variable x is seen as a transformation of the error e, the transformation governed by the value of $\boldsymbol{\theta}$. The key idea is then to *reverse* this relationship, and to interpret $\boldsymbol{\theta}$ as a transformation of e governed by the observed x, so that $\boldsymbol{\theta}$ in a sense "inherits" the probability distribution.

Example B.7. ***(Structural and reference distributions)***. Let $x = \{x_1, \ldots, x_n\}$ be a set of independent measurements with unknown location μ and scale σ. If the errors have a known distribution $p(e)$, the structural equation is

$$x_i = \mu + \sigma e_i, \quad i = 1, \ldots, n$$

and the error distribution is,

$$p(e) = \prod_{i=1}^{n} p(e_i).$$

If $p(e)$ is normal, this structural model may be *reduced* in terms of the sufficient statistics $\bar{x}$ and s^2 to the equations

$$\bar{x} = \mu + \sigma \bar{e}, \qquad s = \sigma s_e$$

and error distributions

$$\bar{e} = z/\sqrt{n}, \quad z \sim \mathrm{N}(z \mid 0, 1)$$
$$s_e = \{w/(n-1)\}^{1/2}, \quad w \sim \chi^2_{n-1}.$$

Reversing the probability relationship in the pivotal functions $(n-1)s^2/\sigma^2 \sim \chi^2_{n-1}$ and $\sqrt{n}(\bar{x} - \mu)/s \sim \mathrm{St}\,(t \mid 0, 1, n-1)$ leads to structural distributions for σ and μ which, as is often the case, coincide with the corresponding reference posterior distributions.

The general formulation of structural inference generalises the affine group structure underlying the last example, and considers a structural equation $x = \theta e$, to be interpreted as the response x generated by some transformation $\theta \in G$ in a group of transformations G, operating on a realised error e, with a completely identified error distribution for e. It is then claimed that $\theta^{-1}(x)$ has the same probability distribution e and, hence, this may be used to provide a *structural* distribution for θ.

Here, the mechanism by which the probability measure on X is transferred to Θ is certainly well-defined in the presence of the group structure central to Fraser's argument. However, the group structure is fundamental and the approach seems to lack general validity and applicability. As Lindley (1969) puts it

> ... Fraser's argument [is] an improvement upon and an extension of Fisher's in the special case where the group structure is present but [one should be] ... suspicious of any argument, ... , that only works in some situations, for inference is surely a whole, and the Poisson distribution [is] not basically different in character from ... the normal.

When the structural argument can be applied it produces answers which are mathematically closely related to Bayesian posterior distributions with "non-informative" priors derived from (group) invariance arguments. In fact, in most examples, the structural distributions are precisely the posterior distributions obtained by using as priors the right Haar measures associated with the structural group, which in turn, are special cases of reference posterior distributions (see Villegas, 1977a, 1981, 1990; Dawid, 1983b, and references therein).

B.3 STYLISED INFERENCE PROBLEMS

B.3.1 Point Estimation

Let $\{p(x \mid \theta), \theta \in \Theta\}$ be a fully specified parametric family of models and suppose that it is desired to calculate from the data x a single value $\tilde{\theta}(x) \in \Theta$ representing the "best estimate" of the unknown parameter θ. This is the so-called *point estimation* problem. Note that, in this formulation, the final answer is an element of Θ, with *no explicit recognition of the uncertainty involved.* Pragmatically, a point estimate of θ may be motivated as being the simplest possible summary of the inferences to be drawn from x about the value of θ: alternatively, one may genuinely require a point estimate as the solution to a decision problem; for example, adjusting a control mechanism, or setting a stock level.

We recall from Section 5.1.5 that, within the Bayesian framework, the problem of point estimation is naturally described as a decision problem where the set of possible answers to the inference problem, $\mathcal{A}$, is the parameter space Θ. Formally, one specifies the loss function $l(\boldsymbol{a}, \boldsymbol{\theta})$ which describes the decision maker's preferences in that context, and chooses as the (*Bayes*) estimate that value $\boldsymbol{\theta}^*(\boldsymbol{x})$ which minimises the posterior expected loss,

$$\int l(\boldsymbol{a}, \boldsymbol{\theta})p(\boldsymbol{\theta} \,|\, \boldsymbol{x})d\boldsymbol{\theta},$$

where

$$p(\boldsymbol{\theta} \,|\, \boldsymbol{x}) \propto p(\boldsymbol{x} \,|\, \boldsymbol{\theta})p(\boldsymbol{\theta}).$$

We have seen (Propositions 5.2 and 5.9) that intuitively natural solutions, such as the mean, mode or median of the posterior distribution of $\boldsymbol{\theta}$, are particular cases of this formulation for appropriately chosen loss functions. We also note that the definition of an optimal Bayesian estimator is constructive, in that it identifies a precise procedure for obtaining the required value.

Classical decision theory ideas can obviously be applied to point estimation viewed as a decision problem. Thus, one may define admissible estimates, minimax estimates, etc., with respect to any particular loss function. From our perspective, the problems and limitations of classical decision theory that we identified in Section B.2.1 carry over to particular applications such as point estimation. Thus, admissible estimators are essentially Bayes estimators, but classical decision theory provides no foundationally justified procedure for choosing among admissible estimators, with—as we noted—the general minimax principle being unpalatable to most statisticians.

The frequentist approach proceeds by defining possible *desiderata* of the long run behaviour of point estimators, and, *using these desiderata as criteria*, proposes methods for obtaining "best" estimators, and identifies conditions under which "good behaviour" will result. The criteria adopted are typically non-constructive.

The likelihood approach proceeds by using the likelihood function to measure the strength with which the possible parameter values are supported by the data. Hence, the optimal estimator is naturally taken to be that $\hat{\boldsymbol{\theta}}$ which maximises the likelihood function. It is worth stressing that this is a constructive criterion, in that the very definition of a maximum likelihood estimator (MLE) determines its method of construction.

Fiducial, pivotal and structural inference approaches all produce "posterior" probability distributions for $\boldsymbol{\theta}$. Hence, their "solution" to the problem of point estimation is essentially that suggested by the Bayesian approach; either to offer as an estimator of $\boldsymbol{\theta}$ some location measure of the probability distribution of $\boldsymbol{\theta}$ or, more formally, to obtain that value of $\boldsymbol{\theta}$ which minimises some specified loss function with respect to such a distribution.

Criteria for Point Estimation

It should be clear from Sections 5.1.4 and B.2.2 that the search for good estimators may safely be limited to those based on *sufficient* statistics, for then, and only then, is one certain to use all the relevant information about the parameter of interest. However, the following two points introduce a note of caution.

(i) Sufficiency is a *global* concept; thus, if $\tilde{\boldsymbol{\theta}}(\boldsymbol{x})$ is sufficient for $\boldsymbol{\theta}$, it does *not* follow that $\tilde{\theta}_i(\boldsymbol{x})$ is sufficient for a component parameter θ_i, even if $\tilde{\theta}_i(\boldsymbol{x})$ is sufficient for θ_i when $\boldsymbol{\theta} - \{\theta_i\}$ is known. For instance, with univariate normal data $(\bar{x}, s^2)$ is jointly sufficient for (μ, σ^2), but $\bar{x}$ is not sufficient for μ, nor is s^2 sufficient for σ^2.

(ii) Sufficiency is a concept *relative* to a model; thus, even a small perturbation to the assumed model may destroy sufficiency. For example $(\bar{x}, s^2)$ is *not* sufficient for (μ, σ) if the true model is $\mathrm{St}\,(x \mid \mu, \sigma, 1000)$ or the mixture form $0.999 \times \mathrm{N}(x \mid \mu, \sigma) + 0.001 \times \mathrm{N}(x \mid 0, 1)$, even though these two models are indeed very "close" to $\mathrm{N}(x \mid \mu, \sigma)$.

The *bias* of an estimator $\tilde{\boldsymbol{\theta}}(\boldsymbol{x})$ is defined to be

$$b(\boldsymbol{\theta}) = \int \hat{\boldsymbol{\theta}}(\boldsymbol{x})\, p(\boldsymbol{x} \mid \boldsymbol{\theta})\, d\boldsymbol{x} - \boldsymbol{\theta}$$

and its *mean squared error* (mse) to be

$$\mathrm{mse}(\tilde{\boldsymbol{\theta}} \mid \boldsymbol{\theta}) = \int \{\tilde{\boldsymbol{\theta}}(\boldsymbol{x}) - \boldsymbol{\theta}\}^2\, p(\boldsymbol{x} \mid \boldsymbol{\theta})\, d\boldsymbol{x} = V(\tilde{\boldsymbol{\theta}} \mid \boldsymbol{\theta}) + \{b(\boldsymbol{\theta})\}^2.$$

From a frequentist point of view it is desired that, in the long run, $\tilde{\boldsymbol{\theta}}$ should be as close to $\boldsymbol{\theta}$ as possible; thus, if quadratic loss is judged to be an appropriate "distance" measure, a frequentist would like an estimator $\tilde{\boldsymbol{\theta}}$ with small $\mathrm{mse}(\tilde{\boldsymbol{\theta}} \mid \boldsymbol{\theta})$ for almost all values of $\boldsymbol{\theta}$. A concept of *relative efficiency* is developed in these terms. An estimator $\tilde{\boldsymbol{\theta}}_1$ is *more efficient* than $\tilde{\boldsymbol{\theta}}_2$, if, for all $\boldsymbol{\theta}$, $\mathrm{mse}(\tilde{\boldsymbol{\theta}}_1 \mid \boldsymbol{\theta}) < \mathrm{mse}(\tilde{\boldsymbol{\theta}}_2 \mid \boldsymbol{\theta})$.

A simple theory is available if attention is restricted to *unbiased* estimators, i.e., estimators such that $b(\boldsymbol{\theta}) = \mathbf{0}$, since then we simply have to minimise $V(\tilde{\boldsymbol{\theta}} \mid \boldsymbol{\theta})$ in this unbiased class. However, although requiring the sampling distribution of $\hat{\boldsymbol{\theta}}$ to be centred at $\boldsymbol{\theta}$ may have some intuitive appeal, there are powerful arguments *against* requiring unbiasedness. Indeed:

(i) In many problems, *there are no unbiased estimators.* For instance, r/n is an unbiased estimator of the parameter θ for a binomial $\mathrm{Bi}(r \mid \theta, n)$ distribution, but there is no unbiased estimator of $\theta^{1/2}$.

(ii) Even when they exist, unbiased estimators may give nonsensical answers, and no theory exists which specifies conditions under which this can be guaranteed not to happen. For example, the (unique) unbiased estimator of the parameter $\theta \in (0, 1)$ of a geometric distribution $p(x \mid \theta) = \theta(1 - \theta)^x$, $x = 0, 1, \ldots$, is $\tilde{\theta}(0) = 1$, $\tilde{\theta}(x) = 0, x = 1, 2, \ldots$; hardly a sensible solution! Similarly (see Ferguson, 1967), if θ is the mean of a Poisson distribution, $\text{Pn}(x \mid \theta) = e^{-\theta}\theta^x/x!$, $x = 0, 1, \ldots$, then the *only* unbiased estimator of $e^{-\theta}$, a quantity which must lie in $(0, 1)$, is 1 if x is even and 0 if it is odd (again, hardly sensible); but—even more ridiculously—the *only* unbiased estimate of $e^{-2\theta}$ is $(-1)^x$, leading to the estimate of a probability as -1 (for all odd x)!

(iii) The unbiasedness requirement violates the likelihood principle, by making the answer dependent on the sampling mechanism. Thus, the unbiased estimator of μ from a $\text{N}(x \mid \mu, \sigma)$ observation is x, but the unbiased estimator from

$$\begin{aligned} p(x \mid \mu, \sigma) &= \text{N}(x \mid \mu, \sigma), \qquad \text{if } x \leq 100 \\ &= \text{N}(x \mid 0, 1) \qquad \text{otherwise} \end{aligned}$$

will be something else. Yet, if one is measuring μ with an instrument which only works for values $x \leq 100$ and obtains $x = 50$, i.e., a valid measurement, it seems inappropriate to make our estimate of μ dependent on the fact that we *might* have obtained an invalid measurement, but did *not*.

(iv) Even from a frequentist perspective, unbiased estimators may well be unappealing if they lead to large mean squared errors, so that an estimator with small bias and small variance may be preferred to one with zero bias but a large variance.

For further discussion of the conflict between Bayes and unbiased estimators, see Bickel and Blackwell (1967). See, also, Wald (1939).

Another frequentist criterion for judging an estimator concerns the asymptotic behaviour of its sampling distribution. If we write $\tilde{\theta}_n = \tilde{\theta}(x_1, \ldots, x_n)$ to make explicit the dependence of the estimator on the sample size, a frequentist would clearly like $\tilde{\theta}_n$ to converge to θ (in some sense) as n increases. An estimator $\tilde{\theta}_n$ is said to be *weakly consistent* if $\tilde{\theta}_n \to \theta$ in probability, and *strongly consistent* if $\tilde{\theta}_n \to \theta$ with probability one. By Chebychev's inequality, a sufficient condition for the weak consistency of unbiased estimators is that $V(\tilde{\theta}_n) \to 0$ as $n \to 0$. Obviously, a consistent estimator is *asymptotically* unbiased.

For discussion on the consistency of Bayes estimators, see, for example, Schwartz (1965), Freedman and Diaconis (1983), de la Horra (1986) and Diaconis and Freedman (1986a, 1986b). For the frequentist properties of Bayes estimators, see Diaconis and Freedman (1983).

"Optimum" Estimators

We have mentioned before that minimising the variance among unbiased estimators is often suggested as a procedure for obtaining "good" estimators. Sometimes, this procedure is even further restricted to *linear* functions; thus, provided $\mu = E(x \mid \theta)$ and $\sigma^2 = V(x \mid \theta)$ exist, $\bar{x}$ is said to be the *best linear unbiased estimator* (BLUE) of μ, in the sense that it has the smallest mse among all linear, unbiased estimators.

It is easy, however, to demonstrate, with appropriate examples, that this is a rather restricted view of optimality, since non-linear estimators may be considerably more efficient. An "absolute" standard by which unbiased estimators may be judged is provided by the Cramer-Rao inequality. Let $\tilde{g} = \tilde{g}(\boldsymbol{x})$ be an unbiased estimator of $g(\theta)$ and define the *efficient score function* $u(\boldsymbol{x} \mid \theta)$ to be

$$u(\boldsymbol{x} \mid \theta) = \frac{\partial}{\partial \theta} \log p(\boldsymbol{x} \mid \theta).$$

Then, under suitable regularity conditions, $E_{x \mid \theta}[u(\boldsymbol{x} \mid \theta)] = 0$,

$$V(\tilde{g} \mid \theta) \geq \frac{[dg(\theta)/d\theta]^2}{I(\theta)},$$

where

$$I(\theta) = E_{x|\theta}\left[u^2(\boldsymbol{x} \mid \theta)\right] = -E_{x|\theta}\left[\frac{\partial^2}{\partial \theta^2} u(\boldsymbol{x} \mid \theta)\right],$$

with equality if, and only if,

$$u(\boldsymbol{x} \mid \theta) = k(\theta)\,\{\tilde{g}(\boldsymbol{x}) - g(\theta)\},$$

where $k(\theta)$ does not depend on $\boldsymbol{x}$, in which case $\tilde{g}$ is said to be a *minimum variance bound* (MVB) estimator of $g(\theta)$. It follows that a minimum variance bound estimator must be sufficient, unbiased, and a linear function of the score function.

We have already stressed that limiting attention to unbiased estimators may not be a good idea in the first place. Moreover, the range of situations where "optimal" unbiased estimators, i.e., the MVB estimators, can be found is rather limited. Indeed, if $\tilde{\theta}$ is sufficient for θ there is a *unique* function $g(\theta)$ for which a MVB exists, namely that described above. For example, if $\boldsymbol{x} = \{x_1, \ldots, x_n\}$ is a random sample from $\text{N}(x \mid 0, \sigma^2)$, $\Sigma x_i^2/n$ is a MVB estimator for σ^2, but no MVB estimator exists for σ!

One might then ask whether it is at least possible to obtain an unbiased estimator with a variance which is lower than that of any other estimator for each θ, even if it does not reach the Cramer-Rao lower bound. Under suitable regularity conditions, the existence of such *uniformly minimum variance* (UMV) estimators can indeed be established. Specifically, Rao (1945) and Blackwell (1947) independently proved that if $\tilde{\theta}(\boldsymbol{x})$ is an estimator of θ and $t = t(\boldsymbol{x})$ is a sufficient statistic for θ, then, given the value of the sufficient statistic t, the conditional expectation of $\tilde{\theta}(\boldsymbol{x})$,

$$\hat{\theta}(t) = E[\tilde{\theta} \mid t] = \int \tilde{\theta}(\boldsymbol{x}) p(\boldsymbol{x} \mid t) d\boldsymbol{x},$$

is an *improved estimator* of θ, in the sense that, for every value of θ, $\text{mse}(\hat{\theta} \mid \theta) \leq \text{mse}(\tilde{\theta} \mid \theta)$, a result which can be generalised to multidimensional problems.

A decision-theoretic consequence of the so-called Rao-Blackwell theorem is that any estimator of θ which is not a function of the sufficient statistic t must be inadmissible. However, as a constructive procedure for obtaining estimators this result is of limited value due the fact that it is usually very difficult to calculate the required conditional expectation.

If $\tilde{\theta}(x)$ is unbiased and complete, and there is a complete sufficient statistic $t = t(x)$,then $\hat{\theta}(t)$ is unbiased, and is the UMV estimator of θ. For example, r/n is the MVB estimator of the parameter θ of a binomial distribution $\mathrm{Bi}(r \mid \theta, n)$, but there is no MVB estimator of θ^2. However, the result may be used to show that $r(r-1)/[n(n-1)]$ is a UMV estimator of θ^2.

MLE estimators are not guaranteed to exist or to be unique; but when they do exist they typically have very good asymptotic properties. Under fairly general conditions, MLE's can be shown to be consistent (hence asymptotically unbiased, even if biased in small samples), asymptotically fully efficient, and asymptotically normal, so that, if $n \to \infty$, the sampling distribution of $\hat{\theta}$ converges to the normal distribution $\mathrm{N}(\hat{\theta} \mid \theta, I(\theta))$ with mean θ and precision $I(\theta)$, the information function.

Bayesian estimators always exist for appropriately chosen loss functions and automatically use all the relevant information in the data. They are typically biased, and have analogous asymptotic properties to maximum likelihood estimators (i.e., from a frequentist perspective they are consistent, asymptotically fully efficient and, under suitable regularity conditions, asymptotically normal). A famous example is the *Pitman estimator* (Pitman (1939), which may be obtained as the posterior mean which corresponds to a uniform prior; see, also, Robert *et al.* (1993).

Both the likelihood and the Bayesian solutions to the point estimation problem automatically define procedures for obtaining them; the frequentist approach does not (expect for special cases like the exponential family). In addition to the MLE approach, other methods of construction include minimum chi-squared, least squares, and the method of moments. However, these methods do not in themselves guarantee any particular properties for the resulting estimators, which usually have to be investigated case by case. Historically, all these construction methods have been used at various times within the frequentist approach to produce candidate "good estimators", which have then been analysed using the criteria described above. Nowadays, partly under the influence of classical decision theory, some frequentist statisticians pragmatically minimise an expected posterior loss to obtain an estimator, whose behaviour they then proceed to study using non-Bayesian criteria.

For an extensive treatment of the topic of point estimation, see Lehmann (1959/1983).

B.3.2 Interval Estimation

Let $\{p(x \mid \boldsymbol{\theta}), \boldsymbol{\theta} \in \Theta\}$ be a fully specified parametric family of models and suppose that it is desired to calculate, from the data x, a *region* $C(x)$ within which the parameter $\boldsymbol{\theta}$ may reasonably be expected to lie. Thus, rather than mapping X

into Θ, as in point estimation, a *subset* of Θ is associated with each value of x, whose elements may be claimed to be supported by the data as "likely" values of the unknown parameter $\boldsymbol{\theta}$. This is the so-called *region estimation* problem; when $\boldsymbol{\theta}$ is one-dimensional, the regions obtained are typically intervals, hence the more standard reference to the *interval estimation problem.*

Region estimates of $\boldsymbol{\theta}$ may be motivated pragmatically as informative simple summaries of the inferences to be drawn from x about the value of $\boldsymbol{\theta}$ or, more formally, as a set of $\boldsymbol{\theta}$ values which may safely be declared to be consistent with the observed data.

We recall from Section 5.1.5 that, within a Bayesian framework, credible regions provide a sensible solution to the problem of region estimation. Indeed, for each α value, $0 < \alpha < 1$, a $100(1-\alpha)\%$ credible region C, i.e., such that

$$\int_C p(\boldsymbol{\theta} \mid \boldsymbol{x}) d\boldsymbol{\theta} = 1 - \alpha,$$

contains the true value of the parameter with (posterior) probability $1 - \alpha$ and, among such regions, those of the smallest size, i.e., the highest posterior density (HPD) regions, suggest themselves as summaries of the inferential content of the posterior distribution. Note that this formulation is equally applicable to *prediction* problems simply by using the corresponding posterior predictive distribution.

Confidence Limits

For $0 < \alpha < 1$ and scalar $\theta \in \Theta \subseteq \Re$, a statistic $\bar{\theta}^{\alpha}(x)$ such that for all θ,

$$\Pr\{\bar{\theta}^{\alpha}(x) \geq \theta \mid \theta\} = 1 - \alpha,$$

and such that if $\alpha_1 > \alpha_2$ then $\bar{\theta}^{\alpha_1} \leq \bar{\theta}^{\alpha_2}$, is called an *upper confidence limit* for θ with *confidence coefficient* $1 - \alpha$. Note that if g is strictly increasing, then $g(\bar{\theta}^{\alpha})$ is an upper confidence limit for $g(\theta)$. The nesting condition is important to avoid inconsistency; see e.g., Plante (1984, 1991).

Given x, the specific interval $(-\infty, \bar{\theta}^{\alpha}(x)]$ is then typically interpreted as a region where, given x, the parameter θ may reasonably be expected to lie. It is crucial however to recognise that the only proper *probability interpretation* of a confidence interval is that, *in the long run*, a proportion $1 - \alpha$ of the $\bar{\theta}^{\alpha}(x)$ values will be larger than θ. Whether or not the *particular* $\bar{\theta}^{\alpha}(x)$ which corresponds to the observed data x is smaller or greater than θ is *entirely uncertain*. One only has the rather dubious "transferred assurance" from the long-run definition.

A *lower confidence* limit $\underline{\theta}_{\alpha}(x)$ is similarly defined as a statistic $\underline{\theta}_{\alpha}(x)$ such that $\Pr\{\underline{\theta}_{\alpha}(x) \leq \theta \mid \theta\} = 1 - \alpha$ with the corresponding nesting property. Combining a lower limit at confidence level $1 - \alpha_1$, with an upper limit at confidence level

$1 - \alpha_2$, we obtain a *two-sided confidence interval* $[\underline{\theta}_{\alpha_1}(\boldsymbol{x}), \bar{\theta}^{\alpha_2}(\boldsymbol{x})]$ at confidence level $1 - \alpha_1 - \alpha_2$, such that, for all θ,

$$\Pr\{\underline{\theta}_{\alpha_1}(\boldsymbol{x}) < \theta < \bar{\theta}^{\alpha_2}(\boldsymbol{x}) \,|\, \theta\} = 1 - \alpha_1 - \alpha_2.$$

For two-sided confidence intervals, a convenient choice is $\alpha_1 = \alpha_2$, which produces *central* confidence intervals based on equal tail-area probabilities. There are, however, other alternatives.

(i) *Shortest confidence intervals.* For fixed $\alpha_1 + \alpha_2 = \alpha$, α_1 and α_2 may be chosen to minimise the expected interval length $E_{\boldsymbol{x} \,|\, \theta}[\bar{\theta}^{\alpha_2}(\boldsymbol{x}) - \underline{\theta}_{\alpha_1}(\boldsymbol{x}) \,|\, \theta]$. It must be realised, however, that shortest intervals for θ do not generally transform to shortest intervals for functions $g(\theta)$. It can be proved that intervals based on the score function

$$u(\boldsymbol{x} \,|\, \theta) = \frac{\partial}{\partial \theta} \log p(\boldsymbol{x} \,|\, \boldsymbol{\theta})$$

have asymptotically minimum expected length; moreover, the fact that u has a sampling distribution which is asymptotically normal $\mathrm{N}(u \,|\, 0, I(\theta))$, with mean 0 and precision $I(\theta)$, may be used to provide approximate confidence intervals for θ.

(ii) *Most selective intervals.* For fixed $\alpha_1 + \alpha_2 = \alpha$, one could try to choose α_1 and α_2 to minimise the probability that the interval contains false values of θ. However, such *uniformly most accurate* intervals are not guaranteed to exist.

It is worth noting that, for a variety of reasons, the *construction* of confidence intervals is by no means immediate.

(i) They typically do not exist for arbitrary confidence levels when the model is discrete.

(ii) There is no general constructive guidance on which particular statistic to choose in constructing the interval.

(iii) There are serious difficulties in incorporating any known restrictions on the parameter space, and no *systematic* procedure exists for incorporating such knowledge in the construction of confidence intervals.

(iv) In multiparameter situations, the construction of simultaneous confidence intervals is rather controversial. It is less than obvious whether one should use the confidence limits associated with individual intervals, or whether one should think of the problem as that of estimating a region for a single vector parameter, or as one of considering the probability that a number of confidence statements are simultaneously correct.

(v) Interval estimation in the presence of nuisance parameters is another controversial topic. Unless appropriate pivotal quantities can be found, the properties of various alternative procedures, typically based on replacing the unknown nuisance parameters by estimates, are generally less than clear.

(vi) Interval estimation of future observations poses yet another set of difficulties. Unless one is able to find a function of the present and future observation whose sampling distribution does not depend on the parameters (and this is not typically the case), one is again limited to *ad hoc* approximations based on substituting estimates for parameters.

But, even in the simplest case where θ is a scalar parameter labelling a continuous model $p(x \mid \theta)$, the *concept* of a confidence interval is open to what many would regard as a rather devastating criticism. Namely the fact that the confidence limits can turn out to be either vacuous or just plain silly *in the light of the observed data.* We give two examples.

(i) In the Fieller-Creasy problem, where the parameter of interest is the ratio of two normal means, there are values $\alpha < 1$ such that, for a subset of possible data with positive probability, the corresponding $1 - \alpha$ confidence interval is the *entire real line.* Solemnly quoting the whole real line as a 95% confidence interval for a real parameter is not a good advertisement for statistics. For Bayesian solutions, see Bernardo (1977) and Raftery and Schweder (1993).

(ii) If x_1 and x_2 are two random observations from a uniform distribution on the interval $(\theta - 0.5, \theta + 0.5)$, and y_1 and y_2 are, respectively, the smaller and the larger of these two observations, then it is easily established that for all θ

$$P\{y_1 < \theta < y_2 \mid \theta\} = 0.5$$

so that (y_1, y_2) provides a 50%. confidence interval. However, if for the observed data it turns out that $y_2 - y_1 \geq 0.5$ then certainly $y_1 < \theta < y_2$, so that we know *for sure* that θ belongs to the interval (y_1, y_2), even though the confidence level of the interval is only 50%.

These examples reflect the inherent difficulty that the frequentist approach to statistics has of being unable to condition on the complete observed data. Conditioning on ancillary statistics, when possible, may mitigate this problem, but it certainly does not solve it and, as discussed in Section B.2.2, it may create others. The reader interested in other blatant counterexamples to the (unconditional) frequentist approach to statistics will find references in the literature under the key-words *relevant subsets*, which refer to subsets of the sample space yielding special information and subverting the "long-run" or "on average" frequentist viewpoint. Two important such references are Robinson (1975) and Jaynes (1976); see, also, Buehler (1959), Basu (1964, 1988), Cornfield (1969), Pierce (1973), Robinson (1979a, 1979b), Casella (1987, 1992), Maatta and Casella (1990) and Goutis and Casella (1991).

As a final point, we should mention that for many of the standard textbook examples of confidence intervals (typically those which can be derived from univariate continuous pivotal quantities), the quoted intervals are *numerically* equal to credible regions of the same level obtained from the corresponding reference posterior distributions. This means that, in these cases, the intuitive interpretation that many users (incorrectly, of course!) tend to give to frequentist intervals of confidence $1-\alpha$, namely that, *given the data*, there is probability $1-\alpha$ that the interval contains the true parameter value, would in fact be correct, if described, instead, as a reference posterior credible interval.

A typical example of this situation is provided by the class of intervals

$$\bar{x} - t^{\alpha}_{n-1} s/\sqrt{n-1} < \mu < \bar{x} + t^{\alpha}_{n-1} s/\sqrt{n-1}, \qquad \alpha > 0$$

for the mean μ of a normal distribution with unknown precision. These are both the "best" confidence intervals for μ, derivable from the sampling distribution of the pivotal quantity $\sqrt{n-1}(\bar{x}-\mu)/s$, and also the credible intervals which correspond to the reference posterior distribution for μ, $\pi(\mu \mid \boldsymbol{x}) = \text{St}(\mu \mid \bar{x}, (n-1)s^{-2}, n-1)$ derived in Example 5.17. Buehler and Feddersen (1963) demonstrated that relevant subsets exist even in this standard case. Indeed, if $\boldsymbol{x} = \{x_1, x_2\}$, then $C = \{x_{\min}, x_{\max})$ is a 50% interval for μ, but if both observations belong to the set

$$R = \{(x_1, x_2); \quad |x_1 - x_2| \geq 4|\bar{x}|/3\}$$

then $\Pr\{C \mid \boldsymbol{x} \in R, \mu, \sigma\} = 0.5181$. Pierce (1973) has shown that similar situations can occur whenever the confidence interval cannot be interpreted as a credible region corresponding to a posterior distribution with respect to a *proper* prior. Note that although this long-term coverage probability is not directly relevant to a Bayesian, the example suggests that special care should be exercised when *interpreting* posterior distributions obtained from *improper* priors.

Casella *et al.* (1993) have proposed, for interval estimation, alternative loss functions to the standard linear functions of volume and coverage probability.

B.3.3 Hypothesis Testing

Let $\{p(\boldsymbol{x} \mid \boldsymbol{\theta}), \boldsymbol{\theta} \in \Theta\}$ be a fully specified parametric family of models, with Θ, partitioned into two disjoint subsets Θ_0 and Θ_1, and suppose that we wish to *decide* whether the unknown $\boldsymbol{\theta}$ lies in Θ_0 or in Θ_1. If H_0 denotes the hypothesis that $\theta \in \Theta_0$ and H_1 the hypothesis that $\theta \in \Theta_1$, we have a decision problem, with only two possible answers to the inference problem, $a_0 \equiv$ accept H_0 or $a_1 \equiv$ accept H_1, where the choice is to be made on the basis of the observed data $\boldsymbol{x}$. This is the so-called problem of *hypothesis testing*. In most such problems, the two hypotheses are not symmetrically treated; the *working* hypothesis H_0 is usually called the *null* hypothesis, while H_1 is referred to as the *alternative* hypothesis. Although the theory can easily be extended to any finite number of alternative hypotheses, we will present our discussion in terms of a single alternative hypothesis.

We recall from Section 6.1 that, within a Bayesian framework, the problem of hypothesis testing, as formulated above, can be appropriately treated using standard decision theoretical methodology; that is, by specifying a prior distribution and an appropriate utility function, and maximising the corresponding posterior expected utility. We also recall that the solution to the decision problem posed generally depends on whether or not the "true" model is *assumed to be included* in the family of analysed models. Assuming the stylised $\mathcal{M}$-closed case, where the true model is assumed to belong to the family $\{p(\boldsymbol{x} \mid \boldsymbol{\theta}), \boldsymbol{\theta} \in \Theta\}$ and the utility structure is simply

$$
\begin{aligned}
u(a_i, \theta) &= 0 \qquad \theta \in \Theta_i, \quad i = 0, 1 \\
&= -l_{ij} \qquad \theta \in \Theta_j, \quad j \neq i,
\end{aligned}
$$

we have seen (Proposition 6.1) that the null hypothesis H_0 should be rejected if, and only if,

$$
B_{01}(\boldsymbol{x}) < \frac{l_{01}}{l_{10}} \frac{p(H_1)}{p(H_0)} \cdot
$$

This corresponds to checking whether the appropriate (integrated) likelihood ratio, or Bayes factor,

$$
B_{01}(\boldsymbol{x}) = \frac{\int_{\Theta_0} p(\boldsymbol{x} \mid \boldsymbol{\theta}) p(\boldsymbol{\theta}) d\boldsymbol{\theta}}{\int_{\Theta_1} p(\boldsymbol{x} \mid \boldsymbol{\theta}) p(\boldsymbol{\theta}) d\boldsymbol{\theta}} \Bigg/ \frac{\int_{\Theta_0} p(\boldsymbol{\theta}) d\boldsymbol{\theta}}{\int_{\Theta_1} p(\boldsymbol{\theta}) d\boldsymbol{\theta}},
$$

is smaller than a cut-off point which depends on the *ratio* l_{01}/l_{10} of the losses incurred, respectively, by accepting a false null and rejecting a true null, and on the *ratio* of the prior probabilities of the hypotheses,

$$
p(H_i) = \int_{\Theta_i} p(\boldsymbol{\theta}) d\boldsymbol{\theta}, \quad i = 0, 1.
$$

From the point of view of classical decision theory, the problem of hypothesis testing is naturally posed in terms of decision rules. Thus, a decision rule for this problem (henceforth called a *test procedure* δ, or simply a *test* δ) is specified in terms of a *critical region* R_δ, defined as the set of $\boldsymbol{x}$ values such that H_0 is rejected whenever $\boldsymbol{x} \in R_\delta$. The most relevant frequentist aspect of such a procedure δ is its *power function*

$$
\text{pow}(\boldsymbol{\theta} \mid \delta) = \Pr\{\boldsymbol{x} \in R_\delta \mid \boldsymbol{\theta}\},
$$

which specifies, as a function of $\boldsymbol{\theta}$, the long-run probability that the test rejects the null hypothesis H_0. Obviously, the ideal power function would be

$$
\begin{aligned}
\text{pow}(\boldsymbol{\theta} \mid \delta) &= 0, \qquad \boldsymbol{\theta} \in \Theta_0 \\
&= 1, \qquad \boldsymbol{\theta} \in \Theta_1
\end{aligned}
$$

although, naturally, one will seldom be able to derive a test procedure with such an ideal power function. For any $\boldsymbol{\theta} \in \Theta_0$, $\text{pow}(\boldsymbol{\theta} \mid \delta)$ is the long-run probability of incorrect rejection of the null hypothesis; frequentist statisticians often specify an upper bound for such probability, which is then called the *level of significance* of the tests to be considered. The *size* of any specific test δ is defined to be

$$\alpha = \sup_{\theta \in \Theta_0} \text{pow}(\boldsymbol{\theta} \mid \delta);$$

thus, to specify a significance level α is to restrict attention to those tests whose size is not larger than α.

Either Θ_0 or Θ_1 may contain just a single value of $\boldsymbol{\theta}$. In this case, the corresponding hypothesis is referred to as a *simple hypothesis*; if Θ_i contains more than one value of $\boldsymbol{\theta}$, then H_i is referred to as a *composite hypothesis*.

For any test procedure δ one may explicitly consider two types of error; rejecting a true null hypothesis, a so-called *error of type 1*, and accepting a false null hypothesis, a so-called *error of type 2*. Let us denote by $\alpha(\delta \mid \boldsymbol{\theta})$ and $\beta(\delta \mid \boldsymbol{\theta})$ the respective probabilities of these two types of error,

$$\begin{aligned} \alpha(\delta \mid \boldsymbol{\theta}) &= \Pr\{\boldsymbol{x} \in R_\delta \mid \boldsymbol{\theta}\} && \text{if } \boldsymbol{\theta} \in \Theta_0, \\ &= 0 && \text{otherwise} \end{aligned}$$

$$\begin{aligned} \beta(\delta \mid \boldsymbol{\theta}) &= \Pr\{\boldsymbol{x} \notin R_\delta \mid \boldsymbol{\theta}\} && \text{if } \boldsymbol{\theta} \in \Theta_1, \\ &= 0 && \text{otherwise.} \end{aligned}$$

It would obviously be desirable to identify tests which keep both error probabilities as small as possible. However, typically, modifying R_δ to reduce one would make the other larger. Hence, one usually tries to minimise some function of the two; for example, a linear combination $a\alpha(\delta \mid \boldsymbol{\theta}) + b\beta(\delta \mid \boldsymbol{\theta})$.

Testing Simple Hypotheses

When *both* H_0 and H_1 are *simple* hypothesis, so that $\alpha(\delta \mid \boldsymbol{\theta}) = \alpha(\delta \mid \boldsymbol{\theta}_0) = \alpha(\delta)$ and $\beta(\delta \mid \boldsymbol{\theta}) = \beta(\delta \mid \boldsymbol{\theta}_1) = \beta(\delta)$, it can be proved that a test which minimises $a\alpha(\delta) + b\beta(\delta)$ should reject H_0 if, and only if,

$$\frac{p(\boldsymbol{x} \mid \boldsymbol{\theta}_0)}{p(\boldsymbol{x} \mid \boldsymbol{\theta}_1)} < \frac{a}{b};$$

i.e., if the likelihood ratio in favour of the null is smaller than the ratio of the weights given to the two kinds of error. This can be seen as a particular case of the Bayesian solution recalled above, and is closely related to the *Neyman-Pearson lemma* (Neyman and Pearson, 1933, 1967) which says that a test which minimises $\beta(\delta)$ subject to $\alpha(\delta) \leq \alpha$ must reject H_0 if, and only if,

$$\frac{p(\boldsymbol{x} \mid \boldsymbol{\theta}_0)}{p(\boldsymbol{x} \mid \boldsymbol{\theta}_1)} < k$$

for some appropriately chosen constant k. It has become standard practice among many frequentist statisticians to choose a significance level α_0 (often "conventional" quantities such as 0.05 or 0.01) and then to find a test procedure which minimises $\beta(\delta)$ among all tests such that $\alpha(\delta) \leq \alpha_0$ (rather than explicitly minimising some combination of the two probabilities of error). The Neyman-Pearson lemma shows explicitly how to derive such a test, but it should be emphasised that this is *not* a sensible procedure. Indeed:

(i) With discrete data one cannot attain a fixed specific size $\alpha(\delta)$ without recourse to auxiliary, *irrelevant randomisation*, whereas minimisation of a linear combination of the form $a\alpha(\delta) + b\beta(\delta)$ can always be achieved. For a Bayesian view on randomisation, see Kadane and Seidenfeld (1986).

(ii) More importantly, by fixing $\alpha(\delta)$ and minimising $\beta(\delta)$ one may find that, with large sample sizes, H_0 is rejected when $p(\boldsymbol{x} \mid H_0)$ is far larger than $p(\boldsymbol{x} \mid H_1)$, due to the fact that the minimising $\beta(\delta)$ may be extremely small compared with the fixed $\alpha(\delta)$. Although this can be avoided by carefully selecting $\alpha(\delta)$ as a decreasing function of the sample size, it seems far more natural to minimise a linear combination $a\alpha(\delta) + b\beta(\delta)$ of the two error probabilities, in which case no difficulties of this type can arise.

Other strategies for the choice of $\alpha(\delta)$ and $\beta(\delta)$ have been proposed. For example, in the $0-1$ loss case, $\alpha(\delta) = \beta(\delta)$ corresponds to the minimax principle. However, it is important to note (see e.g., Lindley, 1972) that minimising a *linear* combination of the two types of error is actually the *only* coherent way of making a choice, in the sense that no other procedure is equivalent to minimising an expected loss.

Composite Alternative Hypotheses

In spite of the difficulties described above, frequentist statisticians have traditionally defined an optimal test δ to be one which minimises $\beta(\delta \mid \boldsymbol{\theta})$ for a fixed significance level α_0. In terms of the power function, this implies deriving a test δ such that

$$\text{pow}(\boldsymbol{\theta} \mid \delta) \leq \alpha_0, \qquad \boldsymbol{\theta} \in \Theta_0$$

and for which $\text{pow}(\boldsymbol{\theta} \mid \delta)$ is as large as possible in Θ_1. A test procedure δ^* is called a *uniformly most powerful* (UMP) test, at *level of significance* α_0, if $\alpha(\delta^* \mid \boldsymbol{\theta}) \leq \alpha_0$ and, for any other δ such that $\alpha(\delta \mid \boldsymbol{\theta}) \leq \alpha_0$,

$$\text{pow}(\boldsymbol{\theta} \mid \delta) \leq \text{pow}(\boldsymbol{\theta} \mid \delta^*), \qquad \text{for all } \boldsymbol{\theta} \in \Theta_1.$$

It can be proved that, when Θ is *one-dimensional*, UMP tests often exist for *one-sided* alternative hypotheses.

A model $\{p(\boldsymbol{x} \mid \theta), \theta \in \Theta \subseteq \Re\}$ is said to have a monotone likelihood ratio in the statistic $t = t(\boldsymbol{x})$ if for all $\theta_1 < \theta_2$, $p(\boldsymbol{x} \mid \theta_2)/p(\boldsymbol{x} \mid \theta_1)$ is an increasing function of t. If $p(\boldsymbol{x} \mid \theta)$ has a monotone likelihood ratio in t and c is a constant such that

$$\Pr\{t \geq c \mid \theta_0\} = \alpha_0,$$

then the test δ which rejects H_0 if $t \geq c$ is a UMP test of the hypothesis $H_0 \equiv \theta \leq \theta_0$ versus the alternative $H_1 \equiv \theta > \theta_0$, at the level of significance α_0. However, UMP tests do not generally exist.

Example B.8. ***(Non-existence of a UMP test).*** If $\boldsymbol{x} = \{x_1, \ldots, x_n\}$ is a random sample from a normal distribution $\mathrm{N}(x \mid \mu, 1)$, then the test δ_1 defined by critical region $R_{\delta_1} = \{\boldsymbol{x};\ \bar{x} - \mu_0 > 1.282/\sqrt{n}\}$ is a UMP test for $H_0 \equiv \mu \leq \mu_0$ versus $H_1 \equiv \mu > \mu_0$, with 0.10 significance level. Similarly, the test δ_2 defined by $R_{\delta_2} = \{\boldsymbol{x};\ \mu_0 - \bar{x} > 1.282/\sqrt{n}\}$ is a UMP test for $H_0 \equiv \mu \geq \mu_0$ versus $H_1 \equiv \mu < \mu_0$, with the same level. Since these critical regions are different, it follows that there is no UMP test for $\mu = \mu_0$ versus $\mu \neq \mu_0$.

The fact, illustrated in the above example, that UMP tests typically do not exist for two-sided alternatives, suggests that a less demanding criterion must be used if one is to define a "best" test among those with a fixed significance level. Since the power function $\mathrm{pow}(\boldsymbol{\theta} \mid \delta)$ describes the probability that the test δ rejects the null, it seems desirable that, when H_0 is true, $\mathrm{pow}(\boldsymbol{\theta} \mid \delta)$ should be smaller in Θ_0 than elsewhere. A test δ is called *unbiased* if for any pair $\boldsymbol{\theta}_0 \in \Theta_0$ and $\boldsymbol{\theta}_1 \in \Theta_1$ it is true that $\mathrm{pow}(\boldsymbol{\theta}_0 \mid \delta) \leq \mathrm{pow}(\boldsymbol{\theta}_1 \mid \delta)$.

Example B.9. ***(Comparative power of different tests).*** If $\boldsymbol{x} = \{x_1, \ldots, x_n\}$ is a random sample from a normal distribution $\mathrm{N}(x \mid \mu, 1)$ then the test δ_3 defined by the region

$$R_{\delta_3} = \{\boldsymbol{x};\ |\,\bar{x} - \mu_0\,| > 1.645/\sqrt{n}\}$$

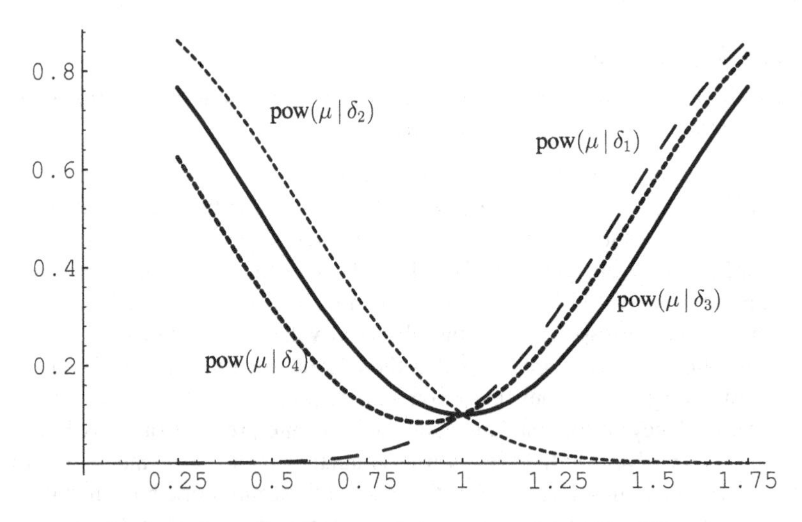

Figure B.2 *Power of tests for the mean of a normal distribution*
$\mu_0 = 1,\ n = 30,\ c_1 = 1.40,\ c_2 = 2.05$

is an unbiased test for $H_0 \equiv \mu = \mu_0$ versus $H_1 \equiv \mu \neq \mu_0$. Figure B.2 compares the power of this test with those defined in Example B.8, and with that of a typical non-symmetric test δ_4 of the same level, which has the critical region

$$R_{\delta_4} = \{\boldsymbol{x};\ \bar{x} - \mu_0 > c_1/\sqrt{n},\ \text{or}\ \mu_0 - \bar{x} > c_2/\sqrt{n}\}$$

for suitably chosen constants $c_1 < c_2$. It seems obvious that δ_4, which is more cautious about accepting values of μ larger than μ_0 than about accepting values of μ smaller than μ_0, should be preferred to the unbiased test δ_3 whenever the consequences of the first class of errors are more serious, or whenever the values of μ smaller than μ_0 are considered to be more likely.

It is clear from Example B.9 above that, even when they exist, unbiased procedures may only be reasonable in special circumstances. We are drawn again to the general comment that, in any decision procedure, prior information and utility preferences should be an integral part of the solution.

Yet another approach to defining a "good" test when UMP tests do not exist is to focus attention on *local power*, by requiring the power function to be maximised in a neighbourhood of the null hypothesis. Under suitable regularity conditions, *locally most powerful tests* may be derived by using the sampling distribution of the efficient score function in a process which is closely related to that described in our discussion of interval estimation. However, the requirement of maximum local power does not say anything about the behaviour of the test in a region of high power and, indeed, locally most powerful tests may be very inappropriate when the true value of $\boldsymbol{\theta}$ is far from Θ_0.

Methodological Discussion

Testing hypotheses using the frequentist methodology described above may be misleading in many respects. In particular:

(i) It should be obvious that the all too frequent practice of simply quoting whether or not a null hypothesis is rejected at a specified significance level α_0 ignores a lot of relevant information. Clearly, if such a test is to be performed, the statistician should report the cut-off point α such that H_0 would not be rejected for any level of significance smaller than α. This value is called the *tail area* or *p-value* corresponding to the observed value of the statistic. An added advantage of this approach is that there is no need to select beforehand an arbitrary significance level. As noted in the case of confidence intervals, there is a tendency on the part of many users to interpret a p-value as implying that the *probability* that H_0 is true is smaller than the p-value. Not only, of course, is this false within the frequentist framework but, in this case, there is, in general, no simple form of reinterpretation which would have a Bayesian justification, so that, even numerically, p-values cannot generally be interpreted as posterior probabilities. For detailed discussions see, for example, Berger (1985a), Berger and Delampady (1987) and Berger and Sellke (1987). See Casella and Berger (1987) for an attempted reconciliation in the case of one-sided tests.

(ii) Another statistical "tradition" related to hypothesis testing consists of declaring an observed value *statistically significant*, implying that there exists statistical evidence which is sufficient to reject the null hypothesis, whenever the corresponding tail area is smaller than a "conventional" value such as 0.05 or 0.01. However, since the classical theory of hypothesis testing does not make any use of a utility function, there is no way to assess formally whether or not the true value of the parameter $\boldsymbol{\theta}$, which may well be *numerically* different from a hypothetical value $\boldsymbol{\theta}_0$, is *significantly* different from $\boldsymbol{\theta}_0$ in the sense of implying any practical difference. Thus, a vote proportion of 34% for a political party is technically different from a proportion of 34.001%, but under most plausible utility functions the difference has no political significance.

(iii) Finally, the mutual inconsistency of frequentist desiderata often makes it impossible, even in the theory's own terms, to identify the most appropriate procedure. For example, if x is a random sample from $\mathrm{N}(x \mid \mu, m\lambda)$ with precision $m\lambda$ determined by a *random* integer m, then m is ancillary and hence, by the conditionality principle, tests on μ or λ should condition on the observed m. Yet, Durbin (1969) showed that, at least asymptotically, unrestricted tests may be uniformly more powerful.

See Chernoff (1951) and Stein (1951) for further arguments against standard hypothesis testing.

B.3.4 Significance Testing

In the previous section, we have reviewed the problem of hypothesis testing where, given a family $\{p(\boldsymbol{x} \mid \boldsymbol{\theta}), \boldsymbol{\theta} \in \Theta\}$, a null hypothesis $H_0 \equiv \boldsymbol{\theta} \in \Theta_0$ is tested *against* (at least) one specific alternative. In this section we shall review the problem of *pure significance tests*, where *only* the null hypothesis $H_0 = \{p(\boldsymbol{x} \mid \boldsymbol{\theta}), \boldsymbol{\theta} \in \Theta_0\}$ has been initially proposed, and it is desired to test whether or not the data $\boldsymbol{x}$ are *compatible* with this hypothesis, without considering specific alternatives. The null hypothesis may be either *simple*, if it completely specifies a density $p(\boldsymbol{x} \mid \boldsymbol{\theta}_0)$, or *composite*.

We recall from Section 6.2 that, within the Bayesian framework, the problem of significance testing, as formulated above, could be solved by embedding the hypothetical model in some larger class $\{p(\boldsymbol{x} \mid \boldsymbol{\theta}), \boldsymbol{\theta} \in \Theta\}$, designed either to contain *actual* alternatives of practical interest, or *formal* alternatives generated by selecting a mathematical neighbourhood of H_0. For any discrepancy measure

$$\delta(\boldsymbol{\theta}) = u\{H_0^c, \boldsymbol{\theta}\} - u\{H_0, \boldsymbol{\theta}\},$$

describing, for each $\boldsymbol{\theta}$, the conditional utility difference, and for any function $\varepsilon_0(\boldsymbol{x})$, describing the additional utility obtained by retaining H_0 because of its special status, we showed that H_0 should be rejected if,

$$t(\boldsymbol{x}) > \varepsilon_0(\boldsymbol{x}),$$

where

$$t(\boldsymbol{x}) = \int_\Theta \delta(\boldsymbol{\theta}) p(\boldsymbol{\theta} \mid \boldsymbol{x}) d\boldsymbol{\theta}$$

is the *expected posterior discrepancy*. In particular, we proposed the logarithmic discrepancy

$$\delta(\boldsymbol{\theta}) = \int_X p(\boldsymbol{x} \mid \boldsymbol{\theta}) \log \frac{p(\boldsymbol{x} \mid \boldsymbol{\theta})}{p(\boldsymbol{x} \mid \boldsymbol{\theta}_0)} d\boldsymbol{x}$$

as a reasonable general discrepancy measure. This (fully Bayesian) procedure could be described as that of selecting a test statistic $t(\boldsymbol{x})$ which is expected to measure the discrepancy between H_0 and the true model, and rejecting H_0 if $t(\boldsymbol{x})$ is larger than some cut-off point $\varepsilon_0(\boldsymbol{x})$ describing the additional utility of keeping H_0 if it were true, due to its special status corresponding to simplicity (Occam's razor), scientific support (or fashion), or whatever.

From a frequentist point of view, a test statistic $t = t(\boldsymbol{x})$ is selected with two requirements in mind.

(i) The sampling distribution of t under the null hypothesis $p(t \mid H_0)$ must be known and, if H_0 is composite, $p(t \mid H_0)$ should be the same for all $\boldsymbol{\theta} \in \Theta_0$.
(ii) The larger the value of t the stronger the evidence of the departure from H_0 of the kind which it is desired to test.

Then, given the data $\boldsymbol{x}$, a *p-value* or *significance level* is calculated as the probability, conditional on H_0, that, in repeated samples, t would exceed the observed value $t(\boldsymbol{x})$, so that p is given by

$$p = \int_{t(\boldsymbol{x})}^{\infty} p(t \mid H_0) dt.$$

Small values of p are regarded as strong evidence that H_0 should be rejected. The result of the analysis is typically reported by stating the p-value and declaring that H_0 should be rejected for all significance levels which are smaller than p.

Comparison with the Bayesian analogue summarised above prompts the following remarks.

(i) The frequentist theory does not generally offer any guidance on the choice of an appropriate test statistic (the *generalised likelihood ratio test*, a disguised Bayes factor seems to be the only proposal). While in the Bayesian analysis

$t(\boldsymbol{x})$ is naturally and constructively derived as an expected measure of discrepancy, the frequentist statistician must, in general, rely on intuition to select t. The absence of declared alternatives even precludes the use of the frequentist optimality criteria used in hypothesis testing.

(ii) Even if a function $t = t(\boldsymbol{x})$ is found which may be regarded as a sensible discrepancy measure, the frequentist statistician needs to determine the unconditional sampling distribution of t under H_0; this may be very difficult, and actually impossible when there are nuisance parameters. Moreover, in the more interesting situation of composite null hypotheses, it is required that $p(t \mid \boldsymbol{\theta})$ be the same for all $\boldsymbol{\theta}$ in Θ_0, which, often, is simply not the case.

(iii) If a measure of the strength of evidence against H_0 is all that is required, the position of the *observed* value of t with respect to its posterior *predictive* distribution $p(t \mid \boldsymbol{x}, H_0)$ under the null hypothesis seems a more reasonable, more relevant answer than quoting the realised p-value. Indeed, the compatibility of $t(\boldsymbol{x})$ with H_0 may be described by quoting the HPD intervals to which it belongs, or may be measured with any proper scoring rule such as $A \log p(t(\boldsymbol{x}) \mid \boldsymbol{x}, H_0) + B$. Thus, in Figure B.3, $t_1(\boldsymbol{x})$ may readily be accepted as compatible with H_0 while $t_2(\boldsymbol{x})$ may not.

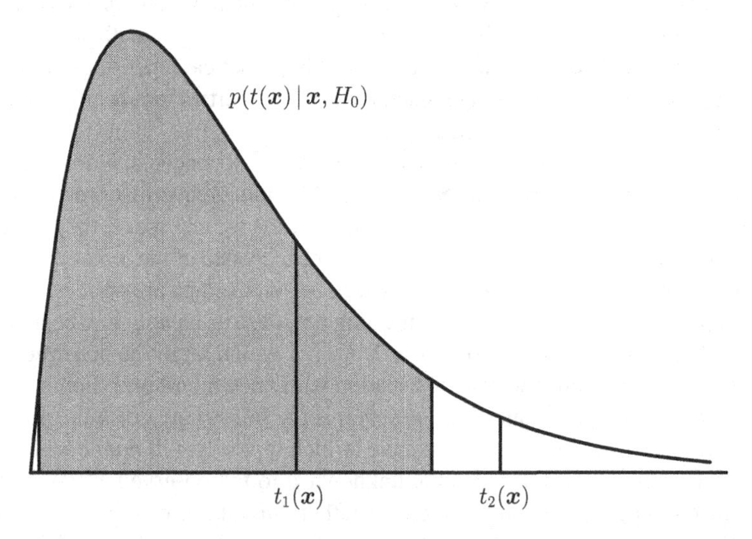

Figure B.3 *Visualising the compatibility of* $t(\boldsymbol{x})$ *with* H_0

(iv) If a *decision* on whether or not to reject H_0 has to be made, this should certainly take into account the advantages of keeping H_0, i.e., defining the cut-off point in terms of utility. We described in Section 6.2 how this may actually be chosen to guarantee a specified significance level, but this is only one possible choice, not necessarily the most appropriate in all circumstances.

We should finally point out that most of the criticisms already made about hypothesis testing are equally applicable to significance testing. Similarly, criticisms made of confidence intervals typically apply to significance testing, since confidence intervals can generally be thought of as consisting of those null values which are not rejected under a significance test.

B.4 COMPARATIVE ISSUES

B.4.1 Conditional and Unconditional Inference

At numerous different points of this Appendix we have emphasised the following essential difference between Bayesian and frequentist statistics: Bayesian statistics directly produces statements about the uncertainty of unknown quantities, either parameters or future observations, conditional on known data; frequentist statistics produces probability statements about hypothetical repetitions of the data conditional on the unknown parameter, and then seeks (indirectly) ways of making this relevant to inferences about the unknown parameters given the observed data. Indeed, the problem at the very heart of the frequentist approach to statistics is that of connecting *aggregate*, *long-run sampling* properties under hypothetical repetitions, to *specific inferences* of a totally different type. Not only may one dispute the existence of the conceptual "*collective*" where these hypothetical repetitions might take place, but the *relevance* of the aggregate, long-run properties for specific inference problems seems, at best, only tangential.

It is useful to distinguish between two very different concepts, *initial precision* and *final precision*, introduced by Savage (1962). Thus, frequentist procedures are designed in terms of their expected behaviour over the sample space; they typically have average characteristics which describe, for each value of the unknown parameters, the "precision" we may *initially* expect, before the data are collected. Thus, for example, one might expect that, in the long run, the true mean μ will be included in 95% of the intervals of the form $\bar{x} \pm 1.96/\sqrt{n}$ which might be constructed by repeated sampling from a normal distribution with known unit precision.

A far more pertinent question, however, is the following: given the observed $\bar{x}$ which derives from the observed sample (which typically will *not* be repeated in any actual practice), how close is the unknown μ to the observed $\bar{x}$? Within the frequentist approach, one must rest on the rather dubious "transferred" properties of the long-run behaviour of the procedure, with no logical possibility of assessing the relevant *final* precision.

Thus, p-values or confidence intervals are largely irrelevant once the sample has been observed, since they are concerned with events which might have occurred, but have not. Indeed, to quote Jeffreys (1939/1961, p. 385)

> ... a hypothesis which may be true may be rejected because it has not predicted observable results which have *not* occurred. This seems a remarkable procedure.

The following example, taken from Welch (1939) further illustrates the difference between initial and final precision.

Example B.10. ***(Initial and final precision).*** Let $\boldsymbol{x} = \{x_1, \ldots, x_n\}$ be a random sample from a uniform distribution over $]\mu - \frac{1}{2}, \mu + \frac{1}{2}[$. It is easily verified that the midrange $\tilde{\mu} = (x_{\min} + x_{\max})/2$ is a very efficient estimator with a sampling variance of the order of $1/n^2$, rather than the usual $1/n$, so that, from *large* samples, we may expect, *on average*, very precise estimates of μ. Suppose however that we obtain a specific large sample with a *small range*; this is, admittedly, unlikely, but nevertheless possible. Given the sample and using a uniform (reference) prior for μ, we can only really claim that $\mu \in]x_{\max} - \frac{1}{2}, x_{\min} + \frac{1}{2}[$ (since the reference posterior distribution is uniform on that interval). Thus, if the *actual data* turn out this way, the final precision of our inferences about μ is bound to be rather poor, no matter how efficient the estimator $\tilde{\mu}$ was *expected* to be.

The need for conditioning on observed data can be partially met in frequentist procedures by conditioning on an ancillary statistic. Indeed, we saw in Examples B.2 and B.3 that it is easy to construct examples where totally unconditional procedures produce ludicrous results. However, as pointed out in our discussion of the conditionality principle, there remain many problems with conditioning on ancillary statistics; they are not easily identifiable, they are not necessarily unique and, moreover, conditioning on an ancillary statistic, can yield a totally uninformative sampling distribution, and can conflict with other frequentist desiderata, such as the search for maximum power in hypothesis testing; see, for example, Basu (1964, 1992) and Cox and Reid (1987). See, also, Berger (1984b).

B.4.2 Nuisance Parameters and Marginalisation

Most realistic probability models make the sampling distribution dependent not only on the unknown quantity of primary interest but also on some other parameters. Thus, the full parameter vector $\boldsymbol{\theta}$ can typically be partitioned into $\boldsymbol{\theta} = \{\boldsymbol{\phi}, \boldsymbol{\lambda}\}$ where $\boldsymbol{\phi}$ is the subvector of interest and $\boldsymbol{\lambda}$ is the complementary subvector of $\boldsymbol{\theta}$, often referred to as the vector of nuisance parameters.

We recall from Section 5.1 that, within a Bayesian framework, the presence of nuisance parameters does not pose any formal, theoretical problems. Indeed, the desired result, namely the (marginal) posterior distribution of the parameter of interest, can simply be written as

$$p(\boldsymbol{\phi} \,|\, \boldsymbol{x}) = \int p(\boldsymbol{\phi}, \boldsymbol{\lambda} \,|\, \boldsymbol{x}) d\boldsymbol{\lambda}$$

where the full posterior $p(\phi, \lambda \mid x)$ is directly obtained from Bayes' theorem.

The situation is very different from a frequentist point of view. Indeed, the problem posed by the presence of nuisance parameters is only satisfactorily solved within a pure frequentist framework in those few cases where the optimality criterion used leads to a procedure which depends on a statistic whose sampling distribution does not depend on the nuisance parameter. Frequentist inferences about the mean of a normal distribution with unknown variance based on the Student-t statistic, whose sampling distribution does not involve the variance, provides the best known example. In general, frequentists are forced to use *approximate* methods, typically based on asymptotic theory. Indeed, some statisticians see this as the main motivation for developing asymptotic results:

> a ... serious difficulty is that the techniques ... for problems with nuisance parameters are of *fairly restricted applicability*. It is, therefore, essential to have widely applicable procedures that in some sense provide good *approximations* when "exact" solutions are not available. ... the central idea being that when the number n of observations is large and errors of estimation correspondingly small, simplifications become available that are not available in general (Cox and Hinkley, 1974, p. 279, our italics)

However, even the domain of "fairly restricted applicability" resulting from reliance on asymptotic methods can be problematic. In an early paper, Neyman and Scott (1948) illustrated such problems by considering models with many nuisance parameters of the type

$$p(\boldsymbol{x} \mid \phi, \boldsymbol{\lambda}) = \prod_{i=1}^{n} p(x_i \mid \phi, \lambda_i),$$

where a new nuisance parameter λ_i is introduced with each observation. Note that such models are not unrealistic: for example, x_i may be a physiological measurement on individual i, which may have a normal distribution with mean λ_i and common variance ϕ, the latter being the parameter of interest. Kiefer and Wolfowitz (1956) and Cox (1975) proposed solutions for this type of problem based on treating the λ_i's as independent observations from some distribution. From a Bayesian viewpoint, this, of course, then becomes a case of hierarchical modelling as discussed in Section 4.6.5.

The only general alternative strategy which has been proposed to avoid resorting to asymptotics when exact methods are not available is to use a modified form of likelihood, (estimated, conditional, or marginal), where the dependence on the nuisance parameters has been reduced or eliminated.

An *estimated likelihood* is obtained by replacing nuisance parameters by (for example) their maximum likelihood estimates. This procedure does not take account of the uncertainty due to the lack of knowledge about the nuisance parameters, and may be misleading both in the precision and in the location associated with inferences about the parameters of interest. For example, in linear regression with many regressors, substitution of the regression coefficients by their mle's leads to an estimate of the variance which is misleadingly precise.

Marginal and *conditional* likelihoods are based on breaking the likelihood function into two factors, either using invariance arguments or conditioning on sufficient statistics for the nuisance parameters. In both cases, one factor provides a likelihood function for the parameter of interest while the other is assumed "to contain no information about the parameter of interest in the absence of knowledge about the nuisance parameter". Key references to this approach are Kalbfleish and Sprott (1970, 1973) and Andersen (1970, 1973).

There are however two main problems with this type of approach.

(i) They are not general and can only be applied under rather specific circumstances.

(ii) They critically depend on the highly controversial notion of a "function not containing relevant information in the absence of knowledge about the nuisance parameters", for which no operational definition has ever been provided.

In the cases where the techniques can be applied, and a consensus seems to exist about this vague information condition, the resulting forms tend to coincide, as one might expect, with the integrated likelihood

$$\int p(\boldsymbol{x} \mid \phi, \boldsymbol{\lambda})\pi(\boldsymbol{\lambda} \mid \phi)d\boldsymbol{\lambda},$$

integrated with respect to the conditional reference prior distribution $\pi(\boldsymbol{\lambda} \mid \phi)$ of the nuisance parameters given the parameter of interest.

Profile likelihood provides a much more refined version of this approach, which often gives answers which closely correspond to Bayesian marginalisation results; the Fieller-Creasy problem concerning the ratio of normal means provides a typical example (see Bernardo, 1977). For further discussion and extensive references, see, for example, Barndorff-Nielsen (1983, 1991), Cox and Reid (1987), Cox (1988) and Fraser and Reid (1989). Another suggestion, closely related to fiducial inference, is the *implied likelihood* (Efron, 1993).

Liseo (1993) shows that reference posterior credible regions have better frequentist coverage properties than those obtained from likelihood methods. For a Bayesian overview of methods for treating nuisance parameters, see Basu (1977), Dawid (1980a), Willing (1988) and Albert (1989).

B.4.3 Approaches to Prediction

The general problem of statistical prediction may be described as that of inferring the values of unknown *observable* variables from current available information. Thus, from data x, usually a random sample $\{x_1, \ldots, x_n\}$, inference statements are desired about, as yet, unobserved data y, often x_{n+1} (the original problem considered by Bayes, 1763, in a binomial setting).

We recall from Section 5.1.3 that, from a Bayesian point of view, with an operationalist concern with modelling uncertainty in terms of observables, Bayes' theorem, in its central role as a coherent learning process about parameters, is just a convenient step in the process of passing from

$$p(\boldsymbol{x}) = \int \prod_{i=1}^{n} p(x_i \mid \boldsymbol{\theta}) p(\boldsymbol{\theta}) d\boldsymbol{\theta}$$

to

$$p(y \mid \boldsymbol{x}) = \int p(y \mid \boldsymbol{\theta}) p(\boldsymbol{\theta} \mid \boldsymbol{x}) d\boldsymbol{\theta}$$

by means of $p(\boldsymbol{\theta} \mid \boldsymbol{x}) \propto p(\boldsymbol{x} \mid \boldsymbol{\theta}) p(\boldsymbol{\theta})$. Since any valid coherent inferential statement about y given x is contained in the *posterior predictive* distribution $p(y \mid x)$, no special theory has to be developed. Of course, the inferential content of the predictive distribution may be appropriately summarised by location or spread measures, respectively providing "estimators" of y, such as the mean and the mode of $p(y \mid x)$, or "interval estimates" of y such as the class of HPD intervals which may be derived from $p(y \mid x)$. Moreover, if one is faced with a decision problem whose utility function $u(a, y)$ involves a future observable, then $p(y \mid x)$ becomes the necessary ingredient in determining the optimal action, a^*, which maximises the appropriate (posterior) expected utility

$$\bar{u}(a \mid \boldsymbol{x}) = \int_Y u(a, y) p(y \mid \boldsymbol{x})\, dy.$$

The range of potential applications of these ideas is extensive.

(i) *Density estimation.* The action space consists of the class of sampling distributions; the predictive distribution, which is the posterior expectation of the sampling distribution, is, for squared error loss, the optimal estimator of the sampling distribution.

(ii) *Calibration.* Two observations (x_{1i}, x_{2i}) are made on a set of n individuals using two different measuring procedures, and it is desired to estimate the measurement y_2 that the second procedure would yield on a new individual, given that the measurement using the first procedure has turned out to be y_1. The solution is a simple exercise in probability calculus leading to the required posterior predictive density $p(y_2 \mid y_1, \boldsymbol{x}_1, \boldsymbol{x}_2)$.

(iii) *Classification.* This is a particular case of the problem of calibration where the x_{2i}'s (and y_2) can only take on a discrete, usually finite, set of values.

(iv) *Regulation.* In contexts analogous to (ii) and (iii), it is desired to select and fix a value of y_1 so that y_2 is as close as possible to a prescribed value. The solution is obtained by minimising the expectation of an appropriate loss function with respect to the predictive distribution $p(y_2 \,|\, y_1, \boldsymbol{x}_1, \boldsymbol{x}_2)$. The particular case of *optimisation* obtains when it is desired to make y_2 as large (or small) as possible.

(v) *Model comparison.* In a setting with alternative models, the latter may be compared in terms of their predictive posterior probabilities (cf. Section 6.1).

(vi) *Model criticism.* The compatibility of a given model with observed data may be assessed by comparing the realised value of a test statistic with its predictive distribution under that model (cf. Section 6.2).

For further details of the systematic use of predictive ideas, the reader is referred, for example, to Roberts (1965), Geisser (1966, 1974, 1980b, 1988) and Zellner (1986b). The books by Aitchison and Dunsmore (1975) and Geisser (1993) contain a wealth of detailed discussion of prediction problems, including those involving decision making. Applications of predictive ideas to classification, calibration, regulation, optimization and smoothing are found, for instance, in Dunsmore (1966, 1968, 1969), Bernardo (1988), Racine-Poon (1988), Klein and Press (1992), Lavine and West (1992) and Zidek and Weerahandi (1992). See, also, Gelfand and Desu (1968) and Amaral-Turkman and Dunsmore (1985).

It is important to recall here (see Section 5.1) that, by virtue of the representation theorems, parameters are limiting forms of observables and, hence, inference about parameters may be seen as a limiting form of predictive inference about observables. Although in practice it is usually convenient to work via parametric models, this point, stressed by de Finetti (1970/1974, 1970/1975) has considerable theoretical importance. Cifarelli and Regazzini (1982), among others, have continued this tradition by trying to develop a completely predictive approach which bypasses entirely the use of parametric models.

We should emphasise again that the all too often adopted naïve solution of prediction based on the "plug-in estimate" form

$$p(y \,|\, \boldsymbol{x}) = \int p(y \,|\, \boldsymbol{\theta}) p(\boldsymbol{\theta} \,|\, \boldsymbol{x}) d\boldsymbol{\theta} \simeq p(y \,|\, \hat{\boldsymbol{\theta}}),$$

effectively replacing the posterior distribution by a degenerate distribution assigning probability one to an estimator of $\boldsymbol{\theta}$, usually the maximum likelihood estimate, is bound to give misleadingly overprecise inference statements about y, since it effectively ignores the uncertainty about $\boldsymbol{\theta}$. The point is illustrated in detail by Aitchison and Dunsmore (1975).

By comparison, the possibilities for frequentist-based prediction are fairly limited. They are essentially limited to producing *tolerance regions*, $R(x)$, designed to guarantee that, in the long run, a proportion p of possible samples x would produce regions $R(x)$ such that

$$\Pr[y \in R(x) \mid \boldsymbol{\theta}] = 1 - \alpha, \qquad \text{for all } \boldsymbol{\theta} \in \Theta,$$

i.e., regions which, for all parameters values, will contain a proportion $1 - \alpha$ of future observations. If this sounds obscure, particularly in comparison with the simple idea of an HPD region from the predictive distribution $p(y \mid x)$, we can but agree! Moreover:

(i) In order to construct a tolerance region it is essential to find a function of y and x with a sampling distribution which does not involve $\boldsymbol{\theta}$, something which is typically only possible in very simple stylised problems.

(ii) The difficulties of "transferring" the long-run aggregate properties of confidence intervals into inference statements conditional on the observed data, are even more acute in a tolerance region setting.

Descriptions of the frequentist approach to prediction are given in Cox (1975), Mathiasen (1979) and Barndorff-Nielsen (1980); Guttman (1970) provides a comparison between frequentist tolerance regions and HPD regions from predictive distributions.

Kalbfleish (1971) was one of the first to examine likelihood methods for prediction. Essentially, with t denoting a sufficient statistic for $\boldsymbol{\theta}$, he proposed computing a predictive distribution of the form

$$p(y \mid t) = \int_{\Theta} p(y \mid \boldsymbol{\theta}) f(\boldsymbol{\theta} \mid t)\, d\boldsymbol{\theta}$$

whenever a fiducial distribution for $\boldsymbol{\theta}$, $f(\boldsymbol{\theta} \mid t)$, can be derived from the sampling distribution of t. Of course, the method is not always applicable; moreover, even when it is, it may lead to inconsistent results when the fiducial distribution is not a Bayes' posterior. For instance, in the discussion which follows Kalbfleish's paper, Lindley points out that if the model is

$$p(x \mid \theta) = \frac{\theta^2}{\theta + 1}(x + 1)e^{-x\theta}, \qquad x > 0, \theta > 0$$

and the method is applied both to obtain directly $p(x_{n+1} \mid x_1, \ldots, x_n)$ and to obtain $p(x_{n+1}, x_n \mid x_1, \ldots, x_{n-1})$ and then $p(x_{n+1} \mid x_1, \ldots, x_n)$ from the joint predictive, one obtains different answers. This is an interesting example of the fact that fiducial distributions do not necessarily have basic coherence properties unless they are equivalent to Bayesian posterior distributions.

Since the late 1970's a variety of more sophisticated "likelihood prediction" methods have been proposed, some sufficiency-based, others relating to profile likelihood ideas. Seminal contributions include those by Hinkley (1979), Lejeune and Faulkenberry (1982), Butler (1986) and Lane and Sudderth (1989). A review is given by Bjrnstad (1990), and a further overview is provided by Geisser (1993).

A more radical approach to prediction is set out in Dawid (1984), who sets out a theory of *prequential analysis.* This is closely related to our view that a model or theory is simply a probability forecasting system, but Dawid's theory is not predicated on such a system necessarily being Bayesian. Instead, the basic ingredients are simply two sequences; one a string of observations, the other a string of probability forecasts. Theoretical developments requiring an extension of the standard Kolmogorov (1933/1950) framework for probability are pursued in Vovk (1993a). See, also, Vovk (1993b) and Vovk and Vyugin (1993). Links with *stochastic complexity* (Solomonoff, 1978; Rissanen, 1987, 1989; Wallace and Freeman, 1987) are reviewed in Dawid (1992).

B.4.4 Aspects of Asymptotics

In most statistical problems, a number of simplifications become available when the sample size becomes sufficiently large. In frequentist statistics, this is often the only way to obtain analytic results. From a Bayesian point of view, such simplifications are *never theoretically necessary*, although, of course, they often make computations easier and sometimes provide valuable analytic insight.

We recall from Section 5.3 that, as the sample size increases, the posterior distribution of the parameter of interest $\boldsymbol{\theta}$ converges to a degenerate distribution which gives probability one to the true parameter value when the parameter of interest is discrete and, under suitable regularity conditions when the parameter of interest is continuous, converges to a normal distribution $\mathrm{N}(\boldsymbol{\theta} \mid \hat{\boldsymbol{\theta}}_n,\ H(\hat{\boldsymbol{\theta}}_n))$, with precision matrix

$$H(\hat{\boldsymbol{\theta}}_n) = \left(-\frac{\partial^2 \log p(\boldsymbol{x} \mid \boldsymbol{\theta})}{\partial \theta_i \partial \theta_j} \right) \Big|_{\boldsymbol{\theta} = \hat{\boldsymbol{\theta}}_n} .$$

The most frequently used asymptotic results in frequentist statistics concern the large sample behaviour of the maximum likelihood estimate $\hat{\boldsymbol{\theta}}_n$ which, under suitable regularity conditions (mathematically usually closely related to those required to guarantee posterior asymptotic normality), may be shown to have an asymptotically normal sampling distribution $\mathrm{N}(\hat{\boldsymbol{\theta}}_n \mid \boldsymbol{\theta}, nI(\boldsymbol{\theta}))$, with precision matrix whose general element is

$$(I(\boldsymbol{\theta}))_{ij} = \int p(x \mid \boldsymbol{\theta}) \left(-\frac{\partial^2 \log p(x \mid \boldsymbol{\theta})}{\partial \theta_i \partial \theta_j} \right) dx.$$

For details, see, for example, LeCam (1956, 1970, 1986), and references therein.

Since it is easily established that, for large n, $H(\hat{\theta}_n)$ converges to $nI(\theta)$, and since, asymptotically, the sampling distribution of $\hat{\theta}_n$ becomes a location model for θ, it follows (Lindley, 1958) that the reference posterior distribution for θ and the asymptotic fiducial distribution of θ based on the sampling distribution of $\hat{\theta}_n$ are asymptotically equivalent. Moreover, the maximum likelihood estimator of θ and the asymptotic confidence intervals based on $\hat{\theta}_n$ will be, respectively, numerically identical to the mode (or the mean) and the HPD intervals based on the reference posterior distribution (or any other posterior distribution based on a reasonably well-behaved prior).

These results explain the fact that, for large samples (relative to the dimensionality of the parametric model component) there are typically very few *numerical* differences between Bayesian inferential statements and frequentist statements based on asymptotic properties. This asymptotic equivalence carries over, of course, to a number of applications. For example:

(i) We showed (Corollary 2 to Proposition 5.17) that if θ is asymptotically normal $\mathrm{N}(\theta \mid \hat{\theta}_n, -L_n''(\hat{\theta}_n))$ then, under appropriate regularity conditions $g(\theta)$ is asymptotically normal

$$\mathrm{N}(g(\theta) \mid g(\hat{\theta}_n), H(\hat{\theta}_n)[g'(\hat{\theta}_n)]^{-2}).$$

The frequentist equivalent (typically derived using the *delta method* for determining the asymptotic distribution of an estimator) is that if $\hat{\theta}_n$ has an asymptotic sampling distribution $\mathrm{N}(\hat{\theta}_n \mid \theta, nI(\theta))$, then $g(\hat{\theta})$ has an asymptotic sampling distribution

$$\mathrm{N}(g(\hat{\theta}_n) \mid g(\theta), nI(\theta)[g'(\theta)]^{-2}).$$

(ii) The predictive distribution $p(y \mid x)$ is asymptotically approached by $p(y \mid \hat{\theta}_n)$.

(iii) The action which maximises the posterior expected utility is, asymptotically, the same as that which maximises $u(a, \hat{\theta}_n)$.

In the fictional world of unlimited data, *numerical* differences between frequentist and Bayesian solutions would tend to disappear with increasing sample size although, even then, differences in *interpretation* would persist. However, in the real world of limited data relative to the (often multiparameter) models required for realism, there is no reason to expect, in general, close coincidence of numerical solutions.

B.4.5 Model Choice Criteria

We have discussed earlier in Sections B.3.3 and B.3.4, the hypothesis and significance testing approaches to parametric hypotheses, but have noted that, in general, no satisfactory exact procedures exist. This may be because of a lack of simplification via sufficiency or invariance arguments, resulting in intractable distributions, or because a procedure cannot be found which has uniformly optimal properties through the range of parameter values under the alternative hypothesis.

A procedure frequently adopted in such situations is the so-called *general maximum likelihood ratio test*, which we describe first for the case of a simple null hypothesis, $\boldsymbol{\theta} = \boldsymbol{\theta}_0 \in \Re^k$ and observations $\boldsymbol{x} = (x_1, \ldots, x_n)$. The procedure is motivated by considering the ratio

$$r(\boldsymbol{x}) = \frac{p(\boldsymbol{x} \mid \boldsymbol{\theta}_0)}{p(\boldsymbol{x} \mid \hat{\boldsymbol{\theta}})},$$

where $\hat{\boldsymbol{\theta}}$ is the maximum likelihood estimate. Intuitively, small values of $r(\boldsymbol{x})$ suggest rejection of the null hypothesis, but using this type of test requires deriving the distribution of $r(\boldsymbol{x})$, which is, in general, not possible. However, a simple asymptotic argument (see, for example, Cox and Hinkley, 1974, Section 9.3) reveals that, for suitable regularity conditions, under the null hypothesis, as $n \to \infty$, $\lambda(\boldsymbol{x}) = -2 \log r(\boldsymbol{x})$ has a limiting χ^2_k distribution.

The procedure is easily extended to the case of a composite null hypothesis $\boldsymbol{\theta} \in \Theta_0 \subseteq \Re^k$. If the alternative hypothesis is $\boldsymbol{\theta} \in \Theta_1$, and we define $\Theta = \Theta_0 \cup \Theta_1$, we consider the ratio

$$r(\boldsymbol{x}) = \frac{\sup_{\boldsymbol{\theta} \in \Theta_0} p(\boldsymbol{x} \mid \boldsymbol{\theta})}{\sup_{\boldsymbol{\theta} \in \Theta} \; p(\boldsymbol{x} \mid \boldsymbol{\theta})}.$$

In this case, asymptotic analysis reveals that, for suitable regularity conditions, $\lambda(\boldsymbol{x}) = -2 \log r(x)$ has a limiting χ^2_d distribution, where d is the difference in dimensionality, $\dim(\Theta) - \dim(\Theta_0)$, of the general and null hypothesis parameter spaces, respectively.

It is interesting to compare this with a widely used Bayesian form of assessment of null and alternative models. Schwarz (1978) shows that, asymptotically,

$$-2 \log B_{01} = \lambda(x) - d \log n,$$

where B_{01} is the Bayes factor (Section 6.1.4).

We see, therefore, that the so-called *Schwarz criterion* for model choice adjusts the $-2 \log r(\boldsymbol{x})$ factor by a $\log n$ multiple of the dimensionality difference.

An earlier proposal for adjusting the general likelihood ratio criterion is that of Akaike (1973, 1974, 1978b, 1987), whose so-called *Akaike Information Criterion* (AIC) takes the form

$$\text{AIC} = \lambda(\boldsymbol{x}) - 2d.$$

See also, Akaike (1978b, 1979) for a Bayesian extension (BIC) of the AIC procedure.

Yet another variant is found in Nelder and Wedderburn (1972), whose suggestion for goodness-of-fit comparisons of general linear models through plotting degrees of freedom against deviance is, in effect, the criterion

$$\lambda(\boldsymbol{x}) - d.$$

These and other related proposals are reviewed from a Bayesian perspective in Smith and Spiegelhalter (1980). See Stone (1977, 1979a) for further discussion and comparison.

Roughly speaking, the Akaike criterion can be derived from a Bayes factor perspective as corresponding to a prior which concentrates on a neighbourhood of the alternative which is close, in an appropriate sense depending on n, to the null. The Schwarz criterion is derived from a Bayes factor perspective through a prior which does not depend on n.

Finally, we note that the prequential theory of Dawid (1984)—see, also, Section B4.3—directly embraces the view that models are simply predictive tools and should be compared on that basis, but does not necessarily use a Bayesian mechanism for such prediction. In Dawid (1992), it is shown that a particular form of so-called prequential assessment, based on the logarithmic scoring rule, leads to a model choice criterion which is asymptotically equivalent to the Schwarz criterion. It is also shown that this approach is essentially equivalent to model choice procedures arising in the stochastic complexity theory of Rissanen (1987).

References

Abramowitz, M. and Stegun, I. A. (1964). *Handbook of Mathematical Functions*. New York: Dover.

Achcar, J. A. and Smith, A. F. M. (1989). Aspects of reparameterisation in approximate Bayesian inference. *Bayesian and Likelihood Methods in Statistics and Econometrics: Essays in Honor of George A. Barnard* (S. Geisser, J. S. Hodges, S. J. Press and A. Zellner, eds.). Amsterdam: North-Holland, 439–452.

Aczel, J. and Pfanzagl, J. (1966). Remarks on the measurement of subjective probability and information. *Metrika* **11**, 91–105.

Aitchison, J. (1964). Bayesian tolerance regions. *J. Roy. Statist. Soc. B* **26**, 161–175.

Aitchison, J. (1966). Expected cover and linear utility tolerance intervals. *J. Roy. Statist. Soc. B* **28**, 57–62.

Aitchison, J. (1968). In discussion of Dempster (1968). *J. Roy. Statist. Soc. B* **30**, 234–237.

Aitchison, J. (1970). *Choice against chance. An Introduction to Statistical Decision Theory*. Reading, MA: Addison-Wesley.

Aitchison, J. and Dunsmore, I. R. (1975). *Statistical Prediction Analysis*. Cambridge: University Press.

Aitken, C. G. G. and Stoney, D. A. (1991). *The Use of Statistics in Forensic Science*. Chichester: Ellis Horwood.

Aitkin, M. (1991). Posterior Bayes factors. *J. Roy. Statist. Soc. B* **53**, 111–142 (with discussion).

Akaike, H. (1973). Information theory and an extension of the maximum likelihood principle. *2nd. Int. Symp. Information Theory*. Budapest: Akademia Kaido, 267-281.

Akaike, H. (1974). A new look at the statistical model identification. *IEEE Trans. Automatic Control* **19**, 716–727.

Akaike, H. (1978a). A new look at the Bayes procedure. *Biometrika* **65**, 53–59.

Akaike, H. (1978b). A Bayesian analysis of the minimum AIC procedure. *Ann. Inst. Statist. Math.* **30**, 9–14.

Akaike, H. (1979). A Bayesian extension of the minimum AIC procedure of autoregressive model fifting. *Biometrika* **66**, 53-59.

Akaike, H. (1980a). The interpretation of improper prior distributions as limits of data dependent proper prior distributions. *J. Roy. Statist. Soc. B* **45**, 46–52.

Akaike, H. (1980b). Likelihood and the Bayes procedure. *Bayesian Statistics* (J. M. Bernardo, M. H. DeGroot, D. V. Lindley and A. F. M. Smith, eds.). Valencia: University Press, 144–166 and 185–203 (with discussion).

Akaike, H. (1987). Factor analysis and the AIC. *Psychometrika* **52**, 317–332.

Albert, J. H. (1989). Nuisance parameters and the use of exploratory graphical methods in Bayesian analysis. *Amer. Statist.* **43**, 191–196.

Albert, J. H. (1990). Algorithms for Bayesian computing using Mathematica. *Computing Science and Statistics: Proceedings of the Symposium on the Interface* (C. Page and R. LePage eds.). Berlin: Springer, 286–290.

Albert, J. H. (1993). Teaching Bayesian statistics using sampling methods and MINITAB. *Amer. Statist.* **47**, 182-191.

Aldous, D. (1985). *Exchangeability and Related Topics*. Berlin: Springer.

Allais, M. (1953). Le comportement de l'homme rational devant le risque: Critique des postulats et axiomes de l'école Amëricaine. *Econometrica* **21**, 503–546.

Allais, M. and Hagen, D. (1979). *Expected Utility Hypotheses and the Allais Paradox*. Dordrecht: Reidel.

Amaral-Turkman, M. A. and Dunsmore, I. R. (1985). Measures of information in the predictive distribution. *Bayesian Statistics 2* (J. M. Bernardo, M. H. DeGroot, D. V. Lindley and A. F. M. Smith, eds.), Amsterdam: North-Holland, 603–612.

Ameen, J. R. M. (1992). Non linear prediction models. *J. Forecasting* **11**, 309–324.

Amster, S. J. (1963). A modified Bayes stopping rule. *Ann. Math. Statist.* **34**, 1404-1413.

Andersen, E. B. (1970). Asymptotic properties of conditional maximum-likelihood estimators. *J. Roy. Statist. Soc. B* **32**, 283–301.

Andersen, E. B. (1973). *Conditional Inference and Models for Measuring*. Copenhagen: Mental Hygiejnisk Foslay.

Anderson, T. W. (1984). *An Introduction to Multivariate Statistical Analysis*. New York: Wiley.

Angers, J.-F. and Berger, J. O. (1991). Robust hierarchical Bayes estimation of exchangeable means. *Canadian J. Statist.* **19**, 39–56.

Anscombe, F. J. (1961). Bayesian statistics. *Amer. Statist.* **15**, 21–24.

Anscombe, F. J. (1963). Sequential medical trials. *J. Amer. Statist. Assoc.* **58**, 365–383.

Anscombe, F. J. (1964a). Some remarks on Bayesian statistics. *Human Judgement and Optimality* (Shelly and Bryan, eds.). New York: Wiley, 155–177.

Anscombe, F. J. (1964b). Normal likelihood functions. *Ann. Inst. Statist. Math.* **16**, 1–41.

Anscombe, F. J. and Aumann, R. J. (1963). A definition of subjective probability. *Ann. Math. Statist.* **34**, 199–205.

Ansley, C. F., Kohn, R. and Wong, C.-M. (1993). Non-parametric spline regression with prior information. *Biometrika* **80**, 75–88.

Antoniak, C. (1974). Mixtures of Dirichlet processes with applications to Bayesian nonparametric problems. *Ann. Statist.* **2**, 1152–1174.

Aoki, M. (1967). *Optimization of Stochastic Systems.* New York: Academic Press.

Arimoto, S. (1970). Bayesian decision rule and quantity of equivocation. *Systems, Computers, Controls* **1**, 17–23.

Arnaiz, G. and Ruíz-Rivas, C. (1986). Outliers in circular data, a Bayesian approach. *Qüestiió* **10**, 1–6.

Arnold, S. F. (1993). Gibbs sampling. *Handbook of Statistics* **9**. *Computational Statistics* (C. R. Rao, ed.). Amsterdam: North-Holland, 599–625.

Arrow, K. J. (1951a). Alternative approaches to theory of choice in risk-taking situations. *Econometrica* **19**, 404–437.

Arrow, K. J. (1951b). *Social Choice and Individual Values.* New York: Wiley

Arrow, K. J. and Raynaud, H. (1987). *Social Choice and Multicriteria Decision Making.* Cambridge, MA: The MIT Press

Ash, R. B. (1972). *Real Analysis and Probability.* New York: Academic Press.

Aumann, R. J. (1987). Correlated equilibrium as an expression of Bayesian rationality. *Econometrica* **55**, 1–18.

Aykaç, A. and Brumat, C. (eds.) (1977). *New Developments in the Applications of Bayesian Methods.* Amsterdam: North-Holland.

Bahadur, R. R. (1954). Sufficiency and statistical decision functions. *Ann. Math. Statist.* **25**, 423–462.

Bailey, R. W. (1992). Distributional identities of Beta and chi-squared variates:a geometrical interpretation. *Amer. Statist.* **46**, 117–120.

Balch, M. and Fishburn, P. C. (1974). Subjective expected utility for conditional primitives. *Essays on Economic Behaviour under Uncertainty* (M. Balch, D. McFadden and S. Wu, eds.). Amsterdam: North-Holland, 45–54.

Bandemer, H. (1977). *Theorie und Anwendung der Optimalen Versuchsplanung.* Berlin: Akademie-Verlag.

Barlow, R. E. (1989). Influence diagrams. *Encyclopedia of Statistical Sciences* **Suppl.** (S. Kotz, N. L. Johnson and C. B. Read, eds.). New York: Wiley, 72–74.

Barlow, R. E. (1991). Introduction to de Finetti (1937). *Breakthroughs in Statistics* **1** (S. Kotz and N. L. Johnson, eds.). Berlin: Springer, 125–133.

Barlow, R. E. and Irony, T. Z. (1992). Foundations of statistical quality control. *Current Issues in Statistical Inference: Essays in Honor of D. Basu.* (M. Ghosh and P. K. Pathak eds.). Hayward, CA: IMS, 99–112.

Barlow, R. E. and Mendel, M. B. (1992). De Finetti-type representations for lifetime distributions. *J. Amer. Statist. Assoc.* **87**, 1116-1123.

Barlow, R. E. and Mendel, M. B. (1994). The operational Bayesian approach. *Aspects of Uncertainty: a Tribute to D. V. Lindley* (P. R. Freeman, and A. F. M. Smith, eds.). Chichester: Wiley, 19–28.

Barlow, R. E., Wechsler, S. and Spizzichino, F. (1988). De Finetti's approach to group decision making. *Bayesian Statistics 3* (J. M. Bernardo, M. H. DeGroot, D. V. Lindley and A. F. M. Smith, eds.). Oxford: University Press, 1–15 (with discussion).

Barnard, G. A. (1949). Statistical inference. *J. Roy. Statist. Soc. B* **11**, 115–149 (with discussion).

Barnard, G. A. (1951). The theory of information. *J. Roy. Statist. Soc. B* **13**, 46–64.

Barnard, G. A. (1952). The frequency justification of certain sequential tests. *Biometrika* **39**, 155–150.

Barnard, G. A. (1958). Thomas Bayes, a biographical note. *Biometrika* **45**, 293–295.

Barnard, G. A. (1963). Some aspects of the fiducial argument. *J. Roy. Statist. Soc. B* **25**, 111-114.

Barnard, G. A. (1967). The use of the likelihood function in statistical practice. *Proc. Fifth Berkeley Symp.* **1** (J. Neyman and E. L. Scott, eds.). Berkeley: Univ. California Press, 27–40.

Barnard, G. A. (1980a). In discussion of Box (1980). *J. Roy. Statist. Soc. A* **143**, 404–406.

Barnard, G. A. (1980b). Pivotal inference and the Bayesian controversy. *Bayesian Statistics* (J. M. Bernardo, M. H. DeGroot, D. V. Lindley and A. F. M. Smith, eds.). Valencia: University Press, 295–318 (with discussion).

Barnard, G. A., Jenkins, G. M. and Winsten, C. B. (1962). Likelihood inference and time series. *J. Roy. Statist. Soc. A* **125**, 321–372 (with discussion).

Barnard, G. A. and Sprott, D. A. (1968). Likelihood. *Encyclopedia of Statistical Sciences* **9** (S. Kotz, N. L. Johnson and C. B. Read, eds.). New York: Wiley, 639–644.

Barndorff-Nielsen, O. E. (1978). *Information and Exponential Families in Statistical Theory*. New York: Wiley.

Barndorff-Nielsen, O. E. (1980). Likelihood prediction. *Symposium Mathematica* **25**, 11–24.

Barndorff-Nielsen, O. E. (1983). On a formula for the distribution of the maximum likelihood estimator. *Biometrika* **70**, 343–365.

Barndorff-Nielsen, O. E. (1991). Likelihood theory. *Statistical Theory and Modelling. In Honour of Sir David Cox* (D. V. Hinkley, N. Reid and E. J. Snell, eds.). London: Chapman and Hall, 232–265.

Barnett, V. (1973/1982). *Comparative Statistical Inference*. Second edition in 1982, Chichester: Wiley.

Barrai, I., Coletti, G. and Di Bacco, M. (eds.) (1992). *Probability and Bayesian Statistics in Medicine and Biology.* Pisa: Giardini.

Bartholomew, D. J. (1965). A comparison of some Bayesian and frequentist inferences. *Biometrika* **52**, 19–35.

Bartholomew, D. J. (1967). Hypothesis testing when the sample size is treated as a random variable. *J. Roy. Statist. Soc. B* **29**, 53–82.

Bartholomew, D. J. (1971). A comparison of Bayesian and frequentist approaches to inferences with prior knowledge. *Foundations of Statistical Inference* (V. P. Godambe and D. A. Sprott, eds.). Toronto: Holt, Rinehart and Winston, 417–434 (with discussion).

Bartholomew, D. J. (1994). Bayes theorem in latent variable modelling. *Aspects of Uncertainty: a Tribute to D. V. Lindley* (P. R. Freeman, and A. F. M. Smith, eds.). Chichester: Wiley, 41–50.

Bartlett, M. (1957). A comment on D. V. Lindley's statistical paradox. *Biometrika* **44**, 533–534.

Basu, D. (1959). The family of ancillary statistics. *Sankhyā A* **21**, 247–256.

Basu, D. (1964). Recovery of ancillary information, *Sankhyā A* **26**, 3–16.

Basu, D. (1969). Role of sufficiency and likelihood principles in survey sampling theory. *Sankhyā A* **31**, 441–454.

Basu, D. (1971). An essay on the logical foundations of survey sampling. *Foundations of Statistical Inference* (V. P. Godambe and D. A. Sprott, eds.). Toronto: Holt, Rinehart and Winston, 203–242 (with discussion).

Basu, D. (1975). Statistical information and likelihood. *Sankhyā A* **37**, 1–71 (with discussion).

Basu, D. (1977). On the elimination of nuisance parameters. *J. Amer. Statist. Assoc.* **72**, 355-366.

Basu, D. (1988). *Statistical Information and Likelihood: a Collection of Critical Essays* (J. K. Ghosh, ed.). Berlin: Springer.

Basu, D. (1992). Learning statistics from counter examples: ancillary statistics. *Bayesian Analysis in Statistics and Econometrics* (P. K. Goel and N. S. Iyengar, eds.). Berlin: Springer, 217–224.

Basu, D. and Pereira, C. (1983). A note on Blackwell sufficiency and a Skibinsky characterization of distributions. *Sankhyā A* **45**, 99–104.

Bauwens, L. (1984). *Bayesian Full Information Analysis of Simultaneous Equation Models Using Integration by Monte Carlo*. Berlin: Springer

Bayarri, M. J. (1981). Inferencia Bayesiana sobre el coeficiente de correlación de una población normal bivariante. *Trab. Estadist.* **32**, 18–31.

Bayarri, M. J. and Berger, J. O. (1994). Applications and limitations of robust Bayesian bounds and type II MLE. *Statistical Decision Theory and Related Topics V* (S. S. Gupta and J. O. Berger, eds.). Berlin: Springer, 121–134.

Bayarri, M. J. and DeGroot, M. H. (1987). Bayesian analysis of selection models. *The Statistician* **36**, 137–146.

Bayarri, M. J. and DeGroot, M. H. (1988). Gaining weight: a Bayesian approach. *Bayesian Statistics 3* (J. M. Bernardo, M. H. DeGroot, D. V. Lindley and A. F. M. Smith, eds.). Oxford: University Press, 25–44 (with discussion).

Bayarri, M. J. and DeGroot, M. H. (1989). Optimal reporting of predictions. *J. Amer. Statist. Assoc.* **84**, 214–222.

Bayarri, M. J. and DeGroot, M. H. (1990). Selection models and selection mechanisms. *Bayesian and Likelihood Methods in Statistics and Econometrics: Essays in Honor of George A. Barnard* (S. Geisser, J. S. Hodges, S. J. Press and A. Zellner, eds.). Amsterdam: North-Holland, 211–227.

Bayarri, M. J. and DeGroot, M. H. (1991). What Bayesians expect of each other. *J. Amer. Statist. Assoc.* **86**, 924–932.

Bayarri, M. J. and DeGroot, M. H. (1992a). A 'BAD' view of weighted distributions and selection models. *Bayesian Statistics 4* (J. M. Bernardo, J. O. Berger, A. P. Dawid and A. F. M. Smith, eds.). Oxford: University Press, 17–33 (with discussion).

Bayarri, M. J. and DeGroot, M. H. (1992b). Difficulties and ambiguities in the definition of a likelihood function. *J. It. Statist. Soc.* **1**, 1–15.

Bayarri, M. J., DeGroot, M. H. and Kadane, J. B. (1988). What is the likelihood function? *Statistical Decision Theory and Related Topics IV* **1** (S. S. Gupta and J. O. Berger, eds.). Berlin: Springer, 3–27.

Bayes, T. (1763). An essay towards solving a problem in the doctrine of chances. Published posthumously in *Phil. Trans. Roy. Soc. London* **53**, 370–418 and **54**, 296–325. Reprinted in *Biometrika* **45** (1958), 293–315, with a biographical note by G. A. Barnard. Reproduced in Press (1989), 185–217.

Becker, G. M., DeGroot, M. H. and Marschak, J. (1963). Stochastic models of choice behavior. *Behavioral Sci.* **8**, 41–55. Reprinted in *Decision Making* (W. L. Edwards and A. Tversky, eds.) Baltimore: Penguin.

Becker, G. M. and McClintock, C. G. (1967). Value: Behavioral decision theory. *Annual Rev. Psychology* **18**, 239–286.

Bellman, R. E. (1957). *Dynamic Programming*. Princeton: University Press.

Berger, J. O. (1979). Multivariate estimation with nonsymmetric loss functions. *Optimizing Methods in Statistics* (J. S. Rustagi, ed.). New York: Academic Press.

Berger, J. O. (1980). A robust generalized Bayes estimator and confidence region for a multivariate normal mean. *Ann. Statist.* **8**, 716–761.

Berger, J. O. (1982). Bayesian robustness and the Stein effect. *J. Amer. Statist. Assoc.* **77**, 358–368.

Berger, J. O. (1984a). The robust Bayesian viewpoint. *Robustness of Bayesian Analysis* (J. B. Kadane, ed.). Amsterdam: North-Holland, 63-144 (with discussion).

Berger, J. O. (1984b). The frequentist viewpoint and conditioning. *Proc. Berkeley Symp. in Honor of Kiefer and Neyman* (L. LeCam and R. Olshen, eds.). Pacific Drove, CA: Wadsworth.

Berger, J. O. (1985a). *Statistical Decision Theory and Bayesian Analysis.* Berlin: Springer.

Berger, J. O. (1985b). In defense of the likelihood principle: axiomatics and coherence. *Bayesian Statistics 2* (J. M. Bernardo, M. H. DeGroot, D. V. Lindley and A. F. M. Smith, eds.), Amsterdam: North-Holland, 33–65, (with discussion).

Berger, J. O. (1986). Bayesian salesmanship. *Bayesian Inference and Decision Techniques: Essays in Honor of Bruno de Finetti* (P. K. Goel and A. Zellner, eds.). Amsterdam: North-Holland, 473–488.

Berger, J. O. (1990). Robust Bayesian analysis: sensitivity to the prior. *J. Statist. Planning and Inference* **25**, 303–328.

Berger, J. O. (1993). The present and future of Bayesian multivariate analysis. *Multivariate Analysis: Future Directions* (C. R. Rao, ed.). Amsterdam: North-Holland, 25–53.

Berger, J. O. (1994). A review of recent developments in robust Bayesian analysis. *Test* **3**, (to appear, with discussion).

Berger, J. O. and Berliner, L. M. (1986). Robust Bayes and empirical Bayes analysis with ϵ-contaminated priors. *Ann. Statist.* **14**, 461–486.

Berger, J. O. and Bernardo, J. M. (1989). Estimating a product of means: Bayesian analysis with reference priors. *J. Amer. Statist. Assoc.* **84**, 200–207.

Berger, J. O. and Bernardo, J. M. (1992a). Ordered group reference priors with applications to a multinomial problem. *Biometrika* **79**, 25–37.

Berger, J. O. and Bernardo, J. M. (1992b). Reference priors in a variance components problem. *Bayesian Analysis in Statistics and Econometrics* (P. K. Goel and N. S. Iyengar, eds.). Berlin: Springer, 323–340.

Berger, J. O. and Bernardo, J. M. (1992c). On the development of reference priors. *Bayesian Statistics 4* (J. M. Bernardo, J. O. Berger, A. P. Dawid and A. F. M. Smith, eds.). Oxford: University Press, 35–60 (with discussion).

Berger, J. O., Bernardo, J. M. and Mendoza, M. (1989). On priors that maximize expected information. *Recent Developments in Statistics and their Applications* (J. P. Klein and J. C. Lee, eds.). Seoul: Freedom Academy, 1–20.

Berger, J. O. and Berry, D. A. (1988). The relevance of stopping rules in statistical inference. *Statistical Decision Theory and Related Topics IV* **1** (S. S. Gupta and J. O. Berger, eds.). Berlin: Springer, 29–72 (with discussion).

Berger, J. O. and DasGupta, A. (1991). *Multivariate Estimation, Bayes, Empirical Bayes and Stein Approaches*. Philadelphia, PA: SIAM.

Berger, J. O. and Delampady, M. (1987). Testing precise hypotheses. *Statist. Sci.* **2**, 317–352 (with discussion).

Berger, J. O. and Fan, T. H. (1991). Behaviour of the posterior distribution and inferences for a normal mean with t prior distributions. *Statistics and Decisions* **10**, 99–120.

Berger, J. O. and Jefferys, W. H. (1992) The application of robust Bayesian analysis to hypothesis testing and Occams razor. *J. It. Statist. Soc.* **1**, 17–32.

Berger, J. O. and Mortera, J. (1991a). Interpreting the stars in precise hypothesis testing. *Internat. Statist. Rev.* **59**, 337–353.

Berger, J. O. and Mortera, J. (1991b). Bayesian analysis with limited communication. *J. Statist. Planning and Inference* **28**, 1–24.

Berger, J. O. and Mortera, J. (1994). Robust Bayesian hypothesis testing in the presence of nuisance parameters. *J. Statist. Planning and Inference* **31**, 357–373.

Berger, J. O. and O'Hagan A. (1988). Ranges of posterior probabilities of unimodal priors with specified quantiles. *Bayesian Statistics 3* (J. M. Bernardo, M. H. DeGroot, D. V. Lindley and A. F. M. Smith, eds.). Oxford: University Press, 45–65 (with discussion).

Berger, J. O. and Robert, C. P. (1990). Subjective hierarchical Bayes estimation of a multivariate normal mean: on the frequentist interface. *Ann. Statist.* **18**, 617–651.

Berger, J. O. and Sellke, T. (1987). Testing a point null hypothesis: the irreconcilability of significance levels and evidence. *J. Amer. Statist. Assoc.* **82**, 112–133 (with discussion).

Berger, J. O. and Srinivasan, C. (1978). Generalized Bayes estimators in multivariate problems. *Ann. Statist.* **6**, 783–801.

Berger, J. O. and Wolpert, R. L. (1984/1988). *The Likelihood Principle*. Second edition in 1988, Hayward, CA: IMS.

Berger, R. L. (1981). A necessary and sufficient condition for reaching a consensus using DeGroot's method. *J. Amer. Statist. Assoc.* **76**, 415–418.

Berk, R. H. (1966). Limiting behaviour of the posterior distributions when the model is incorrect. *Ann. Math. Statist.* **37**, 51–58.

Berk, R. H. (1970). Consistency a posteriori. *Ann. Math. Statist.* **41**, 894–906.

Berliner, L. M. (1987). Bayesian control in mixture models. *Technometrics* **29**, 455–460.

Berliner, L. M. and Goel P. K. (1990). Incorporating partial prior information: ranges of posterior probabilities. *Bayesian and Likelihood Methods in Statistics and Econometrics: Essays in Honor of George A. Barnard* (S. Geisser, J. S. Hodges, S. J. Press and A. Zellner, eds.). Amsterdam: North-Holland, 397–406.

Berliner, L. M. and Hill, B. M. (1988). Bayesian non-parametric survival analysis. *J. Amer. Statist. Assoc.* **83**, 772–782 (with discussion).

Bermúdez. J. D. (1985). On the asymptotic normality of the posterior distribution of the logistic classification model. *Statistics and Decisions* **2**, 301–308.

Bernardo, J. M. (1977). Inferences about the ratio of normal means: a Bayesian approach to the Fieller-Creasy problem. *Recent Developments in Statistics* (J. R. Barra *et al.* eds.). Amsterdam: North-Holland, 345–349.

Bernardo, J. M. (1978a). Una medida de la información útil proporcionada por un experimento. *Rev. Acad. Ciencias Madrid* **72**, 419–440.

Bernardo, J. M. (1978b). Unacceptable implications of the left Haar measure in a standard normal theory inference problem *Trab. Estadist.* **29**, 3–9.

Bernardo, J. M. (1979a). Expected information as expected utility. *Ann. Statist.* **7**, 686–690.

Bernardo, J. M. (1979b). Reference posterior distributions for Bayesian inference. *J. Roy. Statist. Soc. B* **41**, 113–147 (with discussion).

Bernardo, J. M. (1980). A Bayesian analysis of classical hypothesis testing. *Bayesian Statistics* (J. M. Bernardo, M. H. DeGroot, D. V. Lindley and A. F. M. Smith, eds.). Valencia: University Press, 605–647 (with discussion).

Bernardo, J. M. (1981a). Reference decisions. *Symposia Mathematica* **25**, 85–94.

Bernardo, J. M. (1981b). *Bioestadística, una Perspectiva Bayesiana.* Barcelona: Vicens-Vives.

Bernardo, J. M. (1982). Contraste de modelos probabilísticos desde una perspectiva Bayesiana. *Trab. Estadist.* **33**, 16–30.

Bernardo, J. M. (1984). Monitoring the 1982 Spanish socialist victory: a Bayesian analysis. *J. Amer. Statist. Assoc.* **79**, 510–515.

Bernardo, J. M. (1985a). Análisis Bayesiano de los contrastes de hipótesis paramétricos. *Trab. Estadist.* **36**, 45–54.

Bernardo, J. M. (1985b). On a famous problem of induction. *Trab. Estadist.* **36**, 24–30.

Bernardo, J. M. (1988). Bayesian linear probabilistic classification. *Statistical Decision Theory and Related Topics IV* **1** (S. S. Gupta and J. O. Berger, eds.). Berlin: Springer, 151–162.

Bernardo, J. M. (1989). Análisis de datos y métodos Bayesianos. *Historia de la Ciencia Estadística* (S. Ríos, ed.). Madrid: Academia de Ciencias, 87–105.

Bernardo, J. M. (1994). Optimal prediction with hierarchical models: Bayesian clustering. *Aspects of Uncertainty: a Tribute to D. V. Lindley* (P. R. Freeman, and A. F. M. Smith, eds.). Chichester: Wiley, 67–76.

Bernardo, J. M. and Bayarri, M. J. (1985). Bayesian model criticism. *Model Choice* (J.-P. Florens, M. Mouchart, J.-P. Raoult and L. Simar, eds.). Brussels: Pub. Fac. Univ. Saint Louis, 43–59.

Bernardo, J. M., Berger, J. O., Dawid, A. P. and Smith, A. F. M. (eds.) (1992). *Bayesian Statistics 4*. Oxford: University Press.

Bernardo, J. M. and Bermúdez, J. D. (1985). The choice of variables in probabilistic classification. *Bayesian Statistics 2* (J. M. Bernardo, M. H. DeGroot, D. V. Lindley and A. F. M. Smith, eds.), Amsterdam: North-Holland, 67–81 (with discussion).

Bernardo, J. M., DeGroot, M. H., Lindley, D. V. and Smith, A. F. M. (eds.) (1980). *Bayesian Statistics*. Valencia: University Press.

Bernardo, J. M., DeGroot, M. H., Lindley, D. V. and Smith, A. F. M. (eds.) (1985). *Bayesian Statistics 2*. Amsterdam: North-Holland.

Bernardo, J. M., DeGroot, M. H., Lindley, D. V. and Smith, A. F. M. (eds.) (1988). *Bayesian Statistics 3*. Oxford: University Press.

Bernardo, J. M., Ferrándiz, J. R. and Smith, A. F. M. (1985). The foundations of decision theory: an intuitive, operational approach with mathematical extensions. *Theory and Decision* **18**, 127–150.

Bernardo, J. M. and Girón F. J. (1988). A Bayesian analysis of simple mixture problems. *Bayesian Statistics 3* (J. M. Bernardo, M. H. DeGroot, D. V. Lindley and A. F. M. Smith, eds.). Oxford: University Press, 67–88 (with discussion).

Bernardo, J. M. and Girón F. J. (1989). A Bayesian approach to cluster analysis. *Qüestiió* **5**, 97–112.

Bernoulli, D. (1730/1954). Specimen theoriae novae de mensura sortis. *Comment. Acad. Sci. Imp. Petropolitanae* **5**, 175–192. English translation as "Exposition of a new theory on the measurement of risk" in 1954, *Econometrica* **22**, 23–26.

Bernoulli, J. (1713/1899). *Ars Conjectandi*. Basel: Thurnisiorum. Translated into German as *Wahrscheinlichkeitsrechnung*. Leipzig: Engelmann. 1899.

Berry, D. A. (1996). *Statistics, A Bayesian perspective*. Belmont, CA: Duxbury.

Berry, D. A. and Stangl, D. K. (eds.) (1996). *Bayesian Biostatistics*. New York: Marcel Dekker.

Besag, J. (1986). Statistical analysis of dirty pictures. *J. Roy. Statist. Soc. B* **48**, 259–302 (with discussion).

Besag, J. (1989). Towards Bayesian image analysis. *J. Appl. Statist.* **16**, 395–407.

Besag, J. and Green, P. J. (1993). Spatial statistics and Bayesian computation. *J. Roy. Statist. Soc. B* **55**, 25–37.

Bickel, P. J. and Blackwell, D. (1967). A note on Bayes estimates. *Ann. Math. Statist.* **38**, 1907–1911.

Bickel, P. J. and Ghosh, J. K. (1990). A decomposition for the likelihood ratio statistic and the Bartlett correction—a Bayesian argument. *Ann. Statist.* **18**, 1070–1090.

Binder, S. (1978). Bayesian cluster analysis. *Biometrika* **65**, 31–38.

Birnbaum, A. (1962). On the foundations of statistical inference. *J. Amer. Statist. Assoc.* **57**, 269–306.

Birnbaum, A. (1968). Likelihood. *Internat. Encyclopedia of the Social Sciences* **9**, 299-301.

Birnbaum, A. (1969). Concepts of statistical evidence. *Philosophy Science and Methods*. (S. Morgenbesso, P. Suppes and M. White eds.) New York: St. John's Press.

Birnbaum, A. (1972). More on concepts of statistical evidence. *J. Amer. Statist. Assoc.* **67**, 858–861.

Birnbaum, A. (1978). Likelihood. *International Encyclopedia of Statistics* (W. H. Kruskal and J. M. Tanur, eds.). London: Macmillan, 519–522.

Bjrnstad, J. F. (1990). Predictive likelihood: a review. *Statist. Sci.* **5**, 242–265 (with discussion).

Blackwell, D. (1947). Conditional expectation and unbiased sequential estimation. *Ann. Math. Statist.* **18**, 105–110.

Blackwell, D. (1951). Comparison of experiments. *Proc. Second Berkeley Symp.* (J. Neyman ed.). Berkeley: Univ. California Press 93–102,

Blackwell, D. (1953). Equivalent comparison of experiments. *Ann. Math. Statist.* **24**, 265–272.

Blackwell, D. (1988). In discussion of Diaconis (1988). *Bayesian Statistics 3* (J. M. Bernardo, M. H. DeGroot, D. V. Lindley and A. F. M. Smith, eds.). Oxford: University Press, 123-124.

Blackwell, D. and Dubins, L. E. (1962). Merging of opinions with increasing information. *Ann. Math. Statist.* **33**, 882–886.

Blackwell, D. and Girshick, M. A. (1954). *Theory of Games and Statistical Decisions*. New York: Wiley.

Blyth, C. R. (1972). On Simpson's paradox and the sure-thing principle. *J. Amer. Statist. Assoc.* **67**, 364–366.

Blyth, C. R. (1973). Simpson's paradox and mutually favourable events. *J. Amer. Statist. Assoc.* **68**, 746.

Booth, N. B. and Smith, A. F. M. (1976). Batch acceptance schemes based on an autogressive prior. *Biometrika* **63**, 133–136.

Bordley, R. F. (1992). An intransitive expectations based Bayesian variant of prospect theory. *J. Risk and Uncertainty* **5**, 127–144.

Borel, E. (1924/1964). A propos d'un traité de probabilités. *Revue Philosophique* **98**, 321–336. Reprinted in 1980 as "A propos of a treatise on probability" in *Studies in Subjective Probability* (H. E. Kyburg and H. E Smokler, eds.). New York: Dover, 45–60.

Borovcnik, M. (1992). *Stochastik im Wechselspiels von Intuitionen und Mathematik*. Mannheim: BI-Wissenschaftsverlag.

Box, G. E. P. (1980). Sampling and Bayes' inference in scientific modelling. *J. Roy. Statist. Soc. A* **143**, 383–430 (with discussion).

Box, G. E. P. (1983). An apology for ecumenism in statistics. *Science* **151**, 15–84.

Box, G. E. P. (1985). *The Collected Works of G. E. P. Box* (G. C. Tiao, ed.). Pacific Drove, CA: Wadsworth.

Box, G. E. P. and Cox, D. R. (1964). An analysis of transformations. *J. Roy. Statist. Soc. B* **26**, 211–252 (with discussion).

Box, G. E. P. and Hill, W. J. (1967). Discrimination among mechanistic models. *Technometrics* **9**, 57–71.

Box, G. E. P., Leonard, T. and Wu, C.-F. (eds.) (1983). *Scientific Inference, Data Analysis and Robustness*. New York: Academic Press.

Box, G. E. P. and Tiao, G. C. (1962). A further look at robustness via Bayes' theorem. *Biometrika* **49**, 419–432.

Box, G. E. P. and Tiao, G. C. (1964). A note on criterion robustness and inference robustness. *Biometrika* **51**, 169–173.

Box, G. E. P. and Tiao, G. C. (1965). Multiparameter problems from a Bayesian point of view. *Ann. Math. Statist.* **36**, 1468–1482.

Box, G. E. P. and Tiao, G. C. (1968). A Bayesian approach to some outlier problems. *Biometrika* **55**, 119–129.

Box, G. E. P. and Tiao, G. C. (1973). *Bayesian Inference in Statistical Analysis*. Reading, MA: Addison-Wesley.

Boyer, M. and Kihlstrom, R. E. (eds.) (1984). *Bayesian Models in Economic Theory*. Amsterdam: North-Holland.

Breslow, N. (1990). Biostatistics and Bayes. *Statist. Sci.* **5**, 269–298 (with discussion).

Bretthorst, G. L. (1988). *Bayesian Spectrum Analysis and Parameter Estimation*. New York: Springer-Verlag.

Bridgman, P. W. (1927). *The Logic of Modern Physics*. London: Macmillan.

Brier, G. W. (1950). Verification of forecasts expressed in terms of probability. *Month. Weather Rev.* **78**, 1–3.

Brillinger, D. R. (1962). Examples bearing on the definition of fiducial probability, with a bibliography. *Ann. Math. Statist.* **33**, 1349–1355.

Broemeling, L. D. (1985). *Bayesian Analysis of Linear Models*. New York: Marcel Dekker.

Brown, L. D. (1973). Estimation with incompletely specified loss functions. *J. Amer. Statist. Assoc.* **70**, 417–427.

Brown, L. D. (1985). *Foundations of Exponential Families*. Hayward, CA: IMS.

Brown, P. J., Le, N. D. and Zidek, J. V. (1994). Inference for a covariance matrix. *Aspects of Uncertainty: a Tribute to D. V. Lindley* (P. R. Freeman, and A. F. M. Smith, eds.). Chichester: Wiley, 77–92.

Brown, P. J. and Mäkeläinen, T. (1992). Regression, sequential measurements and coherent calibration. *Bayesian Statistics 4* (J. M. Bernardo, J. O. Berger, A. P. Dawid and A. F. M. Smith, eds.). Oxford: University Press, 97–108 (with discussion).

Brown, R. V. (1993). Impersonal probability as an ideal assessment. *J. Risk and Uncertainty* **7**, 215–235.

Brown, R. V. and Lindley, D. V. (1982). Improving judgement by reconciling incoherence. *Theory and Decision* **14**, 113–132.

Brown, R. V. and Lindley, D. V. (1986). Plural analysis: multiple approach to quantitative research. *Theory and Decision* **20**, 133–154.

Brunk, H. D. (1991). Fully coherent inference. *Ann. Statist.* **19**, 830–849.

Buehler, R. J. (1959). Some validity criteria for statistical inference. *Ann. Math. Statist.* **30**, 845–863.

Buehler, R. J. (1971). Measuring information and uncertainty. *Foundations of Statistical Inference* (V. P. Godambe and D. A. Sprott, eds.). Toronto: Holt, Rinehart and Winston, 330–351 (with discussion).

Buehler, R. J. (1976). Coherent preferences. *Ann. Statist.* **4**, 1051–1064.

Buehler, R. J. and Feddersen, A. P. (1963). Note on a conditional property of Student's t. *Ann. Math. Statist.* **34**, 1098–1100.

Bunn, D. J. (1984). *Applied Decision Analysis*. New York: McGraw-Hill

Butler, R. W. (1986). Predictive likelihood inference with applications. *J. Roy. Statist. Soc. B* **48**, 1–38 (with discussion).

Cano, J. A., Hernández, A. and Moreno, E. (1988). On Kolmogorov's partial sufficiency. *Bayesian Statistics 3* (J. M. Bernardo, M. H. DeGroot, D. V. Lindley and A. F. M. Smith, eds.). Oxford: University Press, 553–556.

Carlin, B. P. and Gelfand, A. E. (1991). An iterative Monte Carlo method for nonconjugate Bayesian analysis. *Statist. Computing* **1**, 119-128.

Carlin, B. P. and Polson N. G. (1992). Monte Carlo Bayesian methods for discrete regression models and categorical time series. *Bayesian Statistics 4* (J. M. Bernardo, J. O. Berger, A. P. Dawid and A. F. M. Smith, eds.). Oxford: University Press, 577–586.

Carlin, J. B. and Dempster, A. P. (1989). Sensitivity analysis of seasonal adjustments: empirical case studies. *J. Amer. Statist. Assoc.* **84**, 6–32 (with discussion).

Carnap, R. (1950/1962). *Logical Foundations of Probability*. Chicago: University Press.

Caro, E., Domínguez, J. I., and Girón, F. J. (1984). Compatibilidad del método de DeGroot para llegar a un consenso en la fórmula de Bayes. *Trab. Estadist.* **35**, 139–153.

Casella, G. (1987). Conditionally acceptable recentered set estimators. *Ann. Statist.* **15**, 1364–1371.

Casella, G. (1992). Conditional inference for confidence sets. *Current Issues in Statistical Inference: Essays in Honor of D. Basu.* (M. Ghosh and P. K. Pathak eds.). Hayward, CA: IMS.

Casella, G. and Berger, R. L. (1987). Reconciling Bayesian and frequentist evidence in the one-sided testing problem. *J. Amer. Statist. Assoc.* **82**, 106–135, (with discussion).

Casella, G. and Berger, R. L. (1990). *Statistical Inference*. Pacific Drove, CA: Wadsworth.

Casella, G. and George, E. I. (1992). Explaining the Gibbs sampler. *Amer. Statist.* **46**, 167–174.

Casella, G., Hwang, J. T. G. and Robert, C. P. (1993). A paradox in decision theoretic interval estimation. *Statistica Sinica* **3**, 141–155.

Chaloner, K. (1984). Optimal Bayesian experimental design for linear models. *Ann. Statist.* **12**, 283–300.

Chaloner, K. (1994). Residual analysis and outliers in hierarchical models. *Aspects of Uncertainty: a Tribute to D. V. Lindley* (P. R. Freeman, and A. F. M. Smith, eds.). Chichester: Wiley, 149–157.

Chaloner, K. and Brant, R. (1988). A Bayesian approach to outlier detection and residual analysis. *Biometrika* **75**, 651–659.

Chan, K. S. (1993). Asymptotic behaviour of the Gibbs sampler. *J. Amer. Statist. Assoc.* **88**, 320–326.

Chang, T. and Eaves, D. M. (1990). Reference priors for the orbit of a group model. *Ann. Statist.* **18**, 1595–1614.

Chang, T. and Villegas, C. (1986). On a theorem of Stein relating Bayesian and classical inferences in group models. *Canadian J. Statist.* **14**, 289–296.

Chankong, V. and Haimes, Y. (1982). *Multiobjective Decision Making*. Amsterdam: North-Holland.

Chao, M. T. (1970). The asymptotic behavior of Bayes' estimators. *Manag. Sci.* **41**, 601–608.

Chateaneuf, A. and Jaffray, J. Y. (1984). Archimedean qualitative probabilities. *J. Math. Psychology* **28**, 191–204.

Chen, C. F. (1985). On asymptotic normality of limiting density functions with Bayesian implications. *J. Roy. Statist. Soc. B* **97**. 540–546.

Chernoff, H. (1951). A property of some type A regions. *Ann. Math. Statist.* **22**, 472–474.

Chernoff, H. (1959). Sequential design of experiments. *Ann. Math. Statist.* **30**, 755–770.

Chernoff, H. and Moses, L. E. (1959). *Elementary Decision Theory*. New York: Wiley.

Chow, Y. S. and Teicher, H. (1978/1988). *Probability Theory*. Berlin: Springer. Second edition in 1988. Berlin: Springer.

Chuaqui, R. and Malitz, J. (1983). Preorderings compatible with probability measures. *Trans. Amer. Math. Soc.* **279**, 811–824.

Cifarelli, D. M. (1987). Recent contributions to Bayesian statistics. *Italian Contributions to the Methodology of Statistics* (A. Naddeo, ed.). Padova: Cleub, 483–516.

Cifarelli, D. M. and Muliere, P. (1989). *Statistica Bayesiana*. Pavia: G. Iuculano.

Cifarelli, D. M. and Regazzini, E. (1982). Some considerations about mathematical statistics teaching methodology suggested by the concept of exchangeability. *Exchangeability in Probability and Statistics*. (G. Koch and F. Spizzichino, eds.). Amsterdam: North-Holland, 185–205.

Clarke, B. and Wasserman, L. (1993). Non-informative priors and nuisance parameters. *J. Amer. Statist. Assoc.* **88**, 1427–1432.

Clarke, R. D. (1954). The concept of probability. *J. Inst. Actuaries* **80**, 1–31 (with discussion).

Claroti, C. A. and Lindley, D. V. (eds.) (1988). *Accelerated Life testing and Expert Opinion in Reliability*. Amsterdam: North-Holland.

Clayton, M. K, Geisser, S. and Jennings, D. E. (1986). A comparison of several model selection procedures. *Bayesian Inference and Decision Techniques: Essays in Honor of Bruno de Finetti* (P. K. Goel and A. Zellner, eds.). Amsterdam: North-Holland, 425–442.

Clemen, R. T. (1989). Combining forecasts: a review and annotated bibliography. *Int. J. Forecasting* **5**, 559–583.

Clemen, R. T. (1990). Unanimity and compromise among probability forecasters. *Manag. Sci.* **36**, 767–779.

Clemen, R. T. and Winkler, R. L. (1987). Calibrating and combining precipitation probability forecasts. *Probability and Bayesian Statistics* (R. Viertl, ed.). London: Plenum, 97–110.

Clemen, R. T. and Winkler, R. L. (1993). Aggregating point estimates, a flexible modelling approach. *Manag. Sci.* **39**, 501–515.

Cochrane, J. L. and Zeleny, M. (eds.) (1973). *Multiple Criteria Decision Making*. Columbia, SC: University Press

Conlisk, J. (1993). The utility of gambling. *J. Risk and Uncertainty* **6**, 255–275.

Consonni, G. and Veronese, P. (1987). Coherent distributions and Lindley's paradox. *Probability and Bayesian Statistics* (R. Viertl, ed.). London: Plenum, 111–120.

Consonni, G. and Veronese, P. (1992a). Bayes factors for linear models and improper priors. *Bayesian Statistics 4* (J. M. Bernardo, J. O. Berger, A. P. Dawid and A. F. M. Smith, eds.). Oxford: University Press, 587–594.

Consonni, G. and Veronese, P. (1992b). Conjugate priors for exponential families having quadratic variance functions. *J. Amer. Statist. Assoc.* **87**, 1123–1127.

Cooke, R. M. (1991). *Experts in Uncertainty. Opinion and Subjective Probability in Science*. Oxford: University Press.

Cornfield, J. (1969). The Bayesian outlook and its applications. *Biometrics* **25**, 617–657.

Cowell, R. G. (1992). *BAIES*, a probabilistic expert system shell with qualitative and quantitative learning. *Bayesian Statistics 4* (J. M. Bernardo, J. O. Berger, A. P. Dawid and A. F. M. Smith, eds.). Oxford: University Press, 595–600.

Cox, D. D. (1993). An analysis of Bayesian inference for nonparametric regression. *Ann. Statist.* **21**, 903–923.

Cox, D. R. (1958). Some problems connected with statistical inference. *Ann. Math. Statist.* **29**, 357–372.

Cox, D. R. (1975). Partial likelihood. *Biometrika* **62**, 269–276.

Cox, D. R. (1988). Conditional and asymptotic inference. *Sankhyā A* **50**, 314–337.

Cox, D. R. (1990). Models in statistical analysis. *Statist. Sci.* **5**, 169–174.

Cox, D. R. and Hinkley, D. V. (1974). *Theoretical Statistics*. London: Chapman and Hall.

Cox, D. R. and Reid, N. (1987). Parameter orthogonality and approximate conditional inference. *J. Roy. Statist. Soc. B* **49**, 1–39 (with discussion).

Cox, D. R. and Reid, N. (1992). A note on the difference between profile and modified profile likelihood. *Biometrika* **79**, 408–411.

Cox R. T. (1946). Probability, frequency and expectation. *Amer. J. Physics* **14**, 1–13.

Cox R. T. (1961). *The Algebra of Probable Inference*. Baltimore: Johns Hopkins.

Crowder, M. J. (1988). Asymptotic expansions of posterior expectations, distributions and densities for stochastic processes. *Ann. Inst. Statist. Math.* **40**, 297–309.

Csiszár, I. (1985). An extended maximum entropy principle and a Bayesian justification. *Bayesian Statistics 2* (J. M. Bernardo, M. H. DeGroot, D. V. Lindley and A. F. M. Smith, eds.), Amsterdam: North-Holland, 83–98, (with discussion).

Cuevas, A. and Sanz, P. (1988). On differentiability properties of Bayes operators. *Bayesian Statistics 3* (J. M. Bernardo, M. H. DeGroot, D. V. Lindley and A. F. M. Smith, eds.). Oxford: University Press, 569–577.

Cyert, R. M. and DeGroot, M. H. (1987). *Bayesian Analysis and Uncertainty in Economic Theory*. London: Chapman and Hall.

Daboni, L. and Wedlin, A. (1982). *Statistica. Un'Introduzione all'Impostazione Neo-Bayesiana.* Torino: UTET.

Dalal, S. and Hall, W. J. (1980). On approximating parametric models by nonparametric Bayes models. *Ann. Statist.* **8**, 664–672.

Dalal, S. and Hall, W. J. (1983). Approximating priors by mixtures of natural conjugate priors. *J. Roy. Statist. Soc. B* **45**, 278–286.

Dale, A. I. (1990). Thomas Bayes: some clues to his education. *Statist. and Prob. Letters* **9**, 289–290.

Dale, A. I. (1991). *History of Inverse Probability: From Thomas Bayes to Karl Pearson.* Berlin: Springer.

Darmois, G. (1936). Sur les lois de probabilité à estimation exhaustive. *C. R. Acad. Sci. Paris* **200**, 1265–1266.

DasGupta, A. (1991). Diameter and volume minimizing confidence sets in Bayes and classical problems. *Ann. Statist.* **19**, 1225–1243.

DasGupta, A. and Studden, W. J. (1991). Robust Bayesian experimental designs in normal linear models. *Ann. Statist.* **19**, 1244–1256.

Davison, A. C. (1986). Approximate predictive likelihood. *Biometrika* **73**, 323–332.

Davison, D., Suppes, P. and Siegel, S. (1957). *Decision Making: An Experimental Approach.* Stanford: University Press.

Dawid, A. P. (1970). On the limiting normality of posterior distributions. *Proc. Camb. Phil. Soc.* **67**, 625–633.

Dawid, A. P. (1973). Posterior expectations for large observations. *Biometrika* **60**, 644–666.

Dawid, A. P. (1977). Invariant distributions and analysis of variance models. *Biometrika* **64**, 291–297.

Dawid, A. P. (1978). Extendibility of spherical matrix distributions. *J. Multivariate Analysis* **14**, 559–566.

Dawid, A. P. (1979a). Conditional independence in statistical theory. *J. Roy. Statist. Soc. B* **41**, 1–31, (with discussion).

Dawid, A. P. (1979b). Some misleading arguments involving conditional independence. *J. Roy. Statist. Soc. B* **41**, 249–252.

Dawid, A. P. (1980a). A Bayesian look at nuisance parameters. *Bayesian Statistics* (J. M. Bernardo, M. H. DeGroot, D. V. Lindley and A. F. M. Smith, eds.). Valencia: University Press, 167–203 (with discussion).

Dawid, A. P. (1980b). Conditional independence for statistical operators. *Ann. Statist.* **8**,, 598–617.

Dawid, A. P. (1982a). The well-calibrated Bayesian. *J. Amer. Statist. Assoc.* **77**, 605–613.

Dawid, A. P. (1982b). Intersubjective statistical models. *Exchangeability in Probability and Statistics*. (G. Koch and F. Spizzichino, eds.). Amsterdam: North-Holland, 217–232.

Dawid, A. P. (1983a). Statistical inference. *Encyclopedia of Statistical Sciences* **4** (S. Kotz, N. L. Johnson and C. B. Read, eds.). New York: Wiley, 89–105.

Dawid, A. P. (1983b). Invariant prior distributions. *Encyclopedia of Statistical Sciences* **4** (S. Kotz, N. L. Johnson and C. B. Read, eds.). New York: Wiley, 228–236.

Dawid, A. P. (1984). Statistical theory, the prequential approach. *J. Roy. Statist. Soc. A* **147**, 278–292.

Dawid, A. P. (1986a). Probability forecasting. *Encyclopedia of Statistical Sciences* **7** (S. Kotz, N. L. Johnson and C. B. Read, eds.). New York: Wiley, 210–218.

Dawid, A. P. (1986b). A Bayesian view of statistical modelling. *Bayesian Inference and Decision Techniques: Essays in Honor of Bruno de Finetti* (P. K. Goel and A. Zellner, eds.). Amsterdam: North-Holland. 391–404.

Dawid, A. P. (1988a). The infinite regress and its conjugate analysis. *Bayesian Statistics 3* (J. M. Bernardo, M. H. DeGroot, D. V. Lindley and A. F. M. Smith, eds.). Oxford: University Press, 95–110 (with discussion).

Dawid, A. P. (1988b). Symmetry models and hypotheses for structured data layouts. *J. Roy. Statist. Soc. B* **50**, 1–34 (with discussion).

Dawid, A. P. (1992). Prequential analysis, stochastic complexity and Bayesian inference. *Bayesian Statistics 4* (J. M. Bernardo, J. O. Berger, A. P. Dawid and A. F. M. Smith, eds.). Oxford: University Press, 109–125 (with discussion).

Dawid, A. P. (1994). The island problem: coherent use of identification evidence. *Aspects of Uncertainty: a Tribute to D. V. Lindley* (P. R. Freeman, and A. F. M. Smith, eds.). Chichester: Wiley, 159–170.

Dawid, A. P. and Fang, B. Q. (1992). Conjugate Bayes discrimination with infinitely many variables. *J. Multivariate Analysis* **41**, 27–42.

Dawid, A. P. and Smith, A. F. M. (eds.) (1983). *1982 Conference on Practical Bayesian Statistics*. Special issue, *The Statistician* **32**, Numbers 1 and 2.

Dawid, A. P. and Stone, M. (1972). Expectation consistency of inverse probability distributions. *Biometrika* **59**, 486–489.

Dawid, A. P. and Stone, M. (1973). Expectation consistency and generalised Bayes inference. *Ann. Statist.* **1**, 478–485.

Dawid, A. P., Stone, M. and Zidek, J. V. (1973). Marginalization paradoxes in Bayesian and structural inference. *J. Roy. Statist. Soc. B* **35**, 189–233 (with discussion).

Debreu, G. (1960). Topological methods in cardinal utility. *Mathematical Methods in the Social Sciences* (K. J. Arrow, S. Karlin and P. Suppes, eds.). Stanford: University Press, 16–26.

Deely, J. J. and Lindley, D. V. (1981). Bayes empirical Bayes. *J. Amer. Statist. Assoc.* **76**, 833-841.

de Finetti, B. (1930). Funzione caratteristica di un fenomeno aleatorio. *Mem. Accad. Naz. Lincei* **4**, 86–133.

de Finetti, B. (1937/1964). La prévision: ses lois logiques, ses sources subjectives. *Ann. Inst. H. Poincaré* **7**, 1–68. Reprinted in 1980 as 'Foresight; its logical laws, its subjective sources' in *Studies in Subjective Probability* (H. E. Kyburg and H. E Smokler, eds.). New York: Dover, 93–158.

de Finetti, B. (1938). Sur la condition d'équivalence partielle. *Actualités Scientifiques et Industrielles* **739**. Paris: Herman and Cii. Translated in *Studies in Inductive Logic and Probability* **2** (R. Jeffrey, ed.). Berkeley: Univ. California Press, 193–206.

de Finetti, B. (1951). Recent suggestions for the reconciliation of theories of probability. *Proc. Second Berkeley Symp.* (J. Neyman ed.). Berkeley: Univ. California Press, 217–226.

de Finetti, B. (1961). The Bayesian approach to the rejection of outliers. *Proc. Fourth Berkeley Symp.* **1** (J. Neyman and E. L. Scott, eds.). Berkeley: Univ. California Press, 199–210.

de Finetti, B. (1962). Does it make sense to speak of 'Good Probability Appraisers'? *The Scientist Speculates: An Anthology of Partly-Baked Ideas* (I. J. Good, ed.). New York: Wiley, 257–364. Reprinted in 1972, *Probability, Induction and Statistics* New York: Wiley, 19–23.

de Finetti, B. (1963). La décision et les probabilitiés. *Rev. Roumaine Math. Pures Appl.* **7**, 405–413.

de Finetti, B. (1964). Probabilità subordinate e teoria delle decisioni. *Rendiconti Matematica* **23**, 128–131. Reprinted as 'Conditional probabilities and decision theory' in 1972, *Probability, Induction and Statistics* New York: Wiley, 13–18.

de Finetti, B. (1965). Methods for discriminating levels of partial knowledge concerning a test item. *British J. Math. Statist. Psychol.* **18**, 87–123. Reprinted in 1972, *Probability, Induction and Statistics* New York: Wiley, 25–63.

de Finetti, B. (1967). Logical foundations and measurement of subjective probability. *Acta Psychologica* **34**, 129–145.

de Finetti, B. (1968). Probability: interpretations. *Internat. Encyclopedia of the Social Sciences*, **12**. London: Macmillan, 496–504.

de Finetti, B. (1970/1974). *Teoria delle Probabilità* **1**. Turin: Einaudi. English translation as *Theory of Probability* **1** in 1974, Chichester: Wiley.

de Finetti, B. (1970/1975). *Teoria delle Probabilità* **2**. Turin: Einaudi. English translation as *Theory of Probability* **2** in 1975, Chichester: Wiley.

de Finetti, B. (1972). *Probability, Induction and Statistics*. Chichester: Wiley.

de Finetti, B. (1978). Probability: interpretations. *Internat. Encyclopedia of Statistics* (W. H. Kruskal and J. M. Tanur, eds.) London: Macmillan, 496–505.

de Finetti, B. (1993). *Induction and Probability*. (P. Monori and D. Cocchi, eds.). Bologna: Clueb.

DeGroot, M. H. (1962). Uncertainty, information and sequential experiments. *Ann. Math. Statist.* **33**, 404–419.

DeGroot, M. H. (1963). Some comments on the experimental measurement of utility. *Behavioral Sci.* **8**, 146–149.

DeGroot, M. H. (1970). *Optimal Statistical Decisions*. New York: McGraw-Hill.

DeGroot, M. H. (1973). Doing what comes naturally: interpreting a tail area as a posterior probability or as a likelihood ratio. *J. Amer. Statist. Assoc.* **68**, 966–969.

DeGroot, M. H. (1974). Reaching a consensus. *J. Amer. Statist. Assoc.* **69**, 118–121.

DeGroot, M. H. (1980). Improving predictive distributions. *Bayesian Statistics* (J. M. Bernardo, M. H. DeGroot, D. V. Lindley and A. F. M. Smith, eds.). Valencia: University Press, 385–395 and 415–429 (with discussion).

DeGroot, M. H. (1982). Decision theory. *Encyclopedia of Statistical Sciences* **2** (S. Kotz, N. L. Johnson and C. B. Read, eds.). New York: Wiley, 277–286.

DeGroot, M. H. (1987). *Probability and Statistics*. Reading, MA: Addison-Wesley.

DeGroot, M. H. and Fienberg, S. E. (1982). Assessing probability assessors: calibration and refinement. *Statistical Decision Theory and Related Topics III* **1** (S. S. Gupta and J. O. Berger, eds.). New York: Academic Press, 291–314.

DeGroot, M. H. and Fienberg, S. E. (1983). The comparison and evaluation of forecasters. *The Statistician* **32**, 12–22.

DeGroot, M. H. and Fienberg, S. E. (1986). Comparing probability forecasters: basic binary concepts and multivariate extensions. *Bayesian Inference and Decision Techniques: Essays in Honor of Bruno de Finetti* (P. K. Goel and A. Zellner, eds.). Amsterdam: North-Holland, 247–264.

DeGroot, M. H., Fienberg, S. E. and Kadane, J. B. (eds.) (1986). *Statistics and the Law*. New York: Wiley.

DeGroot, M. H. and Kadane, J. B. (1980) Optimal challenges for selection. *Operations Research* **28**, 952–968.

DeGroot, M. H. and Mortera, J. (1991). Optimal linear opinion pools. *Manag. Sci.* **37**, 546–558.

DeGroot, M. H. and Rao, M. M. (1963). Bayes estimation with convex loss. *Ann. Math. Statist.* **34**, 839–846.

DeGroot, M. H. and Rao, M. M. (1966). Multidimensional information inequalities and prediction. *Multivariate Statistics* (P. R. Krishnaiah, ed.). New York: Academic Press, 287–313.

de la Horra, J. (1986). Convergencia del vector de probabilidad a posterior bajo una distribución predictiva. *Trab. Estadist.* **1**, 3–11.

de la Horra, J. (1987). Generalized estimators: a Bayesian decision theoretic view. *Statistics and Decisions* **5**, 347–352.

de la Horra, J. (1988). Parametric estimation with L_1 distance. *Bayesian Statistics 3* (J. M. Bernardo, M. H. DeGroot, D. V. Lindley and A. F. M. Smith, eds.). Oxford: University Press, 579–583.

de la Horra, J. (1992). Using the prior mean of a nuisance parameter. *Test* **1**, 31–38.

de la Horra, J. and Fernández, C. (1994). Bayesian robustness of credible regions in the presence of nuisance parameters. *Comm. Statist. Theory and Methods* **23**, 689–699.

Delampady, M. (1989). Lower bounds on Bayes factors for interval null hypotheses. *J. Amer. Statist. Assoc.* **84**,120–124.

Delampady, M. and Berger, J. O. (1990). Lower bounds on Bayes factors for multinomial and chi-squared tests of fit. *Ann. Statist.* **18**, 1295–1316.

Delampady, M. and Dey, D. K. (1994). Bayesian robustness for multiparameter problems. *J. Statist. Planning and Inference* **40**, 375–382.

Dellaportas, P. and Smith, A. F. M. (1993). Bayesian inference for generalised linear and proportional hazards models via Gibbs sampling. *Appl. Statist.* **42**, 443–460.

Dellaportas, P. and Wright, D. E. (1992). A numerical integration strategy in Bayesian analysis. *Bayesian Statistics 4* (J. M. Bernardo, J. O. Berger, A. P. Dawid and A. F. M. Smith, eds.). Oxford: University Press, 601–606.

De Morgan, A. (1847). *Formal Logic*. London: Taylor and Walton.

Dempster, A. P. (1967). Upper and lower probabilities induced by a multivalued mapping. *Ann. Math. Statist.* **38**, 325–339.

Dempster, A. P. (1968). A generalization of Bayesian inference. *J. Roy. Statist. Soc. B* **30**, 205–247 (with discussion).

Dempster, A. P. (1975). A subjective look at robustness. *Internat. Statist. Rev.* **46**, 349–374.

Dempster, A. P. (1985). Probability, evidence and judgement. *Bayesian Statistics 2* (J. M. Bernardo, M. H. DeGroot, D. V. Lindley and A. F. M. Smith, eds.), Amsterdam: North-Holland, 119–132 (with discussion).

DeRobertis, L. and Hartigan, J. (1981). Bayesian inference using intervals of measures. *Ann. Statist.* **9**, 235–244.

Devroye, L. (1986). *Non-Uniform Random Variate Generation*. Berlin: Springer.

De Waal, D. J. and Groenewald, P. C. N. (1989). On measuring the amount of information from the data in a Bayesian analysis. *South African Statist. J.* **23**, 23–61 (with discussion).

De Waal, D. J., Groenewald, P. C. N., van Zyl, D. and Zidek, J. (1986). Multi-Bayesian estimation theory. *Statistics and Decisions* **4**, 1–18.

Diaconis, P. (1977). Finite forms of de Finetti's theorem on exchangeability. *Synthese* **36**, 271–281.

Diaconis, P. (1988a). Recent progress on de Finetti's notion of exchangeability. *Bayesian Statistics 3* (J. M. Bernardo, M. H. DeGroot, D. V. Lindley and A. F. M. Smith, eds.). Oxford: University Press, 111-125 (with discussion).

Diaconis, P. (1988b). Bayesian numerical analysis. *Statistical Decision Theory and Related Topics IV* **1** (S. S. Gupta and J. O. Berger, eds.). Berlin: Springer, 163–175.

Diaconis, P., Eaton, M. L. and Lauritzen, S. L. (1992). Finite de Finetti theorems in linear models and multivariate analysis. *Scandinavian J. Statist.* **19**, 289–316.

Diaconis, P. and Freedman, D. (1980a). Finite exchangeable sequences. *Ann. Prob.* **8**, 745–764.

Diaconis, P. and Freedman, D. (1980b). De Finetti generalizations of exchangeability. *Studies Inductive Logic and Probability* (Jeffrey, R. C. ed.). Berkeley: Univ. California Press, 223–249.

Diaconis, P. and Freedman, D. (1983). Frequency properties of Bayes rules. *Scientific Inference, Data Analysis and Robustness* (G. E. P. Box, T. Leonard and C. F. Wu, eds.). New York: Academic Press, 105–116

Diaconis, P. and Freedman, D. (1984). Partial exchangeability and sufficiency. *Statistics: Applications and New Directions* (J. K. Ghosh and J. Roy, eds.). Calcutta: Indian Statist. Institute, 205–236.

Diaconis, P. and Freedman, D. (1986a). On the consistency of Bayes estimates. *Ann. Statist.* **14**, 1–67, (with discussion).

Diaconis, P. and Freedman, D. (1986b). On inconsistent Bayes estimates of location. *Ann. Statist.* **14**, 68–87.

Diaconis, P. and Freedman, D. (1987). A dozen de Finetti-style results in search of a theory. *Ann. Inst. H. Poincaré* **23**, 397–423.

Diaconis, P. and Freedman, D. (1990). Cauchy's equation and de Finetti's theorem. *Scandinavian J. Statist.* **17**, 235–274.

Diaconis, P. and Ylvisaker, D. (1979). Conjugate priors for exponential families. *Ann. Statist.* **7**, 269–281.

Diaconis, P. and Ylvisaker, D. (1985). Quantifying prior opinion. *Bayesian Statistics 2* (J. M. Bernardo, M. H. DeGroot, D. V. Lindley and A. F. M. Smith, eds.), Amsterdam: North-Holland, 133-156 (with discussion).

Diaconis, P. and Zabell, S. L. (1982). Updating subjective probability. *J. Amer. Statist. Assoc.* **77**, 822–830.

Dickey, J. M. (1968). Three multidimensional integral identities with Bayesian applications. *Ann. Math. Statist.* **39**, 1615–1627.

Dickey, J. M. (1969). Smoothing by cheating. *Ann. Math. Statist.* **40**, 1477–1482.

Dickey, J. M. (1971). The weighted likelihood ratio, linear hypotheses on normal location parameteters. *Ann. Math. Statist.* **42**, 204–223.

Dickey, J. M. (1973). Scientific reporting and personal probabilities: Student's hypothesis. *J. Roy. Statist. Soc. B* **35**, 285–305. Reprinted in 1974 in *Studies in Bayesian Econometrics and Statistics: in Honor of Leonard J. Savage* (S. E. Fienberg and A. Zellner, eds.). Amsterdam: North-Holland, 485–511.

Dickey, J. M. (1974). Bayesian alternatives to the F test and least-squares estimate in normal linear model. *Studies in Bayesian Econometrics and Statistics: in Honor of Leonard J. Savage* (S. E. Fienberg and A. Zellner, eds.). Amsterdam: North-Holland, 515–554.

Dickey, J. M. (1976). Approximate posterior distributions. *J. Amer. Statist. Assoc.* **71**, 680–689.

Dickey, J. M. (1977). Is the tail area useful as an approximate Bayes factor? *J. Amer. Statist. Assoc.* **72**, 138–142.

Dickey, J. M. (1980). Beliefs about beliefs, a theory of stochastic assessment of subjective probabilities. *Bayesian Statistics* (J. M. Bernardo, M. H. DeGroot, D. V. Lindley and A. F. M. Smith, eds.). Valencia: University Press, 471–487 and 504–512 (with discussion).

Dickey, J. M. (1982). Conjugate families of distributions. *Encyclopedia of Statistical Sciences* **2** (S. Kotz, N. L. Johnson and C. B. Read, eds.). New York: Wiley, 135–145.

Dickey, J. M. and Freeman, P. R. (1975). Population-distributed personal probabilities. *J. Amer. Statist. Assoc.* **70**, 362–364.

Dickey, J. M. and Kadane, J. B. (1980). Bayesian decision theory and the simplification of models. *Evaluation of Econometric Models* (J. Kwenta and J. Ramsey, eds.). New York: Academic Press, 245–268.

Dickey, J. M. and Lientz, B. P. (1970). The weighted likelihood ratio, sharp hypotheses about chances, the order of a Markov chain. *Ann. Math. Statist.* **41**, 214–226.

Diebolt, J. and Robert, C. P. (1994). Estimation of finite mixture distributions through Bayesian sampling. *J. Roy. Statist. Soc. B* **56**, 363–375.

Doksum, K. A. (1974). Tailfree and neutral random probabilities and their posterior distributions. *Ann. Prob.* **2**, 183–201.

Doksum, K. A. and Lo, A. Y. (1990). Consistent and robust Bayes procedures for location based on partial information. *Ann. Statist.* **18**, 443–453.

Domotor, Z. and Stelzer, J. (1971). Representation of finitely additive semiordered qualitative probability structures. *J. Math. Psychology* **8**, 145-158.

Draper, N. R. and Guttman, I. (1969). The value of prior information. *New Developments in Survey Sampling* (N. L. Johnson and H. Smith Jr., eds.). New York: Wiley.

Drèze, J. H. (1974). Bayesian theory of identification in simultaneous equations models. *Studies in Bayesian Econometrics and Statistics: in Honor of Leonard J. Savage* (S. E. Fienberg and A. Zellner, eds.). Amsterdam: North-Holland, 159–174.

Dubins, L. E. and Savage, L. J. (1965/1976). *How to Gamble if you Must: Inequalities for Stochastic Processes*. New York: McGraw-Hill. Second edition in 1976. New York: Dover.

DuMouchel, W. (1990). Bayesian metaanalysis. *Statistical Methodology in the Pharmaceutical Sciences* (D. A. Berry, ed.). New York: Marcel Dekker, 509–529.

DuMouchel, W. and Harris, J. E. (1983). Bayes methods for combining the results of cancer studies in humans and other species. *J. Amer. Statist. Assoc.* **78**, 293–315.

Duncan, G. and DeGroot, M. H. (1976). A mean squared error approach to optimal design theory. *Proc. 1976 Conf. Information Sciences and Systems*. Baltimore: John Hopkins University Press, 217–221.

Duncan L. R. and Raiffa, H. (1957). *Games and Decisions*. New York: Wiley.

Dunsmore, I. R. (1966). A Bayesian approach to classification. *J. Roy. Statist. Soc. B* **28**, 568–577.

Dunsmore, I. R. (1968). A Bayesian approach to calibration. *J. Roy. Statist. Soc. B* **30**, 396–405.

Dunsmore, I. R. (1969). Regulation and optimization. *J. Roy. Statist. Soc. B* **31**, 160–170.

Durbin, J. (1969). Inferential aspects of the randomness of sample size in survey sampling. *New Developments in Survey Sampling* (N. L. Johnson and H. Smith Jr., eds.). New York: Wiley, 629–651.

Durbin, J. (1970). On Birnbaum's theorem on the relation between sufficiency, conditionality and likelihood. *J. Amer. Statist. Assoc.* **65**, 395–398.

Dykstra, R. L. and Laud, P. (1981). A Bayesian nonparametric approach to reliability. *Ann. Statist.* **9**, 356–367.

Dynkin, E. B. (1953). Klassy ekvivalentnych slucainychy velicin. *Uspechi. Mat. Nauk* **54**, 125–134.

Earman, J. (1990). Bayes' Bayesianism. *Stud. History Philos. Sci.* **21**, 351–370.

Eaton, M. L. (1982). A method for evaluating improper prior distributions. *Statistical Decision Theory and Related Topics III* **1** (S. S. Gupta and J. O. Berger, eds.). New York: Academic Press,

Eaton, M. L. (1992). A statistical diptych: admissible inferences, recurrence of symmetric Markov chains. *Ann. Statist.* **20**, 1147–1179.

Eaves, D. M. (1985). On maximizing the missing information about a hypothesis. *J. Roy. Statist. Soc. B* **47**, 263-266.

Edwards, A. W. F. (1972/1992). *Likelihood.* Cambridge: University Press. Second edition in 1992. Baltimore: John Hopkins University Press.

Edwards, A. W. F. (1974). The history of likelihood. *Internat. Statist. Rev.* **42**, 9–15.

Edwards, W. L. (1954). The theory of decision making. *Psychological Bul.* **51**, 380–417.

Edwards, W. L. (1961). Behavioral decision theory. *Annual Rev. Psychology* **12**, 473–498.

Edwards, W. L., Lindman, H. and Savage, L. J. (1963). Bayesian statistical inference for psychological research. *Psychol. Rev.* **70**, 193–242. Reprinted in *Robustness of Bayesian Analysis* (J. B. Kadane, ed.). Amsterdam: North-Holland, 1984, 1–62.

Edwards, W. L. and Newman, J. R. (1982). *Multiattribute Evaluation.* Beverly Hills, CA: Sage.

Edwards, W. L., Phillips, L. D., Hays, W. L. and Goodman, B. C. (1968). Probability information processing systems: design and evaluation. *IEEE Trans. Systems, Science and Cybernetics* **4**, 248–265.

Edwards, W. L. and Tversky, A. (eds.) (1967). *Decision Making.* Baltimore: Penguin.

Efron, B. (1973). In discussion of Dawid, Stone and Zidek (1973). *J. Roy. Statist. Soc. B* **35**, 219.

Efron, B. (1982). *The Jacknife, the Bootstrap and other Resampling Plans.* Philadelphia, PA: SIAM.

Efron, B. (1986). Why isn't everyone a Bayesian? *Amer. Statist.* **40**, 1–11 (with discussion).

Efron, B. (1993). Bayes and likelihood calculations from confidence intervals. *Biometrika* **80**, 3–26.

Efron, B. and Morris, C. N. (1972). Empirical Bayes estimators on vector observations—an extension of Stein's method. *Biometrika* **59**, 335–347.

Efron, B. and Morris, C. N. (1975). Data analysis using Stein's estimator and its generalisations. *J. Amer. Statist. Assoc.* **70**, 311–319.

Efron, B. and Tibshirani, R. J. (1993). *An Introduction to the Bootstrap.* London: Chapman and Hall.

Eichhorn, W. (1978). *Functional Equations in Economics.* Reading, MA: Addison-Wesley.

Eliashberg, J. and Winkler, R. L. (1981). Risk sharing and group decision making. *Manag. Sci.* **27**, 1121–1235.

El-Krunz, S. M. and Studden, W. J. (1991). Bayesian optimal designs for linear regression models. *Ann. Statist.* **19**, 2183–2208.

Ellsberg, D. (1961). Risk, ambiguity and the Savage axioms. *Quart. J. Econ.* **75**, 643–669.

Erickson, G. J. and Smith, C. R. (eds.) (1988). *Maximum Entropy and Bayesian Methods in Science and Engineering*. (2 volumes). Dordrecht: Kluwer.

Ericson, W. A. (1969a). Subjective Bayesian models in sampling finite populations. *J. Roy. Statist. Soc. B* **31**, 195–233.

Ericson, W. A. (1969b). Subjective Bayesian models in sampling finite populations: stratification. *New Developments in Survey Sampling* (N. L. Johnson and H. Smith Jr., eds.). New York: Wiley, 326–357.

Ericson, W. A. (1988). Bayesian inference in finite populations. *Handbook of Statistics 6. Sampling* (P. R. Krishnaiah and C. R. Rao eds.). Amsterdam: North-Holland, 213–246.

Farrell, R. H. (1964). Estimators of a location parameter in the absolutely continuous case. *Ann. Math. Statist.* **35**, 949–998.

Farrell, R. H. (1968). Towards a theory of generalized Bayes tests. *Ann. Math. Statist.* **39**, 1–22.

Fearn, T. and O'Hagan, A. (eds.) (1993). *1992 Conference on Practical Bayesian Statistics*. Special issue, *The Statistician* **42**, Number 4.

Fedorov, V. V. (1972). *Theory of Optimal Experiments*. New York: Academic Press.

Feller, W. (1950/1968). *An Introduction to Probability Theory and its Applications* **1**. Chichester: Wiley. Third edition in 1968.

Fellner, W. (1965). *Probability and Profits: A Study of Economic Behavior along Bayesian Lines*. Homewood, IL: Irwin.

Felsenstein, K. (1988). Iterative procedures for continuous Bayesian designs. *Bayesian Statistics 3* (J. M. Bernardo, M. H. DeGroot, D. V. Lindley and A. F. M. Smith, eds.). Oxford: University Press, 609–613.

Felsenstein, K. (1992). Optimal Bayesian design for discrimination among rival models. *J. Comp. Statist. and Data Analysis* **14**, 427–436.

Ferguson, T. S. (1967). *Mathematical Statistics: a Decision Theoretic Approach*. New York: Academic Press.

Ferguson, T. S. (1973). A Bayesian analysis of some nonparametric problems. *Ann. Statist.* **1**, 209–230.

Ferguson, T. S. (1974). Prior distributions on spaces of probability measures. *Ann. Statist.* **2**, 615–629.

Ferguson, T. S. (1989). Who solved the secretary problem? *Statist. Sci.* **4**, 282–296 (with discussion).

Ferguson, T. S. and Phadia, E. G. (1979). Bayesian nonparametric estimation based on censored data. *Ann. Statist.* **7**, 163–186.

Ferguson, T. S., Phadia, E. G. and Tiwari, R. C. (1992). Bayesian nonparametric inference. *Current Issues in Statistical Inference: Essays in Honor of D. Basu.* (M. Ghosh and P. K. Pathak eds.). Hayward, CA: IMS, 127–150.

Ferrándiz, J. R. (1982). Una solución Bayesiana a la paradoja de Stein. *Trab. Estadist.* **33**, 31–46.

Ferrándiz, J. R. (1985). Bayesian inference on Mahalanobis distance: an alternative approach to Bayesian model testing. *Bayesian Statistics 2* (J. M. Bernardo, M. H. DeGroot, D. V. Lindley and A. F. M. Smith, eds.), Amsterdam: North-Holland, 645–654.

Ferrándiz, J. R. and Sendra, M. (1982). *Tablas de Bioestadística*. Valencia: University Press.

Fieller, E. C. (1954). Some problems in interval estimation. *J. Roy. Statist. Soc. B* **16**, 186–194 (with discussion).

Fienberg, S. E. and Zellner, A. (eds.) (1974). *Studies in Bayesian Econometrics and Statistics: in Honor of Leonard J. Savage*. Amsterdam: North-Holland.

Fine, T. L. (1973). *Theories of Probability: an Examination of Foundations*. New York: Academic Press.

Fishburn, P. C. (1964). *Decision and Value Theory*. New York: Wiley.

Fishburn, P. C. (1967a). Bounded expected utility. *Ann. Math. Statist.* **38**, 1054–1060.

Fishburn, P. C. (1967b). Preference-based definitions of subjective utility. *Ann. Math. Statist.* **38**, 1605–1617.

Fishburn, P. C. (1968). Utility theory. *Manag. Sci.* **14**, 335–378.

Fishburn, P. C. (1969). A general theory of subjective probability and expected utilities. *Ann. Math. Statist.* **40**, 1419–1429.

Fishburn, P. C. (1970). *Utility Theory for Decision Making*. New York: Wiley.

Fishburn, P. C. (1975). A theory of subjective expected utility with vague preferences. *Theory and Decision* **6**, 287–310.

Fishburn, P. C. (1981). Subjective expected utility: a review of normative theories. *Theory and Decision* **13**, 139–199.

Fishburn, P. C. (1982). *The Foundations of Expected Utility*. Dordrecht: Reidel.

Fishburn, P. C. (1986). The axioms of subjective probability. *Statist. Sci.* **1**, 335–358 (with discussion).

Fishburn, P. C. (1987). *Interprofile Conditions and Impossibility*. London: Harwood.

Fishburn, P. C. (1988a). *Non-linear Preference and Utility Theory*. Baltimore: John Hopkins University Press.

Fishburn, P. C. (1988b). Utility theory. *Encyclopedia of Statistical Sciences* **9** (S. Kotz, N. L. Johnson and C. B. Read, eds.). New York: Wiley, 445–452.

Fisher, R. A. (1915). Frequency distribution of the values of the correlation coefficient in samples from an indefinitely large population. *Biometrika* **10**, 507–521.

Fisher, R. A. (1922). On the mathematical foundations of theoretical statistics. *Phil. Trans. Roy. Soc. London A* **222**, 309–368. Reprinted in *Breakthroughs in Statistics* **1** (S. Kotz and N. L. Johnson, eds.). Berlin: Springer, 1991, 11–44.

Fisher, R. A. (1925). Theory of statistical information. *Proc. Camb. Phil. Soc.* **22**, 700–725.

Fisher, R. A. (1930). Inverse probability. *Proc. Camb. Phil. Soc.* **26**, 528–535.

Fisher, R. A. (1933). The concepts of inverse probability and fiducial probability referring to unknown parameters. *Proc. Roy. Soc. A* **139**, 343–348.

Fisher, R. A. (1935). The fiducial argument in statistical inference. *Ann. Eugenics* **6**, 391–398.

Fisher, R. A. (1939). A note on fiducial inference. *Ann. Statist.* **10**, 383–388.

Fisher, R. A. (1956/1973). *Statistical Methods and Scientific Inference*. Third edition in 1973. Edinburgh: Oliver and Boyd. Reprinted in 1990 whithin *Statistical Methods, Experimental Design, and Scientific Inference* (J. H. Bennet, ed.). Oxford: University Press.

Florens, J.-P. (1978). Mesures à priori et invariance dans une expérience Bayésienne. *Pub. Inst. Statist. Univ. Paris* **23**, 29–55.

Florens, J.-P. (1982). Expériences Bayésiennes invariantes. *Ann. Inst. M. Poincaré* **18**, 309–317.

Florens, J.-P. and Mouchart, M. (1985). Model selection: some remarks from a Bayesian viewpoint. *Model Choice* (Florens, J.-P., Mouchart, M., Raoult J.-P. and Simar, L., eds.). Brussels: Pub. Fac. Univ. Saint Louis, 27–44.

Florens, J.-P. and Mouchart, M. (1986). Exaustivité, ancillarité et identification en statistique bayésienne. *Ann. Econ. Statist.* **4**, 63–93.

Florens, J.-P., Mouchart, M., Raoult J.-P. and Simar, L. (eds.) (1985). *Model Choice*. Brussels: Pub. Fac. Univ. Saint Louis.

Florens, J.-P., Mouchart, M., Raoult, J.-P., Simar, L. and Smith, A. F. M. (eds.) (1983). *Specifying Statistical Models*. Berlin: Springer.

Florens, J.-P., Mouchart, M. and Rolin, J.-M. (1990). *Elements of Bayesian Statistics*. New York: Marcel Dekker.

Florens, J.-P., Mouchart, M. and Rolin, J.-M. (1992). Bayesian analysis of mixtures: some results on exact estimability and identification. *Bayesian Statistics 4* (J. M. Bernardo, J. O. Berger, A. P. Dawid and A. F. M. Smith, eds.). Oxford: University Press, 127–145 (with discussion).

Flournoy, N. and Tsutakawa, R. K. (eds.) (1991). *Statistical Multiple Integration*. Providence: RI: ASA.

Fougère, P. T. (ed.) (1990). *Maximum Entropy and Bayesian Methods*. Dordrecht: Kluwer.

Fraser, D. A. S. (1963). On the sufficiency and likelihood principles. *J. Amer. Statist. Assoc.* **58**, 641–647.

Fraser, D. A. S. (1968). *The Structure of Inference*. New York: Wiley.

Fraser, D. A. S. (1972). Bayes, likelihood or structural. *Ann. Math. Statist.* **43**, 777–790.

Fraser, D. A. S. (1979). *Inference and Linear Models*. New York: McGraw-Hill.

Fraser, D. A. S. and McDunnough, P. (1984). Further remarks on the asymptotic normality of likelihood and conditional analysis. *Canadian J. Statist.* **12**, 183–190.

Fraser, D. A. S. and Reid, N. (1989). Adjustments to profile likelihood. *Biometrika* **76**, 477–488.

Freedman, D. A. (1962). Invariants under mixing which generalize de Finetti's theorem. *Ann. Math. Statist.* **33**, 916–923.

Freedman, D. A. (1963a). Invariants under mixing which generalize de Finetti's theorem: continuous time parameter. *Ann. Math. Statist.* **34**, 1194–1216.

Freedman, D. A. (1963b). On the asymptotic behavior of Bayes estimates in the discrete case. *Ann. Math. Statist.* **34**, 1386–1403.

Freedman, D. A. (1965). On the asymptotic behavior of Bayes estimates in the discrete case II. *Ann. Math. Statist.* **36**, 454–456.

Freedman, D. A. and Diaconis, P. (1983). On inconsistent Bayes estimates in the discrete case. *Ann. Statist.* **11**, 1109–1118.

Freedman, D. A. and Purves, R. A. (1969). Bayes' method for bookies. *Ann. Math. Statist.* **40**, 1117–1186.

Freeman, P. R. (1980). On the number of outliers in data from a linear model. *Bayesian Statistics* (J. M. Bernardo, M. H. DeGroot, D. V. Lindley and A. F. M. Smith, eds.). Valencia: University Press, 349–365 and 370–381 (with discussion).

Freeman, P. R. (1983). The secretary problem and its extensions—a review. *Internat. Statist. Rev.* **51**, 189–206.

Freeman, P. R. and Smith, A. F. M. (eds.) (1994). *Aspects of Uncertainty: a Tribute to D. V. Lindley*. Chichester: Wiley.

French, S. (1980). Updating of beliefs in the light of someone else's opinion. *J. Roy. Statist. Soc. A* **143**, 43–48.

French, S. (1981). Consensus of opinion. *Eur. J. Oper. Res.* **7**, 332–340.

French, S. (1982). On the axiomatisation of subjective probabilities. *Theory and Decision* **14**, 19–33.

French, S. (1985). Group consensus probability distributions: a critical survey. *Bayesian Statistics 2* (J. M. Bernardo, M. H. DeGroot, D. V. Lindley and A. F. M. Smith, eds.), Amsterdam: North-Holland, 183–202 (with discussion).

French, S. (1986). *Decision Theory: an Introduction to the Mathematics of Rationality*. Chichester: Ellis Horwood.

French, S. (ed.) (1989). *Readings in Decision Analysis*. London: Chapman and Hall.

French, S., Hartley, R., Thomas, L. C. and White, D. J (eds.) (1983). *Multiobjective Decision Making*. New York: Academic Press.

Friedman, M. and Savage, L. J. (1948). The utility analysis of choice involving risk. *J. Political Econ.* **56**, 279–304.

Friedman, M. and Savage. L. J. (1952). The expected utility hypothesis and the measurement of utility. *J. Political Econ.* **60**, 463–474.

Fu, J. C. and Kass, R. E. (1988). The exponential rate of convergence of posterior distributions. *Ann. Inst. Statist. Math.* **40**, 683–691.

Gamerman, D. (1992). A dynamic approach to the statistical analysis of point processes. *Biometrika* **79**, 39–50.

Gamerman, D. and Migon, H. S. (1993). Dynamic hierarchical models. *J. Roy. Statist. Soc. B* **55**, 629–642.

Gärdenfors, P. and Sahlin, N.-E. (1988) (eds.) *Decision, Probability, and Utility. Selected Readings*. Cambridge: University Press.

Garthwaite, P. H. and Dickey, J. M. (1992). Elicitation of prior distributions for variable selection problems in regression. *Ann. Statist.* **20**, 1697–1719.

Gatsonis, C. A. (1984). Deriving posterior distributions for a location parameter: a decision-theoretic approach. *Ann. Statist.* **12**, 958–970.

Gatsonis, C. A., Hodges, J. S., Kass, R. E. and Singpurwalla, N. (eds.) (1993). *Case Studies in Bayesian Statistics*. Berlin: Springer.

Gaul, W. and Schader, M. (eds.) (1978). *Data, Expert Knowledge and Decisions*. Berlin: Springer.

Geisser, S. (1964). Posterior odds for multivariate normal classification. *J. Roy. Statist. Soc. B* **26**, 69–76.

Geisser, S. (1966). Predictive discrimination. *Multivariate Analysis* (P. R. Krishnaiah, ed.). New York: Academic Press, 149–163.

Geisser, S. (1971). The inferential use of predictive distributions. *Foundations of Statistical Inference* (V. P. Godambe and D. A. Sprott, eds.). Toronto: Holt, Rinehart and Winston, 456–469.

Geisser, S. (1974). A predictive approach to the random effect model. *Biometrika* **61**, 101–107.

Geisser, S. (1975). The predictive sample reuse method, with applications. *J. Amer. Statist. Assoc.* **70**, 320–328.

Geisser, S. (1979). In discussion of Bernardo (1979b). *J. Roy. Statist. Soc. B* **41**, 136–137.

Geisser, S. (1980a). The contributions of Sir Harold Jeffreys to Bayesian inference. *Bayesian Analysis in Econometrics and Statistics: Essays in Honor of Harold Jeffreys* (A. Zellner, ed.). Amsterdam: North-Holland, 13–20.

Geisser, S. (1980b). A predictivist primer. *Bayesian Analysis in Econometrics and Statistics: Essays in Honor of Harold Jeffreys* (A. Zellner, ed.). Amsterdam: North-Holland, 363–381.

Geisser, S. (1982). Bayesian discrimination. *Handbook of Statistics 2. Classification* (P. R. Krishnaiah and L. N. Kanal eds.). Amsterdam: North-Holland, 101–120.

Geisser, S. (1984). On prior distributions for binary trials. *J. Amer. Statist. Assoc.* **38**, 244–251 (with discussion).

Geisser, S. (1985). On the prediction of observables: a selective update. *Bayesian Statistics 2* (J. M. Bernardo, M. H. DeGroot, D. V. Lindley and A. F. M. Smith, eds.), Amsterdam: North-Holland, 203–230 (with discussion).

Geisser, S. (1986). Predictive analysis. *Encyclopedia of Statistical Sciences* **7** (S. Kotz, N. L. Johnson and C. B. Read, eds.). New York: Wiley, 158–170.

Geisser, S. (1987). Influential observations, diagnostics and discordancy tests. *Appl. Statist.* **14**, 133–142.

Geisser, S. (1988). The future of statistics in retrospect. *Bayesian Statistics 3* (J. M. Bernardo, M. H. DeGroot, D. V. Lindley and A. F. M. Smith, eds.). Oxford: University Press, 147–158 (with discussion).

Geisser, S. (1992). Bayesian perturbation diagnostics and robustness. *Bayesian Analysis in Statistics and Econometrics* (P. K. Goel and N. S. Iyengar, eds.). Berlin: Springer, 289–302.

Geisser, S. (1993). *Predictive Inference: an Introduction*. London: Chapman and Hall.

Geisser, S. and Cornfield, J. (1963). Posterior distributions for multivariate normal parameters. *J. Roy. Statist. Soc. B* **25**, 368–376.

Geisser, S. and Eddy, W. F. (1979). A predictive approach to model selection. *J. Amer. Statist. Assoc.* **74**, 153–160.

Geisser, S., Hodges, J. S., Press, S. J. and Zellner, A. (eds.) (1990). *Bayesian and Likelihood methods in Statistics and Econometrics: Essays in Honor of George A. Barnard.* Amsterdam: North-Holland.

Gelfand, A. E. and Desu, A. (1968). Predictive zero-mean uniform discrimination. *Biometrika* **55**, 519–524.

Gelfand, A. E. and Dey, D. K. (1991). On Bayesian robustness in contaminated classes of priors. *Statistics and Decisions* **9**, 63–80.

Gelfand, A. E., Dey, D. K. and Chang, H. (1992). Model determination using predictive distributions with implementation via sampling-based methods. *Bayesian Statistics 4* (J. M. Bernardo, J. O. Berger, A. P. Dawid and A. F. M. Smith, eds.). Oxford: University Press, 147–167 (with discussion).

Gelfand, A. E., Hills, S. E., Racine-Poon, A. and Smith, A. F. M. (1990). Illustration of Bayesian inference in normal models using Gibbs sampling. *J. Amer. Statist. Assoc.* **85**, 972–985.

Gelfand, A. E. and Smith, A. F. M. (1990). Sampling based approaches to calculating marginal densities. *J. Amer. Statist. Assoc.* **85**, 398–409.

Gelfand, A. E., Smith, A. F. M. and Lee, T.-M. (1992). Bayesian analysis of constrained parameter and truncated data problems using Gibbs sampling. *J. Amer. Statist. Assoc.* **87**, 523–532.

Gelman, A. and Rubin, D. B. (1992a). Inference from iterative simulation using multiple sequences. *Statist. Sci.* **7**, 457–511 (with discussion).

Gelman, A. and Rubin, D. B. (1992b). A single series from the Gibbs sampler provides a false sense of security. *Bayesian Statistics 4* (J. M. Bernardo, J. O. Berger, A. P. Dawid and A. F. M. Smith, eds.). Oxford: University Press, 625–631.

Geman, S. (1988). Experiments in Bayesian image analysis. *Bayesian Statistics 3* (J. M. Bernardo, M. H. DeGroot, D. V. Lindley and A. F. M. Smith, eds.). Oxford: University Press, 159–171 (with discussion).

Geman, S. and Geman, D. (1984). Stochastic relaxation, Gibbs distributions and the Bayesian restoration of images. *IEEE Trans. Patt. Anal. Mach. Intelligence* **6**, 721–740.

Genest, C. (1984a). A characterization theorem for externally Bayesian groups. *Ann. Statist.* **12**, 1100–1105.

Genest, C. (1984b). A conflict between two axioms for combining subjective distributions. *J. Roy. Statist. Soc. B* **46**, 403–405.

Genest, C. and Zidek, J. (1986). Combining probability distributions: a critique and an annotated bibliography. *Statist. Sci.* **1**, 114–148 (with discussion).

George, E. I., Makov, U. E. and Smith, A. F. M. (1993). Conjugate likelihood distributions. *Scandinavian J. Statist.* **20**, 147–156.

George, E. I., Makov, U. E. and Smith, A. F. M. (1994). Bayesian hierarchical analysis for exponential families via Markov chain Monte Carlo. *Aspects of Uncertainty: a Tribute to D. V. Lindley* (P. R. Freeman, and A. F. M. Smith, eds.). Chichester: Wiley, 181–199.

George, E. I. and McCulloch, R. (1993a). Variable selection via Gibbs sampling. *J. Amer. Statist. Assoc.* **88**, 881–889.

George, E. I. and McCulloch, R. (1993b). On obtaining invariant prior distributions. *J. Statist. Planning and Inference* **37**, 169–179.

Geweke, J. (1988). Antithetic acceleration of Monte Carlo integration in Bayesian inference. *J. Econometrics* **38**, 73–90.

Geweke, J. (1989). Bayesian inference in econometric models using Monte Carlo integration. *Econometrica* **57**, 1317–1339.

Geyer, C. J. (1992). Practical Markov chain Monte Carlo. *Statist. Sci.* **7**, 473–511 (with discussion).

Ghosh, J. K., Ghosal, S. and Samanta, T. (1994). Stability and convergence of the posterior in non-regular problems. *Statistical Decision Theory and Related Topics V* (S. S. Gupta and J. O. Berger, eds.). Berlin: Springer, 183–199.

Ghosh, J. K. and Mukerjee, R. (1992). Non-informative priors. *Bayesian Statistics 4* (J. M. Bernardo, J. O. Berger, A. P. Dawid and A. F. M. Smith, eds.). Oxford: University Press, 195–210 (with discussion).

Ghosh, M. (1991). Hierarchical and empirical Bayes sequential estimation. *Handbook of Statistics* **8**. *Statistical Methods in Biological and Medical Sciences* (C. R. Rao and R. Chakraborty, eds.). Amsterdam: North-Holland, 441-458.

Ghosh, M. (1992a). Hierarchical and empirical Bayes multivariate estimation. *Current Issues in Statistical Inference: Essays in Honor of D. Basu.* (M. Ghosh and P. K. Pathak eds.). Hayward, CA: IMS, 151–177.

Ghosh, M. (1992b). Constrained Bayes estimation with application. *J. Amer. Statist. Assoc.* **87**, 533–540.

Ghosh, M. and Pathak, P. K. (eds.) (1992). *Current Issues in Statistical Inference: Essays in Honor of D. Basu.* Hayward, CA: IMS.

Gilardoni, G. L. and Clayton, M. K. (1993). On reaching a consensus using DeGroot's iterative pooling. *Ann. Statist.* **21**, 391–401.

Gilio, A. (1992a). C_0-Coherence and extensions of conditional probabilities. *Bayesian Statistics 4* (J. M. Bernardo, J. O. Berger, A. P. Dawid and A. F. M. Smith, eds.). Oxford: University Press, 633-640.

Gilio, A. (1992b). Incomplete probability assessments in decision analysis. *J. It. Statist. Soc.* **1**, 67–76.

Gilio, A. and Scozzafava, R. (1985). Vague distributions in Bayesian testing of a null hypothesis. *Metron* **43**, 167–174.

Gilks, W. R. (1992). Derivative-free adaptive rejection sampling for Gibbs sampling. *Bayesian Statistics 4* (J. M. Bernardo, J. O. Berger, A. P. Dawid and A. F. M. Smith, eds.). Oxford: University Press, 641–649.

Gilks, W. R., Clayton, D. G., Spiegelhalter, D. J., Best, N. G., McNeil, A. J., Sharples, L. D. and Kirby, A. J. (1993). Modelling complexity: applications of Gibbs sampling in medicine. *J. Roy. Statist. Soc. B* **55**, 39–52 (with discussion).

Gilks, W. R. and Wild, P. (1992). Adaptive rejection sampling for Gibbs sampling. *Appl. Statist.* **41**, 337–348.

Gillies, D. A. (1987). Was Bayes a Bayesian? *Hist. Math.* **14**, 325–346.

Gilliland, D. C., Boyer, J. E. Jr. and Tsao, H. J. (1982). Bayes empirical Bayes: finite parameter case. *Ann. Statist.* **10**, 1277–1282.

Girelli-Bruni, E. (ed.) (1981). *Teoria delle Decisioni in Medicina.* Verona: Bertani.

Girón, F. J., Martínez, L. and Morcillo, C. (1992). A Bayesian justification for the analysis of residuals and inference measures. *Bayesian Statistics 4* (J. M. Bernardo, J. O. Berger, A. P. Dawid and A. F. M. Smith, eds.). Oxford: University Press, 651–660.

Girón, F. J. and Ríos, S. (1980). Quasi-Bayesian behaviour: a more realistic approach to decision making? *Bayesian Statistics* (J. M. Bernardo, M. H. DeGroot, D. V. Lindley and A. F. M. Smith, eds.). Valencia: University Press, 17–38 (with discussion).

Girshick, M. A. and Savage, L. J. (1951). Bayes and minimax estimates for quadratic loss functions. *Proc. Second Berkeley Symp.* (J. Neyman ed.). Berkeley: Univ. California Press, 53–74.

Godambe, V. P. (1969). Some aspects of the theoretical development in survey sampling. *New Developments in Survey Sampling* (N. L. Johnson and H. Smith Jr., eds.). New York: Wiley, 27–58.

Godambe, V. P. (1970). Foundations of survey sampling. *Amer. Statist.* **24**, 33–38.

Godambe, V. P. and Sprott, D. A. (eds.) (1971). *Foundations of Statistical Inference*. Toronto: Holt, Rinehart and Winston.

Goel, P. K. (1983). Information measures and Bayesian hierarchical models. *J. Amer. Statist. Assoc.* **78**, 408–410.

Goel, P. K. (1988). Software for Bayesian analysis: current status and additional needs. *Bayesian Statistics 3* (J. M. Bernardo, M. H. DeGroot, D. V. Lindley and A. F. M. Smith, eds.). Oxford: University Press, 173–188 (with discussion).

Goel, P. K. and DeGroot, M. H. (1979). Comparison of experiments and information measures. *Ann. Statist.* **7**, 1066–1077.

Goel, P. K. and DeGroot, M. H. (1980). Only normal distributions have linear posterior expectations in linear regression. *J. Amer. Statist. Assoc.* **75**, 895–900.

Goel, P. K. and DeGroot, M. H. (1981). Information about hyperparameters in hierarchical models. *J. Amer. Statist. Assoc.* **76**, 140–147.

Goel, P. K., Gulati, C. M and DeGroot, M. H. (1992). Optimal stopping for a non-communicating team. *Bayesian Statistics 4* (J. M. Bernardo, J. O. Berger, A. P. Dawid and A. F. M. Smith, eds.). Oxford: University Press, 211–226 (with discussion).

Goel, P. K. and Iyengar, N. S. (eds.) (1992). *Bayesian Analysis in Statistics and Econometrics*. Berlin: Springer

Goel, P. K. and Zellner, A. (eds.) (1986). *Bayesian Inference and Decision Techniques: Essays in Honor of Bruno de Finetti*. Amsterdam: North-Holland.

Goicoechea, A., Duckstein, L. and Zionts, S. (eds.) (1992). *Multiple Criteria Decision Making*. Berlin: Springer

Goldstein, M. (1981). Revising previsions: a geometric interpretation. *J. Roy. Statist. Soc. B* **43**, 105–130.

Goldstein, M. (1985). Temporal coherence. *Bayesian Statistics 2* (J. M. Bernardo, M. H. DeGroot, D. V. Lindley and A. F. M. Smith, eds.), Amsterdam: North-Holland, 231–248 (with discussion).

Goldstein, M. (1986a). Separating beliefs. *Bayesian Inference and Decision Techniques: Essays in Honor of Bruno de Finetti* (P. K. Goel and A. Zellner, eds.). Amsterdam: North-Holland, 197–215.

Goldstein, M. (1986b). Exchangeable belief structures. *J. Amer. Statist. Assoc.* **81**, 971–976.

Goldstein, M. (1986c). Prevision. *Encyclopedia of Statistical Sciences* **7** (S. Kotz, N. L. Johnson and C. B. Read, eds.). New York: Wiley, 175–176.

Goldstein, M. (1987a). Systematic analysis of limited belief specifications. *The Statistician* **36**, 191–199.

Goldstein, M. (1987b). Can we build a subjectivist statistical package? *Probability and Bayesian Statistics* (R. Viertl, ed.). London: Plenum, 203–217.

Goldstein, M. (1988). The data trajectory. *Bayesian Statistics 3* (J. M. Bernardo, M. H. DeGroot, D. V. Lindley and A. F. M. Smith, eds.). Oxford: University Press, 189–209 (with discussion).

Goldstein, M. (1991). Belief transforms and the comparison of hypothesis. *Ann. Statist.* **19**, 2067–2089.

Goldstein, M. (1994). Revising exchangeable beliefs: subjectivist foundations for the inductive argument. *Aspects of Uncertainty: a Tribute to D. V. Lindley* (P. R. Freeman, and A. F. M. Smith, eds.). Chichester: Wiley, 201–222.

Goldstein, M. and Howard, J. V. (1991). A likelihood paradox. *J. Roy. Statist. Soc. B* **53**, 619–628 (with discussion).

Goldstein, M. and Smith, A. F. M. (1974). Ridge-type estimators for regression analysis. *J. Roy. Statist. Soc. B* **36**, 284–319.

Gómez-Villegas, M. A. and Gómez, E. (1992). Bayes factors in testing precise hypotheses. *Comm. Statist. A* **21**, 1707–1715.

Gómez-Villegas, M. A. and Maín, P. (1992). The influence of prior and likelihood tail behaviour on the posterior distribution. *Bayesian Statistics 4* (J. M. Bernardo, J. O. Berger, A. P. Dawid and A. F. M. Smith, eds.). Oxford: University Press, 661–667.

Good, I. J. (1950). *Probability and the Weighing of Evidence.* London : Griffin; New York: Hafner Press.

Good, I. J. (1952). Rational decisions. *J. Roy. Statist. Soc. B* **14**, 107–114.

Good, I. J. (1959). Kinds of probability. *Science* **127**, 443–447.

Good, I. J. (1960). Weight of evidence, corroboration, explanatory power and the utility of experiments. *J. Roy. Statist. Soc. B* **22**, 319–331.

Good, I. J. (1962). Subjective probability on the measure of a non-measurable set. *Logic Methodology and Philosophy of Science* (E. Nagel, P. Suppes and A. Tarski, eds.). Stanford: University Press, 319–329.

Good, I. J. (1965). *The Estimation of Probabilities. An Essay on Modern Bayesian Methods.* Cambridge, Mass: The MIT Press.

Good, I. J. (1966). A derivation of the probabilistic explanation of information. *J. Roy. Statist. Soc. B* **28**, 578–581.

Good, I. J. (1967). A Bayesian test for multinomial distributions. *J. Roy. Statist. Soc. B* **29**, 399–431.

Good, I. J. (1969). What is the use of a distribution? *Multivariate Analysis* **2** (P. R. Krishnaiah, ed.). New York: Academic Press, 183–203.

Good, I. J. (1971). The probabilistic explication of information, evidence, surprise, causality, explanation and utility. Twenty seven principles of rationality. *Foundations of Statistical Inference* (V. P. Godambe and D. A. Sprott, eds.). Toronto: Holt, Rinehart and Winston, 108–141 (with discussion).

Good, I. J. (1976). The Bayesian influence, or how to sweep subjectivism under the carpet. *Foundations of Probability Theory, Statistical Inference and Statistical Theories of Science* **2** (W. L. Harper and C. A. Hooker eds.). Dordrecht: Reidel, 119–168.

Good, I. J. (1977). Dynamic probability, computer chess and the measurement of knowledge. *Machine Intelligence* **8**, (E. W. Elcock and D. Michie, eds.). Chichester: Ellis Horwood, 139–150. Reprinted in Good (1983), 106–116.

Good, I. J. (1980a). The contributions of Jeffreys to Bayesian statistics. *Bayesian Analysis in Econometrics and Statistics: Essays in Honor of Harold Jeffreys* (A. Zellner, ed.). Amsterdam: North-Holland, 21–34.

Good, I. J. (1980b). Some history of the hierarchical Bayesian mehodology. *Bayesian Statistics* (J. M. Bernardo, M. H. DeGroot, D. V. Lindley and A. F. M. Smith, eds.). Valencia: University Press, 489–519.

Good, I. J. (1982). Degrees of belief. *Encyclopedia of Statistical Sciences* **2** (S. Kotz, N. L. Johnson and C. B. Read, eds.). New York: Wiley, 287–292.

Good, I. J. (1983). *Good Thinking: The Foundations of Probability and its Applications.* Minneapolis: Univ. Minnesota Press.

Good, I. J. (1985). Weight of Evidence: a brief survey. *Bayesian Statistics 2* (J. M. Bernardo, M. H. DeGroot, D. V. Lindley and A. F. M. Smith, eds.), Amsterdam: North-Holland, 249–270 (with discussion).

Good, I. J. (1987). Hierarchical Bayesian and empirical Bayesian methods. *Amer. Statist.* **41**, (with discussion).

Good, I. J. (1988a). Statistical evidence. *Encyclopedia of Statistical Sciences* **8** (S. Kotz, N. L. Johnson and C. B. Read, eds.). New York: Wiley, 651–656.

Good, I. J. (1988b). The interface between statistics and philosophy of science. *Statist. Sci.* **3**, 386–398 (with discussion).

Good, I. J. (1992). The Bayes/non-Bayes compromise: a brief review. *J. Amer. Statist. Assoc.* **87**, 597–606.

Good, I. J. and Gaskins, R. (1971). Non-parametric roughness penalties for probability densities. *Biometrika* **58**, 255–277.

Good, I. J. and Gaskins, R. (1980). Density estimation and bump hunting by the penalized likelihood method, exemplified by scattering and meteorite data. *J. Amer. Statist. Assoc.* **75**, 42–73.

Goodwin, P. and Wright, G. (1991). *Decision Analysis for Management Judgement.* New York: Wiley.

Gordon, N. J. and Smith, A. F. M. (1993). Approximate non-Gaussian Bayesian estimation and modal consistency. *J. Roy. Statist. Soc. B* **55**, 913–918.

Goutis, C. and Casella, G. (1991). Improved invariant confidence intervals for a normal variance. *Ann. Statist.* **19**, 2019–2031.

Grandy, W. T. and Schick, L. H. (eds.) (1991). *Maximum Entropy and Bayesian Methods.* Dordrecht: Kluwer.

Grayson, C. J. (1960). *Decisions under Uncertainty: Drilling Decisions by Oil and Gas Operators.* Harvard, MA: University Press.

Grenander, U. and Miller, M. I. (1994). Representations of knowledge in complex systems. *J. Roy. Statist. Soc. B* **56** (to appear, with discussion).

Grieve, A. P. (1987). Applications of Bayesian software: two examples. *The Statistician* **36**, 283–288.

Gu, C. (1992). Penalized likelihood regression: a Bayesian analysis. *Statistica Sinica* **2**, 255-264.

Gupta, S. S. and Berger, J. O. (eds.) (1988). *Statistical Decision Theory and Related Topics IV* **1**. Berlin: Springer.

Gupta, S. S. and Berger, J. O. (eds.) (1994). *Statistical Decision Theory and Related Topics V*. Berlin: Springer.

Gutiérrez-Peña, E. (1992). Expected logarithmic divergence for exponential families. *Bayesian Statistics 4* (J. M. Bernardo, J. O. Berger, A. P. Dawid and A. F. M. Smith, eds.). Oxford: University Press, 669–674.

Guttman, I. (1970). *Statistical Tolerance Regions: Classical and Bayesian*. London: Griffin.

Guttman, I. and Peña, D. (1988). Outliers and influence. Evaluation by posteriors of parameters in the linear model. *Bayesian Statistics 3* (J. M. Bernardo, M. H. DeGroot, D. V. Lindley and A. F. M. Smith, eds.). Oxford: University Press, 631–640.

Guttman, I. and Peña, D. (1993). A Bayesian look at the question of diagnostics. *Statistica Sinica* **3**, 367–390.

Haag, J. (1924). Sur un problème général de probabilités et ses diverses applications. *Proc. Internat. Congress Math. Toronto*, 659–674.

Hacking, I. (1965). Slightly more realistic personal probability. *Philosophy of Science* **34**, 311–325.

Hacking, I. (1975). *The Emergence of Probability*. Cambridge: University Press.

Hadley, G. (1967). *Introduction to Probability and Statistical Decision Theory*. San Francisco, CA: Holden-Day.

Hald, A. (1968), Bayesian single acceptance plans for continuous prior distributions. *Technometrics* **10**, 667–683.

Haldane, J. B. S. (1931). A note on inverse probability. *Proc. Camb. Phil. Soc.* **28**, 55–61.

Haldane, J. B. S. (1948). The precision of observed values of small frequencies. *Biometrika* **35**, 297–303.

Halmos, P. R. and Savage, L. J. (1949). Application of the Radon-Nikodym theorem to the theory of sufficient statistics. *Ann. Math. Statist.* **20**, 225–241.

Halter, A. N. and Dean, G. W. (1971). *Decisions under Uncertainty*. Cincinnati, OH: South-Western.

Harrison P. J. and West, M. (1987). Practical Bayesian forecasting. *The Statistician* **36**, 115–125.

Harsany, J. (1967). Games with incomplete information played by 'Bayesian' players. *Manag. Sci.* **14**, 159–182, 320–334, 486–502.

Hartigan, J. A. (1964). Invariant prior distributions. *Ann. Math. Statist.* **35**, 836–845.

Hartigan, J. A. (1965). The asymptotically unbiased prior distribution. *Ann. Math. Statist.* **36**, 1137–1152.

Hartigan, J. A. (1966a). Estimation by ranking parameters. *J. Roy. Statist. Soc. B* **28**, 32–44.

Hartigan, J. A. (1966b). Note on the confidence prior of Welch and Peers. *J. Roy. Statist. Soc. B* **28**, 55-56.

Hartigan, J. A. (1967). The likelihood and invariance principles. *J. Roy. Statist. Soc. B* **29**, 533–539.

Hartigan, J. A. (1969). Use of subsample values as typical values. *J. Amer. Statist. Assoc.* **104**, 1003–1317.

Hartigan, J. A. (1971). Similarity and probability. *Foundations of Statistical Inference* (V. P. Godambe and D. A. Sprott, eds.). Toronto: Holt, Rinehart and Winston, 305–313 (with discussion).

Hartigan, J. A. (1975). Necessary and sufficient conditions for asymptotic normality of a statistic and its subsample values. *Ann. Statist.* **3**, 573–580.

Hartigan, J. A. (1983). *Bayes Theory*. Berlin: Springer.

Hastings, W. K. (1970). Monte Carlo sampling methods using Markov chains and their applications. *Biometrika* **57**, 97–109.

Heath, D. L. and Sudderth, W. D. (1972). On a theorem of de Finetti, oddsmaking and game theory. *Amer. Statist.* **43**, 2072–2077.

Heath, D. L. and Sudderth, W. D. (1976). De Finetti's theorem for exchangeable random variables. *Amer. Statist.* **30**, 333–345.

Heath, D. L. and Sudderth, W. D. (1978). On finitely additive priors, coherence and extended admissibility. *Ann. Statist.* **6**, 333–345.

Heath, D. L. and Sudderth, W. D. (1989). Coherent inference from improper priors and from finitely additive priors. *Ann. Statist.* **17**, 907–919.

Hens, T. (1992). A note on Savage's theorem with a finite number of states. *J. Risk and Uncertainty* **5**, 63–71.

Herstein, I. N. and Milnor, J. (1953). An axiomatic approach to measurable utility. *Econometrica* **21**, 291–297.

Hewitt, E. and Savage, L. J. (1955). Symmetric measures on Cartesian products. *Trans. Amer. Math. Soc.* **80**, 470–501.

Hewlett, P. S. and Plackett, R. L. (1979). *The Interpretation of Quantal Responses in Biology*. London: Edward Arnold.

Heyde, C. C. and Johnstone, I. M. (1979). On asymptotic posterior normality for stochastic processes. *J. Roy. Statist. Soc. B* **41**, 184–189.

Hildreth, C. (1963). Bayesian statisticians and remote clients. *Econometrica* **31**, 422–438.

Hill, B. M. (1968). Posterior distributions of percentiles: Bayes' theorem for sampling from a finite population. *J. Amer. Statist. Assoc.* **63**, 677–691.

Hill, B. M. (1969). Foundations of the theory of least squares. *J. Roy. Statist. Soc. B* **31**, 89–97.

Hill, B. M. (1974). On coherence, inadmissibility and inference about many parameters in the theory of least squares. *Studies in Bayesian Econometrics and Statistics: in Honor of Leonard J. Savage* (S. E. Fienberg and A. Zellner, eds.). Amsterdam: North-Holland, 555–584.

Hill, B. M. (1975). A simple general approach to inference about the tail of a distribution. *Ann. Statist.* **3**, 1163–1174.

Hill, B. M. (1980). On finite additivity, non-conglomerability, and statistical paradoxes. *Bayesian Statistics* (J. M. Bernardo, M. H. DeGroot, D. V. Lindley and A. F. M. Smith, eds.). Valencia: University Press, 39–66 (with discussion).

Hill, B. M. (1986). Some subjective Bayesian considerations in the selection of models. *Econometric Reviews* **4**, 191–288.

Hill, B. M. (1987). The validity of the likelihood principle. *Amer. Statist.* **41**, 95–100.

Hill, B. M. (1988). De Finetti's theorem, induction and $A_{(n)}$, or Bayesian nonparametric predictive inference. *Bayesian Statistics 3* (J. M. Bernardo, M. H. DeGroot, D. V. Lindley and A. F. M. Smith, eds.). Oxford: University Press, 211–241 (with discussion).

Hill, B. M. (1990). A theory of Bayesian data analysis. *Bayesian and Likelihood Methods in Statistics and Econometrics: Essays in Honor of George A. Barnard* (S. Geisser, J. S. Hodges, S. J. Press and A. Zellner, eds.). Amsterdam: North-Holland, 49–73.

Hill, B. M. (1992). Bayesian nonparametric prediction and statistical inference. *Bayesian Analysis in Statistics and Econometrics* (P. K. Goel and N. S. Iyengar, eds.). Berlin: Springer, 43–76.

Hill, B. M. (1994). On Steinian shrinkage estimators: the finite/infinite problem and formalism in probability and statistics. *Aspects of Uncertainty: a Tribute to D. V. Lindley* (P. R. Freeman, and A. F. M. Smith, eds.). Chichester: Wiley, 223–260.

Hill, B. M. and Lane, D. (1986). Conglomerability and countable additivity. *Bayesian Inference and Decision Techniques: Essays in Honor of Bruno de Finetti* (P. K. Goel and A. Zellner, eds.). Amsterdam: North-Holland, 45–57.

Hills, S. E. (1987). Reference priors and identifiability problems in non-linear models. *The Statistician* **36**, 235–240.

Hills, S. E. and Smith, A. F. M. (1992). Parametrization issues in Bayesian inference. *Bayesian Statistics 4* (J. M. Bernardo, J. O. Berger, A. P. Dawid and A. F. M. Smith, eds.). Oxford: University Press, 227–246 (with discussion).

Hills, S. E. and Smith, A. F. M. (1993). Diagnostic plots for improved parametrisation in Bayesian inference. *Biometrika* **80**, 61–74.

Hinkelmann, K. (ed.) (1990). *Foundations of Statistics. An International Symposium in Honor of I. J. Good.* Special issue, *J. Statist. Planning and Inference* **25**.

Hinkley, D. V. (1979). Predictive likelihood. *Ann. Statist.* **7**, 718–728.

Hipp, C. (1974). Sufficient statistics and exponential families. *Ann. Statist.* **2**, 1283–1292.

Hjort, N. L. (1990). Nonparametric Bayes estimator based on beta processes in models for life history data. *Ann. Statist.* **18**, 1259–1294.

Hoadley, B. (1970). A Bayesian look at inverse regression. *J. Amer. Statist. Assoc.* **65**, 356–369.

Hodges, J. S. (1987). Uncertainty, policy analysis and statistics. *Statist. Sci.* **2**, 259–291 (with discussion).

Hodges, J. S. (1990). Can/may Bayesians use pure tests of significance? *Bayesian and Likelihood Methods in Statistics and Econometrics: Essays in Honor of George A. Barnard* (S. Geisser, J. S. Hodges, S. J. Press and A. Zellner, eds.). Amsterdam: North-Holland, 75–90.

Hodges, J. S. (1992). Who knows what alternative lurks in the hearts of significance tests? *Bayesian Statistics 4* (J. M. Bernardo, J. O. Berger, A. P. Dawid and A. F. M. Smith, eds.). Oxford: University Press, 247–266 (with discussion).

Hogarth, R. (1975). Cognitive processes and the assessment of subjective probability distributions. *J. Amer. Statist. Assoc.* **70**, 271–294.

Hogarth, R. (1980). *Judgement and Choice*. New York: Wiley

Holland, G. D. (1962). The reverend Thomas Bayes, F.R.S. (1702–1761). *J. Roy. Statist. Soc. A* **125**, 421–461.

Howson, C. and Urbach, P. (1989). *Scientific Reasoning: the Bayesian Approach.* La Salle, IL: Open Court.

Hull, J., Moore, P. G. and Thomas, H. (1973). Utility and its measurement. *J. Roy. Statist. Soc. A* **136**, 226–247.

Huseby, A. B. (1988). Combining opinions in a predictive case. *Bayesian Statistics 3* (J. M. Bernardo, M. H. DeGroot, D. V. Lindley and A. F. M. Smith, eds.). Oxford: University Press, 641–651.

Huzurbazar, V. S. (1976). *Sufficient Statistics*. New York: Marcel Dekker.

Hwang, J. T. (1985). Universal domination and stochastic domination: decision theory under a broad class of loss functions. *Ann. Statist.* **13**, 295–314.

Hwang, J. T. (1988). Stochastic and universal domination. *Encyclopedia of Statistical Sciences* **8** (S. Kotz, N. L. Johnson and C. B. Read, eds.). New York: Wiley, 781–784.

Hylland, A, and Zeckhauser, R. (1981). The impossibility of Bayesian group decision making with separate aggregation of beliefs and values. *Econometrica* **79**, 1321–1336.

Ibragimov, I. A. and Hasminski, R. Z. (1973). On the information in a sample about a parameter. *Proc. 2nd Internat. Symp. Information Theory*. (B. N. Petrov and F. Csaki, eds.), Budapest: Akademiaikiadó, 295–309.

Irony, T. Z. (1992). Bayesian estimation for discrete distributions. *J. Appl. Statist.* **19**, 533–549.

Irony, T. Z. (1993). Information in sampling rules. *J. Statist. Planning and Inference* **36**, 27–38.

Irony, T. Z., Pereira, C. A. de B. and Barlow, R. E. (1992). Bayesian models for quality assurance. *Bayesian Statistics 4* (J. M. Bernardo, J. O. Berger, A. P. Dawid and A. F. M. Smith, eds.). Oxford: University Press, 675–688.

Isaacs, G. L., Christ, D. E., Novick, M. R. and Jackson, P. H. (1974). *Tables for Bayesian Statisticians*. Ames, IO: Iowa University Press.

Iversen, G. R. (1984). *Bayesian Statistical Inference*. Beverly Hills, CA: Sage

Jackson, J. E. (1960). Bibliography on sequential analysis. *J. Amer. Statist. Assoc.* **55**, 561–580.

James, W. and Stein, C. (1961). Estimation with quadratic loss. *Proc. Fourth Berkeley Symp.* **1** (J. Neyman and E. L. Scott, eds.). Berkeley: Univ. California Press, 361–380.

Jaynes, E. T. (1958). *Probability Theory in Science and Engineering*. Dallas: Mobil Oil Co.

Jaynes, E. T. (1968). Prior probabilities. *IEEE Trans. Systems, Science and Cybernetics* **4**, 227–291.

Jaynes, E. T. (1971). The well posed problem. *Foundations of Statistical Inference* (V. P. Godambe and D. A. Sprott, eds.). Toronto: Holt, Rinehart and Winston, 342–356 (with discussion).

Jaynes, E. T. (1976). Confidence intervals vs. Bayesian intervals. *Foundations of Probability Theory, Statistical Inference and Statistical Theories of Science* **2** (W. L. Harper and C. A. Hooker eds.). Dordrecht: Reidel, 175–257 (with discussion).

Jaynes, E. T. (1980). Marginalization and prior probabilities. *Bayesian Analysis in Econometrics and Statistics: Essays in Honor of Harold Jeffreys* (A. Zellner, ed.). Amsterdam: North-Holland, 43–87 (with discussion).

Jaynes, E. T. (1983). *Papers on Probability, Statistics and Statistical Physics*. (R. D. Rosenkrantz, ed.). Dordrecht: Kluwer.

Jaynes, E. T. (1985). Highly informative priors. *Bayesian Statistics 2* (J. M. Bernardo, M. H. DeGroot, D. V. Lindley and A. F. M. Smith, eds.), Amsterdam: North-Holland, 329–359 (with discussion).

Jaynes, E. T. (1986). Some applications and extensions of the de Finetti representation theorem. *Bayesian Inference and Decision Techniques: Essays in Honor of Bruno de Finetti* (P. K. Goel and A. Zellner, eds.). Amsterdam: North-Holland, 31–42.

Jeffrey, R. C. (1965/1983). *The Logic of Decision*. New York: McGraw-Hill. Second edition in 1983. Chicago: University Press.

Jeffrey, R. C. (ed.) (1981). *Studies in Inductive Logic and Probability*. Berkeley: Univ. California Press.

Jeffreys, H. (1931/1973). *Scientific Inference*. Cambridge: University Press. Third edition in 1973, Cambridge: University Press.

Jeffreys, H. (1939/1961). *Theory of Probability*. Oxford: University Press. Third edition in 1961, Oxford: University Press.

Jeffreys, H. (1946). An invariant form for the prior probability in estimation problems. *Proc. Roy. Soc. A* **186**, 453–461.

Jeffreys, H. (1955). The present position in probability theory. *Brit. J. Philos. Sci.* **5**, 275–289.

Jeffreys, H. and Jeffreys, B. S. (1946/1972). *Methods of Mathematical Physics*. Cambridge: University Press. Third edition in 1972, Cambridge: University Press.

Jewell, W. S. (1974). Credible means are exact Bayesian for simple exponential families. *ASTIN Bulletin* **8**, 77–90.

Jewell, W. S. (1988). A heterocedastic hierarchical model. *Bayesian Statistics 3* (J. M. Bernardo, M. H. DeGroot, D. V. Lindley and A. F. M. Smith, eds.). Oxford: University Press, 657–663.

Johnson, N. L. and Kotz, S. (1969). *Discrete Distributions*. New York: Wiley.

Johnson, N. L. and Kotz, S. (1970). *Continuous Univariate Distributions*. New York: Wiley.

Johnson, N. L. and Kotz, S. (1972). *Continuous Multivariate Distributions*. New York: Wiley.

Johnson, R. A. (1967). An asymptotic expansion for posterior distributions. *Ann. Math. Statist.* **38**, 1899–1906.

Johnson, R. A. (1970). Asymptotic expansions associated with posterior distributions. *Ann. Math. Statist.* **41**, 851–864.

Johnson, R. A. and Ladalla, J. N. (1979). The large-sample behaviour of posterior distributions with sampling from muitiparameter exponential family models and allied results. *Sankhyā B* **41**, 169–215.

Johnson, W. and Geisser, S. (1982). Assessing the predictive influence of observations. *Statistics and Probability Essays in Honor of C. R. Rao* (G. Kallianpur, P. K. Krishnaiah and J. K. Ghosh, eds.). Amsterdam: North-Holland, 343–358.

Johnson, W. and Geisser, S. (1983). A predictive view of the detection and characterisation of influential observations in regression analysis. *J. Amer. Statist. Assoc.* **78**, 137–144.

Johnson, W. and Geisser, S. (1985). Estimative influence measures for the multivariate general linear model. *J. Statist. Planning and Inference* **11**, 33–56.

Joshi, V. M. (1983). Likelihood principle. *Encyclopedia of Statistical Sciences* **4** (S. Kotz, N. L. Johnson and C. B. Read, eds.). New York: Wiley, 644–647.

Justice, J. M. (ed.) (1987). *Maximum Entropy and Bayesian Methods in Applied Statistics.* Cambridge: University Press.

Kadane, J. B. (1974). The role of identification in Bayesian theory. *Studies in Bayesian Econometrics and Statistics: in Honor of Leonard J. Savage* (S. E. Fienberg and A. Zellner, eds.). Amsterdam: North-Holland, 175–191.

Kadane, J. B. (1980). Predictive and structural methods for eliciting prior distributions. *Bayesian Analysis in Econometrics and Statistics: Essays in Honor of Harold Jeffreys* (A. Zellner, ed.). Amsterdam: North-Holland, 89–109.

Kadane, J. B. (ed.) (1984). *Robustness of Bayesian Analysis.* Amsterdam: North-Holland.

Kadane, J. B. (1992). Healthy scepticism as an expected utility explanation of the phenomena of Allais and Ellsberg. *Theory and Decision* **32**, 57–64.

Kadane, J. B. (1993). Several Bayesians: a review. *Test* **2**, 1–32 (with discussion).

Kadane, J. B. and Chuang, D. T. (1978). Stable decision problems. *Ann. Statist.* **6**, 1095–1110.

Kadane, J. B. and Larkey, P. (1982). Subjective probability and the theory of games. *Manag. Sci.* **28**, 113–120.

Kadane, J. B. and Larkey, P. (1983). The confusion of is and ought in game theoretic contexts. *Manag. Sci.* **29**, 1365–1379.

Kadane, J. B., Schervish, M. J. and Seidenfeld, T. (1986). Statistical implications of finitely additive probability. *Bayesian Inference and Decision Techniques: Essays in Honor of Bruno de Finetti* (P. K. Goel and A. Zellner, eds.). Amsterdam: North-Holland, 59–76.

Kadane, J. B. and Seidenfeld, T. (1990). Randomization in a Bayesian perspective. *J. Statist. Planning and Inference* **25**, 329–345.

Kadane, J. B. and Seidenfeld, T. (1992). Equilibrium, common knowledge and optimal sequential decisions. *Knowledge, Beliefs and Strategic Information* (C. Bicchini and M. L. Dalla Chiara, eds.). Cambridge: University Press, 27–45.

Kagan, A. M., Linnik, Y. V. and Rao, C. R. (1973). *Characterization Problems in Mathematical Statistics.* New York: Wiley.

Kahneman, D., Slovick, P. and Tversky, A. (eds.) (1982). *Judgement under Uncertainty: Heuristics and Biases.* Cambridge: University Press.

Kahneman, D. and Tversky, A. (1979). Prospect theory: an analysis of decision under risk. *Econometrica* **47**, 263–291.

Kalbfleish, J. G. (1971). Likelihood methods in prediction. *Foundations of Statistical Inference* (V. P. Godambe and D. A. Sprott, eds.). Toronto: Holt, Rinehart and Winston, 372–392 (with discussion).

Kalbfleish, J. G. and Sprott, D. A. (1970). Application of likelihood methods to models involving large number of parameters. *J. Roy. Statist. Soc. B* **32**, 175–208 (with discussion).

Kalbfleish, J. G. and Sprott, D. A. (1973). Marginal and conditional likelihoods. *Sankhyā A* **35**, 311–328.

Kapur, J. M. and Kesavan, H. K. (1992). *Entropy Optimization Principles and Applications.* New York: Academic Press.

Karlin, S. and Rubin, H. (1956). The theory of decision procedures for distributions with monotone likelihood ratio. *Ann. Math. Statist.* **27**, 272–299.

Kashyap, R. L. (1971). Prior probability and uncertainty. *IEEE Trans. Information Theory* **14**, 641–650.

Kashyap, R. L. (1974). Minimax estimation with divergence loss function. *Information Sciences* **7**, 341–364.

Kass, R. E. (1989). The geometry of asymptotic inference. *Statist. Sci.* **4**, 188–234.

Kass, R. E. (1990). Data-translated likelihood and Jeffreys' rule. *Biometrika* **77**, 107–114.

Kass, R. E. and Slate E. H. (1992). Reparametrization and diagnostics of posterior non-normality. *Bayesian Statistics 4* (J. M. Bernardo, J. O. Berger, A. P. Dawid and A. F. M. Smith, eds.). Oxford: University Press, 289–305 (with discussion).

Kass, R. E. and Steffey, D. (1989). Approximate Bayesian inference in conditionally independent hierarchical models (parametric empirical Bayes). *J. Amer. Statist. Assoc.* **84**, 717–726.

Kass, R. E., Tierney, L. and Kadane, J. B. (1988). Asymptotics in Bayesian computation. *Bayesian Statistics 3* (J. M. Bernardo, M. H. DeGroot, D. V. Lindley and A. F. M. Smith, eds.). Oxford: University Press, 261–278, (with discussion).

Kass, R. E., Tierney, L. and Kadane, J. B. (1989a). The validity of posterior expansions based on Laplace's method. *Bayesian and Likelihood Methods in Statistics and Econometrics: Essays in Honor of George A. Barnard* (S. Geisser, J. S. Hodges, S. J. Press and A. Zellner, eds.). Amsterdam: North-Holland, 473–488.

Kass, R. E., Tierney, L. and Kadane, J. B. (1989b). Approximate methods for assessing influence and sensitivity in Bayesian analysis. *Biometrika* **76**, 663–674.

Kass, R. E., Tierney, L. and Kadane, J. B. (1991). Laplace's method in Bayesian analysis. *Statistical Multiple Integration* (N. Flournoy and R. K. Tsutakawa eds.). Providence: RI: ASA, 89-99.

Kass, R. E. and Vaidyanathan, S. (1992). Approximate Bayes factors and orthogonal parameters, with application to testing equality of two binomial proportions. *J. Roy. Statist. Soc. B* **54**, 129–144.

Keeney, R. L. (1992). *Value-Focused Thinking*. Harvard, MA: University Press.

Keeney, R. L. and Raiffa, H. (1976). *Decisions with Multiple Objectives: Preferences and Value Tradeoffs*. New York: Wiley.

Kelly, J. S. (1991). Social choice bibliography. *Social Choice and Welfare* **8**, 97–169.

Kempthorne, P. J. (1986). Decision-theoretic measures of influence in regression. *J. Roy. Statist. Soc. B* **48**, 370–378.

Kestemont, M.-P. (1987). The Kolmogorov distance as comparison measure between parametric and non-parametric Bayesian predictions. *The Statistician* **36**, 259–264.

Keynes, J. M. (1921/1929). *A Treatise on Probability*. London: Macmillan. Second edition in 1929, London: Macmillan. Reprinted in 1962. New York: Harper and Row.

Khintchine, A. I. (1932). Sur les classes d'événements équivalents. *Mat. Sbornik* **39**, 40–43.

Kiefer, J. and Wolfowitz, J. (1956). Consistency of the maximum likelihood estimator in the presence of infinitely many nuisance parameters. *Ann. Math. Statist.* **27**, 887–906.

Kim, K. H. and Roush, F. W. (1987). *Team Theory*. Chichester: Ellis Horwood.

Kimeldorf, G. S. and Wahba, G. (1970). A correspondence between Bayesian estimation in stochastic processes and smoothing by splines. *Ann. Math. Statist.* **41**, 495–502.

Kingman, J. F. C. (1972). On random sequences with spherical symmetry. *Biometrika* **59**, 492–494.

Kingman, J. F. C. and Taylor, S. J. (1966). *Introduction to Measure and Probability*. Cambridge: University Press.

Klein, R. and Press, S. J. (1992). Adaptive Bayesian classification of spatial data. *J. Amer. Statist. Assoc.* **87**, 844–851.

Klein, R. W. and Brown, S. J. (1984). Model selection when there is 'minimal' prior information. *Econometrica* **52**, 1291–1312.

Kleiter, G. D. (1980). *Bayes-Statistik: Grundlagen und Anwendungen*. Berlin: W. de Gruyter.

Kloek, T. and van Dijk, H. K. (1978). Bayesian estimates of system equation parameters: an application of integration by Monte Carlo. *Econometrica* **46**, 1–19.

Klugman, S. A. (1992). *Bayesian Statistics in Actuarial Science, with Emphasis on Credibility*. Dordrecht: Kluwer.

Koch, G. and Spizzichino, F. (eds.) (1982). *Exchangeability in Probability and Statistics*. Amsterdam: North-Holland.

Kogan, N. and Wallace, M. A. (1964). *Risk Taking.: A Study in Cognition and Personality*. Toronto: Holt, Rinehart and Winston.

Kolmogorov, A. N. (1933/1950). *Grundbegriffe der Wahrscheinlichkeitsrechnung*. Berlin: Springer. English translation in 1950 as *Foundations of the Theory of Probability*, New York: Chelsea.

Koopman, B. O. (1940). The axioms and algebra of intuitive probability. *Ann. Math. Statist.* **41**, 269–292.

Koopman, L. H. (1936). On distributions admitting a sufficient statistics. *Trans. Amer. Math. Soc.* **39**, 399–409.

Korsan, R. J. (1992). Decision analytica: an example of Bayesian inference and decision theory using Mathematica. *Economic and Financial Modelling with Mathematica* (H. R. Varian, ed.). Berlin: Springer, 407–458.

Kraft, C., Pratt, J. W. and Seidenberg, A. (1959). Intuitive probability on finite sets. *Ann. Math. Statist.* **30**, 408–419.

Krantz, D. H., Luce, R. D., Suppes, P. and Tversky, A. (1971). *Foundations of Measurement* **1**. New York: Academic Press.

Krasker, W. S. (1984). A note on selecting parametric models in Bayesian inference. *Ann. Statist.* **12**, 751–757.

Küchler, U. and Lauritzen, S. L. (1989). Exponential families, extreme point models, and minimal space-time invariant functions for stochastic processes with stationary and independent increments. *Scandinavian J. Statist.* **15**, 237–261.

Kuhn, T. S. (1962). *The Structure of Scientific Revolutions*. Chicago: University Press.

Kullback, S. (1959/1968). *Information Theory and Statistics*. New York: Wiley. Second edition in 1968, New York: Dover. Reprinted in 1978, Gloucester, MA: Peter Smith.

Kullback, S. and Leibler, R. A. (1951). On information and sufficiency. *Ann. Math. Statist.* **22**, 79–86.

Kyburg, H. E. (1961). *Probability and the Logic of Rational Belief*. Middletown: Wesleyan University Press.

Kyburg, H. E. (1974). *The Logical Foundations of Statistical Inference*. Dordrecht: Reidel.

Kyburg, H. E. and Smokler, H. E. (eds.) (1964/1980). *Studies in Subjective Probability*. Chichester: Wiley. Second edition in 1980, New York: Dover.

Lad, F. and Deely, J. J. (1994). Experimental design from a subjective utilitarian viewpoint. *Aspects of Uncertainty: a Tribute to D. V. Lindley* (P. R. Freeman, and A. F. M. Smith, eds.). Chichester: Wiley, 267–282.

Lad, F., Dickey, J. M. and Rahman, M. A. (1990). The fundamental theorem of prevision. *Statistica* **50**, 19–38.

LaMotte, L. R. (1985). Bayesian linear estimators. *Encyclopedia of Statistical Sciences* **5** (S. Kotz, N. L. Johnson and C. B. Read, eds.). New York: Wiley, 20–22.

Lane, D. A. and Sudderth, W. D. (1983). Coherent and continuous inference. *Ann. Statist.* **11**, 114–120.

Lane, D. A. and Sudderth, W. D. (1984). Coherent predictive inference. *Sankhyā A* **46**, 166–185.

Laplace, P. S. (1774/1986). Mémoire sur la probabilité des causes par les évenements. *Mem. Acad. Sci. Paris* **6**, 621–656. English translation in 1986 as "Memoir on the probability of the causes of events", with an introduction by S. M. Stigler, *Statist. Sci.* **1**, 359–378.

Laplace, P. S. (1812). *Théorie Analytique des Probabilités*. Paris: Courcier. Reprinted as *Oeuvres Complètes de Laplace* **7**, 1878–1912. Paris: Gauthier-Villars.

Laplace, P. S. (1814/1952). *Essai Philosophique sur les Probabilitiés*. Paris: Courcier. The 5th edition (1825) was the last revised by Laplace. English translation in 1952 as *Philosophical Essay on Probabilities*. New York: Dover.

Lauritzen, S. L. (1982). *Statistical Families as Extremal Families*. Aalborg: University Press.

Lauritzen, S. L. (1988). *Extremal Families and Systems of Sufficient Statistics*. Berlin: Springer.

Lauritzen, S. L. and Spiegelhalter, D. J. (1988). Local computations with probabilities on graphical structures, and their application to expert systems. *J. Roy. Statist. Soc. B* **50**, 157–224 (with discussion).

Lavalle, I. H. (1968). On cash equivalents and information evaluation in decisions under uncertainty. *J. Amer. Statist. Assoc.* **63**, 252–290.

Lavalle, I. H. (1970). *An Introduction to Probability, Decision and Inference*. Toronto: Holt, Rinehart and Winston.

Lavalle, I. H. (1978). *Fundamentals of Decision Analysis*. Toronto: Holt, Rinehart and Winston.

Lavine, M. (1991a). Sensitivity in Bayesian statistics: the prior and the likelihood. *J. Amer. Statist. Assoc.* **86**, 396–399.

Lavine, M. (1991b). An approach to robust Bayesian analysis for multidimensional parameter spaces. *J. Amer. Statist. Assoc.* **86**, 400-403.

Lavine, M. (1992a). Some aspects of Polya tree distributions for statistical modelling. *Ann. Statist.* **20**, 1222-1235.

Lavine, M. (1992b). Sensitivity in Bayesian statistics: the prior and the likelihood. *J. Amer. Statist. Assoc.* **86**, 396–399.

Lavine, M. (1994). An approach to evaluating sensitivity in Bayesian regression analysis. *J. Statist. Planning and Inference* **40**, 242–244. (with discussion).

Lavine, M., Wasserman, L. and Wolpert, R. L. (1991). Bayesian inference with specified prior marginals. *J. Amer. Statist. Assoc.* **86**, 964–971.

Lavine, M., Wasserman, L. and Wolpert, R. L. (1993). Linearization of Bayesian robustness problems. *J. Statist. Planning and Inference* **37**, 307–316.

Lavine, M. and West, M. (1992). A Bayesian method for classification and discrimination. *Canadian J. Statist.* **20**, 451–461.

Leamer, E. E. (1978). *Specification Searches: Ad hoc Inference with Nonexperimental Data.* New York: Wiley.

LeCam, L. (1953). On some asymptotic properties of maximum likelihood estimates and related Bayes' estimates. *Univ. California Pub. Statist.* **1**, 277–329.

LeCam, L. (1956). On the asymptotic theory of estimation and testing hypothesis. *Proc. Third Berkeley Symp.* **1** (J. Neyman and E. L. Scott, eds.). Berkeley: Univ. California Press, 129–156.

LeCam, L. (1958). Les propietés asymptotiques de solutions de Bayes. *Pub. Inst. Statist. Univ. Paris* **7**, 17–35.

LeCam, L. (1966). Likelihood functions for large number of independent observations. *Research Papers in Statistics. Festschrift for J. Neyman* (F. N. David, ed.). New York: Wiley, 167–187.

LeCam, L. (1970). On the assumptions used to prove asymptotic normality of maximum likelihood estimates. *Ann. Math. Statist.* **41**, 802–828.

LeCam, L. (1986). *Asymptotic Methods in Statistical Decision Theory.* Berlin: Springer.

Lecoutre, B. (1984). *L'Analyse Bayésienne des Comparaisons.* Lille: Presses Universitaires.

Lee, P. M. (1989). *Bayesian Statistics: an Introduction.* London: Edward Arnold.

Lehmann, E. L. (1959/1983). *Theory of Point Estimation.* Second edition in 1983, New York: Wiley. Reprinted in 1991, Belmont, CA: Wadsworth.

Lehmann, E. L. (1959/1986). *Testing Statistical Hypotheses.* Second edition in 1986, New York: Wiley. Reprinted in 1991, Belmont, CA: Wadsworth.

Lehmann, E. L. (1990). Model specification. *Statist. Sci.* **5**, 160–168.

Lejeune, M. and Faulkenberry, G. D. (1982). A simple predictive density function. *J. Amer. Statist. Assoc.* **87**, 654–657.

Lempers, F. B. (1971). *Posterior Probabilities of Alternative Linear Models.* Rotterdam: University Press.

Lenk, P. J. (1991). Towards a practicable Bayesian nonparametric density estimator. *Biometrika* **78**, 531–543.

Leonard, T. (1973). A Bayesian method for histograms. *Biometrika* **60**, 297–308.

Leonard, T. (1975). Bayesian estimation methods for two-way contingency tables. *J. Roy. Statist. Soc. B* **37**, 23–37.

Leonard, T. (1980). The roles of inductive modelling and coherence in Bayesian statisitcs. *Bayesian Statistics* (J. M. Bernardo, M. H. DeGroot, D. V. Lindley and A. F. M. Smith, eds.). Valencia: University Press, 537–555 and 568–581 (with discussion).

Leonard, T. and Hsu, J. S. J. (1992). Bayesian inference for a covariance matrix. *Ann. Statist.* **20**, 1669-1696.

Leonard, T. and Hsu, J. S. J. (1994). The Bayesian analysis of categorical data: a selective review. *Aspects of Uncertainty: a Tribute to D. V. Lindley* (P. R. Freeman, and A. F. M. Smith, eds.). Chichester: Wiley, 283–310.

Leonard, T., Hsu, J. S. J. and Tsui, K.-W. (1989). Bayesian marginal inference. *J. Amer. Statist. Assoc.* **84**, 1051–1058.

Leonard, T. and Ord, K. (1976). An investigation of the F test procedure as an estimation short-cut. *J. Roy. Statist. Soc. B* **38**, 95–98.

Levine, R. D. and Tribus, M. (eds.) (1978). *The Maximum Entropy Formalism.* Cambridge, MA: The MIT Press.

Ley, E. and Steel, M. F. J. (1992). Bayesian econometrics, conjugate analysis and rejection sampling. *Economic and Financial Modelling with Mathematica* (H. R. Varian, ed.). Berlin: Springer, 344–367.

Lindgren, B. W. (1971). *Elements of Decision Theory*. London: Macmillan.

Lindley, D. V. (1953). Statistical inference. *J. Roy. Statist. Soc. B* **15**, 30–76.

Lindley, D. V. (1956). On a measure of information provided by an experiment. *Ann. Math. Statist.* **27**, 986–1005.

Lindley, D. V. (1957). A statistical paradox. *Biometrika* **44**, 187–192.

Lindley, D. V. (1958). Fiducial distribution and Bayes' Theorem. *J. Roy. Statist. Soc. B* **20**, 102–107.

Lindley, D. V. (1961a). Dynamic programming and decision theory. *Appl. Statist.* **10**, 39–51.

Lindley, D. V. (1961b). The use of prior probability distributions in statistical inference and decision. *Proc. Fourth Berkeley Symp.* **1** (J. Neyman and E. L. Scott, eds.). Berkeley: Univ. California Press, 453–468.

Lindley, D. V. (1964). The Bayesian analysis of contingency tables. *Ann. Math. Statist.* **35**, 1622-1643.

Lindley, D. V. (1965). *Introduction to Probability and Statistics from a Bayesian Viewpoint.* Cambridge: University Press.

Lindley, D. V. (1969). Review of Fraser (1968). *Biometrika* **56**, 453–456.

Lindley, D. V. (1971). The estimation of many parameters. *Foundations of Statistical Inference* (V. P. Godambe and D. A. Sprott, eds.). Toronto: Holt, Rinehart and Winston, 435–453 (with discussion).

Lindley, D. V. (1971/1985). *Making Decisions*. Second edition in 1985, Chichester: Wiley.

Lindley, D. V. (1972). *Bayesian Statistics, a Review*. Philadelphia, PA: SIAM.

Lindley, D. V. (1976). Bayesian Statistics. *Foundations of Probability Theory, Statistical Inference, and Statistical Theories of Science* **2** (W, L. Harper and C. A. Hooker, eds.), Dordrecht: Reidel, 353–363.

Lindley, D. V. (1977). A problem in forensic science. *Biometrika* **44**, 187–192.

Lindley, D. V. (1978). The Bayesian approach. *Scandinavian J. Statist.* **5**, 1–26.

Lindley, D. V. (1980a). Jeffreys's contribution to modern statistical thought. *Bayesian Analysis in Econometrics and Statistics: Essays in Honor of Harold Jeffreys* (A. Zellner, ed.). Amsterdam: North-Holland, 35–39.

Lindley, D. V. (1980b). Approximate Bayesian methods. *Bayesian Statistics* (J. M. Bernardo, M. H. DeGroot, D. V. Lindley and A. F. M. Smith, eds.). Valencia: University Press, 223–245 (with discussion).

Lindley, D. V. (1982a). Scoring rules and the inevitability of probability. *Internat. Statist. Rev.* **50**, 1–26 (with discussion).

Lindley, D. V. (1982b). Bayesian inference. *Encyclopedia of Statistical Sciences* **1** (S. Kotz, N. L. Johnson and C. B. Read, eds.). New York: Wiley, 197–204.

Lindley, D. V. (1982c). Coherence. *Encyclopedia of Statistical Sciences* **2** (S. Kotz, N. L. Johnson and C. B. Read, eds.). New York: Wiley, 29–31.

Lindley, D. V. (1982d). The improvement of probability judgements. *J. Roy. Statist. Soc. A* **145**, 117–126.

Lindley, D. V. (1983). Reconciliation of probability distributions. *Operations Res.* **31**, 866–880.

Lindley, D. V. (1984). The next 50 years. *J. Roy. Statist. Soc. A* **147**, 359–367.

Lindley, D. V. (1985). Reconciliation of discrete probability distributions. *Bayesian Statistics 2* (J. M. Bernardo, M. H. DeGroot, D. V. Lindley and A. F. M. Smith, eds.), Amsterdam: North-Holland, 375–390 (with discussion).

Lindley, D. V. (1986). The reconciliation of decision analyses. *Oper. Research* **14**, 289–295.

Lindley, D. V. (1987). The probability approach to the treatment of uncertainty in artificial intelligence and expert systems. *Statist. Sci.* **2**, 17–44 (with discussion).

Lindley, D. V. (1990). The present position in Bayesian Statistics. *Statist. Sci.* **5**, 44–89 (with discussion).

Lindley, D. V. (1991). Subjective probability, decision analysis and their legal consequences. *J. Roy. Statist. Soc. A* **154**, 83–92.

Lindley, D. V. (1992). Is our view of Bayesian statistics too narrow? *Bayesian Statistics 4* (J. M. Bernardo, J. O. Berger, A. P. Dawid and A. F. M. Smith, eds.). Oxford: University Press, 1–15 (with discussion).

Lindley, D. V. (1993). On the presentation of evidence. *Math. Scientist* **18**, 60–63.

Lindley, D. V. and Deely, J. J. (1993). Optimal allocation of stratified sampling with partial information. *Test* **2**, 147–160.

Lindley, D. V. and Novick, M. R. (1981). The role of exchangeability in inference. *Ann. Statist.* **9**, 45–58.

Lindley, D. V. and Phillips, L. D. (1976). Inference for a Bernoulli process (a Bayesian view). *Amer. Statist.* **30**, 112–119.

Lindley, D. V. and Scott, W. F. (1985). *New Cambridge Elementary Statistical Tables.* Cambridge: University Press.

Lindley, D. V. and Singpurwalla, N. D. (1991). On the evidence needed to reach agreed action between adversaries, with application to acceptance sampling. *J. Amer. Statist. Assoc.* **86**, 933–937.

Lindley, D. V. and Singpurwalla, N. D. (1993). Adversarial life testing. *J. Roy. Statist. Soc. B* **55**, 837–847.

Lindley, D. V. and Smith, A. F. M. (1972). Bayes estimates for the linear model. *J. Roy. Statist. Soc. B* **34**, 1–41 (with discussion).

Lindley, D. V., Tversky, A. and Brown, R. V. (1979). On the reconciliation of probability assessments. *J. Roy. Statist. Soc. A* **142**, 146–180.

Liseo, B. (1993). Elimination of nuisance parameters with reference priors. *Biometrika* **80**, 295–304.

Liseo, B., Petrella, L. and Salinetti, G. (1993). Block unimodality for multivariate Bayesian robustness. *J. It. Statist. Soc.* **2**, 55-71.

Little, R. J. A. and Rubin, D. B. (1987). *Statistical Analysis with Missing Data*. New York: Wiley.

Lo, A. Y. (1984). On a class of Bayesian non-parametric estimates: I. Density estimates. *Ann. Statist.* **12**, 351–357.

Lo, A. Y. (1986). Bayesian statistical inference for sampling a finite population. *Ann. Statist.* **14**, 1226–1233.

Lo, A. Y. (1987). A large sample study of the Bayesian bootstrap. *Ann. Statist.* **15**, 360–375.

Lo, A. Y. (1993). A Bayesian bootstrap for censored data. *Ann. Statist.* **20**, 100–123.

Luce, R. D. (1959). *Individual Choice Behaviour*. New York: Wiley.

Luce, R. D. (1992). When does subjective expected utility fail descriptively? *J. Risk and Uncertainty* **5**, 5–27.

Luce, R. D. and Krantz, D. H. (1971). Conditional expected utility. *Econometrica* **39**, 253–271.

Luce, R. D. and Narens, L. (1978). Qualitative independence in probability theory. *Theory and Decision* **9**, 225–239.

Luce, R. D. and Raiffa, H. (1957). *Games and Decisions. Introduction and Critical Survey*. Chichester: Wiley.

Luce, R. D. and Suppes, P. (1965). Preference, utility and subjective probability. *Handbook of Mathematical Psychology* **3** (R. D. Luce, Bush and Galanter, eds.). New York: Wiley, 249–410.

Lusted, L. B. (1968). *Introduction to Medical Decision Making*. Springfield, IL: Thomas.

Maatta, J. and Casella, G. (1990). Developments in decision theoretic variance estimation. *Statist. Sci.* **5**, 90–120 (with discussion).

Machina, M. (1982). 'Expected utility' analysis without the independence axiom. *Econometrica* **50**, 277–323.

Machina, M. (1987). Choices under uncertainty. Problems solved and unsolved. *J. Econ. Perspectives* **1**, 121–154.

Maín, P. (1988). Prior and posterior tail comparisons. *Bayesian Statistics 3* (J. M. Bernardo, M. H. DeGroot, D. V. Lindley and A. F. M. Smith, eds.). Oxford: University Press, 669–675.

Makov, U. E. (1988). On stochastic approximation and Bayes linear estimators. *Bayesian Statistics 3* (J. M. Bernardo, M. H. DeGroot, D. V. Lindley and A. F. M. Smith, eds.). Oxford: University Press, 697–699.

Mardia, K. V., Kent, J. T. and Walder, A. N. (1992). Statistical shape models in image analysis. *Computer Science and Statistics: Proc. 23rd. Symp. Interface* (E. M. Keramidas, ed.). Fairfax Station: Interface Foundation, 550–557.

Marinell, G. and Seeber, G. (1988). *Angewandte Statistik*. Munich: Oldenbourg Verlag.

Maritz, J. S. and Lwin, T. (1989). *Empirical Bayes Methods*. London: Chapman and Hall.

Marriott, J. M. (1988). Reparametrisation for Bayesian inference in ARMA time series. *Bayesian Statistics 3* (J. M. Bernardo, M. H. DeGroot, D. V. Lindley and A. F. M. Smith, eds.). Oxford: University Press, 701–704.

Marriott, J. M. and Naylor, J. C. (1993). Teaching Bayes on *MINITAB*. *Appl. Statist.* **42**, 223–232.

Marriott, J. M. and Smith, A. F. M. (1992). Reparametrisation aspects of numerical Bayesian methodology for autoregressive moving-average models. *J. Time Series Anal.* **13**, 327–343.

Marschak, J. (1950). Statistical inference in economics: an introduction. *Statistical Inference in Dynamic Economic Models*. New York: Cowles Commission, 1–50.

Marschak, J. and Radner, R. (1972). *Economic Theory of Teams*. New Haven: Yale University Press.

Martin, J. J. (1967). *Bayesian Decision Problems and Markov Chains*. New York: Wiley.

Martz, H. F. and Waller, R. A. (1982). *Bayesian Reliability Analysis*. New York: Wiley

Masreliez, C. J. (1975). Approximate non-Gaussian filtering with linear state and observation relations. *IEEE Trans. Automatic Control* **20**, 107–110.

Mathiasen, P. E. (1979). Prediction functions. *Scandinavian J. Statist.* **6**, 1–21.

Mazloum, R. and Meeden, G. (1987). Using the stepwise Bayes technique to choose between experiments. *Ann. Statist.* **15**, 269–277.

McCarthy, J. (1956). Measurements of the value of information. *Proc. Nat. Acad. Sci. USA* **42**, 654–655.

McCulloch, R. E. (1989). Local model influence. *J. Amer. Statist. Assoc.* **84**, 473–478.

McCulloch, R. E. and Rossi, P. E. (1992). Bayes factors for non-linear hypothesis and likelihood distributions. *Biometrika* **79**, 663–676.

McCulloch, R. E. and Tsay, R. S. (1993). Bayesian inference and prediction for mean and variance shifts in autoregressive time series. *J. Amer. Statist. Assoc.* **88**, 968–978.

Meeden, G. (1990). Admissible contour credible sets. *Statistics and Decisions* **8**, 1–10.

Meeden, G. and Isaacson, D. (1977). Approximate behavior of the posterior distribution for a large observation. *Ann. Statist.* **5**, 899–908.

Meeden, G. and Vardeman, S. (1991). A non-informative Bayesian approach to interval estimation in finite population sampling. *J. Amer. Statist. Assoc.* **86**, 972–986.

Meinhold, R. and Singpurwalla, N. D. (1983). Understanding the Kalman filter. *Amer. Statist.* **37**, 123–127.

Mendel, M. B. (1992). Bayesian parametric models for lifetimes. *Bayesian Statistics 4* (J. M. Bernardo, J. O. Berger, A. P. Dawid and A. F. M. Smith, eds.). Oxford: University Press, 697–705.

Mendoza, M. (1994). Asymptotic posterior normality under transformations. *Test* **3**, 173–180.

Meng X.-L. and Rubin, D. B. (1992). Recent extensions to the EM algorithm. *Bayesian Statistics 4* (J. M. Bernardo, J. O. Berger, A. P. Dawid and A. F. M. Smith, eds.). Oxford: University Press, 307–320 (with discussion).

Merkhofer, M. W. (1987). Quantifying judgemental uncertainty: methodology, experiences and insights. *IEEE Trans. Systems, Science and Cybernetics* **17**. 741–752.

Metropolis, N., Rosenbluth, A. W., Rosenbluth, M. N., Teller, A. H. and Teller, E. (1953). Equation of state calculations by fast computing machines. *J. Chem. Phys.* **21**, 1087–1092.

Meyer, D. L. and Collier, R. O. (eds.) (1970). *Bayesian Statistics*. Itasca, IL: Peacock.

Mills, J. A. (1992). Bayesian prediction tests for structural stability. *J. Econometrics* **52**, 381–388.

Mitchell, T. J. and Beauchamp, T. J. (1988). Bayesian variable selection in linear regression. *J. Amer. Statist. Assoc.* **83**, 1023–1035 (with discussion).

Mitchell, T. J. and Morris, M. D. (1992). Bayesian design and analysis of computer experiments: two examples. *Statistica Sinica* **2**, 359–379.

Mockus, J. (1989). *Bayesian Approach to Global Optimization*. Dordrecht: Kluwer.

Mohammad-Djafari, A. and Demoment, G. (eds.) (1993). *Maximum Entropy and Bayesian Methods*. Dordrecht: Kluwer.

Monahan, J. F. and Boos, D. D. (1992). Proper likelihoods for Bayesian analysis. *Biometrika* **79**, 271–278.

Morales, J. A. (1971). *Bayesian Full Information Structural Analysis*. Berlin: Springer.

Moreno, E. and Cano, J. A. (1989). Testing a point null hypothesis: asymptotic robust Bayesian analysis with respect to priors given on a sub-sigma field. *Internat. Statist. Rev.* **57**, 221-232.

Moreno, E. and Cano, J. A. (1991). Robust Bayesian analysis with ϵ-contaminations partially known. *J. Roy. Statist. Soc. B* **53**, 143–155.

Moreno, E. and Pericchi, L. R. (1992). Bands of probability measures: a robust Bayesian analysis. *Bayesian Statistics 4* (J. M. Bernardo, J. O. Berger, A. P. Dawid and A. F. M. Smith, eds.). Oxford: University Press, 707–713.

Moreno, E. and Pericchi, L. R. (1993). Prior assessments for bands of probability measures: empirical Bayesian analysis. *Test* **2**, 101–110.

Morgan, M. G. and Henrion, M. (1990). *Uncertainty: a Guide to Dealing with Uncertainty in Quantitative Risk and Policy Analysis*. Cambridge: University Press.

Morris, C. N. (1982). Natural exponential families with quadratic variance functions. *Ann. Statist.* **10**, 65–80.

Morris, C. N. (1983). Parametric empirical Bayes inference: theory and applications. *J. Amer. Statist. Assoc.* **8**, 47–59.

Morris, C. N. (1988). Approximating posterior distributions and posterior moments. *Bayesian Statistics 3* (J. M. Bernardo, M. H. DeGroot, D. V. Lindley and A. F. M. Smith, eds.). Oxford: University Press, 327–344 (with discussion).

Morris, P. A. (1974). Decision analysis expert use. *Manag. Sci.* **20**, 1233–1241.

Morris, W. T. (1968). *Management Science, a Bayesian Introduction*. Englewood Cliffs, NJ: Prentice-Hall.

Mortera, J. (1986). Bayesian forecasting. *Metron* **44**, 277-296.

Mosteller, F. and Wallace, D. L. (1964/1984). *Inference and Disputed Authorship: The Federalist*. Reading, MA: Addison-Wesley. Second edition, published in 1984 as *Applied Bayesian and Classical Inference, the Case of the Federalist Papers*. Berlin: Springer.

Mosteller, F. and Youtz, C. (1990). Quantifying probabilistic expressions. *Statist. Sci.* **5**, 2–24 (with discussion).

Mouchart, M. (1976). A note on Bayes' theorem. *Statistica* **36**, 349–357.

Mouchart, M. and Simar, L. (1980). Least squares approximation in Bayesian analysis. *Bayesian Statistics* (J. M. Bernardo, M. H. DeGroot, D. V. Lindley and A. F. M. Smith, eds.). Valencia: University Press, 207–222 and 237–245 (with discussion).

Muirhead, C. R. (1986). Distinguishing outlier types in time series. *J. Roy. Statist. Soc. B* **48**, 39–47.

Mukerjee, R. and Dey, D. K. (1993). Frequentist validity of posterior quantiles in the presence of a nuisance parameter: Higher order asymptotics. *Biometrika* **80**, 499–505.

Murphy, A. H. and Epstein, E. S. (1967). Verification of probabilistic predictions: a brief review. *J. Appl. Meteorology* **6**, 748–755.

Murray, R. G., McKillop, J. H., Bessant, R. G., Hutton, I, Lorimer, A. R. and Lawrie, T. D. V. (1981). Bayesian analysis of stress thallium-201 scintigraphy. *Eur. J. Nucl. Med.* **6**, 201–204.

Myerson, R. B. (1979). An axiomatic derivation of subjective probability, utility and evaluation functions. *Theory and Decision* **11**, 339–352.

Nakamura, Y. (1993). Subjective utility with upper and lower probabilities on finite states. *J. Risk and Uncertainty* **6**, 33–48.

Narens, L. (1976). Utility, uncertainty and trade-off structures. *J. Math. Psychol.* **13**, 296–332.

Nau, R. F. (1992). Indeterminate probabilities on finite sets. *Ann. Statist.* **20**, 1737–1767.

Nau, R. F. and McCardle, K. F. (1990). Coherent behavior in non-cooperative games. *J. Economic Theory* **50**, 242–444.

Naylor, J. C. and Smith, A. F. M. (1982). Applications of a method for the efficient computation of posterior distributions. *Appl. Statist.* **31**, 214–225.

Naylor, J. C. and Smith, A. F. M. (1988). Economic illustrations of novel numerical integration methodology for Bayesian inference. *J. Econometrics* **38**, 103–125.

Nelder, J. A. and Wedderburn, R. W. M. (1972). Generalised linear models. *J. Roy. Statist. Soc. A* **135**, 370–384.

Neyman, J. (1935). Sur un teorema concerte le cosidette statistiche sufficenti. *Giorn. Ist. Ital.* **6**, 320–334.

Neyman, J. and Pearson, E. S. (1933). On the problem of the most efficient tests of statistical hypothesis. *Phil. Trans. Roy. Soc. London A* **231**, 289–337.

Neyman, J. and Pearson, E. S. (1967). *Joint Statistical Papers*. Cambridge: University Press.

Neyman, J. and Scott, E. L. (1948). Consistent estimates based on partially consistent observations. *Econometrica* **16**, 1–32.

Nicolau, A. (1993). Bayesian intervals with good frequentist behaviour in the presence of nuisance parameters. *J. Roy. Statist. Soc. B* **55**, 377–390.

Normand, S.-L. and Tritchler, D. (1992). Parameter updating in a Bayes network. *J. Amer. Statist. Assoc.* **87**, 1109–1115.

Novick, M. R. (1969). Multiparameter Bayesian indifference procedures. *J. Roy. Statist. Soc. B* **31**, 29–64.

Novick, M. R. and Hall, W. K. (1965). A Bayesian indifference procedure. *J. Amer. Statist. Assoc.* **60**, 1104–1117.

Novick, M. R. and Jackson, P. H. (1974). *Statistical Methods for Educational and Psychological Research*. New York: McGraw-Hill.

O'Hagan, A. (1979). On outlier rejection phenomena in Bayes inference. *J. Roy. Statist. Soc. B* **41**, 358–367.

O'Hagan, A. (1981). A moment of indecision. *Biometrika* **68**, 329–330.

O'Hagan, A. (1988a). *Probability: Methods and Measurements*. London: Chapman and Hall.

O'Hagan, A. (1988b). Modelling with heavy tails. *Bayesian Statistics 3* (J. M. Bernardo, M. H. DeGroot, D. V. Lindley and A. F. M. Smith, eds.). Oxford: University Press, 345–359 (with discussion).

O'Hagan, A. (1990). Outliers and credence for location parameter inference. *J. Amer. Statist. Assoc.* **85**, 172–176.

O'Hagan, A. (1991). Bayes-Hermite quadrature. *J. Statist. Planning and Inference* **29**, 245–260.

O'Hagan, A. (1992). Some Bayesian numerical analysis. *Bayesian Statistics 4* (J. M. Bernardo, J. O. Berger, A. P. Dawid and A. F. M. Smith, eds.). Oxford: University Press, 345-363 (with discussion).

O'Hagan, A. (1994a). *Kendall's Advanced Theory of Statistics* **2B**: *Bayesian Inference*. London: Edward Arnold

O'Hagan, A. (1994b). Robust modelling for asset management. *J. Statist. Planning and Inference* **40**, 245–259.

O'Hagan, A. and Berger, J. O. (1988). Ranges of posterior probabilities for quasimodal priors with specified quantiles. *J. Amer. Statist. Assoc.* **83**, 503–508.

O'Hagan, A. and Le, H. (1994). Conflicting information and a class of bivariate heavy-tailed distributions. *Aspects of Uncertainty: a Tribute to D. V. Lindley* (P. R. Freeman, and A. F. M. Smith, eds.). Chichester: Wiley, 311–327.

Oliver, R. M. and Smith, J. Q. (eds.) (1990). *Influence Diagrams, Belief Nets and Decision Analysis*. Chichester: Wiley.

Osiewalski, J. and Steel, M. F. J. (1993). Robust Bayesian inference in l_q-spherical models. *Biometrika* **80**, 456–460.

Osteyee, D. D. B. and Good, I. J. (1974). *Information, Weight of Evidence, the Singularity between Probability Measures and Signal Detection*. Berlin: Springer.

Pack, D. J. (1986a). Posterior distributions. Posterior probabilities. *Encyclopedia of Statistical Sciences* **7** (S. Kotz, N. L. Johnson and C. B. Read, eds.). New York: Wiley, 121–124.

Pack, D. J. (1986b). Prior distributions. *Encyclopedia of Statistical Sciences* **7** (S. Kotz, N. L. Johnson and C. B. Read, eds.). New York: Wiley, 194–196.

Padgett, W. J. and Wei, L. J. (1981). A Bayesian nonparametric estimator of survival probability assuming increasing failure rate. *Comm. Statist. Theory and Methods* **10**, 49–63.

Page, A. N. (ed.) (1968). *Utility Theory: A Book of Readings*. New York: Wiley.

Pardo, L., Taneja, I. J. and Morales, D. (1991). λ-measures of hypoentropy and comparison of experiments: Bayesian approach. *The Statistician* **51**, 173–184.

Parenti, G. (ed.) (1978). *I Fondamenti dell'Inferenza Statistica*. Florence: Università degli Studi.

Parmigiani, G. and Berry, D. A. (1994). Applications of Lindley information to the design of clinical experiments. *Aspects of Uncertainty: a Tribute to D. V. Lindley* (P. R. Freeman, and A. F. M. Smith, eds.). Chichester: Wiley, 329–348.

Pearson, E. S. (1978). *The History of Statistics in the 17th and 18th Centuries*. London: Macmillan.

Peers, H. W. (1965). On confidence points and Bayesian probability points in the case of several parameters. *J. Roy. Statist. Soc. B* **27**, 9–16.

Peers, H. W. (1968). Confidence properties of Bayesian interval estimates. *J. Roy. Statist. Soc. B* **30**, 535–544.

Peirce, C. S. (1878). How to make our ideas clear. *Popular Science Monthly* **12**, 286–302.

Peizer, D. B. and Pratt, J. W. (1968). A normal approximation for binomial, F, beta, and other common related tail probabilities. *J. Amer. Statist. Assoc.* **43**, 24–26.

Peña, D. and Guttman, I. (1993). Comparing probabilistic methods for outlier detection. *Biometrika* **80**, 603–610.

Peña, D. and Tiao, G. C. (1992). Bayesian robustness functions for linear models. *Bayesian Statistics 4* (J. M. Bernardo, J. O. Berger, A. P. Dawid and A. F. M. Smith, eds.). Oxford: University Press, 365–388 (with discussion).

Pereira, C. A. de B. and Lindley, D. V. (1987). Examples questioning the use of partial likelihood. *The Statistician* **37**, 15–20.

Pérez, M. E. and Pericchi, L. R. (1992). Analysis of multistage survey as a hierarchical model. *Bayesian Statistics 4* (J. M. Bernardo, J. O. Berger, A. P. Dawid and A. F. M. Smith, eds.). Oxford: University Press, 723–730.

Pericchi, L. R. (1981). A Bayesian approach to transformations to normality. *Biometrika* **68**, 35–43.

Pericchi, L. R. (1984). An alternative to the standard Bayesian procedure for discrimination between normal linear models. *Biometrika* **71**, 576–586.

Pericchi, L. R. (1993). Personal communication.

Pericchi, L. R. and Nazaret, W. A. (1988). On being imprecise at the higher levels of a hierarchical linear model. *Bayesian Statistics 3* (J. M. Bernardo, M. H. DeGroot, D. V. Lindley and A. F. M. Smith, eds.). Oxford: University Press, 361–375 (with discussion).

Pericchi, L. R. and Pérez, M. E. (1994). Posterior robustness with more than one sampling model. *J. Statist. Planning and Inference* **40**, 279–984.

Pericchi, L. R., Sansó, B. and Smith, A. F. M. (1993). Posterior cumulant relationships in Bayesian inference involving the exponential family. *J. Amer. Statist. Assoc.* **88**, 1419–1426.

Pericchi, L. R. and Smith, A. F. M. (1992). Exact and approximate posterior moments for a normal location parameter. *J. Roy. Statist. Soc. B* **54**, 793–804.

Pericchi, L. R. and Walley, P. (1991). Robust Bayesian credible intervals and prior ignorance. *Internat. Statist. Rev.* **59**, 1–23.

Perks, W. (1947). Some observations on inverse probability, including a new indifference rule. *J. Inst. Actuaries* **73**, 285–334 (with discussion).

Peskun, P. H. (1973). Optimal Monte Carlo sampling using Markov chains. *Biometrika* **60**, 607–612.

Pettit, L. I. (1986). Diagnostics in Bayesian model choice. *The Statistician* **35**, 183–190.

Pettit, L. I. (1992). Bayes factors for outlier models using the device of imaginary observations. *J. Amer. Statist. Assoc.* **87**, 541–545.

Pettit, L. I. and Smith, A. F. M. (1985). Outliers and influential observation in linear models. *Bayesian Statistics 2* (J. M. Bernardo, M. H. DeGroot, D. V. Lindley and A. F. M. Smith, eds.), Amsterdam: North-Holland, 473–494 (with discussion).

Pettit, L. I. and Young, K. S. (1990). Measuring the effect of observations on Bayes factors. *Biometrika* **77**, 455–466.

Pfanzagl, J. (1967). Subjective probability derived from the Morgenstern-von Neumann utility concept. *Essays in Mathematical Economics* (M. Shubik, ed.). Princeton: University Press, 237–251.

Pfanzagl, J. (1968). *Theory of Measurement*. Chichester: Wiley.

Pham-Gia, T. and Turkkan, N. (1992). Sample size determination in Bayesian analysis. *The Statistician* **41**, 389–404.

Phillips, L. D. (1973). *Bayesian Statistics for Social Scientists*. London: Nelson.

Piccinato, L. (1973). Un metodo per determinare distribuzioni iniziali relativamente noninformative. *Metron* **31**, 124–156.

Piccinato, L. (1977). Predictive distributions and non-informative priors. *Trans. 7th. Prague Conf. Information Theory* (M. Uldrich, ed.). Prague: Czech. Acad. Sciences, 399–407.

Piccinato, L. (1986). De Finetti's logic of uncertainty and its impact on statistical thinking and practice. *Bayesian Inference and Decision Techniques: Essays in Honor of Bruno de Finetti* (P. K. Goel and A. Zellner, eds.). Amsterdam: North-Holland, 13–20.

Piccinato, L. (1992). Critical issues in different inferential paradigms. *J. It. Statist. Soc.* **2**, 251–274.

Pierce, D. (1973). On some difficulties in a frequency theory of inference. *Ann. Statist.* **1**, 241–250.

Pilz, J. (1983/1991). *Bayesian Estimation and Experimental Design in Linear Regression Models*. Leipzig: Teubner. Second edition in 1991, Chichester: Wiley.

Pitman E. J. G. (1936). Sufficient statistics and intrinsic accuracy. *Proc. Camb. Phil Soc.* **32**, 567–579.

Pitman E. J. G. (1939). Location and scale parameters. *Biometrika* **36**, 391–421.

Plante, A. (1971). Counter-example and likelihood. *Foundations of Statistical Inference* (V. P. Godambe and D. A. Sprott, eds.). Toronto: Holt, Rinehart and Winston, 357–371 (with discussion).

Plante, A. (1984). A reexamination of Stein's antifiducial example. *Canad. J. Statist.* **12**, 135–141.

Plante, A. (1991). An inclusion-consistent solution to the problem of absurd confidence statements. *Canad. J. Statist.* **19**, 389–397.

Poirier, D. J. (1985). Bayesian hypothesis testing in linear models with continuously induced conjugate priors across hypotheses. *Bayesian Statistics 2* (J. M. Bernardo, M. H. DeGroot, D. V. Lindley and A. F. M. Smith, eds.), Amsterdam: North-Holland, 711–722.

Poirier, D. J. (1993). *Intermediate Statistics and Econometrics: a Comparative Approach*. Cambridge, MA: The MIT Press.

Polasek, W. and Pötzelberger, K. (1988). Robust Bayesian analysis in hierarchical models. *Bayesian Statistics 3* (J. M. Bernardo, M. H. DeGroot, D. V. Lindley and A. F. M. Smith, eds.). Oxford: University Press, 377–394.

Polasek, W. and Pötzelberger, K. (1994). Robust Bayesian methods in simple ANOVA problems. *J. Statist. Planning and Inference* **40**, 295–311.

Pole, A. and West, M. (1989). Reference analysis of the dynamic linear model. *J. Time Series Analysis* **10**, 13–147.

Pole, A., West, M. and Harrison P. J. (1994). *Applied Bayesian Forecasting and Time Series Analysis*. New York: Chapman and Hall.

Pollard, W. E. (1986). *Bayesian Statistics for Evaluation Research: an Introduction*. Beverly Hills, CA: Sage.

Polson, N. G. (1991). A representation of the posterior mean for a location model. *Biometrika* **78**, 426–430.

Polson, N. G. (1992). In discussion of Ghosh and Mukerjee (1992). *Bayesian Statistics 4* (J. M. Bernardo, J. O. Berger, A. P. Dawid and A. F. M. Smith, eds.). Oxford: University Press, 203–205.

Polson, N. G. and Tiao, G. C. (1995). *Bayesian Inference* (2 volumes). Aldershot: Edward Elger.

Poskitt, D. S. (1987). Precision, complexity and Bayesian model determination. *J. Roy. Statist. Soc. B* **49**, 199–208.

Pötzelberger, K. and Polasek, W. (1991). Robust HPD regions in Bayesian regression models. *Econometrica* **59**, 1581–1590.

Pratt, J. W. (1961). Length of confidence intervals. *J. Amer. Statist. Assoc.* **56**, 549–567.

Pratt, J. W. (1964). Risk aversion in the small and in the large. *Econometrica* **32**, 122–136.

Pratt, J. W. (1965). Bayesian interpretation of standard inference statements. *J. Roy. Statist. Soc. B* **27**, 169–203.

Pratt, J. W., Raiffa, H. and Schlaifer, R. (1964). The foundations of decision under uncertainty: an elementary exposition. *J. Amer. Statist. Assoc.* **59**, 353–375.

Pratt, J. W., Raiffa, H. and Schlaifer, R. (1965). *Introduction to Statistical Decision Theory*. New York: McGraw-Hill.

Press, S. J. (1972/1982). *Applied Multivariate Analysis: using Bayesian and Frequentist Methods of Inference*. Second edition in 1982, Melbourne, FL: Krieger.

Press, S. J. (1978). Qualitative controlled feedback for forming group judgements and making decisions. *J. Amer. Statist. Assoc.* **73**, 526–535.

Press, S. J. (1980a). Bayesian Inference in MANOVA. *Handbook of Statistics 1. Analysis of Variance*. (P. R. Krishnaiah, ed.). Amsterdam: North-Holland, 117–132.

Press, S. J. (1980b). Bayesian inference in group judgement formulation and decision making using qualitative controlled feedback. *Bayesian Statistics* (J. M. Bernardo, M. H. DeGroot, D. V. Lindley and A. F. M. Smith, eds.). Valencia: University Press, 383–430 (with discussion).

Press, S. J. (1985a). Multivariate Analysis (Bayesian). *Encyclopedia of Statistical Sciences* **6** (S. Kotz, N. L. Johnson and C. B. Read, eds.). New York: Wiley, 16–20.

Press, S. J. (1985b). Multivariate group assessment of probabilities of nuclear war. *Bayesian Statistics 2* (J. M. Bernardo, M. H. DeGroot, D. V. Lindley and A. F. M. Smith, eds.), Amsterdam: North-Holland, 425–462 (with discussion).

Press, S. J. (1989). *Bayesian Statistics*. New York: Wiley.

Rabena, M. (1998). Deriving reference decisions. *Test* **7**, 161–178.

Racine-Poon, A. (1988). A Bayesian approach to non-linear calibration problems. *J. Amer. Statist. Assoc.* **83**, 650–656.

Racine-Poon, A. (1992). *SAGA*: Sample assisted graphical analysis. *Bayesian Statistics 4* (J. M. Bernardo, J. O. Berger, A. P. Dawid and A. F. M. Smith, eds.). Oxford: University Press, 389–404 (with discussion).

Racine-Poon, A., Grieve, A. P., Flühler, H. and Smith, A. F. M. (1986). Bayesian methods in practice: experiences in the pharmaceutical industry. *Appl. Statist.* **35**, 93–150 (with discussion).

Raftery, A. E. and Lewis, S. M. (1992). How many iterations in the Gibbs sampler? *Bayesian Statistics 4* (J. M. Bernardo, J. O. Berger, A. P. Dawid and A. F. M. Smith, eds.). Oxford: University Press, 763–773.

Raftery, A. E. and Schweder, T. (1993). Inference about the ratio of two parameters, with applications to whale censusing. *Amer. Statist.* **47**, 259–264.

Raiffa, H. (1961). Risk ambiguity and the Savage axioms. Comment. *Quart. J. Econ.* **75**, 690–694.

Raiffa, H. (1968). *Decision Analysis. Introductory Lectures on Choices under Uncertainty.* Reading, MA: Addison-Wesley

Raiffa, H. (1982). *The Art and Science of Negotiation.* Cambridge: University Press

Raiffa, H. and Schlaifer, R. (1961). *Applied Statistical Decision Theory.* Boston: Harvard University.

Ramsey, F. P. (1926). Truth and probability. *The Foundations of Mathematics and Other Logical Essays* (R. B. Braithwaite, ed.). London: Kegan Paul (1931), 156–198. Reprinted in 1980 in *Studies in Subjective Probability* (H. E. Kyburg and H. E Smokler, eds.). New York: Dover, 61–92.

Ramsey, J. O. and Novick, M. R. (1980). PLU robust Bayesian decision theory: point estimation. *J. Amer. Statist. Assoc.* **75**, 901–907.

Randall, C. H. and Foulis, D. J. (1975). A mathematical setting for inductive reasoning. *Foundations of Probability Theory, Statistical Inference, and Statistical Theories of Science* **3** (W. L. Harper and C. A. Hooker, eds.). Dordrecht: Reidel.

Rao, C. R. (1945). Information and accuracy attainable in estimation of statistical parameters. *Bull. Calcutta Math. Soc.* **37**, 81–91.

Regazzini, E. (1983). *Sulle Probabilità Coerenti nel Senso di de Finetti.* Bologna: Clueb.

Regazzini, E. (1987). De Finetti's coherence and statistical inference. *Ann. Statist.* **15**, 845–864.

Regazzini, E. and Petris, G. (1992). Some critical aspects of the use of exchangeability in statistics. *J. It. Statist. Soc.* **1**, 103–130.

Reichenbach, H. (1935). *The Theory of Probability.* Berkeley: Univ. California Press.

Renyi, A. (1955). On a new axiomatic theory of probability. *Acta Math. Acad. Sci. Hungaricae* **6**, 285–335.

Renyi, A. (1961). On measures of entropy and information. *Proc. Fourth Berkeley Symp.* **1** (J. Neyman and E. L. Scott, eds.). Berkeley: Univ. California Press, 547–561.

Renyi, A. (1962/1970). *Wahrscheinlichkeitsrechnung.* Berlin: Deutscher Verlag der Wissenschaften. English translation in 1970 as *Probability Theory.* San Francisco, CA: Holden-Day.

Renyi, A. (1964). On the amount of information concerning an unknown parameter in a sequence of observations. *Pub. Math. Inst. Hung. Acad. Sci.* **9**, 617–624.

Renyi, A. (1966). On the amount of missing information and the Neyman-Pearson lemma. *Research Papers in Statistics. Festschrift for J. Neyman* (F. N. David, ed.). New York: Wiley, 281–288.

Renyi, A. (1967). On some basic problems of statistics from the point of view of information theory. *Proc. Fifth Berkeley Symp.* **1** (J. Neyman and E. L. Scott, eds.). Berkeley: Univ. California Press, 531–543.

Ressel, P. (1985). de Finetti type theorems: an analytical approach. *Ann. Prob.* **13**, 818–922.

Richard, J. F. (1973). *Posterior and Predictive Densities for Simultaneous Equations Models*. Berlin: Springer.

Ríos, D. (1990). *Sensitivity Analysis in Multiobjective Decision Making*. Berlin: Springer.

Ríos, D. (1992). Foundations for a robust theory of decision making: the simple case. *Test* **1**, 69–78.

Ríos, D. and Martín, J. (1994). Robustness issues under precise beliefs and preferences. *J. Statist. Planning and Inference* **40**, 383–389.

Ríos, S. (1977). *Análisis de Decisiones*. Madrid: ICE.

Ríos, S., Ríos, S. Jr. and Ríos, M. J. (1989). *Procesos de Decision Multicriterio*. Madrid: Eudema.

Ripley, B. D. (1987). *Stochastic Simulation*. Chichester: Wiley.

Rissanen, J. (1983). A universal prior for integers and estimation by minimum description length. *Ann. Statist.* **11**, 416–431.

Rissanen, J. (1987). Stochastic complexity. *J. Roy. Statist. Soc. B* **49**, 223-239 and 252-265 (with discussion).

Rissanen, J. (1989). *Stochastic Complexity in Statistical Enquiry*. Singapore: World Scientific.

Ritter, C, and Tanner, M. A. (1992). Facilitating the Gibbs sampler: the Gibbs stopper and the griddy-Gibbs sampler. *J. Amer. Statist. Assoc.* **87**, 861–868.

Rivadulla, A. (1991). *Probabilidad e Inferencia Científica*. Barcelona: Anthropos.

Robbins, H. (1955). An empirical Bayes approach to statistics. *Proc. Third Berkeley Symp.* **1** (J. Neyman and E. L. Scott, eds.). Berkeley: Univ. California Press, 157–164.

Robbins, H. (1964). The empirical Bayes approach to statistical decision problems. *Ann. Math. Statist.* **35**, 1–20.

Robbins, H. (1983). Some thoughts on empirical Bayes estimation. *Ann. Statist.* **1**, 713–723.

Robert, C. P. (1992). *L'Analyse Statistique Bayésienne*. Paris: Economica.

Robert, C. P. (1993). A note on Jeffreys-Lindley paradox. *Statistica Sinica* **3**, 603–608.

Robert, C. P., Hwang, J. T. G. and Strawderman, W. E. (1993). Is Pitman closeness a reasonable criterion? *J. Amer. Statist. Assoc.* **88**, 57–76 (with discussion).

Robert, C. P. and Soubiran, C. (1993). Estimation of a normal mixture model through Gibbs sampling and prior feedback. *Test* **2**, 125–146.

Roberts, F. (1974). Laws of exchange and their applications. *SIAM J. Appl. Math.* **26**, 260–284.

Roberts, F. (1979). *Measurement Theory*. Reading, MA: Addison-Wesley

Roberts, G. O. (1992). Convergence diagnostics of the Gibbs sampler. *Bayesian Statistics 4* (J. M. Bernardo, J. O. Berger, A. P. Dawid and A. F. M. Smith, eds.). Oxford: University Press, 775–782.

Roberts, G. O. and Smith, A. F. M. (1994). Simple conditions for the convergence of the Gibbs sampler and Metropolis-Hastings algorithms. *Stoch. Proc. and their Applic.* **44**, 207–216.

Roberts, H. V. (1963). Risk ambiguity and the Savage axioms. Comment. *Quart. J. Econ.* **77**, 327–342.

Roberts, H. V. (1965). Probabilistic prediction. *J. Amer. Statist. Assoc.* **60**, 50–62.

Roberts, H. V. (1966). *Statistical Inference and Decision.* Chicago: University Press.

Roberts, H. V. (1967). Informative stopping rules and inferences about population size. *J. Amer. Statist. Assoc.* **62**, 763–775.

Roberts, H. V. (1974). Reporting of Bayesian studies. *Studies in Bayesian Econometrics and Statistics: in Honor of Leonard J. Savage* (S. E. Fienberg and A. Zellner, eds.). Amsterdam: North-Holland, 465–483.

Roberts, H. V. (1978). Bayesian inference. *International Encyclopedia of Statistics* (W. H. Kruskal, and J. M. Tanur, eds.). London: Macmillan, 9–16.

Robinson, G. K. (1975). Some counter-examples to the theory of confidence intervals. *Biometrika* **62**, 155–161.

Robinson, G. K. (1979a). Conditional properties of statistical procedures. *Ann. Statist.* **7**, 742–755.

Robinson, G. K. (1979b). Conditional properties of statistical procedures for location and scale parameters. *Ann. Statist.* **7**, 756–771.

Rodríguez, C. C. (1991). From Euclid to entropy. *Maximum Entropy and Bayesian Methods* (W. T. Grandy and L. H. Schick eds.). Dordrecht: Kluwer, 343–348.

Rolin, J.-M. (1983). Non-parametric Bayesian statistics: a stochastic processes approach. *Specifying Statistical Models* (J.-P. Florens *et al.* eds.). Berlin: Springer. 108–133.

Rosenkranz, R. D. (1977). *Inference, Method and Decision. Towards a Bayesian Philosophy of Science.* Dordrecht: Reidel.

Royall, R. M. (1992). The elusive concept of statistical evidence. *Bayesian Statistics 4* (J. M. Bernardo, J. O. Berger, A. P. Dawid and A. F. M. Smith, eds.). Oxford: University Press, 405–418 (with discussion).

Rubin, D. B. (1981). The Bayesian bootstrap. *Ann. Statist.* **9**, 130–134.

Rubin, D. B. (1984). Bayesianly justifiable and relevant frequency calculations for the applied statistician. *Ann. Statist.* **12**, 1151–1172.

Rubin, D. B. (1987). *Multiple Imputation for Non-Response in Surveys.* New York: Wiley

Rubin, D. B. (1988). Using the SIR algorithm to simulate posterior distributions. *Bayesian Statistics 3* (J. M. Bernardo, M. H. DeGroot, D. V. Lindley and A. F. M. Smith, eds.). Oxford: University Press, 395–402 (with discussion).

Rubin, H. (1971). A decision-theoretic approach to the problem of testing a null hypothesis. *Statistical Decision Theory and Related Topics* (S. S. Gupta and J. Yackel, eds.). New York: Academic Press, 103–108.

Rubin, H. (1977). Robust Bayesian estimation. *Statistical Decision Theory and Related Topics II* (S. S. Gupta and D. S. Moore, eds.). New York: Academic Press,

Rubin, H. (1987). A weak system of axioms for 'rational' behaviour and the non-separability of utility from prior. *Statistics and Decisions* **5**, 47–58.

Rubin, H. (1988a). Some results on robustness in testing. *Statistical Decision Theory and Related Topics IV* **1** (S. S. Gupta and J. O. Berger, eds.). Berlin: Springer, 271–278.

Rubin, H. (1988b). Robustness in generalized ridge regression and related topics. *Bayesian Statistics 3* (J. M. Bernardo, M. H. DeGroot, D. V. Lindley and A. F. M. Smith, eds.). Oxford: University Press, 403–410 (with discussion).

Rueda, R. (1992). A Bayesian alternative to parametric hypothesis testing. *Test* **1**, 61-67.

Saaty, T. L. (1980). *The Analytic Hierarchy Process*. New York: McGraw-Hill.

Sacks, J. (1963). Generalized Bayes solutions in estimation problems. *Ann. Math. Statist.* **34**, 787–794.

Salinetti, G. (1994). Stability of Bayesian decisions. *J. Statist. Planning and Inference* , (to appear).

San Martini, A. and Spezzaferri F. (1984). A predictive model selection criterion. *J. Roy. Statist. Soc. B* **46**, 296–303.

Sansó, B. and Pericchi, L. R. (1992). Near ignorance classes of log-concave priors for the location model. *Test* **1**, 39–46.

Särndal C.-E. (1970). A class of explicata for 'information' and 'weight of evidence'. *Internat. Statist. Rev.* **38**, 223–235.

Savage, I. R. (1968). *Statistics: Uncertainty and Behavior*. Boston: Houghton Miffin.

Savage, I. R. (1980). On not being rational. *Bayesian Statistics* (J. M. Bernardo, M. H. DeGroot, D. V. Lindley and A. F. M. Smith, eds.). Valencia: University Press, 321–328 and 339–346 (with discussion).

Savage, L. J. (1954/1972). *The Foundations of Statistics*. New York: Wiley. Second edition in 1972, New York: Dover.

Savage, L. J. (1962) (with others). *The Foundations of Statistical Inference: a Discussion*. London: Methuen.

Savage, L. J. (1961). The foundations of statistics reconsidered. *Proc. Fourth Berkeley Symp.* **1** (J. Neyman and E. L. Scott, eds.). Berkeley: Univ. California Press, 575–586. Reprinted in 1980 in *Studies in Subjective Probability* (H. E. Kyburg and H. E Smokler, eds.). New York: Dover, 175–188.

Savage, L. J. (1970). Reading suggestions for the foundations of statistics. *Amer. Statist.* **24**, 23–27.

Savage, L. J. (1971). Elicitation of personal probabilities and expectations. *J. Amer. Statist. Assoc.* **66**, 781–801. Reprinted in 1974 in *Studies in Bayesian Econometrics and Statistics: in Honor of Leonard J. Savage* (S. E. Fienberg and A. Zellner, eds.). Amsterdam: North-Holland, 111–156.

Savage, L. J. (1981). *The Writings of Leonard Jimmie Savage: a Memorial Collection*. Washington: ASA/IMS.

Savchuk, V. P. (1989). *Bayesovskiye Metodi Statisticheskogo Otsenivaniya*. Moscow: Nauka.

Sawagari, Y., Sunahara, Y. and Nakamizo, T. (1967). *Statistical Decision Theory in Adaptive Control Systems*. New York: Academic Press.

Schervish, M. J., Seidenfeld, T. and Kadane, J. B. (1990). State-dependent utilities. *J. Amer. Statist. Assoc.* **85**, 840–847.

Schervish, M. J., Seidenfeld, T. and Kadane, J. B. (1992). Bayesian analysis of linear models. *Bayesian Statistics 4* (J. M. Bernardo, J. O. Berger, A. P. Dawid and A. F. M. Smith, eds.). Oxford: University Press, 419–434 (with discussion).

Schlaifer, R. (1959). *Probability and Statistics for Business Decisions*. New York: McGraw-Hill.

Schlaifer, R. (1961). *Introduction to Statistics for Business Decisions*. New York: McGraw-Hill.

Schlaifer, R. (1969). *Analysis of Decisions under Uncertainty*. New York: McGraw-Hill.

Schmitt, S. A. (1969). *Measuring Uncertainty: an Elementary Introduction to Bayesian Statistics*. Reading, MA: Addison-Wesley

Schwartz, L. (1965). On Bayes procedures. *Z. Wahr.* **4**, 10–26.

Schwarz, G. (1978). Estimating the dimension of a model. *Ann. Statist.* **6**, 461–464.

Scott, D. (1964). Measurement structures and linear inequalities. *J. Math. Psychology* **1**, 233–247.

Scozzafava, R. (1989). *La Probabilità Soggettiva e le sue Applicazioni*. Milano: Veschi.

Seidenfeld, T. (1979). *Philosophical Problems of Statistical Inference*. Dordrecht: Reidel.

Seidenfeld, T. (1992). R. A. Fisher's fiducial argument and Bayes' theorem. *Statist. Sci.* **7**, 358–368.

Seidenfeld, T., Kadane, J. B. and Schervish, M. J. (1989). On the shared preferences of two Bayesian decision makers. *J. of Psychology* **5**, 225–244.

Seidenfeld, T. and Schervish, M. J. (1983). A conflict between finite additivity and avoiding Dutch book. *Philos. of Science* **50**, 398–412.

Sen, A. K. (1970). *Collective Choice and Social Welfare*. San Francisco, CA: Holden-Day.

Serfling, R. J. (1980). *Approximation Theorems of Mathematical Statistics*. New York: Wiley.

Shafer, G. (1976). *A Mathematical Theory of Evidence*. Princeton: University Press.

Shafer, G. (1982a). Belief functions and parametric models. *J. Roy. Statist. Soc. B* **44**, 322–352 (with discussion).

Shafer, G. (1982b). Lindley's paradox. *J. Amer. Statist. Assoc.* **77**, 325–351 (with discussion).

Shafer, G. (1986). Savage revisited. *Statist. Sci.* **1**, 435–462 (with discussion).

Shafer, G. (1990). The unity and diversity of probability. *Statist. Sci.* **5**, 463–501 (with discussion).

Shannon, C. E. (1948). A mathematical theory of communication. *Bell System Tech. J.* **27** 379–423 and 623–656. Reprinted in *The Mathematical Theory of Communication* (Shannon, C. E. and Weaver, W., 1949). Urbana, IL.: Univ. Illinois Press.

Shao, J. (1989). Monte Carlo approximations in Bayesian decision theory. *J. Amer. Statist. Assoc.* **84**, 727–732.

Shao, J. (1990). Limiting behaviour of Monte Carlo approximation to Bayesian action. *Statistics and Decisions* **8**, 85–99.

Shao, J. (1993). Linear model selection by cross-validation. *J. Amer. Statist. Assoc.* **88**, 486–494.

Shaw, J. E. H. (1988a). A quasi-random approach to integration in Bayesian statistics. *Ann. Statist.* **16**, 895–914.

Shaw, J. E. H. (1988b). Aspects of numerical integration and summarisation. *Bayesian Statistics 3* (J. M. Bernardo, M. H. DeGroot, D. V. Lindley and A. F. M. Smith, eds.). Oxford: University Press, 411–428 (with discussion).

Simon, J. C. (1984). *La Reconnaissance des Formes*. Paris: Masson.

Simpson, E. H. (1951). The interpretation of interaction in contingency tables. *J. Roy. Statist. Soc. B* **13**, 238–241.

Singpurwalla, N. D. and Soyer, R. (1992). Non homogeneous autoregressive processes for tracking (software) reliability growth, and their Bayesian analysis. *J. Roy. Statist. Soc. B* **54**, 145–156.

Singpurwalla, N. D. and Wilson, S. P. (1992). Warranties. *Bayesian Statistics 4* (J. M. Bernardo, J. O. Berger, A. P. Dawid and A. F. M. Smith, eds.). Oxford: University Press, 435–446 (with discussion).

Sivaganesan S. (1991). Sensitivity of some standard Bayesian estimates to prior uncertainty: a comparison. *J. Statist. Planning and Inference* **27**, 85–103.

Sivaganesan S. (1993). Robust Bayesian analysis of the binomial empirical Bayes problems. *Canadian J. Statist.* **21**, 107–119.

Sivaganesan S. and Berger, J. O. (1989). Ranges of posterior measures for priors with unimodal contamination. *Ann. Statist.* **17**, 868–889.

Skene, A. M., Shaw, J. E. H. and Lee, T. D. (1986). Bayesian modelling and sensitivity analysis. *The Statistician* **35**, 281–288.

Skilling, J. (ed.) (1989). *Maximum Entropy and Bayesian Methods*. Dordrecht: Kluwer.

Smith, A. F. M. (1973a). Bayes estimates in one-way and two way models. *Biometrika* **60**, 319–330.

Smith, A. F. M. (1973b). A general Bayesian linear model. *J. Roy. Statist. Soc. B* **35**, 67–75.

Smith, A. F. M. (1978). In discussion of Tanner (1978). *J. Roy. Statist. Soc. A* **141**, 50–51.

Smith, A. F. M. (1981). On random sequences with centred spherical symmetry. *J. Roy. Statist. Soc. B* **43**, 208–209.

Smith, A. F. M. (1983). Bayesian approaches to outliers and robustness. *Specifying Statistical Models* (J.-P. Florens, M. Mouchart, J.-P. Raoult, L. Simar and A. F. M. Smith, eds.). Berlin: Springer, 13–55.

Smith, A. F. M. (1984). Bayesian Statistics. Present position and potential developments: some personal views. *J. Roy. Statist. Soc. A* **147**. 245–259 (with discussion).

Smith, A. F. M. (1986). Some Bayesian thoughts on modeling and model choice. *The Statistician* **35**, 97–102.

Smith, A. F. M. (1988). What should be Bayesian about Bayesian software? *Bayesian Statistics 3* (J. M. Bernardo, M. H. DeGroot, D. V. Lindley and A. F. M. Smith, eds.). Oxford: University Press, 429–435 (with discussion).

Smith, A. F. M. (1991). Bayesian computational methods. *Phil. Trans. Roy. Soc. London A* **337**, 369–386.

Smith, A. F. M. and Dawid, A. P. (eds.) (1987). *1986 Conference on Practical Bayesian Statistics*. Special issue, *The Statistician* **36**, Numbers 2 and 3.

Smith, A. F. M. and Gelfand, A. E. (1992). Bayesian statistics without tears: a sampling-resampling perspective. *Amer. Statist.* **46**, 84–88.

Smith, A. F. M. and Roberts, G. O. (1993). Bayesian computation via the Gibbs sampler and related Markov chain Monte Carlo methods. *J. Roy. Statist. Soc. B* **55**, 3–23 (with discussion).

Smith, A. F. M., Skene, A. M., Shaw, J. E. H. and Naylor, J. C. (1987). Progress with numerical and graphical methods for Bayesian statistics. *The Statistician* **36**, 75–82.

Smith, A. F. M., Skene, A. M., Shaw, J. E. H., Naylor, J. C. and Dransfield, M. (1985). The implementation of the Bayesian paradigm. *Comm. Statist. Theory and Methods* **14**, 1079–1109.

Smith, A. F. M. and Spiegelhalter, D. J. (1980). Bayes factors and choice criteria for linear models. *J. Roy. Statist. Soc. B* **42**, 213–220.

Smith, A. F. M. and Verdinelli, I. (1980). A note on Bayes designs for inference using a hierarchical linear model. *Biometrika* **47**, 613–619.

Smith, C. A. B. (1961). Consistency in statistical inference and decision. *J. Roy. Statist. Soc. B* **23**, 1–37 (with discussion).

Smith, C. A. B. (1965). Personal probability and statistical analysis. *J. Roy. Statist. Soc. A* **128**, 469–499.

Smith, C. R. and Erickson, J. G. (eds.) (1987). *Maximum Entropy and Bayesian Spectral Analysis and Estimation Problems.* Dordrecht: Reidel.

Smith, C. R. and Grandy, W. T. (eds.) (1985). *Maximum Entropy and Bayesian Methods in Inverse Problems.* Dordrecht: Reidel.

Smith, J. Q. (1988a). *Decision Analysis, a Bayesian Approach.* London: Chapman and Hall.

Smith, J. Q. (1988b). Models, optimal decisions and influence diagrams. *Bayesian Statistics 3* (J. M. Bernardo, M. H. DeGroot, D. V. Lindley and A. F. M. Smith, eds.). Oxford: University Press, 765–776.

Smith, J. Q. (1992). A comparison of the characteristics of some Bayesian forecasting models. *Internat. Statist. Rev.* **60**, 75–87.

Smith, J. Q. and Gathercole, R. B. (1986). Principles of interactive forecasteing. *Bayesian Inference and Decision Techniques: Essays in Honor of Bruno de Finetti* (P. K. Goel and A. Zellner, eds.). Amsterdam: North-Holland, 405–423.

Smouse, E. P. (1984). A note on Bayesian least squares inference for finite population models. *J. Amer. Statist. Assoc.* **79**, 390–392.

Solomonoff, R. J. (1978). Complexity based induction systems: comparison and convergence theorems. *IEEE Trans. Information Theory* **24**, 422–432.

Spall, J. C. (ed.) (1988). *Bayesian Analysis of Time Series and Dynamic Models.* New York: Marcel Dekker.

Spall, J. C. and Hill, S. D. (1990). Least informative Bayesian prior distributions for finite samples based on information theory. *IEEE Trans. Automatic Control* **35**, 580–583.

Spall, J. C. and Maryak, J. C. (1992). A feasible Bayesian estimator of quantiles for projectile accuracy for i.d.d. data. *J. Amer. Statist. Assoc.* **87**, 676–681.

Spiegelhalter, D. J. (1987). Probability expert systems in medicine: practical issues in handling uncertainty. *Statist. Sci.* **2**, 25–34 (with discussion).

Spiegelhalter, D. J. and Cowell, R. G. (1992). Learning in probabilistic expert systems. *Bayesian Statistics 4* (J. M. Bernardo, J. O. Berger, A. P. Dawid and A. F. M. Smith, eds.). Oxford: University Press, 447–465 (with discussion).

Spiegelhalter, D. J. and Knill-Jones, R. (1984). Statistical and knowledge-based approaches to clinical decision support systems with application in gastroenterology. *J. Roy. Statist. Soc. A* **147**, 34–77 (with discussion).

Spiegelhalter, D. J. and Smith, A. F. M. (1982). Bayes factors for linear and log-linear models with vague prior information. *J. Roy. Statist. Soc. B* **44**, 377–387.

Stäel von Holstein, C.-A. S. (1970). *Assessment and Evaluation of Subjective Probability Distributions.* Stockholm: School of Economics.

Stäel von Holstein, C.-A. S. and Matheson, J. E. (1979). *A Manual for Encoding Probability Distributions.* Palo Alto: CA.: SRI International.

Steel, M. F. J. (1992). Posterior analysis of restricted seemingly unrelated regression equation models. *Econometric Reviews* **11**, 129–142.

Stein, C. (1951). A property of some tests of composite hypotheses. *Ann. Math. Statist.* **22**, 475–476.

Stein, C. (1956). Inadmissibility of the usual estimation of the mean of a multivariate normal distribution. *Proc. Third Berkeley Symp.* **1** (J. Neyman and E. L. Scott, eds.). Berkeley: Univ. California Press, 197–206.

Stein, C. (1959). An example of wide discrepancy between fiducial and confidence intervals. *Ann. Math. Statist.* **30**, 877–880.

Stein, C. (1962). Confidence sets for the mean of a multivariate normal distribution. *J. Roy. Statist. Soc. B* **24**, 265–296 (with discussion).

Stein, C. (1965). Approximation of improper prior measures by proper probability measures. *Bernoulli, Bayes, Laplace Festschrift.* (J. Neyman and L. LeCam, eds.). Berlin: Springer, 217–240.

Stephens, D. A. and Smith, A. F. M. (1992). Sampling-resampling techniques for the computation of posterior densities in normal means problems. *Test* **1**, 1–18.

Stewart, L. (1979). Multiparameter univariate Bayesian analysis. *J. Amer. Statist. Assoc.* **74**, 684–693.

Stewart, L. (1983). Bayesian analysis using Monte Carlo integration, a powerful methodology for handling some difficult problems. *The Statistician* **32**, 195–200.

Stewart, L. (1985). Multiparameter Bayesian inference using Monte Carlo integration, some techniques for bivariate analysis. *Bayesian Statistics 2* (J. M. Bernardo, M. H. DeGroot, D. V. Lindley and A. F. M. Smith, eds.), Amsterdam: North-Holland, 495–510.

Stewart, L. (1987). Hierarchical Bayesian analysis using Monte Carlo integration: computing posterior distributions when there are many models. *The Statistician* **36**, 211–219.

Stewart, L. and Davis, W. W. (1986). Bayesian posterior distributions over sets of possible models with inferences computed by Monte Carlo integration. *The Statistician* **35**, 175–182.

Stigler, S. M. (1982). Thomas Bayes' Bayesian inference. *J. Roy. Statist. Soc. A* **145**, 250–258.

Stigler, S. M. (1986a). *The History of Statistics.* Harvard, MA: University Press.

Stigler, S. M. (1986b). Laplace's 1774 memoir on inverse probability. *Statist. Sci.* **1**, 359–378.

Stigum, B. P. (1972). Finite state space and expected utility maximization. *Econometrica* **40**, 253–259.

Stone, M. (1959). Application of a measure of information to the design and comparison of experiments. *Ann. Math. Statist.* **30**, 55–70.

Stone, M. (1961). The opinion pool. *Ann. Math. Statist.* **32**, 1339–1342,

Stone, M. (1963). The posterior t distribution. *Ann. Math. Statist.* **34**, 568–573.

Stone, M. (1965). Right Haar measures for convergence in probability to invariant posterior distributions. *Ann. Math. Statist.* **36**, 440–453.

Stone, M. (1970). Necessary and sufficient conditions for convergence in probability to invariant posterior distributions. *Ann. Math. Statist.* **41**, 1939–1953.

Stone, M. (1974). Cross-validatory choice and assessment of statistical predictions. *J. Roy. Statist. Soc. B* **36**, 11–147 (with discussion).

Stone, M. (1976). Strong inconsistency from uniform priors. *J. Amer. Statist. Assoc.* **71**, 114–125 (with discussion).

Stone, M. (1977). An asymptotic equivalence of choice of model by cross-validation and Akaike's criterion. *J. Roy. Statist. Soc. B* **39**, 44–47.

Stone, M. (1979a). Comments on model selection criteria of Akaike and Schwarz. *J. Roy. Statist. Soc. B* **41**, 276–278.

Stone, M. (1979b). Review and analysis of some inconsistencies related to improper distributions and finite additivity. *Proc. 6th Internat. Conf. Logic, Methodology and Philosophy of Science* (L. J. Cohen, J. Los, H. Pfeiffer and K. P. Podewski, eds.). Amsterdam: North-Holland,

Stone, M. (1986). In discussion of Fishburn (1986). *Statist. Sci.* **1**, 356–357.

Stone, M. and Dawid, A. P. (1972). Un-Bayesian implications of improper Bayesian inference in routine statistical problems. *Biometrika* **59**, 369–373.

Stroud, A. H. (1971). *Approximate Calculation of Multiple Integrals*. Englewood Cliffs, NJ: Prentice-Hall

Sudderth, W. D. (1980). Finitely additive priors, coherence and the marginalization paradox. *J. Roy. Statist. Soc. B* **42**, 339–341.

Sugden, R. A. (1985). A Bayesian view of ignorable designs in survey sampling inference. *Bayesian Statistics 2* (J. M. Bernardo, M. H. DeGroot, D. V. Lindley and A. F. M. Smith, eds.), Amsterdam: North-Holland, 751–754.

Suppes, P. (1956). The role of subjective probability in decision making. *Proc. Third Berkeley Symp.* **5** (J. Neyman and E. L. Scott, eds.). Berkeley: Univ. California Press, 61–73.

Suppes, P. (1960). Some open problems in the foundations of subjective probability. *Information and Decision Processes* (Machol, ed.). New York: McGraw-Hill, 162–170.

Suppes, P. (1974). The measurement of belief. *J. Roy. Statist. Soc. B* **36**, 160–175.

Suppes, P. and Walsh, K. (1959). A non-linear model for the experimental measurement of utility. *Behavioral Sci.* **4**, 204–211.

Suppes, P. and Zanotti, M. (1976). Necessary and sufficient conditions for the existence of a unique measure strictly agreeing with a qualitative probability ordering. *J. Philos. Logic* **5**, 431–438.

Suppes, P. and Zanotti, M. (1982). Necessary and sufficient qualitative axioms for conditional probability. *Z. Wahisch. verw. Gebiete* **60**, 163–169.

Susarla V. and van Ryzin, J. (1976). Nonparametric Bayesian estimation of survival curves from incomplete observations. *J. Amer. Statist. Assoc.* **71**, 897–902.

Sweeting, T. J. (1984). On the choice of prior distributions for the Box-Cox transformed linear model. *Biometrika* **71**, 127–134.

Sweeting, T. J. (1985). Consistent prior distributions for transformed models. *Bayesian Statistics 2* (J. M. Bernardo, M. H. DeGroot, D. V. Lindley and A. F. M. Smith, eds.), Amsterdam: North-Holland, 755–762.

Sweeting, T. J. (1992). On asymptotic posterior normality in the multiparameter case. *Bayesian Statistics 4* (J. M. Bernardo, J. O. Berger, A. P. Dawid and A. F. M. Smith, eds.). Oxford: University Press, 825–835.

Sweeting, T. J. and Adekola, A. D. (1987). Asymptotic posterior normality for stochastic processes revisited *J. Roy. Statist. Soc. B* **49**, 215–222.

Tanner, M. A. (1991). *Tools for Statistical Inference: Observed Data and Data Augmentation Methods*. Berlin: Springer.

Tanner, M. A. and Wong, W. H. (1987). The calculation of posterior distributions by data augmentation. *J. Amer. Statist. Assoc.* **82**, 582–548 (with discussion).

Teichroew, D. (1965). A history of distribution sampling prior to the era of the computer and its relevance to simulation. *J. Amer. Statist. Assoc.* **60**, 27–49.

Thatcher, A. R. (1964). Relationships between Bayesian and confidence limits for prediction *J. Roy. Statist. Soc. B* **26**, 126–210.

Thomas, A., Spiegelhalter, D. J. and Gilks, W. R. (1992). *BUGS*, a program to perform Bayesian inference using Gibbs sampling. *Bayesian Statistics 4* (J. M. Bernardo, J. O. Berger, A. P. Dawid and A. F. M. Smith, eds.). Oxford: University Press, 837–842.

Thorburn, D. (1986). A Bayesian approach to density estimation. *Biometrika* **73**, 65–75.

Tiao, G. C. and Box, G. E. P. (1974). Some comments on 'Bayes' estimators. *Studies in Bayesian Econometrics and Statistics: in Honor of Leonard J. Savage* (S. E. Fienberg and A. Zellner, eds.). Amsterdam: North-Holland, 620–626.

Tibshirani, R. (1989). Noninformative priors for one parameter of many. *Biometrika* **76**, 604–608.

Tierney, L. (1990). *LISP-STAT. An Object Oriented Environment for Statistical Computing and Dynamic Graphics*. Chichester: Wiley.

Tierney, L. (1992). Exploring posterior distributions using Markov chains. *Computer Science and Statistics: Proc. 23rd. Symp. Interface* (E. M. Keramidas, ed.). Fairfax Station: Interface Foundation, 563–570.

Tierney, L. and Kadane, J. B. (1986). Accurate approximations for posterior moments and marginal densities. *J. Amer. Statist. Assoc.* **81**, 82–86.

Tierney, L., Kass, R. E. and Kadane, J. B. (1987). Interactive Bayesian analysis using accurate asymptotic approximations. *Computer Science and Statistics: 19th Symposium on the Interface* (R. Heiberger, ed.). Alexandria, VA: ASA, 15–21.

Tierney, L., Kass, R. E. and Kadane, J. B. (1989a). Fully exponential Laplace approximations to expectations and variances of nonpositive functions. *J. Amer. Statist. Assoc.* **84**, 710–716.

Tierney, L., Kass, R. E. and Kadane, J. B. (1989b). Approximate marginal densities of nonlinear functions. *Biometrika* **76**, 425–433.

Titterington, D. M., Smith, A. F. M. and Makov, U. E. (1985). *Statistical Analysis of Finite Mixture Distributions*. Chichester: Wiley.

Torgesen, E. N. (1981). Measures of information based on comparison with total information and with total ignorance. *Ann. Statist.* **9**, 638–657.

Trader, R. L. (1989). Thomas Bayes. *Encyclopedia of Statistical Sciences* **suppl.** (S. Kotz, N. L. Johnson and C. B. Read, eds.). New York: Wiley, 14–17.

Tribus, M. (1969). *Rational Descriptions, Decisions and Designs*. New York: Pergamon.

van der Merwe, A. J. and van der Merwe, C. A. (1992). Empirical and hierarchical Bayes estimation in multivariate regression models. *Bayesian Statistics 4* (J. M. Bernardo, J. O. Berger, A. P. Dawid and A. F. M. Smith, eds.). Oxford: University Press, 843–850.

van Dijk, H. K., Hop, J. P. and Louter, A. S. (1987). An algorithm for the computation of posterior moments and densities using simple importance sampling. *The Statistician* **36**, 83–90.

van Dijk, H. K. and Kloek, T. (1983). Monte Carlo analysis of skew posterior distributions: An illustrative econometric example. *The Statistician* **32**, 216–223.

van Dijk, H. K. and Kloek, T. (1985). Experiments with some alternatives for simple importance sampling in Monte Carlo integration. *Bayesian Statistics 2* (J. M. Bernardo, M. H. DeGroot, D. V. Lindley and A. F. M. Smith, eds.), Amsterdam: North-Holland, 511–530 (with discussion).

Venn, J. (1886). *The Logic of Chance*. London: MacMillan. Reprinted in 1963, New York: Chelsea.

Verbraak, H. L. F. (1990). *The Logic of Objective Bayesianism*. The Hague: CIP-DATA.

Verdinelli, I. (1992). Advances in Bayesian experimental design. *Bayesian Statistics 4* (J. M. Bernardo, J. O. Berger, A. P. Dawid and A. F. M. Smith, eds.). Oxford: University Press, 467–481, (with discussion).

Verdinelli, I. and Kadane, J. B. (1992). Bayesian designs for maximizing information and outcome. *J. Amer. Statist. Assoc.* **87**, 510–515.

Verdinelli, I. and Wasserman, L. (1991). Bayesian analysis of outlier problems using the Gibbs sampler. *Statist. Computing* **1**, 135–139.

Viertl, R. (ed.) (1987). *Probability and Bayesian Statistics*. London: Plenum.

Villegas, C. (1964). On qualitative σ-algebras. *Ann. Math. Statist.* **35**, 1787–1796.

Villegas, C. (1969). On the a priori distribution of the covariance matrix. *Ann. Math. Statist.* **40**, 1098–1099.

Villegas, C. (1971). On Haar priors. *Foundations of Statistical Inference* (V. P. Godambe and D. A. Sprott, eds.). Toronto: Holt, Rinehart and Winston, 409–414 (with discussion).

Villegas, C. (1977a). On the representation of ignorance. *J. Amer. Statist. Assoc.* **72**, 651–654.

Villegas, C. (1977b). Inner statistical inference. *J. Amer. Statist. Assoc.* **72**, 453–458.

Villegas, C. (1981). Inner statistical inference II. *Ann. Statist.* **9**, 768–776.

Villegas, C. (1990). Bayesian inference in models with euclidean structures. *J. Amer. Statist. Assoc.* **85**, 1159–1164.

von Mises, R. (1928). *Probability, Statistics and Truth*. Reprinted in 1957, London: Macmillan.

von Neumann, J. and Morgenstern, O. (1944/1953). *Theory of Games and Economic Behaviour*. 3rd. edition in 1953. Princeton: University Press.

Vovk, V. G. (1993a). A logic of probability, with applications to the foundation of statistics. *J. Roy. Statist. Soc. B* **55**, 317–352 (with discussion).

Vovk, V. G. (1993b). Forecasting point and continuous processes: prequential analysis. *Test* **2**, 189–217.

Vovk, V. G. and Vyugin V. V. (1993). On the empirical validity of the Bayesian method. *J. Roy. Statist. Soc. B* **55**, 253–266.

Wahba, G. (1978). Improper priors, spline smoothing and the problems of guarding against model errors in regression. *J. Roy. Statist. Soc. B* **40**, 364–372.

Wahba, G. (1983). Bayesian confidence intervals for the cross-validated smoothing spline. *J. Roy. Statist. Soc. B* **45**, 133–150.

Wahba, G. (1988). Partial and interaction spline models. *Bayesian Statistics 3* (J. M. Bernardo, M. H. DeGroot, D. V. Lindley and A. F. M. Smith, eds.). Oxford: University Press, 479–491 (with discussion).

Wakefield, J. C., Gelfand, A. E. and Smith, A. F. M. (1991). Efficient generation of random variates via the ratio-of-uniforms method. *Statistics and Computing* **1**, 129–133.

Wald, A. (1939). Contributions to the theory of statistical estimation and testing hypothesis. *Ann. Math. Statist.* **10**, 299–326.

Wald, A. (1947). *Sequential Analysis*. New York: Wiley.

Wald, A. (1950). *Statistical Decision Functions*. New York: Wiley.

Walker, A. M. (1969). On the asymptotic behaviour of posterior distributions. *J. Roy. Statist. Soc. B* **31**, 80–88.

Wallace, C. S. and Freeman, P. R. (1987). Estimation and inference by compact coding. *J. Roy. Statist. Soc. B* **49**, 240–260 (with discussion).

Walley, P. (1987). Belief function representations of statistical evidence. *Ann. Statist.* **15**, 1439–1465.

Walley, P. (1991). *Statistical Reasoning with Imprecise Probabilities*. London: Chapman and Hall.

Walley, P. and Fine, T. L. (1979). Varieties of modal (classificatory) and comparative probability. *Synthese* **41**, 321–374.

Wallsten, T. S. (1974). The psychological concept of subjective probability: a measurement theoretic view. *The Concept of Probability in Psychological Experiments* (C.-A. S. Stäel von Holstein, ed.). Dordrecht: Reidel, 49–72.

Wasserman, L. (1989). A robust Bayesian interpretation of likelihood regions. *Ann. Statist.* **17**, 1387–1393.

Wasserman, L. (1990a). Belief functions and statistical inference. *Canadian J. Statist.* **18**, 183–196.

Wasserman, L. (1990b). Prior envelopes based on belief functions. *Ann. Statist.* **18**, 454–464.

Wasserman, L. (1992a). Recent methodological advances in robust Bayesian inference. *Bayesian Statistics 4* (J. M. Bernardo, J. O. Berger, A. P. Dawid and A. F. M. Smith, eds.). Oxford: University Press, 483–502 (with discussion).

Wasserman, L. (1992b). Invariance properties of density ratio priors. *Ann. Statist.* **20**, 2177–2182.

Wasserman, L. and Kadane, J. B. (1990). Bayes' theorem for Choquet capacities. *Ann. Statist.* **18**, 1328–1339.

Wasserman, L. and Kadane, J. B. (1992a). Computing bounds on expectations. *J. Amer. Statist. Assoc.* **87**, 516–522.

Wasserman, L. and Kadane, J. B. (1992b). Symmetric upper probabilities. *Ann. Statist.* **20**, 1720–1736.

Wechsler, S. (1993). Exchangeability and predictivism. *Erkenntnis* **38**, 343–350.

Weerahandi, S. and Zidek, J. V. (1981). Multi-Bayesian statistical decision theory. *J. Roy. Statist. Soc. A* **144**, 85–93.

Weerahandi, S. and Zidek, J. V. (1983). Elements of multi-Bayesian decision theory. *Ann. Statist.* **11**, 1032–1046.

Weiss, R. E. and Cook, R. D. (1992). A graphical case statistic for assessing posterior inference. *Biometrika* **79**, 51–55.

Welch, B. L. (1939). On confidence limits and sufficiency, with particular reference to parameters of location. *Ann. Math. Statist.* **10**, 58–69.

Welch, B. L. (1965). On comparisons between confidence point procedures in the case of a single parameter. *J. Roy. Statist. Soc. B* **27**, 1–8.

Welch, B. L. and Peers, H. W. (1963). On formulae for confidence points based on intervals of weighted likelihoods. *J. Roy. Statist. Soc. B* **25**, 318–329.

West, M. (1981). Robust sequential approximate Bayesian estimation. *J. Roy. Statist. Soc. B* **43**, 157–166.

West, M. (1984). Outlier models and prior distributions in linear regression. *J. Roy. Statist. Soc. B* **46**, 431–439.

West, M. (1985). Generalized linear models: scale parameters, outlier accomodation and prior distributions. *Bayesian Statistics 2* (J. M. Bernardo, M. H. DeGroot, D. V. Lindley and A. F. M. Smith, eds.), Amsterdam: North-Holland, 531–557 (with discussion).

West, M. (1986). Bayesian model monitoring. *J. Roy. Statist. Soc. B* **48**, 70–78.

West, M. (1988). Modelling expert opinion. *Bayesian Statistics 3* (J. M. Bernardo, M. H. DeGroot, D. V. Lindley and A. F. M. Smith, eds.). Oxford: University Press, 493–508 (with discussion).

West, M. (1992a). Modelling agent forecast distributions. *J. Roy. Statist. Soc. B* **54**, 553–568.

West, M. (1992b). Modelling with mixtures. *Bayesian Statistics 4* (J. M. Bernardo, J. O. Berger, A. P. Dawid and A. F. M. Smith, eds.). Oxford: University Press, 503–524 (with discussion).

West, M. and Crosse, J. (1992). Modelling probabilistic agent opinion. *J. Roy. Statist. Soc. B* **54**, 285–299.

West, M. and Harrison, P. J. (1986). Monitoring and adaptation in Bayesian forecasting models. *J. Amer. Statist. Assoc.* **81**, 741–750.

West, M. and Harrison, P. J. (1989). *Bayesian Forecasting and Dynamic Models*. Berlin: Springer.

West, M. and Migon, H. S. (1985). Dynamic generalised linear models and Bayesian forecasting. *J. Amer. Statist. Assoc.* **80**, 73–83.

West, M., Mueller, P. and Escobar, M. D. (1994). Hierarchical priors and mixture models with applications in regression and density estimation. *Aspects of Uncertainty: a Tribute to D. V. Lindley* (P. R. Freeman, and A. F. M. Smith, eds.). Chichester: Wiley, 363–386.

Wetherill, G. B. (1961). Bayesian sequential analysis. *Biometrika* **48**, 281–292.

Wetherill, G. B. (1966). *Sequential Methods in Statistics*. New York: Wiley.

Wetherill, G. B. and Campling, G. E. G. (1966). The decision theory approach to sampling inspection. *J. Roy. Statist. Soc. B* **28**, 381–416.

White, D. J. (1976a). *Fundamentals of Decision Theory*. Amsterdam: North-Holland.

White, D. J. (1976b). *A Decision Methodology*. Chichester: Wiley.

White, D. J. and Bowen, K. C. (eds.) (1975). *The Role and Effectiveness of Theories of Decision in Practice*. London: Hodder and Stoughton.

Whittle, P. (1958). On the smoothing of probability density functions. *J. Roy. Statist. Soc. B* **20**, 334–343.

Whittle, P. (1976). *Probability*. Chichester: Wiley.

Wichmann, D. (1990). *Bayes-Statistik*. Mannheim: BI-Wissenschaftsverlag.

Wiener, N. (1948). *Cybernetics*. Cambridge, Mass.: The MIT Press. Reprinted in 1961.

Wilkinson, G. N. (1977). On resolving the controversy in statistical inference. *J. Roy. Statist. Soc. B* **39**, 119–171 (with discussion).

Wilks, S. S. (1962). *Mathematical Statistics*. New York: Wiley.

Willing, R. (1988). Information contained in nuisance parameters. *Bayesian Statistics 3* (J. M. Bernardo, M. H. DeGroot, D. V. Lindley and A. F. M. Smith, eds.). Oxford: University Press, 801–805.

Wilson, J. (1986). Subjective probabilities and the prisoners' dilemma. *Manag. Sci.* **32**, 45–55.

Wilson, R. B. (1968). On the theory of syndicates. *Econometrica* **36**, 119–132.

Winkler, R. L. (1967a). The assessment of prior distributions in Bayesian analysis. *J. Amer. Statist. Assoc.* **62**, 776–800.

Winkler, R. L. (1967b). The quantification of judgement; some methodological suggestions. *J. Amer. Statist. Assoc.* **62**, 1105–1120.

Winkler, R. L. (1968). The consensus of subjective probability distributions. *Manag. Sci.* **15**, 861–875.

Winkler, R. L. (1972). *Introduction to Bayesian Inference and Decision*. Toronto: Holt, Rinehart and Winston.

Winkler, R. L. (1980). Prior information, predictive distributions and Bayesian model building. *Bayesian Analysis in Econometrics and Statistics: Essays in Honor of Harold Jeffreys* (A. Zellner, ed.). Amsterdam: North-Holland,

Winkler, R. L. (1981). Combining probability distributions from dependent information sources. *Manag. Sci.* **27**, 479–488.

Witmer, J. A. (1986). Bayesian multistage decision problems. *Ann. Statist.* **14**, 283–297.

Wolpert, R. L. (1991). Monte Carlo importance sampling in Bayesian statistics. *Statistical Multiple Integration* (N. Flournoy, and R. K. Tsutakawa, (eds.). Providence: RI: ASA,

Wolpert, R. L. and Warren-Hicks, W. J. (1992). Bayesian hierarchical logistic models for combining field and laboratory survival data. *Bayesian Statistics 4* (J. M. Bernardo, J. O. Berger, A. P. Dawid and A. F. M. Smith, eds.). Oxford: University Press, 525–546 (with discussion).

Wong, W. H. and Li, B. (1992). Laplace expansion for posterior densities of nonlinear functions of parameters. *Biometrika* **79**, 393–398.

Wooff, D. A. (1992). [B/D] works. *Bayesian Statistics 4* (J. M. Bernardo, J. O. Berger, A. P. Dawid and A. F. M. Smith, eds.). Oxford: University Press, 851–859.

Wright, D. E. (1986). A note on the construction of highest posterior density intervals. *Appl. Statist.* **35**, 49–53.

Wrinch, D. H. and Jeffreys, H. (1919). On some aspects of the theory of probability. *Phil. Mag. Ser. 6*, **38**, 715–731.

Wrinch, D. H. and Jeffreys, H. (1921). On certain fundamental principles of scientific inquiry. *Phil. Magazine 6*, **42**, 363–390; **45**, 368–374.

Yaglom, A. M. and Yaglom, I. M. (1960/1983). *Verojatnost i Informacija*. Moscow: Nauka. English translation in 1983 as *Probability and Information*. Dordrecht: Reidel.

Ye, K. (1993). Reference priors when the stopping rule depends on the parameter of interest. *J. Amer. Statist. Assoc.* **88**, 360–363.

Ye, K. and Berger, J. O. (1991). Non-informative priors for inferences in exponential regression models. *Biometrika* **78**, 645–656.

Yilmaz, M. R. (1992). An information-expectation framework for decision under uncertainty. *J. Multi-Criteria Dec. Analysis* **1**, 65–80.

Young, S. C. and Smith, J. Q. (1991). Deriving and analysing optimal strategies in Bayesian models of games. *Manag. Sci.* **37**, 559–571.

Yu, P. L. (1985). *Multiple Criteria Decision Making*. London: Plenum.

Zellner, A. (1971). *An Introduction to Bayesian Inference in Econometrics*. New York: Wiley. Reprinted in 1987, Melbourne, FL: Krieger.

Zellner, A. (1977). Maximal data information prior distributions. *New Developments in the Applications of Bayesian Methods* (A. Aykaç and C. Brumat, eds.). Amsterdam: North-Holland, 211–232.

Zellner, A. (ed.) (1980). *Bayesian Analysis in Econometrics and Statistics: Essays in Honor of Harold Jeffreys*. Amsterdam: North-Holland.

Zellner, A. (1984). Posterior odds ratios for regression hypothesis: general considerations and some specific results. *Basic Issues in Econometrics* (A. Zellner, ed.). Chicago: University Press, 275–305.

Zellner, A. (1985). Bayesian econometrics. *Econometrica* **53**, 253–269.

Zellner, A. (1986a). On assessing prior distibutions and Bayesian regression analysis with g-prior distributions. *Bayesian Inference and Decision Techniques: Essays in Honor of Bruno de Finetti* (P. K. Goel and A. Zellner, eds.). Amsterdam: North-Holland, 233–243.

Zellner, A. (1986b). Bayesian estimation and prediction using asymmetric loss functions. *J. Amer. Statist. Assoc.* **81**, 446–451.

Zellner, A. (1987). Bayesian inference. *The New Palgrave: a Dictionary of Economics* **1** (J. Eatwell, M. Milgate and P. Newman, eds.). London: Macmillan, 208–218.

Zellner, A. (1988a). A Bayesian era. *Bayesian Statistics 3* (J. M. Bernardo, M. H. DeGroot, D. V. Lindley and A. F. M. Smith, eds.). Oxford: University Press, 509–516.

Zellner, A. (1988b). Optimal information processing and Bayes' theorem. *Amer. Statist.* **42**, 278–284 (with discussion).

Zellner, A. (1988c). Bayesian analysis in econometrics. *J. Econometrics* **37**, 27–50.

Zellner, A. (1991). Bayesian methods and entropy in economics and econometrics. *Maximum Entropy and Bayesian Methods* (W. T. Grandy and L. H. Schick eds.). Dordrecht: Kluwer, 17–31.

Zellner, A. and Siow, A. (1980). Posterior odds ratios for selected regression hypothesis. *Bayesian Statistics* (J. M. Bernardo, M. H. DeGroot, D. V. Lindley and A. F. M. Smith, eds.). Valencia: University Press, 585–603 and 618–647 (with discussion).

Zidek, J. (1969). A representation of Bayes invariant procedures in terms of Haar measure. *Ann. Inst. Statist. Math.* **21**, 291–308.

Zidek, J. and Weerahandi, S. (1992). Bayesian predictive inference for samples from smooth processes. *Bayesian Statistics 4* (J. M. Bernardo, J. O. Berger, A. P. Dawid and A. F. M. Smith, eds.). Oxford: University Press, 547–563 (with discussion).

Subject Index

Author Index

In an attempt to signal the contributions of authors who otherwise are subsumed anonymously in the "*et al.*" of multi-author papers, we have included page references for all authors of such papers, even though only the first author's name actually appears on the page.

WILEY SERIES IN PROBABILITY AND STATISTICS

ESTABLISHED BY WALTER A. SHEWHART AND SAMUEL S. WILKS

Probability and Statistics Section

*ANDERSON · The Statistical Analysis of Time Series
ARNOLD, BALAKRISHNAN, and NAGARAJA · A First Course in Order Statistics
ARNOLD, BALAKRISHNAN, and NAGARAJA · Records
BACCELLI, COHEN, OLSDER, and QUADRAT · Synchronization and Linearity: An Algebra for Discrete Event Systems
BARNETT · Comparative Statistical Inference, *Third Edition*
BASILEVSKY · Statistical Factor Analysis and Related Methods: Theory and Applications
BERNARDO and SMITH · Bayesian Theory
BILLINGSLEY · Convergence of Probability Measures
BOROVKOV · Asymptotic Methods in Queuing Theory
BOROVKOV · Ergodicity and Stability of Stochastic Processes
BRANDT, FRANKEN, and LISEK · Stationary Stochastic Models
CAINES · Linear Stochastic Systems
CAIROLI and DALANG · Sequential Stochastic Optimization
CONSTANTINE · Combinatorial Theory and Statistical Design
COOK · Regression Graphics
COVER and THOMAS · Elements of Information Theory
CSÖRGŐ and HORVÁTH · Weighted Approximation in Probability Statistics
CSÖRGŐ and HORVÁTH · Limit Theorems in Change Point Analysis
DETTE and STUDDEN · The Theory of Canonical Moments with Applications in Statistics, Probability, and Analysis
*DOOB · Stochastic Processes
DUPUIS and ELLIS · A Weak Convergence Approach to the Theory of Large Deviations
ETHIER and KURTZ · Markov Processes: Characterization and Convergence
FELLER · An Introduction to Probability Theory and Its Applications, Volume 1, *Third Edition*, Revised; Volume II, *Second Edition*
FULLER · Introduction to Statistical Time Series, *Second Edition*
FULLER · Measurement Error Models
GHOSH, MUKHOPADHYAY, and SEN · Sequential Estimation
GIFI · Nonlinear Multivariate Analysis
GUTTORP · Statistical Inference for Branching Processes
HALL · Introduction to the Theory of Coverage Processes

*Now available in a lower priced paperback edition in the Wiley Classics Library.

Probability and Statistics Section (Continued)

HAMPEL · Robust Statistics: The Approach Based on Influence Functions
HANNAN and DEISTLER · The Statistical Theory of Linear Systems
HUBER · Robust Statistics
HUŠKOVÁ, BERAN, and DUPAČ · Collected Works of Jaroslav Hájek—With Commentary
IMAN and CONOVER · A Modern Approach to Statistics
JUREK and MASON · Operator-Limit Distributions in Probability Theory
KASS and VOS · The Geometrical Foundations of Asymptotic Inference
KAUFMAN and ROUSSEEUW · Finding Groups in Data: An Introduction to Cluster Analysis
KELLY · Probability, Statistics, and Optimization
KENDALL, BARDEN, CARNE and LE · Shape and Shape Theory
LINDVALL · Lectures on the Coupling Method
McFADDEN · Management of Data in Clinical Trials
MANTON, WOODBURY, and TOLLEY · Statistical Applications Using Fuzzy Sets
MARDIA and JUPP · Directional Statistics
MORGENTHALER and TUKEY · Configural Polysampling: A Route to Practical Robustness
MUIRHEAD · Aspects of Multivariate Statistical Theory
OLIVER and SMITH · Influence Diagrams, Belief Nets, and Decision Analysis
*PARZEN · Modern Probability Theory and Its Applications
PRESS · Bayesian Statistics: Principles, Models, and Applications
PUKELSHEIM · Optimal Experimental Design
RAO · Asymptotic Theory of Statistical Inference
RAO · Linear Statistical Inference and Its Applications, *Second Edition*
RAO and SHANBHAG · Choquet-Deny Type Functional Equations with Applications to Stochastic Models
ROBERTSON, WRIGHT, and DYKSTRA · Order Restricted Statistical Inference
ROGERS and WILLIAMS · Diffusions, Markov Processes, and Martingales, Volume I: Foundations, *Second Edition*, Volume II: Itô Calculus
RUBINSTEIN and SHAPIRO · Discrete Event Systems: Sensitivity Analysis and Stochastic Optimization by the Score Function Method
RUZSA and SZEKELEY · Algebraic Probability Theory
SCHEFFÉ · The Analysis of Variance
SEBER · Linear Regression Analysis
SEBER · Multivariate Observations
SEBER and WILD · Nonlinear Regression
SERFLING · Approximation Theorems of Mathematical Statistics
SHORACK and WELLNER · Empirical Processes with Applications to Statistics
SMALL and McLEISH · Hilbert Space Methods in Probability and Statistical Inference
STAPLETON · Linear Statistical Models
STAUDTE and SHEATHER · Robust Estimation and Testing
STOYANOV · Counterexamples in Probability
TANAKA · Time Series Analysis: Nonstationary and Noninvertible Distribution Theory
THOMPSON and SEBER · Adaptive Sampling
WELSH · Aspects of Statistical Inference
WHITTAKER · Graphical Models in Applied Multivariate Statistics
YANG · The Construction Theory of Denumerable Markov Processes

---

*Now available in a lower priced paperback edition in the Wiley Classics Library.

Applied Probability and Statistics Section

ABRAHAM and LEDOLTER · Statistical Methods for Forecasting
AGRESTI · Analysis of Ordinal Categorical Data
AGRESTI · Categorical Data Analysis
ANDERSON, AUQUIER, HAUCH, OAKES, VANDAELE, and WEISBERG · Statistical Methods for Comparative Studies
ARMITAGE and DAVID (editors) · Advances in Biometry
*ARTHANARI and DODGE · Mathematical Programming in Statistics
ASMUSSEN · Applied Probability and Queues
*BAILEY · The Elements of Stochastic Processes with Applications to the Natural Sciences
BARNETT and LEWIS · Outliers in Statistical Data, *Third Edition*
BARTHOLOMEW, FORBES, and McLEAN · Statistical Techniques for Manpower Planning, *Second Edition*
BATES and WATTS · Nonlinear Regression Analysis and Its Applications
BECHHOFER, SANTNER, and GOLDSMAN · Design and Analysis of Experiments for Statistical Selection, Screening, and Multiple Comparisons
BELSLEY · Conditioning Diagnostics: Collinearity and Weak Data in Regression
BELSLEY, KUH, and WELSCH · Regression Diagnostics: Identifying Influential Data and Sources of Collinearity
BHAT · Elements of Applied Stochastic Processes, *Second Edition*
BHATTACHARYA and WAYMIRE · Stochastic Processes with Applications
BIRKES and DODGE · Alternative Methods of Regression
BLOOMFIELD · Fourier Analysis of Time Series: An Introduction
BOLLEN · Structural Equations with Latent Variables
BOULEAU · Numerical Methods for Stochastic Processes
BOX · Bayesian Inference in Statistical Analysis
BOX and DRAPER · Empirical Model-Building and Response Surfaces
BOX and DRAPER · Evolutionary Operation: A Statistical Method for Process Improvement
BUCKLEW · Large Deviation Techniques in Decision, Simulation, and Estimation
BUNKE and BUNKE · Nonlinear Regression, Functional Relations, and Robust Methods: Statistical Methods of Model Building
CHATTERJEE and HADI · Sensitivity Analysis in Linear Regression
CHERNICK · Bootstrap Methods: A Practitioner's Guide
CHILÈS and DELFINER · Geostatistics: Modeling Spatial Uncertainty
CHOW and LIU · Design and Analysis of Clinical Trials
CLARKE and DISNEY · Probability and Random Processes: A First Course with Applications, *Second Edition*
*COCHRAN and COX · Experimental Designs, *Second Edition*
CONOVER · Practical Nonparametric Statistics, *Second Edition*
CORNELL · Experiments with Mixtures, Designs, Models, and the Analysis of Mixture Data, *Second Edition*
*COX · Planning of Experiments
CRESSIE · Statistics for Spatial Data, *Revised Edition*
DANIEL · Applications of Statistics to Industrial Experimentation
DANIEL · Biostatistics: A Foundation for Analysis in the Health Sciences, *Sixth Edition*
DAVID · Order Statistics, *Second Edition*
*DEGROOT, FIENBERG, and KADANE · Statistics and the Law
*DODGE · Alternative Methods of Regression
DOWDY and WEARDEN · Statistics for Research, *Second Edition*
DRYDEN and MARDIA · Statistical Shape Analysis

---

*Now available in a lower priced paperback edition in the Wiley Classics Library.

Applied Probability and Statistics (Continued)

DUNN and CLARK · Applied Statistics: Analysis of Variance and Regression, *Second Edition*
ELANDT-JOHNSON and JOHNSON · Survival Models and Data Analysis
EVANS, PEACOCK, and HASTINGS · Statistical Distributions, *Second Edition*
FLEISS · The Design and Analysis of Clinical Experiments
FLEISS · Statistical Methods for Rates and Proportions, *Second Edition*
FLEMING and HARRINGTON · Counting Processes and Survival Analysis
GALLANT · Nonlinear Statistical Models
GLASSERMAN and YAO · Monotone Structure in Discrete-Event Systems
GNANADESIKAN · Methods for Statistical Data Analysis of Multivariate Observations. *Second Edition*
GOLDSTEIN and LEWIS · Assessment: Problems, Development, and Statistical Issues
GREENWOOD and NIKULIN · A Guide to Chi-squared Testing
*HAHN · Statistical Models in Engineering
HAHN and MEEKER · Statistical Intervals: A Guide for Practitioners
HAND · Construction and Assessment of Classification Rules
HAND · Discrimination and Classification
HEDAYAT and SINHA · Design and Inference in Finite Population Sampling
HEIBERGER · Computation for the Analysis of Designed Experiments
HINKELMAN and KEMPTHORNE · Design and Analysis of Experiments, Volume 1: Introduction to Experimental Design
HOAGLIN, MOSTELLER, and TUKEY · Exploratory Approach to Analysis of Variance
HOAGLIN, MOSTELLER, and TUKEY · Exploring Data Tables, Trends and Shapes
HOAGLIN, MOSTELLER, and TUKEY · Understanding Robust and Exploratory Data Analysis
HOCHBERG and TAMHANE · Multiple Comparison Procedures
HOCKING · Methods and Applications of Linear Models: Regression and the Analysis of Variables
HOGG and KLUGMAN · Loss Distributions
HOLLANDER and WOLFE · Nonparametric Statistical Methods
HOSMER and LEMESHOW · Applied Logistic Regression
HØYLAND and RAUSAND · System Reliability Theory: Models and Statistical Methods
HUBERTY · Applied Discriminant Analysis
HUNT and KENNEDY · Financial Derivatives in Theory and Practice
JACKSON · A User's Guide to Principal Components
JOHN · Statistical Methods in Engineering and Quality Assurance
JOHNSON · Multivariate Statistical Simulation
JOHNSON & KOTZ · Distributions in Statistics

Continuous Multivarious Distributions

JOHNSON, KOTZ, and BALAKRISHNAN · Continuous Univariate Distributions, Volume 1, *Second Edition*
JOHNSON, KOTZ, and BALAKRISHNAN · Continuous Univariate Distributions, Volume 2, *Second Edition*
JOHNSON, KOTZ, and BALAKRISHNAN · Discrete Multivariate Distributions
JOHNSON, KOTZ, and KEMP · Univariate Discrete Distributions, *Second Edition*
JUREČKOVÁ and SEN · Robust Statistical Procedures: Asymptotics and Interrelations
KADANE · Bayesian Methods and Ethics in a Clinical Trial Design
KADANE and SCHUM · A Probabilistic Analysis of the Sacco and Vanzetti Evidence
KALBFLEISCH and PRENTICE · The Statistical Analysis of Failure Time Data
KELLY · Reversibility and Stochastic Networks

---

*Now available in a lower priced paperback edition in the Wiley Classics Library.

Applied Probability and Statistics (Continued)

KHURI, MATHEW, and SINHA · Statistical Tests for Mixed Linear Models
KLUGMAN, PANJER, and WILLMOT · Loss Models: From Data to Decisions
KLUGMAN, PANJER, and WILLMOT · Solutions Manual to Accompany Loss Models: From Data to Decisions
KOVALENKO, KUZNETZOV, and PEGG · Mathematical Theory of Reliability of Time-Dependent Systems with Practical Applications
LAD · Operational Subjective Statistical Methods: A Mathematical, Philosophical, and Historical Introduction
LANGE, RYAN, BILLARD, BRILLINGER, CONQUEST, and GREENHOUSE · Case Studies in Biometry
LAWLESS · Statistical Models and Methods for Lifetime Data
LEE · Statistical Methods for Survival Data Analysis, *Second Edition*
LEPAGE and BILLARD · Exploring the Limits of Bootstrap
LINHART and ZUCCHINI · Model Selection
LITTLE and RUBIN · Statistical Analysis with Missing Data
MAGNUS and NEUDECKER · Matrix Differential Calculus with Applications in Statistics and Econometrics, *Revised Edition*
MALLER and ZHOU · Survival Analysis with Long Term Survivors
MANN, SCHAFER, and SINGPURWALLA · Methods for Statistical Analysis of Reliability and Life Data
McLACHLAN and KRISHNAN · The EM Algorithm and Extensions
McLACHLAN · Discriminant Analysis and Statistical Pattern Recognition
McNEIL · Epidemiological Research Methods
MEEKER and ESCOBAR · Statistical Methods for Reliability Data
MILLER · Survival Analysis
MONTGOMERY and PECK · Introduction to Linear Regression Analysis, *Second Edition*
MYERS and MONTGOMERY · Response Surface Methodology: Process and Product in Optimization Using Designed Experiments
NELSON · Accelerated Testing, Statistical Models, Test Plans, and Data Analyses
NELSON · Applied Life Data Analysis
OCHI · Applied Probability and Stochastic Processes in Engineering and Physical Sciences
OKABE, BOOTS, and SUGIHARA · Spatial Tesselations: Concepts and Applications of Voronoi Diagrams
PANKRATZ · Forecasting with Dynamic Regression Models
PANKRATZ · Forecasting with Univariate Box–Jenkins Models: Concepts and Cases
PIANTADOSI · Clinical Trials: A Methodologic Perspective
PORT · Theoretical Probability for Applications
PUTERMAN · Markov Decision Processes: Discrete Stochastic Dynamic Programming
RACHEV · Probability Metrics and the Stability of Stochastic Models
RÉNYI · A Diary on Information Theory
RIPLEY · Spatial Statistics
RIPLEY · Stochastic Simulation
ROLSKI, SCHMIDLI, SCHMIDT and TEUGELS · Stochastic Processes for Insurance and Finance
ROUSSEEUW and LEROY · Robust Regression and Outlier Detection
RUBIN · Multiple Imputation for Nonresponse in Surveys
RUBINSTEIN · Simulation and the Monte Carlo Method
RUBINSTEIN and MELAMED · Modern Simulation and Modeling
RYAN · Statistical Methods for Quality Improvement
SCHUSS · Theory and Applications of Stochastic Differential Equations

---

*Now available in a lower priced paperback edition in the Wiley Classics Library.

Applied Probability and Statistics (Continued)

SCOTT · Multivariate Density Estimation: Theory, Practice, and Visualization
*SEARLE · Linear Models
SEARLE · Linear Models for Unbalanced Data
SEARLE, CASELLA, and McCULLOCH · Variance Components
STOYAN, KENDALL, and MECKE · Stochastic Geometry and Its Applications, *Second Edition*
STOYAN and STOYAN · Fractals, Random Shapes, and Point Fields: Methods of Geometrical Statistics
THOMPSON · Empirical Model Building
THOMPSON · Sampling
THOMPSON · Simulation: A Modeler's Approach
TIJMS · Stochastic Modeling and Analysis: A Computational Approach
TIJMS · Stochastic Models: An Algorithmic Approach
TITTERINGTON, SMITH, and MARKOV · Statistical Analysis of Finite Mixture Distributions
UPTON and FINGLETON · Spatial Data Analysis by Example, Volume 1: Point Pattern and Quantitative Data
UPTON and FINGLETON · Spatial Data Analysis by Example, Volume II: Categorical and Directional Data
VAN RIJKEVORSEL and DE LEEUW · Component and Correspondence Analysis
VIDAKOVIC · Statistical Modeling by Wavelets
WEISBERG · Applied Linear Regression, *Second Edition*
WESTFALL and YOUNG · Resampling-Based Multiple Testing: Examples and Methods for *p*-Value Adjustment
WHITTLE · Systems in Stochastic Equilibrium
WOODING · Planning Pharmaceutical Clinical Trials: Basic Statistical Principles
WOOLSON · Statistical Methods for the Analysis of Biomedical Data
*ZEELNER · An Introduction to Bayesian Inference in Econometrics

Texts and References Section

AGRESTI · An Introduction to Categorical Data Analysis
ANDERSON · An Introduction to Multivariate Statistical Analysis, *Second Edition*
ANDERSON and LOYNES · The Teaching of Practical Statistics
ARMITAGE and COLTON · Encyclopedia of Biostatistics. 6 Volume set
BARTOSZYNSKI and NIEWIADOMSKA-BUGAJ · Probability and Statistical Inference
BERRY, CHALONER, and GEWEKE · Bayesian Analysis in Statistics and Econometrics: Essays in Honor of Arnold Zellner
BHATTACHARYA and JOHNSON · Statistical Concepts and Methods
BILLINGSLEY · Probability and Measure, *Second Edition*
BOX · R. A. Fisher, the Life of a Scientist
BOX, HUNTER, and HUNTER · Statistics for Experimenters: An Introduction to Design, Data Analysis, and Model Building
BOX and LUCEÑO · Statistical Control by Monitoring and Feedback Adjustment
BROWN and HOLLANDER · Statistics: A Biomedical Introduction
CHATTERJEE and PRICE · Regression Analysis by Example, *Second Edition*
COOK and WEISBERG · An Introduction to Regression Graphics
COOK and WEISBERG · Applied Regression Including Computing and Graphics
COX · A Handbook of Introductory Statistical Methods
DILLON and GOLDSTEIN · Multivariate Analysis: Methods and Applications

*Now available in a lower priced paperback edition in the Wiley Classics Library.

Texts and References Section (Continued)

DODGE and ROMIG · Sampling Inspection Tables, *Second Edition*
DRAPER and SMITH · Applied Regression Analysis, *Third Edition*
DUDEWICZ and MISHRA · Modern Mathematical Statistics
DUNN · Basic Statistics: A Primer for the Biomedical Sciences, *Second Edition*
FISHER and VAN BELLE · Biostatistics: A Methodology for the Health Sciences
FREEMAN and SMITH · Aspects of Uncertainty: A Tribute to D. V. Lindley
GROSS and HARRIS · Fundamentals of Queueing Theory, *Third Edition*
HALD · A History of Probability and Statistics and their Applications Before 1750
HALD · A History of Mathematical Statistics from 1750 to 1930
HELLER · MACSYMA for Statisticians
HOEL · Introduction to Mathematical Statistics, *Fifth Edition*
HOLLANDER and WOLFE · Nonparametric Statistical Methods, *Second Edition*
HOSMER and LEMESHOW · Applied Survival Analysis: Regression Modeling of Time to Event Data
JOHNSON and BALAKRISHNAN · Advances in the Theory and Practice of Statistics: A Volume in Honor of Samuel Kotz
JOHNSON and KOTZ (editors) · Leading Personalities in Statistical Sciences: From the Seventeenth Century to the Present
JUDGE, GRIFFITHS, HILL, LÜTKEPOHL, and LEE · The Theory and Practice of Econometrics, *Second Edition*
KHURI · Advanced Calculus with Applications in Statistics
KOTZ and JOHNSON (editors) · Encyclopedia of Statistical Sciences. Volumes 1 to 9 with Index
KOTZ and JOHNSON (editors) · Encyclopedia of Statistical Sciences: Supplement Volume
KOTZ, REED, and BANKS (editors) · Encyclopedia of Statistical Sciences: Update Volume 1
KOTZ, REED, and BANKS (editors) · Encyclopedia of Statistical Sciences: Update Volume 2
LAMPERTI · Probability: A Survey of the Mathematical Theory, *Second Edition*
LARSON · Introduction to Probability Theory and Statistical Inference, *Third Edition*
LE · Applied Categorical Data Analysis
LE · Applied Survival Analysis
MALLOWS · Design, Data, and Analysis by Some Friends of Cuthbert Daniel
MARDIA · The Art of Statistical Science: A Tribute to G. S. Watson
MASON, GUNST, and HESS · Statistical Design and Analysis of Experiments with Applications to Engineering and Science
MURRAY · X-STAT 2.0 Statistical Experimentation, Design Data Analysis, and Nonlinear Optimization
PURI, VILAPLANA, and WERTZ · New Perspectives in Theoretical and Applied Statistics
RENCHER · Methods of Multivariate Analysis
RENCHER · Multivariate Statistical Inference with Applications
ROSS · Introduction to Probability and Statistics for Engineers and Scientists
ROHATGI · An Introduction to Probability Theory and Mathematical Statistics
RYAN · Modern Regression Methods
SCHOTT · Matrix Analysis for Statistics
SEARLE · Matrix Algebra Useful for Statistics
STYAN · The Collected Papers of T. W. Anderson: 1943–1985
TIERNEY · LISP-STAT: An Object-Oriented Environment for Statistical Computing and Dynamic Graphics
WONNACOTT and WONNACOTT · Econometrics, *Second Edition*

---

*Now available in a lower priced paperback edition in the Wiley Classics Library.

WILEY SERIES IN PROBABILITY AND STATISTICS
ESTABLISHED BY WALTER A. SHEWHART AND SAMUEL S. WILKS

Editors
Robert M. Groves, Graham Kalton, J. N. K. Rao, Norbert Schwarz, Christopher Skinner

Survey Methodology Section

BIEMER, GROVES, LYBERG, MATHIOWETZ, and SUDMAN · Measurement Errors in Surveys
COCHRAN · Sampling Techniques, *Third Edition*
COUPER, BAKER, BETHLEHEM, CLARK, MARTIN, NICHOLLS and O'REILLY (editors) · Computer Assisted Survey Information Collection
COX, BINDER, CHINNAPPA, CHRISTIANSON, COLLEDGE, and KOTT (editors) · Business Survey Methods
*DEMING · Sample Design in Business Research
DILLMAN · Mail and Telephone Surveys: The Total Design Method
GROVES · Survey Errors and Survey Costs
GROVES and COUPER · Nonresponse in Household Interview Surveys
GROVES, BIEMER, LYBERG, MASSEY, NICHOLLS, and WAKSBERG · Telephone Survey Methodology
*HANSEN, HURWITZ, and MADOW · Sample Survey Methods and Theory, Volume I: Methods and Applications
*HANSEN, HURWITZ, and MADOW · Sample Survey Methods and Theory, Volume II: Theory
KASPRZYK, DUNCAN, KALTON, and SINGH · Panel Surveys
KISH · Statistical Design for Research
*KISH · Survey Sampling
LESSLER and KALSBEEK · Nonsampling Error in Surveys
LEVY and LEMESHOW · Sampling of Populations: Methods and Applications
LYBERG, BIEMER, COLLINS, de LEEUW, DIPPO, SCHWARZ, TREWIN (editors) · Survey Measurement and Process Quality
SIRKEN, HERRMANN, SCHECHTER, SCHWARZ, TANUR and TOURNANGEAU (editors) · Cognition and Survey Research
SKINNER, HOLT and SMITH · Analysis of Complex Surveys

*Now available in a lower priced paperback edition in the Wiley Classics Library.